Methods in Enzymology

Volume 345
G Protein Pathways
Part C
Effector Mechanisms

METHODS IN ENZYMOLOGY

EDITORS-IN-CHIEF

John N. Abelson   Melvin I. Simon

DIVISION OF BIOLOGY
CALIFORNIA INSTITUTE OF TECHNOLOGY
PASADENA, CALIFORNIA

FOUNDING EDITORS

Sidney P. Colowick and Nathan O. Kaplan

Methods in Enzymology

Volume 345

G Protein Pathways

Part C
Effector Mechanisms

EDITED BY

Ravi Iyengar

MOUNT SINAI SCHOOL OF MEDICINE
NEW YORK, NEW YORK

John D. Hildebrandt

MEDICAL UNIVERSITY OF SOUTH CAROLINA
CHARLESTON, SOUTH CAROLINA

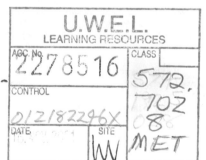

ACADEMIC PRESS

San Diego London Boston New York Sydney Tokyo Toronto

Academic Press
A Harcourt Science and Technology Company
525 B Street, Suite 1900, San Diego, California 92101-4495, USA
http://www.academicpress.com

Academic Press
Harcourt Place, 32 Jamestown Road, London NW1 7BY, UK
http://www.academicpress.com

International Standard Book Number: 0-12-182246-X

PRINTED IN THE UNITED STATES OF AMERICA
01 02 03 04 05 06 07 SB 9 8 7 6 5 4 3 2 1

Table of Contents

CONTRIBUTORS TO VOLUME 345 xi

PREFACE xvii

VOLUME IN SERIES xxi

Section I. Modeling Intracellular Signaling Pathways

1. Use of Kinetikit and GENESIS for Modeling Signaling Pathways   UPINDER S. BHALLA   3

Section II. Phosphodiesterases

2. Assays of G Protein/cGMP–Phosphodiesterase Interactions   ALEXEY E. GRANOVSKY AND NIKOLAI O. ARTEMYEV   27

3. Assaying G Protein–Phosphodiesterase Interactions in Sensory Systems   KOSEI MORIYAMA, MANJIRI M. BAKRE, FAROOQ AHMED, NANCY SPICKOFSKY, MARIANNA MAX, AND ROBERT F. MARGOLSKEE   37

Section III. Calcium and Potassium Channels

4. Studies of Endogenous G-Protein-Mediated Pathways in Neurons by Whole-Cell Electrophysiology   ARUN ANANTHARAM AND MARÍA A. DIVERSÉ-PIERLUISSI   51

5. Biochemical Approaches to Study Interaction of Calcium Channels with RGS12 in Primary Neuronal Cultures   ARUN ANANTHARAM AND MARÍA A. DIVERSÉ-PIERLUISSI   60

6. Assaying Phosphatidylinositol Bisphosphate Regulation of Potassium Channels   TIBOR ROHÁCS, COELI LOPES, TOORAJ MIRSHAHI, TAIHAO JIN, HAILIN ZHANG, AND DIOMEDES E. LOGOTHETIS   71

Section IV. Adenylyl Cyclases

7. Purification of Soluble Adenylyl Cyclase   JOCHEN BUCK,
                                                 MEEGHAN L. SINCLAIR, AND
                                                 LONNY R. LEVIN   95

8. Calcium-Sensitive Adenylyl Cyclase/Aequorin DERMOT M. F. COOPER   105
   Chimeras as Sensitive Probes for Discrete Modes
   of Elevation of Cytosolic Calcium

9. Kinetic Analysis of the Action of P-Site Analogs CARMEN W. DESSAUER   112

10. Expression, Purification, and Assay of Cytoso- MARK E. HATLEY,
    lic (Catalytic) Domains of Membrane-Bound   ALFRED G. GILMAN, AND
    Mammalian Adenylyl Cyclases   ROGER K. SUNAHARA   127

11. Identification of Putative Direct Effectors for J. DEDRICK JORDAN AND
    $G\alpha_o$ Using Yeast Two-Hybrid Method   RAVI IYENGAR   140

12. Identification of Transmembrane Adenylyl Cy- MARTIN J. CANN AND
    clase Isoforms   LONNY R. LEVIN   150

13. Functional Analyses of Type V Adenylyl Cyclase TARUN B. PATEL,
                                                 CLAUS WITTPOTH,
                                                 ANN J. BARBIER,
                                                 YINGES YIGZAW, AND
                                                 KLAUS SCHOLICH   160

14. Photoaffinity Labeling of Adenylyl Cyclase   MICHAEL K. SIEVERT,
                                                 GÜLHAN PILLI, YU LIU,
                                                 ELIZABETH M. SUTKOWSKI,
                                                 KENNETH B. SEAMON, AND
                                                 ARNOLD E. RUOHO   188

15. Crystallization of Complex between Soluble Do- JOHN J. G. TESMER,
    mains of Adenylyl Cyclase and Activated $G_s\alpha$ ROGER K. SUNAHARA,
                                                 DAVID A. FANCY,
                                                 ALFRED G. GILMAN, AND
                                                 STEPHEN R. SPRANG   198

16. Generation of Adenylyl Cyclase Knockout Mice SCOTT T. WONG AND
                                                 DANIEL R. STORM   206

17. Construction of Soluble Adenylyl Cyclase from SHUI-ZHONG YAN AND
    Human Membrane-Bound Type 7 Adenylyl   WEI-JEN TANG   231
    Cyclase

18. Genetic Selection of Regulatory Mutants of Mam- PETER CLAPP,
    malian Adenylyl Cyclases   AUSTIN B. CAPPER, AND
                                                 RONALD TAUSSIG   241

Section V. Phospholipases and Lipid-Derived Products

19. Expression and Characterization of Rat Brain ZHI XIE, HA KUN KIM, AND
    Phospholipase D   JOHN H. EXTON   255

20. G-Protein-Coupled Receptor Regulation of Phos-   GUANGWEI DU,
    pholipase D   ANDREW J. MORRIS,
                                                     VICKI A. SCIORRA, AND
                                                     MICHAEL A. FROHMAN   265

21. Analysis and Quantitation of Ceramide   PAOLA SIGNORELLI AND
                                                     YUSUF A. HANNUN   275

22. Assays for Phospholipase D Reaction Products   YUHUAN XIE AND
                                                     KATHRYN E. MEIER   294

23. Determination of Strength and Specificity of   SUZANNE SCARLATA   306
    Membrane-Bound G Protein–Phospholipase C
    Association Using Fluorescence Spectroscopy

24. Assays and Characterization of Mammalian Phos-   XUEJUN JIANG,
    phatidylinositol 4,5-Bisphosphate-Sensitive   STEPHEN GUTOWSKI,
    Phospholipase D   WILLIAM D. SINGER, AND
                                                     PAUL C. STERNWEIS   328

25. Characterization and Purification of Phosphatidyl- RUDIGER WOSCHOLSKI   335
    inositol Trisphosphate 5-Phosphatase from Rat
    Brain Tissues

Section VI. Small GTP-Binding Proteins

26. Assay of Cdc42, Rac, and Rho GTPase Activation VALERIE BENARD AND
    by Affinity Methods   GARY M. BOKOCH   349

27. Assays of ADP-Ribosylation Factor Function   JUN KUAI AND
                                                     RICHARD A. KAHN   359

28. Functional Characterization of p115 RhoGEF   CLARK WELLS,
                                                     XUEJUN JIANG,
                                                     STEPHEN GUTOWSKI, AND
                                                     PAUL C. STERNWEIS   371

29. Nonisotopic Methods for Detecting Activation   KENDALL D. CAREY AND
    of Small G Proteins   PHILIP J. S. STORK   383

30. BIG1 and BIG2: Brefeldin A-Inhibited Guanine   GUSTAVO PACHECO-RODRIGUEZ,
    Nucleotide-Exchange Proteins for ADP-   JOEL MOSS, AND
    Ribosylation Factors   MARTHA VAUGHAN   397

31. Functional Interaction of $G\alpha_{13}$ with p115RhoGEF JUNHAO MAO AND
    Determined with Transcriptional Reporter   DIANQING WU   404
    System

Section VII. Protein Kinases and Phosphatases

32. Analysis of c-Jun N-Terminal Kinase Regulation J. PERRY HALL AND
    and Function   ROGER J. DAVIS   413

33. Double-Label Confocal Microscopy of Phospho-   MARIA GRAZIA GIOVANNINI   426
    rylated Protein Kinases Involved in Long-Term
    Potentiation

34. Regulation of Mitogen-Activated Protein Kinases   MARIO CHIARIELLO AND
    by G-Protein-Coupled Receptors   J. SILVIO GUTKIND   437

35. Analysis of Protein Kinase B/Akt   MICHELLE M. HILL AND
                                                        BRIAN A. HEMMINGS   448

36. Direct Stimulation of Bruton's Tyrosine Kinase by   WILLIAM E. LOWRY,
    G Protein α Subunits   YONG-CHAO MA,
                                                        SVETLANA CVEJIC, AND
                                                        XIN-YUN HUANG   464

37. Isozyme-Specific Inhibitors and Activators of   DEBORAH SCHECHTMAN AND
    Protein Kinase C   DARIA MOCHLY-ROSEN   470

38. Assay of Raf-1 Activity   JÜRGEN MÜLLER AND
                                                        DEBORAH K. MORRISON   490

39. Analyzing Protein Kinase C Activation   ALEXANDRA C. NEWTON   499

40. Assays for Protein-Tyrosine Phosphatases   DANIEL F. MCCAIN AND
                                                        ZHONG-YIN ZHANG   507

41. Differential Display of mRNAs Regulated by   HIRONORI EDAMATSU,
    G-Protein Signaling   YOSHITO KAZIRO, AND
                                                        HIROSHI ITOH   521

42. Gene Profiling of Transgenic Mice with Tar-   HSIEN-YU WANG,
    geted Expression of Activated Heterotrimeric   XIAOSONG SONG,
    G Protein α Subunits Using DNA Microarray   XI-PING HUANG, AND
                                                        JIANGCHUAN TAO   529

43. Retroviral Vectors Applied to Gene Regulation   T. J. MURPHY,
    Studies   GRACE K. PAVLATH,
                                                        XIAOFEI WANG,
                                                        VALERIE BOSS,
                                                        KAREN L. ABBOTT,
                                                        AARON M. ROBIDA,
                                                        JIM NICHOLS, KAIMING XU,
                                                        MICHELLE L. ELLINGTON,
                                                        AND JAMES R. LOSS II   539

44. Overexpression of Tightly Regulated Proteins: PRAHLAD T. RAM   551
    Protein Phosphatase 2A Overexpression in NIH
    3T3 Cells

45. Monitoring G-Protein-Coupled Receptor Signal-   TONY YUEN,
    ing with DNA Microarrays and Real-Time Poly-   WEN ZHANG,
    merase Chain Reaction   BARBARA J. EBERSOLE, AND
                                                        STUART C. SEALFON   556

46. cAMP Response Element-Mediated Gene Expression in Transgenic Reporter Gene Mouse Strain   KARL OBRIETAN, SOREN IMPEY, AND DANIEL R. STORM   570

47. Functional Genomic Search of G-Protein-Coupled Receptors Using Microarrays with Normalized cDNA Library   SUSUMU KATSUMA, SATOSHI SHIOJIMA, AKIRA HIRASAWA, YASUHITO SUZUKI, HIROSHI IKAWA, KAZUCHIKA TAKAGAKI, YOSHINORI KAMINISHI, MASATOSHI MURAI, TADAAKI OHGI, JUNICHI YANO, AND GOZOH TSUJIMOTO   585

AUTHOR INDEX 601

SUBJECT INDEX 635

Contributors to Volume 345

Article numbers are in parentheses following the names of contributors.
Affiliations listed are current.

KAREN L. ABBOTT (43), *Department of Biochemistry and Molecular Biology, University of Georgia, Athens, Georgia 30602*

FAROOQ AHMED (3), *Department of Physiology and Biophysics, Howard Hughes Medical Institute, Mount Sinai School of Medicine, New York, New York 10029*

ARUN ANANTHARAM (4, 5), *Department of Pharmacology, Mount Sinai School of Medicine, New York, New York 10029*

NIKOLAI O. ARTEMYEV (2), *Department of Physiology and Biophysics, University of Iowa College of Medicine, Iowa City, Iowa 52242*

MANJIRI M. BAKRE (3), *Cancer Center, University of California at San Diego, La Jolla, California 92037*

ANN J. BARBIER (13), *Hoechst Marion Roussel, Inc., CNS In Vitro Group, Bridgewater, New Jersey 08807*

VALERIE BENARD (26), *UPRESA 5018-CNRS, 31054 Toulouse cédex, France*

UPINDER S. BHALLA (1), *GKVK Campus, National Centre for Biological Sciences, Bangalore 560065, India*

GARY M. BOKOCH (26), *Departments of Immunology and Cell Biology, The Scripps Research Institute, La Jolla, California 92037*

VALERIE BOSS (43), *Department of Pharmacology, Emory University School of Medicine, Atlanta, Georgia 30322*

JOCHEN BUCK (7), *Department of Pharmacology, Joan and Sanford I. Weill Medical College of Cornell University, New York, New York 10021*

MARTIN J. CANN (12), *Department of Biological Sciences, University of Durham, Durham DH1 3LE, United Kingdom*

AUSTIN B. CAPPER (18), *Department of Biological Chemistry, University of Michigan Medical School, Ann Arbor, Michigan 48109*

KENDALL D. CAREY (29), *Vollum Institute L-474, Department of Cell and Developmental Biology, Oregon Health and Sciences University, Portland, Oregon 97201*

MARIO CHIARIELLO (34), *Oral and Pharyngeal Cancer Branch, National Institute of Dental and Craniofacial Research, National Institutes of Health, Bethesda, Maryland 20892*

PETER CLAPP (18), *Department of Biological Chemistry, University of Michigan Medical School, Ann Arbor, Michigan 48109*

DERMOT M. F. COOPER (8), *Department of Pharmacology, University of Colorado Medical School, Denver, Colorado 80262*

SVETLANA CVEJIC (36), *Department of Physiology, Cornell University Medical College, New York, New York 10021*

ROGER J. DAVIS (32), *Howard Hughes Medical Institute, Program in Molecular Medicine, University of Massachusetts Medical School, Worcester, Massachusetts 01605*

CARMEN W. DESSAUER (9), *Department of Integrative Biology and Pharmacology, University of Texas—Houston Medical School, Houston, Texas 77030*

MARÍA A. DIVERSÉ-PIERLUISSI (4, 5), *Department of Pharmacology, Mount Sinai School of Medicine, New York, New York 10029*

GUANYWEI DU (20), *Department of Pharmacology and Center for Developmental Genetics, State University of New York, Stony Brook, New York 11794*

BARBARA J. EBERSOLE (45), *Department of Neurology, Mount Sinai School of Medicine, New York, New York 10029*

HIRONORI EDAMATSU (41), *Department of Molecular and Cellular Biology, Kobe University Graduate School of Medicine, Chuo-ku, Kobe 650-0017, Japan*

MICHELLE L. ELLINGTON (43), *Department of Pharmacology, Emory University School of Medicine, Atlanta, Georgia 30322*

JOHN H. EXTON (19), *Department of Molecular Physiology and Biophysics, Howard Hughes Medical Institute, Vanderbilt University School of Medicine, Nashville, Tennessee 37232*

DAVID A. FANCY (15), *Department of Biochemistry, University of Texas Southwestern Medical Center, Dallas, Texas 75390*

MICHAEL A. FROHMAN (20), *Department of Pharmacology and Center for Developmental Genetics, State University of New York, Stony Brook, New York 11794*

ALFRED G. GILMAN (10, 15), *Department of Pharmacology, University of Texas Southwestern Medical Center, Dallas, Texas 75390*

MARIA GRAZIA GIOVANNINI (33), *Dipartimento di Farmacologia Preclinica e Clinica "Mario Aiazzi Mancini," Universitá di Firenze, 50139 Firenze, Italy*

ALEXEY E. GRANOVSKY (2), *Department of Physiology and Biophysics, University of Iowa College of Medicine, Iowa City, Iowa 52242*

J. SILVIO GUTKIND (34), *Oral and Pharyngeal Cancer Branch, National Institute of Dental and Craniofacial Research, National Institutes of Health, Bethesda, Maryland 20892*

STEPHEN GUTOWSKI (24, 28), *Department of Pharmacology, University of Texas Southwestern Medical Center, Dallas, Texas 75390*

J. PERRY HALL (32), *Howard Hughes Medical Institute, Program in Molecular Medicine, University of Massachusetts Medical School, Worcester, Massachusetts 01605*

YUSUF A. HANNUN (21), *Department of Biochemistry and Molecular Biology, Medical University of South Carolina, Charleston, South Carolina 29425*

MARK E. HATLEY (10), *Department of Pharmacology, University of Texas Southwestern Medical Center, Dallas, Texas 75390*

BRIAN A. HEMMINGS (35), *Friedrich Miescher Institute, CH-4058 Basel, Switzerland*

MICHELLE M. HILL (35), *Friedrich Miescher Institute, CH-4058 Basel, Switzerland*

AKIRA HIRASAWA (47), *Department of Molecular, Cell Pharmacology, National Children's Medical Research Center, Setagaya-ku, Tokyo 154-8509, Japan*

XIN-YUN HUANG (36), *Department of Physiology, Cornell University Medical College, New York, New York 10021*

XI-PING HUANG (42), *Department of Molecular Pharmacology, Diabetes and Metabolic Diseases Research Program, University Medical Center, State University of New York, Stony Brook, New York 11794*

HIROSHI IKAWA (47), *Department of Molecular, Cell Pharmacology, National Children's Medical Research Center, Setagaya-ku, Tokyo 154-8509, Japan*

SOREN IMPEY (46), *Department of Pharmacology, University of Washington, Seattle, Washington 98195*

HIROSHI ITOH (41), *Division of Signal Transduction, Graduate School of Biological Sciences, Nara Institute of Science and Technology, Ikoma, Nara 630-0101, Japan*

RAVI IYENGAR (11), *Department of Pharmacology, Mount Sinai School of Medicine, New York, New York 10029*

XUEJUN JIANG (24, 28), *Department of Pharmacology, University of Texas Southwestern Medical Center, Dallas, Texas 75390*

TAIHAO JIN (6), *Department of Physiology and Biophysics, Mount Sinai School of Medicine, New York, New York 10029*

J. DEDRICK JORDAN (11), *Department of Pharmacology, Mount Sinai School of Medicine, New York, New York 10029*

RICHARD A. KAHN (27), *Department of Biochemistry, Emory University School of Medicine, Atlanta, Georgia 30322*

YOSHINORI KAMINISHI (47), *Research Laboratories, Nippon Shinyaku Co. Ltd., Tsukuba, Ibaraki 305-0003, Japan*

SUSUMU KATSUMA (47), *Department of Molecular, Cell Pharmacology, National Children's Medical Research Center, Setagaya-ku, Tokyo 154-8509, Japan*

YOSHITO KAZIRO (41), *Sanyo Gakuen University and College, Okayama-shi, Okayama 703-8501, Japan*

JUN KUAI (27), *The Genetics Institute, Cambridge, Massachusetts 02140*

HA KUN KIM (19), *Department of Molecular Physiology and Biophysics, Howard Hughes Medical Institute, Vanderbilt University School of Medicine, Nashville, Tennessee 37232*

LONNY R. LEVIN (7, 12), *Department of Pharmacology, Joan and Sanford I. Weill Medical College of Cornell University, New York, New York 10021*

YU LIU (14), *GlaxoSmithKline, Philadelphia, Pennsylvania 19102*

DIOMEDES E. LOGOTHETIS (6), *Department of Physiology and Biophysics, Mount Sinai School of Medicine, New York, New York 10029*

COELI LOPES (6), *Department of Physiology and Biophysics, Mount Sinai School of Medicine, New York, New York 10029*

JAMES R. LOSS II (43), *Department of Pharmacology, Emory University School of Medicine, Atlanta, Georgia 30322*

WILLIAM E. LOWRY (36), *Department of Physiology, Cornell University Medical College, New York, New York 10021*

YONG-CHAO MA (36), *Department of Physiology, Cornell University Medical College, New York, New York 10021*

JUNHAO MAO (31), *Department of Genetics and Developmental Biology, University of Connecticut, Farmington, Connecticut 06030*

ROBERT F. MARGOLSKEE (3), *Department of Physiology and Biophysics, Howard Hughes Medical Institute, Mount Sinai School of Medicine, New York, New York 10029*

MARIANNA MAX (3), *Department of Physiology and Biophysics, Mount Sinai School of Medicine, New York, New York 10029*

DANIEL F. MCCAIN (40), *Department of Biochemistry, Albert Einstein College of Medicine, Bronx, New York 10461*

KATHRYN E. MEIER (22), *Department of Pharmacology, Medical University of South Carolina, Charleston, South Carolina 29425*

TOORAJ MIRSHAHI (6), *Department of Physiology and Biophysics, Mount Sinai School of Medicine, New York, New York 10029*

DARIA MOCHLY-ROSEN (37), *Department of Molecular Pharmacology, Stanford University School of Medicine, Stanford, California 94305*

KOSEI MORIYAMA (3), *Section of Internal Medicine, Department of Medicine, Fukuoko Dental College, Sawara-ku, Fukuoka 814-0193, Japan*

ANDREW J. MORRIS (20), *Department of Cell and Developmental Biology, University of North Carolina at Chapel Hill, Chapel Hill, North Carolina 27599*

DEBORAH K. MORRISON (38), *Regulation of Cell Growth Laboratory, Center for Cancer Research, National Cancer Institute, Frederick, Maryland 21702*

JOEL MOSS (30), *Pulmonary-Critical Care Medicine Branch, National Heart, Lung, and Blood Institute, National Institutes of Health, Bethesda, Maryland 20892*

JÜRGEN MÜLLER (38), *Regulation of Cell Growth Laboratory, National Cancer Institute-Frederick Cancer Research and Development Center, Frederick, Maryland 21702*

MASATOSHI MURAI (47), *Research Laboratories, Nippon Shinyaku Co. Ltd., Tsukuba, Ibaraki 305-0003, Japan*

T. J. MURPHY (43), *Department of Pharmacology, Emory University School of Medicine, Atlanta, Georgia 30322*

ALEXANDRA C. NEWTON (39), *Department of Pharmacology, University of California at San Diego, La Jolla, California 92093*

JIM NICHOLS (43), *Department of Biology, Abilene Christian University, Abilene, Texas 79699*

KARL OBRIETAN (46), *Neuroscience Department, Ohio State University, Columbus, Ohio 43210*

TADAAKI OHGI (47), *Research Laboratories, Nippon Shinyaku Co. Ltd., Tsukuba, Ibaraki 305-0003, Japan*

GUSTAVO PACHECO-RODRIGUEZ (30), *Pulmonary-Critical Care Medicine Branch, National Heart, Lung, and Blood Institute, National Institutes of Health, Bethesda, Maryland 20892*

TARUN B. PATEL (13), *Department of Pharmacology and Vascular Biology Center, University of Tennessee, The Health Science Center, Memphis, Tennessee 38163*

GRACE K. PAVLATH (43), *Department of Pharmacology, Emory University School of Medicine, Atlanta, Georgia 30322*

GÜLHAN PILLI (14), *Eczacibaşi, Istanbul, Turkey*

PRAHLAD T. RAM (44), *Department of Pharmacology, Mount Sinai School of Medicine, New York, New York 10029*

AARON M. ROBIDA (43), *Department of Pharmacology, Emory University School of Medicine, Atlanta, Georgia 30322*

TIBOR ROHÁCS (6), *Department of Physiology and Biophysics, Mount Sinai School of Medicine, New York, New York 10029*

ARNOLD E. RUOHO (14), *Department of Pharmacology, University of Wisconsin Medical School, Madison, Wisconsin 53706*

SUZANNE SCARLATA (23), *Department of Physiology and Biophysics, State University of New York, Stony Brook, New York 11794*

DEBORAH SCHECHTMAN (37), *Department of Molecular Pharmacology, Stanford University School of Medicine, Stanford, California 94305*

KLAUS SCHOLICH (13), *Department of Pharmacology, University of Tennessee, The Health Science Center, Memphis, Tennessee 38163*

VICKI A. SCIORRA (20), *Department of Pharmacology, State University of New York, Stony Brook, New York 11794*

STUART C. SEALFON (45), *Department of Neurology, Fishberg Research Center for Neurobiology, Mount Sinai School of Medicine, New York, New York 10029*

KENNETH B. SEAMON (14), *Immunex Corporation, Seattle, Washington 98101*

SATOSHI SHIOJIMA (47), *Department of Molecular, Cell Pharmacology, National Children's Medical Research Center, Tokyo 154-8509, Japan*

MICHAEL K. SIEVERT (14), *Department of Pharmacology, University of Wisconsin Medical School, Madison, Wisconsin 53706*

PAOLA SIGNORELLI (21), *Department of Biochemistry and Molecular Biology, Medical University of South Carolina, Charleston, South Carolina 29425*

MEEGHAN L. SINCLAIR (7), *Department of Pharmacology, Joan and Sanford I. Weill Medical College of Cornell University, New York, New York 10021*

WILLIAM D. SINGER (24), *Department of Pharmacology, University of Texas Southwestern Medical Center, Dallas, Texas 75390*

XIAOSONG SONG (42), *Department of Molecular Pharmacology, Diabetes and Metabolic Diseases Research Center, University Medical Center, State University of New York, Stony Brook, New York 11794*

NANCY SPICKOFSKY (3), *Office of Regional Counsel-Region 2, U.S. Environmental Protection Agency, New York, New York 10007*

STEPHEN R. SPRANG (15), *Department of Biochemistry, Howard Hughes Medical Institute, University of Texas Southwestern Medical Center, Dallas, Texas 75390*

PAUL C. STERNWEIS (24, 28), *Department of Pharmacology, University of Texas Southwestern Medical Center, Dallas, Texas 75390*

PHILIP J. S. STORK (29), *Vollum Institute L-474, Oregon Health and Sciences University, Portland, Oregon 97201*

DANIEL R. STORM (16, 46), *Department of Pharmacology, University of Washington School of Medicine, Seattle, Washington 98195*

ROGER K. SUNAHARA (10, 15), *Department of Pharmacology, University of Michigan Medical School, Ann Arbor, Michigan 48109*

ELIZABETH M. SUTKOWSKI (14), *Food and Drug Administration, Division of Vaccines and Related Products Applications, Rockville, Maryland 20852*

YASUHITO SUZUKI (47), *Department of Molecular, Cell Pharmacology, National Children's Medical Research Center, Setagaya-ku, Tokyo 154-8509, Japan*

KAZUCHIKA TAKAGAKI (47), *Research Laboratories, Nippon Shinyaku Co. Ltd., Ibaraki 305-0003, Japan*

WEI-JEN TANG (17), *Ben May Institute for Cancer Research and Department of Neurobiology, Pharmacology, and Physiology, The University of Chicago, Chicago, Illinois 60637*

JIANGCHUAN TAO (42), *Department of Molecular Pharmacology, Diabetes and Metabolic Diseases Research Center, University Medical Center, State University of New York, Stony Brook, New York 11794*

RONALD TAUSSIG (18), *Department of Pharmacology, University of Texas Southwestern Medical Center, Dallas, Texas 75390*

JOHN J. G. TESMER (15), *Department of Chemistry and Biochemistry, University of Texas, Austin, Texas 78712*

GOZOH TSUJIMOTO (47), *Department of Molecular, Cell Pharmacology, National Children's Medical Research Center, Setagaya-ku, Tokyo 154-8509, Japan*

MARTHA VAUGHAN (30), *Pulmonary-Critical Care Medicine Branch, National Heart, Lung, and Blood Institute, National Institutes of Health, Bethesda, Maryland 20892*

HSIEN-YU WANG (42), *Department of Physiology and Biophysics, Diabetes and Metabolic Diseases Research Center, University Medical Center, State University of New York, Stony Brook, New York 11794*

XIAOFEI WANG (43), *Department of Immunology, Scripps Research Institute, La Jolla, California 92037*

CLARK WELLS (28), *Department of Pharmacology, University of Texas Southwestern Medical Center, Dallas, Texas 75390*

CLAUS WITTPOTH (13), *MelTec GmbH, Leipziger Str. 44, Germany*

SCOTT T. WONG (16), *Department of Pharmacology, University of Washington School of Medicine, Seattle, Washington 98195*

RUDIGER WOSCHOLSKI (25), *Department of Biology and Biochemistry, Wolfson Laboratories, Imperial College of Science, Technology, and Medicine, London SW7 2AY, United Kingdom*

DIANQING WU (31), *Department of Genetics and Developmental Biology, University of Connecticut, Farmington, Connecticut 06030*

YUHUAN XIE (22), *Department of Pharmacology, Medical University of South Carolina, Charleston, South Carolina 29425*

ZHI XIE (19), *Department of Molecular Physiology and Biophysics, Howard Hughes Medical Institute, Vanderbilt University School of Medicine, Nashville, Tennessee 37232*

KAIMING XU (43), *Department of Pharmacology, Emory University School of Medicine, Atlanta, Georgia 30322*

SHUI-ZHONG YAN (17), *Ben May Institute for Cancer Research, The University of Chicago, Chicago, Illinois 60637*

JUNICHI YANO (47), *Research Laboratories, Nippon Shinyaku Co. Ltd., Tsukuba, Ibaraki 305-0003, Japan*

YINGES YIGZAW (13), *Department of Pharmacology, University of Tennessee, The Health Science Center, Memphis, Tennessee 38163*

TONY YUEN (45), *Department of Neurology, Mount Sinai School of Medicine, New York, New York 10029*

HAILIN ZHANG (6), *Department of Physiology and Biophysics, Mount Sinai School of Medicine, New York, New York 10029*

WEN ZHANG (45), *Department of Physiology, Mount Sinai School of Medicine, New York, New York 10029*

ZHONG-YIN ZHANG (40), *Department of Molecular Pharmacology, Albert Einstein College of Medicine, Bronx, New York 10461*

Preface

The heterotrimeric G proteins are the central component of one of the primary mechanisms used by eukaryotic cells to receive, interpret, and respond to extracellular signals. Many of the basic concepts associated with the entire range of fields collectively referred to in terms such as "signal transduction" originate from the pioneering work of Sutherland, Krebs and Fischer, Rodbell and Gilman, Greengard, and others on systems that are primary examples of G protein signaling pathways. The study of these proteins has been and remains at the forefront of research on cell signaling mechanisms. The G Protein Pathways volumes (343, 344, and 345) of *Methods in Enzymology* have come about as part of a continuing attempt to use the methods developed in studying the G protein signaling pathways as a resource both within this field and throughout the signal transduction field.

Several volumes of this series have been devoted in whole or in part to approaches for studying the heterotrimeric G proteins. Volumes 109 and 195 were the earliest to devote substantial parts to G protein-mediated signaling systems. In 1994 Volumes 237 and 238 comprehensively covered this field. The continued growth, the ever increasing impact of this field, and the continued evolution of approaches and questions generated by research related to G proteins have led inevitably to the need for a new and comprehensive treatment of the approaches used to study these proteins. Each volume of G Protein Pathways brings together varied topics and approaches to this central theme.

Very early in the development of the concepts of G protein signaling mechanisms, Rodbell and Birnbaumer recognized at least three components of these signaling systems. They compared them to a receiver, a transducer, and an amplifier. The receiver was the receptor for an extracellular signal. The transducer referred to those mechanisms and components required for converting an extracellular signal into an intracellular response. The amplifier was synonymous with the effector enzymes that generate the beginning of the intracellular signal. Over the years these ideas have evolved in many ways. We now know an immense amount about the receptors and their great range of diversity. Through the work of Gilman, along with his associates and contemporaries, the transducer component turned out to be nearly synonymous with the heterotrimeric G proteins themselves. Nevertheless, the complexity of this component of the system continues to become more and more apparent with the recognition of the diversity of these proteins, the many ways they interact, and the increasing number of regulatory influences on their function. The key concepts associated with the effector enzymes, such as adenylyl cyclase that produces the intracellular "second messenger" cAMP, have been broadened substantially to include other enzymes, ion channels, and the components of other signaling systems that form an interacting network of systems inside cells.

The organization of Volumes 343, 344, and 345 is still conveniently centered on these three components of the G proteins signaling pathway: receptor, G protein, and effector. The evolution of the field, however, inevitably left a mark on the form of these volumes. So, for example, receptors in Volume 343 and G proteins in Volume 344 no longer stand alone as individual components in these volumes, but share space with other directly interacting proteins that influence their function. In addition, Volume 345 addresses, more generally, effector mechanisms and forms a bridge between the many different cell regulatory mechanisms that cooperate to control cellular function. Thus, there are chapters that include methods related to small GTP binding proteins, ion channels, gene regulation, and novel signaling compounds.

As we learn more about G protein signaling systems, we acquire an ever increasing appreciation of their complexity. In the previous volumes on G proteins it was already evident that there were many different isoforms of each of the three heterotrimeric G protein subunits. Initial analysis of the Human Genome Project suggests some constraints on the number of members of these proteins with 27 α subunit isoforms, 5 β subunits, and 13 γ subunits. It is interesting that in recent years we had nearly accounted for all or most of the β and γ subunit isoforms, but that there are nearly twice as many potential α subunits as the ones we currently understand. These recent discoveries, along with the recognition of the existence of between 600 and 700 G protein-coupled receptors in the human genome, may place limits on the complexity of the G protein signaling system itself. These potential limits, however, are balanced by the immense number of possible combinations of interactions that can be generated from all these components, by the possible variation of all of these proteins at levels after their genomic structure, and by our continual discovery of additional interacting components of the system.

Perhaps one of the really substantial gains in our understanding of this system since the earlier volumes is the increasing recognition of the role of accessory proteins in G protein signaling pathways. Those proteins recognized nearly 20 years ago that work at the level of the receptor continue to grow and have a prominent place in Volume 343. One of the real breakthroughs though has been the rapid development of our knowledge of accessory proteins that interact with the G proteins themselves. Prominent among these are the RGS (regulators of G protein signaling) proteins that act as GTPase activating proteins for selective G protein α subunits. A fairly substantial section of these volumes is devoted to these proteins. One apparent aside related to the RGS proteins though, is that as much as we have rapidly learned about them, there is much more yet to be learned, because there is wide variation in the structure of these proteins outside their G protein interaction sites. These proteins likely have many different stories yet to be developed based on their interactions with the G proteins, perhaps mediating functions that we do not yet know about. The RGS proteins are not the end of the interacting proteins either, however, with the description of additional G protein-interacting proteins such as the AGS (activators of G protein signaling) proteins. These are likely a heterogeneous group of proteins with several different mechanisms of interaction and roles in G protein signaling mechanisms.

The range of topics covered in these volumes turned out to be quite large. This is a result of the wide range of approaches that creative scientists can develop to gain an understanding of a complex and rapidly evolving field. There are several chapters that provide a theoretical basis for the analysis and interpretation of data, several chapters on the application of modeling techniques at several different levels, and chapters on structural biology approaches, classical biochemical techniques juxtaposed to protein engineering, molecular biology, gene targeting strategies addressing physiological questions, and DNA array approaches to evaluating the effects of pathway activation. In all likelihood, the G protein signaling field will continue to be one that moves at the forefront of scientific approaches to studying events at the interface between biochemistry and molecular biology, on the one hand, and physiology and cell biology, on the other. Thus, it is our hope that these volumes will serve a scientific readership beyond those that study G proteins per se, or even those that study cell signaling mechanisms. We would hope that the approaches and techniques described here would hold relevance for those large number of scientists involved in many different kinds of projects that address the interface between the molecular/biochemical world and the cell/tissue/organism world.

We owe a tremendous debt of gratitude to our colleagues who so readily contributed chapters to these volumes. Truly, without their so willing participation this work would not have evolved into as substantial and comprehensive a work as it ultimately became. We would also like to thank Ms. Shirley Light for her support, encouragement, and patience throughout this long process.

JOHN D. HILDEBRANDT
RAVI IYENGAR

METHODS IN ENZYMOLOGY

VOLUME I. Preparation and Assay of Enzymes
Edited by SIDNEY P. COLOWICK AND NATHAN O. KAPLAN

VOLUME II. Preparation and Assay of Enzymes
Edited by SIDNEY P. COLOWICK AND NATHAN O. KAPLAN

VOLUME III. Preparation and Assay of Substrates
Edited by SIDNEY P. COLOWICK AND NATHAN O. KAPLAN

VOLUME IV. Special Techniques for the Enzymologist
Edited by SIDNEY P. COLOWICK AND NATHAN O. KAPLAN

VOLUME V. Preparation and Assay of Enzymes
Edited by SIDNEY P. COLOWICK AND NATHAN O. KAPLAN

VOLUME VI. Preparation and Assay of Enzymes (*Continued*)
Preparation and Assay of Substrates
Special Techniques
Edited by SIDNEY P. COLOWICK AND NATHAN O. KAPLAN

VOLUME VII. Cumulative Subject Index
Edited by SIDNEY P. COLOWICK AND NATHAN O. KAPLAN

VOLUME VIII. Complex Carbohydrates
Edited by ELIZABETH F. NEUFELD AND VICTOR GINSBURG

VOLUME IX. Carbohydrate Metabolism
Edited by WILLIS A. WOOD

VOLUME X. Oxidation and Phosphorylation
Edited by RONALD W. ESTABROOK AND MAYNARD E. PULLMAN

VOLUME XI. Enzyme Structure
Edited by C. H. W. HIRS

VOLUME XII. Nucleic Acids (Parts A and B)
Edited by LAWRENCE GROSSMAN AND KIVIE MOLDAVE

VOLUME XIII. Citric Acid Cycle
Edited by J. M. LOWENSTEIN

VOLUME XIV. Lipids
Edited by J. M. LOWENSTEIN

VOLUME XV. Steroids and Terpenoids
Edited by RAYMOND B. CLAYTON

VOLUME XVI. Fast Reactions
Edited by KENNETH KUSTIN

VOLUME XVII. Metabolism of Amino Acids and Amines (Parts A and B)
Edited by HERBERT TABOR AND CELIA WHITE TABOR

VOLUME XVIII. Vitamins and Coenzymes (Parts A, B, and C)
Edited by DONALD B. MCCORMICK AND LEMUEL D. WRIGHT

VOLUME XIX. Proteolytic Enzymes
Edited by GERTRUDE E. PERLMANN AND LASZLO LORAND

VOLUME XX. Nucleic Acids and Protein Synthesis (Part C)
Edited by KIVIE MOLDAVE AND LAWRENCE GROSSMAN

VOLUME XXI. Nucleic Acids (Part D)
Edited by LAWRENCE GROSSMAN AND KIVIE MOLDAVE

VOLUME XXII. Enzyme Purification and Related Techniques
Edited by WILLIAM B. JAKOBY

VOLUME XXIII. Photosynthesis (Part A)
Edited by ANTHONY SAN PIETRO

VOLUME XXIV. Photosynthesis and Nitrogen Fixation (Part B)
Edited by ANTHONY SAN PIETRO

VOLUME XXV. Enzyme Structure (Part B)
Edited by C. H. W. HIRS AND SERGE N. TIMASHEFF

VOLUME XXVI. Enzyme Structure (Part C)
Edited by C. H. W. HIRS AND SERGE N. TIMASHEFF

VOLUME XXVII. Enzyme Structure (Part D)
Edited by C. H. W. HIRS AND SERGE N. TIMASHEFF

VOLUME XXVIII. Complex Carbohydrates (Part B)
Edited by VICTOR GINSBURG

VOLUME XXIX. Nucleic Acids and Protein Synthesis (Part E)
Edited by LAWRENCE GROSSMAN AND KIVIE MOLDAVE

VOLUME XXX. Nucleic Acids and Protein Synthesis (Part F)
Edited by KIVIE MOLDAVE AND LAWRENCE GROSSMAN

VOLUME XXXI. Biomembranes (Part A)
Edited by SIDNEY FLEISCHER AND LESTER PACKER

VOLUME XXXII. Biomembranes (Part B)
Edited by SIDNEY FLEISCHER AND LESTER PACKER

VOLUME XXXIII. Cumulative Subject Index Volumes I-XXX
Edited by MARTHA G. DENNIS AND EDWARD A. DENNIS

VOLUME XXXIV. Affinity Techniques (Enzyme Purification: Part B)
Edited by WILLIAM B. JAKOBY AND MEIR WILCHEK

VOLUME XXXV. Lipids (Part B)
Edited by JOHN M. LOWENSTEIN

VOLUME XXXVI. Hormone Action (Part A: Steroid Hormones)
Edited by BERT W. O'MALLEY AND JOEL G. HARDMAN

VOLUME XXXVII. Hormone Action (Part B: Peptide Hormones)
Edited by BERT W. O'MALLEY AND JOEL G. HARDMAN

VOLUME XXXVIII. Hormone Action (Part C: Cyclic Nucleotides)
Edited by JOEL G. HARDMAN AND BERT W. O'MALLEY

VOLUME XXXIX. Hormone Action (Part D: Isolated Cells, Tissues, and Organ Systems)
Edited by JOEL G. HARDMAN AND BERT W. O'MALLEY

VOLUME XL. Hormone Action (Part E: Nuclear Structure and Function)
Edited by BERT W. O'MALLEY AND JOEL G. HARDMAN

VOLUME XLI. Carbohydrate Metabolism (Part B)
Edited by W. A. WOOD

VOLUME XLII. Carbohydrate Metabolism (Part C)
Edited by W. A. WOOD

VOLUME XLIII. Antibiotics
Edited by JOHN H. HASH

VOLUME XLIV. Immobilized Enzymes
Edited by KLAUS MOSBACH

VOLUME XLV. Proteolytic Enzymes (Part B)
Edited by LASZLO LORAND

VOLUME XLVI. Affinity Labeling
Edited by WILLIAM B. JAKOBY AND MEIR WILCHEK

VOLUME XLVII. Enzyme Structure (Part E)
Edited by C. H. W. HIRS AND SERGE N. TIMASHEFF

VOLUME XLVIII. Enzyme Structure (Part F)
Edited by C. H. W. HIRS AND SERGE N. TIMASHEFF

VOLUME XLIX. Enzyme Structure (Part G)
Edited by C. H. W. HIRS AND SERGE N. TIMASHEFF

VOLUME L. Complex Carbohydrates (Part C)
Edited by VICTOR GINSBURG

VOLUME LI. Purine and Pyrimidine Nucleotide Metabolism
Edited by PATRICIA A. HOFFEE AND MARY ELLEN JONES

VOLUME LII. Biomembranes (Part C: Biological Oxidations)
Edited by SIDNEY FLEISCHER AND LESTER PACKER

VOLUME LIII. Biomembranes (Part D: Biological Oxidations)
Edited by SIDNEY FLEISCHER AND LESTER PACKER

VOLUME LIV. Biomembranes (Part E: Biological Oxidations)
Edited by SIDNEY FLEISCHER AND LESTER PACKER

VOLUME LV. Biomembranes (Part F: Bioenergetics)
Edited by SIDNEY FLEISCHER AND LESTER PACKER

VOLUME LVI. Biomembranes (Part G: Bioenergetics)
Edited by SIDNEY FLEISCHER AND LESTER PACKER

VOLUME LVII. Bioluminescence and Chemiluminescence
Edited by MARLENE A. DELUCA

VOLUME LVIII. Cell Culture
Edited by WILLIAM B. JAKOBY AND IRA PASTAN

VOLUME LIX. Nucleic Acids and Protein Synthesis (Part G)
Edited by KIVIE MOLDAVE AND LAWRENCE GROSSMAN

VOLUME LX. Nucleic Acids and Protein Synthesis (Part H)
Edited by KIVIE MOLDAVE AND LAWRENCE GROSSMAN

VOLUME 61. Enzyme Structure (Part H)
Edited by C. H. W. HIRS AND SERGE N. TIMASHEFF

VOLUME 62. Vitamins and Coenzymes (Part D)
Edited by DONALD B. MCCORMICK AND LEMUEL D. WRIGHT

VOLUME 63. Enzyme Kinetics and Mechanism (Part A: Initial Rate and Inhibitor Methods)
Edited by DANIEL L. PURICH

VOLUME 64. Enzyme Kinetics and Mechanism (Part B: Isotopic Probes and Complex Enzyme Systems)
Edited by DANIEL L. PURICH

VOLUME 65. Nucleic Acids (Part I)
Edited by LAWRENCE GROSSMAN AND KIVIE MOLDAVE

VOLUME 66. Vitamins and Coenzymes (Part E)
Edited by DONALD B. MCCORMICK AND LEMUEL D. WRIGHT

VOLUME 67. Vitamins and Coenzymes (Part F)
Edited by DONALD B. MCCORMICK AND LEMUEL D. WRIGHT

VOLUME 68. Recombinant DNA
Edited by RAY WU

VOLUME 69. Photosynthesis and Nitrogen Fixation (Part C)
Edited by ANTHONY SAN PIETRO

VOLUME 70. Immunochemical Techniques (Part A)
Edited by HELEN VAN VUNAKIS AND JOHN J. LANGONE

VOLUME 71. Lipids (Part C)
Edited by JOHN M. LOWENSTEIN

VOLUME 72. Lipids (Part D)
Edited by JOHN M. LOWENSTEIN

VOLUME 73. Immunochemical Techniques (Part B)
Edited by JOHN J. LANGONE AND HELEN VAN VUNAKIS

VOLUME 74. Immunochemical Techniques (Part C)
Edited by JOHN J. LANGONE AND HELEN VAN VUNAKIS

VOLUME 75. Cumulative Subject Index Volumes XXXI, XXXII, XXXIV–LX
Edited by EDWARD A. DENNIS AND MARTHA G. DENNIS

VOLUME 76. Hemoglobins
Edited by ERALDO ANTONINI, LUIGI ROSSI-BERNARDI, AND EMILIA CHIANCONE

VOLUME 77. Detoxication and Drug Metabolism
Edited by WILLIAM B. JAKOBY

VOLUME 78. Interferons (Part A)
Edited by SIDNEY PESTKA

VOLUME 79. Interferons (Part B)
Edited by SIDNEY PESTKA

VOLUME 80. Proteolytic Enzymes (Part C)
Edited by LASZLO LORAND

VOLUME 81. Biomembranes (Part H: Visual Pigments and Purple Membranes, I)
Edited by LESTER PACKER

VOLUME 82. Structural and Contractile Proteins (Part A: Extracellular Matrix)
Edited by LEON W. CUNNINGHAM AND DIXIE W. FREDERIKSEN

VOLUME 83. Complex Carbohydrates (Part D)
Edited by VICTOR GINSBURG

VOLUME 84. Immunochemical Techniques (Part D: Selected Immunoassays)
Edited by JOHN J. LANGONE AND HELEN VAN VUNAKIS

VOLUME 85. Structural and Contractile Proteins (Part B: The Contractile Apparatus and the Cytoskeleton)
Edited by DIXIE W. FREDERIKSEN AND LEON W. CUNNINGHAM

VOLUME 86. Prostaglandins and Arachidonate Metabolites
Edited by WILLIAM E. M. LANDS AND WILLIAM L. SMITH

VOLUME 87. Enzyme Kinetics and Mechanism (Part C: Intermediates, Stereochemistry, and Rate Studies)
Edited by DANIEL L. PURICH

VOLUME 88. Biomembranes (Part I: Visual Pigments and Purple Membranes, II)
Edited by LESTER PACKER

VOLUME 89. Carbohydrate Metabolism (Part D)
Edited by WILLIS A. WOOD

VOLUME 90. Carbohydrate Metabolism (Part E)
Edited by WILLIS A. WOOD

VOLUME 91. Enzyme Structure (Part I)
Edited by C. H. W. HIRS AND SERGE N. TIMASHEFF

VOLUME 92. Immunochemical Techniques (Part E: Monoclonal Antibodies and General Immunoassay Methods)
Edited by JOHN J. LANGONE AND HELEN VAN VUNAKIS

VOLUME 93. Immunochemical Techniques (Part F: Conventional Antibodies, Fc Receptors, and Cytotoxicity)
Edited by JOHN J. LANGONE AND HELEN VAN VUNAKIS

VOLUME 94. Polyamines
Edited by HERBERT TABOR AND CELIA WHITE TABOR

VOLUME 95. Cumulative Subject Index Volumes 61–74, 76–80
Edited by EDWARD A. DENNIS AND MARTHA G. DENNIS

VOLUME 96. Biomembranes [Part J: Membrane Biogenesis: Assembly and Targeting (General Methods; Eukaryotes)]
Edited by SIDNEY FLEISCHER AND BECCA FLEISCHER

VOLUME 97. Biomembranes [Part K: Membrane Biogenesis: Assembly and Targeting (Prokaryotes, Mitochondria, and Chloroplasts)]
Edited by SIDNEY FLEISCHER AND BECCA FLEISCHER

VOLUME 98. Biomembranes (Part L: Membrane Biogenesis: Processing and Recycling)
Edited by SIDNEY FLEISCHER AND BECCA FLEISCHER

VOLUME 99. Hormone Action (Part F: Protein Kinases)
Edited by JACKIE D. CORBIN AND JOEL G. HARDMAN

VOLUME 100. Recombinant DNA (Part B)
Edited by RAY WU, LAWRENCE GROSSMAN, AND KIVIE MOLDAVE

VOLUME 101. Recombinant DNA (Part C)
Edited by RAY WU, LAWRENCE GROSSMAN, AND KIVIE MOLDAVE

VOLUME 102. Hormone Action (Part G: Calmodulin and Calcium-Binding Proteins)
Edited by ANTHONY R. MEANS AND BERT W. O'MALLEY

VOLUME 103. Hormone Action (Part H: Neuroendocrine Peptides)
Edited by P. MICHAEL CONN

VOLUME 104. Enzyme Purification and Related Techniques (Part C)
Edited by WILLIAM B. JAKOBY

VOLUME 105. Oxygen Radicals in Biological Systems
Edited by LESTER PACKER

VOLUME 106. Posttranslational Modifications (Part A)
Edited by FINN WOLD AND KIVIE MOLDAVE

VOLUME 107. Posttranslational Modifications (Part B)
Edited by FINN WOLD AND KIVIE MOLDAVE

VOLUME 108. Immunochemical Techniques (Part G: Separation and Characterization of Lymphoid Cells)
Edited by GIOVANNI DI SABATO, JOHN J. LANGONE, AND HELEN VAN VUNAKIS

VOLUME 109. Hormone Action (Part I: Peptide Hormones)
Edited by LUTZ BIRNBAUMER AND BERT W. O'MALLEY

VOLUME 110. Steroids and Isoprenoids (Part A)
Edited by JOHN H. LAW AND HANS C. RILLING

VOLUME 111. Steroids and Isoprenoids (Part B)
Edited by JOHN H. LAW AND HANS C. RILLING

VOLUME 112. Drug and Enzyme Targeting (Part A)
Edited by KENNETH J. WIDDER AND RALPH GREEN

VOLUME 113. Glutamate, Glutamine, Glutathione, and Related Compounds
Edited by ALTON MEISTER

VOLUME 114. Diffraction Methods for Biological Macromolecules (Part A)
Edited by HAROLD W. WYCKOFF, C. H. W. HIRS, AND SERGE N. TIMASHEFF

VOLUME 115. Diffraction Methods for Biological Macromolecules (Part B)
Edited by HAROLD W. WYCKOFF, C. H. W. HIRS, AND SERGE N. TIMASHEFF

VOLUME 116. Immunochemical Techniques (Part H: Effectors and Mediators of Lymphoid Cell Functions)
Edited by GIOVANNI DI SABATO, JOHN J. LANGONE, AND HELEN VAN VUNAKIS

VOLUME 117. Enzyme Structure (Part J)
Edited by C. H. W. HIRS AND SERGE N. TIMASHEFF

VOLUME 118. Plant Molecular Biology
Edited by ARTHUR WEISSBACH AND HERBERT WEISSBACH

VOLUME 119. Interferons (Part C)
Edited by SIDNEY PESTKA

VOLUME 120. Cumulative Subject Index Volumes 81–94, 96–101

VOLUME 121. Immunochemical Techniques (Part I: Hybridoma Technology and Monoclonal Antibodies)
Edited by JOHN J. LANGONE AND HELEN VAN VUNAKIS

VOLUME 122. Vitamins and Coenzymes (Part G)
Edited by FRANK CHYTIL AND DONALD B. MCCORMICK

VOLUME 123. Vitamins and Coenzymes (Part H)
Edited by FRANK CHYTIL AND DONALD B. MCCORMICK

VOLUME 124. Hormone Action (Part J: Neuroendocrine Peptides)
Edited by P. MICHAEL CONN

VOLUME 125. Biomembranes (Part M: Transport in Bacteria, Mitochondria, and Chloroplasts: General Approaches and Transport Systems)
Edited by SIDNEY FLEISCHER AND BECCA FLEISCHER

VOLUME 126. Biomembranes (Part N: Transport in Bacteria, Mitochondria, and Chloroplasts: Protonmotive Force)
Edited by SIDNEY FLEISCHER AND BECCA FLEISCHER

VOLUME 127. Biomembranes (Part O: Protons and Water: Structure and Translocation)
Edited by LESTER PACKER

VOLUME 128. Plasma Lipoproteins (Part A: Preparation, Structure, and Molecular Biology)
Edited by JERE P. SEGREST AND JOHN J. ALBERS

VOLUME 129. Plasma Lipoproteins (Part B: Characterization, Cell Biology, and Metabolism)
Edited by JOHN J. ALBERS AND JERE P. SEGREST

VOLUME 130. Enzyme Structure (Part K)
Edited by C. H. W. HIRS AND SERGE N. TIMASHEFF

VOLUME 131. Enzyme Structure (Part L)
Edited by C. H. W. HIRS AND SERGE N. TIMASHEFF

VOLUME 132. Immunochemical Techniques (Part J: Phagocytosis and Cell-Mediated Cytotoxicity)
Edited by GIOVANNI DI SABATO AND JOHANNES EVERSE

VOLUME 133. Bioluminescence and Chemiluminescence (Part B)
Edited by MARLENE DELUCA AND WILLIAM D. MCELROY

VOLUME 134. Structural and Contractile Proteins (Part C: The Contractile Apparatus and the Cytoskeleton)
Edited by RICHARD B. VALLEE

VOLUME 135. Immobilized Enzymes and Cells (Part B)
Edited by KLAUS MOSBACH

VOLUME 136. Immobilized Enzymes and Cells (Part C)
Edited by KLAUS MOSBACH

VOLUME 137. Immobilized Enzymes and Cells (Part D)
Edited by KLAUS MOSBACH

VOLUME 138. Complex Carbohydrates (Part E)
Edited by VICTOR GINSBURG

VOLUME 139. Cellular Regulators (Part A: Calcium- and Calmodulin-Binding Proteins)
Edited by ANTHONY R. MEANS AND P. MICHAEL CONN

VOLUME 140. Cumulative Subject Index Volumes 102–119, 121–134

VOLUME 141. Cellular Regulators (Part B: Calcium and Lipids)
Edited by P. MICHAEL CONN AND ANTHONY R. MEANS

VOLUME 142. Metabolism of Aromatic Amino Acids and Amines
Edited by SEYMOUR KAUFMAN

VOLUME 143. Sulfur and Sulfur Amino Acids
Edited by WILLIAM B. JAKOBY AND OWEN GRIFFITH

VOLUME 144. Structural and Contractile Proteins (Part D: Extracellular Matrix)
Edited by LEON W. CUNNINGHAM

VOLUME 145. Structural and Contractile Proteins (Part E: Extracellular Matrix)
Edited by LEON W. CUNNINGHAM

VOLUME 146. Peptide Growth Factors (Part A)
Edited by DAVID BARNES AND DAVID A. SIRBASKU

VOLUME 147. Peptide Growth Factors (Part B)
Edited by DAVID BARNES AND DAVID A. SIRBASKU

VOLUME 148. Plant Cell Membranes
Edited by LESTER PACKER AND ROLAND DOUCE

VOLUME 149. Drug and Enzyme Targeting (Part B)
Edited by RALPH GREEN AND KENNETH J. WIDDER

VOLUME 150. Immunochemical Techniques (Part K: *In Vitro* Models of B and T
Cell Functions and Lymphoid Cell Receptors)
Edited by GIOVANNI DI SABATO

VOLUME 151. Molecular Genetics of Mammalian Cells
Edited by MICHAEL M. GOTTESMAN

VOLUME 152. Guide to Molecular Cloning Techniques
Edited by SHELBY L. BERGER AND ALAN R. KIMMEL

VOLUME 153. Recombinant DNA (Part D)
Edited by RAY WU AND LAWRENCE GROSSMAN

VOLUME 154. Recombinant DNA (Part E)
Edited by RAY WU AND LAWRENCE GROSSMAN

VOLUME 155. Recombinant DNA (Part F)
Edited by RAY WU

VOLUME 156. Biomembranes (Part P: ATP-Driven Pumps and Related Transport:
The Na, K-Pump)
Edited by SIDNEY FLEISCHER AND BECCA FLEISCHER

VOLUME 157. Biomembranes (Part Q: ATP-Driven Pumps and Related Transport:
Calcium, Proton, and Potassium Pumps)
Edited by SIDNEY FLEISCHER AND BECCA FLEISCHER

VOLUME 158. Metalloproteins (Part A)
Edited by JAMES F. RIORDAN AND BERT L. VALLEE

VOLUME 159. Initiation and Termination of Cyclic Nucleotide Action
Edited by JACKIE D. CORBIN AND ROGER A. JOHNSON

VOLUME 160. Biomass (Part A: Cellulose and Hemicellulose)
Edited by WILLIS A. WOOD AND SCOTT T. KELLOGG

VOLUME 161. Biomass (Part B: Lignin, Pectin, and Chitin)
Edited by WILLIS A. WOOD AND SCOTT T. KELLOGG

VOLUME 162. Immunochemical Techniques (Part L: Chemotaxis and Inflammation)
Edited by GIOVANNI DI SABATO

VOLUME 163. Immunochemical Techniques (Part M: Chemotaxis and Inflammation)
Edited by GIOVANNI DI SABATO

VOLUME 164. Ribosomes
Edited by HARRY F. NOLLER, JR., AND KIVIE MOLDAVE

VOLUME 165. Microbial Toxins: Tools for Enzymology
Edited by SIDNEY HARSHMAN

VOLUME 166. Branched-Chain Amino Acids
Edited by ROBERT HARRIS AND JOHN R. SOKATCH

VOLUME 167. Cyanobacteria
Edited by LESTER PACKER AND ALEXANDER N. GLAZER

VOLUME 168. Hormone Action (Part K: Neuroendocrine Peptides)
Edited by P. MICHAEL CONN

VOLUME 169. Platelets: Receptors, Adhesion, Secretion (Part A)
Edited by JACEK HAWIGER

VOLUME 170. Nucleosomes
Edited by PAUL M. WASSARMAN AND ROGER D. KORNBERG

VOLUME 171. Biomembranes (Part R: Transport Theory: Cells and Model Membranes)
Edited by SIDNEY FLEISCHER AND BECCA FLEISCHER

VOLUME 172. Biomembranes (Part S: Transport: Membrane Isolation and Characterization)
Edited by SIDNEY FLEISCHER AND BECCA FLEISCHER

VOLUME 173. Biomembranes [Part T: Cellular and Subcellular Transport: Eukaryotic (Nonepithelial) Cells]
Edited by SIDNEY FLEISCHER AND BECCA FLEISCHER

VOLUME 174. Biomembranes [Part U: Cellular and Subcellular Transport: Eukaryotic (Nonepithelial) Cells]
Edited by SIDNEY FLEISCHER AND BECCA FLEISCHER

VOLUME 175. Cumulative Subject Index Volumes 135–139, 141–167

VOLUME 176. Nuclear Magnetic Resonance (Part A: Spectral Techniques and Dynamics)
Edited by NORMAN J. OPPENHEIMER AND THOMAS L. JAMES

VOLUME 177. Nuclear Magnetic Resonance (Part B: Structure and Mechanism)
Edited by NORMAN J. OPPENHEIMER AND THOMAS L. JAMES

VOLUME 178. Antibodies, Antigens, and Molecular Mimicry
Edited by JOHN J. LANGONE

VOLUME 179. Complex Carbohydrates (Part F)
Edited by VICTOR GINSBURG

VOLUME 180. RNA Processing (Part A: General Methods)
Edited by JAMES E. DAHLBERG AND JOHN N. ABELSON

VOLUME 181. RNA Processing (Part B: Specific Methods)
Edited by JAMES E. DAHLBERG AND JOHN N. ABELSON

VOLUME 182. Guide to Protein Purification
Edited by MURRAY P. DEUTSCHER

VOLUME 183. Molecular Evolution: Computer Analysis of Protein and Nucleic
Acid Sequences
Edited by RUSSELL F. DOOLITTLE

VOLUME 184. Avidin-Biotin Technology
Edited by MEIR WILCHEK AND EDWARD A. BAYER

VOLUME 185. Gene Expression Technology
Edited by DAVID V. GOEDDEL

VOLUME 186. Oxygen Radicals in Biological Systems (Part B: Oxygen Radicals
and Antioxidants)
Edited by LESTER PACKER AND ALEXANDER N. GLAZER

VOLUME 187. Arachidonate Related Lipid Mediators
Edited by ROBERT C. MURPHY AND FRANK A. FITZPATRICK

VOLUME 188. Hydrocarbons and Methylotrophy
Edited by MARY E. LIDSTROM

VOLUME 189. Retinoids (Part A: Molecular and Metabolic Aspects)
Edited by LESTER PACKER

VOLUME 190. Retinoids (Part B: Cell Differentiation and Clinical Applications)
Edited by LESTER PACKER

VOLUME 191. Biomembranes (Part V: Cellular and Subcellular Transport:
Epithelial Cells)
Edited by SIDNEY FLEISCHER AND BECCA FLEISCHER

VOLUME 192. Biomembranes (Part W: Cellular and Subcellular Transport:
Epithelial Cells)
Edited by SIDNEY FLEISCHER AND BECCA FLEISCHER

VOLUME 193. Mass Spectrometry
Edited by JAMES A. MCCLOSKEY

VOLUME 194. Guide to Yeast Genetics and Molecular Biology
Edited by CHRISTINE GUTHRIE AND GERALD R. FINK

VOLUME 195. Adenylyl Cyclase, G Proteins, and Guanylyl Cyclase
Edited by ROGER A. JOHNSON AND JACKIE D. CORBIN

VOLUME 196. Molecular Motors and the Cytoskeleton
Edited by RICHARD B. VALLEE

VOLUME 197. Phospholipases
Edited by EDWARD A. DENNIS

VOLUME 198. Peptide Growth Factors (Part C)
Edited by DAVID BARNES, J. P. MATHER, AND GORDON H. SATO

VOLUME 199. Cumulative Subject Index Volumes 168–174, 176–194

VOLUME 200. Protein Phosphorylation (Part A: Protein Kinases: Assays,
Purification, Antibodies, Functional Analysis, Cloning, and Expression)
Edited by TONY HUNTER AND BARTHOLOMEW M. SEFTON

VOLUME 201. Protein Phosphorylation (Part B: Analysis of Protein
Phosphorylation, Protein Kinase Inhibitors, and Protein Phosphatases)
Edited by TONY HUNTER AND BARTHOLOMEW M. SEFTON

VOLUME 202. Molecular Design and Modeling: Concepts and Applications
(Part A: Proteins, Peptides, and Enzymes)
Edited by JOHN J. LANGONE

VOLUME 203. Molecular Design and Modeling: Concepts and Applications
(Part B: Antibodies and Antigens, Nucleic Acids, Polysaccharides, and Drugs)
Edited by JOHN J. LANGONE

VOLUME 204. Bacterial Genetic Systems
Edited by JEFFREY H. MILLER

VOLUME 205. Metallobiochemistry (Part B: Metallothionein and Related
Molecules)
Edited by JAMES F. RIORDAN AND BERT L. VALLEE

VOLUME 206. Cytochrome P450
Edited by MICHAEL R. WATERMAN AND ERIC F. JOHNSON

VOLUME 207. Ion Channels
Edited by BERNARDO RUDY AND LINDA E. IVERSON

VOLUME 208. Protein–DNA Interactions
Edited by ROBERT T. SAUER

VOLUME 209. Phospholipid Biosynthesis
Edited by EDWARD A. DENNIS AND DENNIS E. VANCE

VOLUME 210. Numerical Computer Methods
Edited by LUDWIG BRAND AND MICHAEL L. JOHNSON

VOLUME 211. DNA Structures (Part A: Synthesis and Physical Analysis of DNA)
Edited by DAVID M. J. LILLEY AND JAMES E. DAHLBERG

VOLUME 212. DNA Structures (Part B: Chemical and Electrophoretic Analysis of
DNA)
Edited by DAVID M. J. LILLEY AND JAMES E. DAHLBERG

VOLUME 213. Carotenoids (Part A: Chemistry, Separation, Quantitation, and Antioxidation)
Edited by LESTER PACKER

VOLUME 214. Carotenoids (Part B: Metabolism, Genetics, and Biosynthesis)
Edited by LESTER PACKER

VOLUME 215. Platelets: Receptors, Adhesion, Secretion (Part B)
Edited by JACEK J. HAWIGER

VOLUME 216. Recombinant DNA (Part G)
Edited by RAY WU

VOLUME 217. Recombinant DNA (Part H)
Edited by RAY WU

VOLUME 218. Recombinant DNA (Part I)
Edited by RAY WU

VOLUME 219. Reconstitution of Intracellular Transport
Edited by JAMES E. ROTHMAN

VOLUME 220. Membrane Fusion Techniques (Part A)
Edited by NEJAT DÜZGUÜNES

VOLUME 221. Membrane Fusion Techniques (Part B)
Edited by NEJAT DÜZGÜNES

VOLUME 222. Proteolytic Enzymes in Coagulation, Fibrinolysis, and Complement Activation (Part A: Mammalian Blood Coagulation Factors and Inhibitors)
Edited by LASZLO LORAND AND KENNETH G. MANN

VOLUME 223. Proteolytic Enzymes in Coagulation, Fibrinolysis, and Complement Activation (Part B: Complement Activation, Fibrinolysis, and Nonmammalian Blood Coagulation Factors)
Edited by LASZLO LORAND AND KENNETH G. MANN

VOLUME 224. Molecular Evolution: Producing the Biochemical Data
Edited by ELIZABETH ANNE ZIMMER, THOMAS J. WHITE, REBECCA L. CANN, AND ALLAN C. WILSON

VOLUME 225. Guide to Techniques in Mouse Development
Edited by PAUL M. WASSARMAN AND MELVIN L. DEPAMPHILIS

VOLUME 226. Metallobiochemistry (Part C: Spectroscopic and Physical Methods for Probing Metal Ion Environments in Metalloenzymes and Metalloproteins)
Edited by JAMES F. RIORDAN AND BERT L. VALLEE

VOLUME 227. Metallobiochemistry (Part D: Physical and Spectroscopic Methods for Probing Metal Ion Environments in Metalloproteins)
Edited by JAMES F. RIORDAN AND BERT L. VALLEE

VOLUME 228. Aqueous Two-Phase Systems
Edited by HARRY WALTER AND GÖTE JOHANSSON

VOLUME 229. Cumulative Subject Index Volumes 195–198, 200–227

VOLUME 230. Guide to Techniques in Glycobiology
Edited by WILLIAM J. LENNARZ AND GERALD W. HART

VOLUME 231. Hemoglobins (Part B: Biochemical and Analytical Methods)
Edited by JOHANNES EVERSE, KIM D. VANDEGRIFF, AND ROBERT M. WINSLOW

VOLUME 232. Hemoglobins (Part C: Biophysical Methods)
Edited by JOHANNES EVERSE, KIM D. VANDEGRIFF, AND ROBERT M. WINSLOW

VOLUME 233. Oxygen Radicals in Biological Systems (Part C)
Edited by LESTER PACKER

VOLUME 234. Oxygen Radicals in Biological Systems (Part D)
Edited by LESTER PACKER

VOLUME 235. Bacterial Pathogenesis (Part A: Identification and Regulation of Virulence Factors)
Edited by VIRGINIA L. CLARK AND PATRIK M. BAVOIL

VOLUME 236. Bacterial Pathogenesis (Part B: Integration of Pathogenic Bacteria with Host Cells)
Edited by VIRGINIA L. CLARK AND PATRIK M. BAVOIL

VOLUME 237. Heterotrimeric G Proteins
Edited by RAVI IYENGAR

VOLUME 238. Heterotrimeric G-Protein Effectors
Edited by RAVI IYENGAR

VOLUME 239. Nuclear Magnetic Resonance (Part C)
Edited by THOMAS L. JAMES AND NORMAN J. OPPENHEIMER

VOLUME 240. Numerical Computer Methods (Part B)
Edited by MICHAEL L. JOHNSON AND LUDWIG BRAND

VOLUME 241. Retroviral Proteases
Edited by LAWRENCE C. KUO AND JULES A. SHAFER

VOLUME 242. Neoglycoconjugates (Part A)
Edited by Y. C. LEE AND REIKO T. LEE

VOLUME 243. Inorganic Microbial Sulfur Metabolism
Edited by HARRY D. PECK, JR., AND JEAN LeGALL

VOLUME 244. Proteolytic Enzymes: Serine and Cysteine Peptidases
Edited by ALAN J. BARRETT

VOLUME 245. Extracellular Matrix Components
Edited by E. RUOSLAHTI AND E. ENGVALL

VOLUME 246. Biochemical Spectroscopy
Edited by KENNETH SAUER

VOLUME 247. Neoglycoconjugates (Part B: Biomedical Applications)
Edited by Y. C. LEE AND REIKO T. LEE

VOLUME 248. Proteolytic Enzymes: Aspartic and Metallo Peptidases
Edited by ALAN J. BARRETT

VOLUME 249. Enzyme Kinetics and Mechanism (Part D: Developments in Enzyme Dynamics)
Edited by DANIEL L. PURICH

VOLUME 250. Lipid Modifications of Proteins
Edited by PATRICK J. CASEY AND JANICE E. BUSS

VOLUME 251. Biothiols (Part A: Monothiols and Dithiols, Protein Thiols, and Thiyl Radicals)
Edited by LESTER PACKER

VOLUME 252. Biothiols (Part B: Glutathione and Thioredoxin; Thiols in Signal Transduction and Gene Regulation)
Edited by LESTER PACKER

VOLUME 253. Adhesion of Microbial Pathogens
Edited by RON J. DOYLE AND ITZHAK OFEK

VOLUME 254. Oncogene Techniques
Edited by PETER K. VOGT AND INDER M. VERMA

VOLUME 255. Small GTPases and Their Regulators (Part A: Ras Family)
Edited by W. E. BALCH, CHANNING J. DER, AND ALAN HALL

VOLUME 256. Small GTPases and Their Regulators (Part B: Rho Family)
Edited by W. E. BALCH, CHANNING J. DER, AND ALAN HALL

VOLUME 257. Small GTPases and Their Regulators (Part C: Proteins Involved in Transport)
Edited by W. E. BALCH, CHANNING J. DER, AND ALAN HALL

VOLUME 258. Redox-Active Amino Acids in Biology
Edited by JUDITH P. KLINMAN

VOLUME 259. Energetics of Biological Macromolecules
Edited by MICHAEL L. JOHNSON AND GARY K. ACKERS

VOLUME 260. Mitochondrial Biogenesis and Genetics (Part A)
Edited by GIUSEPPE M. ATTARDI AND ANNE CHOMYN

VOLUME 261. Nuclear Magnetic Resonance and Nucleic Acids
Edited by THOMAS L. JAMES

VOLUME 262. DNA Replication
Edited by JUDITH L. CAMPBELL

VOLUME 263. Plasma Lipoproteins (Part C: Quantitation)
Edited by WILLIAM A. BRADLEY, SANDRA H. GIANTURCO, AND JERE P. SEGREST

VOLUME 264. Mitochondrial Biogenesis and Genetics (Part B)
Edited by GIUSEPPE M. ATTARDI AND ANNE CHOMYN

VOLUME 265. Cumulative Subject Index Volumes 228, 230–262

VOLUME 266. Computer Methods for Macromolecular Sequence Analysis
Edited by RUSSELL F. DOOLITTLE

VOLUME 267. Combinatorial Chemistry
Edited by JOHN N. ABELSON

VOLUME 268. Nitric Oxide (Part A: Sources and Detection of NO; NO Synthase)
Edited by LESTER PACKER

VOLUME 269. Nitric Oxide (Part B: Physiological and Pathological Processes)
Edited by LESTER PACKER

VOLUME 270. High Resolution Separation and Analysis of Biological Macro-molecules (Part A: Fundamentals)
Edited by BARRY L. KARGER AND WILLIAM S. HANCOCK

VOLUME 271. High Resolution Separation and Analysis of Biological Macro-molecules (Part B: Applications)
Edited by BARRY L. KARGER AND WILLIAM S. HANCOCK

VOLUME 272. Cytochrome P450 (Part B)
Edited by ERIC F. JOHNSON AND MICHAEL R. WATERMAN

VOLUME 273. RNA Polymerase and Associated Factors (Part A)
Edited by SANKAR ADHYA

VOLUME 274. RNA Polymerase and Associated Factors (Part B)
Edited by SANKAR ADHYA

VOLUME 275. Viral Polymerases and Related Proteins
Edited by LAWRENCE C. KUO, DAVID B. OLSEN, AND STEVEN S. CARROLL

VOLUME 276. Macromolecular Crystallography (Part A)
Edited by CHARLES W. CARTER, JR., AND ROBERT M. SWEET

VOLUME 277. Macromolecular Crystallography (Part B)
Edited by CHARLES W. CARTER, JR., AND ROBERT M. SWEET

VOLUME 278. Fluorescence Spectroscopy
Edited by LUDWIG BRAND AND MICHAEL L. JOHNSON

VOLUME 279. Vitamins and Coenzymes (Part I)
Edited by DONALD B. MCCORMICK, JOHN W. SUTTIE, AND CONRAD WAGNER

VOLUME 280. Vitamins and Coenzymes (Part J)
Edited by DONALD B. MCCORMICK, JOHN W. SUTTIE, AND CONRAD WAGNER

VOLUME 281. Vitamins and Coenzymes (Part K)
Edited by DONALD B. MCCORMICK, JOHN W. SUTTIE, AND CONRAD WAGNER

VOLUME 282. Vitamins and Coenzymes (Part L)
Edited by DONALD B. MCCORMICK, JOHN W. SUTTIE, AND CONRAD WAGNER

VOLUME 283. Cell Cycle Control
Edited by WILLIAM G. DUNPHY

VOLUME 284. Lipases (Part A: Biotechnology)
Edited by BYRON RUBIN AND EDWARD A. DENNIS

VOLUME 285. Cumulative Subject Index Volumes 263, 264, 266–284, 286–289

VOLUME 286. Lipases (Part B: Enzyme Characterization and Utilization)
Edited by BYRON RUBIN AND EDWARD A. DENNIS

VOLUME 287. Chemokines
Edited by RICHARD HORUK

VOLUME 288. Chemokine Receptors
Edited by RICHARD HORUK

VOLUME 289. Solid Phase Peptide Synthesis
Edited by GREGG B. FIELDS

VOLUME 290. Molecular Chaperones
Edited by GEORGE H. LORIMER AND THOMAS BALDWIN

VOLUME 291. Caged Compounds
Edited by GERARD MARRIOTT

VOLUME 292. ABC Transporters: Biochemical, Cellular, and Molecular Aspects
Edited by SURESH V. AMBUDKAR AND MICHAEL M. GOTTESMAN

VOLUME 293. Ion Channels (Part B)
Edited by P. MICHAEL CONN

VOLUME 294. Ion Channels (Part C)
Edited by P. MICHAEL CONN

VOLUME 295. Energetics of Biological Macromolecules (Part B)
Edited by GARY K. ACKERS AND MICHAEL L. JOHNSON

VOLUME 296. Neurotransmitter Transporters
Edited by SUSAN G. AMARA

VOLUME 297. Photosynthesis: Molecular Biology of Energy Capture
Edited by LEE MCINTOSH

VOLUME 298. Molecular Motors and the Cytoskeleton (Part B)
Edited by RICHARD B. VALLEE

VOLUME 299. Oxidants and Antioxidants (Part A)
Edited by LESTER PACKER

VOLUME 300. Oxidants and Antioxidants (Part B)
Edited by LESTER PACKER

VOLUME 301. Nitric Oxide: Biological and Antioxidant Activities (Part C)
Edited by LESTER PACKER

VOLUME 302. Green Fluorescent Protein
Edited by P. MICHAEL CONN

VOLUME 303. cDNA Preparation and Display
Edited by SHERMAN M. WEISSMAN

VOLUME 304. Chromatin
Edited by PAUL M. WASSARMAN AND ALAN P. WOLFFE

VOLUME 305. Bioluminescence and Chemiluminescence (Part C)
Edited by THOMAS O. BALDWIN AND MIRIAM M. ZIEGLER

VOLUME 306. Expression of Recombinant Genes in Eukaryotic Systems
Edited by JOSEPH C. GLORIOSO AND MARTIN C. SCHMIDT

VOLUME 307. Confocal Microscopy
Edited by P. MICHAEL CONN

VOLUME 308. Enzyme Kinetics and Mechanism (Part E: Energetics of Enzyme Catalysis)
Edited by DANIEL L. PURICH AND VERN L. SCHRAMM

VOLUME 309. Amyloid, Prions, and Other Protein Aggregates
Edited by RONALD WETZEL

VOLUME 310. Biofilms
Edited by RON J. DOYLE

VOLUME 311. Sphingolipid Metabolism and Cell Signaling (Part A)
Edited by ALFRED H. MERRILL, JR., AND YUSUF A. HANNUN

VOLUME 312. Sphingolipid Metabolism and Cell Signaling (Part B)
Edited by ALFRED H. MERRILL, JR., AND YUSUF A. HANNUN

VOLUME 313. Antisense Technology (Part A: General Methods, Methods of Delivery, and RNA Studies)
Edited by M. IAN PHILLIPS

VOLUME 314. Antisense Technology (Part B: Applications)
Edited by M. IAN PHILLIPS

VOLUME 315. Vertebrate Phototransduction and the Visual Cycle (Part A)
Edited by KRZYSZTOF PALCZEWSKI

VOLUME 316. Vertebrate Phototransduction and the Visual Cycle (Part B)
Edited by KRZYSZTOF PALCZEWSKI

VOLUME 317. RNA–Ligand Interactions (Part A: Structural Biology Methods)
Edited by DANIEL W. CELANDER AND JOHN N. ABELSON

VOLUME 318. RNA–Ligand Interactions (Part B: Molecular Biology Methods)
Edited by DANIEL W. CELANDER AND JOHN N. ABELSON

VOLUME 319. Singlet Oxygen, UV-A, and Ozone
Edited by LESTER PACKER AND HELMUT SIES

VOLUME 320. Cumulative Subject Index Volumes 290–319

VOLUME 321. Numerical Computer Methods (Part C)
Edited by MICHAEL L. JOHNSON AND LUDWIG BRAND

VOLUME 322. Apoptosis
Edited by JOHN C. REED

VOLUME 323. Energetics of Biological Macromolecules (Part C)
Edited by MICHAEL L. JOHNSON AND GARY K. ACKERS

VOLUME 324. Branched-Chain Amino Acids (Part B)
Edited by ROBERT A. HARRIS AND JOHN R. SOKATCH

VOLUME 325. Regulators and Effectors of Small GTPases (Part D: Rho Family)
Edited by W. E. BALCH, CHANNING J. DER, AND ALAN HALL

VOLUME 326. Applications of Chimeric Genes and Hybrid Proteins (Part A: Gene
Expression and Protein Purification)
Edited by JEREMY THORNER, SCOTT D. EMR, AND JOHN N. ABELSON

VOLUME 327. Applications of Chimeric Genes and Hybrid Proteins (Part B: Cell
Biology and Physiology)
Edited by JEREMY THORNER, SCOTT D. EMR, AND JOHN N. ABELSON

VOLUME 328. Applications of Chimeric Genes and Hybrid Proteins (Part C: Protein-
Protein Interactions and Genomics)
Edited by JEREMY THORNER, SCOTT D. EMR, AND JOHN N. ABELSON

VOLUME 329. Regulators and Effectors of Small GTPases (Part E: GTPases
Involved in Vesicular Traffic)
Edited by W. E. BALCH, CHANNING J. DER, AND ALAN HALL

VOLUME 330. Hyperthermophilic Enzymes (Part A)
Edited by MICHAEL W. W. ADAMS AND ROBERT M. KELLY

VOLUME 331. Hyperthermophilic Enzymes (Part B)
Edited by MICHAEL W. W. ADAMS AND ROBERT M. KELLY

VOLUME 332. Regulators and Effectors of Small GTPases (Part F: Ras Family I)
Edited by W. E. BALCH, CHANNING J. DER, AND ALAN HALL

VOLUME 333. Regulators and Effectors of Small GTPases (Part G: Ras Family II)
Edited by W. E. BALCH, CHANNING J. DER, AND ALAN HALL

VOLUME 334. Hyperthermophilic Enzymes (Part C)
Edited by MICHAEL W. W. ADAMS AND ROBERT M. KELLY

VOLUME 335. Flavonoids and Other Polyphenols
Edited by LESTER PACKER

VOLUME 336. Microbial Growth in Biofilms (Part A: Developmental and Molecular
Biological Aspects)
Edited by RON J. DOYLE

VOLUME 337. Microbial Growth in Biofilms (Part B: Special Environments and
Physicochemical Aspects)
Edited by RON J. DOYLE

VOLUME 338. Nuclear Magnetic Resonance of Biological Macromolecules (Part A)
Edited by THOMAS L. JAMES, VOLKER DÖTSCH, AND ULI SCHMITZ

VOLUME 339. Nuclear Magnetic Resonance of Biological Macromolecules (Part B)
Edited by THOMAS L. JAMES, VOLKER DÖTSCH, AND ULI SCHMITZ

VOLUME 340. Drug–Nucleic Acid Interactions
Edited by JONATHAN B. CHAIRES AND MICHAEL J. WARING

VOLUME 341. Ribonucleases (Part A)
Edited by ALLEN W. NICHOLSON

VOLUME 342. Ribonucleases (Part B)
Edited by ALLEN W. NICHOLSON

VOLUME 343. G Protein Pathways (Part A: Receptors) (in preparation)
Edited by RAVI IYENGAR AND JOHN D. HILDEBRANDT

VOLUME 344. G Protein Pathways (Part B: G Proteins and Their Regulators) (in preparation)
Edited by RAVI IYENGAR AND JOHN D. HILDEBRANDT

VOLUME 345. G Protein Pathways (Part C: Effector Mechanisms)
Edited by RAVI IYENGAR AND JOHN D. HILDEBRANDT

VOLUME 346. Gene Therapy Methods (in preparation)
Edited by M. IAN PHILLIPS

VOLUME 347. Protein Sensors and Reactive Oxygen Species (Part A: Selenoproteins and Thioredoxin) (in preparation)
Edited by HELMUT SIES AND LESTER PACKER

VOLUME 348. Protein Sensors and Reactive Oxygen Species (Part B: Thiol Enzymes and Proteins) (in preparation)
Edited by HELMUT SIES AND LESTER PACKER

VOLUME 349. Superoxide Dismutase (in preparation)
Edited by LESTER PACKER

Section I

Modeling Intracellular Signaling Pathways

[1] Use of Kinetikit and GENESIS for Modeling Signaling Pathways

By UPINDER S. BHALLA

Introduction

Computer modeling is a way of doing experiments *in silico,* to bridge the gap between precise theories of simplified systems and the complexity of experimental biology. Modeling harnesses computer power to scale theoretical calculations up from equations describing individual reactions, to more lifelike systems involving thousands of interacting molecules. The start and end point of this process is experiment: from data to predictions to further data. The ideal of the modeling process is to gain insights into system function along the way.

Modeling has tended to be the preserve of the mathematician or computer scientist, but its greatest value is in the hands of the experimentalist. Modern simulation systems seek to reach their users in two ways: first, by providing a simple graphical interface suitable for nonprogrammers, and second, by tying closely into advances in web-based databases to tap into better data. There are currently several systems that strive toward this goal in the field of signaling pathways and genetic networks. These include Vcell,[1] MCell,[2] Dbsolve,[3] Jarnac,[4] GENESIS/Kinetikit,[5] and others.

Several levels of detail can be considered in models of signaling pathways. In increasing order of complexity and realism, these are (1) logical/Boolean models of signaling and genetic cascades, which treat activity as either on or off and regard signaling interactions as Boolean operations, (2) the "well-stirred cell" approach, which assumes perfect diffusional access to all components, and models signaling pathways in terms of chemical kinetics, (3) the diffusional approach, which treats everything in terms of reaction–diffusion systems, and (4) the Monte Carlo approach, which considers the stochastic interactions and movements of individual molecules.

The limiting factor in all these approaches is data. Computational requirements also increase steeply with increasing realism. The classic test tube biochemical experiment remains the staple data source for most models. It is still unusual for sufficient experimental information to be available to justify diffusional or

---

[1] http://www.nrcam.uchc.edu.

[2] http://www.mcell.cnl.salk.edu.

[3] http://websites.ntl.com/~igor.goryanin/.

[4] http://members.tripod.co.uk/sauro/biotech.htm, http://www.fssc.demon.co.uk.

[5] U. S. Bhalla, *In* "The Book of GENESIS" (J. M. Bower and D. Beeman, eds.), 2nd Ed. Springer-Verlag, Berlin, 1998. http://www.bbb.caltech.edu/GENESIS, http://www.ncbs.res.in/downloads.

stochastic models, although it is clear that such factors are indeed important in biology.

This article is based on the "well-stirred" approach, with compartmentalization as a slight elaboration. We use G-protein-coupled receptor signaling through the cAMP pathway as an example. We work through the process of determining reaction mechanisms and parameters, incorporating them into a functioning model, and assessing its behavior. In developing a research simulation, these stages would occur at a very fine grain: The user might add a single reaction at a time, test out the model for various parameters, and then study which behaves most realistically. Here we go through the simulation process in the same sequence, but at a much coarser level because we must describe an entire signaling pathway. It may be helpful to build the model while reading the chapter, and skip back and forth through the stages described in the article to get a better feel for the overall modeling process.

Reaction Mechanisms in Signaling Pathways

Data Sources

Block diagrams are the basal level information required to set up a model of a signaling pathway. If all else fails, a block diagram can be transcribed into equivalent reactions, which can then be fitted to the kinetic data as described below. Information of the form "A is an upstream activator of B" comes from many sources, typically from tissue culture, genetic, and molecular biology methods. More detailed mechanistic data rely on pharmacology and test tube biochemistry. Such data can provide details even beyond what a typical model might require.[6]

The generic block diagram for a signaling pathway specifies inputs, outputs, and regulators for constituent pathways (Fig. 1A). The starting point for devising a model is to expand the block diagram to specify it precisely in terms of individual reactions. The guiding rule for this is to choose the simplest set of reactions that embody the known interactions. As is clear from the expanded version of the block diagram into a precise reaction scheme, there is an alarming proliferation of molecules and reactions even in this highly simplified system (Fig. 1B–D). This complexity is the central problem in developing realistic models of signaling pathways. Kinetikit is therefore designed specifically to provide a friendly interface for managing complex reactions and large amounts of data.

We use the example of G-protein-mediated signal transduction to describe how to create and connect up the reaction and enzyme steps. Subsequent sections consider parameterization, running models, and analysis. In actual simulation development these steps take place iteratively and at a much finer grain. It would be

[6] S. O. Døskeland and D. Øgreid, *J. Biol. Chem.* **259**, 2291 (1984).

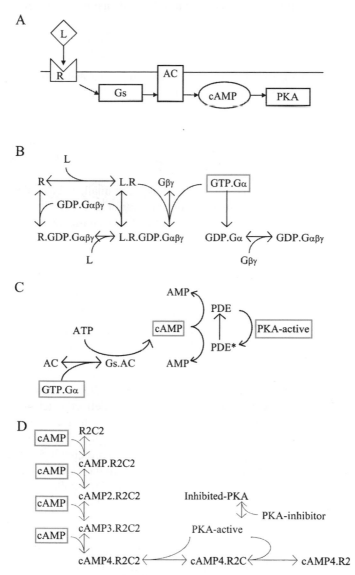

FIG. 1. cAMP signaling pathway example. (A) Block diagram. (B) Reaction details for receptor–G protein module. (C) Reaction details for AC–PDE module. (D) Reaction details for PKA module. In (B–D) some molecules may be represented more than once for illustrative reasons, but they are modeled as a single pool. The molecules highlighted within a box are common to more than one module and are used when merging modules to form the composite simulation. L, Ligand; R, receptor; Gs, stimulatory G protein; AC, adenylyl cyclase; cAMP, cyclic AMP; PKA, protein kinase A; PDE, phosphodiesterase; PDE*, phosphorylated phosphodiesterase; R2C2, PKA holoenzyme [two regulatory (R) and two catalytic (C) subunits].

typical to add a reaction at a time and test its effects, then go on to another reaction, and so on.

Software and Hardware

GENESIS and Kinetikit are freely available from a number of Internet sites as well as on the CD-ROM available as part of the GENESIS reference textbook.[5] At the time of writing, the current versions are GENESIS 2.2 and Kinetikit 7. The examples in this article refer to Kinetikit 7, but earlier versions are functionally equivalent except for the unit conversions. Kinetikit 7 will run on GENESIS 2.1 as well. The model script files used in the examples in this article are available at the same GENESIS sites.[5]

The computational demands of kinetic models are small, and a modern PC can run all but the largest models hundreds of times faster than real time. GENESIS runs on a wide range of Unix machines, including PCs running the freely available Unix clone, Linux.

Setting Up Reaction Mechanisms for Simulation

To illustrate, we start off with our example of receptor–G protein signaling (Fig. 1). Conversion of the block diagram into reactions should proceed in a modular manner. Modules are typically well-defined signaling molecules or pathways, each individually constrained by experimental data. In this pathway, there are experiments describing ligand–receptor interactions[7,8] and the closely coupled receptor–G protein interactions.[9] These collectively describe the receptor–G protein interaction nicely, and it is a clear conceptual unit. The next item in the block diagram is adenylyl cyclase (AC), which is also clearly defined in terms of its rates and interactions.[10] The block diagram can thus be reduced to a set of modules that can each be individually tackled to build up a library of signaling components. In our example, we consider receptor–G protein, AC, and protein kinase A (PKA) as distinct modules. In addition to the basics of setting up the reaction scheme, each module illustrates modeling design decisions and potential pitfalls. The integration of multiple modules is described in Save, Restore and Merging (see p. 18).

Receptor–G-Protein Module. It is relatively easy to describe receptor–ligand binding as a simple reversible reaction. The coupling of the receptor to the heterotrimeric G protein is more complex: does the coupling happen before ligand binding or after? It turns out that both reactions are observed, and in fact the reaction details are complex.[9] This example confines itself to a loop of reactions as in Fig. 1B. Care

---

[7] R. S. Kent, A. De Lean, and R. J. Lefkowitz, *Mol. Pharmacol.* **17,** 14 (1980).

[8] U. Gether, S. Lin, and B. K. Kobilka, *J. Biol. Chem.* **270,** 28268 (1995).

[9] P. Samama, S. Cotecchia, T. Costa, and R. J. Lefkowitz, *J. Biol. Chem.* **268,** 4625 (1993).

[10] J. P. Pieroni, O. Jacobowitz, J. Chen, and R. Iyengar, *Curr. Opin. Neurobiol.* **3,** 345 (1993).

must be exercised to ensure that the product of K_{eq} around the loop is unity, because there is no net free energy change around the loop. Further steps of this module are dissociation of the heterotrimer, binding of GTP, the hydrolysis of GTP, and the reassociation to form the GDP-bound trimer. Many design decisions arise at this stage. In this example we choose to ignore GTP metabolism, as we can treat it fairly accurately as a well-buffered pool. We also simplify the details of the process of GTP GDP exchange and release of $GTP \cdot G\alpha$ (GTP.Ga in Kinetikit; GTP–G protein α subunit complex). Another simplification pertains to the activation of the downstream pathway (in this case AC). This is at first sight straightforward: it only involves binding of activated stimulatory G protein (G_s; Gs in Kinetikit) to AC to obtain an active $G_s \cdot AC$ (Gs.Ac) complex. The obvious enzyme activity of the complex is cyclization of ATP to form cAMP, as described below. The other, often ignored activity is the hydrolysis of GTP by the G protein. Because AC levels are low it may be reasonable to ignore this. On the other hand, results show that several G-protein–activated molecules also act as GTPase-activating proteins (GAPs), which enhance the rate of hydrolysis of GTP by the bound G protein.[11] This apparently simple, and easily ignored step, turns out to be potentially important for signal flow.

Adenylyl Cyclase Module. In designing the AC module we again can choose to ignore the metabolism of ATP. In our example the cyclase is treated as a two-state enzyme: it is either off or, in the G_s-bound form, it is on. It is possible to add in further regulatory interactions [inhibitory G protein (G_i; Gi in Kinetikit), phosphorylation, calmodulin (CaM) binding, etc.], depending on the AC subtype. Indeed, if we wished to go into details for subtype interactions we could consider some 15 or so isoforms of AC.[10] This is a modeling decision that requires judicious evaluation of likely interactions and the details of the specific system being modeled.

There are at least two ways to treat isoforms: the explicit approach is to model each AC variant separately and treat its interactions as distinct. This is more accurate but more tedious, and the proliferation of similar reaction steps increases the need for care in constructing the model. For many purposes it is adequate to lump various isoforms together and treat them as an "averaged molecule." Fortunately for this example, we are interested only in the linear pathway of activation of AC by G proteins, and can ignore the rest.

Protein Kinase A Module. PKA is one of the best constrained enzymes in the signaling literature.[6] It is fairly straightforward to construct the reaction sequence on the basis of these experiments, but here we encounter a different design decision: should we include all the available experimental data? For example, it is known that the sharpness of PKA activation is increased by the presence of an

[11] K. Scholich, J. B. Mullenix, C. Wittpoth, H. M. Poppleton, S. C. Pierre, M. A. Lindorfer, J. C. Garrison, and T. B. Patel, *Science* **283,** 1328 (1999).

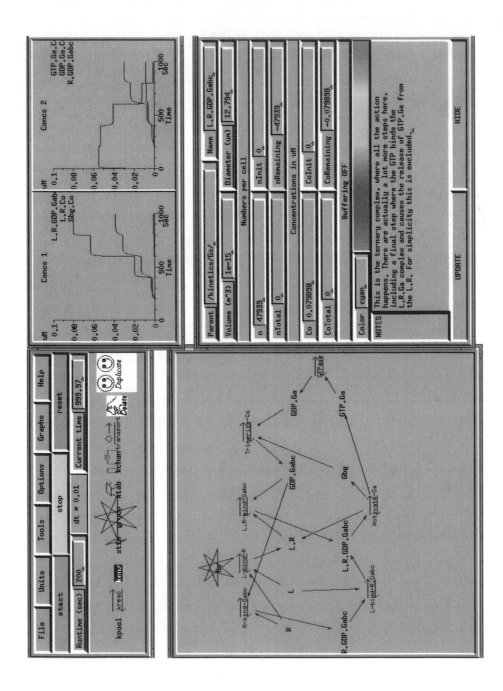

inhibitor molecule, which stoichiometrically blocks the kinase until all the inhibitor is bound. Should this interaction be included? Here we have chosen to do so, at the cost of introducing a further free parameter into the model.

Regulatory and Housekeeping Functions. We have already bypassed several sources of complexity in the model by treating highly regulated molecules such as GTP and ATP as buffered to a fixed value. A general principle that we cannot so easily bypass is that every enzyme should have a counteracting enzyme. In the case of the cyclase, we must provide a phosphodiesterase (PDE) for removing the cAMP. This mundane task assumes signaling importance because the PDE itself can be regulated and further activated on phosphorylation by PKA. (There are other PDEs, which we do not discuss here.) Similarly, PKA will need a counteracting protein phosphatase to work on its substrates. These balancing or housekeeping molecules can, of course, be modeled in their full glory as highly regulated enzymes. A simpler approach that can often be justified is to treat them as a fixed-rate back reaction. In our example mechanism we have done both: modeled the PDE as an enzyme that itself is regulated by PKA, and modeled the phosphatase as a simple irreversible back reaction.

Molecules

All reactions are set up in Kinetikit by clicking and dragging objects from the menu bar into the edit window (Fig. 2). The creation of an object (a pool, reaction, enzyme, or other simulation module) is accompanied by the appearance of a dialog box for editing the object. The dialog box allows complete specification of object parameters including name, kinetic parameters, and display attributes such as color. At this stage of dealing with reaction mechanisms we need only change the name from the default. For this example, we would create distinct molecular pools by dragging the Pool icon from the menu bar into the edit window and assign their names (Table I). For now we can ignore the concentration terms, which are discussed in Volume and Concentration Units (below).

Reactions

Reaction objects are also created by dragging Reaction Arrow icons from the menu bar into the edit window. The bare reaction object is represented by

---

FIG. 2. Kinetikit interface layout. The top row of buttons includes the menu options, followed by several buttons for controlling the running of the model. The menu bar with the library of prototypes is below this. The edit window is the large window to the bottom left and is shown displaying the reaction scheme for the receptor–G protein module. The graph windows at upper right show the results of a simulation with step increases in ligand levels. The pool parameter window at lower right displays parameters for the $L \cdot R \cdot GDP \cdot G\alpha\beta\gamma$ (L.R.GDP.Gabc) pool.

TABLE I
MOLECULAR POOLS[a]

Receptor–G protein module		AC module		PKA module	
Name	Initial concentration (μM)	Name	Initial concentration (μM)	Name	Initial concentration (μM)
L	Variable	AC	0.015	R2C2	0.5
R	0.1	Gs.AC	0	cAMP.R2C2	0
GDP.Gabc	1	ATP	5000 (buffered)	cAMP2.R2C2	0
R.GDP.Gabc	0	cAMP	0	cAMP3.R2C2	0
L.R	0	AMP	1000 (buffered)	cAMP4.R2C2	0
L.R.GDP.Gabc	0	cAMP-PDE	0.5	cAMP4.R2C	0
GTP.Ga	0	cAMP-PDE*	0	cAMP4.R2	0
Gbg	0			PKA-active	0
GDP.Ga	0			PKA-inhibitor	0.25
				inhibited-PKA	0

[a] L, Ligand; R, receptor; AC, adenylyl cylase; R2C2, PKA holoenzyme [two regulatory (R) and two catalytic (C) subunits]; PKA, protein kinase A; PDE, phosphodiesterase; PDE*, phosphorylated phosphodiesterase.

a reversible arrow symbol, and only two parameters are needed to specify it: the forward and backward rates (K_f and K_b, respectively; Table II). The reaction arrow symbol defines a reaction object, but does not set up its substrates or products. This is because Kinetikit maintains a distinction between the objects (pools, reactions, etc.) and their interactions. Once the substrate pools, reactions, and products have been set up, the reaction can be completed by dragging the substrate pool onto the reaction symbol and then dragging the reaction symbol onto the product pool. In our example, to complete the binding of ligand to receptor we must first drag the Ligand (L) icon to the L-bind-R icon (a green arrow appears) and then drag the L-bind-R icon onto the receptor (R) icon (another green arrow appears). The assignment of reaction rates is described in Setting Reaction Kinetics (below).

Enzymes

Enzyme objects in Kinetikit should be regarded as enzyme activities of pre-existing molecules. They can be created only on existing molecular pools. This is done by dragging the Enzyme icon from the menu bar onto an already defined molecular pool in the edit window. Any number of enzyme activities can be set up on a single pool. For example, a single protein kinase might have a large number of substrates, each with different kinetics. Each of the substrates would therefore

TABLE II
REACTIONS[a]

	Receptor–G protein module			AC module			PKA module	
Name	K_f (1/sec/μM)	K_b (1/sec)	Name	K_f (1/sec/μM)	K_b (1/sec)	Name	K_f (1/sec/μM)	K_b (1/sec)
L-bind-R	0.1	0.1	Dephosph-PDE	0.1 (1/sec)	0	cAMP-bind-B1	54	33
L-bind-R. Gabc	5	0.1	G_s-bind-AC	500	1	cAMP-bind-B2	54	33
R-bind-Gabc	0.002	0.1				cAMP-bind-A1	75	110
L.R-bind-Gabc	0.1	0.1				cAMP-bind-A2	75	32.5
Activate-Gs	1 (1/sec)	0				Release-C1	60 (1/sec)	18 (1/sec/μM)
GTPase	0.0667 (1/sec)	0				Release-C2	60 (1/sec)	18 (1/sec/μM)
Trimerize-Gs	6	0				inhibit-PKA	60	1

[a] K_f, Forward rate; K_b, backward rate; B, cAMP binding site B; A, cAMP binding site A; C, catalytic subunit of PKA. See also Table I.

TABLE III
ENZYMES

Parent molecule	Symbol in model	K_m (μM)	V_{max} (1/sec)	$k_2 : k_3$ ratio
Gs.AC	cyclase	20	18	4
cAMP–PDE	PDE	19.84	10	4
cAMP–PDE*	PDE*	19.84	20	4
PKA-active	phosph-PDE	7.5	9	4

interact with a different enzyme activity object. This makes it possible to specify different substrates with different reaction rates. It does not, of course, imply that the actual structure of the enzyme contains multiple sites. As with the reaction objects, the interactions between substrates and products of the enzyme are set up by click-and-drag operations within the edit window. To set up the enzymatic reaction, the substrate is dragged onto the enzyme activity, and the enzyme site is dragged onto the product. Enzyme rate assignment is discussed in Setting Enzyme Constants (below). The only enzymes we have to consider in this example are AC, PKA, and cAMP–PDE (Table III).

Groups

A useful organizational feature of Kinetikit is that closely coupled sets of reactions and molecules can be placed on "groups." Each signaling module is typically represented by a distinct group. A group is created in the same way as any other simulation object, by dragging it from the menu window to the reaction edit window. Assignment of reactions and pools to a group is done by dragging them onto the Group icon. Once grouped, the objects in a signaling module can be saved, displayed, moved, deleted, and duplicated as a unit. In our example there are three groups: for the receptor–Gs, for AC, and for PKA.

Modeling Regulation

In modeling terms there are two principal approaches to describing enzyme regulation. The first (illustrated here by AC) is to regard the enzyme as either on or off. In this situation the only enzyme site required is on the activated form of the enzyme. Activation level here is determined purely by the amount of enzyme in the activated state. The second approach is to consider all possible forms of activation and represent each by an individual enzyme site with different activity. For example, PDE has a basal activity, but can also be stimulated further by phosphorylation. An enzyme with complex regulation might need several such enzyme sites. If the enzyme has multiple substrates the number of potential sites grows rapidly.

There are various philosophies regarding reaction regulation. Kinetikit is built around the assumption that a given molecular species has a fixed interaction rate with any other molecule. Regulation is therefore regarded as the formation of a different molecule that has its own interaction rate. This is perhaps more correct, but does lead to a proliferation of reaction sites as mentioned above. A different and commonly used approach is to vary the reaction rates themselves. Kinetikit permits this, but grudgingly. This and other rarely used options are available under the Options menu item. In some situations (e.g., voltage-gated channels) this formalism is more appropriate.

Correcting Errors

Objects can be renamed and their parameters changed at any time. The objects themselves can be removed (deleted) by dragging them onto the dinosaur (Delete) icon in the menu bar. Interactions between objects obviously vanish when one of the objects is deleted. Interactions can also be removed by repeating the click-and-drag that created the interaction.

Parameters

Parameter specification is the most difficult part of biological modeling. Signaling pathway models can draw on a large body of experimental work, particularly on the biochemistry of signaling molecules, and so the bench work will often have been taken care of. A careful reading of the methods and conditions and a good deal of informed judgment are required to extract useful parameters from typical articles. Most published kinetic data are not in a form that can be plugged directly into a model. It is necessary, therefore, to work backward by simulating the experimental conditions, comparing results with the measurement, and adjusting parameters until they match. In this section we examine common strategies for setting parameters for various kinds of signaling reactions.

Volume and Concentration Units

GENESIS internally models all reactions in terms of numbers of interacting molecules. This approach is independent of local volume and localization issues, but has the drawback that the units are unfamiliar. Kinetikit therefore provides conversions to commonly used units. These are selected through the Units menu item. All reaction pools are initialized with the volume specified here, which defaults to $1 \times 10^{-15} \text{m}^3$, corresponding to a cube with 10-μm sides or a sphere with a diameter of 12.41 μm. These defaults can be set to other values if desired. Concentration and time units default to micromolar (μM) and seconds, respectively. We retain all these defaults for our example.

Setting Pool Concentrations

The Pool parameter dialog can now be filled in for each of the molecules in the simulation. Note that it provides its own local volume, which can differ from the default setting. The Pool parameter window displays concentrations both in terms of numbers of molecules, and also in terms of the selected concentration units. The most important fields in the parameter window are current concentration (Co) and initial concentration (CoInit). The counterparts of these values in terms of numbers of molecules are n and nInit. Co and n are calculated by the simulator, and are not normally set by the user. Most pools in the model are formed as intermediates in the signaling pathway, and should have a CoInit of zero. Only a few key reactants will need to have this value assigned. For example, the concentration of the unbound receptor, the unstimulated AC, and the GDP-bound $G\alpha\beta\gamma$ trimer should be set up to known total values. Provided that the key reactants have the correct total levels, and the rates are reasonable, the system will settle to a reasonable steady state. Metabolites such as ATP, which are tightly regulated, can be "buffered" to a known level by using the Buffering toggle. This is a facility that should be used with care, because it implies an infinite source (or sink) of the molecule. Enzymes are an interesting case, because their concentrations *in vivo* are usually estimated by successive purifications and activity assays. Thus an impure enzyme will overestimate the concentration but underestimate the activity in proportion. In modeling terms, the total cellular enzyme activity may be correct even if the levels are not. Unfortunately, the regulation of the enzyme is likely to depend on its exact levels. In our example, the rate of binding of Gs to AC depends on exact enzyme levels, and would be affected by such an error.

Setting Reaction Kinetics

By far the most common experimental approach to quantitation of reaction parameters is the concentration–effect curve. This, in conjunction with time course information, is enough to tightly constrain most binding reactions. If a perfect match to the theoretical reaction is assumed,

$$A + B \underset{k_b}{\overset{k_f}{\rightleftharpoons}} C \tag{1}$$

the half-maximum of the curve defines the K_d for binding. K_d is related to the rate constants by

$$K_d = k_b/k_f \tag{2}$$

Because the time course is approximately

$$\tau \sim\, = 1/(k_b + k_f) \qquad \text{(at fixed B)} \tag{3}$$

we could in principle solve for k_f and k_b. In practice, the binding curve is usually more complex and the time course not as well defined. Here the calculated values of k_f and k_b could be used as a starting point for modeling the reaction, and progressive refinement could be done by using the model itself to match the observed curves. This process of refinement is also useful when the experimental input is several steps removed from the measured response. In our example, we might have data on the formation of cAMP as a function of ligand concentration. If we can approximate some of the steps as linear responses, the remainder can be constrained by the results. In this system we might be able to treat cAMP response as proportional to the number of free Gs molecules, and use this to improve our estimates for rates in the G protein module.

Some care is required in working out units, because this depends on the order of reaction. The k_f and k_b dialogs use units of numbers of molecules and seconds, and do not require scaling when additional substrates or products are added to the reaction. K_f and K_b are somewhat friendlier and use the concentration and time units specified by the user. These do, however, depend on the order of reaction, and will be rescaled if this is changed. K_d is also calculated in terms of user-specified units. This may need some care in relating to the experimental value if one or more reactants is assumed fixed. The variable τ (tau) is in user-specified time units, and is really only applicable for first-order reactions.

Setting Enzyme Constants

Enzyme interactions in Kinetikit follow the classic Michaelis–Menten irreversible scheme:

$$\text{Sub} + \text{Enz} \underset{k_2}{\overset{k_1}{\rightleftharpoons}} \text{Sub} \cdot \text{Enz} \xrightarrow{k_3} \text{Prd} + \text{Enz} \tag{4}$$

(where Sub is substrate, Enz is enzyme, and Prd is product). The enzyme object is therefore really just a shortcut way of implementing two reactions and a molecular pool (the Sub$\cdot$Enz complex). Enzyme reactions require three parameters: k_1, k_2, and k_3. Most experiments report the two parameters for Michaelis–Menten enzymes, that is, K_m and V_{max}. These are related as follows:

$$K_m = (k_2 + k_3)/k_1 \tag{5}$$

$$V_{max} = k_3 \quad \text{(allowing for enzyme levels)} \tag{6}$$

These two Michaelis–Menten parameters do not fully constrain the requisite three rate parameters. The third parameter may be fixed by assuming a fixed ratio of k_2 to k_3. It turns out that most models are insensitive to this ratio. Then,

$$k_3 = V_{max} \tag{7}$$

$$k_2 = \text{ratio} \cdot V_{max} \tag{8}$$

$$k_1 = [(1 + \text{ratio}) \cdot V_{max}]/K_m \tag{9}$$

On the basis of the fact that typical enzymes do not accumulate much enzyme–substrate complex, a ratio of 4 is usually satisfactory. Either of these alternative representations of enzyme rates can be used in the enzyme dialog box to set up rates.

Further Constraints

One benefit of using detailed reaction schemes for representing biochemical reactions is that experimental manipulations can usually be faithfully replicated. This proves useful in constraining parameters on the basis of indirect data. For example, it is common to have data for steady state levels of various reactants but no direct rate information. In the example there were useful numbers of this kind for numbers of GTP.Ga complexes after stimulation.[12] This was used to improve estimates for rates of binding of the L.R complex to the GDP · $G\alpha\beta\gamma$ (GDP.Gabc in Kinetikit) complex. A simulation of the reaction system could predict reactant levels for given reaction rates. By iteration it is possible to work backward to obtain rates that fit the observed levels.

Parameters versus Mechanisms

In any reasonably complex signaling network, situations will probably occur in which the obvious reaction mechanisms do not appear to yield the observed responses. At some point parameter adjustment must give way to a reformulation of the reaction scheme. This is usually a tedious operation, as all reaction rates may need recalculation. Nevertheless, it is critical to the iterative process of model development. It closes the cycle involving specification of rate, concentration, and mechanistic details. This process is likely to involve many cycles: 50 versions of models for a single pathway are common.

Modeling

Modeling begins as soon as there are two or more pools and reactions to simulate. The process of modeling includes running, displaying, saving, and other aspects of controlling the simulation.

Running Simulations

There are three appropriately colored buttons just under the menu bar: Start, Stop, and Reset (Fig. 2). The Start button runs the simulation. The Stop button halts the simulation cleanly while it is running, so that it can be resumed by another click of the start button. The Reset button reinitializes the simulation and sets the time back to zero, losing any information from the previous run. The user specifies the desired duration of the simulation in the Runtime dialog box. The current simulation time is continually updated in the Current Time dialog box.

[12] S. R. Post, R. Hilal-Dandan, K. Urusawa, L. L. Brunton, and P. A. Insel, *Biochem. J.* **311,** 75 (1995).

Displaying Simulation Output

The current level of any pool can be checked by double-clicking on the pool. This will pop up the Pool parameter window. The current pool level (n as well as Co) updates during the course of the simulation.

It is usually more useful to display the levels of several pools at a time, using plots. Plots, as the name implies, generate time course plots of reactant levels. These are set up by clicking on any reactant in the edit window, and dragging it to a graph window. This causes the appearance of the name of the plot in the graph window. Plotting will not start until the Reset button has been clicked. The plot name can be dragged to the Delete icon in the menu window to delete it. Double-clicking of the plot name opens a plot editor window offering various self-explanatory options. One of the most important is the Save to File option, which dumps all the data points in the plot into a text file with time and concentration value pairs on each line. Several global plot options are available in the Graphs menu item. These include the Overlay facility for overlaying new runs on previous plots for comparison.

Plots of long simulations can consume a great deal of memory. The number of data points is equal to simulation duration divided by time step for the plot. Therefore it is advisable to use a fairly long time step for the plots. It is usually necessary to update the plot only every 1 sec or so, unless the reactant is varying exceptionally rapidly. This is controlled by setting the plot time step in the Options menu.

Time Steps and Accuracy

GENESIS/Kinetikit uses the exponential Euler method, which is well suited to handling differential equations that exhibit an exponential time course.[13] Kinetic equations are of this form. The method is explicit, and so it requires some care in selecting integration time steps to ensure accuracy. In general, a time step 10 times faster than the fastest rates in the system should be accurate. The time step is specified in the Options menu. A quick way of selecting a time step is to successively halve time steps and compare output plots for convergence. The process is then repeated with longer time steps until a good balance between accuracy and speed is attained. Given that typical reaction parameters have fairly large error bars, a numerical accuracy of 1% should be adequate.

Many systems have rapid initial kinetics followed by much slower, prolonged responses. Furthermore, stimulus delivery in the simulation can be instantaneous, which is difficult to handle accurately. A simple solution to this is to set a fine time step for initial transients, and then to switch over to a slower time step for the smoother part of the simulation. This feature can be set up in the Options menu. The initial time steps can be selected by using the same convergence criterion

---

[13] R. J. MacGregor, "Neural and Brain Modeling." Academic Press, San Diego, California, 1987.

described above. This procedure is a crude but effective version of more complex, variable time step methods.

Volumes and Units

All computations in the simulator actually use numbers of interacting particles. This makes it possible to specify different volumes for pools (e.g., in distinct cellular compartments) without incurring scaling errors. Therefore, all concentration terms are actually scaled by the volume for the purposes of calculation.

The default volume for reactions is $10^{-15} m^3$, or a 10-μm cube. A diameter for a sphere can also be used. These terms can be set in the Units menu item. All subsequently created pools will be set to this volume on creation. The volume can be edited as required within the Pool edit box. Note that enzymes sites too have a volume term that applies to the enzyme–substrate complex. It is always set equal to the volume of the parent enzyme molecule. The Units menu also provides for altering the units used for display and data entry purposes.

It is instructive to monitor the value of n (number of molecules per cell) in the Pool parameter window. In small cellular compartments, several pools may exist in single-molecule quantities. The continuous approximation used here breaks down under such situations, and stochastic calculations must be used. This is one of the key areas for future simulator developments.

Given a preferred set of concentration and time units, rate constants are scaled from the more fundamental but less friendly *#/cell/second* units used internally. The dimensions of rate terms depend on the order of the reaction and care is needed to match them with the experimental conditions so that this dimension is handled correctly.

Save, Restore, and Merging

One of the most important operations in modeling is frequent saving of the model state, with annotations. The File menu option is used to store the complete current state of the model and interface (with the exception of plotted data points). If a signaling module is encapsulated within a group, it can also be saved as an option in the Group parameter window. The data file in both cases is a standard GENESIS script file. This has two implications. First, the simulation can be resumed simply by entering

genesis <filename>

Second, for more sophisticated users, complex batch jobs using the model can be set up by using the GENESIS scripting language and simply including the simulation file name. If the simulation is included with the option nox it will not display the interface, and only load in the numerical part of the model:

include <filename> nox

Previously stored models can also be opened from the GENESIS command line. If Kinetikit is not yet running, the simulator will automatically load it and then open the model.

genesis> include model.g // loads in model file model.g

There is a special situation when Kinetikit is already running with a previously loaded model. In such situations the system will attempt to "merge" the two models. In doing so it always loads up the most recent model in its entirety. If there is overlap with an existing model, it does not duplicate existing pools but it does change their parameters to that of the most recent model. Similarly, existing linkages between pools, reactions, and enzymes are not duplicated. This behavior is designed to allow merging of interacting signaling pathways. In our example, there are three signaling modules that interact as follows. First, the GTP.Ga pool from the Gs module binds to AC to produce the Gs.AC active enzyme. Second, cAMP from the AC module binds to and activates PKA in several steps. Third, PKA itself phosphorylates the PDE from the AC module in a negative feedback loop. These molecules are highlighted in Fig. 1B–D. The merging facility in Kinetikit allows all these interactions to be set up by simply loading in the three modules in sequence. To do this, the only requirement is that the signaling molecule involved in an interaction between two modules must be defined (with the same name) in both modules. In our example, GTP.Ga is produced by the Gs module and is also defined in the AC module as an input molecule. The second-messenger cAMP was present as an end product in the model of AC, and as a regulator of PKA in the PKA model. PKA itself phosphorylated the PDE present in the AC model. The following sequence of commands was used to merge the three modules:

genesis>include pka.g

genesis>include ac.g

genesis>include gs.g

Note that the parameter values are those of the most recently loaded module. This usually does not matter. In rare cases, there is a shared pool or reaction whose parameter values differ between modules. This (inadvisable) situation can lead to a dependence on order of loading. The graphical layout of objects in the edit window is also determined by the most recently loaded module, and this is often a more practical reason for choosing a particular sequence of module loading.

Miscellaneous Features

There are a wide range of advanced features in Kinetikit that are beyond the scope of this chapter, but which are documented in detail in its Help menu. A brief road map of some of the capabilities of the options is provided in Table IV.

TABLE IV
KINETIKIT FEATURES

Feature	Location	Function
kpool	Menu bar	Pool of reactants
kreac	Menu bar	Reversible reaction
kenz	Menu bar	Michaelis–Menten enzyme
stim	Menu bar	Pulse stimuli
group	Menu bar	Hierarchy and organization of models
xtab	Menu bar	Smooth stimuli
kchan	Menu bar	Membrane channel
Transport	Menu bar	Molecular transport with delay
Delete	Menu bar	Delete objects
Duplicate	Menu bar	Duplicate objects
Postscript	Tools menu	Specify PostScript options. Control-P in the edit window will generate PostScript output according to these settings
Plot	Tools menu	Options for graphs
Compare models	Tools menu	Compare model parameters
Tabulate models	Tools menu	Generate text table of model parameters
Timestep control	Options menu	Set time steps for simulation and display
Reaction options	Options menu	Numerous options for special reaction situations
Graph axis limits	Graphs menu	Globally set graph axis limits
Show more graphs	Graphs menu	Display additional graph windows
Plot overlay	Graphs menu	Toggle retention of previous plots after reset
Save all plots	Graphs menu	Save data points from all plots in text file

Interpreting Models

Detail

One of the most difficult aspects of modeling is in deciding how much detail to include. Even in this small example, we have had to make many decisions about simplifying and excluding interactions. These decisions determine how we interpret model results. For example, our model explicitly considers the association of G protein subunits. If we were now to include a second $G\alpha$ that bound the same $\beta\gamma$ subunits, the simulation would calculate the rate at which free $\beta\gamma$ subunits bound to the first or second $G\alpha$. Thus we would be justified in using our model to interpret $\beta\gamma$-mediated cross-talk between the two $G\alpha$ subtypes.

A different aspect of detail, which again we have encountered, is that every molecule seems to interact with several others. In our example, we chose to stop the reaction cascade at PKA, just at the point where we would have had to deal with half a dozen different substrates. Where should the line be drawn in terms of the scope of the model? One answer is to explicitly model every interaction that is part of a measurement that the model seeks to replicate. In the simulator we can readily monitor PKA activity by plotting the concentration of its active form. In an experiment, we would have had to put in a test substrate. It would therefore be more accurate to model this substrate explicitly as well. By doing so the model itself would incorporate the effects of enzyme saturation, etc., and the output would be directly equivalent to the experimental measurement of phosphorylation. Similar issues of scope crop up when considering isoforms. In our model we have ignored the 10 or so G protein and AC isoforms, let alone the hundreds of G-protein-coupled receptors. There is a happy medium level of detail where we have enough information to reasonably represent the system but not so much that interpretation becomes difficult.

Reliability

The error bars in model output are at least as large as those of the model parameters. We can, however, be fairly confident that they are not much larger if the model is able to quantitatively predict experimental outcomes. In our example, we could assess the reliability of most of the model by examining the time course of cAMP stimulation by agonist. This is not a result that the model has been fine-tuned to predict, and so a reasonable match to experiment means that our model mechanism and parameters are probably reasonable as well. Good experimental articles frequently overconstrain a model. That is, they conduct more than the minimal number of experiments required to specify the model parameters. If this is the case then the model can be set up on the basis of the first few experiments, and tested for prediction of the remainder.

Robustness

Biological systems are usually robust. Signaling pathways reflect this by usually having a wide operating range over which their quantitative behavior may change moderately, but their qualitative behavior does not. In modeling terms, it should be possible to vary the rates, concentrations, etc., of the model over a range comparable to the experimental error, without affecting the basic model properties. A factor of 2 above and below the basal level is common. A single "fragile" parameter might be interpreted as a sensitive regulator of the system. Several such parameters usually indicate that the model mechanisms need re-working. An analysis of robustness (or parameter sensitivity analysis) proceeds by varying the selected parameters in turn and measuring a key response of the system as a whole. In the GENESIS/Kinetikit context, this is usually done by writing a GENESIS script program that loads in the model, systematically adjusts the selected parameters, runs the simulation and monitors the response. In the cAMP signaling example, we might choose to vary such parameters as receptor affinity, GTP exchange rates, levels of key molecules, and so on. The obvious response to monitor would be PKA activation as a function of ligand concentration.

Complexity

How should results from a model that itself is too complex to intuitively comprehend be interpreted? The modeling process itself frequently gives rise to insights at the abstract level, well above the tedious details of parameter specification. The process of decomposing the example block diagrams into detailed reaction schemes will probably have suggested to the modeler that certain reactions could be sensitive points of regulation. GTP exchange and hydrolysis might be such points of regulation in the cAMP pathway. Modeling can be applied more concretely, as a counterpart to experiment. In our example, we could ask mechanistic questions: does the receptor bind the G protein before or after the ligand? The model would predict quantitatively different responses, which could be tested. We could also consider quantitative issues: what would happen if AC increased the GTPase activity of $G\alpha_s$? In each case the model encapsulates all the complex detail of reaction mechanisms and rates, and outputs a simple, experimentally verifiable quantity.

Emergent Properties

One of the most important functions of a model is to predict something that the user does not expect. There are unlikely to be many surprises in a pathway as well established as cAMP signaling. Despite this, it would be an instructive exercise to work out whether the modeled pathway could generate oscillations or bistable behavior. On theoretical grounds this might occur if the PKA phosphorylated the receptor and inhibited or activated it, respectively. This behavior would be emergent in the sense that it would be almost impossible to predict the actual outcome without doing an experiment, whether *in vivo* or *in silico*. It would also be a response of a qualitatively different (and more interesting) kind than the conventional linear flow of signaling information.

The above-described process of setting up a model is really designed to make sure it behaves as expected in situations that are experimentally well understood. It follows that the best way to obtain interesting predictions from the model is to try it out in unknown territory: complex interactions, novel signaling contexts, or simply in systems where the role of the pathway is uncertain. All of these situations are likely to be true as experiments and models scale up to describe biology more completely. At this point the model truly goes from descriptive to predictive.

Concluding Remarks

There are many levels of analysis of cellular signaling, but the starting point remains the definition of reactions, rates, and concentrations. Biology poses many interesting problems at finer levels of detail such as cytoskeletal or single-molecule interactions. Nevertheless, there is an immediate flood of data at the bulk, nondiffusive level of detail we have considered. Modeling in the age of proteomics will have to contend with single proteomics experiments that can provide a thousand data points, and genomics assays that yield a hundred times that. Kinetikit and this article represent an early evolutionary step toward systems that combine the scalability of databases and the predictive and interpretive capacity of models.

Acknowledgment

This work was supported by the National Centre for Biological Sciences.

Section II

Phosphodiesterases

[2] Assays of G Protein/cGMP–Phosphodiesterase Interactions

By ALEXEY E. GRANOVSKY and NIKOLAI O. ARTEMYEV

Introduction

Activation of cGMP phosphodiesterase (PDE) by the photoreceptor G protein, transducin (G_t), is a key event in the vertebrate visual transduction cascade. In the dark, the GDP-bound transducin α subunit ($G_t\alpha$) is complexed with $G_t\beta\gamma$. Photoexcitation of the visual receptor rhodopsin induces a GDP/GTP exchange on $G_t\alpha$ and dissociation of the GTP-bound $G_t\alpha$. $G_t\alpha$GTP rapidly activates PDE, leading to a drop in the intracellular concentration of cGMP in photoreceptor cells. The rod photoreceptor PDE is composed of two homologous catalytic subunits ($P\alpha\beta$) that are kept inactive in the dark via the association with two identical inhibitory γ subunits ($P\gamma$). $G_t\alpha$GTP binds to the $P\gamma$ subunits and displaces them from the $P\alpha\beta$ catalytic sites, thus producing an active enzyme.[1–3]

The assays of cGMP hydrolysis by activated PDE have been a major tool for monitoring the interaction between transducin and PDE, and initial studies mainly relied on such assays.[4,5] Purified $G_t\alpha$ complexed with the nonhydrolyzable analog of GTP, GTPγS, stimulates PDE in solution ineffectively, and the PDE activation assay requires relatively large concentrations of $G_t\alpha$.[5] The presence of rod outer segment (ROS) membranes or lipid vesicles significantly enhances the effectiveness of PDE stimulation because of the formation of an active membrane-bound complex between $G_t\alpha$ and PDE and allows an increase in assay sensitivity.[6,7] However, membrane-supported activation of PDE by transducin is a complex process that depends on a number of factors such as type and concentration of membranes (vesicles), binding of PDE, binding of $G_t\alpha$ to membranes, and the intact state of lipid modifications on $P\alpha\beta$ and $G_t\alpha$. Furthermore, PDE activation is not a direct monitor of transducin binding to the enzyme. This is evident from studies of $G_t\alpha$ mutants that bind but fail to activate PDE,[8] or interact with the effector weakly without causing enzyme activation.[9] Therefore, detailed analysis of the interface

[1] M. Chabre and P. Deterre, *Eur. J. Biochem.* **179**, 255 (1989).
[2] S. Yarfitz and J. B. Hurley, *J. Biol. Chem.* **269**, 14329 (1994).
[3] L. Stryer, *Proc. Natl. Acad. Sci. U.S.A.* **93**, 557 (1996).
[4] B. K.-K. Fung, J. B. Hurley, and L. Stryer, *Proc. Natl. Acad. Sci. U.S.A.* **78**, 152 (1981).
[5] T. G. Wensel and L. Stryer, *Proteins Struct. Funct. Genet.* **1**, 90 (1986).
[6] A. Clerc and N. Bennett, *J. Biol. Chem.* **267**, 6620 (1992).
[7] J. A. Malinski and T. G. Wensel, *Biochemistry* **31**, 9502 (1992).
[8] Q. Li and R. A. Cerione, *J. Biol. Chem.* **272**, 21673 (1997).
[9] M. Natochin, A. E. Granovsky, and N. O. Artemyev, *J. Biol. Chem.* **273**, 21808 (1998).

and mechanism of transducin–PDE interaction required new, simple, and direct methods for G protein–effector binding. Several assays based on different types of fluoresence measurements have been subsequently developed.[10–13] All of them have taken advantage of the fact that within the holo-PDE complex, $P\alpha\beta\gamma_2$, the $P\gamma$ subunits constitute the primary, if not exclusive, target for $G_t\alpha$ binding. Direct binding between $G_t\alpha$ and $P\gamma$ can be measured by changes in intrinsic tryptophan fluorescence of the proteins.[10] The drawback of this method includes the need to separately record the fluorescence spectra of $G_t\alpha$ and $P\gamma$, and analyze the summed individual spectra in comparison with the spectra of mixtures of $G_t\alpha$ and $P\gamma$. Moreover, the changes in tryptophan fluorescence on $P\gamma$ binding are different for $G_t\alpha$GDP and $G_t\alpha$GTPγS.[10] Complicating the analysis, binding of $P\gamma$ to $G_t\alpha$GTPγS leads to a blue shift in the emission spectrum rather than to an increase in the fluorescence quantum yield.[10] Another approach is the use of resonance energy transfer between fluorescently labeled $P\gamma$ and $G_t\alpha$.[11] A disadvantage of this method is the relatively nonselective labeling of $G_t\alpha$ at several sites, using cysteine-reactive probes,[14] whereas a lysine-selective modification may affect GTPase activity and the ability of $G_t\alpha$ to undergo the GTP-induced conformational change.[15] One of the simplest and most sensitive assays to monitor the interaction between $G_t\alpha$ and $P\gamma$ utilizes $P\gamma$ labeled with environmentally sensitive fluorescent probes.[12,13] In this chapter we describe such a fluorescence assay as well as a practical PDE activation assay for analysis of the properties of transducin mutants designed to delineate the $G_t\alpha$ effector interface.

Activation of Holo-PDE by $G_t\alpha$ in a Reconstituted System

All the major components of the vertebrate visual transduction cascade can be readily purified from ROS by standard techniques and then reconstituted for functional analysis. Individual proteins can be substituted for their mutants to analyze effects of the mutation on the cascade. Activation of holo-PDE in the reconstituted system with transducin and ROS disk membranes has been widely used to study transducin–effector interaction. Several techniques are commonly used to measure PDE activity. Hydrolysis of cGMP results in the generation of protons, and the kinetics of pH change during PDE activation can be monitored with a pH

[10] A. Otto-Bruc, B. Antonny, T. Minh Vuong, P. Chardin, and M. Chabre, *Biochemistry* **32**, 8636 (1993).

[11] J. W. Erickson, R. Mittal, and R. A. Cerione, *Biochemistry* **34**, 8693 (1995).

[12] N. O. Artemyev, H. M. Rarick, J. S. Mills, N. P. Skiba, and H. E. Hamm, *J. Biol. Chem.* **267**, 25067 (1992).

[13] N. O. Artemyev, *Biochemistry* **36**, 4188 (1997).

[14] Y.-K. Ho and B. K.-K. Fung, *J. Biol. Chem.* **259**, 6694 (1984).

[15] V. N. Hingorani and Y.-K. Ho, *Biochemistry* **26**, 1633 (1987).

microelectrode.[16] This method is uniquely suited for analysis of photoreceptor PDEs because they have high catalytic rates. However, this assay requires relatively high concentrations of PDE and is not practical for analysis of a large number of samples. A detailed protocol and applications for the proton evolution PDE assay have been previously described.[17] A simple and economical method to measure PDE activity is based on generation and quantification of inorganic phosphate.[18] Inorganic phosphate is released from the product of cGMP hydrolysis, guanosine 5'-monophosphate (GMP), using 5'-nucleotidase or alkaline phosphatase. Application of this assay is limited by the potential presence or release of inorganic phosphate in the samples that is not caused by PDE activity. The PDE assay utilizing [³H]cGMP is the most common and sensitive method by which to analyze PDE activity.[19,20] It allows a study not only of photoreceptor PDEs, but also of enzymes from other PDE families with relatively low catalytic rates.

Assay of Phosphodiesterase Activity Using [³H]cGMP

The PDE assay is a two-step procedure. During the first step PDE converts [³H]cGMP to 5'-[³H]GMP. In the second step, alkaline phosphatase hydrolyzes 5'-[³H]GMP, producing [³H]guanosine and inorganic phosphate. [³H]Guanosine is separated from cGMP by using anion-exchange resin that binds phosphate-containing nucleotides.

To test the activation of PDE by $G_t\alpha$ or its mutants, the holoenzyme is reconstituted with urea-washed ROS membranes (uROS) and $G_t\beta\gamma$ in the presence of GTPγS. For these experiments, holo-PDE can be extracted from ROS membranes and purified as previously described.[21] uROS, $G_t\alpha$GDP, and $G_t\beta\gamma$ are isolated according to proved protocols.[22–24] The reaction mixtures (50-μl total volume) contain PDE (0.2 nM), varying concentrations of $G_t\alpha$GDP or its mutants (0–2 μM), $G_t\beta\gamma$ (2 μM), and uROS membranes (10 μM rhodopsin) in 20 mM Tris-HCl (pH 8.0) buffer, 100 mM NaCl, and 5 mM MgSO₄. They are prepared by mixing the proteins with a 10× assay buffer. Holo-PDE is excluded from control samples. GTPγS (10 μM) is added to the mixture, which is then incubated for 10 min at 25°. Alternatively, $G_t\alpha$GTPγS can be used in the PDE activation assay. $G_t\alpha$GTPγS is extracted from ROS, using GTPγS, and purified on Blue-Sepharose CL-6B as

[16] P. A. Liebman and A. T. Evanczuk, *Methods Enzymol.* **81,** 532 (1982).
[17] M. W. Kaplan and K. Palczewski, *in* "Methods in Neurosciences" (P. A. Hargrave, ed.), p. 205. Academic Press, San Diego, California, 1993.
[18] P. G. Gillespie and J. A. Beavo, *Mol. Pharmacol.* **36,** 773 (1989).
[19] W. J. Thompson and M. M. Appleman, *Biochemistry* **10,** 311 (1971).
[20] R. L. Kincaid and V. C. Manganiello, *Methods Enzymol.* **159,** 457 (1988).
[21] N. O. Artemyev and H. E. Hamm, *Biochem. J.* **283,** 273 (1992).
[22] G. Yamanaka, F. Eckstein, and L. Stryer, *Biochemistry* **24,** 8094 (1985).
[23] A. Yamazaki, M. Tatsumi, and M. W. Bitensky, *Methods Enzymol.* **159,** 702 (1988).
[24] C. Kleuss, M. Pallast, S. Brendel, W. Rosenthal, and G. Schultz, *J. Chromatogr.* **407,** 281 (1987).

described by Kleuss *et al.*[24] A 10× stock solution of [³H]cGMP is prepared by mixing 20 μl (20 μCi) of [³H]cGMP (specific activity, 11.2 Ci/mmol) (Amersham Pharmacia Biotech, Piscataway, NJ) with 25 μl of unlabeled cGMP (40 mM) and 955 μl of doubly distilled H$_2$O. cGMP hydrolysis is initiated with the addition of 5 μl of the [³H]cGMP stock solution (final concentration, 100 μM cGMP) and allowed to proceed for 10 min at 37°. The reaction is stopped by heating the samples for 2 min at 100°. Samples are cooled and treated with 0.1 unit of bacterial alkaline phosphatase (Sigma, St. Louis, MO) for 10 min at 37°. AG 1-X2 anion-exchange resin (Bio-Rad, Hercules, CA) (1 ml of a 20% bed volume suspension) is added and the samples are incubated for 10 min with periodic vortex mixing. Aliquots (0.6 ml) are withdrawn from the supernatants containing [³H]guanosine, mixed with 5 ml of 3a70B scintillation cocktail (Research Products International, Mount Prospect, IL), and counted with a Beckman (Fullerton, CA) LS 6500 liquid scintillation counter.

Procedures for Studies of G$_t\alpha$–Pγ Interaction Using Scanning Fluorescence Labeling of Pγ

Site-Directed Mutagenesis of Pγ Designed for Fluorescent Labeling

A cysteine residue represents an excellent site for introduction of a fluorescent probe to study G$_t\alpha$–Pγ interactions. Cysteine residues can be selectively labeled with a large number of commercially available SH-reactive probes. Furthermore, Pγ contains only one cysteine residue, Cys68, which is localized within the known G$_t\alpha$–Pγ interaction site. If an environmentally sensitive probe is localized near or within the site of interaction between two proteins, it is more likely to lead to a change in the probe fluorescence in the protein–protein complex. In addition to labeling the wild-type Pγ at Cys68, Pγ mutants can be designed to place a single cysteine at any desired position in the polypeptide chain. This allows not only identification of an optimal location for the fluorescence probe, but also a mapping of Pγ sites of interaction with G$_t\alpha$.

The pET-11a-Pγ expression vector has been widely used for Pγ mutagenesis.[25,26] First, PγCys68 must be substituted to introduce a cysteine residue at a selected position. A substitution of Cys68 for serine can be made by using an *Nco*I restriction site (codons for Pro[69] Trp[70]) (Fig. 1). Polymerase chain reaction (PCR)-based mutagenesis is carried out with a forward primer that contains an *Nde*I site and a reverse primer that contains an *Nco*I site and the mutation. The *Nde*I/*Nco*I-digested PCR fragment is then inserted into the *Nde*I/*Nco*I fragment of the Pγ

[25] N. P. Skiba, N. O. Artemyev, and H. E. Hamm, *J. Biol. Chem.* **270,** 13210 (1995).
[26] V. Z. Slepak, N. O. Artemyev, Y. Zhu, C. L. Dumke, L. Sabacan, J. Sondek, H. E. Hamm, M. D. Bownds, and V. Y. Arshavsky, *J. Biol. Chem.* **270,** 14319 (1995).

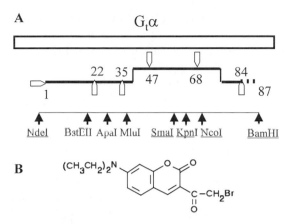

FIG. 1. (A) Schematic representation of BC-labeled Pγ mutants and their interactions with $G_t\alpha$. Restriction sites utilized in the cloning of Pγ mutants for fluorescence labeling are indicated (sites unique for the pET-11a vector are underlined). (B) Chemical structure of 3-(bromoacetyl)-7-diethylaminocoumarin (BC).

expression vector. The PγCys68Ser mutant gene serves as a basic construct for introduction of cysteine. In addition to the *Nco*I site, unique restriction sites *Nde*I, *Sma*I, *Kpn*I, and *Bam*HI allow the placement of cysteine at any position within the N terminus and the C-terminal half of Pγ, using PCR-directed mutagenesis (Fig. 1). A different procedure is required to incorporate a cysteine residue near or within the polycationic region of Pγ, Pγ-24–45, which is a region involved in binding to $G_t\alpha$.[12,27,28] The *Bst*EII, *Apa*I, and *Mlu*I restriction sites are not unique (Fig. 1). A potential strategy includes partial digestion of the pET-11a-Pγ vector with one of these enzymes. The linearized plasmid is isolated and subsequently cut at one of the unique restriction sites, *Nde*I, *Sma*I, *Kpn*I, *Nco*I, or *Bam*HI (Fig. 1). A plasmid DNA of appropriate size is separated and used for subcloning of a PCR-generated DNA fragment carrying a mutation and flanked with the chosen restriction sites. Alternatively, the Pγ cDNA can be subcloned from the pET-11a vector into a vector that does not contain *Bst*EII, *Apa*I, and *Mlu*I sites. Afterward, mutations can be generated by PCR-directed mutagenesis as outlined above. Because the *Bst*EII, *Apa*I, and *Mlu*I sites are in proximity to each other, mutations can also be made by using two annealed complementary mutant oligonucleotides with protruding ends that are compatible with selected restriction enzymes.

PCR are performed with a RoboCycler PCR amplifier (Stratagene, La Jolla, CA). Sizes of PCR fragments for Pγ mutants usually vary from 150 to 250 bp. The

[27] V. M. Lipkin, I. L. Dumler, K. G. Muradov, N. O. Artemyev, and R. N. Etingof, *FEBS Lett.* **234**, 287 (1988).
[28] R. L. Brown, *Biochemistry* **31**, 5918 (1992).

PCR mixtures contain 10–50 ng of pET-11a-PγCys68Ser as a template, 500 ng of each 5' and 3' primer, 0.2 mM dNTP mix, 5 μl of 10× PCR buffer, 2.5 units of *Taq* polymerase (Perkin-Elmer, Norwalk, CT), and doubly distilled H_2O to a final volume of 50 μl. After a 2-min dwell at 95°, 30 cycles of PCR are performed as follows: 94° for 50 sec, 56° for 50 sec, and 72° for 35 sec. The PCR products are separated on a 1.5% (w/v) agarose gel and purified with a Qiagen (Valencia, CA) gel extraction kit. The yields of PCR products are generally 2–4 μg. If several restriction sites (enzymes) are suitable for the mutagenesis, the preference is given to the enzymes that are optimally active in the same restriction buffer. The purified PCR product (1 μg) is incubated with 10 units of each restriction enzyme for 3–4 hr to ensure complete digestion. If simultaneous digestion is not possible, a PCR fragment is first cut with one enzyme, precipitated with 0.7 volume of 2-propanol, dissolved in doubly distilled H_2O, and digested with a second enzyme. An agarose gel-purified DNA fragment is subcloned into the pET-11a-Pγ vector digested with appropriate restriction enzymes. Becuase the sizes of excised fragments are small (150–250 bp), it is difficult to adequately separate fully and incompletely digested (linearized) plasmid. To avoid a high yield of colonies with the original plasmid, the digested vector is additionally treated with shrimp alkaline phosphatase (SAP) (1 unit/0.5 μg DNA, 1 hr, 37°) (U.S. Biochemical, Cleveland, OH) before ligation with a PCR fragment. In control experiments the digested and SAP-treated vector is tested, using ligation in the absence of the PCR fragment.

For the introduction of mutations by a pair of annealed complementary mutant oligonucleotides equimolar amounts (0.2 nmol/10 μl) of two oligonucleotides are mixed and incubated at 95° for 1 min. Afterward, the mix is incubated at room temperature for 20 min. The pET-11a-Pγ vector is prepared by cutting with selected restriction enzymes, separation on an agarose gel, and purification with a Qiagen gel extraction kit. The oligonucleotide duplex and the vector are mixed at a molar ratio of 20 : 1 and ligated overnight at 25°. The control ligation mix contains no duplex oligonucleotide.

Expression and Purification of Pγ or Its Mutants

BL21(DE3) cells carrying the pET-11a-Pγ plasmid or mutant plasmid are grown at 37° overnight in LB medium (1 liter: 10 g of tryptone, 5 g of yeast extract, 5 g of NaCl, 80 μl of 12.5 M NaOH) containing ampicillin (50 μg/ml). The overnight culture is diluted (1 : 250) with 2× TY medium (1 liter: 16 g of tryptone, 10 g of yeast extract, 5 g of NaCl) containing ampicillin (50 μg/ml) and is grown at 37° until it reaches an OD_{600} of ~0.8. The incubation temperature is then reduced to 30° and Pγ expression is induced by addition of 0.5 mM isopropylthiogalactoside (IPTG). The induction is allowed to proceed for 3–4 hr. Afterward, the cells are spun down and washed twice with 50 mM Tris-HCl buffer (pH 7.5) and resuspended in 50 mM Tris-HCl buffer (pH 7.5) containing 20 mM

NaCl, 5 mM EDTA, 1 mM dithiothreitol (DTT), and the protease inhibitors phenyl-methylsulfonyl fluoride (PMSF, 1 mM) and pepstatin A (20 μg/ml). The cells are disrupted by sonication with 30-sec pulses (4 min total sonication time), using a model 550 Sonic Dismembrator (Fisher, Pittsburgh, PA), followed by centrifuga-tion at 100,000g for 30 min at 4°. The supernatant is loaded on a SP Fast Flow Sepharose column (Amersham Pharmacia Biotech) equilibrated with 50 mM Tris-HCl buffer (pH 7.5) containing 20 mM NaCl, 5 mM EDTA, and 1 mM DTT. The bound proteins are eluted at a flow rate of 0.5 ml/min, using a 20–400 mM NaCl gradient. Pγ elutes at $\sim$300 mM NaCl. Additional purification of Pγ is achieved by reversed-phase high-performance liquid chromatography (HPLC) on a C$_4$ column (214TP54; Vydac, Hesperia, CA) with a 0–80% gradient of ace-tonitrile in 0.1% (v/v) trifluoroacetic acid–H$_2$O. Pγ elutes at $\sim$45% acetonitrile. Purified Pγ is lyophilized, dissolved in 20 mM HEPES buffer (pH 7.5), and stored at $-80°$ until use. This procedure yields up to 20 mg of >95% pure Pγ or mutant per liter of culture.

Labeling of Pγ or Pγ Mutants with Fluorescent Probe
3-(Bromoacetyl)-7-diethylaminocoumarin

Two environmentally sensitive probes have been extensively used for labeling Pγ, Lucifer Yellow vinyl sulfone (4-amino-N-[3-(vinylsulfonyl)phenyl]naphthali-mide 3,6-disulfonate) (LY)[12] and 3-(bromoacetyl)-7-diethylaminocoumarin (BC) (Molecular Probes, Eugene, OR).[13] For studies of the effector interface of G$_t\alpha$, la-beling with BC offers several advantages over the use of LY. Binding of G$_t\alpha$GTPγS or G$_t\alpha$GDP · AlF$_4^-$ to PγLY and to PγBC results in maximal fluorescence increases (F/F_0) of $\sim$2- to 3- and 6- to 8-fold, respectively (Fig. 2).[12,13] Furthermore, the affinities of these G$_t\alpha$ forms for PγBC (2–4 nM)[13] are significantly higher than those for PγLY (36 nM).[12] The inactive, GDP-bound G$_t\alpha$ also displays a stronger interaction with PγBC (K_d of 75 nM)[13] as compared with PγLY (K_d of 880 nM).[12]

To label Pγ or its mutants containing a single cysteine with BC, a stock solution of the probe (5 mM) is freshly prepared by dissolving BC in N',N-dimethylform-amide in a light-protected tube. Typically, a 2- to 3-fold molar excess of BC is added from the stock solution to 100–200 μM Pγ or mutant in 20 mM HEPES buffer (pH 7.5). The mixture is incubated for 30 min at room temperature and the reaction is stopped by addition of 2-mercaptoethanol (5 mM final concentration). The BC-labeled Pγ or mutant is then passed through a PD-10 column (Amersham Pharmacia Biotech) equilibrated with 20 mM HEPES buffer (pH 7.5) containing 100 mM NaCl to remove excess BC. The final purification step includes reversed-phase HPLC on a protein C$_4$ column (214TP54; Vydac), using a 0–80% gradient of acetonitrile–0.1% (v/v) trifluoroacetic acid to separate labeled Pγ from the unlabeled protein. Using an ε_{445} of 53,000 for BC, the molar ratio of BC to Pγ in the purified preparations is typically greater than 0.8 mol/mol.

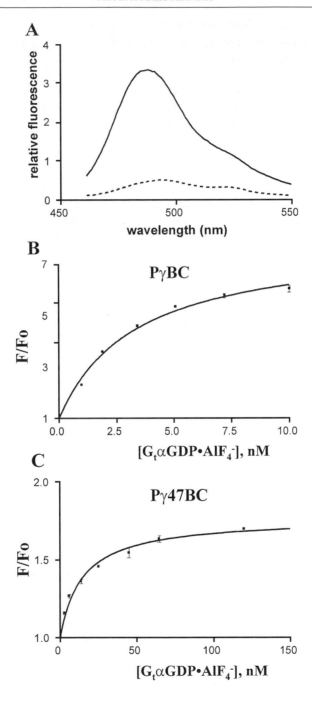

Fluorescence Assay of $G_t\alpha$ Binding to $P\gamma BC$

The assay is based on monitoring the fluorescence change resulting from binding of $G_t\alpha$ to BC-labeled $P\gamma$ or $P\gamma$ mutants. The fluorescence measurements are performed on a F-2000 fluorescence spectrophotometer (Hitachi Instruments, San Jose, CA) or on an AB2 fluorescence spectrophotometer (Spectronic Instruments, Rochester, NY) in 1 ml of 20 mM HEPES buffer (pH 7.5) containing 100 mM NaCl and 4 mM MgSO$_4$. Fluorescence of BC labeled $P\gamma$ (10 nM) is monitored at equilibrium before and after additions of increasing concentrations of $G_t\alpha$GDP or $G_t\alpha$GTPγS, using excitation at 445 nm and emission at 495 nm. When the AlF$_4^-$ activated conformation of $G_t\alpha$ is examined, AlCl$_3$ and NaF are added to the assay buffer before measurements to final concentrations of 30 μM and 10 mM, respectively. The measurements are taken 1–2 min after addition of $G_t\alpha$GDP, allowing it to assume the active conformation.

The K_d values are calculated by fitting the data to Eq. (1):

$$\frac{F}{F_0} = 1 + \frac{[(F/F_0)_{max} - 1]X}{K_d + X} \tag{1}$$

where F_0 is the basal fluorescence of $P\gamma BC$, F is the fluorescence after addition of $G_t\alpha$, $(F/F_0)_{max}$ is the maximal relative increase of fluorescence, and X is the concentration of free $G_t\alpha$. The X value is determined as $[G_t\alpha]_{total} - [G_t\alpha]_{bound} = [G_t\alpha]_{total} - [P\gamma BC](F - F_0)/(F_{max} - F_0)$; where $[P\gamma BC]$ is the concentration of $P\gamma BC$ in the assay, typically 10 nM.

Unlabeled $P\gamma$ or $P\gamma$ mutants can be used to compete with $P\gamma BC$ for the binding to $G_t\alpha$. This competition leads to dissociation of $P\gamma BC$ from $G_t\alpha$ and a decrease in the BC fluorescence. The EC$_{50}$ values can be calculated by fitting the data to the one-site competition equation with variable slope:

$$\frac{F}{F_0} = 1 + \frac{(F/F_0)_{max} - 1}{1 + 10^{(X - \log EC_{50})H}} \tag{2}$$

where X is the concentration of $P\gamma$ or $P\gamma$ mutant, and H is the Hill slope. Fitting of the data is performed with nonlinear least-squares criteria, using Graph-Pad (San Diego, CA) Prizm software. The dissociation constants ($K_{1/2}$) for the

---

FIG. 2. (A) The emission spectra of $P\gamma BC$ (10 nM) alone and in the presence of 100 nM $G_t\alpha$GDP are recorded on an AB2 fluorescence spectrophotometer (Spectronic Instruments) in a stirred cuvette with 1 ml of 20 mM HEPES buffer (pH 7.5) containing 100 mM NaCl, 4 mM MgSO$_4$, 30 μM AlCl$_3$, and 10 mM NaF, using excitation at 445 nm. (B and C) Binding of $G_t\alpha$GDP · AlF$_4^-$ to $P\gamma BC$ and $P\gamma 47BC$. The relative increase in fluorescence (F/F_0) of $P\gamma BC$ (B) and $P\gamma 47BC$ (C) (10 nM each) was determined after addition of increased concentrations of $G_t\alpha$GDP · AlF$_4^-$. The binding curve characteristics are as follows: (B) $K_d = 3.3 \pm 0.3$ nM, maximum $F/F_0 = 7$; (C) $K_d = 14 \pm 1$ nM, maximum $F/F_0 = 1.8$.

Pγ/G$_t\alpha$ interaction can be calculated from the EC$_{50}$ values, using the Cheng and Prusoff equation[29] describing competitive displacement, $K_{1/2} = \text{EC}_{50}/(1 + [\text{P}\gamma\text{BC}]/K_d)$, where EC$_{50}$ is the concentration of Pγ or mutant that reduces the relative fluorescence increase by 50% [from Eq. (2)], [PγBC] is the total concentration of PγBC, and K_d is the dissociation constant for the PγBC/G$_t\alpha$ complex [from Eq. (1)]. If the K_d and $K_{1/2}$ values are comparable to the total concentration of G$_t\alpha$ in the assay, then equations derived by Linden[30] are appropriate for the calculation of $K_{1/2}$.

Application of PγBC Mutants to Mapping G$_t\alpha$-Binding Sites on Pγ

Pγ mutants, each with a single cysteine residue evenly spaced along the Pγ polypeptide chain, have been generated[31] with the pET-11a-PγCys68Ser construct and the mutagenesis procedures outlined in this chapter. Pγ residues at positions 2, 22, 35, 47, and 84–87 have been substituted by cysteine, which is then labeled with BC (Fig. 1). Binding of activated G$_t\alpha$ to the BC-labeled wild-type Pγ (Pγ68BC) and to Pγ47BC leads to a maximal fluorescence enhancement of 7-fold and $\sim$1.8-fold, respectively (Fig. 2). Fluorescence enhancement of Pγ labeled at Cys68 supports the data implicating the region surrounding this residue in the Pγ/G$_t\alpha$ interaction.[10,25] Fluorescence increase of Pγ47BC in the presence of G$_t\alpha$ correlates well with the finding that the triple mutation of Pγ residues K41, K44, and K45 eliminates the ability of G$_t\alpha$ to activate PDE.[28] We found that G$_t\alpha$ does not affect the fluorescence of Pγ2BC, Pγ22BC, Pγ35BC, and Pγ1-83BC, suggesting that either the corresponding regions do not participate in the interaction with transducin, or the probe is oriented away from the contact site. The agreement between existing biochemical evidence and the results from the scanning fluorescence labeling indicates that this approach can be used for mapping protein–protein interactions. This analysis also demonstrates that Cys68 in Pγ naturally represents an optimal site for introduction of fluorescent probes to study the Pγ/G$_t\alpha$ interaction.

Application of PγBC-Binding Assay for Identification of Effector Residues of Transducin

Our studies of delineation of the effector interface of transducin[9,32] illustrate the utility of the fluorescence binding assay using PγBC. In these studies a transducin-like chimeric G$_t\alpha$/G$_i\alpha$ protein, allowing functional expression in

[29] Y.-C. Cheng and W. H. Prusoff, *Biochem. Pharmacol.* **22,** 3099 (1973).

[30] J. Linden, *J. Cyclic Nucleotide Res.* **8,** 163 (1982).

[31] A. E. Granovsky, R. McEntaffer, and N. O. Artemyev, *Cell Biochem. Biophys.* **28,** 115 (1998).

[32] M. Natochin, A. E. Granovsky, K. G. Muradov, and N. O. Artemyev, *J. Biol. Chem.* **274,** 7865 (1999).

Escherichia coli, has been used as a template for mutational analysis of candidate effector regions of $G_t\alpha$. The procedures for site-directed mutagenesis, expression, purification, and evaluation of correct folding of chimeric transducin mutants have been summarized.[33] The analysis of $P\gamma BC$ binding to the GDP-bound and AIF_4^--activated conformations of chimeric $G_t\alpha$ mutants revealed two groups of effector residues on $G_t\alpha$.[9] Switch II residues Arg201, Arg204, Trp207, and Ile208 are essential for the conformation-dependent $G_t\alpha$–effector interaction, whereas residues His244 and Asn247 in the α_3 helix are responsible for the conformation-independent effector-specific interaction.

Acknowledgments

This study was supported by NIH Grant EY-10843.

[33] M. Natochin and N. O. Artemyev, *Methods Enzymol.* **315,** 539 (2000).

[3] Assaying G Protein–Phosphodiesterase Interactions in Sensory Systems

By KOSEI MORIYAMA, MANJIRI M. BAKRE, FAROOQ AHMED,
NANCY SPICKOFSKY, MARIANNA MAX, and ROBERT F. MARGOLSKEE

Introduction

Regulation of retinal phosphodiesterase (PDE) by the transducins is well known, and essential to signal transduction in both rod and cone photoreceptor cells of the retina. Data suggest that G proteins may regulate other PDE isoforms in nonretinal sensory systems (e.g., gustatory,[1] pineal,[2] and parietal eye[3]) and nonsensory cells (retinal bipolar neurons[4] and hepatocytes[5]). We have isolated PDEs from retinal, gustatory, and pineal tissues and tested them for interactions with multiple G protein α subunits (e.g., rod and cone α-transducins, α-gustducin, and $G\alpha_i$) using recombinant G protein subunits, as well as wild-type and substituted peptides corresponding to the so-called effector interaction region of the

[1] L. Ruiz-Avila, S. K. McLaughlin, D. Wildman, P. J. McKinnon, A. Robichon, N. Spickofsky, and R. F. Margolskee, *Nature (London)* **376,** 80 (1995).
[2] F. A. Ahmed and M. Max, unpublished observations (2000).
[3] W. H. Xiong, E. C. Solessio, and K. W. Yau, *Nat Neurosci.* **1,** 359 (1998).
[4] S. Nawy and C. E. Jahr, *Nature (London)* **346,** 269 (1990).
[5] M. Robles-Flores, G. Allende, E. Pina, and J. A. Garcia-Sainz, *Biochem. J.* **312,** 763 (1995).

0076-6879/02 $35.00

G protein α subunit. In this chapter we describe methods for preparing PDEs from sensory tissues and monitoring G protein–PDE interactions using peptides and recombinant proteins.

Methods

Preparation and Characterization of Phosphodiesterases from Sensory Tissues

Preparation of Gustatory Phosphodiesterases

Reagents. Leupeptin and aprotinin are obtained from Boehringer Mannheim Biochemicals (Indianapolis, IN). The DEAE anion-exchange column is from Bio-Rad (Hercules, CA). Peptides are synthesized with a free N terminus and amidated C terminus by the solid-phase method, using an Applied Biosystems 431A synthesizer (PE Biosystems, Foster City, CA). All other reagents are purchased from Sigma (St. Louis, MO).

Preparation of Taste Extract. One to 2 g of circumvallate taste papillae or fungiform papillae from fresh bovine tongues (obtained from a local abattoir) is dissected and homogenized with a Polytron homogenizer (Brinkmann, Westbury, NY) in a buffer containing 20 mM Tris-HCl (pH 7.5), 1 mM EDTA, 1 mM dithiothreitol (DTT), 10% (v/v) glycerol, leupeptin (10 μg/ml), aprotinin (10 μg/ml), 100 μM 4-(2-aminoethyl)-benzene sulfonyl fluoride (AEBSF), and pepstatin (10 μg/ml). The homogenate is spun at 2500g for 10 min at 4°, and the resulting supernatant is then spun at 100,000g for 30 min at 4°. The supernatant of the second spin is checked for PDE activity and fractionated by anion-exchange chromatography.

Anion-Exchange Chromatography. The supernatant prepared from taste papillae is loaded at 1 ml/min on a DEAE-10 anion-exchange column that is pre-equilibrated with buffer A [20 mM Tris-HCl (pH 7.5), 2 mM EDTA, 20 mM 2-mercaptoethanol]. The column is washed for 15 min or until protein in the column wash buffer (buffer A) is no longer detected. The bound protein activity is eluted at a flow rate of 2 ml/min over 45 min with a linear gradient of NaCl (0–1 M) in buffer A. Fractions are collected at 4° and assayed for PDE activity. Fractions can be stored in buffer A at 4° for 7–10 days without significant loss of activity.

Preparation of Bovine Retinal Phosphodiesterase

Urea-washed bovine retinal rod outer segment (ROS) membranes are prepared as described previously.[6–8] Membrane-associated rod PDE (PDE6) is isolated by taking the ROS membrane fraction and subjecting it to a series of hypotonic washes, followed by purification of PDE on DE52 and G-100 chromatography columns.[9] The retinal PDE is typically found to be ~80% pure, as described previously.[10]

---

[6] D. S. Papermaster and W. J. Dreyer, *Biochemistry* **13**, 2438 (1974).
[7] H. G. Smith, Jr., G. W. Stubbs, and B. J. Litman, *Exp. Eye Res.* **20**, 211 (1975).
[8] J. H. McDowell and H. Kuhn, *Biochemistry* **16**, 4054 (1977).

Preparation of Chicken Pineal Phosphodiesterase

Preparation of Pineal Extract. Thirty pineals from 2-week-old chicks (Charles River SPAFAS, Wilmington, MA) are dissected and homogenized in a buffer containing 10 mM Tris-HCl (pH 7.5), 1 mM EDTA, 1 mM DTT, 10% (v/v) glycerol, and 1× complete EDTA-free protease inhibitors (Roche, Nutley, NJ). Homogenization is carried out with a Polytron homogenizer. The homogenate is first spun at 3000g for 10 min at 4° and the resulting supernatant is spun at 100,000g for 60 min at 4°. The supernatant of the second spin is checked for PDE activity and fractionated by anion-exchange chromatography.

Anion-Exchange Chromatography. The supernatant prepared from pineal tissue is loaded at 0.5 ml/min on a DEAE-2 anion exchange column (Bio-Rad) that is preequilibrated with buffer A. The column is washed for 20 min or until the protein in the column wash buffer (buffer A) is no longer detected. The bound protein activity is eluted over 15 min at a flow rate of 1 ml/min with a linear gradient of NaCl (0–1 M) in buffer A. Fractions (0.5 ml) are collected at 4° and assayed for PDE activity.

Preparation and Purification of Recombinant Gα Subunits and PDE6γ

Plasmid Construction and in Vitro Transcription and Translation

The synthetic gene of bovine rod α-transducin[11] is cloned into pBluescript II KS+ (Stratagene, La Jolla, CA) to generate pBSGtα. Mutagenesis is done by polymerase chain reaction (PCR), using substituted oligonucleotide primers and *Pfu*I DNA polymerase (Stratagene). The synthetic gene for PDE6γ is amplified by PCR, using pLcIIFXSG[12] as the template; the product is cloned into pCMV-SPORT (GIBCO, Ronkonkoma, NY), which has SP6 and cytomegaloverus (CMV) promoters upstream of the coding region. This construct is designated pCMVPγ. [35S]Methionine-labeled proteins are produced with the TNT-coupled reticulocyte lysate system (Promega, Madison, WI).

Sf9 Cell Culture and Purification of Recombinant Hexahistidine-Tagged α-Transducin

Generation of Recombinant Constructs. The coding region of α-transducin is excised from pBSGtα and cloned into pFastBacHTc (GIBCO) to generate pFastGtα. The N-terminal tag of the expressed protein is MSYYHHHHHHDYDI-PTTENLYFQ*GA-M, where the asterisk indicates the cleavage site of tobacco etch virus protease. The final "M" in the tag corresponds to the natural methionine start

---

[9] W. Baehr, E. A. Morita, R. J. Swanson, and M. L. Applebury, *J. Biol. Chem.* **257,** 6452 (1982).

[10] N. Spickofsky, A. Robichon, W. Danho, D. Fry, D. Greeley, B. Graves, V. Madison, and R. F. Margolskee, *Nat. Struct. Biol.* **11,** 771 (1994).

[11] T. P. Sakmar and H. G. Khorana, *Nucleic Acids Res.* **16,** 6361 (1988).

[12] R. L. Brown and L. Stryer, *Proc. Natl. Acad. Sci. U.S.A.* **86,** 4922 (1989).

codon of α-transducin. Generation of composite bacmids and baculovirus stocks utilizes the Bac-to-Bac kit (GIBCO): Transposition of the coding region into the virus genome is done by transfecting *Escherichia coli* DH10Bac with wild-type and mutant pFastGtα plasmids. Recombinant bacmid DNA is obtained by antibiotic selection with kanamycin, gentamicin, and tetracycline, and blue–white color selection with isopropylthiogalactoside (IPTG) and Bluo-gal (GIBCO).

Production and Purification of Proteins. Sf9 (spodoptera frugiperda ovary) cells cultured in Sf900 medium (GIBCO) with 7% (v/v) fetal bovine serum are transfected with the viral DNAs; the recombinant baculoviruses are harvested, and the titers determined. For large preparations of the hexahistidine (His$_6$)–α-transducin, Sf9 cells are grown in 1000 ml of culture medium in 2.8-liter glass flasks (Pyrex 4420) and infected with virus at a multiplicity of infection (MOI) of 10. Two days after infection, cells are collected by centrifugation (750 g) for 10 min at 4°, and suspended in 50 ml of cell lysis buffer [50 mM Na–HEPES (pH 8.0), 0.1 mM EDTA, 3 mM MgCl$_2$, 10 mM 2-mercaptoethanol, 100 mM NaCl, 10 μM GDP, leupeptin (30 μg/ml), pepstatin A (1 μg/ml), aprotinin (1 μg/ml)].[13] All remaining procedures are carried out at 4°, unless specified otherwise. Cells are lysed with a Potter–Elvehjem homogenizer, and centrifuged at 750 g for 10 min to remove intact cells and nuclei. The supernatant is centrifuged at 100,000g for 30 min, collected, and diluted 5-fold with buffer KA [20 mM Na–HEPES (pH 8.0), 100 mM NaCl, 1 mM MgCl$_2$, 10 mM 2-mercaptoethanol, 10 μM GDP], and then loaded onto a 10-ml DEAE-Sephacel column that is preequilibrated with buffer KA. The column is washed and the proteins eluted with a linear gradient (10–250 mM NaCl) in buffer KA. A portion of each fraction is dotted onto a nitrocellulose membrane and α-transducin is detected by a monoclonal antibody (TF15; American Qualex, San Clemente, CA) and ECL (Amersham, Arlington Heights, IL). The peak fraction is mixed with 400 μl of Ni-NTA resin (Qiagen, Valencia, CA), which is preequilibrated with Ni-NTA wash buffer [50 mM NaH$_2$PO$_4$ (pH 8.0), 300 mM NaCl, 20 mM imidazole], and then rotated for 1 hr at 4°. The resin is washed four times with Ni-NTA wash buffer (10 ml) at 25°. The protein is eluted with 1 ml of elution buffer [50 mM NaPO$_4$ (pH 8.0), 300 mM NaCl, 150 mM imidazole] and dialyzed against storage buffer [10 mM Na–HEPES (pH 7.6), 2 mM MgCl$_2$, 1 mM DTT, 50% (v/v) glycerin]. Samples are analyzed by silver staining[14] and/or immunoblotting[15] with the TF15 monoclonal antibody after electrophoresis through a 12% (w/v) nongradient sodium dodecyl sulfate (SDS)–polyacrylamide gel. The concentrations of wild-type and alanine-substituted His$_6$–α-transducin proteins are determined by spectrophotometric means, using bovine serum albumin as a standard, and then normalized by the addition of storage buffer. The typical yield from 1 liter of Sf9 cells varies form 50 to 100 μg of α-transducin protein.

[13] T. Kozasa and A. G. Gilman, *J. Biol. Chem.* **270**, 1734 (1995).

[14] W. Wray, T. Boulikas, V. P. Wray, and R. Hancock, *Anal. Biochem.* **118**, 197 (1981).

[15] E. Harlow and D. Lane, "Using Antibodies: A Laboratory Manual." Cold Spring Harbor Laboratory Press, Cold Spring Harbor, New York, 1999.

Phosphodiesterase Assays

Peptide-Stimulated Phosphodiesterase Assay

Selection of Peptides. Peptides (12- to 22-mers) derived from the α_4 helix and adjacent α_4/β_6 loop of rod α-transducin [amino acids (aa) 293–314] activate retinal PDE (PDE6), PDEs derived from bovine taste tissue,[1,16] and chick pineal.[2] Homologous peptides derived from cone α-transducin and α-gustducin also activate PDE6, while peptides from α_i do not, identifying candidate residues within this region important to PDE6 activation.[10] Alanine-substituted and truncated peptides can be used to identify candidate residues required for PDE activation. The peptides provide a rapid "first pass" at identifying residues involved in contacting PDE or otherwise important to the G protein–PDE interaction. However, follow-up studies with recombinant proteins are strongly advised, as results with the substituted peptides may differ from those obtained with recombinant proteins.

Reagents. Crude α-transducin peptides are purified by reversed-phase high-performance liquid chromatography (HPLC) on a 2.3 × 30 cm Bondapak C_{18} column (ES Industries, West Berlin, NJ). The peptides are eluted with a linear gradient (3 hr) of 5–65% (w/v) acetonitrile containing 0.1% (w/v) trifluoroacetic acid at a flow rate of 8 ml/min with detection at 215 nm. The purified peptides are dissolved in distilled water as 1 mM solutions and then stored at $-20°$ before use. [8,5-^{3}H]cGMP and [2,8-^{3}H]cAMP are purchased from NEN-Research Products (Boston, MA).

ASSAY PROCEDURES: PHOSPHATE RELEASE ASSAY. Colorimetric monitoring of phosphate release, determined in triplicate, is used to determine peptide-induced PDE activity.[17] The reaction is done in 90 μl of PDE assay buffer [40 mM Tris (pH 7.5), bovine serum albumin (BSA, 0.5 mg/ml), and 10 mM MgCl$_2$] containing retinal PDE (5 nM) and cGMP (2 mM). The reaction is carried out for 10 min at 30°; during the last 2 min, 10 μl of *Ophiophagus hannah* snake venom (1 mg/ml; Sigma) is added to each tube. The released phosphate is detected by adding an equal volume of color mix [ascorbic acid (100 mg/ml), ammonium molybdate (10 mg/ml), 0.2 N H$_2$SO$_4$]. The absorbance of the wells of the microtiter plate is determined at 750 nm. The mean value ± SEM is plotted. The maximal level of peptide-stimulated PDE activity is determined by comparison with the positive control peptide (aa 293 to 314 of α-transducin).

ASSAY PROCEDURES: PHOSPHODIESTERASE ASSAY USING ^{3}H-LABELED cNMPs AND DEAE-SEPHADEX. The assays are performed according to established procedures.[18] All assays are performed at 30° with either 1 μM cGMP or cAMP as substrate.

---

[16] M. M. Bakre, J. G. Glick, S. Rybalkin, M. Max, J. Beavo, and R. F. Margolskee, submitted (2001).

[17] P. G. Gillespie and J. A. Beavo, *Mol. Pharmacol.* **36,** 773 (1989).

[18] W. K. Sonnenburg, S. D. Rybalkin, K. E. Bornfeldt, K. S. Kwak, I. G. Rybalkina, and J. A. Beavo, *Methods* **14,** 3 (1998).

Recombinant G Protein-Stimulated Phosphodiesterase Assays

ASSAY PROCEDURES: PHOSPHODIESTERASE ASSAY USING ACID ALUMINA METHOD. The PDE-stimulable activity of His_6–α-transducin expressed in Sf9 cells is determined by monitoring [^{3}H]cGMP degradation[19] after acid alumina column separation.[20] Holo-PDE (2 pM) and α-transducin (2.5 nM) are incubated in PDE buffer (50 μl), with or without NaF (0.5 mM) and $AlCl_3$ (2.5 μM) for 15 min at 30°. The reaction is started by adding PDE assay buffer [40 mM Tris (pH 7.5), bovine serum albumin (0.5 mg/ml), and 10 mM $MgCl_2$] (50 μl) containing cGMP (6 μM), with [^{3}H]cGMP (specific activity, $\sim$1000 cpm/pmol), and then the reaction is continued for 15 min at 30°. This reaction is terminated by boiling for 2 min in a heat block, and then the tube is placed on ice. Snake venom nucleotidase (20 μl of a 1-mg/ml solution) is added to each tube, incubated for 20 min at 30°, and boiled for 2 min. The 5'-nucleotidase reaction generates nucleosides, which are separated with the alumina column. The sample is applied onto a dry 0.7-g type 1 acidic alumina column (Sigma) and washed with 0.005 N HCl (3 ml), which is collected into a scintillation vial, mixed with 13 ml of BetaMax (ICN, Costa Mess, CA), and counted.

Monitoring Activity of Recombinant G Proteins

Limited Proteolysis Conformational Assay

We determine the activation state of the alanine-substituted α-transducin proteins by the characteristic cleavage pattern generated on protease treatment. For AIF_4^--induced activation, 0.6 μl of *in vitro* translation mix is combined with 20 μl of trypsin assay buffer [25 mM Tris (pH 7.5), 2 mM $MgCl_2$, 50 mM NaCl, and 5 mM DTT] plus GDP (100 μM) and transferred to a tube containing separate 1-μl drops of AlF_3 (100–2000 μM) and NaF (10–200 mM) (the final concentration of AIF_4^- is 5–100 μM). After 3 min, 1 μl of tolylsulfonyl phenylalanyl chloromethyl ketone (TPCK)-treated trypsin (0.5 mg/ml) is added and the reaction is incubated for 45 min at 25°. The proteolysis reaction is terminated by the addition of 1.5 μl of soybean trypsin inhibitor (3 mg/ml). For the activation assay with rhodopsin, translation mix (0.6 μl) is mixed with 0.2 μl of α-transducin (2 pmol) and incubated for 10 min. Trypsin assay buffer (20 μl) containing 2 mM GDP and 5 μM GTPγS is added, and the mixture is transferred to a tube containing 2 μl of urea-washed ROS (the final rhodopsin concentration is 0.0125–1.25 mM). If basal activation of the G protein is high, the concentration and ratio of GDP to GTPγS are adjusted as necessary (either decrease the GTPγS concentration, or increase the GDP-to-GTPγS ratio). For the data presented herein we always used 2 mM GDP and 5 μM GTPγS. The fragments are separated by SDS–polyacrylamide gel electrophoresis (PAGE),

[19] S. H. Francis and J. D. Corbin, *Methods Enzymol.* **159,** 722 (1988).
[20] R. Alvarez and D. V. Daniels, *Anal. Biochem.* **203,** 76 (1992).

using a 4–20% gradient gel. The electrophoresed polypeptides are transferred to a nitrocellulose membrane, which is exposed overnight to a PhosphorImage screen. The counts of the specific bands are quantified with the ImageQuant software of the Molecular Dynamics PhosphorImage system.

GTPγS-Binding, GTPase, and GTPase Acceleration Activity Assay

In vitro-translated wild-type and mutated α-transducins are used for these assays. For GTPγS binding, 2 μM [γ^{35} S]GTP (specific activity, 1000 cpm/pmol), 0.5 μl of urea-washed ROS (0.5 μM rhodopsin), 2 pmol of retinal α-transducin, and 2 μl of reticulocytes containing 4 ng of *in vitro*-translated α-transducin are mixed in GTPγS buffer [20 mM Na–HEPES (pH 7.5), 100 mM NaCl, 10 mM MgCl$_2$, 1 μM GDP, 2 mM DTT]. The reaction is started by the addition of α-transducin, and terminated by the addition of 500 μl of ice-cold stopping buffer [20 mM Tris (pH 8.0), 100 mM NaCl, 25 mM MgCl$_2$, 2 mM DTT], and then the proteins are bound to a glass fiber filter and washed with the stopping buffer and the bound [γ^{35}-S]GTP is counted in scintillation liquid.

Results

Fractionation of α-Transducin-Stimulated Phosphodiesterases Present in Taste Tissue

Taste tissue is known to be rich in PDE activity[21–23] and regulation of cNMPs has been implicated in sweet as well as bitter signal transduction.[24] Earlier studies showed that α-transducin and/or α-gustducin activate PDE6 and taste tissue-expressed PDEs of previously unknown subtype[1]; the taste PDE activity has now been shown to be two isoforms of PDE1A.[16] To identify which PDEs in taste tissue are regulated by transducin/gustducin, the PDE isoforms expressed in bovine taste tissue were resolved by anion-exchange chromatography. Each fraction was assayed for cGMP and cAMP PDE activity in the absence (basal activity) or presence of the α-transducin peptide (peptide 1).

A typical fractionation profile of cGMP PDE activity revealed four major peaks of activity (Fig. 1). The first peak (fractions 6–7) along with fractions 11–13 of the second peak had low basal activity that was stimulated strongly by the α-transducin peptide. The third peak (fractions 13–15) showed greatest activity when stimulated by the α-transducin peptide, but also displayed significant basal activity, indicating the presence of PDEs that are not stimulated by the peptide, as well as peptide-stimulated PDE. The fourth peak (fractions 21–23) was not stimulated above the

[21] S. Price, *Nature (London)* **241**, 54 (1973).

[22] H. Nomura, *Chem. Senses Flav.* **3**, 319 (1978).

[23] P. Kurihara, *FEBS Lett.* **27**, 279 (1972).

[24] T. A. Gilbertson, S. Damak, and R. F. Margolskee, *Curr. Opin. Neurobiol.* **10**, 519 (2000).

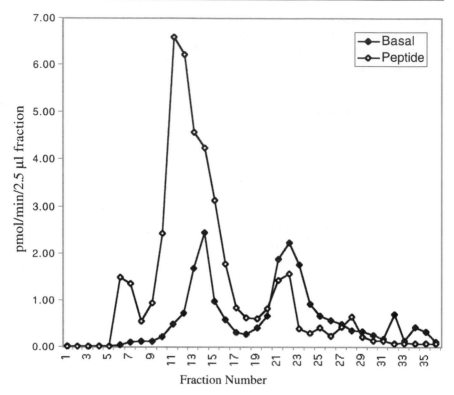

FIG. 1. Fractionation of α-transducin-stimulated PDEs present in taste tissue. PDEs present in the cytoplasm of bovine circumvallate papillae were resolved by DEAE anion-exchange chromatography: proteins were eluted from the DEAE column by a 0–1 M NaCl gradient in buffer A. Assays of PDE activity of each fraction were performed as described (see Methods), using 1 μM cGMP as substrate in the absence of the α-transducin peptide (basal activity, solid diamonds), or in the presence of 100 μM α-transducin peptide (open diamonds).

basal level by the α-transducin peptide. The α-transducin peptide-responsive PDEs present in the first and the second peaks have been identified as PDE1A isoforms.[16]

Identification of G Protein α-Subunit Residues Involved in Phosphodiesterase Activation

Peptide 1, corresponding to aa 293–314 of α-transducin, was previously shown to activate retinal PDE (PDE6)[10,25] and bovine taste PDE (PDE1A).[1,16] To identify the core region of the peptide (and hence of α-transducin) required for activation of PDE6 and PDE1A, we synthesized a series of peptides with N- and C-terminal truncations (Fig. 2). All of the N-terminally truncated peptides induced full activation

[25] H. M. Rarick, N. O. Artemyev, and H. E. Hamm, *Science* **256,** 1031 (1992).

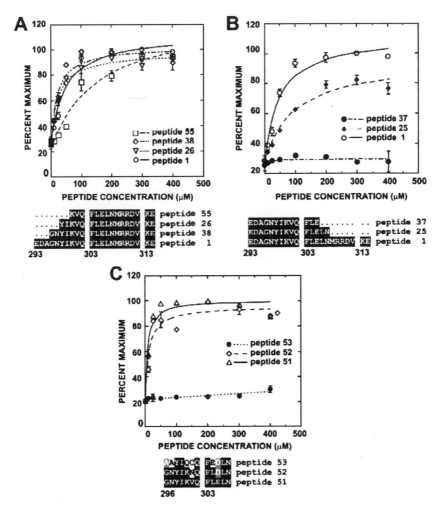

FIG. 2. The minimal region of α-transducin-derived peptides required to activate PDE6 was determined. Peptides at the indicated concentrations were added to 5 n*M* PDE6 in isotonic buffer, cGMP was then added, and PDE activity was measured by the phosphate release assay. Peptide-stimulated PDE activity is expressed as a percentage of stimulation with the control peptide (peptide 1) at 400 n*M*. (*A*) The control peptide (peptide 1, aa 293–314 of α-transducin) and the three N-terminally truncated α-transducin peptides (peptides 26, 38, and 55) all fully activated PDE. (*B*) An α-transducin peptide (peptide 25, aa 293–307) with a C-terminal truncation had ~80% of the activity of control peptide 1. An α-transducin peptide (peptide 37, aa 293–305) with a more extensive truncation at the C terminus did not activate PDE6. (*C*) Peptides derived from the PDE6 activation region of α-transducin (peptide 51, aa 296–307) and the corresponding regions of α-gustducin (peptide 52) and α_{i-3} (peptide 53) were tested for the ability to activate PDE6. The α-transducin and α-gustducin peptides fully activated PDE6, while the α_{i-3} peptide did not.

of PDE6 comparable to that achieved with the peptide 1 22-mer (Fig 2A), indicating that the N-terminal border is distal to aa 300. The C-terminally deleted 15-mer peptide corresponding to aa 293–307 (peptide 25) fully activated PDE6 (Fig 2B), while further truncation led to a nonactivating peptide (peptide 37, aa 293–305), placing the C-terminal border between aa 305 and aa 307. We infer from these results that the core region of α-transducin required for activation of PDE6 is aa 300–307. Peptide 25 fully activated PDE1A to the same extent as did the 22-mer α-transducin peptide (peptide 1) (data not shown), suggesting similar requirements for α-transducin activation of PDE6 and PDE1A.

Within this core region α-transducin differs from the related α-gustducin and α_{i-3} by two or four amino acids, respectively. To determine if these α-transducin-related α subunits would also activate PDE we tested peptides from these α subunits that corresponded to the α-transducin core region. Our results show that the 12-mer peptides derived from α-transducin (peptide 51, aa 296–307) and α-gustducin (peptide 52) fully activated PDE6, while the corresponding peptide from α_{i-3} did not. These results suggest that the two amino acid differences between α-transducin and α-gustducin in this region do not affect their ability to activate PDE6, but apparent critical differences between α-transducin and α_{i-3} in this core region (e.g., K300Q, V301C, and L304E) determine the differential ability of these two G protein α subunits to activate PDE6. Studies using peptides with single alanine substitutions in this region to activate PDE6 also suggest that positions 300, 301, and 304 are important to the α-transducin–PDE6 interaction.[26] However, results with alanine-substituted recombinant α-transducin protein yielded rather different results (see below).

Use of Recombinant α-Transducin Mutants to Activate Phosphodiesterase

Data from truncated and substituted peptides provide an indication of those G-protein α-subunit residues involved in activating PDE. To confirm and/or extend these results, recombinant proteins with specific substitutions should be used. Before using recombinantly expressed G protein α subunits of wild-type and mutant transducin (or other G proteins) in PDE assays, it is imperative to determine that the recombinant proteins fold properly and are capable of activation. Recombinant wild-type and mutant proteins were expressed *in vitro,* and then biochemically assayed for the ability to interact with rhodospin. ^{35}S-labeled wild-type or mutant α-transducin proteins expressed in rabbit reticulocyte lysate were subjected to the trypsin protection assay to determine which mutant proteins folded properly and could undergo GTP-dependent activation on stimulation by rhodopsin. The cleavage pattern derived from the trypsin protection assay distinguishes unfolded from GDP- or GTP-bound forms of α-transducin.[1] Unfolded or partially folded α-transducin is cleaved by trypsin at many sites to generate multiple small fragments.

[26] K. Moriyama, E. Mehler, D. Fry, H. Weinstein, M. Max, and R. F. Margolskee, submitted (2001).

The inherent structure of properly folded GDP-bound α-transducin renders it inaccessible to most of the potential trypsin recognition sites, and on digestion with trypsin leads to generation of two doublet sets of fragments with a relative molecular mass of 23/21 and 17/12 kDa.

Cleavage of the N-terminal His$_6$ tag with tobacco etch virus protease proved unnecessary: baculovirus-expressed α-transducin protein with or without this N-terminal tag displayed the same trypsin digestion pattern, showed an equivalent activation by AlF (based on conformational shift in trypsin activation assays), and displayed the equivalent ability to activate PDE6 (data not shown). Furthermore, the protein with the N terminus removed by protease could form heterotrimers (on the addition of retinal $\beta\gamma$ subunits) which were activated by rhodopsin. Trypsin cleavage of AlF$_4^-$- or rhodopsin-activated wild-type α-transducin and four of the mutants with single alanine substitutions in the core region (K300A, V301A, Q302A, and L306A) generated the 33/31-kDa bands diagnostic of the active (GTP-bound) form of α-transducin (data not shown). In the absence of rhodopsin, or at low concentrations of rhodopsin, these four mutants, like wild type, generated 23/21- and 17/21-kDa doublets diagnostic of GDP-bound α-transducin. In contrast, F303A-substituted α-transducin was completely degraded by trypsin treatment, indicating that this substituted protein is not properly folded. Similar results were obtained after AlF$_4^-$ stimulation of F303A (data not shown).

Position L306 of α-Transducin Required for Phosphodiesterase Activation

The four properly folding alanine substitution mutants (K300A, V301A, Q302A, and L306A) were tested for the ability to activate PDE6. Because PDE activation assays require more protein than can be generated practically by *in vitro* translation, we expressed wild-type and mutant α-transducin in Sf9 cells, using recombinant baculoviruses with N-terminal His$_6$ tags. The nickel column-purified proteins (50 to 150 $\mu g/\mu l$) were of $\geq 80\%$ purity, determined by silver staining of SDS–polyacrylamide gels. On activation with AlF$_4^-$ (0.5mM NaF and 2.5 mM AlCl$_3$) virtually 100% of the wild-type or mutant α-transducin proteins were activated as determined by the trypsin conformational assay, which generated a shift from the predominant 17/12-kDa doublet bands to predominantly 33/31-kDa doublets (data not shown).

We titrated the amount of AlF$_4^-$ such that it fully activated wild-type and mutant His$_6$–α-transducin proteins, but did not interfere significantly with the PDE assay (see Methods for details). AlF$_4^-$-activated wild-type transducin and three mutants (K300A, V301A, and Q302A) were identical in their activation of PDE6 (compare with PDE6 plus AlF$_4^-$ minus transducin) (Fig. 3). In the absence of AlF$_4^-$, neither wild-type transducin nor the three mutant proteins stimulated PDE6 above baseline. Stimulation of wild-type and these three α-transducin mutants by AlF$_4^-$ resulted in an ~5-fold increase in PDE6 activity (Fig. 3). In contrast, AlF$_4^-$-treated L306A

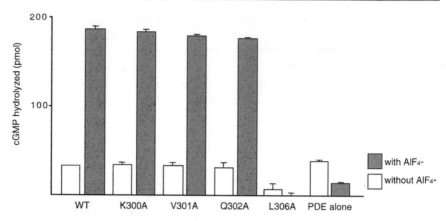

FIG. 3. Single alanine mutations were introduced into the PDE6 interaction region of recombinant His6-tagged α-transducin. Wild-type and mutant His6-tagged α-transducin proteins were activated by AlF$_4^-$, and then tested for their ability to activate PDE6. PDE activity was determined by counting [³H]GMP generated. Wild-type and three of the mutant His6–α-transducin proteins (K300A, V301A, and Q302A) fully activated PDE6, while the L306 mutant did not.

mutant α-transducin did not stimulate PDE6 activity above baseline, demonstrating the importance of L306 in α-transducin-mediated activation of PDE6.

Conclusions

Methods have been presented to study the interaction of G protein α subunits and PDEs. Peptides corresponding to the effector interaction region (aa 293–314) of α-transducin serve as a convenient surrogate for α-transducin protein, which is difficult to obtain in quantity by recombinant means. Truncated and substituted peptides can be used in "first-pass" experiments to identify specific amino acids that are potentially important to G protein–PDE interactions. Follow-up studies with recombinant G protein α subunits are recommended to confirm suspected contacts and determine the specificity of G protein α subunits involved in PDE activation.

Acknowledgments

R.F.M. is an Associate Investigator of the Howard Hughes Medical Institute. This research was supported by NIH Grants DC03055, DC03155, DC03481 (R.F.M.), and MH57241 (M.M.), and by the Human Frontier Science Program via a long-term fellowship grant (K.M.).

Section III

Calcium and Potassium Channels

[4] Studies of Endogenous G-Protein-Mediated Pathways in Neurons by Whole-Cell Electrophysiology

By ARUN ANANTHARAM and MARÍA A. DIVERSÉ-PIERLUISSI

Introduction

High voltage-activated Ca^{2+} channels are inhibited by a wide range of transmitters and hormones.[1,2] The pathways producing such inhibition are heterogeneous and can be dissected into biophysically distinct mechanisms. Perhaps most common is a voltage-dependent effect, but a variety of voltage-independent inhibitory mechanisms have also been described in many cells. Calcium channels are well-known targets for inhibition by receptor–G protein pathways (Fig. 1), and multiple forms of inhibition have been described.[2–4]

One of the best-characterized systems for the study of G-protein-mediated modulation of voltage-dependent calcium channels is the embryonic chick dorsal root ganglion (DRG) neuron. These neurons express only one type of voltage-dependent calcium channel,[5] ω-conotoxin GVIA-sensitive N-type calcium current. This system displays multiple mechanisms of G-protein-mediated modulation of channel function (Fig. 2). These include the $G\alpha$-mediated pathway, and $G\beta\gamma$-mediated, tyrosine kinase-mediated, serine-threonine kinase-mediated, and direct G protein regulation of channel function.[6,7]

The main focus of this article is the study and characterization of G-protein-mediated pathways that modulate voltage-dependent calcium channels in neuronal primary cultures. By monitoring the calcium current, receptor–G protein signaling events can be studied at the single-cell level with a time resolution impossible to obtain by conventional biochemical methods. The electrophysiological recording approach also allows the study of cells with minimal disruption of the molecular environment. This is important because regulation of receptor-mediated signals depends not only on receptor subtype but also on receptor density and cellular environment. Experiments performed in heterologous systems, where

[1] B. Hille, *Trends Neurosci.* **17,** 531 (1994).

[2] K. Wickman and D. E. Clapham, *Physiol. Rev.* **75,** 865 (1995).

[3] S. R. Ikeda and K. Dunlap, *Adv. Second Messenger Phosphoprotein Res.* **33,** 131 (1999).

[4] S. Herlitze, D. E. Garcia, K. Mackie, B. Hille, T. Scheuer, and W. A. Catterall, *Nature (London)* **380,** 258 (1996).

[5] D. H. Cox and K. Dunlap, *J. Neurosci.* **12,** 906 (1992).

[6] M. A. Diversé-Pierluissi, P. K. Goldsmith, and K. Dunlap, *Neuron* **14,** 191 (1995).

[7] M. A. Diversé-Pierluissi, A. E. Remmers, R. Neubig, and K. Dunlap, *Proc. Natl. Acad. Sci. U.S.A.* **94,** 5417 (1997).

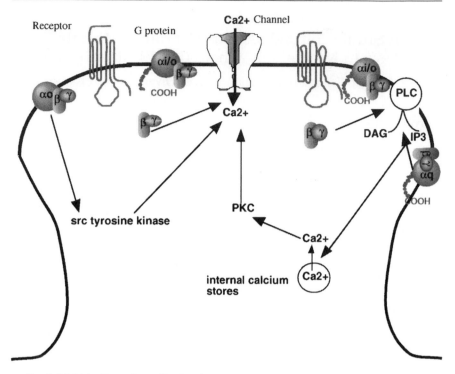

FIG. 1. Multiple G-protein-mediated pathways converge in the modulation of voltage-dependent calcium channels. DAG, Diacylglycerol; PLC, phospholipase C; IP$_3$, inositol triphosphate; PKC, protein kinase C.

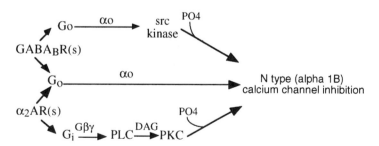

FIG. 2. G-protein-mediated pathways involved in the inhibition of N-type calcium channels in embryonic chick dorsal root ganglion neurons. Activation of GABA$_B$ and α_2-adrenergic receptors mediates different signaling pathways that result in the inhibition of voltage-dependent calcium channels.

the receptor is overexpressed, isolate one receptor subtype in abundance but in a foreign environment, where it may couple to signaling molecules different from its preferred interaction sites in its native environment. The regulatory properties of the receptor might also be altered as the onset and kinetics of desensitization depend on the density of receptors and signaling molecules. The methods presented in this article have been restricted to those using the endogenous cellular machinery. These protocols measure acute changes in signaling without exogenous gene expression. The reader is referred to reconstitution experiments of calcium channel modulation for methods for heterologous expression of calcium channels.[8–10]

General Protocols

Neuronal Primary Cultures

Embryonic chick sensory neurons are grown in culture as described in [5] in this volume.[11] Dorsal root ganglia are dissected from 11- to 12-day-old embryos. Cells are plated at a density of ~50,000 cells per collagen-coated 35-mm tissue culture dish and studied between 1 and 3 days *in vitro*.

Solutions for Electrophysiology

The external solution contains (in mM) NaCl 133, CaCl$_2$ 1, MgCl$_2$ 0.8, HEPES 25, NaOH 12.5, glucose 5, tetraethylammonium chloride 10 (to block K$^+$ channels), tetrodotoxin 0.3 μM (to block voltage-dependent Na$^+$ channels) (pH 7.4). The internal recording solution contains (in mM) CsCl 150, 1,2-bis(*o*-aminophenoxy)ethane-*N,N,N',N'*-tetraacetic acid (BAPTA) 5, MgATP 5, and HEPES 10 (pH 7.2).

Electrophysiology

Whole-cell recordings are performed as previously described.[12] For extracellular application, agents are diluted into standard extracellular saline and applied via a wide-bore (140-μm i.d.) "sewer pipe" that exchanges solutions with subsecond kinetics. For the experiments presented in this article calcium current has been corrected for rundown by measuring calcium current as a function of time in control cells without transmitter. Cells used for experiments exhibit a rundown of the current of less than 1%/min.

[8] D. A. Ewald, I. H. Pang, P. C. Sternweis, and R. J. Miller, *Neuron* **2**, 1185 (1989).

[9] L. P. Jones, P. G. Patil, T. P. Snutch, and D. T. Yue, *J. Physiol.* **498**, 601 (1997).

[10] S. W. Jeong and S. R. Ikeda, *Proc. Natl. Acad. Sci. U.S.A.* **97**, 907 (2000).

[11] A. Ananthram and M. A. Diversé-Pierluissi, *Methods Enzymol.* **345**, [5] 2001 (this volume)

[12] O. P. Hamill, A. Marty, E. Neher, B. Sakmann, and F. J. Sigworth, *Pflugers Arch.* **39**, 85 (1981).

If the reader is not familiar with the technique of whole-cell electrophysiology, details of the technique and equipment have been described by Hescheler in this series.[13]

Data Analysis

Data are filtered at 3 kHz, acquired at 10–20 kHz, and analyzed with Pulse-Fit (HEKA, Lambrecht/Pfalz, Germany) and IGOR (WaveMetrics, Lake Oswego, OR) on a Macintosh computer. Strong depolarizing conditioning pulses (to 80 mV) that precede test pulses (to 0 mV) reverse transmitter-induced voltage dependent components. Such conditioning pulses have no effect on control currents recorded in the absence of transmitter. During the application of the transmitter, test pulse currents measured before and after the conditioning pulse are subtracted to yield the voltage-dependent inhibitory component. Test pulses measured after the conditioning pulse are subtracted from control currents (measured in the absence of transmitter) to yield the voltage-independent component. Integration of these two currents gives measurements of total charge entry carried by the two modulatory components.

Tools with Which Study G-Protein-Mediated Signaling, Using Whole-Cell Patch Clamp in Primary Neuronal Cultures

Calcium channels are known targets for modulation by a wide range of transmitters and hormones. When studying transmitter-mediated modulation of calcium current there are some basic questions to be addressed: (1) Is the response mediated by a G protein? (2) Which G protein is the mediator? (3) Is the modulation produced by direct G protein binding to the channel or are there downstream mediators?

Is the Response Mediated by a G Protein?

The α subunit of the G protein binds to GDP in its basal state and binds to GTP in its active form, dissociating from the $\beta\gamma$ subunits.[14] Responses mediated by the activation of a G-protein-coupled receptor can be regulated by infusion of analogs into the cytoplasm via the recording pipette.[15–17] GDPβS (1–100 μM) prevents G-protein-mediated inhibition of calcium channels. GTP analogs such as guanosine 5'-O-(3-thiophosphate) (GTPγS) and 5'-guanosine imidophosphate (GMP-PNP) are used in the range of 1–500 μM.

[13] J. Hescheler, *Methods Enzymol.* **238,** 365 (1994).
[14] A. G. Gilman, *Annu. Rev. Biochem.* **56,** 615 (1994).
[15] R. H. Scott and A. C. Dolphin, *Neurosci. Lett.* **69,** 59 (1986).
[16] H. Kasai and T. Aosaki, *Pflugers Arch.* **414,** 145 (1989).
[17] M. R. Plummer, D. E. Logothetis, and P. Hess, *Neuron* **2,** 1453 (1989).

Which G Protein Is the Mediator?

Some of the tools used to identify the G protein are (1) ADP-ribosylating bacterial toxins, (2) antibodies raised against G proteins, (3) recombinant G protein subunits, and (4) peptides based on the sequence of the C terminus of the G protein α subunit.

Bacterial Toxins. Cholera toxin ADP-ribosylates an arginine residue of the α subunit of G proteins of the stimulatory (G_s) and olfactory-specific (G_{olf}) families. This modification activates the α subunit permanently. Typically the cells are incubated for 24 hr at concentrations ranging from 100 ng/ml to 1 μg/ml. Shorter incubation intervals ranging from 4 to 8 hr can be used if higher concentrations of toxin are used.

Pertussis toxin ADP-ribosylates members of the (inhibitory) G_i/G_o families in a cysteine residue proximal to the C terminus of the α subunit, uncoupling the G protein from the receptor. Concentrations ranging from 10 to 100 ng/ml are routinely used.

Antibodies. Transient suppression of transmitter-mediated inhibition of calcium channels can be achieved by microinjection of antibodies that selectively recognize different G protein α subunits. Many of these antibodies are raised against decapeptides derived from the C-terminus region of the G protein α subunit. Stock solutions ranging from 200 ng/ml to 20 mg/ml are used in this kind of approach. We routinely inject the antibodies with a fluorescent marker to allow subsequent identification of the cells. A wide range of antibodies is commercially available. We have been more successful in experiments in which affinity-purified antibodies are used. Figure 3 shows the results of experiments in which antibodies raised against $G\alpha_o$ blocked γ-aminobutyric acid receptor B ($GABA_B$)-mediated inhibition of the N-type calcium channels in chick DRG neurons. Control IgG had not effect on modulation. In these kind of experiments the effect of the antibody on the basal calcium current before transmitter application, and on the current–voltage relationship before and during transmitter application, is measured to discard any agonist-independent effect of the antibody on the calcium channel.

Recombinant G Protein Subunits. Recombinant G protein $\beta\gamma$ subunits have been expressed in baculovirus-infected Sf9 (*Spodeptera frugiperda* ovary) cells (kind gift of J. Garrison, University of Virginia, Charlottesville, CA). We have tested a matrix of different G protein $\beta\gamma$ subunits introduced into the cell body through the recording pipette.[18] Concentrations of G protein $\beta\gamma$ from 60 pM to 20 nM induced inhibition of calcium current. The degree of selectivity observed

[18] M. Diversé-Pierluissi, W. E. McIntire, C. S. Myung, M. A. Lindorfer, J. C. Garrison, M. F. Goy, and K. Dunlap, *J. Biol. Chem.* **275**, 28380 (2000).

GABA
(100 μM)

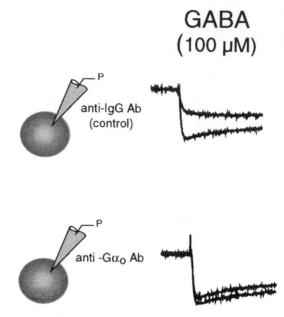

FIG. 3. Blockade of GABA$_B$-mediated inhibition of calcium current by antibodies. Dorsal root ganglion neurons were injected with an internal solution containing IgG or anti-Gα_o (200 ng/ml; Santa Cruz Biotechnologies, Santa Cruz, CA). The solution contained 0.1% (w/v) fluorescein–dextran to allow subsequent visualization by epifluorescence optics. Whole-cell electrophysiological recordings were performed 1 hr after injection. Currents were evoked by a 10-msec voltage step from −80 to 0 mV every 10 sec.

at the single-cell level is high, as Gβ_1-containing dimers inhibited N-type calcium current whereas dimers containing Gβ_2 were without effect.[19] The effect of recombinant subunits in many cases was maximal at 2 nM. Figure 4 shows the effect of G$\beta_1\gamma_2$ on calcium current as a function of time.

Constitutive active and dominant negative forms of the G protein α subunits have been successfully used by many investigators.[19–21]

Peptides. Peptides containing the sequence of the α subunits have been used to disrupt receptor–G protein coupling. These peptides are usually 10 or 11 amino acids (aa) long and can be introduced through the recording pipette.

[19] R. Mattera, M. P. Graziano, A. Yatani, Z. Zhou, R. Graf, J. Codina, L. Birnbaumer, A. G. Gilman, and A. M. Brown, *Science* **10,** 804 (1989).

[20] M. A. Diversé-Pierluissi, A. E. Remmers, R. R. Neubig, and K. Dunlap, *Proc. Natl. Acad. Sci. U.S.A.* **94,** 5417 (1997).

[21] S. Herlitze, G. H. Hockerman, T. Scheuer, and W. A. Catterall, *Proc. Natl. Acad. Sci. U.S.A.* **94,** 1512 (1997).

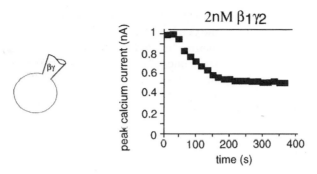

FIG. 4. G$\beta\gamma$ produces inhibition of calcium current. Peak calcium current was measured in cells exposed to 20 nM G$\beta_1\gamma_2$. Currents were normalized to the maximum peak current measured after obtaining whole-cell access, averaged, and plotted as a function of time. The recombinant subunit was introduced through the recording pipette.

Introduction of Reagents into Cytosolic Environment

Introduction of Reagents through Recording Pipette

One of the advantages of whole-cell electrophysiology is that reagents can be introduced into the cellular environment through the recording pipette. Peptides, small molecules (<60 kDa), and pharmacological reagents can be included in the internal recording solution. Experimental and theoretical observations have identified factors that determine the rate of exchange between the recording pipette solution and the cytosol, such as the diffusion coefficient of the reagent, pipette access resistance, and cell volume.

In our experiments with embryonic chick DRG neurons the variability in cell volume is minimized by using cells with a diameter of 30–40 μm. The pipette resistance was 1–2.5 MΩ. The diameter of pipettes with this resistance is large, causing the efficiency of seal formation to be low. Slow seal formation is observed when the internal solution contains a high protein concentration (micromolar range). Once whole-cell access is gained by rupturing the membrane, the capacitive transients should be fast (<1 msec). If they become slower during equilibration, apply more suction to improve cell access. If this problem persists, the exchange will be poor. In this case the investigator cannot obtain reliable data on the time course of the effects of the reagent and negative data become difficult to interpret. It is advisable to discard such cells.

As the internal solution containing the reagent equilibrates with the cytosolic environment, the cell will become granular in appearance under phase-contrast optics. To monitor the rate of exchange the internal solution can contain 0.1% (w/v) fluoresceinated dextran (Molecular Probes, Engene, OR), which will allow visualization under epifluorescence optics. In our experimental system it takes 3–5 min for the soma to become completely fluorescent.

The investigator must pay particular attention to the choice of vehicle or buffer used in the experiments. Some organic solvents such as chloroform and dimethyl sulfoxide (DMSO) have adverse effects on the membrane at concentrations greater than 0.1% (v/v). The same problem arises with hydrophobic proteins that need detergent in their buffer solution. In the case of the G protein $\beta\gamma$ subunits, we use a final concentration of 0.06% (w/v) 3-[(3-cholamidopropyl)-dimethyl-ammonio]-1-propanesulfonate (CHAPS).

Guanine Nucleotide Analogs. Hydrolysis-resistant GTP analogs and recombinant G protein subunits can be introduced and their effects monitored by measuring peak calcium current or charge entry as a function of time. G proteins can be directly activated by the hydrolysis-resistant GTP analog GTPγS. GTPγS on its own can produce inhibition of the calcium current as it exchanges with the cytosolic environment. No further reduction in current is observed when the transmitter is applied after the full effect of the nucleotide has been achieved. If the transmitter is applied before the onset of nucleotide-induced calcium current inhibition (about 1 min of whole-cell access), the presence of the nucleotide potentiates the transmitter-induced inhibition. The magnitude of inhibition observed in this case was the same as the maximal inhibition induced by 100 μM GTPγS alone.

Peptide Inhibitors of Protein Kinases. Inhibitory peptides of protein kinases such as the pseudosubstrate peptide of protein kinase C(19–36) or the Src kinase SH2/SH3 domain peptide can also be introduced by this approach. During equilibration the calcium current should be monitored to detect any transmitter-independent effect on the current. Rundown of the current should be <1%/min. During the equilibration period the current–voltage relationship should become stable. Transmitter can be applied once the current is stable.

Introduction of Recombinant Protein by Microinjection

To evaluate whether G protein-coupled receptor kinases (GRKs) play a role in the desensitization of GABA$_B$ receptor-mediated inhibition of N-type calcium current in DRG neurons, we injected individual neurons with recombinant GRKs (expressed in and purified from baculovirus-infected Sf9 cells, a kind gift of R. J. Lefkowitz, Howard Hughes Medical Institute, Durham, NC). An automated Eppendorf 5246 microinjector with a 5171 micromanipulator is used to introduce the recombinant GRKs into sensory neuron cell bodies. Standardized injection pressures (30–50 HPa) and time (0.3 sec) are used to minimize variability in injection volumes (which are estimated to be ∼0.5 nl).

1. The GRKs are added to the pipette at a concentration of 10 ng/μl and are pressure injected into the cells in the presence of fluorescein–dextran.

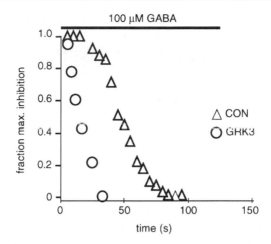

FIG. 5. Recombinant GRK3 accelerates the rate of desensitization of GABA$_B$-mediated inhibition of calcium current. Dorsal root ganglion neurons were injected with an internal solution containing recombinant GRK3 (10 ng/ml). The solution contained 0.1% (w/v) fluorescein–dextran to allow subsequent visualization by epifluorescence optics. Whole-cell electrophysiological recordings were performed 1 hr after injection. Currents were evoked by a 10-msec voltage step from −80 to 0 mV every 10 sec. The percentage of remaining GABA-induced inhibition was measured as a function of time.

2. After a 60-min postinjection incubation period, cells are identified by their fluorescence, and calcium currents are studied under whole-cell recording.

3. Currents are evoked every 5 sec throughout the experiment and responses to 100 μM GABA (in the presence of bicuculline to block GABA$_A$ receptors) are compared with those evoked in control, uninjected cells.

None of the GRKs tested had any effect on basal calcium current amplitude (1.3 ± 0.3 nA in control cells and 1.15 ± 0.4 nA in GRK-injected cells, $n = 14$). Furthermore, the maximal inhibition produced by GABA was unaffected by the presence of the GRKs.

GRK3, however, enhanced the rate of desensitization of the GABA-mediated effect (Fig. 5). No alterations in the rate of desensitization were observed in cells injected with equivalent concentrations of GRK1 (rhodopsin kinase), GRK2 (βARK1), or GRK5. As with GRK3, these other GRKs were without effect on control calcium current amplitude or maximal inhibition produced by the transmitter.

Acknowledgments

This work was supported by grants from the National Institutes of Health (NINDS), the New York City Speaker's Fund for Biomedical Research, and a Hirschl Trust Fund Career Award.

[5] Biochemical Approaches to Study Interaction of Calcium Channels with RGS12 in Primary Neuronal Cultures

By ARUN ANANTHARAM and MARÍA A. DIVERSÉ-PIERLUISSI

Introduction

Regulation of the timing of voltage-dependent calcium channel activity is traditionally thought to involve heterotrimeric G-protein-signaling pathways.[1-3] However, it is now being realized that these complex timing events are controlled by interactions between components of several signaling pathways and the cytoskeletal structure that serves as a scaffold element for interactions between the signaling components and the effector, the calcium channel. γ-Aminobutyric acid type B (GABA$_B$) receptors inhibit N-type calcium channels through Gα_o-mediated activation of a tyrosine kinase pathway.[4] Data from our laboratory suggest that the timing of GABA-induced inhibition is a complex process that involves several signals. The α (pore-forming) subunit of the calcium channel is tyrosine phosphorylated by Src kinase. This phosphorylation makes the channel a target for the binding of a GTPase-activating protein, RGS12. The phosphotyrosine-binding (PTB) domain of RGS12 binds to the tyrosine-phosphorylated channel, altering the kinetics of the termination of the response.[5] Here we describe the biochemical protocols used to determine the phosphorylation of the channel and its interaction with RGS12. The methods used in our laboratory have been optimized for the detection of the endogenous channel and interacting proteins in primary neuronal cultures without the necessity of overexpression.

Materials

Dorsal Root Ganglion Dissociation

Dorsal root ganglia (DRG) are dissected from 10- to 12-day-old chick embryos incubated at 99.5°F.[6] Eggs are obtained from Spafas (Charles River, Wilmington,

---

[1] B. Hille, *Trends Neurosci.* **17**, 531 (1994).

[2] K. Wickman and D. E. Clapham, *Physiol. Rev.* **75**, 865 (1995).

[3] S. R. Ikeda and K. Dunlap, *Adv. Second Messenger Phosphoprotein Res.* **33**, 131 (1999).

[4] M. A. Diversé-Pierluissi, A. E. Remmers, R. Neubig, and K. Dunlap, *Proc. Natl. Acad. Sci. U.S.A.* **94**, 5417 (1997).

[5] M. L. Schiff, D. P. Siderovski, D. J. Jordan, G. Brothers, B. Snow, L. DeVries, D. F. Ortiz, and M. A. Diversé-Pierluissi, *Nature (London)* **408**, 723 (2000).

[6] D. R. Canfield and K. Dunlap, *Br. J. Pharmacol.* **82**, 557 (1984).

MA). Puck's saline solution contains 150 mM NaCl, 5 mM KCl, 1.2 mM Na$_2$HO$_4$, 4 mM KH$_2$PO$_4$, and phenol red (approximately 0.001%, w/v) in deionized H$_2$O; adjust to pH 7.4 with 1 M NaOH. The 1% (w/v) glucose solution is prepared in H$_2$O. Add the glucose solution to Puck's saline to a final concentration of 0.1% (v/v) just before use. All reagents are obtained from Sigma (St. Louis, MO). The forceps used for dissection are Dumont No. 5 forceps obtained from FST (Foster City, CA).

Trituration

Collagenase A is dissolved in H$_2$O to make a stock of 1% (w/v), filtered with a 0.22-μm pore size syringe filter, and set aside in 250-μl aliquots at $-20°$. It is obtained from Boehringer Mannheim/Roche (Indianapolis, IN). The DRG medium is prepared from the following reagents: 83 ml of Dulbecco's modified Eagle's medium (DMEM, low glucose, with L-glutamine, sodium pyruvate, and pyridoxine hydrochloride), 10 ml of heat-inactivated horse serum, 5 ml of chick embryo extract, 1 ml of nerve growth factor, and 1 ml of penicillin–streptomycin/glutamine. Nerve growth factor (murine, 7S) is prepared by dissolving 10 μg in DMEM to a final concentration of 1 μg/ml. Warm the medium to 37° before use. All the reagents are obtained from GIBCO (Rockville, MD).

Plating Dorsal Root Ganglia

Collagen, type VII from rat tail, is dissolved in 1 mM acetic acid to a final concentration of 3 mg/ml. It is obtained from Sigma. The 35-mm Falcon culture dishes are obtained from Fisher (Pittsburgh, PA). To prepare the dishes for plating cells, spread one drop of collagen on each dish that will be used. Then allow the plates to dry (at least 2 hr) before use.

Application of Transmitter to Dorsal Root Ganglion Neurons

The DRG external solution contains 1 mM CaCl$_2$, 133 mM NaCl, 0.8 mM MgCl$_2$, 10 mM tetraethylammonium chloride (TEA-Cl), 25 mM HEPES, 12.5 mM NaOH, 5 mM dextrose, and 3 μM tetrodotoxin in deionized H$_2$O; adjust the solution to pH 7.4, and then filter sterilize with a Nalgene filter unit. Tetrodotoxin (Ttx) blocks voltage-dependent sodium channels and TEA-Cl is used to block voltage-dependent potassium channels. The working concentrations of agonist (GABA), and GABA$_B$ receptor antagonist (bicuculline), are 100 μM, in external solution. Lysis buffer A contains nonidet P-40 (NP-40; 1%, v/v), 1 mM sodium orthovanadate, 1 mM EDTA, calpain inhibitor I (17 μg/ml), calpain inhibitor II (7 μg/ml), pepstatin (10 μg/ml), soybean trypsin inhibitor (10 μg/ml), aprotonin (2 μg/ml), Pefabloc SC (1 mg/ml), leupeptin (10 μg/ml), in phosphate-buffered saline (PBS). Lysis buffer B contains 50 mM N-(2-hydroxyethyl) piperazine-N'-(2-ethanesulfonic acid) (HEPES, pH 7.4), 5 mM EDTA, 1 mM dithiothreitol (DTT), calpain inhibitor I (17 μg/ml), calpain inhibitor II (7 μg/ml), pepstatin

(10 μg/ml), soybean trypsin inhibitor (10 μg/ml), aprotinin (2 μg/ml), Pefabloc SC (1 mg/ml), leupeptin (10 μg/ml), in PBS. The Ttx is obtained from Calbiochem (La Jolla, CA); the other reagents for the external solution are obtained from Sigma. GABA and bicuculline are obtained from RBI (Natick, MA). The filter unit is obtained from Fisher. For lysis buffer A, the detergent NP-40 is obtained from Pierce (Rockford, IL); all other reagents are obtained from Boehringer Mannheim/Roche.

Immunoprecipitation

Immunoprecipitation wash buffer contains sodium azide (0.1%, w/v), bovine serum albumin (BSA; 0.1%, v/v), NP-40 (0.5%, v/v), Pefabloc SC (1 mg/ml), pepstatin (10 μg/ml), leupeptin (10 μg/ml), soybean trypsin inhibitor (10 μg/ml), calpain inhibitor I (17 μg/ml), calpain inhibitor II (7 μg/ml), and aprotinin (2 μg/ml). The sodium azide is obtained from Sigma, the BSA and NP-40 from Pierce. The remaining reagents are obtained from Boehringer Mannheim/Roche.

Detection of Phosphorylated Calcium Channel and RGS12

Separating and stacking gels are prepared according to Laemmli,[7] and with Bio-Rad (Hercules, CA) reagents. For a spacer thickness of 1.00 mm, the total monomer solution volume is 10 ml for the stacking portion, and 25 ml for the separating portion. The 7.5% separating gel is prepared with 6.25 ml of acrylamide–bisacrylamide (30% T, 2.67 C stock), 12.125 ml of deionized water, 6.25 ml of 1.5 M Tris-HCl (pH 8.8), 250 μl of 10% (w/v) sodium dodecyl sulfate (SDS), 125 μl of 10% (w/v) ammonium persulfate, and 12.5 μl of N,N,N',N'-tetramethylethylenediamine (TEMED). The 4% stacking gel is prepared with 1.3 ml of acrylamide–bisacrylamide (30% T, 2.67 C stock), 6.1 ml of deionized water, 2.5 ml of 0.5 M Tris-HCl (pH 6.8), 100 μl of 10% (w/v) SDS, 50 μl of 10% (w/v) ammonium persulfate, and 10 μl of TEMED. The running and transfer buffers are obtained as 10× stocks from Bio-Rad. The gel is run on a Protean II xi cell system with the Power Pac 300 power supply, also obtained from Bio-Rad. The protein marker is obtained from Amersham Pharmacia Biotech (Piscataway, NJ).

Antibodies

Anti sera directed against the N-terminal PDZ/PTB tandem [amino acids (aa) 1–440] and RGS box (aa 664–885) of human and rat RGS12 (kindly provided by D. Siderovski, University of North Carolina, Chapel Hill, NC) are generated and affinity purified by previously described methods.[8] Affinity-purified anti-α_{1B}

[7] U. K. Laemmli, *Nature* (*London*) **227**, 680 (1970).
[8] B. E. Snow, R. A. Hall, A. M. Krumins, G. M. Brothers, D. Bouchard, C. A. Brothers, S. Chung, J. Mangion, A. G. Gilman, R. J. Lefkowitz, and D. P. Siderovski, *J. Biol. Chem.* **273**, 17749 (1998).

is purchased from Alomone Laboratories (Jerusalem, Israel), and monoclonal antibodies against phosphotyrosine, 4G10, are purchased from Upstate Biotechnology (Lake Placid, NY). Antibodies against glutathione S-transferase (GST) are obtained from Santa Cruz Research Antibodies (Santa Cruz, CA). Antibodies against thioredoxin are obtained from Invitrogen (Carlsbad, CA). Horseradish peroxidase-conjugated mouse and rabbit IgG secondary antibodies are purchased from Pierce.

Proteins

Thioredoxin-tagged rat and human RGS12 PDZ/PTB proteins are a kind gift from D. Siderovski. GST-tagged PTB domain of Numb is kindly provided by J. McGlade (Hospital for Sick Children, Toronto, Canada).

Protocols

Dorsal Root Ganglion Isolation

To isolate DRG,[9] open the egg at the broad end with spatula forceps and discard the shell lid. Remove the embryo via the limbs and place it in a large dissection dish. Immediately decapitate, grasping the chick with spatula forceps. Switching to fine forceps, lay the embryo on its back and spread its limbs. With a clipping motion, cut through the sternum from the rostral end. Open the first layers of skin and spread the skin and ribs laterally. Now move the dissection under the light microscope. Wash the cavity with Puck's–glucose medium. Next, remove the intestines and wash the cavity. Remove the lungs after detaching them from the abdominal base and wash the cavity. Remove the kidneys and wash the cavity. Remove the blood vessels and wash the cavity. Remove the superficial nerves and wash the cavity. Remove the connective tissue on either side of the spinal cord and wash the cavity. Using the finest forceps, detach the DRG from peripheral nerves and remove them by cutting their attachments to the spinal cord. Place the DRG in Puck's–glucose medium in an untreated 35-mm culture dish.

Triturating Dorsal Root Ganglia

Add 250 μl of collagenase A [1% (w/v) solution] to 9 ml of Puck's–glucose solution in a sterile 15-ml tube (Corning, Acton, MA). Transfer dissociated DRG from the 35-mm dish to the Puck's–glucose plus collagenase A medium, using a glass Pasteur pipette (in a minimal volume). Incubate at 37° for 15 min. This should be an adequate time for the connective tissue holding the bundled DRG together to begin to deteriorate. Meanwhile, set aside two more 15-ml tubes, each

[9] R. Bellairs, "The Atlas of Chick Development." Academic Press, San Diego, California, 1999.

with 5 ml of the Puck's–glucose medium. Preheat these to 37° in an incubator. They will be used for washing away the collagenase from the DRG, because the dishes on which the DRG will ultimately be plated are treated with collagen. The rest of the procedure is performed under a hood.

Washes. Run the end of a sterile, autoclaved glass Pasteur pipette over a gas flame. Allow it to cool, and then transfer the DRG from the first 15-ml Corning tube (containing the Puck's–glucose plus collagenase A medium) to the next (containing 5 ml of the Puck's–glucose medium only), using a Pasteur pipette. Take care, when transferring DRG, not to draw cells past the narrow portion of the Pasteur pipette. They do have a tendency to stick to the sides of the glass. After the DRG have settled to the bottom of the tube, repeat the procedure, transferring them to the third 15-ml tube (containing 5 ml of the Puck's–glucose medium only). Finally, transfer the DRG from the third 15-ml tube, to a fourth containing 2 ml of DRG medium, preheated to 37°. Again, allow the DRG to settle to the bottom of the tube. Next, fire polish the tip of a glass Pasteur pipette until its diameter is approximately one-third of the original tip diameter. Allow it to cool. Then triturate the DRG by pipetting up and down approximately 15 times. Continue to triturate until the bundled DRG can no longer be seen; only a cloudy pink solution of medium should be seen.

Plating Dorsal Root Ganglia

After triturating, bring up the volume of triturated DRG in complete medium so that approximately 2 ml per collagen-treated 35-mm dish can be plated (the dissection yield is approximately two embryos per dish, depending on DRG viability, dissection efficiency, etc.). After pipetting the appropriate volume of medium into each culture dish, gently move it up and down, and left to right, to disperse the DRG. Return the culture dishes to a 37° incubator.

Feeding Dorsal Root Ganglia

The day after plating the DRG, and every 2 days subsequent to that, feed the cells. Gently remove the lid of a 35-mm dish and place it face down on the surface of the sterile hood. Prop the dish up against the lid so it is at a 20–30° angle to horizontal. Being careful not to touch the bottom of the dish, remove 1 ml of the DRG medium; then place the dish flat with the surface of the hood. Add 1 ml of new DRG medium to the culture by pipetting gently against the side of the dish. Return the 35-mm dishes to the incubator.

Application of Transmitter to Dorsal Root Ganglion Neurons

The lysis buffer that will be used to harvest the stimulated cells should be kept on ice. The control solution (with 100 μM bicuculline, an antagonist of GABA$_A$

receptors) and transmitter-containing solution (with $100 \mu M$ GABA in the presence of bicuculline) should be equilibrated to $25°$, or room temperature.

Using a Pipetman, remove the DRG medium from the first 35-mm culture dish. Immediately add 1 ml of the control solution to the dish and start a stopwatch. (Have a Pipetman in each hand so that external solution can be added and removed without any delay. Timing of exposure to control or transmitter-containing solution is critical to the success of this procedure. After 1 min, remove the control solution, and again add fresh control solution, or GABA-containing solution, to the dish. After 40 sec, remove either the control or stimulating solution, and add 1 ml of lysis buffer. If a membrane preparation will be subsequently performed on the cells, use the buffer without detergent (lysis buffer B) to avoid solubilizing the proteins in that fraction. Then immediately collect the cells with a cell scraper and place the dish on ice. This 1 ml of lysed cells will be used as the buffer for lysing all subsequent cultures, so that the cells are pooled in a constant volume from one dish to the next. At the final 35-mm culture dish, transfer the lysed cells to an appropriately labeled Eppendorf tube.

Preparation of Membrane-Associated Fractions

The cells are homogenized with a mechanized pellet pestle homogenizer (Kontes, Vineland, NJ), using about 20 strokes. The homogenates are then centrifuged at $1000g$ for 10 min at $4°$. Transfer the supernatants from this spin to fresh prechilled Eppendorf tubes and spin again at $100,000g$ for 1 hr. Drain off the supernatant and resuspend the pellet in ice-cold lysis buffer A (with detergent).

Immunoprecipitation of Tyrosine-Phosphorylated Proteins

It is important to note, before beginning, that performing all steps at $4°$ will reduce the protein degradation that can markedly impact the yield of this procedure. For immunoprecipitation of tyrosine-phosphorylated proteins, 500,000 cells from each experimental condition are used. After cells are exposed for 40 sec to buffer $\pm 100 \mu M$ GABA (containing $100 \mu M$ bicuculline to block $GABA_A$ receptors), and collected in ice-cold lysis buffer, gently mix the suspension on a rotating mixer at $4°$ for 15 min. Homogenize the lysates with a mechanized pellet pestle homogenizer. Shear the tissue by running the lysates through syringe needles of progressively smaller sizes, for instance, from 21 to 27 gauge. Centrifuge the lysates at $15,000g$ for 30 min. This will effectively remove any unsolubilized debris. Transfer the supernatant to a fresh, prechilled Eppendorf tube. At this point, the protein concentration of the suspension can be measured by a Bradford (Bio-Rad) or bicinchonisic acid (BCA; Pierce) assay. Then bring 1 mg of cell lysate up to 1 ml with ice-cold lysis buffer A. Incubate it with 20 μl of untreated protein A/G–agarose beads at $4°$ for 1 hr to preclear the lysates. Preclearing the lysate will help to reduce the nonspecific binding of other proteins to the agarose beads when used again later in the immunoprecipitation. Then spin the samples down at 2500g for 5 min to pellet the beads. Remove the supernatant and incubate it in the appropriate primary antibody for 8 hr at 4°. Add 40 μl of protein A/G–agarose to capture the immunocomplex and allow the suspension to incubate at 4° for 4 hr. Collect the immunocomplex by spinning the samples down at 2500g for 5 min. Then wash the pelleted beads three times in ice-cold wash buffer. This should be done in a volume equivalent to 10 times the bed volume of the beads. Care should be taken to avoid disrupting the beads during the washes; trimming the ends off the pipette tips guards against this possibility. Finally, resuspend the beads in 50 μl of 2× Laemmli sample buffer (Bio-Rad). Boil for 5 min to separate the immunocomplex from the beads. Pellet the beads by spinning them down at 10,000g for 5 min. Transfer the supernatant to a fresh tube.

Immunoprecipitation of α Subunit of Calcium Channel

For immunoprecipitation of the calcium channel pore-forming subunit α_{1B}, 1×10^6 cells are lysed. The membrane fraction is then homogenized and pre-cleared, using the protocol described above. The precipitating antibody, in this case, is the N-type calcium channel antibody (Alomone Laboratories) raised in rabbit against CNB1. This corresponds to the loop I–II region of rat brain voltage-gated calcium channels.[10] After preclearing the lysates, add anti-N-type calcium channel antibody to a final dilution of 1 : 50 and incubate the suspension at 4° for 8 hr. Then add 40 μl of protein A/G–agarose beads to capture the immunocomplex and mix for 2 hr at 4°. Collect the immunocomplex by spinning down at 2500g for 5 min. Wash the pelleted beads three times in ice-cold wash buffer. This should be done in a volume equivalent to 10 times the bed volume of the beads. Again, care should be taken here to avoid disrupting the bead complex. Finally, resuspend the beads in 50 μl of 2× Laemmli sample buffer, and boil for 5 min to separate the immunocomplex from the protein A/G–agarose beads. Pellet the beads by spinning down at 10,000g for 5 min. Transfer the supernatant to a fresh tube.

Detection of Phosphorylated Calcium Channel and RGS12

Detection of proteins associated with the α subunit of the calcium channel is carried out by SDS–polyacrylamide gel electrophoresis (PAGE) and then probing for the complex with the antibody of interest. After the samples are loaded, run the gel at 150 V for approximately 30 min (or until the separating gel is reached), and then at 24 mA for 4 hr, until the dye front runs off.

Transfer of the proteins from a polyacrylamide gel to a nitrocellulose membrane is performed in the cold room for 8 hr at 100 mA. After transfer of the proteins from

---

[10] R. E. Westenbroek, *Neuron* **9**, 1099 (1992).

the gel to the membrane, block for 1 hr at 25° (room temperature) or overnight at 4°, using 5% (w/v) milk-PBS-T, pH 7.4 [0.05% (v/v) Tween 20; Pierce]. Then incubate the membrane in primary antibody diluted 1 : 2000 in blocking buffer overnight. The next day, wash the membrane three times in PBS-T [pH 7.4, 0.05% (v/v) Tween], and then incubate for 1 hr in the appropriate horseradish peroxidase-conjugated secondary antibody against primary (Pierce) at a dilution of 1 : 5000. Then wash three more times in PBS-T, and once in PBS; each wash should be 5 min in duration. Bound protein is then detected by enhanced chemiluminescence (Amersham Pharmacia Biotech).

Experimental Data

To test whether the α_1 subunit of the N-type calcium channel has been precipitated, one can perform an immunoblot of the anti-α_{1B} precipitate. A pan-α antibody (Alomone Laboratories) that recognizes all the α_1 subunits of voltage-dependent calcium channels can be used for immunoblotting. The pattern of the bands detected is shown in Fig. 1.

When precipitation with the anti-α_{1B} antibody is performed with lysates from GABA-treated cells and precipitates are immunoblotted with anti-phosphotyrosine (4G10), we detect a protein band of an apparent molecular mass of 240 kDa (Fig. 2). This would be in agreement with the reported range in size of the α subunit of the N-type calcium channel (α_{1B}) between 170 to 260 kDa.[11-12] The pattern of the blotting is also similar to that observed in previous experiments[4] where ^{32}P was used. But if the cells are treated with a tyrosine kinase inhibitor such as genistein, this protein cannot be detected. The data, therefore, indicate that the 240-kDa band represents the α subunit of the N-type calcium channel, and that it is phosphorylated when GABA is applied.

To test whether the N-type calcium channel, which is tyrosine phosphorylated by GABA stimulation, and the phosphotyrosine-binding domain of RGS12 interact *in vivo*, we precipitated RGS12 with the anti-rRGS12-PTB antibody, and subsequently probed for the channel with anti-α_{1B} (Fig. 2). In cells that are exposed to GABA, a band is detected at approximately 240 kDa, indicating that the calcium channel does, in fact, interact with RGS12. Furthermore, we found the interaction of RGS12 with the channel to be time dependent.[4] This was determined by first applying 100 μM GABA to cultured DRG for 0, 20, and 60 sec, and then performing an immunoprecipitation with anti-α_{1B}. Finally, we probed for RGS12 with the RGS12 PTB antibody. During the time at which the inhibition by GABA

[11] M. K. Ahlijanian, J. Striessnig, and W. A. Catterall, *J. Biol. Chem.* **266**, 20192 (1991).
[12] M. W. McEnery, A. M. Snowman, A. H. Sharp, M. E. Adams, and S. H. Snyder, *Proc. Natl. Acad. Sci. U.S.A.* **88**, 11095 (1991).

IB: anti-pan alpha 1

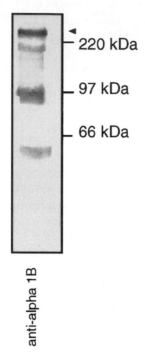

anti-alpha 1B

FIG. 1. Immunoprecipitation of α_{1B} subunit of chick DRG neurons, using anti-pan-α. Immunoblotting performed with anti-α_{1B}.

is maximal (i.e., 20 sec), little of the 187-kDa band corresponding to the reported size of RGS12[13] can be visualized. During desensitization, it appears that more of the RGS12 coprecipitates with the calcium channel, as by 60 sec, the band corresponding to RGS12 becomes distinct.

Overlay Assay (Far Western)

After finding that RGS12 and the N-type calcium channel interact *in vivo,* we were interested to see whether this interaction was direct. The overlay, or "far Western," was helpful to us in resolving this question. After transferring proteins from an SDS–polyacrylamide gel to the nitrocellulose membrane, block for 1 hr in milk-PBS-T. Then incubate the membrane in blocking buffer containing the test protein to a final concentration of 500 ng/ml at 25° or at room temperature

[13] B. E. Snow, L. Antonio, S. Suggs, H. B. Gutstein, and D. P. Siderovski, *Biochem. Biophys. Res. Commun.* **233,** 770 (1997).

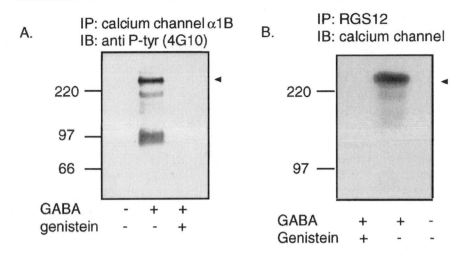

FIG. 2. (A) Calcium channel precipitated with anti-α_{1B}. Channel phosphorylation in GABA-treated cells was detected with anti-phosphotyrosine (4G10). Genistein, a tyrosine kinase inhibitor, prevents channel phosphorylation. (B) RGS12 and phosphorylated calcium channel coprecipitate in GABA-treated cells.

for 4 hr. Ideally, this will encourage the direct association of proteins, which can subsequently be assayed for using an antibody to the protein that was overlaid. Wash the membrane three times in milk-PBS-T before incubating it overnight at 4° in an antibody (directed against the overlaid protein) diluted 1 : 2000 in blocking buffer. Wash the membrane three more times, and incubate it for 1 hr at 25° in the appropriate horseradish peroxidase-conjugated secondary antibody (Pierce) diluted 1 : 5000 in blocking buffer. Then wash the membrane three times in PBS-T and once in PBS. Bound protein is detected by enhanced chemiluminescence (Amersham Pharmacia Biotech).

Experimental Data

By utilizing the overlay assay described above, we were able to show that the interaction between RGS12 and the calcium channel is direct.[4] We first performed a precipitation of the calcium channel from control and GABA-treated DRG neurons with anti-α_{1B}, and overlaid the nitrocellulose membrane to which the precipitates had been transferred with the PTB domain of RGS12 (human RGS12, aa 220–366). Immunoblotting for the RGS12 PTB reveals a 240-kDa band, corresponding to the α subunit of the N-type calcium channel (Fig. 3). This is an effect that is also selectively blocked by the tyrosine kinase inhibitor genistein, demonstrating that there is, in fact, a direct interaction between RGS12 and the tyrosine kinase-phosphorylated calcium channel. To test for the specificity of the calcium channel

IP: anti-alpha 1B

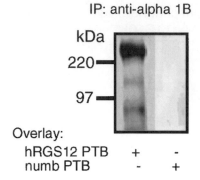

Overlay:
hRGS12 PTB   +   -
numb PTB   -   +

FIG. 3. Overlay assay of anti-α_{1B} precipitate with human RGS12 PTB and Numb PTB. The interaction of RGS12 and calcium channel exhibits selectivity.

for the phosphotyrosine-binding domain of RGS12, we also overlaid our anti-α_{1B} precipitate from GABA-treated cells with the *Drosophila* cell fate determinant Numb.[14-18] Numb has a phosphotyrosine-binding domain through which it is able to form complexes with other proteins at several phosphotyrosine-containing motifs.[19-23] In our experiments, however, PTB–Numb failed to show a direct association with the channel (Fig. 3). This indicated that the interaction of the PTB domain RGS12 with the calcium channel exhibits selectivity.

Acknowledgments

This work was supported by a grant from the National Institutes of Health (NINDS 37443) and by a Hirschl Trust Fund Career Award.

[14] T. Uemura, S. Shepherd, L. Ackerman, L. Y. Jan, and Y. N. Jan, *Cell* **58**, 349 (1989).

[15] M. Park, L. E. Yaich, and R. Bodmer, *Mech. Dev.* **75**, 117 (1998).

[16] E. P. Spana and C. Q. Doe, *Neuron* **17**, 21 (1996).

[17] R. Brewster and R. Bodmer, *Development* **121**, 2823 (1995).

[18] G. M. Ruiz and M. Bate, *Development* **124**, 4857 (1997).

[19] C. T. Chien, S. Wang, M. Rothenberg, L. Y. Jan, and Y. N. Jan, *Mol. Cell. Biol.* **18**, 598 (1998).

[20] B. Lu, M. Rothenberg, L. Y. Jan, and Y. N. Jan, *Cell* **95**, 225 (1998).

[21] M. Guo, L. Y. Jan, and Y. N. Jan, *Neuron* **17**, 27 (1996).

[22] W. Zhong, J. N. Feder, M. M. Jiang, L. Y. Jan, and Y. N. Jan, *Neuron* **17**, 43 (1996).

[23] S. C. Li, Z. Songyang, S. J. Vincent, C. Zwahlen, S. Wiley, L. Cantley, L. E. Kay, J. Forman-Kay, and T. Pawson, *Proc. Natl. Acad. Sci. U.S.A.* **94**, 7204 (1997).

[6] Assaying Phosphatidylinositol Bisphosphate Regulation of Potassium Channels

By Tibor Rohács, Coeli Lopes, Tooraj Mirshahi, Taihao Jin, Hailin Zhang, and Diomedes E. Logothetis

Introduction

Inwardly rectifying K$^+$ (Kir) channels are important regulators of the resting membrane potential and cell excitability. It appears that a common requirement for activity of these channels is the presence of phosphatidylinositol bisphosphate (PIP$_2$) in the inner leaflet of the plasma membrane. Accumulating evidence suggests that direct interactions of the negatively charged head group of PIP$_2$ with positively charged residues of members of this family of ion channels may be of fundamental importance to channel gating. In this article, we describe some of the functional and imaging assays that we have utilized to study channel–PIP$_2$ interactions and their role in regulating channel activity.

Activation of K$^+$ Channels by Different Forms of Phosphatidylinositol Bisphosphate

We have studied PIP$_2$ regulation of several Kir channels. The hallmark of this family of channels, related to their dependence on PIP$_2$, is that they show rundown of their activity on patch excision. Hydrolyzable ATP, in most cases and under certain conditions, prevents this effect.[1,2] Direct application of PIP$_2$ to inside-out membrane patches reactivates Kir channels in a manner similar to ATP reactivation. This suggests that ATP exerts its effect by providing the substrate for phosphorylation of phosphatidylinositol (PI) to phosphatidylinositol phosphate (PIP) and PIP$_2$ by lipid kinases present in the patch.[3,4] Conversely, the rundown is probably due to the breakdown of PIP$_2$ by endogenous lipid phosphatases (see below).

With regard to the location of the phosphate groups on the inositol ring, the plasma membrane of mammalian cells contains three forms of PIP$_2$: PI(4,5)P$_2$, PI(3,4)P$_2$, and PI(3,5)P$_2$. The chemical structure of phosphatidylinositol is shown in Fig. 1C. The most abundant form of PIP$_2$ is PI(4,5)P$_2$. PI(4,5)P$_2$ is a substrate for two important signaling enzymes, phospholipase C (PLC) and phosphatidylinositol 3-kinase (PI3 kinase). In addition to serving as a substrate, PI(4,5)P$_2$ itself

[1] J. L. Sui, K. W. Chan, and D. E. Logothetis, *J. Gen. Physiol.* **108**, 381 (1996).

[2] D. W. Hilgemann, *Annu. Rev. Physiol.* **59**, 193 (1997).

[3] J. L. Sui, J. Petit Jacques, and D. E. Logothetis, *Proc. Natl. Acad. Sci. U.S.A.* **95**, 1307 (1998).

[4] C. L. Huang, S. Feng, and D. W. Hilgemann, *Nature (London)* **391**, 803 (1998).

plays an important role in several processes, such as organization of the cytoskeleton and vesicular transport.[5] PIP_2 also regulates several ion channels and transporters, including inwardly rectifying K^+ channels.[2] $PI(3,4)P_2$ and phosphatidylinositol 3,4,5-trisphosphate $[PI(3,4,5)P_3]$ are generated by PI3 kinase and play similar, although not identical, roles.[6] The generation and function of $PI(3,5)P_2$ are less clear.[6] PIP_2 found in native cell membranes contains a mixture of different acyl chains, the most abundant being arachidonyl in position 2 and stearyl in position 3 (AASt). Commercially available preparations of $PI(4,5)P_2$ are usually purified from bovine brain (AASt). We have used mostly $PI(4,5)P_2$ from Boehringer Mannheim/Roche (Indianapolis, IN). The four double bonds of the arachidonyl chain can easily become oxidized, and therefore most companies synthesize dipalmitoyl (diP)-$PI(4,5)P_2$, which is much more stable. As AASt-$PI(3,4)P_2$, -$PI(3,5)P_2$, and -$PI(3,4,5)P_2$ are much less abundant in biological samples, they have not been commercially available. Synthetic, dipalmitoyl versions, however, are available for all three head group bisphosphate combinations, as well as for $PI(3,4,5)P_3$, from several suppliers [Echelon (Salt Lake City, UT; www.echelon-inc.com), Matreya (Pleasant Gap, PA; www.matreya.com), and Calbiochem (La Jolla, CA; www.calbiochem.com)]. Short-chain synthetic dibutanoyl (diC4)- or dioctanoyl (diC8)-PIP_2 are also available in every head group combination, from Echelon and also from Matreya.

We have used the *Xenopus* oocyte system to express Kir channels and assay the effects of PIP_2 on their activity. We found that Kir3.1/3.4 (GIRK1/GIRK4 G-protein-coupled, inwardly rectifying K^+ subunits 1/4) channels can be activated to a similar extent by any of the PIP_2 head group variants and PIP_3.[7] Similar results have been obtained with GIRK1/GIRK2 channels,[8] Kir6.2/SWR1 (K_{ATP}) channels (T. Rohács and D. E. Logothetis, unpublished results, 2000). In marked contrast to these channels, Kir2.1 (IRK1) currents are specifically activated by the $PI(4,5)P_2$ stereoisomer.[7] These studies were performed with synthetic diP or diC8 analogs.

We also tested the efficiency of the short-acyl chain analogs of PIP_2. DiC4-PIP_2 in our hands was not able to activate any of the channels tested, when applied at 25 μM.[7] DiC8-$PI(4,5)P_2$, on the other hand, activated all the channels that we tested, to different extents. IRK1 channels, which are very sensitive to PIP_2, are almost fully activated by diC8[7] (see also Fig. 1). DiC8-PIP_2 is required in higher

FIG. 1. (A) Representative macropatch measurement of the dose response of IRK1 channel activity to diC8-$PI(4,5)P_2$. The beginning of the trace shows the cell-attached current from an oocyte injected with IRK1 cRNA. The holding potential is -80 mV, a short step to 0 mV and then to $+80$ mV are, also shown, to test for inward rectification. The establishment of the inside-out (i/o) configuration is indicated by the arrow. The applications of 0.2 mM $MgCl_2$, various concentrations of diC8-$PI(4,5)P_2$, and 2.5 μM AASt-$PI(4,5)P_2$ are shown by the horizontal bars. (B) Summary of seven measurements. Data are expressed as a percentage of the response induced by 25 μM diC8-PIP_2. (C) Schematic of phosphatidylinositol, with the numbering of the different OH groups as well as the different acyl chains of the commercially available PIP_2 analogs.

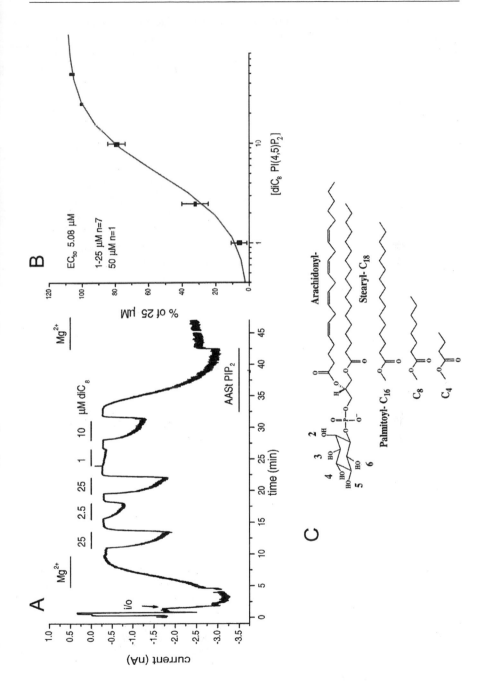

concentrations to activate IRK1 currents: 25 μM was still less active than 2.5–5 μM AASt or diP. The other marked difference between diC_8 and the longer chain forms is in the kinetics of current activation.[7] DiC_8 exerts its effect much more quickly than the long-chain forms. The effect of DiC_8-PIP_2 is also quickly reversible, unlike the effect of the long-chain PIP_2 analogs[7] (see also Fig. 1). This property makes it a useful tool for studying PIP_2 activation of ion channels, as it can be applied to the same patch and removed from it several times, unlike the long-chain forms. Thus different head group combinations of diC_8-PIP_2 can be tested in the same patch, an experiment that is not possible with the longer acyl chain PIP_2 forms (T. Rohács and D. E. Logothetis, unpublished results, 2000).

It is also possible to construct dose–response curves for diC_8-$PI(4,5)P_2$. Figure 1A shows IRK1 currents in response to various diC_8-$PI(4,5)P_2$ doses from a representative macropatch. Application of 0.2 mM $MgCl_2$ accelerates current rundown, whereas different concentrations of diC_8-$PI(4,5)P_2$ stimulate K^+ currents reversibly in a dose-dependent manner. At the end of the experiment, 2.5 μM AASt-$PI(4,5)P_2$ is applied for maximal activation. Figure 1B shows a summary of such measurements as a dose–response curve expressed as a percentage of the response induced by 25 μM diC_8-PIP_2. The 50% effective concentration (EC_{50}) for diC_8-$PI(4,5)P_2$ is ~5 μM. The Hill coefficient of 1.337 indicates that more than one molecule of PIP_2 is needed to activate the channel, which is not surprising given the tetrameric structure of these channels.

The explanation for the kinetic difference in current activation between the short- and long-chain analogs probably lies in their physicochemical properties, specifically, in their different critical micellar concentrations. The long-chain forms mostly exist in solutions as micellar or vesicular aggregates,[9] with low concentrations of free monomers. When these lipids are applied to the patch, channel activation takes several minutes to fully develop,[7] probably because of the slow fusion of these vesicles with the membrane, or the slow incorporation of lipid monomers because of their low concentration. Once these phospholipids are incorporated in the membrane, they come off slowly, as evidenced by slow deactivation after cessation of perfusion of the patch with these lipids (Fig. 1A). The short-chain analogs, on the other hand, are more water soluble, and their free monomer concentration is higher, and consequently they incorporate in the membrane at a relatively faster rate. However, discontinuing perfusion results in fast deactivation, as these lipids quickly diffuse out of the membrane (Fig. 1A).

[5] J. J. Hsuan, S. Minogue, and M. dos Santos, *Adv. Cancer Res.* **74,** 167 (1998).

[6] D. A. Fruman, R. E. Meyers, and L. C. Cantley, *Annu. Rev. Biochem.* **67,** 481 (1998).

[7] T. Rohács, J. Chen, G. D. Prestwich, and D. E. Logothetis, *J. Biol. Chem.* **274,** 36065 (1999).

[8] I. H. M. Ho and R. Murrell-Lagnado, *J. Physiol. Lond.* **520,** 645 (1999).

[9] L. A. Flanagan, C. C. Cunningham, J. Chen, G. D. Prestwich, K. S. Kosik, and P. A. Janmey, *Biophys. J.* **73,** 1440 (1997).

Macropatch Recordings in Xenopus Oocytes

Recombinant ion channels can be conveniently overexpressed in *Xenopus* oocytes by injecting cRNA. Details of procedures for maintaining oocytes,[10] preparing RNA,[11] injection,[12] and macropatch recording have been previously published.[13,14] In this article, we focus on aspects of recording inwardly rectifying K$^+$ channel activity in this system, as it pertains to the study of the effects of phosphoinositides.

Macropatch experiments with IRK1 show high functional expression. For IRK1 channels, injection of 0.5–1 ng of cRNA per oocyte is usually sufficient, and measurements can be performed the next day, with an even higher expression level 2 days after injection. G-protein-sensitive GIRK1/GIRK4 channels show lower levels of expression than IRK1, and thus we usually inject 2–4 ng of each subunit. A convenient alternative is to use GIRK4$_{S143T}$ (GIRK4*), which is active by itself and does not require the coinjection of GIRK1 cRNA.[15] The characteristics of this mutant are similar to those of GIRK1/GIRK4, and it has been extensively used in this laboratory to study both PIP$_2$ and G$\beta\gamma$ regulation of homomeric GIRK channels.[15–18] Because expression levels of this mutant are usually higher than those of GIRK1/GIRK4, we inject 1–2 ng and record 2–3 days after injection.

To gain access to the plasma membrane, we remove the vitelline membrane of the *Xenopus* oocytes before seal formation, using fine forceps. Many authors recommend the use of hyperosmotic shrinking solution to facilitate separation of the vitelline layer from the plasma membrane. As this requires an extra step and waiting period, and devitellinization can be carried out without shrinking solution, we skip this step. We record inward K$^+$ currents in symmetrical K$^+$-based solutions at a holding potential of -80 mV. The pipette solution usually contains (in mM) 96 KCl, 1.8 CaCl$_2$, 1 MgCl$_2$, 1 NaCl, and 10 HEPES, pH 7.4. The bath solution contains (in mM) 96 KCl, 5 EGTA, and 10 HEPES, pH 7.4. We pull microelectrodes from WPI-K borosilicate glass (World Precision Instruments, Sarasota, FL) using a Sutter (Novato, CA) P-97 microelectrode puller. We use patch pipettes with tip diameters up to 30–40 μm, giving a resistance of about 0.5 MΩ.

[10] A. L. Goldin, *Methods Enzymol.* **207**, 266 (1992).

[11] A. L. Goldin and K. Sumikawa, *Methods Enzymol.* **207**, 279 (1992).

[12] H. Soreq and S. Seidman, *Methods Enzymol.* **207**, 225 (1992).

[13] D. W. Hilgemann and C. C. Lu, *Methods Enzymol.* **293**, 267 (1998).

[14] D. W. Hilgemann, *In* "Single-Channel Recording" (B. Sakmann and E. Neher, eds.), p. 307. Plenum Press, New York, 1995.

[15] M. Vivaudou, K. W. Chan, J. L. Sui, L. Y. Jan, E. Reuveny, and D. E. Logothetis, *J. Biol. Chem.* **272**, 31553 (1997).

[16] H. Zhang, C. He, X. Yan, T. Mirshahi, and D. E. Logothetis, *Nat. Cell Biol.* **1**, 183 (1999).

[17] C. He, H. Zhang, T. Mirshahi, and D. E. Logothetis, *J. Biol. Chem.* **274**, 12517 (1999).

[18] E. Kobrinsky, T. Mirshahi, H. Zhang, T. Jin, and D. E. Logothetis, *Nat. Cell Biol.* **2**, 507 (2000).

Seal formation with macropatch pipettes usually takes several minutes. Gentle suction, with the holding potential at -80 mV, usually expedites seal formation. The formation of a seal with gigaohm resistance is usually not necessary, provided the current level is in the nanoampere range. We usually consider 100 MΩ to be sufficient to start an experiment. Once the inside-out configuration is established, the patch is usually stable for up to 1 hr and the channels can be reliably activated with PIP$_2$ even at the end of the experiment (Fig. 1). This stability is much more difficult to achieve with single-channel patches, where the patch usually becomes noisy within 20 min. Also, PIP$_2$ activates the channels less reliably in small patches at the end of a long experiment.

Handling of Lipids

We disperse all the long acyl chain forms in water at a stock concentration of 0.5 mM. We sonicate the stock solution for 30 min on ice, in a Sonic 300 V/T (Imaging Products International, Simi Valley, CA) sonic dismembrator at 90% power. After sonication, we divide the lipids in 25-μl aliquots and keep them at $-80°$. One aliquot is enough to prepare 5 ml of 2.5 μM working solution. When preparing the working solution, we wash the tube containing the stock solution several times with the working solution, as phosphoinositides could adhere to plastic surfaces. Some users recommend fast freezing of the solution in liquid N$_2$ to prevent adherence to the plastic wall. The working solution containing PIP$_2$ is sonicated again at the beginning of the experimental day for 10–30 min on ice. Once an aliquot is thawed, we use it on that day, and discard the unused material. Keeping the working solution at room temperature on the experimental day has not altered the efficiency of the lipid in our hands. Although the double bonds of the arachidonyl side chain may become oxidized in air, the procedure we described does not alter phospholipid efficiency. An alternative way reported by others for preparing solutions of phosphoinositides is to store them in chloroform–methanol (19 : 1, v/v), transfer an aliquot to a siliconized glass tube, and before the experiments evaporate the organic solvent, add the buffer, and sonicate the lipid.[19]

We prepare the short acyl chain PIP$_2$ analogs, which are fully water soluble, as 2.5 mM stock solutions and keep them aliquoted at $-80°$. They do not require sonication. Again, each aliqout is used only for 1 day.

Divalent cations in aqueous solution have been shown to induce aggregation of small PIP$_2$ micelles into larger multilamellar structures.[9] We had difficulties activating the channels with PIP$_2$ in the presence of Mg^{2+}, which is probably due to the aggregation of the small micelles. It is advisable to leave out divalent cations from solutions containing PIP$_2$.

---

[19] V. D. Lupu, E. Kaznacheyeva, U. M. Krishna, J. R. Falck, and I. Bezprozvanny, *J. Biol. Chem.* **273,** 14067 (1998).

Effect of Phosphatidylinositol Bisphosphate on Single-Channel Kinetics of GIRK1/GIRK4

In addition to the analysis of macroscopic currents, the effects of PIP$_2$ on channel activity can also be studied at the single-channel level. Records with a single channel in the patch provide the most reliable kinetic information about the residence of the channel in the closed and open states. However, obtaining patches containing only a single channel, which exhibits stable activity over a long enough period of time that it becomes highly unlikely that a second channel is present in the patch, is not a trivial exercise. Useful kinetic information can be obtained from patches containing two or more channels by focusing the analysis on channel mean residence in the open state and inferring the mean residence in the closed state. Figure 2 shows such an example with the GIRK1/GIRK4 unitary currents. The total activity (NP_o), mean open time (MT_o), and opening frequency NF_o are plotted from an inside-out patch excised from a *Xenopus* oocyte expressing these channels. In control solutions without MgATP, patches show low NP_o and short $MT_o < 1$ msec. Perfusion of Na$^+$ at this stage of the experiment is without effect. Perfusion of PIP$_2$ caused moderate increases in MT_o and NP_o. Perfusion of Na$^+$ after PIP$_2$ exposure dramatically increases channel activity without any further significant effect on MT_o. This effect of Na$^+$ results from an increase in NF_o and because the mean open time is not increased, it reflects a decrease in the mean closed time of the channel. This can be clearly appreciated by considering the relationship of open probability to the mean times the channel spends in the open and closed states:

$$P_o = T_o/(T_o + T_c)$$

where P_o, T_o, and T_c stand for channel open probability, mean open time, and mean closed time, respectively. It has been previously shown[8,16] that Na$^+$ causes channel activation by enhancing channel–PIP$_2$ interaction. Thus an enhancement in channel–PIP$_2$ interaction by Na$^+$ may increase channel activity by decreasing channel mean closed time. Kinetic analysis with a single channel in the patch has been reported in the case of the K$_{ATP}$ channel. The results from this study agree with the conclusion that higher PIP$_2$ levels in the membrane increase channel activity by decreasing channel mean closed time, whereas channel open time kinetics are not affected.[20] The assertion that mean open time is not affected in K$_{ATP}$ channels implicitly assumes the presence of some PIP$_2$ in the membrane, giving rise to activity that can be further increased by higher levels of PIP$_2$.

[20] D. Enkvetchakul, G. Loussouarn, E. N. Makhina, S. L. Shyng, and C. G. Nichols, *Biophys. J.* **78**, 2334 (2000).

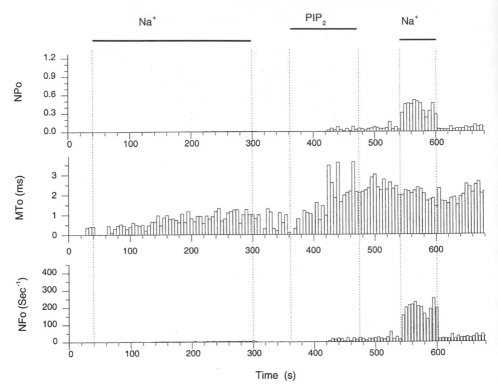

FIG. 2. Effect of PIP_2 on the patch total activity (NP_o), channel mean open time (MT_o), and mean frequency of opening (NF_o) of heteromeric GIRK1/GIRK4 channels coexpressed in *Xenopus* oocytes. At a low PIP_2 level, after the patch has been exposed to control solution, channels have low activity and short mean open time, $MT_o < 1$ msec. Intracellular application of sodium at this PIP_2 level has no effect on channel MT_o and cannot bring channels to their normal high activity. Application of PIP_2 moderately increases channel MT_o and NP_o. Application of Na^+ at this high PIP_2 level dramatically increases NP_o with little effect on channel MT_o, which indicates that enhancing channel–PIP_2 interaction by Na^+ increases channel open probability mainly by decreasing the mean closed time.

Single-Channel Recording in Xenopus Oocytes

For detailed single-channel kinetic studies, particularly evaluation of closed-time kinetics, it is necessary to record activity ideally from a single channel in the patch. To achieve this, oocytes are carefully arranged with the animal pole facing up so that the cRNA is injected just in the center of the dark hemisphere. Knowledge of the location of injection, the amount of cRNA injected, and the time allowed for expression is crucial for obtaining the desired channel density in the membrane, and increasing the probability of obtaining a patch that contains a single active channel. Regarding the location of microinjection, there is an inverse relationship between channel expression and the distance from the site of injection.[21] Thus, it is possible to adjust the distance of the recorded patch from the site of injection, depending on the number of channels present in the patch at each attempted recording. For example, 0.2 ng of GIRK4* cRNA injected together with the same amount of cRNA for human muscarinic 2 receptor (hM2) receptor usually gives optimal expression levels on the next day of injection. cRNA concentrations are estimated from two successive dilutions, which are electrophoresed in parallel on formaldehyde gels and compared with known concentrations of RNA marker (GIBCO, Grand Island, NY).

The entire vitelline membrane of the oocyte is removed before attempting patch clamp, in order to sample patches from different areas from the point of cRNA injection of the oocyte. The single-channel activity is recorded in the cell-attached or inside-out patch configurations,[22,23] using an Axopatch 200A amplifier (Axon Instruments, Foster City, CA). The pipette solution includes (in mM) 96 KCl, 1 MgCl$_2$, and 10 HEPES, pH 7.4. The bath solution includes (in mM) 96 KCl, 5 EGTA, and 10 HEPES, pH 7.4. *Xenopus* oocytes also contain stretch-activated non-selective cation channels. These channels can make single-channel measurements difficult, when they are present in the patch. Stretch-activated channels can be inhibited by 100 μM GdCl$_3$ in the pipette solution.[3] Gd in the pipette does not alter the properties of the inward rectifiers we have studied. GdCl$_3$ stock should be kept in aliquots at $-20°$. We usually prepare a new stock every 4 months. Older stocks may lose activity. For macropatch measurements that require higher expression levels for convenient monitoring of macroscopic currents, the activity of stretch-activated channels is usually negligible compared with the expressed channels, and thus use of Gd is optional. For single-channel measurements, however, use of Gd is recommended. Microelectrodes used in the experiments are pulled from WPI-K borosilicate glass to give a resistance of 1–20 MΩ. Adjusting the pipette size is another important way of increasing the probability of obtaining a single-channel patch. All experiments are performed at room temperature (20–22°). Recordings are performed at a membrane potential of -80 mV. Single-channel currents are filtered at 1–2 KHz with a six-pole low-pass Bessel filter, sampled at 5–10 KHz, and stored directly on hard disk through a DIGIDATA 1200 interface (Axon Instruments). pCLAMP (version 6.01; Axon Instruments) software (complemented by our own analysis routine, as discussed below) is used for data acquisition.

Single-Channel Data Analysis

We analyze single-channel records with pCLAMP software and some of our own programs to complement this software package. The baseline drift is carefully removed with Clampfit8 before idealization with Fetchan.

[21] A. N. Lopatin, E. N. Makhina, and C. G. Nichols, *Biophys. J.* **74,** 2159 (1998).
[22] O. P. Hamill, A. Marty, E. Neher, B. Sakmann, and F. J. Sigworth, *Pflugers Arch.* **391,** 85 (1981).
[23] C. Methfessel, V. Witzemann, T. Takahashi, M. Mishina, S. Numa, and B. Sakmann, *Pflugers Arch.* **407,** 577 (1986).

Fitting dwell-time histograms with exponential functions is a widely used approach to study single-channel kinetics, and relevant programs are commercially available, such as pStat of pClamp. This method, however, is not the ideal approach to study single-channel kinetics as a function of time, because (1) it is not applicable to records exhibiting several multiple openings and (2) it requires channel activity to be in steady state. A method for studying single-channel kinetics parameters as a function of time, and hence for studying the dependence of single-channel kinetics under various experimental conditions, has been previously reported.[1] Here we provide, a detailed description of this method.

In this method, a record is divided into time bins (usually 5 or 10 sec in length), and from the data in each bin, three single-channel kinetic parameters are calculated. The first parameter is the probability that any of the channels in the patch is in the open state, NP_o, where N is the number of the functional channels in the patch and P_o is the open probability of an individual channel. The second parameter is the mean open time of channels, MT_o (to be distinguished from τ_o, which represents the mean open time of a single channel). The third parameter is the open frequency of all of the channels in the patch, NF_o, where N is the number of channels in the patch and F_o is the open frequency of an individual channel. Note that only two of these three parameters are independent, that is, it is possible to compute any one of the three parameters from the other two. Our program reads the event list file, which is the output file of the idealization program (Fetchan) of pClamp. In this event list file, the following parameters, among others, are recorded: (1) the start time of an event, (2) the level of the event, that is, the number of channels opening simultaneously, and (3) the length of the event. Assuming that t_i, L_i, and T_i are the start time, the level, and the length of the ith event of an event list file, respectively, then the three single-channel kinetic parameters, NP_o, MT_o, and NF_o in a time bin can be calculated from the following equations:

$$NP_o = \frac{\sum_{t_i \in \text{bin}} L_i T_i}{T_{\text{bin}}} \tag{1}$$

$$NF_o = \frac{\sum_{t_i \in \text{bin}} f(i)(L_i - L_{i-1})}{T_{\text{bin}}} \tag{2}$$

$$MT_o = \frac{\sum_{t_i \in \text{bin}} L_i T_i}{\sum_{t_i \in \text{bin}} f(i)(L_i - L_{i-1})} \tag{3}$$

where

$$f(i) = \begin{bmatrix} 1 & \text{when } L_i > L_{i-1} \\ 0 & \text{when } L_i < L_{i-1} \end{bmatrix} \tag{4}$$

Equation (1) gives NP_o as the fraction of the time bin in which any of the channels in the patch are in the open state. In Eq. (2), the numerator represents the total number of channel transitions from the closed state to the open state, during the time bin, which gives the frequency of channel opening when divided by the bin size. The factor $f(i)$ is set to zero when the level of the previous event is higher than the present event (i.e., some channels undergo transitions from open state to closed state) to avoid the problem of double counting that has been previously pointed out.[1] Equation (3) gives the mean channel open time during the time bin as the ratio of the total time in which channels are in open state to the total number of channel opening events. Our program outputs these three parameters as a function of time, at the midpoints of each time bin. MT_o calculated by Eq. (3) is independent of the channel number, given that the event level, that is, the number of channels that open simultaneously, is determined accurately. In practice, however, the accuracy of the detection of the event level decreases as the number of channels in the patch increases. Thus, to utilize this method, we usually avoid records containing simultaneous openings of more than three channels. We also avoid records that do not reach baseline for a period longer than several seconds. Thus, it is important to obtain patches that contain a small number of channels, using the approaches discussed above.

Assessment of Channel–Phosphatidylinositol Bisphosphate Interaction, Using Phosphatidylinositol Bisphosphate Antibody

Activation of Kir channels by PIP$_2$ shows that these channels are PIP$_2$ dependent, and suggest that they may directly interact with PIP$_2$. Biochemical studies indeed demonstrated this direct interaction. ^{3}H-Labeled PIP$_2$ binds to purified glutathione S-transferase (GST) fusion proteins of the C terminus of IRK1 channels and this binding can be blocked by unlabeled PIP$_2$ in a dose-dependent manner (EC$_{50}$ of 0.5 μM), but not by PIP or phosphatidylcholine (PC).[4]

Another way to demonstrate channel sensitivity to PIP$_2$ is by means of a monoclonal antibody to PIP$_2$ (PIP$_2$ Ab). When applied to the intracellular side of the plasma membrane, PIP$_2$ Ab inhibits activity presumably by shielding the acidic phosphate groups from basic amino acids of the channel protein.[3,4,16] Much of the work using PIP$_2$ Ab has been performed in studies of channel–PIP$_2$ interactions of Kir channels.[3,4,8,16,24] The sensitivity of Kir channel activity to inhibition by PIP$_2$ Ab correlates well with the *in vitro* biochemical binding strength between Kir channel fragments and PIP$_2$ as well as with the rate of channel activation by PIP$_2$.[4] Thus, for IRK1, which interacts more strongly with PIP$_2$ than GIRK4*, PIP$_2$ Ab will take a longer time to inhibit channel activity[3,4,16] (see also Fig. 3A). The time course of channel inhibition by PIP$_2$ Ab can be fitted with exponential or

---

[24] H.-H. Liou, S.-S. Zhou, and C.-L. Huang, *Proc. Natl. Acad. Sci. U.S.A.* **96,** 5820 (1999).

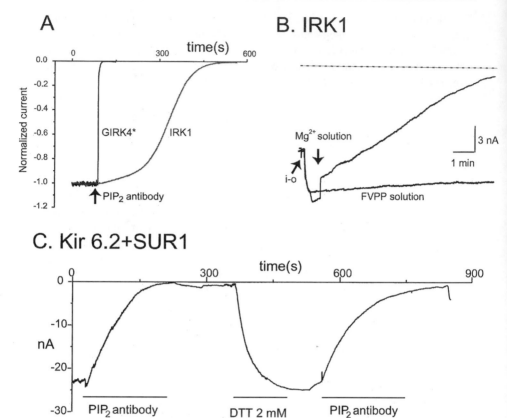

FIG. 3. (A) PIP_2 Ab inhibits K^+ channel currents from two inwardly rectifying K channels, GIRK4* and IRK1. IRK1 cRNA (0.5 ng) and GIRK4* cRNA (2 ng) were injected into each *Xenopus* oocyte. A macropatch was obtained and held at −80 mV. The current traces are from inside-out patches bathed in FVPP solution. Currents are normalized on the basis of their maximal amplitude before the inhibition. PIP_2 Ab (200 times dilution from the stock solution) is applied at the time indicated by the arrow. (B) Comparison of channel activity rundown in inside-out patches. K^+ current through the IRK1 channel was measured at −80 mV. A macropatch was formed and then excised into an inside-out patch (i–o) indicated by the arrow. The current rundown from one patch in the FVPP solution is compared with that from a different patch of comparable current after changing from FVPP solution to a Mg^{2+}-containing solution (arrow). (C) DTT reverses K^+ channel inhibition by PIP_2 Ab. Kir6.2 and SUR1 (2 ng of RNA each per oocyte), which form the K_{ATP} channel, were injected into the *Xenopus* oocyte. An inside-out patch was held at −80 mV and bathed in FVPP solution. PIP_2 Ab and DTT were each applied for the period of time indicated by the bars.

Boltzmann functions. Properties of the channel–PIP_2 interactions can be inferred from such fits (e.g., cooperativity), but for the most part a convenient time marker, such as the time needed for half-maximal (T_{50}) inhibition of channel activity by PIP_2 Ab,[4,16] is sufficient to describe these data.

To obtain an accurate assessment of the kinetics of blockage, channel activity ought to be stable during the period of time in which PIP$_2$ Ab is applied. One common feature of Kir channels is that their channel activity runs down in inside-out patches.[3,4,8,16,24] Strong evidence suggests that the rundown of channel activity is caused by the gradual dephosphorylation of PIP$_2$ in the membrane. This evidence consists first of the observation that PIP$_2$ can reactivate channel activity after its complete rundown.[3,4,16] Second, divalent cations that activate lipid phosphatases, like Mg^{2+}, increase channel rundown, while reagents that block lipid phosphatases slow down or prevent channel rundown[4,8,16] (Fig. 3B). Finally, MgATP can also reactivate run-down channels, presumably by supplying substrate for lipid kinases that resynthesize PIP$_2$ from PI and PIP. This reactivation can be blocked by PIP$_2$ Ab.[3,4,16] The time course of the rundown of channel activity, like the time course of the inhibition by PIP$_2$ Ab, is a good indicator of the strength of the channel–PIP$_2$ interactions, stronger channel–PIP$_2$ interactions yielding slower channel rundown.

To prevent channel rundown in PIP$_2$ Ab experiments, an intracellular solution containing phosphatase inhibitors such as fluoride, vanadate, and pyrophosphate (FVPP solution) can be used.[4,16,24] FVPP solutions we have used contain (in mM) 100 KCl, 20 HEPES, 5 EDTA, 5 NaF, 0.1 Na$_3$VO$_4$, and 10 Na$_4$P$_2$O$_7$, pH 7.4. Although rare, irreversible channel rundown can occur, even in the presence of FVPP solution. It also must be noted here that the FVPP solution described above contains ~45 mM Na$^+$. Some channels, such as GIRK2 or GIRK4, are Na$^+$ sensitive[3,8,16] and thus activity of these channels in the presence of FVPP solution is at least partially due to the presence of Na$^+$. If the Na$^+$ content in the FVPP solution is of concern, potassium salts of phosphatase inhibitors can be used.

We use PIP$_2$ Ab from PerSeptive Biosystems (Framingham, MA) (now available by Assay Designs, Ann Arbor, MI). The PIP$_2$ Ab is dissolved in 0.5 ml of water, divided into aliquots (25 μl), and stored at $-20°$. A 40- to 200-fold dilution of the stock solution has proved effective in blocking completely Kir channel activity[4,16] (Fig. 3A and C). We have observed fairly large variations in the inhibitory effect of PIP$_2$ Ab depending on the batch of PIP$_2$ Ab as well as the batch of oocytes.

Inhibition of Kir channel activity by PIP$_2$ Ab is normally irreversible, even after long washout periods. Furthermore, application of additional PIP$_2$ to Kir channels that have already been inhibited by PIP$_2$ Ab does not reactivate the channels.[4,16,24] The mechanism underlying this persistent inhibition of activity by the PIP$_2$ Ab is not clear, but could be due to the formation of a complex between PIP$_2$ and PIP$_2$ Ab localized in a microdomain.[24] Consistent with this notion, application of reducing agents such as dithiothreitol (DTT), or glutathione, after PIP$_2$ Ab application reversed the inhibition[16,24] (Fig. 3C), probably by reducing the disulfide bonds within the PIP$_2$ Ab molecules[24] and dissociating the PIP$_2$ Ab–PIP$_2$ complexes. After reversal by DTT or glutathione, channels can be inhibited once again by PIP$_2$ Ab.[16,24] Thus, DTT and glutathione are useful tools, because they enable multiple applications of the PIP_2 Ab in the same patch and allow comparisons of PIP_2 Ab inhibition before and after treatments that alter channel–PIP_2 interactions. The reversal by DTT of Kir channel inhibition by PIP_2 Ab is concentration dependent, being complete at 2 mM DTT. Figure 3C shows an example of PIP_2 Ab inhibition and its reversal by DTT. In our hands, DTT (up to 20 mM) proved completely ineffective in reversing PIP_2 Ab inhibition of IRK1.

Polylysine Block

We have searched for other inhibitors of Kir channel activity that could be used in a similar manner to PIP_2 Ab to assess the strength of channel–PIP_2 interactions. Direct application of polylysine to the cytoplasmic side of the membrane inhibits the current of Kir channels. Polylysine with its positively charged residues is thought to compete with the channel for binding to the negatively charged head groups of PIP_2. The stronger the channel interactions with PIP_2, the slower are the kinetics of polylysine block (Fig. 4A). The kinetics of polylysine inhibition can be used to identify residues important for the channel interaction with PIP_2. When critical residues are mutated, channel–PIP_2 interactions are weakened and polylysine blockage occurs faster. In the experiment shown in Fig. 4B, a modified version of IRK1 (IRK1J) was used, in which internal cysteines sensitive to sulfhydryl-modifying reagents have been mutated.[25] For IRK1, we have found two phases of current inhibition by polylysine. The first phase is "instantaneous" and independent of channel interactions with PIP_2. The second phase is slower and PIP_2 dependent, and its T_{50} values of current inhibition correlate well with those obtained from PIP_2 Ab experiments (Fig. 4B). Polylysine of various average molecular weights can be purchased from Sigma (St. Louis, MO). We use a 300-μg/ml concentration of polylysine with an average molecular mass of 7 kDa. We found that polylysine of lower molecular weights resulted in block showing only the fast, PIP_2-independent kinetic phase of IRK1 currents. At this concentration inhibition proceeds at a reasonable rate for channels that bind PIP_2 tightly, such as the Kir1.1 (ROMK1-rat outer medulla K^+ subunit 1) channel ($T_{50} \approx 200$ sec), and is fast for channels that bind weakly, such as GIRK4* ($T_{50} \approx 2$ sec). For channels other than IRK1, we have not experimented with polylysine of molecular mass less than 7 kDa. Polylysine inhibition can be partially reversed by washout (Fig. 4C). A more complete reversal can be obtained by the application of heparin (200 μg/ml; Sigma). Although repeated polylysine applications in the same patch yield reproducible inhibition kinetics, results become inconsistent when heparin is used between applications. Moreover, if FVPP solution (containing 0.1–3 mM Na_3VO_4 and 10 mM $Na_2H_2P_2O_7$) is added to prevent rundown, a precipitate will

[25] T. Lu, B. Nguyen, X. Zhang, and J. Yang, *Neuron* **22,** 571 (1999).

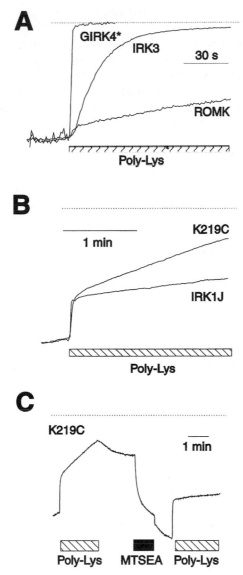

FIG. 4. PIP$_2$-dependent inhibition of Kir channels. Different inhibition kinetics reflect underlying differences in the strength of channel–PIP$_2$ interactions. Normalized currents measured in inside-out macropatches at -100 mV are shown. Polylysine is applied as indicated. The dotted line denotes zero current level. (A) Polylysine inhibition of ROMK1, GIRK4*, and IRK3 channels. (B) Polylysine inhibition of IRK1 mutants. Normalized currents are measured for the control (IRK1J) and K219C mutant channels. The modified channel IRK1J has all native MTSEA-reacting cysteines mutated. (C) Effect of MTSEA application on polylysine inhibition. Current was measured as in (B) for the K219C mutant channel. Note the faster block before exposure to MTSEA, the increase in current levels during MTSEA application, and the slow block by polylysine after MTSEA is covalently bound to the channel. Current levels increased even further after wash of MTSEA. MTSEA is applied as indicated.

appear in the solution over time. To avoid this problem, solutions either must be made immediately before use (less than 1 min) or a modified phosphatase inhibitor solution must be used (no Na_3VO_4 and 2 mM $Na_2H_2P_2O_7$). The low cost of polylysine makes this compound an inexpensive alternative to PIP_2 Ab for measurements of PIP_2 interactions with the channel. Yet, polylysine is not as specific for PIP_2 as PIP_2 Ab. Neither PIP_2 Ab nor polylysine is an ideal reagent for these measurements, as none of these reagents are fully and consistently reversible. Use of small charged peptides (as small as seven amino acids) provides reversible channel–PIP_2 inhibition (D. Hilgemann, personal communication, 2000), but again specificity is not ideal. Thus, the search for a universal, readily reversible, and specific inhibitor of Kir channel–PIP_2 interactions continues.

Cysteine Substitutions

The combination of scanning cysteine accessibility mutagenesis (SCAM) and covalent chemical modification with reagents of different size and charge has been widely used to investigate the architecture of the pore of potassium channels.[25] To identify possible direct interaction sites with negatively charged PIP_2, it is possible to systematically mutate the positively charged residues in both N and C termini to cysteines. Mutant channels can then be probed with a positively charged, water-soluble, sulfhydryl-specific reagent, which forms a covalent bond with accessible sulfhydryl groups. If the channel interaction with PIP_2 is weakened by neutralizing the native positive charge through cysteine mutation and strengthened on application of the sulfhydryl reagent, it is inferred that the original residue is a direct PIP_2-interacting site (Fig. 4B and C).

A prerequisite for employing SCAM is the ability to clearly distinguish the response of the cysteine mutant channels from the wild-type channels. If the wild-type channel reacts with the sulfhydryl reagent applied to the cytoplasmic face of the channel, it is necessary to identify and mutate the reactive endogenous cysteine residues. The resulting nonreactive channel can then be used as a control. The charged residues may then be mutated to cysteines in the background of this channel.

We systematically mutated the positively charged residues in the nonreactive IRK1 mutant, IRK1J[25] (kindly provided by J. Yang, Biological Sciences, Columbia University, New York, NY). As a cysteine-modifying reagent we used methane thiosulfonate-ethylammonium (MTSEA, 2.5 mM; Toronto Research Chemicals, North York, ON, Canada) because a cysteine covalently bound to MTSEA has a similar side chain length and the same$–NH_3^+$ head group as lysine. Consistently, we found that the strengthening of PIP_2 interaction observed after MTSEA application was more pronounced with mutants in which cysteine replaced a lysine residue. These results suggest that PIP_2 interactions with the channel are complex and that the PIP_2–binding site is specific. The current level increased dramatically when

MTSEA reacted with some of the mutant channels studied. These results further confirm that the decrease in current levels observed in the PIP$_2$ mutants can be reversed once the positive charge is replaced (by MTSEA), reestablishing the PIP$_2$-interacting site.

Monitoring Effects of Phosphatidylinositol Bisphosphate Hydrolysis on Channel Activity

Xenopus oocytes possess an endogenous Ca^{2+}-activated Cl^- current that can be used to monitor PIP$_2$ hydrolysis in two-electrode voltage–clamp studies. Agonists that lead to PLC activation cause hydrolysis of PIP$_2$, generating inositol 1,4,5-trisphosphate (IP$_3$) and diacylglycerol (DAG). IP$_3$ causes release of Ca^{2+} from intracellular stores, which in turn activates Cl^- currents. Hydrolysis of PIP$_2$ could inhibit channels whose activity is stimulated by interactions with PIP$_2$, provided that PIP$_2$ levels become limiting for channel activity. Such inhibition should correlate with the activation of Ca^{2+}-activated Cl^- channels, which signifies PIP$_2$ hydrolysis. Thus, *Xenopus* oocytes are a convenient system to monitor the effects of PIP$_2$ hydrolysis on the activity of heterologously expressed PIP$_2$-sensitive ion channels. We have systematically compared three Kir members and have found them to have different sensitivities to PIP$_2$. GIRK channels experience weak interactions with PIP$_2$ compared with the IRK1 channel.[4,16] From the kinetics of PIP$_2$ activation or PIP$_2$ Ab inhibition,[25a] we found the IRK3–PIP$_2$ interaction strength to be intermediate between those of GIRK4* and IRK1. Thus, the order of the strength of interaction with PIP$_2$ for these three channels is as follows: GIRK4 < IRK3 < IRK1. We also compared the inhibition of GIRK4*, IRK3, and IRK1 by muscarinic receptor (M$_1$) activation. The M$_1$ receptor is known to couple to PLC through $G\alpha_q$. Figure 5 shows that GIRK4*, IRK3, and IRK1 are inhibited by M$_1$ activation to different levels. The extent of the inhibition of channel activity correlates well with the relative strength of interaction with PIP$_2$. Thus GIRK4*, which has the weakest interaction with PIP$_2$, is inhibited most, while IRK1, which has the strongest interaction with PIP$_2$, is not significantly inhibited by activation of M$_1$. IRK$_3$, with an intermediate strength of interaction with PIP$_2$, is inhibited significantly but less than GIRK4*.

The experiments shown in Fig. 5 were performed by the two-electrode voltage–clamp technique and the oocytes were bathed in a high K^+ (96 mM) solution. Currents were measured at membrane potentials of -80 mV (inward) and $+80$ mV (outward) and are shown in Fig. 5A, B, and C. M$_1$ was activated by 5 μM acetylcholine. The Ca^{2+}-activated Cl^- currents, which are outwardly rectifying and show transient activation, were more pronounced at $+80$ mV, whereas inwardly rectifying K^+ currents were more pronounced at -80 mV. The activation of Ca^{2+}-activated Cl^- currents correlates well with the start of inhibition of GIRK4*

[25a] H. Zhang and D. E. Logothetis, in preparation (2001).

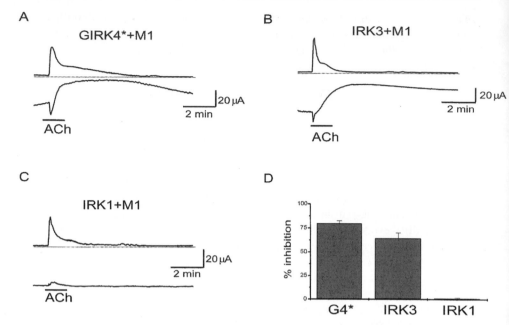

FIG. 5. M$_1$ activation inhibits GIRK4* (A) and IRK3 (B) but not IRK1 (C) currents. Three nanograms of M$_1$ cRNA was coinjected into *Xenopus* oocytes with either 2 ng of GIRK4* (A), 1 ng of IRK3 (B), or 0.5 ng of IRK1 (C) cRNA. The currents were recorded 2 days after injection, using the two-electrode voltage clamp technique. M$_1$ receptors were activated by 5 μM acetylcholine applied to the bath. The currents were monitored at -80 mV (the trace below the dashed zero-current line) and $+80$ mV (the trace above the dashed zero-current line). Summary inhibition of the activity of each channel type is shown in (D).

(Fig. 5A) and IRK3 (Fig. 5B) currents. To calculate percentage inhibition, currents were measured after the inhibition had reached steady state levels, usually after 2 min.

Imaging Phosphatidylinositol Bisphosphate Hydrolysis

Inositol phospholipids play dynamic roles in many physiological processes. Therefore, dynamic measurements of phospholipid turnover and concentration changes provide a wealth of information about the signaling pathways that use these molecules. The development of fluorescent lipid-binding molecules to study phospholipids in living cells has provided tools to achieve real-time monitoring of dynamic changes in these lipids. Specifically, pleckstrin homology (PH) domains from several proteins tagged with fluorescent labels, such as the green fluorescent protein (GFP), have been used extensively for this purpose.[26–28] We have used the PH domain of PLCδ tagged with GFP at the N terminus (GFP–PH) to measure in real time the hydrolysis of PIP$_2$ in the membrane.[18] To date many PH domains have been identified and characterized; however, few of them show specificity for different forms of phosphoinositides.[29,30] The PH domain of PLCδ is unique in that it specifically binds PI(4,5)P$_2$ and not other phosphoinositides.[29,30] Several groups have used this construct to measure PI turnover in the membrane.[26,27,31] Here, we discuss experimental details for using GFP–PH.

cDNA Construct and Expression in Cos-1 Cells

The construct that we have used to monitor PIP$_2$ hydrolysis was kindly provided by T. Meyer (Stanford University, Stanford, CA). Details of this construct have been described elsewhere.[26] The PH domain of PLCδ (amino acids 1–175) was subcloned in frame to the C terminus of GFP. We have subcloned this construct into the expression vector pcDNA3 (Invitrogen, Carlsbad, CA) for optimal expression in mammalian cell lines.

We have also used *Xenopus* oocytes to monitor PIP$_2$ hydrolysis with GFP–PH. We used albino oocytes for confocal imaging because the pigment in normal oocytes occludes laser penetration into the oocyte. Nevertheless, even in albino oocytes, the laser in a confocal microscope cannot penetrate more than 100 μm into the cell. Because the diameter of a normal oocyte is ~1–1.2 mm, confocal microscopy gives information only about the outer one-tenth of the oocyte. Still, we have observed translocation of GFP–PH into the cytoplasm on PLC activation. However, because of their transparency and ease of handling on a confocal microscope, mammalian cells are the preferred expression system.

We use Cos-1 cells, which were transfected with GFP–PH as well as any additional constructs required for the specific experiment. The cells are grown in Dulbecco's Modified Eagle's medium (DMEM) supplemented with 10% (v/v) fetal bovine serum and are plated on special 40-mm coverslips (Bioptechs, Butler, PA), inside a 60-mm tissue culture dish. These coverslips are specifically made for a chamber (Bioptechs) mounted on an inverted Leica TCS-1 confocal microscope used in these experiments. For transfections, we use the Effectene reagent (Qiagen, Valencia, CA), which yields good transfection efficiency in Cos-1 cells (10–25%). GFP-PH expresses at high levels in Cos-1 cells 48 hr after transfection.

---

[26] T. P. Stauffer, S. Ahn, and T. Meyer, *Curr. Biol.* **8**, 343 (1998).

[27] P. Varnai and T. Balla, *J. Cell Biol.* **143**, 501 (1998).

[28] E. G. Tall, I. Spector, S. N. Pentyala, I. Bitter, and M. J. Rebecchi, *Curr. Biol.* **10**, 743 (2000).

[29] L. E. Rameh, A. Arvindsson, K. L. Carraway, A. D. Couvillon, G. Rathbun, A. Crompton, B. Van Renterghem, M. P. Czech, K. S. Ravichandran, S. J. Burakoff, D. S. Wang, C. S. Chen, and L. C. Cantley, *J. Biol. Chem.* **272**, 22059 (1997).

[30] S. A. Dowler, R. G. Currie, D. Campbell, M. Deak, G. Kular, C. P. Downes, and D. R. Alessi, *Biochem. J.* **351**, 19 (2000).

[31] K. Hirose, S. Kadowaki, M. Tanabe, H. Takeshima, and M. Iino, *Science* **284**, 1527 (1999).

Expressed GFP can be easily detected by epifluorescence mainly in the membrane, with a weak signal in the cytoplasmic region (Fig. 6, control). Some intracellular membranes may also display fluorescence, but this is generally negligible compared with the plasma membrane signal. It has been reported that GFP–PH may destabilize cytoskeletal elements,[32] which manifests itself in rounding of the cells. However, in our hands, many cells are morphologically normal after 48 hr of expression. Furthermore, rounded cells that express the protein display normal physiological responses as determined experimentally.

Confocal Microscopy

On the day of the experiment, we remove the coverslip from the tissue culture dish and mount it at the bottom of the chamber. Another coverslip is used as the top of the chamber and the perfusing solution fills the gap between the two coverslips. Input and output ports on the two side of the chamber allow perfusion by a peristaltic pump. Alternatively, an open chamber can be used.

To monitor PIP_2 hydrolysis, we coexpress the human muscarinic receptor type 1 (hM_1) with GFP–PH. Images are acquired on a Leica TCS-1 confocal microscope, using the fluorescein isothiocyanate (FITC) channel and either a $\times40$, $\times63$, or $\times100$ oil immersion objective. Acetylcholine (ACh) is perfused with a pump. The perfusion is stopped during image acquisition to prevent cell movement in the z direction. Application of ACh ($5 \mu M$) causes the translocation of GFP–PH from the membrane into the cytoplasm, which is reversed on washout. Images are obtained every 15–60 sec. The TCS image acquisition software can also be set up to record images at fixed intervals. We acquire three control images before a 3-min application of ACh, which is then washed out for 3–5 min. The frequency of image acquisition is higher when the drug is first applied and in the beginning of the washout period, in order to monitor the kinetics of GFP–PH translocation more closely.

Figure 6 shows images from a representative cell expressing GFP–PH and hM_1. Under control conditions, the signal is concentrated in the cell membrane. Application of ACh causes translocation of the signal to the cytoplasm, which can then be reversed by removing the agonist. Pretreatment with the PLC inhibitor U73122 significantly reduces the translocation of GFP–PH to the cytoplasm on activation of hM_1, indicating that this process is mediated by PLC.[18,26]

Quantification of GFP–PH Translocation

Although the response to hM_1 is clear in cells expressing GFP–PH, it is necessary to quantify this response in order to compare different treatment groups. Quantification is also needed to analyze the kinetics of both the response and the recovery. All images acquired by the confocal microscope are saved as TIFF

---

[32] D. Raucher, T. Stauffer, W. Chen, K. Shen, S. Guo, J. D. York, M. P. Sheetz, and T. Meyer, *Cell.* **100,** 221 (2000).

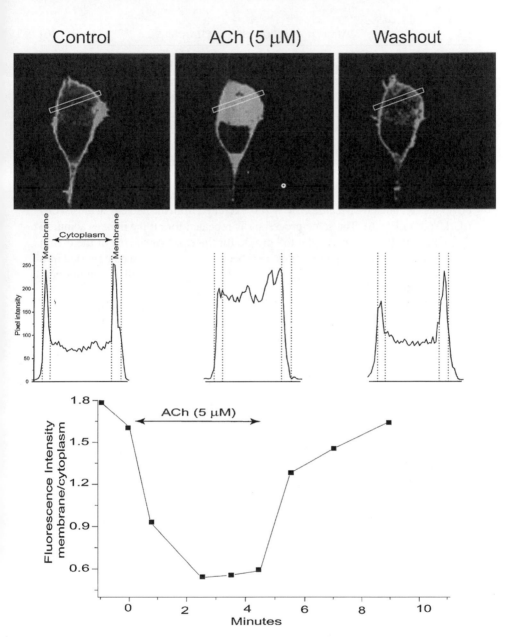

FIG. 6. GFP–PH as a tool for PIP$_2$ imaging. *Top:* Confocal images from Cos-1 cells expressing GFP–PH and hM$_1$ receptors. Under control conditions, the GFP–PH is localized mainly to the membrane. Application of 5 μM ACh leads to translocation of the signal to the cytoplasm, which is reversed on washout. *Middle:* Plot profiles of the images shown above. Each profile represents the intensity along the length of the rectangular box in the respective cell. Two membrane regions and the cytoplasmic region are indicated in one of the plots. *Bottom:* Response of the membrane-to-cytoplasm fluorescence ratio to the application and withdrawal of 5 μM ACh.

files for further analysis. We use Scion (Frederick, MD) imaging software for image analysis, which can be downloaded from the manufacturer web site at no cost (www.scioncorp.com). The program is available for both Macintosh and PC platforms and is based on the NIH Image software for the Macintosh. For analysis, we draw a narrow rectangular box across the cell (Fig. 6, top) and plot the pixel intensity profile for the cross-section. Under control conditions this profile consists of two peaks separated by a wide valley, representing the membrane and cytoplasmic regions, respectively (Fig. 6, middle left). The values for each point in the profile are exported as a text file. It is important to choose an area that does not cover the cell nucleus, because GFP–PH is excluded from the nucleus in these cells (dark spot in Fig. 6). The same procedure is repeated for all the images acquired from the cell. It is crucial to plot the profile for the exact same area of the cell in each image. Once all the plot profiles have been written, they can be imported into a spreadsheet program for further analysis. The plot profile will give numerical values to each point based on the intensity in the image. Several data points at the beginning and end of the profile represent the membrane regions of the cell. We select a fixed number of points and assign them as membrane regions in control. This is relatively easy given the large difference in the intensity of the membrane regions and their neighboring regions (outside the cell and cytoplasmic). The same procedure is repeated for all images acquired. We then average the values for the membrane regions and the cytoplasmic regions, defined as all points between two membranes, to obtain intracellular fluorescence intensity and membrane intensity. We plot the ratio of membrane to cytoplasmic intensity as a function of time (Fig. 6, bottom). This graph gives information not only about the extent but also the kinetics of the GFP–PH translocation. Thus, we have been able to show that there is good correlation between the kinetics of GFP-PH translocation from the membrane into the cytoplasm and desensitization of GIRK channels after activation of M_1 receptors. This observation strengthens the hypothesis that PIP_2 hydrolysis leads to GIRK channel desensitization.[18]

Acknowledgments

We thank Drs. Donald Hilgemann and Massimo Sassaroli for critical comments on the manuscript. This work was supported by Grants HL54185 and HL59949 from the National Institutes of Health to D.E.L., T.M., and T.R. were supported by postdoctoral fellowships from the Revson Foundation and NRSA postdoctoral fellowships from the NIH. We are grateful to Xixin Yan for oocyte preparation.

Section IV

Adenylyl Cyclases

[7] Purification of Soluble Adenylyl Cyclase

By JOCHEN BUCK, MEEGHAN L. SINCLAIR, and LONNY R. LEVIN

Cyclic AMP (cAMP), the nearly universal second messenger, is synthesized by a broad family of adenylyl cyclases (ACs). Mammals have at least nine isoforms of transmembrane-spanning adenylyl cyclases (tmACs). These differ in their expression patterns, and their catalytic activities are differentially regulated by G proteins and other signaling molecules (reviewed in other articles in this volume).

Mammals also possess a cytosolic form of AC. Originally detected solely in soluble extracts derived from mammalian testis,[1] the catalytic domains of soluble AC (sAC) are more closely related to (cyano)bacterial adenylyl cyclases than to the hormone-responsive, G-protein-regulated mammalian tmACs.[2] sAC activity can be biochemically distinguished from the activities of other mammalian cyclases because unlike tmACs, soluble AC activity is insensitive to G protein regulation and forskolin stimulation,[2,3] and it displays approximately 10-fold lower affinity for substrate, ATP ($K_m \sim 1$ mM),[2,4–6] than the tmACs ($K_m \sim 100$ μM).[7] Our laboratories used these properties to develop the following purification scheme for soluble AC activity. Starting with 950 rat testes, we purified sufficient material to isolate and sequence three peptides that enabled us to clone the sAC gene.[2]

Pilot Studies

We confirmed the presence of Mn^{2+}-dependent activity in extracts from frozen rat testis (Fig. 1). Consistent with earlier reports and in contrast to the AC activity in membranes, the soluble AC activity was 10 times more active in the presence of Mn^{2+} than Mg^{2+}. Therefore, subsequent AC assays were performed in the presence of 5 mM $MnCl_2$ in place of $MgCl_2$. In addition, we confirmed the previously reported K_m for ATP of this soluble AC activity to be ~ 1 mM; thus, our assay conditions included 5 mM ATP instead of the more standard 1 mM ATP used in *in vitro* assays of tmAC activity.

[1] T. Braun and R. F. Dods, *Proc. Natl. Acad. Sci. U.S.A.* **72,** 1097 (1975).
[2] J. Buck, M. L. Sinclair, L. Schapal, M. J. Cann, and L. R. Levin, *Proc. Natl. Acad. Sci. U.S.A.* **96,** 79 (1999).
[3] T. Braun, H. Frank, R. Dods, and S. Sepsenwol, *Biochim. Biophys. Acta* **481,** 227 (1977).
[4] E. J. Neer, *J. Biol. Chem.* **253,** 5808 (1978).
[5] J. O. Gordeladze and V. Hansson, *Mol. Cell. Endocrinol.* **23,** 125 (1981).
[6] T. Braun, *Methods Enzymol.* **195,** 130 (1991).
[7] R. A. Johnson, R. Alvarez, and Y. Salomon, *Methods Enzymol.* **238,** 31 (1994).

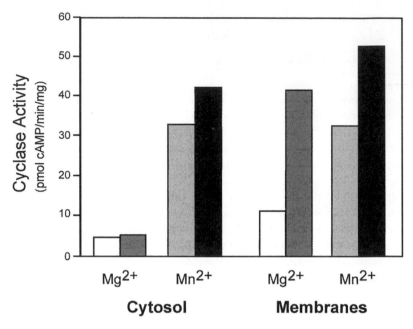

FIG. 1. Cyclase activity in cytosol and membrane fractions from rat testes. Equal percentages of particulate and soluble fractions were assayed in the presence of 5 mM MgCl$_2$ or 5 mM MnCl$_2$ in the presence or absence of 100 μM forskolin. Activity is represented as picomoles of cAMP produced per minute per milligram of protein. Open columns indicate activity in the presence of 5 mM MgCl$_2$; hatched columns indicate activity in the presence of 5 mM MgCl$_2$ plus 100 μM forskolin; gray columns indicate activity in 5 mM MnCl$_2$; solid columns indicate activity assayed in 5 mM MnCl$_2$ plus 100 μM forskolin.

Pilot studies indicated that anion-exchange and gel-filtration chromatographic principles as well as the reactive dyes Reactive Red 120–agarose and Reactive Green 19–agarose could be used for the purification of soluble AC activity (data not shown). Anion exchange and gel filtration had been used previously by other investigators.[5,6] Contrary to previous reports,[8] we were unable to use ATP-affinity chromatography for sAC purification. In our hands, sAC did not bind to three of the most commonly used ATP matrices.

Using gel-filtration chromatography, about 90% of sAC activity eluted as a single peak with an apparent size of approximately 50 kDa (data not shown); this matched several published reports concerning the mass of the enzyme.[5,6] Interestingly, approximately 10% of the total sAC activity eluted with an apparent size > 100 kDa. However, we did not further characterize this peak.

[8] J. Kawabe, Y. Toya, C. Schwencke, N. Oka, T. Ebina, and Y. Ishakawa, *J. Biol. Chem.* **271**, 20132 (1996).

Soluble Adenylyl Cyclase Purification from 950 Rat Testes

sAC is purified from 950 rat testes by sequential column chromatography. Frozen rat testes are purchased from Pel-Freeze Biologicals (Rogers, AR).

Cyclase Assay

The in vitro adenylyl cyclase assay is performed as described previously,[9,10] except that the standard assay conditions for sAC activity include 5 mM MnCl$_2$ in place of MgCl$_2$ and 5 mM [α-^{32}P]ATP (specific activity, 4×10^4 cpm/nmol).

Preparation of Cytosolic Extracts from 950 Rat Testes

Rat testes are homogenized manually in batches of 50 testes in approximately 200 ml of 20 mM Tris-HCl, pH 7.5, in the presence of 1 mM dithiothreitol (DTT) and protease inhibitors [phenylmethylsulfonyl fluoride (PMSF), leupeptin, and aprotinin from Sigma, St. Louis, MO]. The homogenate is aliquoted into 30-ml portions, and each is sonicated on ice twice with 20-sec intervals using a microtip sonicator (output, 13 W). After debris and nuclei are removed by low-speed centrifugation (3000g for 10 min at 4°), a high-speed supernatant (> 100,000g for 60 min at 4°) is prepared. The supernatant corresponding to extract from 50 testes is dialyzed overnight at 4° against 20 mM Tris (pH 7.5), 1 mM DTT (Spectra/Por dialysis tubing; Spectrum, Rancho Dominguez, CA). Cytosol is prepared for subsequent chromatographic separation by filtering through blotting paper VWR 238 (Ahlstrom Filtration, Mount Holly Springs, PA).

Anion Exchange (DE-52)

The total starting cytosolic material (52 g) is purified (as 19 equal portions consisting of 50 testes each) over diethylaminoethyl (DE-52) cellulose anion-exchange columns, a weak anion-exchange matrix. DE-52 cellulose anion-exchange resin is from Whatman (Clifton, NJ; 80-ml column bed volume). Buffer is 20 mM Tris-HCl, pH 7.5, in the presence of 1 mM DTT and protease inhibitors; a linear gradient of 0–0.3 M NaCl; a flow rate of 1.5 ml/min; 600-ml total volume. All detectable sAC activity is bound, and is eluted between 0.20 and 0.24 M NaCl (Fig. 2). Eluting column fractions that contain the peak of soluble AC activity and a high activity-to-protein ratio (fractions 44–49) are used for the next step in the purification.

Gel Filtration

sAC activity recovered from DE-52 (4 g of total protein is divided into 10 equal aliquots consisting of 8 ml each) is separated with an Ultrogel AcA54

---

[9] L. R. Levin, P. L. Han, P. M. Hwang, P. G. Feinstein, R. L. Davis, and R. R. Reed, Cell **68,** 479 (1992).

[10] L.R. Levin and R. R. Reed, J. Biol. Chem. **270,** 7573 (1995).

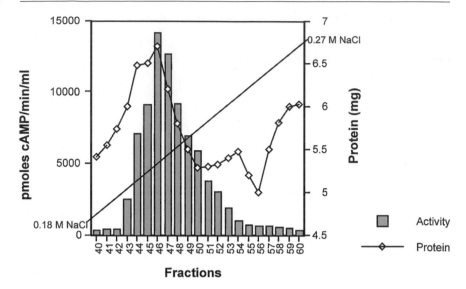

Fractions

Fig. 2. sAC elution profile from representative DE-52 cellulose anion-exchange column. In this particular example fractions 44–49 were combined and further purified over a gel-filtration AcA54 column. The solid line indicates the [NaCl] gradient.

gel-filtration column. The Ultrogel AcA54 gel-filtration matrix is from Pharmacia-LKB, Piscataway, NJ) (column, 4 × 100 cm); the buffer is 20 mM Tris-HCl, pH 7.5, in the presence of 1 mM DTT and protease inhibitors; the flow rate is 1.0 ml/min; the total volume is 500 ml. sAC activity reproducibly elutes in a single peak from these columns with an apparent mass of 50–60 kDa (Fig. 3). Activity-containing fractions pooled from the gel-filtration columns are chosen to maximize the yield of soluble AC activity and avoid the large contaminating protein peak that elutes preceding the peak of this activity.

The 19 DE-52 columns, followed by 10 gel-filtration columns, are run over four successive weeks in order to amass the material for the ensuing steps in the purification. We are able to concentrate and freeze the bioactive fractions from the gel-filtration columns with little subsequent loss after thawing. The following steps are performed with pooled material.

Reactive Dye Column (Red 120–Agarose)

sAC peak fractions are pooled (total of 1 g of protein), applied to a single Reactive Red 120–agarose column, and eluted with a linear salt gradient (0–1.0 M NaCl) (Fig. 4). The reactive Red 120–agarose (Sigma, St. Louis, MO) has a 50-ml bed volume; a buffer consisting of 20 mM Tris-HCl (pH 7.5), 1 mM DTT; a linear gradient of 0–1.0 M NaCl; a flow rate of 2 ml/min; a 600-ml total volume.

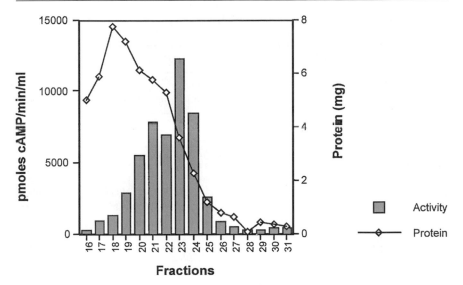

FIG. 3. sAC elution profile from a representative gel-filtration AcA54 column. In this particular example fractions 20–24 were combined, concentrated, frozen, and ultimately loaded onto a Reactive Red 120–agarose column.

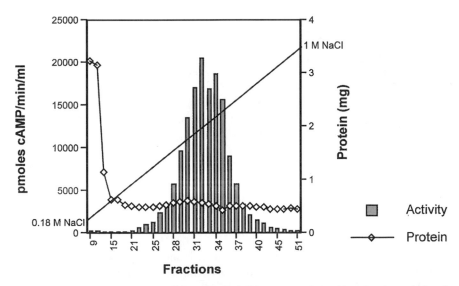

FIG. 4. sAC elution profile from a Reactive Red 120–agarose column. Note that the majority of contaminating protein elutes before the peak of sAC activity. Fractions 31–38 were pooled and applied to the Source Q anion-exchange column. The solid line indicates the [NaCl] gradient for these fractions.

TABLE I
PURIFICATION OF SOLUBLE ADENYLYL CYCLASE FROM 950 TESTES[a]

| Step | Protein (mg) | AC activity | | Enrichment (fold) |
		Total units (nmol/min)	Specific activity [(nmol/min/mg) × 100]	
Cytosol	51,900	2,400	4.6	1
Preparative DE-52	4,015	2,900	74	16
Gel filtration AcA54	1,074	2,100	200	42
Reactive Red	66	1,200	1,900	410
Source Q	8.8	1,100	12,000	2,600
Reactive Green	1.8	380	21,000	4,700
Semipreparative QA (pH 7.4)	0.6	310	49,000	11,000
Analytical QA (pH 6.8)				
Fraction 18	0.003	90	3,000,000	650,000
Fraction 19	0.010	92	920,000	200,000

[a] Total activity (nmol/min), specific activity (nmol/min/mg), and enrichment (fold) of sAC for each step of the purification. The total yield was approximately 8%; each of the final peak fractions yielded approximately 4% of the starting material. [By permission from J. Buck, M. L. Sinclair, L. Schapal, M. J. Cann, and L. R. Levin, *Proc. Natl. Acad. Sci. U.S.A.* **96,** 79 (1999).]

The majority of sAC activity elutes between 0.45 and 0.55 M NaCl, which is after the major peak of protein. This chromatographic step yields almost a 10-fold enrichment in sAC specific activity (summarized in Table I).

Anion Exchange (Source Q)

Active fractions from the Reactive Red 120–agarose columns are pooled, dialyzed versus 20 mM Tris-HCl (pH 7.5), 1 mM DTT, and applied to a Source Q anion-exchange column (Fig. 5). The Source Q anion-exchange resin is from Pharmacia (Piscataway, NJ) and the column has a 15-ml bed volume; the buffer is 20 mM Tris-HCl (pH 7.5), 1 mM DTT; the gradient is 0–0.5 M NaCl; the flow rate is 0.5 ml/min; the total volume is 150 ml. Activity elutes between 0.15 and 0.2 M NaCl. Once again, during this chromatographic separation a large peak of contaminating protein elutes just before the sAC bioactive peak.

Reactive Dye Column (Green 19–Agarose)

Active fractions are pooled, concentrated and dialyzed against 20 mM Tris-HCl (pH 7.5), 1 mM DTT with an Ultrafree-15 centrifugal filter device (Millipore, Bedford, MA), and applied to a reactive Green 19–agarose column. The Reactive Green 19–agarose resin is from Sigma. The column has a 9-ml bed volume, the buffer is 20 mM Tris-HCl (pH 7.5), 1 mM DTT; the linear gradient

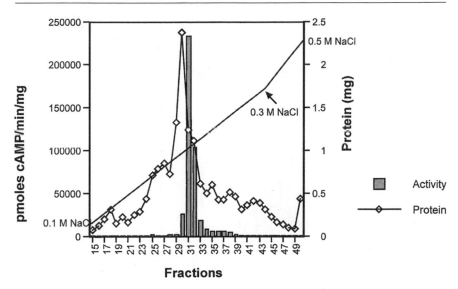

FIG. 5. sAC elution profile from Source Q column. The majority of contaminating protein eluted just prior (fraction 30) to fractions 31–33, which were pooled, concentrated, and loaded onto the Reactive Green 19-agarose column. The solid line indicates the [NaCl] gradient for these fractions.

is 0–0.5 M NaCl; the flow rate is 0.6 ml/min; the total volume is 80 ml. Activity elutes between 0.40 and 0.50 M NaCl in a linear gradient (0.1–1.0 M NaCl) just between two protein peaks (Fig. 6).

High-Performance Liquid Chromatography Columns

Fractions from Reactive Green 19–agarose columns containing peak sAC activity (1.8 mg) are combined, concentrated and dialyzed against 20 mM Tris-HCl (pH 7.4), 1 mM DTT with an Ultrafree-15 centrifugal filter device (Millipore), and loaded onto a semipreparative HydroCell QA 1000 high-performance liquid chromatography (HPLC) anion-exchange column (Fig. 7, top). The semipreparative HydroCell QA 1000 HPLC anion-exchange column is from Biochrom (Terre Haute, IN) (50 × 4.6 mm). It is run in 20 mM Tris-HCl (pH 7.4), 1 mM DTT; the linear gradient is 0–0.3 M NaCl over 30 min, at a flow rate of 2 ml/min. The peak sAC fractions are then loaded onto an analytical QA 1000 HPLC anion-exchange column (HydroCell, 150 × 2.3 mm; Biochrom), which is run in 20 mM Tris-HCl (pH 6.8), 1 mM DTT, with a linear gradient of 0–0.1 M NaCl over 25 min, a flow rate of 1.5 ml/min, and 0.5 ml/fraction. sAC activity elutes from the semipreparative QA1000 HPLC column between 0.04 and 0.06 M NaCl and between 0.07 and 0.10 M NaCl from the subsequent analytical QA 1000

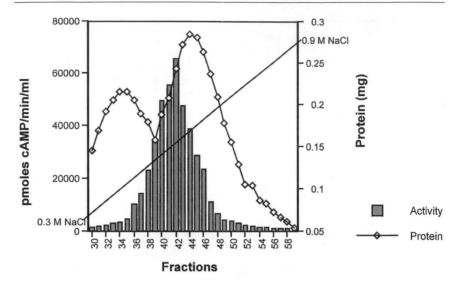

FIG. 6. sAC elution profile from Reactive Green 19–agarose Column. The majority of sAC protein eluted between two protein peaks from the Reactive Green 19–agarose column. Fractions 39–44 containing sAC bioactivity were pooled and applied to the HydroCell QA 1000 HPLC column. The solid line indicates the [NaCl] gradient for these fractions.

HPLC column. This final chromatographic step achieves a greater than 60-fold enrichment even though it uses the same QA anion-exchange matrix as the HPLC column preceding it. By varying the buffer pH (pH 7.4 for the semipreparative QA vs. pH 6.8 for the analytical QA), sAC activity elutes before the majority of contaminating proteins during this final chromatographic separation (Fig. 7, bottom).

A silver-stained 12% (w/v) sodium dodecyl sulfate (SDS)–polyacrylamide gel of selected fractions from the final chromatographic step of the purification reveals two protein bands (of approximately 48 and 62 kDa) whose intensities reflect the relative soluble AC enzymatic activity (Fig. 8). Active protein fractions are separated on SDS–polyacrylamide gels, stained with Coomassie Blue G-250, and both 48- and 62-kDa bands are excised. The proteins (<5 μg) are subjected to tryptic digestion followed by HPLC separation and Edman degradation sequencing at the Rockefeller University Protein/DNA Technology Center.[11,12] The amino acid sequences of six tryptic peptides from the 62-kDa polypeptide reveal it to be unrelated to sAC, whereas the amino acid sequences of three tryptic peptides derived from the 48-kDa candidate polypeptide have allowed us to clone the sAC cDNA.[2] Table I shows the results from each chromatographic step of the

[11] J. Fernandez, M. Demott, D. Atherton, and S. M. Mische, *Anal. Biochem.* **201,** 255 (1992).
[12] J. Fernandez, L. Andrews, and S. M. Mische, *Anal. Biochem.* **218,** 112 (1994).

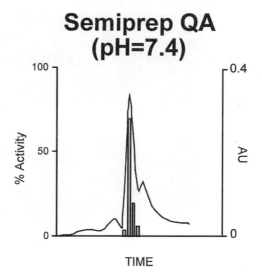

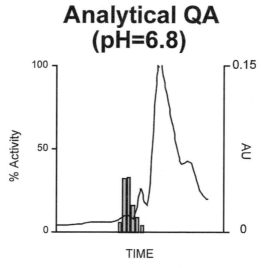

FIG. 7. Diagram of chromatograms from last two QA HPLC columns. Solid line indicates protein eluting over time in an [NaCl] gradient (not shown). Columns indicate activity of fractions. By varying the buffer pH (pH 7.4 for the semipreparative QA vs pH 6.8 for the analytical QA), sAC activity eluted before the majority of contaminating proteins from the last column.

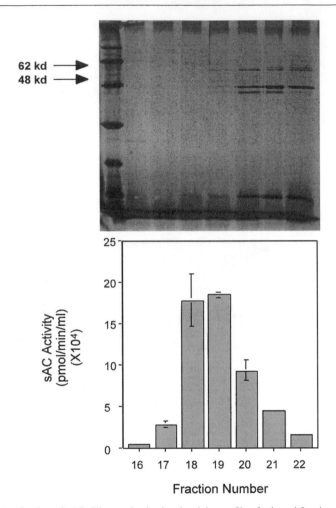

FIG. 8. Purification of sAC. Silver-stained gel and activity profile of selected fractions from Analytical QA column chromatography. Ten-microliter aliquots of selected fractions were separated on a 12% (w/v) SDS–polyacrylamide gel and silver stained. The first lane is low molecular weight silver stain marker, and each subsequent lane corresponds to the fraction assayed for sAC activity shown beneath it. Columns represent the average sAC activity in duplicate assays for each fraction. Error bars indicate the standard deviation from the mean. [By permission from J. Buck, M. L. Sinclair, L. Schapel, M. J. Cann, and L. R. Levin, *Proc. Natl. Acad. Sci. U.S.A.* **96,** 79 (1999).]

purification. Starting with more than 50 g of total cytosolic protein we end up with approximately 13 μg in the final bioactive fractions. sAC specific activity is 460 nmol/min/mg in the cytosolic extracts from the 950 testes and is enriched 10^6-fold to approximately 4.0×10^8 nmol/min/mg in the final fractions from the Analytical QA HPLC column.

Conclusion

Molecular isolation of the sAC gene revealed this form of mammalian adenylyl cyclase to be more similar to (cyano)bacterial adenylyl cyclases than to the more widely studied, hormone-responsive, G-protein-regulated mammalian tmACs.[2] Subsequent to its cloning, our joint laboratory heterologously expressed the sAC cDNA and confirmed that the recombinant protein exhibits all the biochemical hallmarks of testicular soluble adenylyl cyclase.[2] Using the sAC cDNA, we demonstrated that although sAC is widely expressed, its highest levels are found in spermatogenic cells[13]; thus, possibly explaining its previous biochemical detection only in testis.[4] More recent work from our laboratory revealed that sAC is also responsible for the related particulate activity[1,14] identified in mammalian spermatozoa (data not shown).

Because the regulation of sAC differs from that of tmACs and sAC protein is not restricted to the plasma membrane, it defines a distinct mechanism for generating cAMP in mammalian cells. Future studies of sAC will place it into the context of the mammalian tmACs, and will provide a greater understanding of the cellular role of cAMP. Twenty-five years after its first discovery by Braun and Dods,[1] we now have the molecular tools to fully dissect this ubiquitous, ancient signaling pathway.

[13] M. L. Sinclair, X. Y. Wang, M. Mattia, M. Conti, J. Buck, D. J. Wolgemuth, and L. R. Levin, *Mol. Reprod. Dev.* **56**, 6 (2000).

[14] S. Adamo, M. Conti, R. Geremia, and V. Monesi, *Biochem. Biophys. Res. Commun.* **97**, 607 (1980).

[8] Calcium-Sensitive Adenylyl Cyclase/Aequorin Chimeras as Sensitive Probes for Discrete Modes of Elevation of Cytosolic Calcium

By Dermot M. F. Cooper

The fact that Ca^{2+}-sensitive adenylyl cyclases respond to discrete, physiological elevations in cytosolic Ca^{2+} ([Ca^{2+}]$_i$) provides a means for coordinating the actions of the two major second-messenger signaling systems.[1] It turns out that the Ca^{2+}-sensitive adenylyl cyclases are extremely discriminating in the source of the Ca^{2+} to which they respond. In intact nonexcitable cells, they are regulated only by the entry of Ca^{2+} that is triggered by depletion of internal stores, so-called capacitative Ca^{2+} entry (CCE).[2] Store depletion can be triggered by hormones

[1] D. M. F. Cooper, N. Mons, and J. W. Karpen, *Nature (London)* **374**, 421 (1995).

[2] J. W. Putney, Jr., *Adv. Second Mess. Phosphoprot. Res.* **26**, 143 (1992).

coupled to phosphoinositidase C, by inhibition of the Ca^{2+}-reuptake pumps of the endoplasmic reticula (ER), or by ionophore insertion in the ER membranes. However, when CCE is triggered, it regulates Ca^{2+}-sensitive adenylyl cyclases, whether these occur in the cells naturally[3–5] or after heterologous expression.[6,7] Release of Ca^{2+} from internal stores by inositol trisphosphate, thapsigargin, or ionomycin is without effect on these adenylyl cyclases.[6,8] (Actually, transfected AC8 can be stimulated by hormone-stimulated Ca^{2+} release to 15% of the extent achieved by CCE.[6] This may reflect an overexpression of this species in a domain of the cell, where cyclase is not normally expressed and that is accessible to release from stores, e.g., the ER.) Remarkably, in excitable cells, where CCE is dwarfed by voltage-gated Ca^{2+} entry, CCE still robustly regulates adenylyl cyclase.[8a] Indications that the adenylyl cyclases and the sites of CCE must be colocalized to a microdomain of the plasma membrane came from the use of Ca^{2+} chelators as well as by cell fractionation. Loading of cells with the fast chelator 1,2-bis (o-aminophenoxy)ethane -N,N,N',N'-tetraacetic acid (BAPTA) attenuated the regulation of the endogenous adenylyl cyclase in C6-2B glioma cells, whereas the slower chelator ethylene glycol-bis(β-aminoethyl ether)-N,N,N',N'-tetraacetic acid (EGTA), at apparently equivalent intracellular concentrations, was without effect.[3] Adenylyl cyclases have been found to be essentially localized to cholesterol-rich domains in order to retain their susceptibility to CCE.[9] These and more recent data[10] demonstrate that the Ca^{2+}-sensitive adenylyl cyclases are effectively discrete sensors of CCE, a property that could potentially be exploited to monitor this entry of Ca^{2+} (CCE) distinct from other modes of elevating $[Ca^{2+}]_i$.

Shimomura and colleagues first purified the photoprotein aequorin from the jellyfish *Aequorea victoria* and established the Ca^{2+} dependence of its luminescence.[11] The protein was subsequently found to respond to a wide range of concentrations of Ca^{2+}.[12] Cobbold and co-workers exploited this broad dynamic range of aequorin to measure $[Ca^{2+}]_i$ in intact cells by microinjecting the protein, first into oocytes and later into other mammalian cells.[13] This approach showed not only a rapid elevation in Ca^{2+} on hormone or other stimulation, but also that the response

---

[3] K. A. Fagan, N. Mons, and D. M. F. Cooper, *J. Biol. Chem.* **273**, 9297 (1998).

[4] M. M. Burnay, M. B. Vallotton, A. M. Capponi, and M. F. Rossier, *Biochem. J.* **330**, 21 (1998).

[5] E. L. Watson, Z. Wu, K. L. Jacobson, D. R. Storm, J. C. Singh, and S. M. Ott, *Am. J. Physiol.* **274**, C557 (1998).

[6] K. A. Fagan, R. Mahey, and D. M. F. Cooper, *J. Biol. Chem.* **271**, 12438 (1996).

[7] T. J. Shuttleworth and J. L. Thompson, *J. Biol. Chem.* **274**, 31174 (1999).

[8] M. Chiono, R. Mahey, G. Tate, and D. M. F. Cooper, *J. Biol. Chem.* **270**, 1149 (1995).

[8a] K. A. Fagan, R. A. Graf, S. Tolman, J. Schaack, and D. M. F. Cooper, *J. Biol. Chem.* **275**, 40187 (2000).

[9] K. A. Fagan, K. E. Smith, and D. M. F. Cooper, *J. Biol. Chem.* **275**, 26530 (2000).

[10] C. Gu and D. M. F. Cooper, *J. Biol. Chem.* **275**, 6980 (2000).

[11] O. Shimomura, F. H. Johnson, and Y. Saiga, *J. Cell. Comp. Physiol.* **59**, 223 (1962).

[12] D. G. Allen, J. R. Blinks, and F. G. Prendergast, *Science* **195**, 996 (1977).

[13] P. H. Cobbold, K. S. Cuthbertson, M. H. Goyns, and V. Rice, *J. Cell. Sci.* **61**, 123 (1983).

of many cells was to elevate [Ca^{2+}]$_i$ in a pulsatile or oscillatory manner.[14] The cloning of the aequorin cDNA[15,16] permitted the transient expression of the protein, which circumvented the technically demanding step of microinjection and other strategies.[17] Transient and stable expression of aequorin verified that the protein could be expressed in the cytosol and that it reported sensitive and robust measures of changes in [Ca^{2+}]$_i$.[18] A major conceptual breakthrough came with the targeting of aequorin to discrete domains of the cell by placing an organelle-targeting sequence upstream (or downstream) of the aequorin-coding region. This allowed measurement of [Ca^{2+}] in the ER, nuclear, and mitochondrial environments.[19–22] This development led us to consider the possibility that an adenylyl cyclase–aequorin chimera might be a unique sensor of CCE. Consequently, we constructed a full-length adenylyl cyclase (AC6)–aequorin chimera.[23] (Since that time other recombinant aequorin constructs have been designed to measure [Ca^{2+}] near gap junctions,[24] secretory granules,[25] and sites of secretion.[26]) Here we describe some of the issues involved in validating this and related constructs and their application.

Constructs

AC6, AC8, and AC2 High- and Low-Affinity Aequorin Chimeras

Chimeras are made between Ca^{2+}-inhibitable AC6, Ca^{2+}-insensitive AC2, and Ca^{2+}-stimulable AC8 and aequorin. The three different adenylyl cyclases are chosen to assess whether cyclases detect different [Ca^{2+}] on the basis of their intrinsic responsiveness (or not) to various modes (physiological and nonphysiological) of elevating [Ca^{2+}]$_i$. Two aequorin variants are used. The wild-type aequorin responds to [Ca^{2+}] ranging from 0.1 to 10 μM. As a result of our preliminary studies

[14] M. J. Berridge, P. H. Cobbold, and K. S. Cuthbertson, *Phil. Trans. R. Soc. Lond. B. Biol. Sci.* **320**, 325 (1988).

[15] S. Inouye, M. Noguchi, Y. Sakaki, Y. Takagi, T. Miyata, S. Iwanaga, T. Miyata, and F. I. Tsuji, *Proc. Natl. Acad. Sci. U.S.A.* **82**, 3154 (1985).

[16] D. Prasher, R. O. McCann, and M. J. Cormier, *Biochem. Biophys. Res. Commun.* **126**, 1259 (1985).

[17] A. K. Campbell, R. L. Dormer, and M. B. Hallett, *Cell Calcium* **6**, 69 (1985).

[18] D. Button and M. Brownstein, *Cell Calcium* **14**, 663 (1993).

[19] M. Brini, M. Murgia, L. Pasti, D. Picard, T. Pozzan, and R. Rizzuto, *EMBO J.* **12**, 4813 (1993).

[20] J. M. Kendall, R. L. Dormer, and A. K. Campbell, *Biochem. Biophys. Res. Commun.* **189**, 1008 (1992).

[21] R. Rizzuto, A. W. Simpson, M. Brini, and T. Pozzan, *Nature (London)* **358**, 325 (1992).

[22] G. A. Rutter, P. Burnett, R. Rizzuto, M. Brini, M. Murgia, T. Pozzan, J. M. Tavare, and R. M. Denton, *Proc. Natl. Acad. Sci. U.S.A.* **93**, 5489 (1996).

[23] Y. Nakahashi, E. Nelson, K. Fagan, E. Gonzales, J. L. Guillou, and D. M. F. Cooper, *J. Biol. Chem.* **272**, 18093 (1997).

[24] C. H. George, J. M. Kendall, A. K. Campbell, and W. H. Evans, *J. Biol. Chem.* **273**, 29822 (1998).

[25] A. E. Pouli, N. Karagenc, C. Wasmeier, J. C. Hutton, N. Bright, S. Arden, J. G. Schofield, and G. A. Rutter, *Biochem. J.* **330**, 1399 (1998).

[26] R. Marsault, M. Murgia, T. Pozzan, and R. Rizzuto, *EMBO J.* **16**, 1575 (1997).

A.
ACVI cDNA   **HA1-tagged Aequorin cDNA**

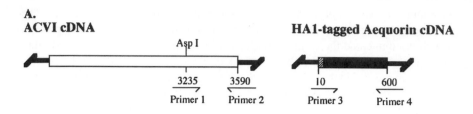

B.
ACVI and HA1-tagged Aequorin chimeric cDNA

FIG. 1. Construction of the AC6–aequorin chimeric cDNA. (A) Two overlapping cDNA fragments were generated in the first-stage PCRs. The coding regions of the AC6 cDNA and the HA1-tagged aequorin cDNA are indicated. The primer sequences are indicated by the arrows. (B) The second-stage PCR generated the chimeric AC6 and HA1-tagged aequorin cDNA, which was then manipulated and ligated back to the full-length AC6. The coding region of the AC6 cDNA, the HA1 tag cDNA, and aequorin cDNA are indicated as open, hatched, and black boxes, respectively. [Reprinted from Y. Nakahashi, E. Nelson, K. Fagan, E. Gonzales, J. L. Guillou, and D. M. F. Cooper, *J. Biol. Chem.* **272,** 18093 (1997), with permission.]

and to expand the potential detection range of the chimeras, a mutated form of aequorin is also employed. Wild-type aequorin possesses three E-F hands. If one of these is mutated (Asp-119 to alanine) the affinity is decreased such that the protein responds to $[Ca^{2+}]$ ranging from 1 to 100 μM.[27] Consequently, chimeras are constructed between adenylyl cyclases and both forms of aequorin.

Construction of Chimeras

Substitution of aequorin at its N terminal does not affect its activity, whereas substitution at the C terminal destroys activity.[28] Furthermore, we have assumed that the N terminal of adenylyl cyclase might play an important targeting role. Consequently, we first decided to place the hemagglutinin 1 (HA1)-tagged aequorin cDNA downstream of the adenylyl cyclase. A two-step polymerase chain reaction (PCR) strategy is used (see Fig. 1). In the first step, two fragments are generated from two separate PCRs. One represents the C-terminal region of the cyclase in frame with the N terminal of the HA1-tagged aequorin, using both the cyclase and aequorin cDNAs as templates; the other fragment overlaps with the N terminal of the HA1-tagged aequorin and encodes the entire aequorin sequence including its 3′ insertion site. A second PCR is then performed, using both fragments as templates

[27] M. Montero, M. Brini, R. Marsault, J. Alvarez, R. Sitia, T. Pozzan, and R. Rizzuto, *EMBO J.* **14,** 5467 (1995).

[28] M. Brini, R. Marsault, C. Bastianutto, J. Alvarez, T. Pozzan, and R. Rizzuto, *J. Biol. Chem.* **270,** 9896 (1995).

and the "outermost" primers from the first PCRs for amplification. Constructs are confirmed by sequencing. For the low-affinity construct, the fragment of aequorin cDNA including the mutated amino acid is switched for the native sequence, using restriction enzymes.

Validation of Chimeras

Retention of Ca^{2+} Sensitivity by Cyclase Component

To establish that the adenylyl cyclase activity is not compromised by the presence of a large aequorin molecule at its C terminal, Ca^{2+} dose–response curves are generated in *in vitro* assays of plasma membranes prepared from cells transfected with adenylyl cyclase–aequorin chimeras. In the case of AC6, the Ca^{2+} sensitivity and overall activity were not detectably different between the chimera and the native AC6.[23] Similar assays should be performed for AC8. Because AC2 is not a Ca^{2+}-sensitive species, only overall activity can be assessed.

Retention of Ca^{2+} Sensitivity by Aequorin Part of Chimera

Because cyclases are large molecules we consider it advisable to verify that the activity of the aequorin component of the chimera is not compromised by the presence of the cyclase. In the case of the AC6–aequorin chimera we again made membranes from cells transfected with AC6–aequorin and performed *in vitro* assessments of the Ca^{2+} sensitivity. This was indistinguishable from that of native acquorin. A similar experiment has been performed with AC8–aequorin, with the same result. Perhaps it is to be expected that if the aequorin can be tolerated by its chimeric partner with no change in activity, an effect on the activity of the aequorin would also not be anticipated.

Appropriate Targeting to Regions of Capacitative Ca^{2+} Entry

The only way we know currently of trying to target adenylyl cyclase–aequorin chimeras appropriately is to use the entire cyclase sequence. Because Ca^{2+}-sensitive adenylyl cyclases, when expressed heterologously, remain exclusively sensitive to CCE, full-length adenylyl cyclase–aequorin chimeras might also be targeted appropriately, unless the aequorin at the C terminal disrupts the targeting. Some time in the future, when the appropriate targeting domains are identified, shorter constructs can be employed. Meanwhile, a functional assay is an extremely robust confirmation of appropriate targeting. The assay that we apply to the Ca^{2+}-sensitive adenylyl cyclase–aequorins is to trigger CCE by passively depleting the Ca^{2+} stores by treatment with thapsigargin in the absence of extracellular Ca^{2+} and subsequently adding adenylyl cyclase stimulants along with progressively increasing [Ca^{2+}]$_{ex}$ and looking for anticipated effects on cAMP accumulation. For instance, in the case of AC6, inhibitory effects of CCE on transfected AC6–aequorin chimera are retained as in the wild type.[23] Inhibition by CCE confirms that the chimera is expressed in the same selective domain of the plasma membrane as the native AC6 and is in close apposition with CCE channels. Of course, this criterion cannot be applied to Ca^{2+}-insensitive adenylyl cyclases. Indeed, one of the reasons behind making Ca^{2+}-sensitive and Ca^{2+}-insensitive adenylyl cyclase–aequorin chimeras would be to determine whether they were exposed to different $[Ca^{2+}]_i$ under a variety of circumstances.

Measurements

Details of Luminescence Measurements

Aequorin measurements are carried out 48 hr after transfection. The *in vitro* $[Ca^{2+}]$ calibration curves are determined by exposing lysates of cells transfected with either the cytosolic aequorin (cytAEQ)- or chimera-carrying plasmids to solutions of EGTA-buffered $[Ca^{2+}]$ as follows. Cells are resuspended in a lysis buffer containing 100 mM HEPES, 0.2 mM EGTA, 0.2% (w/v) bovine serum albumin (BSA), 0.3 mM phenylmethylsulfonyl fluoride (PMSF), protease inhibitor cocktail, pH 7.0, and lysed through a 25-gauge syringe. Cell membranes are separated from cell lysate by centrifugation [20 min, 4°, 12,000 rpm; Sorvall (Newtown, CT) SS34]. The supernatant (cell lysate) is utilized for *in vitro* calibration of cytAEQ and cell pellets (cell membranes) are utilized for *in vitro* calibration of chimerae. Apoaequorin is reconstituted with 5 μM coelenterazine (Molecular Probes, Eugene, OR) for 2–2.5 hr at 4°, in the presence of 1% (v/v) 2-mercaptoethanol. Cell lysate or cell membranes are diluted into an intracellular-like buffer containing the following final concentrations: 130 mM KCl, 10 mM NaCl, 1 mM $MgSO_4$, 0.5 mM K_2HPO_4, 1 mM ATP, 0.2% (w/v) BSA, 100 mM HEPES, pH 7.0 at 22°, supplemented with 0.1 mM EGTA and various amounts of total $CaCl_2$. Ca^{2+} buffers of various concentrations are injected into the chamber and the light emission is recorded. For *in vivo* measurements transfected cells are washed and loaded with 5 μM coelenterazine (Molecular Probes) in minimum essential medium (MEM) supplemented with 1% (v/v) fetal calf serum and 0.5 mM EGTA for 2 hr at 37° in 5% (v/v) CO_2 in the dark. Cells are resuspended in nominally Ca^{2+}-free Krebs buffer at pH 7.4 supplemented with 75 μM Ca^{2+} and 0.2 mM EGTA is added immediately before measurements. Aequorin luminescence is sometimes measured in a Berthold (Bad Wildbad, Germany) Lumat LB 9501 luminometer and collected in ASCII format on an 80486 microprocessor. Standard luminometers such as that marketed by Lumat achieve rapid mixing because of the automated addition of solutions by powerful pumps, which can be a highly desirable property when studying cytosolic proteins. However, the effect of such powerful mixing on intact cells can be rapid lysis. Another drawback of the manufactured luminometers is that rarely can successive additions be made. Consequently, home-made devices are much more desirable when studying biological phenomena in intact cells. The device developed by P. Cobbold[29] has been widely used and is most appropriate for continuous monitoring of luminescent signals

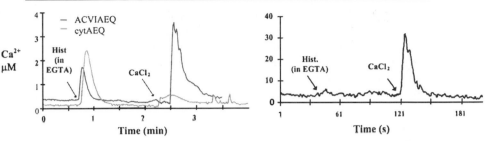

FIG. 2. Measurement of [Ca^{2+}]$_i$ rises with aequorin constructs. HeLa cells transfected with cytosolic aequorin, AC6/high-affinity aequorin, and AC6/low-affinity aequorin cDNA were plated on coverslips and placed on the stage of a warmed chamber apposed to a cooled photomultiplier tube. The chamber was perfused with histamine in calcium-free medium, until external calcium was added, as indicated. Light output was collected, digitized, and normalized according to the L_{max} of each run.[28] *Left:* Cytosolic and AC6/high-affinity aequorin. *Right:* AC6/low-affinity aequorin. Note the peak CCE of 30 μM [Ca^{2+}]$_i$ reported by the low-affinity mutated construct in response to CCE.

during sequential additions. In this setup, cells plated on coverslips on a heated block are directly apposed to (although insulated from) a cooled photomultiplier tube. The chamber containing the coverslip is perfused constantly, which permits the continuous introduction of a range of agonists or different Ca^{2+} solutions, etc. The L_{max} is found by integrating a continuous recording of aequorin-mediated light emission from transfected cells in the presence of 0.2% (v/v) Triton X-100 and 10 mM Ca^{2+} until the light emission returns to basal levels and subsequent injections of Triton X-100 and Ca^{2+} produce no further increases in light emission. This back-calculation strategy also corrects for any variability in the degree of expression of the proteins. Luminescence values determined *in vivo* are transformed into [Ca^{2+}]$_i$ values by using an algorithm based on the reordering of an equation described by Allen *et al.*[12] The light output from transfected cells loaded in the absence of coelenterazine or from mock-transfected cells loaded with coelenterazine is not significantly higher than background.

Detection of [Ca^{2+}]$_i$ Rises with Aequorin Constructs

Because in a device like that devised by Cobbold solutions are perfused through plastic tubing, the retention by the tubing of compounds such as thapsigargin or ionomycin is a consideration. Alternatives such as hydrophilic neurotransmitters or 8-bromo-A23187 avoid the problems of persistent accumulation or leaking of the hydrophobic agents and the ambiguities and uncertainties associated with their use. Transfected cells are treated with a neurotransmitter that releases internal stores of Ca^{2+}, histamine in the case of HeLa cells. When this measurement is performed in the absence of internal Ca^{2+}, release is measured. On the introduction of Ca^{2+}

[29] P. H. Cobbold and J. A. C. Lee, *In* "Cellular Calcium: A Practical Approach" (J. G. McCormack and P. H.Cobbold, eds.), pp. 55–81. IRL Press, Oxford.

to the bathing medium, store-operated Ca^{2+} entry is observed (Fig. 2). In such an experiment the $[Ca^{2+}]_i$ measured by cytosolic aequorin can be compared with that reported by a high-affinity aequorin and a low-affinity aequorin-cyclase chimera (Fig. 2). In the example shown, it can be seen that cytosolic aequorin reports $[Ca^{2+}]_i$ values that would be expected to be reported by Fura-2. However, the chimeras report much higher $[Ca^{2+}]_i$ in response to CCE than in response to release. This observation is consistent with what might be expected, given the close association of adenylyl cyclases with CCE sites and their lack of regulation by Ca^{2+} released from stores. Thus, it seems reasonable to conclude that such constructs could be useful tools for observing selective pools of $[Ca^{2+}]_i$ under a variety of experimental conditions.

[9] Kinetic Analysis of the Action of P-Site Analogs

By CARMEN W. DESSAUER

Overview

P-site inhibitors are adenine or adenosine derivatives that inhibit adenylyl cyclase activity (reviewed in Dessauer et al.[1]). P-site inhibitors are not competitive with adenylyl cyclase activators such as forskolin or $G_s\alpha$,[2] although they are much more potent inhibitors of the activated versus the basal state of the enzyme.[2-5] Londos and Wolff designated these molecules for their strict structural requirement for an intact purine ring.[6] Representative P-site reagents, ordered by potency, include 2',5'-dideoxyadenosine 3'-tetraphosphate $>$ 2',5'-dideoxy-3'-ATP $>$ 2',5'-dideoxy-3'-ADP $>$ 2',5'-dideoxy-3'-AMP $>$ 2'-deoxy-3'-AMP $>$ 3'-AMP $>$ 2'-deoxyadenosine $>$ adenosine.[7-10]

---

[1] C. W. Dessauer, J. J. Tesmer, S. R. Sprang, and A. G. Gilman, *Trends Pharmacol. Sci.* **20,** 205 (1999).

[2] V. A. Florio and E. M. Ross, *Mol. Pharmacol.* **24,** 195 (1983).

[3] C. Londos and M. S. Preston, *J. Biol. Chem.* **252,** 5957 (1977).

[4] R. A. Johnson, W. Saur, and K. H. Jakobs, *J. Biol. Chem.* **254,** 1094 (1979).

[5] J. Wolff, C. Londos, and D. M. F. Cooper, *Adv. Cyclic Nucleotide Res.* **14,** 199 (1981).

[6] C. Londos and J. Wolff, *Proc. Natl. Acad. Sci. U.S.A.* **74,** 5482 (1977).

[7] R. A. Johnson, S.-M. H. Yeung, D. Stübner, M. Bushfield, and I. Shoshani, *Mol. Pharmacol.* **35,** 681 (1989).

[8] L. Desaubry, I. Shoshani, and R. A. Johnson, *J. Biol. Chem.* **271,** 2380 (1996).

[9] L. Desaubry and R. A. Johnson, *J. Biol. Chem.* **273,** 24972 (1998).

[10] R. A. Johnson, L. Desaubry, G. Bianchi, I. Shoshani, E. Lyons, Jr., R. Taussig, P. A. Watson, J. J. Cali, J. Krupinski, J. P Pieroni, and R. Iyengar, *J. Biol. Chem.* **272,** 8962 (1997).

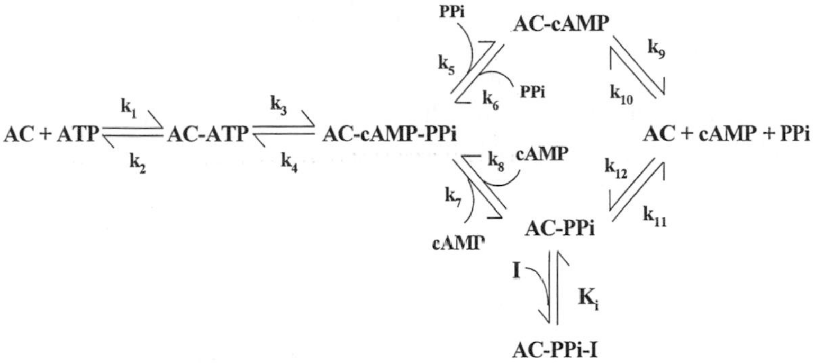

FIG. 1. Mechanism of P-site-mediated inhibition of adenylyl cyclase-catalyzed reaction. The species AC and I represent the adenylyl cyclase enzyme and the P-site inhibitor $2'$-deoxyadenosine, respectively. Estimates of rate constants for the forward and reverse reaction for the cytoplasmic domains (VC_1 and IIC_2) can be found in Dessauer and Gilman; see Ref. 11. The release of cyclic AMP before the release of PP_i is highly preferred in this scheme, allowing a buildup of AC–PP_i to which inhibitor binds.

It is now clear that the requirement for an intact adenine ring reflects the fact that P-site inhibitors bind within the same enzyme pocket as substrate ATP.[11-13] The mechanism of action of P-site inhibitors fall into two broad classes: those that display a strict dead-end mechanism and those that show a mixed inhibition pattern.[11,14-16] The dead-end mechanism is due to the formation of a stable complex containing enzyme, pyrophosphate, and P-site inhibitor. The key to this mechanism is the slow release of pyrophosphate from adenylyl cyclase (AC) after the hydrolysis of ATP.[11] Pyrophosphate (PP_i) is required for binding this type of P-site inhibitor, and the generation of an AC–PP_i complex is a key step in this mechanism (Fig. 1). The estimated rate constants for $G_s\alpha$-stimulated adenylyl cyclase also explain another trait of P-site inhibitors. There are two, approximately equal (60 sec^{-1}), rate-limiting steps for the generation of cyclic AMP: k_3, the rate of ATP breakdown to cAMP and PP_i; and k_{11}, the rate of PP_i release. However, as

[11] C. W. Dessauer and A. G. Gilman, *J. Biol. Chem.* **272**, 27787 (1997).

[12] J. J. G. Tesmer, R. K. Sunahara, A. G. Gilman, and S. R. Sprang, *Science* **278**, 1907 (1997).

[13] J. J. Tesmer, R. K. Sunahara, R. A. Johnson, G. Gosselin, A. G. Gilman, and S. R. Sprang, *Science* **285**, 756 (1999).

[14] M. S. Wolin, Adenylate Cyclase from *Brevibacterium liquefaciens*: A Kinetic Characterization of the Catalytic Mechanism and Regulation of the Enzyme by Uncomplexed Mg^{2+} and Pyruvate. Ph.D. thesis. Yale University, Ithaca, New York, 1981.

[15] V. A. Florio, Inhibition of the Catalytic Unit of Mammalian Adenylate Cyclase by Adenosine and Adenosine Analogs. Ph.D. Thesis. University of Virginia, Charlottesville, Virginia, 1983.

[16] J. J. G. Tesmer, C. W. Dessauer, R. K. Sunahara, L. D. Murray, R. A. Johnson, A. G. Gilman, and S. R. Sprang, *Biochemistry* **39**, 14464 (2000).

activity increases, k_3 also increases, leaving k_{11} as the key rate-limiting step and allowing a buildup of AC–PP_i to which inhibitor binds. Therefore, as activity of adenylyl cyclase increases, P-site inhibitors become more potent inhibitors. Most likely, all P-site molecules that either lack a $3'$-phosphate or contain only a single $3'$-phosphate will fall into this category.

The second major class of P-site inhibitors includes those with $3'$-polyphosphates such as $2',5'$-dideoxy-$3'$-ATP. These molecules contain a "built-in" pyrophosphate moiety that allows for multiple binding states and noncompetitive inhibition patterns.[8,9,16,17] The inhibitor $2',5'$-dideoxy-$3'$-ATP can bind directly to adenylyl cyclase in the absence of PP_i; however, it may also bind to the AC–PP_i complex.[16] Crystal structures of this inhibitor bound to adenylyl cyclase overlap fairly well with structures of bound dead-end inhibitors such as $2'$-deoxy-$3'$-AMP in the presence of PP_i.[16]

Many of the studies that led to our understanding of P-site mechanisms stem from kinetic and binding analysis of P-site inhibitors, using enzyme preparations containing the cytoplasmic domains of adenylyl cyclase. The conserved first and second cytoplasmic domains of adenylyl cyclase (C_1 and C_2) can be individually expressed in *Escherichia coli*.[18,19] Although devoid of enzymatic activity on their own, the simple mixing of these domains reconstitutes full adenylyl cyclase activity, including synergistic activation by $G_s\alpha$ and forskolin and inhibition by P-site inhibitors and $G_i\alpha$.[18–20] This "soluble" system is described in detail in [17] in this volume.[20a] However, this system has several distinct advantages over studies using membranes containing the full-length adenylyl cyclase. Kinetic studies can, of course, be performed with both systems; however, use of the purified enzymes avoids potential complications from ATPases, phosphodiesterases, or other metabolic enzymes. This is particularly important when examining the reverse reaction or inhibition patterns in the presence of PP_i, which rapidly breaks down in most membrane preparations. The largest advantage of the soluble system is the quantity and purity obtained for these proteins. This has allowed for the first time the ability to determine the stoichiometry of ligand and activator binding[11,21,22] and the solution of crystal structures for adenylyl cyclase with a number of bound inhibitors.[12,13,16] These individual domains have been invaluable tools in understanding the structural and kinetic mechanisms of adenylyl cyclase and the inhibition of P-site molecules.

[17] L. Desaubry, I. Shoshani, and R. A. Johnson, *J. Biol. Chem.* **271,** 14028 (1996).

[18] S. Z. Yan, D. Hahn, Z. H. Huang, and W.-J. Tang, *J. Biol. Chem.* **271,** 10941 (1996).

[19] R. E. Whisnant, A. G. Gilman, and C. W. Dessauer, *Proc. Natl. Acad. Sci. U.S.A.* **93,** 6621 (1996).

[20] C. W. Dessauer, J. J. G. Tesmer, S. R. Sprang, and A. G. Gilman, *J. Biol. Chem.* **273,** 25831 (1998).

[20a] S.-Z. Yan and W.-J. Tang, *Methods Enzymol.* **345,** [17] 2001 (this volume).

[21] R. K. Sunahara, C. W. Dessauer, R. E. Whisnant, C. Kleuss, and A. G. Gilman, *J. Biol. Chem.* **272,** 22265 (1997).

[22] C. W. Dessauer, T. T. Scully, and A. G. Gilman, *J. Biol. Chem.* **272,** 22272 (1997).

Kinetic analysis of the action of P-site analogs on adenylyl cyclase activity is in essence the examination of any inhibitor of enzyme action. Therefore, the classic types of kinetic experiments used to determine the mechanism of P-site inhibitors are mentioned, but those techniques that are unique to adenylyl cyclase and this class of inhibitors are emphasized.

Kinetic Analysis of P-Site Inhibition

Inhibition of Cyclic AMP Formation: Forward Reaction

There are many wonderful books that describe in detail the principles of enzyme kinetics, including Segel's "Enzyme Kinetics,"[23] which serves as a comprehensive examination of most enzyme kinetic mechanisms. The classic definitions for inhibitors are competitive, noncompetitive, or uncompetitive. A competitive inhibitor is one that combines with free enzyme in a manner that prevents substrate binding. Typically competitive inhibitors and substrate bind at the same site and are considered mutually exclusive. A good example for adenylyl cyclase is the substrate analog $Ap(CH_2)pp$, which cannot be utilized as a substrate and behaves as a true competitive inhibitor.[22] More recently, the molecule β-L-2',3'-dideoxy-5'-ATP has been shown to be a competitive inhibitor of adenylyl cyclase with tight affinity for the enzyme.[13,24,25] Uncompetitive inhibitors bind to the enzyme only after the addition of substrate; therefore both K_m and V_{max} are equally effected. P-site inhibitors such as 2'-deoxyadenosine or 2'-deoxy-3'-AMP display uncompetitive inhibition in the presence of $G_s\alpha$ and Mg^{2+}.[11,26,27] This is due to their ability to bind tightly to the $AC-PP_i$ complex but not to the free enzyme alone.[11] Noncompetitive inhibitors are defined as decreasing V_{max} but with no effect on K_m. They have no effect on substrate binding but bind both to the free enzyme and to the enzyme–substrate complex. However, irreversible inhibitors will also show noncompetitive kinetic patterns due to a decrease in V_{max}. A mixed inhibition pattern occurs if the affinity of the inhibitor for AC versus $AC-PP_i$ is not identical. Many P-site inhibitors in the presence of Mn^{2+} will display noncompetitive inhibition.[3,9,17,28,29] P-site inhibitors containing 3'-polyphosphates can bind to the free enzyme and most likely to the $AC-PP_i$ complex as well, explaining their noncompetitive or mixed inhibition patterns.[16] Although 2'-deoxyadenosine

[23] I. H. Segel (ed.), "Enzyme Kinetics," Vol. I. John Wiley & Sons, New York, 1975.

[24] R. A. Johnson, I. Shoshani, C. Dessauer, and G. Gosselin, *Nucleosides Nucleotides* **18**, 839 (1999).

[25] I. Shoshani, V. Boudou, C. Pierra, G. Gosselin, and R. A. Johnson, *J. Biol. Chem.* **274**, 34735 (1999).

[26] R. A. Johnson and I. Shoshani, *J. Biol. Chem.* **265**, 11595 (1990).

[27] C. W. Dessauer and A. G. Gilman, *J. Biol. Chem.* **271**, 16967 (1996).

[28] I. Weinryb and I. M. Michel, *Biochim. Biophys. Acta* **334**, 218 (1974).

[29] J. Wolff, C. Londos, and G. H. Cook, *Arch. Biochem. Biophys.* **191**, 161 (1978).

and 2'-deoxy-3'-AMP display noncompetitive or mixed inhibition patterns in the presence of Mn^{2+}, they do not compete with substrate analogs for binding to the free enzyme in the presence of Mn^{2+} or Mg^{2+}.[11,22] This paradox emphasizes the need for multiple approaches in determining the mechanism of these molecules.

Designing Kinetic Assays. The basic adenylyl cyclase assay is given below. A detailed review of this assay is given by Johnson and Salomon.[30]

Basic Reagents

Ro 20-1724	100 μM
EDTA (pH 7.0)	0.6 mM
Bovine serum albumin (BSA)	100 μg/ml
HEPES (pH 8.0)	50 mM
Dipotassium phosphoenolpyruvate	3 mM
Pyruvate kinase	10 μg/ml
[α-^{32}P]ATP	1×10^6 cpm
MgCl$_2$ (this varies, depending on assay)	5 mM

Unlabeled metal–ATP (varies from 2 μM to 2 mM depending on protein)
P-site inhibitor (varies depending on type of inhibitor)
Adenylyl cyclase enzyme and appropriate activator

The final volume is 50–100 μl. The assay is usually started with the addition of adenylyl cyclase enzyme and proceeds for 5–10 min at 30° and then stopped with 0.8 ml of 0.25% (w/v) sodium dodecyl sulfate (SDS), 5 mM ATP, 0.175 mM cAMP. Approximately 10,000 cpm of [^{3}H]cAMP (100 μl) is then added to monitor recovery from sequential chromatography on Dowex and alumina columns.[30,31] For reactions using purified activators and domains of adenylyl cyclase, the phosphodiesterase inhibitor (Ro 20-1724) and the ATP regeneration system (dipotassium phosphoenolpyruvate and pyruvate kinase) can often be eliminated.

Determination of K_m, V_{max}, and kinetic patterns is achieved by varying substrate ATP at different inhibitor concentrations in the presence of constant enzyme and free metal. Concentration ranges should be sufficient to cover the entire binding curves for both substrate and inhibitor. It is important to make sure that less that 10% of substrate ATP is consumed for all ATP concentrations during the reactions and that free metal concentrations are maintained by varying metal–ATP concentrations. ATP concentrations will vary depending on the protein, the metal, and the activator. Generally metal–ATP ranges of 2–200 μM will cover most conditions for native adenylyl cyclase preparations, and ranges of 20–2000 μM will cover most activation conditions for purified C$_1$ and C$_2$ domains. In addition, the

[30] R. A. Johnson and Y. Salomon, *Methods Enzymol.* **195,** 3 (1991).
[31] Y. Salomon, C. Londos, and M. Rodbell, *Anal. Biochem.* **58,** 541 (1974).

time and temperature of the reaction should be chosen such that enzyme activity is completely linear over the course of the reaction.

The appropriate amount of enzyme in these reactions to meet the above-described requirements will completely depend on the activity of the membrane preparation or the purified proteins. Membranes prepared from tissues or from Sf9 (*Spodoptera frugiperda* ovary) cells overexpressing an individual isoform of adenylyl cyclase can require 5–100 μg of protein measured over 4–30 min of time.[30] For determination of K_m values, the appropriate amount of enzyme must be carefully balanced to obtain a sufficient signal at high concentrations of ATP but not to consume more than 10% of the substrate at the lowest ATP concentrations. Reactions utilizing the individual soluble domains of adenylyl cyclase have an added consideration in that the concentration of one domain (usually C_2) must be present at sufficient concentrations to drive interactions with the limiting domain (often C_1). Concentrations of 1–2 μM C_2 are therefore required in these reactions (see Refs. 18, 19, and 21).

Analysis of Multiple Inhibitors. Inhibition patterns are classically plotted as double-reciprocal plots of 1/ATP versus 1/velocity, where the x intercept is $1/K_m$ and the y intercept is $1/V_{max}$. However, analysis of multiple inhibitor interactions is provided by Dixon plots of 1/velocity versus the concentration of one inhibitor at different fixed concentrations of the other inhibitor.[23] The substrate concentration in these experiments remains constant. This analysis can be key in determining whether two inhibitors compete with one another for the same binding site (i.e., cyclic AMP and P-site inhibitors) or whether they are synergistic with one another (i.e., pyrophosphate and 2′-deoxy-3′-AMP). If two inhibitors are mutually exclusive, the family of lines generated will be parallel with one another. A family of intersecting lines indicates that the binding of one inhibitor depends on the concentration of the other. If the two inhibitors are synergistic, the intersection of these curves should lie above the x axis.

Measurement of ATP Synthesis: Reverse Reaction

Because all P-site inhibitors act at least in part as dead-end inhibitors (i.e., they form a tight complex with the product pyrophosphate in the active site), analysis of the reverse reaction can be highly informative. The reaction catalyzed by adenylyl cyclase is readily reversible and actually favors ATP synthesis under standard conditions of 1 M concentrations of reactants and products.[32,33]

ATP synthesis can be measured in a variety of ways. One method is the use of a coupled enzyme system that converts ATP to NADPH, using the enzymes hexokinase and glucose-6-phosphate dehydrogenase (Glu6-P DH) (Fig. 2).[11] This

[32] O. Hayaishi, P. Greengard, and S. P. Colowick, *J. Biol. Chem.* **246**, 5840 (1971).

[33] K. Takai, Y. Kurashina, C. Suzuki, H. Okamoto, and A. Ueki, *J. Biol. Chem.* **246**, 5843 (1971).

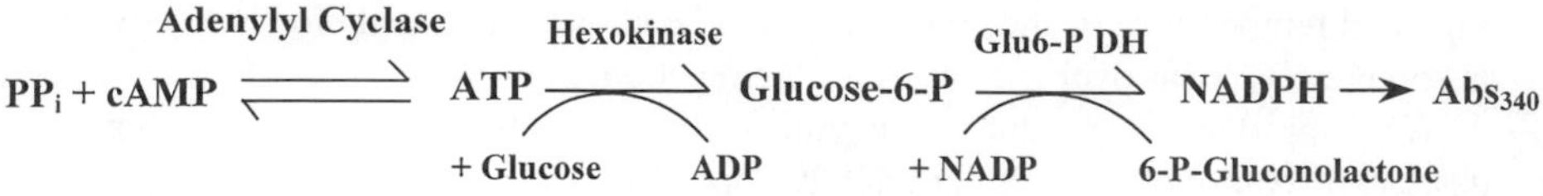

$$\text{PP}_i + \text{cAMP} \overset{\textbf{Adenylyl Cyclase}}{\rightleftharpoons} \text{ATP} \overset{\textbf{Hexokinase}}{\longrightarrow} \text{Glucose-6-P} \overset{\textbf{Glu6-P DH}}{\longrightarrow} \text{NADPH} \longrightarrow \text{Abs}_{340}$$

$$\quad\quad\quad\quad\quad\quad\quad\quad \textbf{+ Glucose} \quad \textbf{ADP} \quad\quad \textbf{+ NADP} \quad \textbf{6-P-Gluconolactone}$$

FIG. 2. Reaction scheme for the measurement of ATP synthesis by a coupled enzyme reaction. The conversion of ATP to NADPH is catalyzed by the enzymes hexokinase and glucose-6-phosphate dehydrogenase (Glu6-P DH) and is measured by an increase in absorbance at 340 nm.

coupled enzyme reaction will rapidly remove ATP from the reaction so that the forward reaction (cyclic AMP synthesis) is minimized.

Coupled Reaction Mixture

Note: Final concentrations in the total reaction are given.

HEPES (pH 8.0)	20 mM
Glucose	50 mM
NADP	0.8 mM
Free MgCl$_2$	3 mM
Hexokinase	1.7 units
Glu6-P DH	0.3 unit

Hexokinase (Boehringer Mannheim, Indianapolis, IN) and glucose-6-phosphate dehydrogenase (Sigma, St. Louis, MO) are obtained as ammonium sulfate slurries and centrifuged (10 min), and the pellets are resuspended in 20 mM HEPES (pH 8.0) before use. These enzymes are >90% pure as judged by Coomassie blue staining. The concentrations of hexokinase, Glu6-P DH, glucose, and NADP were empirically determined to maximize velocities, using $G_s\alpha$- and forskolin-activated cytoplasmic domains from type V and type II adenylyl cyclase (VC$_1$ and IIC$_2$). Under these conditions, the reaction velocities are completely linear with increasing amounts of VC$_1$.

The reactions are set up as follows (500-μl total volume): the coupled reaction mixture (5× stock; 100 μl), cyclic AMP (0.25–20 mM; 100 μl), P-site inhibitors (50 μl), and MgCl$_2$ (0.12–4 mM; 50 μl) are added to methacrylate disposable cuvettes and mixed. MgCl$_2$ is used to balance the amount of pyrophosphate added. Pyrophosphate (100 μl) is added last to avoid precipitation problems, but it is still important to examine the cuvettes for signs of precipitation before the start of the reaction and after data collection. These components are prewarmed at 30° for 3–5 min before addition of adenylyl cyclase and activator (0.4 μM VC$_1$, 2 μM IIC$_2$, 1 μM G$_s\alpha$–GTPγS; 100 μl) to start the reaction. The absorbance changes at 340 nm are measured for 15 min in a Beckman (Fullerton, CA) DU650 spectrophotometer with a six-position, temperature-controlled cuvette holder. Data are collected every 20 sec and typically the range from 1.5 to 10 min is used to determine velocities. Optical densities of greater than 1.5 are excluded from analysis.

Reaction velocities are calculated from the linear increase in absorbance at 340 nm resulting from the reduction of NADP (Fig. 2). The increase in absorbance in the absence of adenylyl cyclase is subtracted as background (typically <0.006/min). We usually express velocities as micromoles of product produced per minute per milligram of protein. Therefore, to convert the absorbance changes to our standard velocity:

$$[\text{Rate}\,(d\text{Abs}/\min) - \text{background rate}](1/\varepsilon)(\ell)(1/E_0)(1/V)(1000)$$
$$= \mu\text{mol}/\min \cdot \text{mg}$$

where ε is the extinction coefficient for NADP $= 6.3\,\text{m}M^{-1}\text{cm}^{-1}$; ℓ is the path length (1 cm); E_0 is the enzyme concentration (mg); and V is the volume of the reaction (0.5×10^{-3} liter).

Potential Problems. Because of the slower rate of the reverse reaction compared with that of the forward reaction ($\sim$1/30th), much larger concentrations of enzyme are required, making this reaction nearly impossible with the membrane-bound enzyme. In addition, most plasma membrane preparations of adenylyl cyclase contain additional ATPases that complicate the analysis; therefore, only purified proteins should be used.

It is also necessary to determine whether the P-site inhibitors used in the studies will have any effect on hexokinase or Glu6-P DH activities in the absence of adenylyl cyclase, using known amounts of ATP as substrates. Although no effect was observed with 2'-deoxyadenosine, many of the new 3'-polyphosphate inhibitors may slightly inhibit the activity of hexokinase, requiring greater concentrations of hexokinase in these reactions. The largest problem with these reactions and many of the binding experiments described below stems from the precipitation problems associated with pyrophosphate. Mn^{2+} cannot be used in these reactions because of the relatively high concentrations of pyrophosphate and its propensity to precipitate as Mn–PP$_i$. Free Mg^{2+} concentrations should be kept below 4 mM to avoid similar problems.

Reversible Binding Measurements

Direct tests of any kinetic model are obtained by binding measurements of inhibitors to the enzyme. Two methods have proved particularly useful: equilibrium dialysis and filter binding. Filter binding works best with tight-binding ligands so that the required washes do not release too much of the bound inhibitor. This technique is useful for rapidly screening a number of conditions for binding, and for providing estimates of K_d, and on and off rates. Equilibrium dialysis works well even for relatively weak-binding inhibitors and provides accurate values of association constants and the number of ligands bound per molecule of enzyme. However, the number of binding sites is averaged over all the binding sites of the protein population; they do not distinguish between different sites on the protein. Therefore, if a ligand can bind to two distinct sites on an enzyme but only a single molecule is bound at any given time, the N number will appear as one, the same as if only a single site exists (differences in affinity may help to distinguish these phenomena). The disadvantage of this technique is the time to reach equilibrium ($\sim$24 hr at 4°) and the amount of protein usually required.

To obtain a dissociation constant with either dialysis or filter binding, there is no need to have an accurate estimate of enzyme concentration. However, to obtain a stoichiometry of binding, it is crucial not only to know the protein concentration but also how much of the protein is active. This is the main reason that most of these techniques have utilized the soluble domains of adenylyl cyclase rather than crude membranes.

Equilibrium Dialysis

Equilibrium dialysis chambers were originally purchased from Hoefer (San Francisco, CA). These units contain eight reaction wells per Teflon disk with a total volume of 60 μl/side. Chambers are separated by dialysis membrane with a cutoff of 14 kDa, sealed between two O rings. A thin coat of silicone grease is applied to the O rings before use. Although the Hoefer units are highly recommended, they are no longer commercially available; a good alternative is the individual units from Sialomed (Harvard Apparatus, Holliston, MA). The Sialomed units are individual chambers but do have the advantage of not using O rings; however, the smaller pieces of dialysis membranes can be more difficult to handle. Before ever adding protein to the system, the time required to reach equilibrium should be determined for each ligand. This can be easily accomplished by adding radiolabeled ligand to one chamber and buffer alone to the opposite chamber. Samples (10 μl) can be removed from each side at various time intervals until an equal amount of ligand is present on each side. This simple test will also detect potential problems with dialysis membranes that have been stretched or otherwise mishandled and no longer allow for rapid dialysis. Experiments should be run for several hours longer than the minimum time required to reach equilibrium in the absence of proteins. For ^{3}H-radiolabeled ligands it is important to periodically lyophilize the sample to remove any [3]H_2O.

Once the dialysis chambers are assembled, equal volumes of sample are injected into each chamber containing buffer A, varying concentrations of ligand, and whatever additional reagents that may be required for binding.[11,16,22] Chamber A also includes 18 μM VC_1, IIC_2, and GTPγS–$G_s\alpha$ (nonradioactive); the opposite chamber contains protein storage buffer in lieu of proteins in a final volume of 50 μl (Fig. 3). Buffer A contains 20 mM Na–HEPES (pH 8.0), 2.5 mM $MgCl_2$, 1 mM

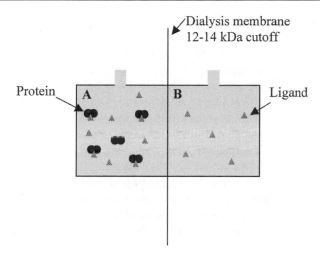

FIG. 3. Equilibrium dialysis setup. Depicted is a reaction at equilibrium where protein components are sequestered in chamber A and the free ligand has reached equilibrium. A 10- to 14-kDa dialysis membrane separates the two chambers.

dithiothreitol (DTT), 75 mM NaCl, and 2'-[^{3}H]deoxyadenosine (0.6 μCi/reaction; ICN, Costa Mesa, CA). Samples are dialyzed for 24 hr at 4° with constant rotation. A small bubble in each chamber helps ensure proper mixing of the samples. It is not advisable to run equilibrium dialysis at room temperature because of the time required to reach equilibrium and the instability of adenylyl cyclase. Duplicate 15-μl aliquots from each chamber are then counted by liquid scintillation spectrometry. A sample of buffer A is counted to determine the total number of counts per minute (cpm) in each well and then the specific activity (cpm/pmol) must be calculated for each concentration of ligand. The concentration of ligand in chambers A and B is simply determined by the counts per minute per microliter from these chambers divided by the specific activity. The concentration of bound ligand is calculated by subtracting the concentration of ligand in chamber B (free ligand) from the concentration of ligand in chamber A (bound plus free ligand).

We typically start with label on both sides to speed the time to reach full equilibrium in the presence of protein. However, it is necessary to also run one set of duplicate samples in which the labeled ligand starts out entirely in chamber B or chamber A. This will also help determine whether the samples come to full equilibrium with added protein present.

The concentrations of ligand used will vary depending on the affinity of the enzyme for each ligand. In the case of 2'-deoxyadenosine ($K_d = 40$ μM for VC$_1$ and IIC$_2$), 10–250 μM concentrations are used to give a specific activity of 1000–40 cpm/pmol.[11] Pyrophosphate (2.5 mM) is also required for

2'-deoxyadenosine binding. Sodium chloride or other salts are important to reduce an asymmetric distribution of ionic charges due to the concentration of macromolecule on one side of the membrane (see Donnan effect[34]). This is also particularly important with a charged ligand to reduce nonspecific binding. The laboratory of R. Johnson (Department of Physiology and Biophysics, Health Sciences Center, State University of New York, Stony Brook, NY) has developed a number of ^{32}P-labeled inhibitors with significantly greater affinity for adenylyl cyclase in the absence of pyrophosphate.[13,16,25] For example, 2',5'-dideoxy-3'-[^{32}P]ATP has an affinity for the soluble domains (VC_1 and IIC_2) at 2.4 μM in the presence of $G_s\alpha$ and Mg^{2+} and 0.2 μM in the presence of $G_s\alpha$ and Mn^{2+}.[16] Because pyrophosphate is not required, binding reactions can be performed easily in the presence of Mn^{2+}. The higher ligand affinities mean that significantly less enzyme (2–4 μM) and unlabeled ligand (0.1–12 μM) are required for these reactions. It is important to note, however, that when using the soluble adenylyl cyclase system, enzyme concentrations must be kept above 2 μM for at least one subunit in the presence of $G_s\alpha$ or forskolin because of the affinity of the two domains for one another (~1 μM). This is also true for any of the kinetic or filter-binding techniques.

Filter Binding

Another way to analyze ligand binding is by using a filter-binding method. This can be used with pure protein or crude membranes and requires less protein than equilibrium dialysis. The affinity of any ligand can be analyzed by competition with unlabeled ligand. Off rates for relatively tight-binding ligands can also be measured easily by this method. This technique can be performed at both room temperature and 4°.

The setup includes a 96-well vacuum manifold unit (Millipore, Bedford, MA), recommended to analyze small sample volumes and for rapid filtration. The 96-well membrane plates are HA, 0.45-μm pore size cellulose ester plates. Larger filter-binding apparatuses can, of course, be used but have slower filtration times and require larger reaction sizes (100 μl) and wash volumes (1 ml).

A typical binding reaction examining 2',5'-dideoxy-3'-ATP binding to the soluble C_1 and C_2 domains contains the following: 20 mM HEPES (pH 8.0), 8 mM $MgCl_2$, 1 mM EDTA, 2 mM DTT, 2 μM 2',5'-dideoxy-3'-ATP, and 2',5'-dideoxy-3'-[^{32}P]ATP (1500 cpm/pmol) (C. W. Dessauer, unpublished conditions, 2000). The binding reaction is started with addition of proteins consisting of 3–8 μM VC_1, IIC_2, $G_s\alpha$, and 100 μM forskolin in a final volume of 25 μl. The reactions are incubated for 10 min at either room temperature or on ice. The filters are prerinsed with 100 μl of wash buffer [20 mM HEPES (pH 8.0), 2 mM $MgCl_2$,

[34] C. R. Cantor and P. R. Schimmel, "Biophysical Chemistry: The Behavior of Biological Macromolecules." W. H. Freeman, New York, 1942.

1 mM EDTA, and 100 mM NaCl]. The vacuum is then shut off and 100 μl of buffer is added to the wells. The reaction is stopped by addition of the sample to 100 μl of buffer in the wells and rapidly filtered. The samples are then washed with an additional 200 μl of buffer. Total radioligand is determined by spotting an aliquot of the ^{32}P-containing mixture on dry filter membranes. Filter membranes are dried in a vacuum desiccator or at room temperature and the bound ligand is quantified by liquid scintillation counting. If a 96-well scintillation counter (Packard, Downers Grove, IL) is available, a bottom plate can be snapped on to the 96-well filter plate and 25 μl of Microscint 20 scintillation fluid added to each filter well and directly counted. Otherwise the individual membranes can be cut out and placed in scintillation vials with scintillation fluid. This latter method, although tedious, works well. The background for these reactions is usually 200–350 cpm with a total of 20,000 cpm in the reaction. We typically obtain more consistent results when both the preincubation and filtration are performed at 4° with cold wash buffer. However, these reactions can be performed at room temperature as shown in Fig. 4A. It is nearly impossible not to have some variability in the washes of each point; therefore, Fig. 4 displays all the data points for a single experiment and the range of values obtained before averaging duplicates. The system described above works well for screening binding conditions or competition experiments and can be assembled in 96-well plates for rapid processing of multiple samples. Timed experiments such as on-rate and off-rate measurements (Fig. 4) can also be performed with a similar setup but require two sets of hands to filter if short time points are analyzed. Off-rate measurements contain the same binding reaction and preincubation as described earlier, but filtration starts at time 0 with addition of a large excess of unlabeled ligand (or buffer as a negative control in Fig. 4A). Small aliquots are then removed from the reaction (in duplicate) and filtered as described above.

This same basic protocol can also be used to examine ligand binding to membranes containing adenylyl cyclase. The specific activity of the ligand needs to be high (> 100 cpm/fmol), and approximately 3–6 μg of membranes is used in the reaction. For the 96-well filtration plates, high concentrations of membranes can clog the filter and significantly slow filtration times. This is greatly improved by using a detergent-solubilized preparation. A similar protocol has been successfully used with the competitive inhibitor β-L-2′,3′-dideoxy-5′-[β-^{32}P]ATP ($K_d \sim$ 30 nM).[25] In this case, nitrocellulose filters were used with a wash buffer consisting of 20 mM TEA-HCl (pH 7.5) and 5 mM MnCl$_2$. For β-L-2′,3′-dideoxy-5′-[β-^{32}P]ATP, the type and pH of the buffering system are important to reduce severe nonspecific binding to the nitrocellulose membranes. The binding reaction consisted of 50 mM TEA-HCl (pH 7.5), 5 mM MnCl$_2$, 1 mM 3-isobutyl-1-methylxanthine, 2 mM creatine phosphate, 10 μg of creatine kinase, 100 μM forskolin, and 0.1–10 μM β-L-2′,3′-dideoxy-5′-[β^{32}P]ATP (80 cpm/fmol) in a final volume of 100 μl. The

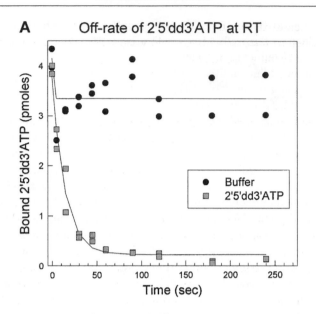

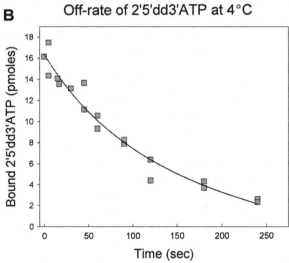

FIG. 4. Off-rate measurements for $2',5'$-dideoxy-$3'$-ATP binding to $G_s\alpha$- and forskolin-activated adenylyl cyclase. (A) $2',5'$-Dideoxy-$3'$-[^{32}P]ATP ($2~\mu M$) was preincubated with $4~\mu M$ VC$_1$, IIC$_2$, and GTPγS-$G_s\alpha$ and $100~\mu M$ forskolin at room temperature for 10 min. At the end of the incubation duplicate aliquots were removed and rapidly filtered with room temperature wash buffer (zero point). An equal volume of buffer alone or buffer containing $250~\mu M$ unlabeled $2',5'$-dideoxy-$3'$-ATP was then added to the reaction (time 0), and duplicate aliquots were filtered at the indicated times. (B) Off-rate measurements were performed exactly as in (A), except that the preincubation and washes were performed at $4°$.

reaction is incubated with 6 μg of detergent-dispersed rat brain extract at 30°
for 15 min before filtering with ice-cold wash buffer. The filters are dried and
quantified by Cerenkov counting.

Data Analysis

The most important rule for modeling a large amount of kinetic data is that
all measurements must be obtained under identical conditions. Of course, this
is nearly impossible, especially when using different measurements to obtain K_d
values and rate constants. For example, 5–10 mM Mg^{2+} is traditionally used to
maximize activities of cyclic AMP synthesis, and care must be taken to maintain
this concentration of free Mg^{2+} on addition of charged nucleotides. However, in
the reverse reaction it is nearly impossible to run these reactions in the presence
of this high Mg^{2+} concentration because of the precipitation problems associated
with pyrophosphate. In addition, equilibrium dialysis measurements performed at
4° will not provide the same K_d as estimates from enzyme reactions performed at
30°. Finally, with the soluble domains it is crucial that all experiments maintain
one domain at sufficient concentrations to drive the interaction with the opposite
domain. If not, analysis could be greatly complicated by changes in the affinity of
these two domains in the presence of inhibitors and assay conditions.

Despite these caveats, it is possible to write equations that describe a particular
mechanistic model and determine the fit of data to this equation and model. By far
the most difficult part of this process is in deriving the velocity equation that depicts
a model. If the enzyme obeys Michaelis–Menten kinetics, then the derivation is
usually fairly straightforward (see Segel[23] for examples). This assumes that the
system is in rapid equilibrium, and is the most direct method for deriving velocity
equations in the absence of any prior knowledge of the relative magnitudes of the
rate constants. If the experimental data fit this model, then it is safe to assume that
the simplest mechanism to explain the system has been found. If not, then more
complex steady state equations must be derived. For example, if k_3 and k_2 are of
the same order of magnitude as for adenylyl cyclase (Fig. 1), the concentration
of the intermediate AC–ATP no longer depends solely on the concentration of
AC and ATP as predicted by the equilibrium constant K values for Michaelis–
Menten kinetics. For this and other reasons, steady state equations were derived
for the adenylyl cyclase reaction. Solution of steady state equations is a matter of
simultaneously solving a series of linear equations and usually involves little more
than basic algebra. For example, the scheme of

$$E + S \underset{k_{-1}}{\overset{k_1}{\rightleftharpoons}} ES \underset{k_{-2}}{\overset{k_2}{\rightleftharpoons}} E + P$$

is given by the equation $v = k_2[ES] - k_{-2}[E][P]$. However, until we solve for [ES]
and [E] this equation is relatively useless. The solution of [E] is straightforward

because the total enzyme concentration $[E_t] = [E] + [ES]$. By taking into account the formation and breakdown of ES, we can show that

$$k_1[E][S] + k_2[E][P] = k_2[ES] + k_{-1}[ES]$$

and solve for [ES]. By putting our solutions for [E] and [ES] back into our original equation, we can derive a steady state equation that is in terms of the starting concentrations for our enzyme, substrate, product, and individual rate constants. It is also possible to solve these equations by grouping methods such as the King–Altman method. The equations can then be entered into a plotting program such as Sigma Plot (via the User-Defined Regression analysis) and used to simultaneously fit data from a number of different kinetic experiments varying substrate, inhibitor, and even product concentrations. All concentrations should be converted to molar units, including the velocity. The quality of the fit and thus the validity of the model is best judged by visual examination of the fit to the data and analysis of statistical parameters.

For complicated mechanistic schemes the solution of steady state equations can be labor intensive. A number of programs currently available can help describe a model in mathematical terms and then solve the equation in terms of velocity as described above or solve the differential equations that describe the relative concentration of various reactant intermediates over time. Examples include Dynafit (Biokin[35]), Berkeley Madonna (R. Macey and G. Oster, University of California, Berkeley, CA), and GENESIS.[36] GENESIS was created for neural modeling but has been used by R. Iyengar and co-workers[37] to simulate signaling networks and more extensively by others to describe calcium dynamics over time. Berkeley Madonna has features similar to GENESIS but does include a chemical reactions window that uses general enzyme terms and allows reactions to be easily entered. The most appropriate of these for the applications discussed above is Dynafit, which can be downloaded free off the internet. It is fairly easy to use with a large number of examples to guide the user through the process (the aminopeptidase example is recommended). This program also allows the user to fit new data and test various models. However, all of these programs will require a significant investment in time to accomplish even basic tasks. Regardless of the methods used to model data, the rewards can be significant, providing insight into the inner workings of an enzyme.

[35] P. Kuzmic, *Anal. Biochem.* **237**, 260 (1996).
[36] J. M. Bower and D. Beeman, "The Book of GENESIS: Exploring Realistic Neural Models with GEneral NEural SImulation System." Springer-Verlag, Berlin, 1998.
[37] G. Weng, U. S. Bhalla, and R. Iyengar, *Science* **284**, 92 (1999).

[10] Expression, Purification, and Assay of Cytosolic (Catalytic) Domains of Membrane-Bound Mammalian Adenylyl Cyclases

By MARK E. HATLEY, ALFRED G. GILMAN, and ROGER K. SUNAHARA

Introduction

Hormonal regulation of intracellular concentrations of cyclic AMP is exerted primarily by modulation of the catalytic activities of the membrane-bound forms of adenylyl cyclase. To date, nine isoforms of the membrane-bound mammalian enzyme and one cytosolic protein have been identified.[1-3] All membrane-bound mammalian adenylyl cyclases are activated by the α subunit of the stimulatory G protein, G_s. With the exception of type IX, all forms are also activated by forskolin, a diterpene. Only types I, V, and VI are inhibited by the α subunit of the inhibitory G protein G_i, but all isoforms are inhibited by a group of adenosine analogs called P-site inhibitors.[1,4]

Membrane-bound adenylyl cyclases are composed of two transmembrane domains (each likely containing six membrane-spanning helices) separated by a large cytosolic loop (C1 domain). In addition, the proteins contain a second cytosolic domain at their carboxyl terminus (C2 domain). This topological arrangement is reminiscent of those of the multidrug resistance transporters and the cystic fibrosis conductance regulators.[5] Each of the cytosolic domains has similar amino acid sequences (approximately 200 residues), and these sequences are also conserved among both mammalian adenylyl cyclases and guanylyl cyclases. Indeed, it is these sequences that represent the catalytic site of the purine nucleoside triphosphate cyclases.[6]

After expression in bacteria, the two cytosolic domains of adenylyl cyclase together display most of the properties of the membrane-bound parent molecules, including the capacity to be regulated by G protein α subunits, inhibited by P-site inhibitors, and activated by forskolin.[7-11] The atomic structure of the catalytic domain of adenylyl cyclase bound to activators $G_s\alpha$ and forskolin has been

[1] R. K. Sunahara, C. W. Dessauer, and A. G. Gilman, *Annu. Rev. Pharmacol. Toxicol.* **36**, 461 (1996).

[2] M. J. Smit and R. Iyengar, *Adv. Second Messenger Phosphoprotein Res.* **32**, 1 (1998).

[3] J. Buck, M. L. Sinclair, L. Schapal, M. J. Cann, and L. R. Levin, *Proc. Natl. Acad. Sci. U.S.A.* **96**, 79 (1999).

[4] W.-J. Tang and J. H. Hurley, *Mol. Pharmacol.* **54**, 231 (1998).

[5] J. Krupinski, F. Coussen, H. A. Bakalyar, W.-J. Tang, P. G. Feinstein, K. Orth, R. R. Reed, and A. G. Gilman, *Science* **244**, 1558 (1989).

[6] W.-J. Tang and A. G. Gilman, *Science* **268**, 1769 (1995).

[7] S.-Z. Yan, D. Hahn, Z.-H. Huang, and W.-J. Tang, *J. Biol. Chem.* **271**, 10941 (1996).

determined.[12,13] Structures of adenylyl cyclase bound with various substrate and product inhibitors have also been delineated.[14] Together with the results of site-directed mutagenesis and kinetic analysis, these structures have provided detailed insight into the catalytic mechanism of adenylyl cyclase.[14,15] Notably, two metal ions are required for cyclic AMP synthesis, similar to the reactions catalyzed by DNA polymerases, reverse transcriptases, and RNA spliceosomes.[16]

This article describes the construction of bacterial expression vectors encoding the cytosolic domains of adenylyl cyclase, the methods for expression and purification of these proteins, and the conditions for assessing their collaborative catalytic activity.

Construction of Expression Vectors

Tang and Gilman[6] first noted that the isolated cytoplasmic domains of adenylyl cyclase displayed regulated catalytic activity. They described the construction, expression, and characterization of a linked chimera between the C1 domain of type I adenylyl cyclase and the C2 domain of the type II enzyme. The molecule was subsequently purified and extensively characterized,[17] revealing properties that were remarkably similar to those of the purified, detergent-extracted parent form of the enzyme.[18-19] It was later determined that the individual domains can be expressed separately and regulated adenylyl cyclase activity reconstituted by simple mixture of the two proteins.[7,9,10] Both the linked and nonlinked forms of the enzyme described initially suffered from problems of protein instability, apparently during translation and folding in the bacterial cell. The problem was identified as the expression of the C1 domain from adenylyl cyclase type I (ACI), which accumulates to concentrations of only 20 μg/liter of culture; by contrast, the C2 domain of ACII accumulates to levels well in excess of 10 mg/liter culture. To circumvent these problems, we investigated expression of C1 domains from other isoforms of adenylyl cyclase.

[8] R. K. Sunahara, C. W. Dessauer, R. E. Whisnant, C. K. Kleuss, and A. G. Gilman, *J. Biol. Chem.* **272**, 22265 (1997).

[9] R. E. Whisnant, C. W. Dessauer, and A. G. Gilman, *Proc. Natl. Acad. Sci. U.S.A.* **93**, 6621 (1996).

[10] K. Scholich, A. J. Barbier, J. B. Mullenix, and T. B. Patel, *Proc. Natl. Acad. Sci. U.S.A.* **97**, 2915 (1997).

[11] C. W. Dessauer, J. J. Tesmer, S. R. Sprang, and A. G. Gilman, *J. Biol. Chem.* **273**, 25831 (1999).

[12] J. J. G. Tesmer, R. K. Sunahara, A. G. Gilman, and S. R. Sprang, *Science* **278**, 1907 (1997).

[13] G. Zhang, Y. Liu, A. E. Ruoho, and J. H. Hurley, *Nature (London)* **386**, 247 (1997).

[14] J. J. G. Tesmer, R. K. Sunahara, R. A. Johnson, G. Gosselin, A. G. Gilman, and S. R. Sprang, *Science* **285**, 6298 (1999).

[15] C. W. Dessauer and A. G. Gilman, *J. Biol. Chem.* **272**, 27787 (1997).

[16] T. A. Steitz, *Nature (London)* **391**, 231 (1998).

[17] C. W. Dessauer and A. G. Gilman, *J. Biol. Chem.* **271**, 16967 (1996).

[18] R. Taussig, L. M. Quarmby, and A. G. Gilman, *J. Biol. Chem.* **268**, 9 (1993).

[19] T. Pfeuffer and H. Metzger, *FEBS Lett.* **146**, 369 (1982).

C1 Domain from Type V Adenylyl Cyclase

The first large cytoplasmic domain of membrane-bound adenylyl cyclases is composed of a highly conserved 200-amino acid catalytic domain (C1a), followed by another stretch of 120–150 amino acids that has a regulatory function (C1b domain). We attempted to express in *Escherichia coli* various polypeptides derived from the C1 domain of six of the isoforms of mammalian adenylyl cyclase; only those from the type I[7,9] and type V[8,10] enzymes yielded significant amounts of active protein. We restrict discussion here to the best of these systems—the C1a domain of type V adenylyl cyclase from *Canis familiaris*. For simplicity, residue numbers are those of Ishikawa *et al.*[20]

Amino- and carboxyl-terminal deletions of the C1 domain are generated by standard techniques for manipulation of recombinant DNA, and the accumulation of these proteins in *E. coli* is assessed (data not shown). Although a number of constructs have been studied, a core of residues, starting from Met-364 and ending with His-591, appear optimal in terms of both protein expression and adenylyl cyclase activity.[8] A 740-bp DNA fragment encoding these amino acid residues is generated by the polymerase chain reaction (PCR). The 3′ oligonucleotide contains sequences encoding a FLAG epitope (Eastman Kodak, Rochester, NY), a stop codon, and a *Hin*dIII restriction enzyme site (sequences available on request). The resulting fragment is digested with *Nco*I and *Hin*dIII and subsequently inserted into the bacterial expression vector H6-pQE60[21] (Fig. 1A). This modified version of pQE60 (Qiagen, Valencia, CA) contains a hexahistidine (His6) tag that may be positioned at the amino terminus of proteins rather than at the carboxyl terminus. We refer to the expressed protein, H6-VC1(591)FLAG, as VC1 for the remainder of the article.

Shorter versions of VC1 have been expressed, purified to homogeneity, and utilized for X-ray crystallographic purposes. A description of the construction and purification of these proteins is found in [15] in this volume.[22] We suggest the use of H6-VC1(591)FLAG for other purposes, because it is the most stable of all of the truncation/deletion constructs of VC1 tested (data not shown).

C2 Domain from Type II Adenylyl Cyclase

The carboxyl-terminal cytoplasmic domain of type II adenylyl cyclase from rat can be expressed luxuriantly as a recombinant protein. Although the protein is susceptible to proteolysis when synthesized in *E. coli,* many of fragments are fully active. We routinely use a carboxyl-terminally hexahistidine-tagged version of

---

[20] Y. Ishikawa, K. Shuichi, L. Chen, N. J. Halnon, J.-I. Kawabe, and C. Homcy, *J. Biol. Chem.* **267,** 13553 (1992).

[21] E. Lee, M. E. Linder, and A. G. Gilman, *Methods Enzymol.* **237,** 146 (1994).

[22] J. J. G. Tesmer, R. K. Sunahara, D. A. Fancy, A. G. Gilman, and S. R. Sprang, *Methods Enzymol.* **345,** [15] 2001 (this volume).

A   **B**

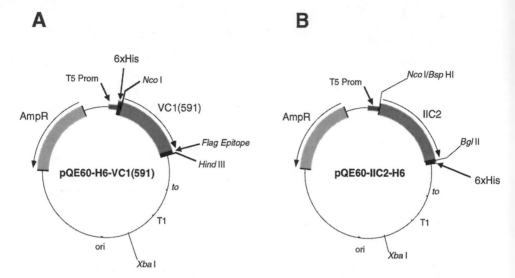

FIG. 1. Plasmid maps of (A) pQE60-H6-VC1(591) and (B) pQE60-IIC2-H6. Complementary DNA fragments encoding the adenylyl cyclase domains were generated by PCR as described in text. The pQE bacterial expression system (Qiagen) utilizes the T5 promoter together with a *lac* operator. The distance between the *Eco*RI/RBS (ribosomal binding sequence) and the initiating methionine has been optimized to facilitate strong regulation of transcription, provided that ample levels of LacI are produced in host cells. For expression in BL21(DE3) (Novagen), we suggest cotransformation with pREP4, which directs expression of *lacI* and kanamycin resistance.

IIC2 that includes residues 822–1090 from rat type II adenylyl cyclase. Oligonucleotide primers containing a *Bsp*HI restriction enzyme site (encoding a methionine) and a *Bgl*II 3′ site (for the C terminus) are used to amplify IIC2 (822–1090) cDNA by PCR.[9] This fragment is subsequently inserted in the bacterial expression vector pQE60 at the *Nco*I and *Bgl*II sites (Fig. 1B). The predicted molecular mass of this tagged protein is approximately 31.8 kDa. However, electron spray mass spectrometry has revealed that the major species that accumulates in *E. coli* has an M_r of 28,258, consistent with a protein containing Met-847 to Ser-1090.[9] The discrepancy between the predicted and determined molecular weights may be a result of translation initiation from an internal methionine (Met-847), or a result of proteolysis. We prefer the former possibility because few proteases cut between a methionine and serine (Ser-846). We refer to the major protein product of this DNA construct (IIC2-Met-847 to Ser-1090) as IIC2 for the remainder of the article.

   Shorter, nontagged versions of IIC2 have been isolated and purified to homogeneity and have served as optimal sources of protein for X-ray crystallography. A description of the construction and purification of these proteins is found in [15] in this volume.[22]

Expression of Adenylyl Cyclase Cytosolic Domains

Each of the two cytosolic domains of mammalian adenylyl cyclase is expressed in *E. coli* separately and purified to homogeneity. The bacterial strain BL21(DE3) is cotransformed with one of the expression plasmids encoding a cyclase domain (described above) and pREP4 (Qiagen). The pREP4 plasmid allows constitutive expression of *lacI* repressor under kanamycin selection. These colonies are used to inoculate 100 ml of Luria broth (LB) containing a 50-μg/ml concentration of both ampicillin and kanamycin (overnight starter culture). The expression conditions for VC1 and IIC2 differ and are detailed below.

The VC1 starter culture is used to inoculate 12 liters of T7 medium[21] containing a 50-μg/ml concentration of both ampicillin and kanamycin. These cultures are shaken at 30° until the OD_{600} is 1.1. Synthesis of VC1 is induced with 30 μM isopropyl-β-D-thiogalactopyranoside (IPTG), and cultures are incubated at room temperature for 4 hr. VC1 aggregates and is found in inclusion bodies after longer incubation times or at higher temperatures. The cells are then collected by centrifugation and stored at −80°.

The IIC2 starter culture is used to inoculate 1 liter of LB medium. The culture is shaken at 30° until the OD_{600} is 0.6. IIC2 synthesis is induced with 30 μM IPTG, and cultures are incubated at room temperature for 16 hr. Cells are collected by centrifugation and stored at −80°.

Purification of VC1

Preparation of Lysates. All the purification steps are performed at 4°. The frozen cells are suspended in 800 ml (1/15 culture volume) of lysis buffer [50 mM Tris-HCl (pH 8.0), 120 mM NaCl, 1 mM 2-mercaptoethanol (2-ME), and a mixture of protease inhibitors]. The mixture of protease inhibitors used throughout this work contains 34 μg/ml each of L-tosylamido-2-phenylethyl chloromethyl ketone, 1-chloro-3-tosylamido-7-amino-2-heptanone, and phenylmethylsulfonyl fluoride and 3 μg/ml each of leupeptin and lima bean trypsin inhibitor. Cells are lysed by addition of lysozyme (0.4 mg/ml) and incubated for 40 min on ice while stirring. The slurry is treated with DNase I (8 μg/ml) and $MgCl_2$ (1 mM) for an additional 30 min. Cellular debris and membranes are removed by centrifugation at 100,000g for 35 min at 4°; the supernatants are collected. The pellets are snap frozen in liquid N_2 and subjected to lysis again by thawing, suspension in 400 ml (1/30 culture volume) of lysis buffer, and treatment with DNase I as described above. The supernatant, clarified by ultracentrifugation, is pooled with the first supernatant and applied to a TALON column (Clontech, Palo Alto, CA).

TALON Column Chromatography. A 5-ml cobalt metal chelate column (TALON; Clontech) is equilibrated with 50 ml of lysis buffer before application of the pooled supernatants at a slow rate (0.5 ml/min). The column is washed with 20 volumes of buffer A [50 mM Tris-HCl (pH 8.0), 500 mM NaCl, 1 mM 2-ME,

TABLE I
PURIFICATION OF VC1 FROM *Escherichia coli*[a]

Pool	Concentration (mg/ml)	Volume (ml)	Yield (mg)	Specific activity (μmol/min/mg)	Total activity (μmol/min)	Recovery (%)	Purification (-fold)
Clarified lysate	4.4	1200	5280	0.3	1584	100	—
TALON	2.4	10	24	25.6	614	39	85
Mono Q	9.6	0.775	7.4	65	481	30	217

[a] Pools from the lysates, the TALON column, and the Mono Q column were assayed for forskolin-stimulated adenylyl cyclase activity. Peak fractions were selected on the basis of their specific adenylyl cyclase activity. The VC1 concentration in the assay of the TALON and Mono Q pools was 2 nM. For the lysate, 5.6 ng was used. Fractions were assayed in a total volume of 100 μl in the presence of 2 μM IIC2, 100 μM forskolin, and 1 mM [^{32}P]ATP for 10 min at 30°.

and protease inhibitors], followed by 10 volumes of buffer B [50 mM Tris-HCl (pH 8.0), 50 mM NaCl, 1 mM 2-ME, and protease inhibitors]. VC1 is then eluted in 2.5-ml fractions of buffer B containing 100 mM imidazole. The peak fractions are pooled and diluted 5-fold with buffer C [20 mM HEPES (pH 8.0), 2 mM MgCl$_2$, 1 mM EDTA, 2 mM dithiothreitol (DTT), and protease inhibitors]. VC1 specific activity in the pooled peak is increased 80- to 100-fold compared with the applied supernatant (see Table I).

Ion-Exchange Fast Protein Liquid Chromatography: Mono Q HR 5/5. VC1 is applied to a Mono Q HR 5/5 column in buffer C and eluted from the column with buffer C and a shallow gradient of 0 to 300 mM NaCl over 25 column volumes, followed by a steep gradient to 1 M NaCl. The UV$_{280}$ absorbance tracing and salt gradient are shown in Fig. 2A. VC1 elutes as a single major peak of activity (at approximately 180 mM NaCl), coincident with a large peak in the UV$_{280}$ absorbance tracing. The peak fractions are pooled, concentrated, and exchanged into buffer C with 50 mM NaCl, using a Centricon 10 (spin concentrator, 10,000 M_r cutoff; Millipore, Bedford, MA). Typical yields from this purification vary between 5 and 10 mg of VC1 per 10-liter culture. The lysate and the pooled fractions from the TALON and Mono Q columns are resolved by sodium dodecyl sulfate–polyacrylamide gel electrophoresis (SDS–PAGE) and visualized by staining with Coomassie blue (Fig. 2B). The pools are assayed for adenylyl cyclase activity in the presence of 100 μM forskolin. A summary of the results is shown in Table I.

Purification of IIC2

Preparation of Lysates. The lysis procedure for bacteria containing IIC2 is similar to that described above. The frozen pellets are suspended in 100 ml of lysis buffer. The cells are lysed with lysozyme and treated with DNase I, and the

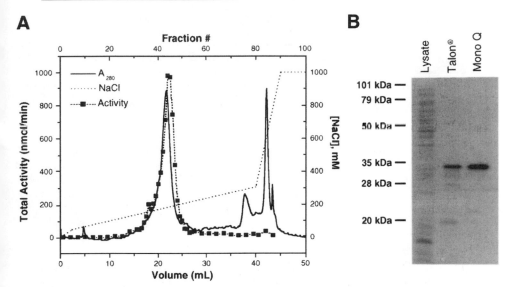

FIG. 2. Purification of VC1 by Mono Q ion-exchange FPLC. (A) The TALON column pool was applied to a Mono Q HR 5/5 FPLC column and eluted with an NaCl gradient as described in text. VC1 was eluted with a shallow linear gradient of 50–300 mM NaCl (37.5 ml; dotted line, right y axis). Also shown is the UV$_{280}$ absorbance tracing of the eluate (solid line). Aliquots (diluted as necessary) from every other 0.5-ml fraction were assayed for adenylyl cyclase activity in the presence of 1 μM IIC2, 100 μM forskolin, and 1 mM [^{32}P]ATP for 10 min at 30° (solid squares, dashed line). Lower than usual concentrations of IIC2 were used to screen column fractions. (B) SDS–PAGE analysis of the lysate, TALON column pool, and Mono Q pool. Five micrograms of lysate and 2 μg of the TALON and Mono Q pools were resolved on a 15% (w/v) SDS–polyacrylamide gel and visualized by staining with Coomassie blue. Molecular weight markers are indicated.

lysate is clarified by centrifugation. The supernatant is applied to a 2-ml column of nitrilotriacetic acid–nickel ion–agarose (Ni^{2+}–NTA; Qiagen), and the pellets are lysed again as described above.

Ni^{2+}–NTA Column Chromatography. The column is equilibrated with 20 ml of lysis buffer before application of the supernatant. The column is washed with 10 column volumes of buffer D [50 mM Tris-HCl (pH 8.0), 500 mM NaCl, 2 mM DTT, and protease inhibitors] followed by 10 column volumes of buffer E [50 mM Tris-HCl (pH 8.0), 50 mM NaCl, 2 mM DTT, and protease inhibitors] containing 5 mM imidazole. The column is then eluted with 1-ml volumes of buffer E containing 150 mM imidazole. The peak fractions are pooled and diluted 7-fold with buffer C.

Ion-Exchange Fast Protein Liquid Chromatography: Mono Q HR 5/5. The Ni^{2+}–NTA pool containing IIC2 is applied to a Mono Q HR 5/5 column in buffer C and eluted by the protocol described above for VC1 (Fig. 3A). The major peak of IIC2 activity elutes at 100 mM NaCl and corresponds to the UV absorbance. The

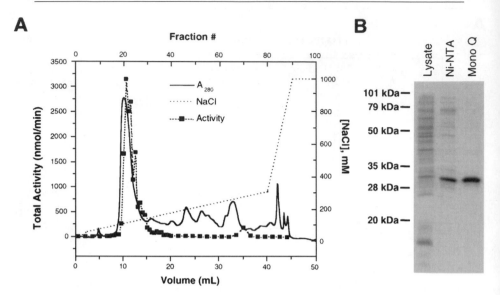

FIG. 3. Purification of IIC2 by Mono Q ion-exchange FPLC. (A) The Ni^{2+}–NTA pool was applied to a Mono Q HR 5/5 FPLC column and eluted with an NaCl gradient as described in text. IIC2 was eluted with a shallow linear gradient of 50–300 mM NaCl (37.5 ml; dotted line, right y axis). Also shown is the UV_{280} absorbance tracing of the eluate (solid line). Aliquots (diluted as necessary) from every other 0.5-ml fraction were assayed for adenylyl cyclase activity in the presence of 1 μM VC1, 100 mM forskolin, and 1 mM [^{32}P]ATP for 10 min at 30° (solid squares, dashed line). Lower than usual concentrations of VC1 were used to screen column fractions. (B) SDS–PAGE analysis of the lysate, Ni^{2+}–NTA column pool, and Mono Q pool. Five micrograms of lysate and 2 μg of the Ni^{2+}–NTA and Mono Q pools were resolved on a 15% (w/v) SDS–polyacrylamide gel and visualized by staining with Coomassie blue. Molecular weight markers are indicated.

typical protein yield varies between 15 and 40 mg/liter of culture. The lysate and the pools from the Ni^{2+}–NTA and Mono Q columns are resolved by SDS–PAGE and stained with Coomassie blue (Fig. 3B). The IIC2 activities of the pools are determined in the presence of 100 μM forskolin (Table II).

Assays of Soluble Domains

The activity of adenylyl cyclase is determined by measuring the conversion of [α-^{32}P]ATP to [^{32}P]cAMP. The assay is initiated by combining the labeled substrate with the mixed soluble protein domains and terminated by addition of sodium dodecyl sulfate [2% (w/v), final] and unlabeled cAMP (1.75 mM, final) and ATP (50 mM, final). The procedure for isolation of [^{32}P]cAMP is identical to that used to assay the membrane-bound form of adenylyl cyclase.[23,24]

[23] M. D. Smigel, *J. Biol. Chem.* **261,** 1976 (1986).
[24] Y. Saloman, C. Londos, and M. Rodbell, *Anal. Biochem.* **58,** 541 (1974).

TABLE II
PURIFICATION OF IIC2 FROM *Escherichia coli*[a]

Pool	Concentration (mg/ml)	Volume (ml)	Yield (mg)	Specific activity (μmol/min/mg)	Total activity (μmol/min)	Recovery (%)	Purification (-fold)
Clarified lysate	4.5	150	675	2	1350	100	—
Ni^{2+}–NTA	4.9	6	29.4	40	1176	87	20
Mono Q	10	1.4	14	67	938	69	34

[a] Pools from the lysates, the Ni^{2+}–NTA column, and the Mono Q column were assayed for forskolin-stimulated adenylyl cyclase activity. Peak fractions were selected on the basis of their specific adenylyl cyclase activity. The IIC2 concentration in the assay of the Ni^{2+}–NTA and Mono Q pools was 2 nM. For the lysate, 5.6 ng was assayed. Fractions were assayed in a total volume of 100 μl in the presence of 2 μM VC1, 100 μM forskolin, and 1 mM [^{32}P]ATP for 10 min at 30°.

In general, no ATP-regenerating system is required, provided that the proteins are pure (i.e., no contaminating ATPase activity). Forskolin (Calbiochem, La Jolla, CA) is typically prepared as a 10 mM solution in either ethanol or dimethyl sulfoxide (DMSO). GTPγS-activated G$_s\alpha$ is produced by incubating purified G$_s\alpha$ (native sequence or hexahistidine tagged; see Graziano *et al.*[25]) with a 5-fold molar excess of [^{35}S]GTPγS (200–500 cpm/pmol of GTPγS) in buffer [20 mM HEPES (pH 8.0), 2 mM MgCl$_2$, 1 mM EDTA, and 2 mM DTT] for 30–45 min at 30°. Free [^{35}S]GTPγS is removed by gel filtration on a Sephadex G-50 (Pharmacia, Piscataway, NJ) spin column. The concentration of G$_s\alpha$–GTPγS is determined on the basis of the amount of bound [^{35}S]GTPγS.[25]

Concentrations of C1 and C2 Domains

Synthesis of cyclic AMP requires functional interaction between the C1 and C2 domains. VC1 and IIC2 exist in isolation as homodimers but form a heterodimer when mixed in the presence of activators of adenylyl cyclase.[8] When a fixed low concentration of one domain is mixed with increasing concentrations of the other, adenylyl cyclase activity increases in a concentration-dependent and saturable manner (Fig. 4). We thus assay limiting concentrations of one domain in the presence of saturating concentrations of the other. When measuring adenylyl cyclase activity with limiting concentrations of VC1 (e.g., 2 nM), at least 10 μM IIC2 should be present (Fig. 4A). Similarly, saturating concentrations of VC1 should be present (at least 1 μM) if the concentration of IIC2 is limiting (Fig. 4B). In addition, the affinity of the two domains for each other is dependent on the identity of the activator. For example, the concentration of IIC2 necessary for maximal activity is

[25] M. P. Graziano, M. Freissmuth, and A. G. Gilman, *J. Biol. Chem.* **264**, 409 (1989).

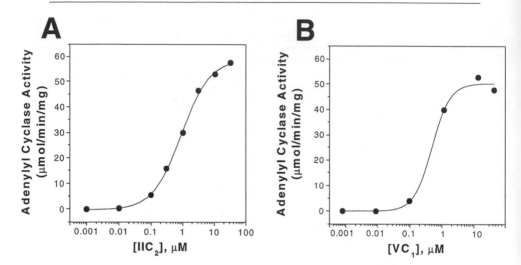

FIG. 4. Concentration-dependent activity of the adenylyl cyclase domains. (A) Forskolin-stimulated adenylyl cyclase activity was measured in the presence of a fixed concentration of VC1 (2 nM) and increasing concentrations of IIC2. (B) Forskolin-stimulated adenylyl cyclase activity was measured in the presence of a fixed concentration of IIC2 (2 nM) and increasing concentrations of VC1. Adenylyl cyclase activity in (A) and (B) was measured in the presence of 100 μM forskolin and 1 mM [^{32}P]ATP for 10 min at 30°.

high in the absence of an activator ($>>10$ μM) and is reduced in the presence of either forskolin or $G_s\alpha$–GTPγS (~ 10 μM) or both activators together (~ 1 μM) (Fig. 5). It should be noted that the saturating concentrations differ slightly depending on which domain is limiting (Fig. 4A vs. 4B). We speculate that differences in the K_d for heterodimerization (and hence the maximal concentrations required for the assay) arise from differences in the K_d for homodimerization of each domain. Data from Fig. 4 would suggest that the homodimerization constant for IIC2 is higher than from VC1. This would imply that higher concentrations of VC1 are required if IIC2 were constant and limiting in the assay, versus the concentration of IIC2 required if VC1 were limiting.

Other Assay Conditions

The adenylyl cyclase activity of the reconstituted cytoplasmic domains is constant for at least 30 min if the assay temperature is 30° or below (Fig. 6A). Thermal lability of the proteins is evident if assays are performed at 37°. The pH optimum for the assay is 8.0 (Fig. 6B), in good agreement with results obtained with native adenylyl cyclases.[26]

[26] G. G. Hammes and M. Rodbell, *Proc. Natl. Acad. Sci. U.S.A.* **73,** 1189 (1976).

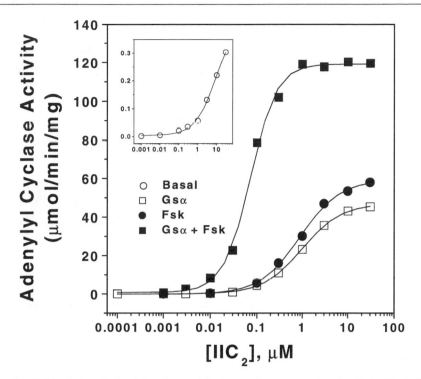

FIG. 5. Stimulation of adenylyl cyclase activity by regulators affects domain affinity. Forskolin (Fsk)-stimulated (100 μM, solid circles), $G_s\alpha$–GTPγS-stimulated (400 nM, open squares), or forskolin- plus $G_s\alpha$–GTPγS-stimulated (solid squares) adenylyl cyclase activity was measured in the presence of a fixed concentration of VC1 (2 nM) and increasing concentrations of IIC2. Measurements of basal enzymatic activity were made with 30 nM VC1. Adenylyl cyclase activity was measured in the presence of 1 mM [^{32}P]ATP for 10 min at 30°.

Adenylyl and guanylyl cyclases both require divalent cations for catalysis.[27,28] X-ray crystallographic studies of the nucleotide-bound catalytic core of adenylyl cyclase reveal two binding sites for Mg^{2+} (sites A and B), both of which are associated with the substrate ATP.[14] In contrast to the membrane-bound proteins, where the effect of Mg^{2+} is relatively constant over a wide range of concentrations (3–100 mM),[29] the soluble domains display a relatively sharp optimum at 3–5 mM (Fig. 6C). The catalytic activity of most isoforms of adenylyl cyclase is also enhanced by Mn^{2+}.[30] Crystallographic studies have identified one of the Mg^{2+}-binding sites (site B) as the site of action of Mn^{2+}.

[27] D. L. Garbers and R. A. Johnson, *J. Biol. Chem.* **250**, 8449 (1975).
[28] D. L. Garbers, E. L. Dyer, and J. G. Hardman, *J. Biol. Chem.* **250**, 382 (1975).
[29] G. Zimmerman, D. Zhou, and R. Taussig, *J. Biol. Chem.* **273**, 19650 (1999).
[30] E. W. Sutherland, T. W. Rall, and T. Menon, *J. Biol. Chem.* **237**, 1220 (1962).

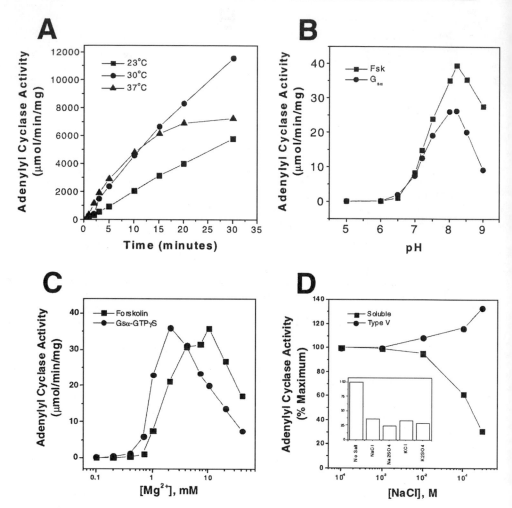

FIG. 6. Effects of time, temperature, pH, Mg^{2+}, and salt on adenylyl cyclase activity. (A) Forskolin-stimulated (100 μM) adenylyl cyclase activity was measured at various times at 23° (solid squares), 30° (solid circles), and 37° (solid triangles). (B) Adenylyl cyclase activity was measured at various values of pH with either 100 μM forskolin or 400 nM $G_s\alpha$–GTPγS. The buffers used in the assay (100 mM) were sodium citrate (pH 5.0), MES (pH 6.0 and 6.5), Na–HEPES (pH 7 to 8.2), and Tris-HCl (pH 8.5 and 9.0). (C) Forskolin (100 μM)-stimulated adenylyl cyclase activity was measured with varying concentrations of MgCl$_2$ and 1 mM [^{32}P]ATP for 10 min at 30°. (D) Inhibition of the $G_s\alpha$–GTPγS (200 nM)-stimulated adenylyl cyclase activity of the soluble domains of adenylyl cyclase by NaCl. Also shown is the effect of NaCl on forskolin (100 μM)-stimulated type V adenylyl cyclase activity in membranes. *Inset:* Inhibition of $G_s\alpha$–GTPγS (200 nM)-stimulated adenylyl cyclase activity of the soluble domains by various salts (300 mM). The data were normalized to the adenylyl cyclase activity observed in the absence of added salt.

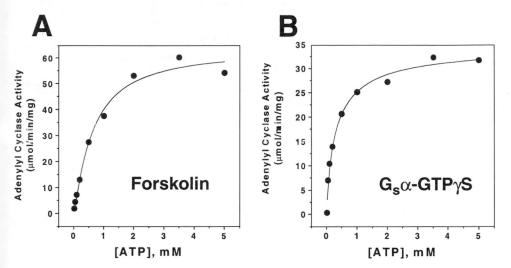

FIG. 7. K_m and V_{max} determinations. The adenylyl cyclase activity of the cytosolic domains was assessed with varying substrate (ATP) concentrations in the presence of either (A) forskolin (100 μM) or (B) $G_s\alpha$–GTPγS (400 nM). Activity was measured with 2 nM VC1, 10 μM IIC2, and 5 mM MgCl$_2$ for 10 min at 30°.

Unlike the activity of membrane-bound adenylyl cyclases, the activity of the soluble domains is sensitive to the concentration of salt. The $G_s\alpha$-stimulated activity of the soluble domains is inhibited substantially by 100 mM NaCl (Fig. 6D); forskolin-stimulated activity is inhibited similarly (not shown). This is in contrast to the behavior of membrane-bound type V adenylyl cyclase, where NaCl has a slight potentiating effect. Other salts behave similarly.

Kinetic Constants

The forskolin-stimulated adenylyl cyclase formed by mixture of VC1 and IIC2 utilizes ATP with a K_m of approximately 0.6 mM; the V_{max} is 60 μmol/min/mg (Fig. 7A). The corresponding constants in the presence of $G_s\alpha$–GTPγS are 0.3 mM and 35 μmol/min/mg, respectively. The K_m for MgATP is thus higher than that observed with membrane-bound adenylyl cyclases (K_m ~40 μM under similar conditions).[31] Care must be taken to avoid consumption of significant amounts of substrate because of the high specific activity of the reconstituted enzyme and the elevated K_m for ATP.

[31] W.-J. Tang, M. Stanzel, and A. G. Gilman, *Biochemistry* **34**, 14563 (1995).

Summary

The identification and isolation of the soluble catalytic domains of adenylyl cyclase have provided investigators with useful reagents for the study of these enzymes. They have permitted detailed mechanistic investigation of the actions of forskolin, $G_s\alpha$, and the inhibitory G protein, $G_i\alpha$.[11] They have served as critical reagents for the development of plausible models of the catalytic mechanism of the enzyme. They have enabled X-ray crystallographic analysis of adenylyl cyclase; this technique was not approachable with the small quantities of the membrane-bound enzyme available previously. The information obtained by using the soluble domains of adenylyl cyclase has provided templates for description of the behavior of many forms of purine nucleotide cyclases from a variety of species. We now appreciate both adenylyl cyclases and guanylyl cyclases as dimeric enzymes with a 2-fold symmetrical domain arrangement (or pseudosymmetrical in the case of heterodimerization). The active sites are located at the interface between the two domains, both of which contribute binding surfaces.

Acknowledgments

We thank Michelle Clark and Julie Collins for excellent technical assistance. This work was supported by grants from the Medical Research Council of Canada (to R.K.S.) and the National Institute of General Medical Sciences (GM34497 to A.G.G.) and by the Raymond and Ellen Willie Distinguished Chair in Molecular Neuropharmacology (A.G.G.).

[11] Identification of Putative Direct Effectors for $G\alpha_o$, Using Yeast Two-Hybrid Method

By J. Dedrick Jordan and Ravi Iyengar

Introduction

Heterotrimeric G proteins enable the cell to receive and process extracellular signals by linking the G-protein-coupled receptors to intracellular signaling networks.[1] When a ligand binds to the receptor a conformational change is induced in the α subunit of the heterotrimeric complex such that the bound GDP is exchanged for GTP. This exchange "activates" the α subunit, which releases the $\beta\gamma$ complex resulting in two independent signaling components. The GTP-bound form of the α subunit is able to bind to and regulate downstream signaling

[1] A. G. Gilman, *Annu. Rev. Biochem.* **56,** 615 (1987).

components. The identification of direct binding partners for the α subunits has enabled the building of complex signaling networks that control many aspect of cellular function.[2]

To identify effectors for the α subunit of the G_o heterotrimeric G protein, a yeast two-hybrid strategy was utilized.[3,4] The constitutively activated form of the α_o subunit was used in the screen. This form is constitutively in the activated state because of a mutation that abolishes the intrinsic GTPase activity and therefore the α subunit is bound preferentially to GTP versus GDP. Because this is the activated state of the G protein α subunit it should preferentially interact with direct downstream signaling molecules. We used an embryonic day 12 chick dorsal root ganglion (DRG) library that we had constructed to carry out the screen.[5] Here we describe the methods, depicted as a flow chart in Fig. 1, used to identify putative effectors for the $G\alpha_o$ subunit.

Materials

The yeast two-hybrid system is available from Life Technologies (Rockville, MD) as the ProQuest two-hybrid system.[4] All media including YPD (yeast extract–peptone–dextrose) and SD (synthetic defined) dropout reagents are obtained from Clontech (Palo Alto, CA). Primers are synthesized by MWG Biotech (High Point, NC).

Small-Scale Yeast Transformation

To perform the two-hybrid control reactions the α_o^* bait and library plasmid must be introduced into the yeast strain MaV203.[6,7] This is done by transforming the yeast strain by the lithium acetate method.[8] The yeast strain MaV203 is streaked onto a YPD plate and incubated at 30° for 48 hr. Several colonies of yeast are suspended in 100 μl of sterile 0.9% (w/v) saline solution and spread onto a fresh YPD plate in a 2×2 cm patch. The plate is incubated overnight at 30°. The patch of yeast is then scraped into several milliliters of 0.9% (w/v) saline and

[2] J. D. Jordan, E. M. Landau, and R. Iyengar, *Cell* **103,** 193 (2000).

[3] S. Fields and O. Song, *Nature (London)* **340,** 245 (1989).

[4] M. Vidal, *in* "The Yeast Two-Hybrid System" (P. L. Bartel and S. Fields, eds.), p. 109. Oxford University Press, New York, 1997.

[5] J. D. Jordan, K. D. Carey, P. J. Stork, and R. Iyengar, *J. Biol. Chem.* **274,** 21507 (1999).

[6] M. Vidal, P. Braun, E. Chen, J. D. Boeke, and E. Harlow, *Proc. Natl. Acad. Sci. U.S.A.* **93,** 10321 (1996).

[7] M. Vidal, R. K. Brachmann, A. Fattaey, E. Harlow, and J. D. Boeke, *Proc. Natl. Acad. Sci. U.S.A.* **93,** 10315 (1996).

[8] H. Ito, Y. Fukuda, K. Murata, and A. Kimura, *J. Bacteriol.* **153,** 163 (1983).

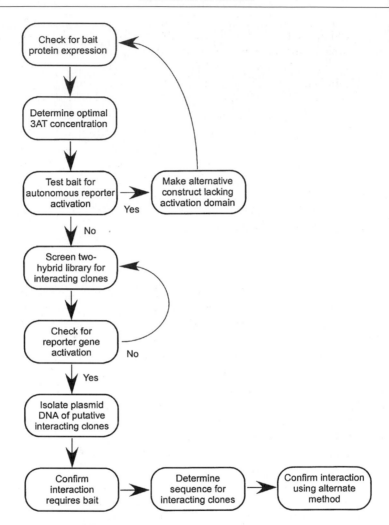

FIG. 1. Flow chart depicting the two-hybrid screening procedure.

completely resuspended. Enough of this suspension is then added to 50 ml of YPD liquid medium in a 250-ml flask until the OD_{600} is approximately 0.09–0.11. The culture is incubated at 30° in an orbital shaker at 225 rpm until the OD_{600} is 0.4. The culture is then centrifuged at 3000g for 5 min at room temperature to pellet the cells. The supernatant is removed; the pellet is resuspended in 20 ml of sterile 0.9% (w/v) saline and again centrifuged at 3000g for 5 min at room temperature. The supernatant is then removed and the pellet is resuspended in 10 ml of TE–LiAc [10 mM Tris-HCl (pH 7.5), 1 mM EDTA, and 0.1 M lithium acetate] and again centrifuged at 3000g for 5 min at room temperature. The supernatant is removed and the pellet is resuspended in 200 μl of TE–LiAc. For each small-scale transformation the following are combined in a sterile microcentrifuge tube: 50 μl of cells, 50 ng of each plasmid DNA, and 5 μl of carrier DNA (freshly boiled for 5 min followed by incubation on ice for 5 min) (Clontech) and mixed by gentle pipetting. Three hundred microliters of polyethylene glycol (PEG)–LiAc [40% (w/v) PEG 3350, filter sterilized; 0.1 M lithium acetate] is added to each tube and mixed by gentle inversion. The tubes are incubated in a 30° water bath for 30 min, followed by heat shock in a 42° water bath for 15 min. The tubes are then centrifuged at 5000g for 1 min at room temperature to pellet the cells. The supernatant is removed and the pellet is resuspended in 0.5 ml of 0.9% (w/v) saline. One hundred microliters of the transformation mixture is then plated on the appropriate dropout medium and incubated for 2–3 days at 30°. Resultant colonies contain both the pDBLeu and pPC86 plasmids.

Construction of $G\alpha_o^*$ Two-Hybrid Bait

The constitutively activated α subunit of G_o (rat species) is amplified by polymerase chain reaction (PCR), using *Pfu* polymerase and the following primers: 5′-ACGTGTCGACCATGGGATGTACTCTGAGC-3′ and 5′-AGCTTAGCGGC-CGCTCAGTACAAGCCACAGCC-3′. These primers encode a *Sal*I restriction site and a *Not*I restriction site, respectively. The PCR product and two-hybrid bait vector pDBLeu (Life Technologies) are digested with *Sal*I and *Not*I and gel purified. The α subunit is subcloned into the bait vector and sequence verified. To verify that the α_o^* subunit is expressed, the MaV203 yeast strain (Life Technologies) is transformed with the empty library vector pPC86 plus either pDBLeu or pDBLeu-α_o^*, using the lithium acetate method mentioned above. Once individual colonies are obtained one is resuspended in 1 ml of SD/−Leu−Trp dropout medium and grown at 30°. The yeast is grown in dropout medium until an OD_{600} of 0.6 and harvested by centrifugation at 5000 g at room temperature for 5 min. Yeast are then lysed in 100 μl of sodium dodecyl sulfate (SDS) protein sample buffer containing dithiothreitol (DTT). Approximately 50 μg of protein is then separated by SDS–polyacrylamide gel electrophoresis (PAGE) and immunoblotted with an α_o-specific antibody (Santa Cruz Biotechnology, Santa Cruz, CA).

Yeast Two-Hybrid Controls

Once the α_o^* construct expression is verified it is necessary to perform control reactions to determine the amount of leakiness of the histidine (His) reporter as well as whether there is any autonomous activation of the other reporter genes. For these controls yeast are transformed with either the empty pDBLeu vector or pDBLeu-α_o^* plus the empty library vector pPC86, plated on SD/−Leu−Trp dropout medium, and grown for 2–3 days at 30°. To determine the correct 3-amino-1, 2,

4-triazole (3-AT) concentration for library screening in the SD/−Leu−Trp−His medium it is necessary to do a titration curve with several concentrations of 3-AT (Sigma, St. Louis, MO)(3-AT suppresses background growth on −His dropout medium). A series of dropout plates are made containing increasing concentrations of 3-AT from 0 to 100 mM. A single colony containing pPC86 and either pDBLeu or pDBLeu-α_0^* is picked and streaked onto each of these plates and incubated for 2–3 days. After that time the plates are examined for growth and the lowest concentration that gives no growth is used from this point on when making SD/−Leu−Trp−His dropout plates. For the screening of α_0^* interacting proteins the concentration used is 25 mM 3-AT.

It is also necessary to determine whether the expression of α_0^* is able to autonomously activate gene expression of β-galactosidase (β-Gal). The same plate from the 3-AT screening is used to streak an SD/−Leu−Trp plate, using three colonies of both the control and α_0^*, each also containing the empty pPC86 vector. This plate is incubated for 3 days at 30°. For the β-Gal assay a filter assay is used. A nitrocellulose filter (MSI, Westboro, MA) is placed onto a YPD plate. The plate containing the streaked colonies is then replica plated onto the filter, which has been placed onto the YPD plate. The plate containing the filter is then incubated for 18–20 hr at 30°. The filter is then removed from the plate and air dried for 2–3 min. The β-galactosidase assay reagent is prepared by dissolving 5 mg of 5-bromo-5-chloro-3-indolyl-β-D-galactopyranoside (IPTG; Sigma) in 100 μl of dimethyl formamide (DMF). The X-Gal solution is then dissolved in 5 ml of assay buffer (60 mM Na$_2$HPO$_4$, 40 mM NaH$_2$PO$_4$, 10 mM KCl, and 1 mM MgSO$_4$, pH 7.0) plus 30 μl of 2-mercaptoethanol. Two 15-cm Whatman (Clifton, NJ) filters are then placed into a 150-mm petri dish and soaked with the assay reagent. Once the filter is dry it is submerged in liquid nitrogen for 30 sec and then placed colony side up onto the X-Gal reagent-soaked Whatman filter. The dish is then incubated at 30° and monitored intermittently for 24 hr. The presence of blue colonies indicates that the β-Gal reporter is activated as depicted in Fig. 2.

Library Screening

Once it has been determined that there is no autonomous reporter gene activation the library screening can begin. A chick neuronal library has already been prepared and cloned into the yeast *GAL4* activation domain plasmid pPC86. Premade competent yeast strain MaV203 (Life Technologies) is used. The PEG–LiAc solution is thawed out at room temperature and mixed well. The competent cells are thawed by placing them in a 30° water bath for 1 min. Five hundred microliters of cells is aliquoted into a 15-cm sterile tube. Ten micrograms of the pDBLeu-α_0^* and 10 μg of the pPC86 library are added to the cells and mixed by swirling followed by the addition of 1.5 ml of the PEG–LiAc solution. The tube is swirled gently

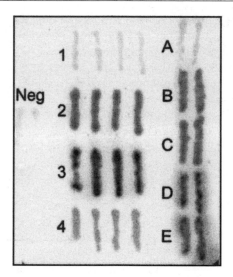

FIG. 2. β-Galactosidase filter assay. Yeast were transferred to a nitrocellulose membrane and grown on a YPD plate for 18 hr at 30°. The membrane was submerged in liquid nitrogen and then placed onto a Whatman filter that had been soaked in X-Gal assay solution. The membrane was incubated at 30° for 24 hr to allow color development. On the left is the negative control, which is the pDBLeu vector only. On the right (labeled A–E) are the two-hybrid controls included with the ProQuest two-hybrid kit. A is a negative control whereas B–E are positive controls, with B encoding a weak interaction, C a medium interaction, D a strong interaction, and E a very strong interaction. Numbered 1–4 are independent putative interacting clones obtained with α_0^* as a bait. The variability in interaction strength among the different interacting proteins can be seen in the experimental results.

to mix and then placed in a 30° water bath for 30 min. After the incubation 88 μl of sterile dimethyl sulfoxide (DMSO) is added to the cells and swirled to mix. The cells are heat shocked by placing the tube in a 42° water bath for 20 min with occasional swirling to mix the cells. The tube is centrifuged at 600 g for 5 min at room temperature to pellet the cells. The supernatant is aspirated and the pellet is resuspended in 15 ml of 0.9% (w/v) sterile saline. Four hundred-microliter aliquots are then plated on approximately 35–40 plates containing 150 mm SD/−Leu−Trp−His plus 25 mM 3-AT. The plates are incubated for 2 days at 30° followed by replica cleaning. The plates are then incubated for an additional 3 days at 30°. Colonies that grow to more than 1 mm in diameter are considered positive clones.

Characterization of Positive Clones

Once positive clones have been obtained it is necessary to characterize the reporter gene activation. Clones that grow on the initial library screen are streaked onto fresh SD/−Leu−Trp plates and grown at 30° for 2 days. Three individual colonies for each positive clone are then streaked onto an SD/−Leu−Trp plate and incubated for 18–20 hr. The streaked plate is then replica plated onto a YPD plate containing a nitrocellulose membrane as well as an SD/−Leu−Trp−His plus 3-AT and an SD/−Leu−Trp−Ura plate. These three plates are then incubated for 24 hr at 30°. The nitrocellulose membrane is then used for a β-galactosidase assay according to the above-mentioned protocol. The other two plates are replica cleaned and incubated for an additional 48 hr. After this time the plates are analyzed for growth on the dropout medium and compared with the β-galactosidase filter assay. Positive clones should indicate a positive β-galactosidase assay, growth on the SD/−Leu−Trp−His plus 3-AT medium, and growth on the SD/−Leu−Trp−Ura dropout medium. Those that fulfill at least two of the characteristics are further evaluated.

Isolation of Plasmid DNA Encoding Putative Interacting Proteins

It is necessary to isolate the plasmid DNA specific for the pPC86 library clones obtained in the library screening.[9] A colony of each candidate clone is inoculated into 2 ml of SD/−Leu−Trp and grown overnight until saturation at 30° with shaking. The yeast cells are then pelleted by centrifugation in a microcentrifuge tube at 13,000g for 2 min at room temperature. The medium is removed and the pellet is resuspended in 100 μl of lysis buffer [2% (w/v) SDS, 0.2 N NaOH] and incubated for 10 min at room temperature. One-half milliliter of TE [10 mM Tris-HCl (pH 7.5), 1 mM EDTA] is added and mixed gently followed by the addition of 60 μl of 3 M sodium acetate. After mixing, 650 μl of Tris-buffered phenol–isoamyl alcohol (25 : 24 : 1, v/v/v) is added. The solution is vortexed for 1 min followed by centrifugation at 13,000g for 2 min at room temperature. The upper aqueous phase is transferred to a new microcentrifuge tube. The phenol extraction is repeated two more times as described above. To the aqueous phase, 650 μl of 2-propanol is added and mixed by repeated inversions followed by incubation for 15 min at −20°. The tube is then centrifuged at 13,000g for 10 min at 4°. The supernatant is removed and the pellet is rinsed with 200 μl of 70% (v/v) ethanol followed by centrifugation at 13,000g for 10 min at 4°. The supernatant is again discarded and the pellet is allowed to air dry for 15–20 min. Once all residual ethanol has evaporated the pellet is resuspended in 15 μl of sterile doubly distilled H_2O. One microliter of the DNA solution is then used to electroporate 40 μl of electrocompetent cells and plated on an LB plus ampicillin (100 μg/ml) plate. Resultant clones should harbor the pPC86 library plasmid encoding the putative interacting protein.

[9] J. Polaina and A. C. Adam, *Nucleic Acids Res.* **19,** 5443 (1991).

Confirmation of Putative Interacting Proteins

Once candidate plasmids have been isolated it is necessary to reintroduce them into the yeast strain to determine whether the reporter gene activation is dependent on the interaction with the α_o^* bait. This is done by introducing the pPC86 candidate plasmid with either the pDBLeu vector only or the pDBLeu-α_o^* construct. The transformation is done using the small-scale transformation. The yeast is plated onto SD/−Leu−Trp and independent clones are isolated. These clones are then tested for their ability to activate the transcription of the reporter genes, as was done in the characterization of positive clones. Clones that activate reporter transcription only in the presence of the pDBLeu-α_o^* are considered to encode interacting proteins.

Specificity of Interactions

Once it has been shown that the interaction is dependent on the presence of the α_o^* subunit further experiments can be done to determine the specificity of the interaction. The first set of experiments is to determine whether the interaction is specific for the activated form of the α_o subunit that is mimicked by the constitutively activated form of α_o. This is done by transforming the yeast strain with the pPC86 library clone with either the wild-type or constitutively activated form of α_o in the pDBLeu vector and testing for reporter activation. Most proteins identified that interact with α_o^* are specific for the activated form versus the wild-type form, as would be expected for effector proteins.[5] An example of one such interacting clone is shown in Fig. 3.

It is also informative to determine the specificity for the different α subunit families. This can be accomplished by making pDBLeu constructs containing the following constitutively activated forms of the following α subunits: α_{i2}, α_o, α_z, α_s, α_q, and α_{12}. These α subunits are then transformed with and without the putative interacting clone and reporter assays are performed to determine the strength of the interaction. Putative interacting proteins identified thus far for α_o^* have been specific for the $\alpha_{i/o}$ family.[5]

Two-Hybrid Clone Identification and Independent Confirmation

Once clones have been confirmed by the reintroduction of bait and prey plasmids into the yeast it is necessary to determine the sequence identification of the clones. This can be done by sequencing the minipreparation plasmid DNA obtained after the electroporation and isolation step in the above-described confirmation assay. The sequencing primer used is within the *GAL4* activation domain about 100 bp upstream of the *Sal*I insert site. The primer sequence used

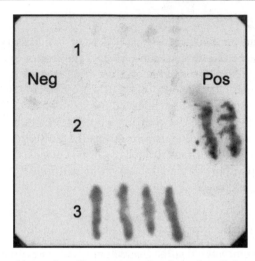

FIG. 3. Specificity of interactions. Yeast were transferred to a nitrocellulose membrane and grown on a YPD plate for 18 hr at 30°. The membrane was submerged in liquid nitrogen and then placed onto a Whatman filter that had been soaked in X-Gal assay solution. The membrane was incubated at 30° for 24 hr to allow color development. On the left is the negative control, which is the pDBLeu vector only. On the right is a positive control. Colony 1 is the library clone plus the empty pDBLeu vector, colony 2 is the library clone plus pDBLeu-α_o (wild type), and colony 3 is the library clone plus pDBLeu-α_o^*. As can be seen by the color development the interaction is selective for the constitutively activated form of Gα_o.

is 5′-AACGCGTTTGGAATCACT-3′ and standard automated sequencing is performed. Once sequence data are obtained the reading frame is in-frame with the *GAL4* activation domain and can be translated as such. The translated sequence is then used for conventional sequence analysis, using BLAST algorithms.

It is still necessary to confirm independently that the predicted interaction occurs. This can be done by several methods. The bait and prey can be swapped and two-hybrid assays performed. In addition, assays can be performed in mammalian cells to validate the interactions. One way of using a mammalian system is to perform coimmunoprecipitations of the bait and prey in the 293T cell line.[5] If antibodies are not readily available tagged constructs can be assembled and used for immunoprecipitations and immunoblotting. A second way of validating the interaction in a mammalian system is by using a glutathione *S*-transferase (GST) pull-down assay. This can be performed by tagging the bait with a GST tag, expressing it in bacteria, and affinity purifying it on a GST column. The GST–bait construct can then be mixed with 293T cell lysate exogenously expressing the prey construct and allowed to incubate overnight. The resultant complexes can then be isolated with GST beads, separated by SDS–PAGE, and immunoblotted for the interacting proteins.

TABLE I
RESULTS OF $G\alpha_o^*$ YEAST TWO-HYBRID SCREEN

$G\alpha_o$-interacting protein	GenBank accession number
Rap1GAP	AF151966
GRIN3	AF151734
GzGAP	AF151967
RGS-17	AF151968
Paladin	X99384

Results

We screened an embryonic day 12 chick DRG library with the constitutively activated form of $G\alpha_o$. The system that we used is now available from Life Technologies as the ProQuest two-hybrid system. This system has the benefit of having few false positives when used with the MaV203 yeast strain. In our screen we had only one false positive after screening 5×10^6 clones. The positive clones isolated are listed in Table I. Several of these clones were isolated independently of our laboratory, using both yeast two-hybrid as well as phage display techniques.[10-12]

Concluding Remarks

With the identification of additional $G\alpha$-interacting proteins the role that G proteins play in signal transduction will inevitably increase, as will the complexity of signaling in general. The yeast two-hybrid method aids in this process by providing a powerful tool to help elucidate the roles that G proteins play in signaling by enabling the identification of novel G protein-interacting proteins.

Acknowledgments

Research in our laboratories is supported by NIH Grants GM-54508 and DK-38761 to R.I. J.D.J. is supported by the MSTP and currently by an individual predoctoral fellowship (DA-05798).

[10] L. T. Chen, A. G. Gilman, and T. Kozasa, *J. Biol. Chem.* **274,** 26931 (1999).
[11] N. Mochizuki, Y. Ohba, E. Kiyokawa, T. Kurata, T. Murakami, T. Ozaki, A. Kitabatake, K. Nagashima, and M. Matsuda, *Nature (London)* **400,** 891 (1999).
[12] J. Meng, J. L. Glick, P. Polakis, and P. J. Casey, *J. Biol. Chem.* **274,** 36663 (1999).

[12] Identification of Transmembrane Adenylyl Cyclase Isoforms

By MARTIN J. CANN and LONNY R. LEVIN

Introduction

Cyclic AMP (cAMP), the second messenger ubiquitous throughout the animal and bacterial kingdoms, is generated from ATP by adenylyl cyclases (ACs). Mammals express two distinct classes of AC. One class, represented by the mammalian soluble adenylyl cyclase (sAC), is evolutionarily related to cyanobacterial cyclases[1] and appears to represent the more ancient of the mammalian ACs. The second class is the more widely studied, G-protein-responsive, transmembrane adenylyl cyclases (tmAC).

Molecular cloning and database analysis have so far identified a single isoform of the cytosolic form in mammalian species. Among other eukaryotes, a single related cyclase, AcrA, has thus far been identified in *Dictyostelium*.[2] The limited number of known members of this family restricts methods for finding related cyclases to low-stringency cloning and database searches.

In contrast to sAC, molecular cloning has revealed the existence of at least nine distinct tmAC isoforms in mammals. Characterized by 12 membrane-spanning domains and 2 homologous catalytic domains, tmAC isoforms have been cloned by a variety of methods including low-stringency hybridization and polymerase chain reaction (PCR) using degenerate primers recognizing conserved sequence motifs (reviewed in Premont[3]). The nine characterized mammalian tmAC isoforms differ in their patterns of expression and regulatory properties (reviewed in [10] in this volume[3a]), and gene knockout studies are underway to elucidate their individual physiological significance.[3b]

The cAMP signaling pathway has been studied in a number of genetically tractable model systems, but understanding the physiological significance of the many mammalian tmAC isoforms necessitates an organism displaying similar complexity and diversity of isoforms.[4] Among nonmammalian eukaryotic model systems, tmAC-related isoforms have been identified in *Dictyostelium, Caenorhabditis elegans, Drosophila melanogaster,* and *Xenopus*. Each of these organisms

---

[1] J. Buck, M. L. Sinclair, L. Schapal, M. J. Cann, and L. R. Levin, *Proc. Natl. Acad. Sci. U.S.A.* **96**, 79 (1999).

[2] F. Soderbom, C. Anjard, N. Iranfar, D. Fuller, and W. F. Loomis, *Development* **126**, 5463 (1999).

[3] R. T. Premont, *Methods Enzymol.* **238**, 116 (1994).

[3a] M. E. Hatley, A. G. Gilman, and R. K. Sunahara, *Methods Enzymol.* **345**, [10] 2001 (this volume).

[3b] S. T. Wong and D. R. Storm, *Methods Enzymol.* **345**, [16] 2001 (this volume).

[4] M. J. Cann and L. R. Levin, *Adv. Second Messenger Phosphoprotein Res.* **32**, 121 (1998).

offers various advantages, which make them attractive for studying the molecular basis of the cAMP signaling pathway. A challenge for studying tmAC function in model organisms is in identifying the number of isoforms they express and relating those isoforms to the known mammalian isozymes. Furthermore, molecular and genetic studies of model organisms have identified tmAC-like molecules that have not been identified in mammals. It remains unclear whether these novel tmAC forms are unique to *Dictyostelium*[5] or *Drosophila*,[6,7] or whether they have yet to be found in mammalian systems. In this article we describe methods to identify new tmAC genes and to study the complete tmAC repertoire in an organism.

Cloning Transmembrane Adenylyl Cyclase Genes

The degenerate PCR method described by Premont[3] is still relevant for identifying novel tmAC genes and for determining the tmAC profile of an organism. However, assessing whether all the tmAC-related sequences within a source organism have been identified presents a problem. Once a number of tmAC genes have been identified in the source organism, novel or rarely expressed tmACs may be difficult to identify among abundant or preferentially amplified "known" genes. To streamline the effort required to identify previously unknown or rare tmAC sequences, we developed a protocol to quickly select unknown sequences from a mixture of PCR products.

When we started our analysis of *Drosophila* tmAC (DAC) diversity, Rutabaga/ DAC1 was the only identified insect AC.[8] We also had genomic fragments believed to correspond to three independent DACs isolated by low-stringency hybridization to mammalian tmAC1 and tmAC2.[8] The DAC genes represented on these genomic fragments were subsequently cloned by library screening and renamed DAC76E (cytologically mapped to band 76E on the *Drosophila* third chromosome),[4] DACXA (previously reported to be AC-62D), and DACXB (cytologically mapped to band 34A).[6]

As first described by Premont,[3] alignment of Rutabaga/DAC1 with the known mammalian ACs revealed at least four short stretches (each consisting of approximately five to seven amino acids) conserved in all known metazoan cyclases. Two of these blocks of absolutely conserved residues were in the first catalytic domain, C_{1a}, separated from each other by 70 amino acids, and the other two were in C_{2a} separated by approximately the same distance. Using fully degenerate PCR primers for the sequence in C_{1a}, we identified a piece of a novel *Drosophila* tmAC

---

[5] G. S. Pitt, N. Milona, J. Borleis, K. C. Lin, R. R. Reed, and P. N. Devreotes, *Cell* **69**, 305 (1992).

[6] M. J. Cann, E. Chung, and L. R. Levin, *Dev. Genes Evol.* **210**, 200 (2000).

[7] M. J. Cann and L. R. Levin, *Dev. Genes Evol.* **210**, 34 (2000).

[8] L. R. Levin, P. L. Han, P. M. Hwang, P. G. Feinstein, R. L. Davis, and R. R. Reed, *Cell* **68**, 479 (1992).

subsequently named DAC39E,[9] and using degenerate C_{2a} primers, we identified a piece of the *Drosophila* tmAC subsequently named DAC78C.[7]

The DAC39E and DAC78C fragments were isolated during our initial screen of reverse transcription PCR (RT-PCR)-generated fragments, which consisted of simply picking and sequencing individually cloned products. These represented the only new DAC genes identified after sequencing approximately 72 RT-PCR products from both primer pairs; all other RT-PCR products sequenced represented previously identified DACs. To expand our search without having to sequence hundreds of cloned RT-PCR products, we developed the strategy outlined in Fig. 1. In this method, RT-PCR products from each primer pair were TA cloned *en masse* into pCR2.1 (Invitrogen, Carlsbad, CA). Transforming this ligation into supercompetent bacteria resulted in a large number of colonies, each containing an independently derived RT-PCR product. A small amount of cells from each colony containing an insert* was screened by performing a second PCR amplification with vector primers flanking the insert. In this step, it is important to use flanking primers to minimize cross-hybridization to the probe during subsequent Southern blotting. Using our standard medium-sized agarose gel boxes, up to 80 PCR-amplified inserts were loaded onto a single agarose gel, Southern blotted, and hybridized at high stringency to a mixed probe containing the analogous regions of all known DAC genes.[†] Any PCR products not recognized by the mixed probe were candidates for novel DAC genes. The colonies corresponding to these candidate inserts were grown, miniprepped, and sequenced. Using this strategy, we screened approximately 400 independently derived RT-PCR products from both primer pairs (C_{1a} and C_{2a}), and we identified a rare RT-PCR product that corresponded to DAC9[10] with minimal DNA sequencing.

The sequences of full-length cDNA clones corresponding to DACXA and DACXB revealed that these cyclases were sufficiently diverged from other metazoan tmACs such that they would not be recognized by our C_{1a} and C_{2a} primer pairs.[6] The quick screen was repeated, using a newly designed set of degenerate PCR primers specific for DACX-like cyclases, and two additional members of this subfamily, DACXC and DACXD, were identified.[6]

Genomic Resources

New tmAC genes can also be identified from genomic resources via database searches. Considerable effort is currently being made to sequence the genomes

---

* Insert-containing colonies can be determined by "blue–white selection" according to the manufacturer instructions.

† At that time, known DAC genes included Rutabaga/DAC1, DAC78C, DAC39E, DAC76E, DACXA, and DACXB.

[9] V. Iourgenko and L. R. Levin, *Biochim. Biophys. Acta* **1495,** 125 (2000).

[10] V. Iourgenko, B. Kliot, M. J. Cann, and L. R. Levin, *FEBS Lett.* **413,** 104 (1997).

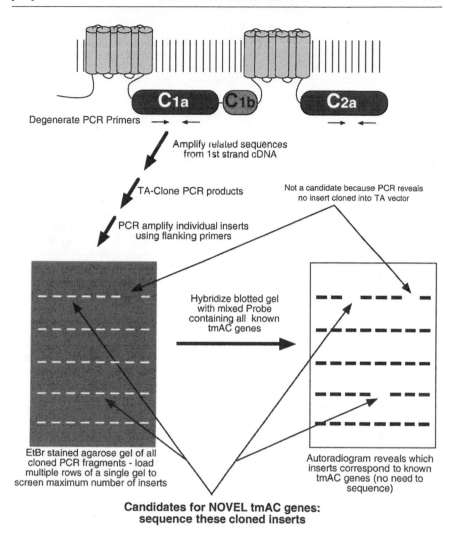

FIG. 1. Screen for selecting new tmAC genes.

of a number of model organisms, and completed and annotated genomes are now available for two widely used members of the animal kingdom, *C. elegans* and *D. melanogaster*. In addition, there are accumulating databases of short cDNA segments, known as expressed sequence tags (ESTs), from a wide variety of organisms, that can be easily searched. Annotated genome resources considerably ease the identification of gene family repertoires; however, the ability to identify complete gene families from genomic resources is only as good as the annotation and the methods used to search.

TABLE I
TRANSMEMBRANE ADENYLYL CYCLASES FROM *Drosophila* GENOME[a]

GadFly	Gene	Map	Transcript	Exons	Length
CG5983	DACXC	2L	CT18649	12	4,148
CG17174	DACXB	2L	CT38106	7	3,326
CG17176	DACXA	2L	CT38108	8	3,686
CG17178	DACXE	2L	CT38110	10	3,920
CG1506	DAC39E	2L	CT3791	19	15,369
CG5712	DACXD	3L	CT6304	10	5,195
CG8970	Unknown	3L	CT25794	23	8,200
CG12243	Unknown	3L	CT14186	23	20,620
CG7978	DAC76E	3L	CT23940	19	38,594
CG10564	DAC78C	3L	CT29620	24	35,481
CG9533	Rutabaga	X	CT26958	16	11,932
CG9210	DAC9	X	CT26306	5	10,714

[a] *GadFly* and *transcript* annotations correspond to the GadFly accession numbers for the presumed amino acid and cDNA sequences, respectively. *Gene* presents the corresponding gene isolated by this laboratory (Unknown indicates the presumed gene was not previously identified). *Map* corresponds to the chromosome on which the gene is found, *Exons* denotes the number of presumed exons in the gene, and *Length* indicates the presumed size of the locus, in base pairs.

The annotated *C. elegans* genome can be accessed via the World Wide Web (WWW) at http://www.sanger.ac.uk/Projects/C_elegans/. Using this database we identified three tmACs. Two of the three had been previously published[11,12] whereas the remaining sequence was novel. The annotated *Drosophila* genome can be accessed via the WWW, using GadFly (http://www.flybase.org/annot/). Twelve putative tmACs were identified in this database (Table I). Ten of these tmACs had previously been identified by methods described above whereas the remaining two (CG8970 and CG12243) represented novel sequences.

As stated above, when identifying new tmACs it is important to determine how they relate to the already characterized isoforms. Relationships between tmACs can be determined on the basis of biochemical properties, but even before isolating the full-length cDNA or developing conditions for heterologous expression and enzyme purification, a great deal can be learned by determining the evolutionary relatedness among the known tmAC isoforms. For such analysis, all that is required is the sequence of the tmAC isoform.

Phylogenetic Analysis

Phylogenetic analysis is a powerful tool that enables the study of relationships among sequences and is an essential step in inferring possible similarities between

[11] A. J. Berger, A. C. Hart, and J. M. Kaplan, *J. Neurosci.* **18,** 2871 (1998).
[12] H. C. Korswagen, A. M. van der Linden, and R. H. Plasterk, *EMBO J.* **17,** 5059 (1998).

newly identified tmACs and those previously cloned. There are a number of analysis methods based on different algorithms that represent different assumptions made for inferring relatedness. We used PHYLIP (*Phylogeny Inference Package*), which is a freeware multiplatform computer package that contains tools covering most of the commonly used methods.[13] PHYLIP software and documentation can be obtained through the WWW at http://www.ibb.waw.pl/docs/PHYLIPdoc/.

The different regions of the tmAC gene, for example, transmembrane and catalytic domains, may be under differing evolutionary pressures. Therefore, to simplify and speed analysis we examined a fusion of the first and second catalytic domains. This also negated the effects of the differing lengths of N and C termini that may erroneously skew phylogenetic relationships. DNA and amino acid sequences corresponding to the C1a–C2a concatamers for mammalian tmAC types 1–9 and all the identified *Drosophila* and *C. elegans* tmACs were aligned using CLUSTAL W in DNAStar. A freeware program for aligning sequences using the CLUSTAL W algorithm can also be obtained via the internet at http://www2.ebi.ac.uk/clustalw. An outgroup was also included in the alignment. An outgroup is a related sequence that is not a member of the group under study and provides a reference point when performing the phylogenetic analysis. Selecting a sequence to use as an outgroup is a subjective process. We used a tmAC from *Dictyostelium* because it is structurally and functionally related, but its sequence is sufficiently diverged to define a separate subfamily. The CLUSTAL W sequence alignment was reformatted for use by the PHYLIP software, using READSEQ obtained via anonymous ftp from D. Gilbert at the University of Indiana (Bloomington, IN; ftp.bio.indiana.edu/molbio/readseq).

There are three main classes of phylogenetic analysis methods; parsimony, maximum likelihood, and distance. PHYLIP offers software for analysis of both DNA and amino acid sequences, using all three methods. PHYLIP also includes statistical analysis of obtained phylogenies (SEQBOOT), which is important for judging the validity of any inferred relationships. In our example, 100 data sets for each amino acid and DNA sequence alignment were analyzed by SEQBOOT.

Figure 2 shows a flow diagram that traces typical paths that can be taken in PHYLIP version 3.5 to obtain phylogenies, using the three main methods for DNA or amino acid sequences. A more detailed treatment of the PHYLIP package is provided by Retief.[14]

Parsimony Analysis

Parsimony is a character-based analysis (i.e., the program examines the sequence directly) that considers only those changes found in at least two sequences

---

[13] J. Felsenstein, "PHYLIP—Phylogeny Inference Package." Department of Genetics, University of Washington, Seattle, Washington, 1993.

[14] J. D. Retief, *In* "Bioinformatics: Methods and Protocols" (S. Misener and S. A. Krawetz, eds.), Vol. 132. Humana Press, Totowa, New Jersey, 2000.

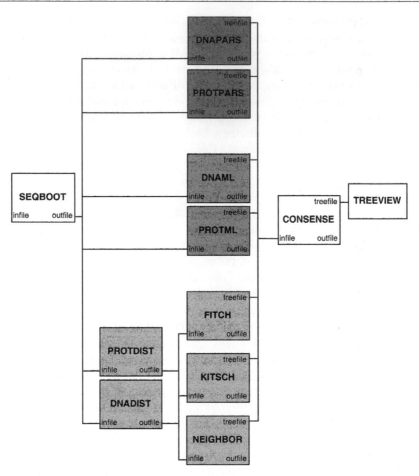

FIG. 2. Flow diagram demonstrating the combinations of programs for generating phylogenies by Parsimony analysis (darkest gray shading), Maximum Likelihood analysis (medium gray), and Distance analysis (lightest gray). Each box has the name of the computer program placed centrally. The required name of the input file is indicated in the lower left-hand corner of each box. The names of the output files generated are indicated at the right-hand side of each box. The diagram moves from left to right and indicates which files must be carried over into the next program. Output files carried over to the next program must be renamed according to the required input for the next box. [Adapted from Ref. 14.]

(those that are considered "informative" sites). Parsimony for DNA and protein is analyzed by the programs DNAPARS and PROTPARS, respectively. In both cases, the output file is copied from SEQBOOT into INFILE. Each data set is examined by using the jumble option to randomize the input order of sequences and remove any inherent bias this may introduce. The output file, called TREEFILE is then copied into INFILE to run CONSENSE. CONSENSE generates a consensus phylogenetic tree from all data sets and applies bootstrap values to each node of the tree. The bootstrap value refers to the number of times, out of the total number of data sets, that the sequences to the right of that node are grouped together. Hence, the higher the bootstrap value, the more reliable the grouping. We subsequently generated phylogenetic trees from these data by using TreeView,[15] a phylogeny drawing package (available at http://taxonomy.zoology.gla.ac.uk/rod/treeview.html).

Maximum Likelihood Analysis

Maximum likelihood is also a character-based analysis, but it considers both informative and noninformative sites. Likelihood analysis is performed by DNAML and PROTML for DNA and amino acid sequences, respectively. The programs are used in an essentially similar fashion to DNAPARS and PROTPARS except for the additional option G which is an optimization procedure for reexamining branch placement. As described above, data sets are run through CONSENSE and examined by using TreeView.

Distance Analysis

Distance methods utilize a single set of programs for DNA and protein sequence analysis because they rely on a pregenerated distance matrix. These matrices are first constructed with the SEQBOOT output in PROTDIST and DNADIST for amino acid and DNA sequences, respectively. FITCH, KITSCH, and NEIGHBOR are similar distance tree-building methods that use the output of PROTDIST/DNADIST and differ slightly in the assumptions used for tree building. Outputs from these programs are run through CONSENSE and TreeView as described above.

Transmembrane Adenylyl Cyclase Phylogenies

To illustrate the differences between tree-building methods, we show two different phylogenetic trees of metazoan tmAC isoforms (Fig. 3) constructed by FITCH (distance method) and PROTPARS (parsimony method). The overall appearances of the trees are similar, and the consistencies between the two trees support previous arrangements of tmAC isoforms into a smaller number of subfamilies. Phylogenetic assignment into one of these subfamilies may provide clues to the biochemical properties of newly identified tmACs and may reveal functional relationships after genetic analysis of tmAC isoforms in model organisms.

The similarities between the two trees generated for tmACs confirm that the *Drosophila* Rutabaga tmAC is an ortholog of mammalian tmAC1[8]; *Drosophila* DAC76E and *C. elegans* ACY-2 each represent single orthologs of the mammalian

---

[15] R. D. Page, *Comput. Appl. Biosci.* **12,** 357 (1996).

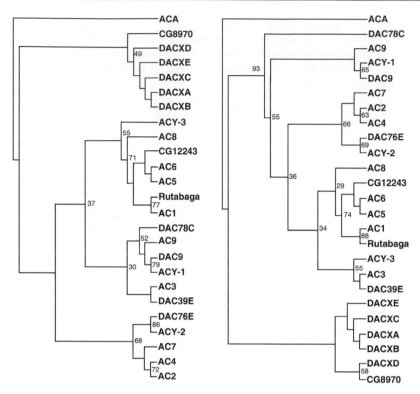

FIG. 3. Two representative phylogenetic trees, generated by FITCH (*left*) and by PROTPARS (*right*), obtained from a CLUSTAL W alignment of the amino acid sequences of a concatamer of C_{1a} and C_{2a} catalytic domains. Bootstrap values were obtained from 100 data sets. Values over 95 are omitted for clarity. Sequences used are bovine AC1 (AC1), rat AC2 (AC2), rat AC3 (AC3), rat AC4 (AC4), rat AC5 (AC5), rat AC6 (AC6), mouse AC7 (AC7), rat AC8 (AC8), mouse AC9 (AC9), *Drosophila* DACXA, DACXB, DACXC, DACXD, DAC78C, DAC13E, DAC39E, DAC76E, and Rutabaga, *C. elegans* ACY-1 and ACY-2, presumed *Drosophila* tmACs CG8970 and CG12243, presumed *C. elegans* tmAC ACY-3, and *Dictyostelium* ACA as an outgroup.

tmAC2, -4, and -7 subfamily[4]; DAC9 and ACY-1 are orthologs of mammalian tmAC9[10]; DAC39E is an ortholog of mammalian tmAC3[9]; and the DACX cyclases define a distinct subfamily of tmAC isoforms.[6] Also, as revealed by both tree-building methods, we can now conclude that the novel *Drosophila* tmAC CG12243 obtained from GadFly appears to represent an ortholog of mammalian tmAC5 and -6, and the sequence CG8970 is an additional member of the DACX subfamily.

Interestingly, there are discrepancies between the two methods. ACY-3 and DAC78C are differentially placed by the two tree-building methods, with low bootstrap values assigned to those placements. Such inconsistency supports our previous conclusion that DAC78C defines a novel tmAC isoform.[6] These relationships

TABLE II
PHYLOGENETIC RELATIONSHIPS BETWEEN TRANSMEMBRANE
ADENYLYL CYCLASES[a]

Mammals	Drosophila	Caenorhabditis elegans
tmAC1	Rutabaga	
tmAC2, -4, and -7	DAC76E	ACY-2
tmAC3	DAC39E	
tmAC5 and -6	CG12243	
tmAC8		
tmAC9	DAC9	ACY-1
	DACXA-E and	
?	CG8970	
?	DAC78C	
?		ACY-3

[a] Presumed phylogenetic relationships between the cloned mammalian tmACs and those tmACs identified by cloning and genomic analysis in *Drosophila* and *C. elegans*. Question marks indicate that in the absence of complete genome information, it is unclear whether orthologous tmACs are yet to be identified.

are summarized in Table II, and they demonstrate the importance of investigating multiple possible tree-building methods before assigning novel tmACs to a particular subfamily.

Conclusions

Phylogenetic examination reveals that molecular cloning and genome analysis have identified orthologs for all mammalian isoforms except tmAC8 in the model systems, *Drosophila* and *C. elegans*. Such analysis confirms the validity of genetic study in these organisms for elucidating physiological significance of individual tmAC isoforms. It is also clear from this analysis that model organisms can express a greater variety of tmACs than mammals, and it is important to bear this in mind when investigating tmAC diversity in model organisms.

A combination of traditional cloning methods, high-throughput PCR analysis, and genomic analysis permits the identification of complete tmAC repertoires in a number of organisms. When coupled with phylogenetic analysis and genetics, such studies will reveal common and dissimilar elements of cAMP signaling throughout the animal kingdom.

Acknowledgments

The authors thank Ben Kliot, Eugene Chung, and Vadim Iourgenko for technical assistance, and Jochen Buck and Mark Wuttke for critical reading of the manuscript. This work was supported by NIH Grant GM52891 (L.R.L.).

[13] Functional Analyses of Type V Adenylyl Cyclase

By Tarun B. Patel, Claus Wittpoth, Ann J. Barbier,
Yinges Yigzaw, and Klaus Scholich

Type V and VI forms of adenylyl cyclase (AC) are closely related not only in their amino acid sequences but also in their regulatory properties (reviewed in Sunahara *et al.*[1] and Iyengar[2]). In brief, both these isoforms of AC are stimulated by the α subunit of stimulatory G protein ($G_s\alpha$) and are inhibited by the α subunit of inhibitory G protein ($G_i\alpha$) as well as directly by calcium. In this article, we describe the methods to generate the full-length and engineered, soluble forms of canine type V adenylyl cyclase to investigate the regions on this enzyme that interact with each other and modulate the actions of $G_s\alpha$. In addition, we describe the use of chimeric forms of adenylyl cyclase consisting of domains derived from bovine type I (ACI) and type V (ACV) enzymes to investigate the domains on these enzymes that interact with $G\beta\gamma$ and $G_i\alpha$, respectively; ACI, but not ACV, is inhibited by $G\beta\gamma$ (reviewed in Sunahara *et al.*[1] and Iyengar[2]). Moreover, we describe the methods utilized to uncover two additional functions of ACV, namely its GTPase-activating protein (GAP) activity against $G_s\alpha$ and its ability to enhance receptor-mediated GTP–GDP exchange on α subunits of G_s and G_i.

Given the aforementioned framework, this article is divided into the following major sections: expression of full-length and two soluble forms of ACV; use of full-length and soluble forms of ACV to elucidate intramolecular interactions; use of chimeric forms of ACV and ACI to identify $G_i\alpha$- and $G\beta\gamma$-interacting sites; GAP activity of ACV; and guanine nucleotide exchange-enhancing activity of ACV.

I. Expression of Full-Length and Two Soluble Forms of Type V Adenylyl Cyclase

A. Introduction

A common feature shared by the various isoforms of AC is that the predicted structure for all isoforms is similar. Hence the AC molecule traverses the cell membrane 12 times and has 2 major cytosolic domains (referred to as C1 and C2, bracketed in Fig. 1A).[1–3] The C1a and C2a subdomains within the two major cytosolic domains share some degree of similarity with each other and also show homology with the catalytic domain of guanylyl cyclases.[3] Notably, the C2b region is present only in ACI, ACII, ACIII, and ACVIII[1]; the slightly shorter

---

[1] R. K. Sunahara, C. W. Dessauer, and A. G. Gilman, *Annu. Rev. Pharmacol. Toxicol.* **36,** 461 (1996).
[2] R. Iyengar, *FASEB J.* **7,** 768 (1993).
[3] W.-J. Tang and A. G. Gilman, *Cell* **70,** 869 (1992).

A

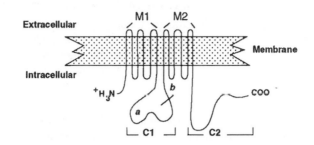

C1- Domain of ACV :

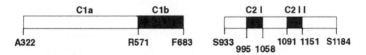

C2 -Domain of ACV:

B

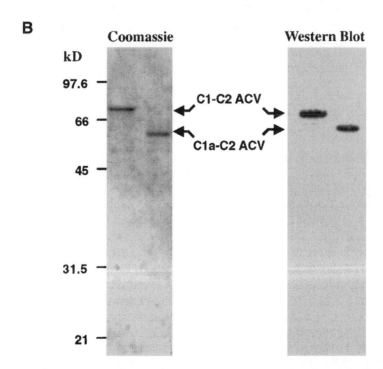

FIG. 1. Schematic representation of ACV, its major subdomains, and immunoblot of soluble forms of C1–C2 and C1a–C2 forms of ACV. (A) Schematic of the full-length ACV. The locations of the major cytosolic regions C1 and C2 are shown in the context of the whole molecule. The M1 and M2 regions, which span the membrane six times each, are also denoted. The lower panel depicts the

ACV does not have a C2b subdomain. Studies of naturally occurring, membrane-bound mammalian adenylyl cyclases have been difficult for the following reasons. First, because the regulatory properties of the different isoforms are varied[1-3] and second, because most mammalian cells express a mixture of isoforms, it is difficult to study a given isoform in isolation. Although this problem can be circumvented by expression of the desired AC isoform in insect cell lines such as *Spodoptera frugiperda*-9 (Sf9), the amount of protein that is expressed is low. Second, the membrane-bound adenylyl cyclases are extremely sensitive to detergents and, therefore, the purification of active enzymes has been difficult. Furthermore, when the active, full-length enzyme has been purified, it has been obtained only in small amounts and then the protein does not store well. Given these limitations, attempts have been made to express the protein subunits and determine whether the enzyme activity can be reconstituted. Interestingly, none of the two halves of the ACI, ACII, or ACV molecule when expressed alone (i.e., M1C1 or M2C2 in Fig. 1A) exhibit AC activity.[4,5] However, coexpression of the two halves of the ACI, ACII, or ACV molecule[4,5] or the expression of C1a and C2a domains from ACI and ACII, respectively, joined by a linker[6,7] demonstrate activity that can be stimulated by forskolin and GTP-bound $G_s\alpha$.[4-7]

Given these findings, our objective was to engineer soluble forms of nonchimeric AC in which both the C1 and C2 regions were derived from type V AC. We also compared the soluble forms of mammalian AC with the full-length ACV. In this section, we describe the methods we used to obtain the full-length canine ACV in Sf9 cells and the soluble counterparts of this enzyme in bacteria.

B. Full-Length Type V Adenylyl Cyclase

Plasmid Construction and Generation of Baculovirus. cDNA encoding canine ACV was provided by Y. Ishikawa, who originally cloned the enzyme.[8] To express the full-length enzyme using the baculovirus expression system, ACV cDNA in pCDNA3[9] is treated with *Pvu*II and *Eco*RI to excise the 3' region of the cDNA. This *Pvu*II–*Eco*RI fragment is ligated into pBlueBacHis2A shuttle vector to generate a construct that we have designated 7.pBlueBacHis2A. The *Bam*HI–*Xba*I fragment from ACV pCDNA3 is excised and inserted into the *Bam*HI–*Xho*I sites of pBlueBacHis2B; the *Xba*I site on ACV and the *Xho*I site on pCDNA3 are blunted to permit ligation. This latter construct is utilized to excise an *Eco*RI–*Eco*RI fragment corresponding to the 5' end of the ACV cDNA. This fragment is then cloned into

---

amino acid residues that define the C1a, C1b, C2I, and C2II subdomains within the C1 and C2 regions. (B) Coomassie-stained gel and Western analysis of the two purified soluble forms of ACV, C1–C2 and C1a–C2. The two soluble forms of ACV were expressed in *E. coli* TP2000 as described in text. Supernatants from cell lysates were applied to Ni–NTA and Mono Q chromatography. Each purified protein (0.5 μg) was separated on 10% (w/v) acrylamide gels and Western analysis was performed with Anti-Xpress antibody, which recognizes the epitope tag in the N terminus of each protein.

the *Eco*RI site of 7.pBlueBacHis2A. This ligation results in the insertion of the full-length ACV into the plasmid pBlueBacHis2A (ACV.pBlueBacHis2A). Although this construct is suitable for generation of recombinant virus we have opted to follow the procedure described below because of the convenience of using the Baculogold system (PharMingen, Carlsbad, CA). The *Eco*RV–*Hin*dIII fragment from this latter construct containing the ACV cDNA in-frame with a hexahistidyl (His$_6$) tag at the 5' end is then cloned into the baculovirus shuttle vector p2Bac. To obtain recombinant baculovirus, the shuttle vector is used in conjunction with Baculogold (PharMingen) to infect Sf9 cells; Sf9 cells are cultured at 27° in serum-free growth medium (GIBCO-BRL Gaithersburg, MD). After three rounds of amplification, the recombinant virus is selected and used to infect Sf9 cell cultures in 75-cm^2 flasks. The expression of ACV is monitored at various times after infection by measuring the basal and forskolin-stimulated AC activities in uninfected and infected cells. Typically, cells expressing ACV have forskolin-stimulated AC activities that are at least four times higher than control cells. Once the conditions for a specific pool of virus are identified, larger amounts of cells are infected and membranes are harvested by the method of Kassis and Fishman.[10]

C. C1–C2 and C1a–C2 Forms of Type V Adenylyl Cyclase

1. Plasmid Construction. Both these forms of soluble ACV are modeled on the chimeric soluble adenylyl cyclase comprising the C1a region of ACI linked to the C2a region of type II AC described by Tang and Gilman.[6] Figure 1A schematically represents the location of the C1, C1a, and C2 domains of ACV in the context of the full-length protein. The expression vector pTrcHisB (Invitrogen, San Diego, CA) is used for expression of both forms of hexahistidyl-tagged soluble adenylyl cyclases in *Escherichia coli*. First, the C2 region (amino acid residues 933–1184 of canine type V adenylyl cyclase) is amplified, using ACV in pCDNA3 (Invitrogen) as a template. The primers used for amplification introduce a *Bgl*II site at the 5' end of the polymerase chain reaction (PCR) product and a stop codon followed by a *Hin*dIII restriction site at the 3' end (primer sequences: 5'-ATATATAGATCTTCCACCGCCCGCCTCGAC and 5'-ATATATAAGCTTCTAACTGAGCGGGGG). The PCR product is cloned into the *Bgl*II and *Hin*dIII sites in pTrcHisB. The resulting plasmid is cut by *Bam*HI and

[4] W.-J. Tang, M. Stanzel, and A. G. Gilman, *Biochemistry* **34,** 14563 (1995).

[5] S. Katsushika, J.-I. Kawabe, C. J. Homcy, and Y. Ishikawa, *J. Biol. Chem.* **268,** 2273 (1993).

[6] W.-J. Tang and A. G. Gilman, *Science* **268,** 1769 (1995).

[7] C. W. Dessauer and A. G. Gilman, *J. Biol. Chem.* **271,** 16967 (1996).

[8] Y. Ishikawa, K. Shuichi, L. Chen, N. J. Halnon, J.-I. Kawabe, and C. Homcy, *J. Biol. Chem.* **267,** 13553 (1992).

[9] Z. Chen, H. S. Nield, H. Sun, A. Barbier, and T. B. Patel, *J. Biol. Chem.* **270,** 27525 (1995).

[10] S. Kassis and P. A. Fishman, *J. Biol. Chem.* **257,** 5312 (1984).

*Bgl*II and a PCR product encoding either the C1a region (amino acid residues 964–1713) or the whole C1 region (amino acid residues 964–2049) is inserted. The oligonucleotides used to generate the C1a and C1 regions introduce a unique *Bam*HI restriction site at the 5′ end and a unique *Bgl*II site and a 14-amino acid linker between C1 (or C1a) and the C2 region at the 3′ end. The 5′ oligonucleotide sequence is 5′-ATATATGGATCCGCCTGAGGTCTCCCAG and the 3′ oligonucleotide sequences are 5′-ATATATAGATCTCATACCTCCAGCTGCAG-CTGGAGGCATTCCACCTGCAGCTGCCCGCTTCTGGGTGCAGCGCAG for C1, and 5′-ATATATAGATCTCATACCTCCAGCTGCAGCTGGAGGCATTCCA-CCTGCAGCTGCGAATCGGTCATCCACCTGCTT for C1a. All clones are sequenced to confirm in-frame ligations and correctness of the sequences.

2. *Expression of Soluble Adenylyl Cyclases. Escherichia coli* strain TP2000, which does not express any endogenous adenylyl cyclase activity,[11,12] is used for expression of both C1–C2 and C1a–C2 forms of ACV. In our experience, the best expression of proteins is obtained when bacteria are freshly transformed. Bacterial cells transformed with a plasmid encoding C1–C2 ACV or C1a–C2 ACV are grown in Luria broth containing ampicillin (100 μg/ml) at 37° until they reach an $OD_{600\ nm}$ of 0.4. Isopropyl-B-D-thiogalactopyranoside (IPTG) is then added to a final concentration of 0.1 mM and the cells are incubated for 15 hr at 23° before they are pelleted by centrifugation (6000g for 5 min at 4°) and stored at −80°. Frozen cells are thawed in 1/100 culture volume of medium containing 50 mM Tris-HCl (pH 8.0), 100 mM NaCl, 1 mM 2-mercaptoethanol, 1 mM benzamidine, and 20 μg/ml each of aprotinin, leupeptin, and soybean trypsin inhibitor*. Lysozyme is added to a final concentration of 0.1 mg/ml and the cells are incubated on ice for 30 min. MgCl$_2$ and DNase are then added to a final concentration of 5.0 mM and 0.2 mg/ml, respectively. After 5 min of incubation on ice, the lysed cells are centrifuged at 27,000g for 30 min at 4° and the protein concentration in the supernatant is determined by using the Bradford reagent (Bio-Rad Hercules, CA) and bovine serum albumin (BSA) as standard. Typically the pellet of a 2 liter culture of bacteria will receive 20 ml of the lysis buffer and yield a final supernatant volume of approximately 23 ml. To confirm expression of the appropriate proteins, the proteins in the supernatant are separated by sodium dodecyl sulfate–polyacrylamide gel electrophoresis (SDS–PAGE) [7.5% (w/v) polyacrylamide gels] and transferred onto nitrocellulose, and immunoblots are performed with anti-Xpress antibody (Invitrogen) and the enhanced chemiluminescence (ECL) system (Pierce, Rockford, IL); the proteins contain an anti-Xpress epitope immediately after the hexahistidyl tag. The supernatant expressing C1–C2 ACV or C1a–C2 ACV is

---

* The buffer described is for purification of C1a–C2 and C1–C2 forms of ACV. When supernatants of bacterial cell lysates are used only for activity measurements, the 100 mM NaCl and 1 mM 2-mercaptoethanol are substituted with 1 mM EDTA.

[11] A. Roy and A. Danchin, *Mol. Gen. Genet.* **188,** 465 (1982).

[12] A. Beauve, B. Boeston, M. Crasnier, A. Danchin, and F. O'Gara, *J. Bacteriol.* **172,** 2614 (1990).

either aliquoted and stored at $-80°$ or used immediately for purification of the recombinant protein.[†] The supernatants from cells expressing either C1–C2 ACV or C1a–C2 ACV can be used to measure adenylyl cyclase activity, which is stimulated by both $G_s\alpha$ as well as forskolin and inhibited by $G_i\alpha$.[13]

D. Purification of C1–C2 and C1a–C2 Forms of Type V Adenylyl Cyclase

The bacterial supernatant containing C1–C2 and C1a–C2 ACV is incubated with metal affinity chromatography resin (TALON; Clontech, Palo Alto, CA), which is washed with 50 mM Tris-HCl (pH 8.0), 100 mM NaCl, and 1 mM 2-mercaptoethanol. Typically, 1 ml of TALON resin is used for each 25 ml of the supernatant (125 mg of total protein) and the mixture is incubated for 1 hr at $4°$ with gentle rolling. The resin is then loaded on a 1×5 cm Econo column (Bio-Rad) and washed with 20 column volumes of buffer containing 50 mM Tris (pH 8.0), 500 mM NaCl, and 1 mM 2-mercaptoethanol; the flow rate is maintained at 1.2 ml/min. Thereafter, the column is eluted with 100 mM imidazole in 50 mM Tris-HCl (pH 8.0), 100 mM NaCl, and 1 mM 2-mercaptoethanol. The flow rate is 1.2 ml/min and soluble cyclases elute between 2 and 6 ml from a column that has 4 ml of the resin. Note that storage of the soluble adenylyl cyclases with imidazole for even short times inactivates the enzymes; this is also true for the C1 and C2 domains because the mixture of the two does not reconstitute enzyme activity if stored in imidazole. Therefore, it is imperative to remove the imidazole rapidly. For this purpose, the imidazole eluate from the metal affinity column is immediately applied to a Mono Q 5/5 fast protein liquid chromatography (FPLC) column (Pharmacia, Piscataway, NJ) and washed extensively with 20–30 column volumes of buffer containing 50 mM Tris-HCl (pH 8.0), 100 mM NaCl, 1 mM EDTA, and 1 mM dithiothreitol (DTT), the Mono Q column is equilibrated at a flow rate of 0.5 ml/min with the same buffer. The recombinant proteins are eluted from the column with a 150–450 mM NaCl gradient over 60 min in the buffer used for column equilibration and wash. The fractions (1 ml) are analyzed on a 10% (w/v) SDS–polyacrylamide gel, using Coomassie staining, and peak fractions are pooled and concentrated. For concentration, Centricon 50 (Amicon, Danvers, MA) spin columns are used according to the manufacturer instructions. At the same time, the buffer is exchanged with 50 mM Tris (pH 8.0), 1 mM DTT, 1 mM EDTA, and before storage 5% (v/v) glycerol is added. The soluble enzymes are stored at $-80°$, optimum storage concentration is ≥ 1 mg of protein/ml. Starting with an 8-liter culture of bacteria, this procedure would typically result in 800 μg of C1a–C2 and C1–C2 proteins. A Coomassie-stained gel showing the purified

---

[†] If the supernatants are stored at $-80°$, they must be used within 30 days to ensure that the soluble forms of ACV are still active.

[13] K. Scholich, A. J. Barbier, J. B. Mullenix, and T. B. Patel, *Proc. Natl. Acad. Sci. U.S.A.* **94,** 2916 (1997).

proteins is presented in Fig. 1B. An identical gel is subjected to Western analysis with anti-Xpress epitope antibody (Invitrogen) and is shown in Fig. 1B.

E. Adenylyl Cyclase Activity Measurements

AC activity is measured on the basis of the principles of separating cAMP by column chromatography[14] and by methods previously described from our laboratory.[15,16] AC assays are performed in the presence of 5 mM Mg$_2$Cl$_2$. However, the concentration of Mg^{2+} is kept at 1 mM in assays involving use of S49 cyc$^-$ membranes.

II. Elucidation of Intramolecular Interactions in Adenylyl Cyclase

A. Introduction

In this section, we describe the three methods that identify the intramolecular interaction in ACV that modulates activation by G$_s\alpha$. The methods described here use the full-length and the two engineered, recombinant, soluble forms of ACV described above and elsewhere,[13] as well as two other approaches, namely the two-hybrid yeast assay and competition by peptides corresponding to sequences in ACV. These findings, which are fully described elsewhere,[17] demonstrate that the C1b region of ACV interacts with a 10-amino acid region within the C2 domain of the enzyme and that this interaction modulates the stimulation of enzyme activity by G$_s\alpha$. Moreover, these findings provide a plausible model for the biphasic G$_s\alpha$ concentration response that is observed for ACV and ACVI.[17,18] According to this model[17] a high- and low-affinity site for G$_s\alpha$ may be present on ACV and ACVI and, at least in the case of ACV, the interaction between the C1b region and its C2 domain may alter the affinity of the low-affinity site. Although the crystal structure data of Tesmer et al.[19] demonstrate that a chimeric soluble AC comprising the C1a region of ACV and the C2 domain of ACII binds one molecule of G$_s\alpha$, it should be noted that the C1b region of ACV is not included in this structure, and also that the C1a of ACV was not studied in the context of the C2 region of ACV. Therefore, how the intramolecular interaction of C1b region of ACV with its C2 domain alters the intermolecular interactions with G$_s\alpha$ is not entirely clear. Nevertheless, because of the intramolecular interaction within the ACV molecule, the C1–C2 ACV more closely resembles the full-length enzyme in its regulation.

[14] Y. Salomon, C. Londos, and M. Rodbell, Anal. Biochem. **54**, 541 (1974).

[15] B. G. Nair, H. M. Rashed, and T. B. Patel, Biochem. J. **264**, 563 (1989).

[16] H. Sun, J. M. Seyer, and T. B. Patel, Proc. Natl. Acad. Sci. U.S.A. **92**, 2229 (1995).

[17] K. Scholich, C. Wittpoth, A. J. Barbier, J. B. Mullenix, and T. B. Patel, Proc. Natl. Acad. Sci. U.S.A. **94**, 9602 (1997).

[18] A. Harry, Y. Chen, R. Magnusson, R. Iyengar, and G. Weng, J. Biol. Chem. **272**, 19017 (1997).

[19] J. J. G. Tesmer, R. K. Sunahara, A. G. Gilman, and S. R. Sprang, Science **278**, 1907 (1997).

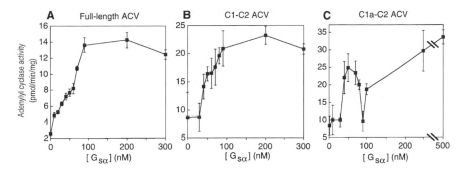

FIG. 2. Activation of the full-length and soluble forms of ACV by increasing concentrations of $G_s\alpha^*$. (A) Stimulation of the full-length ACV in Sf9 cell membranes by varying concentrations of $G_s\alpha^*$. (B) Stimulation of the C1–C2 soluble form of ACV in the presence of various $G_s\alpha^*$ concentrations. (C) Stimulation of the C1a–C2 soluble form of ACV by varying concentrations of $G_s\alpha^*$. Membranes of Sf9 cells (20 μg of protein) or supernatants of lysates (20 μg of protein) from bacteria expressing either the C1–C2 or C1a–C2 form of ACV were assayed for AC activity in the presence of various concentrations of $G_s\alpha^*$ as described in text. Activities are presented as the means $\pm$ SEM of three experiments. [These data have been published elsewhere; see Ref. 17.]

B. Use of Full-Length and Soluble Forms of Type V Adenylyl Cyclase

When the full-length and two soluble forms (C1–C2 and C1a–C2) of ACV are tested for stimulation of enzyme activity in response to different concentrations of the GTPase-deficient, constitutively active mutant (Q213L) form of $G_s\alpha$ ($G_s\alpha^*$), an interesting difference is observed. As illustrated by data in Fig. 2A, the activity of the full-length ACV expressed in Sf9 cells is stimulated by $G_s\alpha^*$ in a concentration-dependent manner. However, the $G_s\alpha^*$ concentration–response curve is biphasic with an inflexion at approximately 50 nM $G_s\alpha^*$ (Fig. 2A). A similar biphasic stimulation of $G_s\alpha^*$-mediated AC activity is also observed when the C1–C2 form of soluble ACV is used in the $G_s\alpha^*$ concentration–response curves (Fig. 2B). From the shapes of the $G_s\alpha^*$ concentration–response curves observed with full-length and C1–C2 forms of ACV it would appear that two saturation curves for $G_s\alpha^*$ are superimposed and that there are probably two $G_s\alpha^*$-interacting sites on these enzymes. One site that has a higher affinity for $G_s\alpha^*$ apparently begins to become saturated at $\sim$50–60 nM concentrations of $G_s\alpha$ (inflexion in Fig. 2A and B). The other, low-affinity site requires higher concentrations of $G_s\alpha^*$ and appears to be saturated at an $\sim$200 nM concentration of $G_s\alpha^*$ (Fig. 2A and B). Using the type VI AC (ACVI), which is homologous to ACV, the Iyengar laboratory has also presented similar evidence.[18]

This notion is further supported by studies employing the shorter C1a–C2 form of ACV, in which the C1b region comprising 112 amino acids in the C terminus of the C1 domain is missing (Fig. 1A). As shown in Fig. 2C, at concentrations of $G_s\alpha^*$ up to 60 nM, activity of the C1a–C2 form of ACV is stimulated.

However, concentrations of $G_s\alpha^*$ between 60 and 80 nM do not activate the enzyme (Fig. 2C). Further increases in the $G_s\alpha^*$ concentration ($\geq$100 nM) elevate activity of the C1a–C2 form of ACV (Fig. 2C). Assuming saturation of the high-affinity site for $G_s\alpha^*$ at the inflexion observed in Fig. 2A and B ($\sim$50 nM $G_s\alpha$) and the peak observed with C1a–C2 ACV at 50 nM $G_s\alpha^*$, the calculated apparent 50% effective concentration (EC$_{50}$) values of $G_s\alpha^*$ for the high-affinity site on all three forms of ACV are similar (39 $\pm$ 14 nM for full length; 41.7 $\pm$ 6 nM for C1–C2; 44.3 $\pm$ 5.6 nM for C1a–C2). On the other hand, the apparent EC$_{50}$ values of $G_s\alpha^*$ for the low-affinity site on the C1a–C2 form of ACV is markedly higher than for the full-length or C1–C2 ACV (cf. 128.3 $\pm$ 14.4 nM for C1a–C2 vs. 72.3 $\pm$ 5 nM for C1–C2 and 63 $\pm$ 9 nM for full-length enzyme). These data suggest that the C1b region, which is missing in the C1a–C2 form of ACV, is important for modulating the affinity of the second, low-affinity, site on ACV for $G_s\alpha^*$. Thus, in the absence of the C1b region, as seen with the C1a–C2 ACV, the affinity of the second site for $G_s\alpha^*$ is decreased and the two saturation curves are separated by a trough (Fig. 2C). The decrease in $G_s\alpha^*$-mediated stimulation of activity of the C1a–C2 form at concentrations of the G protein between 50 and 80 nM (Fig. 2C) may be due to $G_s\alpha^*$ making contact at two loci on ACV that form the high-affinity site for the G protein and is elaborated on in a previous publication.[17]

Importantly, however, the data in Fig. 2 also demonstrate that the C1–C2, and not the C1a–C2, form of soluble adenylyl cyclase more closely resembles the full-length enzyme. Therefore, the longer C1–C2 ACV is the preferred form to substitute for the full-length enzyme in experimental systems.

C. Yeast Two-Hybrid Assay

Because data in Fig. 2 indicate that the C1b region of ACV modulates the ability of $G_s\alpha^*$ to stimulate activity, using the yeast two-hybrid assay, we have investigated whether the C1b region of ACV is involved in interactions with $G_s\alpha^*$ or with another region of ACV that modulates the actions of $G_s\alpha^*$. Essentially, by the yeast two-hybrid assay, we do not observe any interactions between the C1b region of ACV and $G_s\alpha^*$. Therefore, only the intramolecular interactions of C1b are described below.

For the purpose of yeast two-hybrid assay, the Matchmaker kit (Clontech) is used. Employing cDNA encoding the full-length ACV as template and *Bam*HI-tagged (5$'$) and *Sal*I-tagged (3$'$) primers corresponding to nucleotides 1861–1875 and 2178–2199, the C1b domain of ACV (amino acids 572–683) is generated by PCR. The *Bam*HI and *Sal*I sites in the 5$'$ and 3$'$ primers, respectively, facilitate the directional cloning of cDNA encoding the C1b region into the plasmids pGBT9 and pGAD424. Likewise, cDNA encoding the C2 domain (amino acids 933–1184; nucleotides 2797–3555) of ACV is amplified by PCR, using ACV cDNA as template, and cloned into the *Bam*HI and *Sal*I sites of the plasmids pGBT9 and pGAD424. The subdomains of C2, C2I (amino acids 995–1058; nucleotides 2985–3174) and C2II (amino acids 1091–1151; nucleotides 3271–3453), are amplified by PCR to introduce an *Eco*RI site at the 5′ terminus and a *Sal*I site at the 3′ terminus. Both subdomains are cloned into the *Eco*RI and *Sal*I sites of plasmids pGAD424 and pGBT9. All constructs are sequenced to confirm the correct sequences and reading frames. The plasmids pGAD424 and pGBT9 contain the *GAL4* activation domain and binding domains, respectively, and expression of proteins is under the control of the yeast alcohol dehydrogenase promoter.

The HF7c yeast strain provided in the Matchmaker kit (Clontech) is transformed with the pGAD424 and pGBT9 constructs. Growth conditions, media, and transformation protocols are performed according to the manufacturer instructions. Transformed yeast cells are grown on plates containing either medium devoid of L-leucine and L-tryptophan (Leu$^-$/Trp$^-$) or medium in which L-histidine as well as L-leucine and L-tryptophan have been omitted (Leu$^-$/Trp$^-$/His$^-$). The plates are incubated at 30° for 3 days. Several of the colonies from transformants are then individually streaked out onto new plates containing the corresponding medium. To verify the inferences from data in Fig. 2, the expression of *lacZ* gene (β-galactosidase) is monitored. After overnight growth of at least six individual colonies of the various transformants in liquid Trp$^-$/Leu$^-$/His$^-$ medium (3 ml), the density of cells is determined by measuring the OD$_{600 nm}$. The cultures are each diluted to an OD$_{600 nm}$ of 0.3 and allowed to grow for an additional 3 hr in complete liquid medium. Thereafter, β-galactosidase activity in equal numbers of cells (determined by OD$_{600 nm}$) is assayed by the chemiluminescence assay[‡] of Jain and Magrath[20] as described in our previous publications.[17,21] To permit comparisons, HF7c cells are transformed with the various plasmid constructs at the same time and simultaneously assayed for growth and β-galactosidase activity. Activity of β-galactosidase is corrected for cell number.

Control HF7c yeast cells transformed with C2 in either pGBT9 or pGAD424 and the corresponding other plasmid without any cDNA insert do not grow in His$^-$ medium and, therefore, β-galactosidase activity in these cells cannot be monitored (not shown). However, control cells transformed with pGAD424-C1b and pGBT9 alone show enough growth on His$^-$ medium to allow β-galactosidase activity measurements that are not above background (Fig. 3). On the other hand, cells transformed with pGBT9-C2 and pGAD424-C1b show a significantly higher level of β-galactosidase activity (Fig. 3).

[‡] The widely used replica filter assay, using X-Gal as substrate to monitor β-galactosidase activity, is not sensitive enough to observe these interactions.

[20] V. K. Jain and I. T. Magrath, *Anal. Biochem.* **199,** 119 (1991).

[21] H. Sun, Z. Chen, H. Poppleton, K. Scholich, J. Mullenix, G. J. Weipz, D. L. Fulgham, P. J. Bertics, and T. B. Patel, *J. Biol. Chem.* **272,** 5413 (1997).

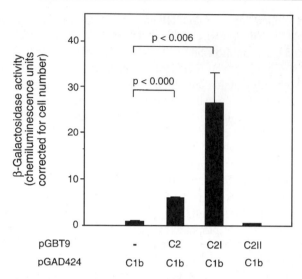

FIG. 3. Interactions between the C1b and C2 or C2I domain of ACV. Expression of β-galactosidase activity in HF7c cells cotransformed with plasmids pGBT9 and pGAD424 either containing the indicated cDNAs corresponding to regions within ACV or devoid of any cDNA insert (denoted by "–"). After growth of the various transformants on medium devoid of L-histidine for 3 days, β-galactosidase activity in six colonies each of the transformants was measured after growing cells for 24 hr in medium devoid of L-histidine followed by incubation for 3 hr in complete medium. Equal numbers of cells as monitored by $OD_{600\ nm}$ were used to assay β-galactosidase activity. Student unpaired t test analyses were employed to assess the significance of differences shown. [These data have been published elsewhere; see Ref. 17.]

To delineate more precisely the region(s) in the C2 domain that interact with the C1b region, using the two-hybrid assay, we have tested the ability of the C1b region to interact with two subdomains (C2I and C2II; see Fig. 1) of C2, which are highly conserved among all known isoforms of mammalian AC. Transformants carrying the plasmids pGBT9-C2I and pGAD424-C1b show robust cell growth on His⁻ medium (not shown) and significant β-galactosidase activity over background, indicating interaction between the C2I region and the C1b domain (Fig. 3). In contrast, no significant β-galactosidase activity above background can be monitored for the C2II region in any of the combination of plasmids (see, e.g., pGBT9-C2II and pGAD424-C1b; (Fig. 3). It is noteworthy that by both criteria, growth on His⁻ medium and β-galactosidase activity, the interaction between the C2 or C2I region of ACV and C1b is observed only when the C1b domain is expressed as a fusion protein with the *GAL4* activation domain. This observation is not unique for the proteins being tested in this system. Similar phenomena have previously been reported by us[21] and others (reviewed in Fields and Sternglanz[22]) and underscore the necessity to test the interacting proteins as chimeras of both the binding and activation domain of *GAL4*. Nonetheless, the findings with the two-hybrid assay suggest that the C1b domain of ACV interacts with the C2 region and, more specifically, a 64-amino acid (C2I) subdomain of C2.

D. Use of Peptides

Because the yeast two-hybrid data suggest that the C1b region of ACV interacts with its C2I subdomain and because the lack of the C1b region in C1a-C2 ACV altered the $G_s\alpha^*$ concentration–response curve (Fig. 2), we postulate that disruption of the C1b–C2I interaction would modulate stimulation of enzyme activity by $G_s\alpha$. To test this postulate, peptides corresponding to the C2I region are synthesized and used in activity assays with the full-length and two soluble forms of ACV. The rationale in these experiments is that if a given peptide interferes with the C1b–C2I interaction then the $G_s\alpha^*$ concentration–response curve with either the full-length ACV or C1–C2 ACV should resemble that observed with the C1a–C2 ACV in Fig. 2. Moreover, the peptide should not alter the profile of the $G_s\alpha^*$ concentration–response curve observed with C1a–C2 ACV. The following four peptides corresponding to the C2I region of ACV are used: peptide 1 (P1) corresponding to residues 1013–1033 (sequence NNEGVECLRVLNEI-IADFDEI), peptide 2 (P2) corresponding to amino acids 1042–1058 (sequence LEKIKTIGSTYMAASGL), peptide 5 (P5) corresponding to the N terminus of P2 (sequence LEKIKTIGST), and peptide 4 (P4) corresponding to the C terminus of P2 (sequence YMAASGLNDS).

None of the peptides, at concentrations up to 100 μM, alter basal or forskolin-stimulated activity of either the full-length or two soluble forms of ACV (not shown). However, when the activity of C1–C2 ACV is stimulated $G_s\alpha^*$, peptides 2 and 5, but not 1 and 4, decrease AC activity such that maximal inhibition is observed at 3 and 10 μM concentrations of P2 and P5, respectively (see Scholich *et al.*[13] for details). Thereafter, experiments have been performed in which the AC activity of the full-length and soluble forms of ACV is monitored in the presence of maximally effective concentrations of P2 or P5 and varying concentrations of $G_s\alpha^*$. Controls are performed under similar conditions in the presence of peptide P4 or P1, which do not alter AC activity. In the presence of either P2 (not shown) or P5 (Fig. 4A and B) the profile of the $G_s\alpha^*$ concentration–response curve with either C1–C2 ACV or the full-length ACV is altered to resemble the profile observed with C1a–C2 ACV in the absence of any peptides (see Fig. 2C). Notably, peptide P4 does not alter the $G_s\alpha^*$ concentration–response profile of C1–C2 ACV (not shown) or the full-length

[22] S. Fields and R. Sternglanz, *Trends Genet.* **10,** 286 (1994).

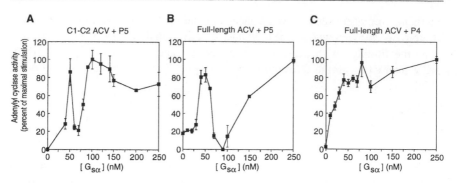

FIG. 4. Peptide P5 converts the profile of $G_s\alpha$ concentration–response curves of C1–C2 and full-length ACV to mimic the effects of $G_s\alpha^*$ on C1a–C2 ACV. (A) Effect of peptide 5 (10 μM) on the ability of different $G_s\alpha^*$ concentrations to stimulate the C1–C2 soluble form of ACV in supernatants of bacterial cell lysates (20 μg of protein). (B) Stimulation of the full-length ACV in Sf9 cell membranes (20 μg of protein) by the indicated varying concentrations of $G_s\alpha^*$, in the presence of 10 μM P5. (C) Same as (B) except 10 μM peptide 4 was used. To facilitate comparisons between different forms of ACV, AC activities are presented as a percentage of maximal activity measured. Each value is the mean $\pm$ SEM of three determinations. [These data have been published elsewhere; see Ref. 17.]

ACV (Fig. 4C). Moreover, none of the peptides alter the $G_s\alpha^*$ concentration–response curves observed with C1a–C2 ACV (see Scholich et al.[13] for details). These data, along with other similar experiments performed in the presence of forskolin and varying concentrations of $G_s\alpha^*$ (see Scholich et al.[13]) as well as the finding that the C1b region of ACV interacts with its C2I domain (Fig. 3), permit the following conclusions. First, peptides P2 and P5 alter the profile of the $G_s\alpha^*$ concentration–response curves of the full-length and C1–C2 ACV by disrupting an interaction of their C1b regions with the C2I domain, and this renders the two enzymes similar to their C1a–C2 counterpart. Second, because the N terminus of P2 (peptide P5) but not the C terminus of P2 (peptide P4) alters the profile of the $G_s\alpha^*$ concentration–response curves, it may be concluded that the C1b region of ACV interacts with a 10-amino acid region (L^{1042}–T^{1051}) in the C2 domain. Third, the intramolecular interactions of the C1b region with these 10 amino acids in the C2 domain modulate stimulatory actions of $G_s\alpha^*$ on ACV.

The cumulative data obtained from the three approaches, namely, utilizing different forms of the soluble ACV, the two-hybrid yeast assay, and peptides, also permit us to hypothesize about a two-site model for $G_s\alpha^*$ interactions with ACV. Such a model would certainly explain the biphasic concentration–response curve that is observed for ACV (see Ref. 17 and Fig. 2) and ACVI.[18] Perhaps the C1b interaction with the C2I domain is unique for the two isoforms of the enzyme (ACV and ACVI), which are closely related to each other, and it is entirely possible that in AC isoforms where the $G_s\alpha$ concentration–response curve is not biphasic[18] only one $G_s\alpha$ interaction site is present on the molecule as demonstrated by the crystal structure of the C1a domain of ACV and the C2 region of ACII.[19]

III. Use of Chimeric Forms of Type V and Type I Adenylyl Cyclase to Identify $G_i\alpha$- and $G\beta\gamma$-Interacting Sites

A. Introduction

One of the unique features of ACV is that this enzyme is inhibited by $G_i\alpha$ (reviewed in Sunahara et al.[1] and Iyengar[2]). Likewise, the type I adenylyl cyclase (ACI) is unique in that its activity is inhibited by $G\beta\gamma$ subunits (reviewed in Sunahara et al.[1] and Iyengar[2]). Interestingly, ACI activity is also stimulated by Ca^{2+} plus calmodulin (CaM) (reviewed in Sunahara et al.[1] and Iyengar[2] and the region on ACI that is necessary to observe stimulation of enzyme activity is the C1b region.[23–25] Moreover, it has previously been demonstrated that ACI activity can be inhibited by $G_i\alpha$ when the enzyme is stimulated by Ca^{2+}/CaM, but not when its activity is enhanced by $G_s\alpha$.[26] Given these unique features of ACV and ACI, we describe here the method of using individual and chimeric domains of these enzymes to identify $G\beta\gamma$- and $G_i\alpha_1$-interacting regions. Because the C1 and C2 domains of AC isoforms are sufficient to observe enzymatic activity[6,13,17] and because these domains do not have to be linked to each other to reconstitute enzyme activity[27,28] our methods involved the use of wild-type and chimeric C1 regions of ACI and ACV mixed with their C2 domains. Figure 5A illustrates the regions of ACI and ACV that were used, the chimeric C1 domains, and the nomenclature that is utilized here to refer to the different soluble AC forms that were derived by mixing the subunits.[§]

B. Generation of cDNA Constructs Encoding Wild-Type and Chimeric Adenylyl Cyclase Domains

All constructs are created in the plasmid pTrcHisB (Invitrogen). The full-length cDNA encoding bovine ACI was provided by A. G. Gilman (Department

---

[§] Because the AC isoforms used in this section are derived by mixing AC subunits, in our nomenclature, a period is placed between the subunits. The only exception to this rule is the IC1a. VC2 chimera, which is linked by the artificial linker described in Section I.

[23] Z. L. Wu, S. T. Wong, and D. R. Storm, *J. Biol. Chem.* **168**, 23766 (1993).

[24] T. Vorherr, L. Knopfel, F. Hofmann, S. Mollner, T. Pfeuffer, and E. Carafoli, *Biochemistry* **32**, 6081 (1993).

[25] L. Levin and R. R. Reed, *J. Biol. Chem.* **270**, 7573 (1995).

[26] C. Wittpoth, K. Scholich, Y. Yigzaw, T. Stringfield, and T. B. Patel, *Proc. Natl. Acad. Sci. U.S.A.* **96**, 9551 (1999).

[27] R. K. Sunahara, C. W. Dessauer, R. E. Whisnant, C. Kleuss, and A. G. Gilman, *J. Biol. Chem.* **272**, 22265 (1997).

[28] C. W. Dessauer and A. G. Gilman, *J. Biol. Chem.* **271**, 16967 (1996).

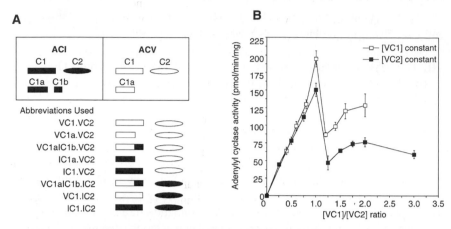

FIG. 5. Reconstitution of enzyme activity from mixtures of soluble subdomains of ACV and ACI. (A) Schematic representation of the various forms of soluble adenylyl cyclases derived by mixing the cytosolic C1 or C1a or chimeric C1 regions with C2 domains of either ACV or ACI. The various domains of ACI (solid) and ACV (open) are represented in the boxes. The amino acid residues encompassing the domains in bovine ACI are as follows: C1, 236–607; C1a, 236–471; C1b, 472–607; C2, 809–1133. In canine ACV the domains shown comprise the following amino acid residues: C1, 322–683; C1a, 322–571; C2, 933–1184. In the list of abbreviations the roman numeral preceding the C1 and C2 domains and their subregions denotes the AC isoform from which that particular region is derived. (B) Titration profile of the reconstituted enzyme activity of VC1 and VC2. Increasing amounts of lysates containing one of the subdomains were mixed with 10 μg (total protein) of the complementary subdomain and stimulated with $G_s\alpha^*$ (100 nM) and forskolin (100 μM). As shown, the optimal activity was achieved at a concentration ratio of 1 : 1 for either subdomain.

of Pharmacology, University of Texas Southwestern medical Center, Dallas, Tx). Complementary DNAs encoding the C1a [IC1a; amino acid (aa) 236–471], C1 (IC1; aa 236–607), and C2 (IC2; aa 809–1133) regions of ACI are obtained by PCR, using ACI cDNA as template and the following primers:

IC1a, primer A:

5′-ATATATGGATCCGGCTGAGCGCGCCCAG-3′

primer B:

5′-ATATATAGCGCTATGAGTTTTCAGAAAACTGTTCCTCTC-3′

IC1, primer C:

5′-ATATATGGATCCGGCTGAGCGCGCCCAG-3′

primer D:

5′- ATATATAAGCTTCTAGTCCTGAAGCTGGTGGTACTTTCGCTCTCG-3′

IC2, primer E:

5′-ATATATAGATCTGTCAAGCTGCGGCTG-3′

primer F:

5′-ATATATAAGCTTCTAAGCCTCCTTCCCAGAGGC-3′

The PCR products are cloned into the *Bam*HI–*Hin*dIII sites (IC1 and IC1a regions) and *Bgl*II–*Hin*dIII sites (IC2 domain) of the plasmid pTrcHisB.

To generate the chimeric C1 domain (VC1aIC1b) consisting of the C1a region (aa 322–571) from ACV and the C1b region (aa 472–607) of ACI, unique *Eco*47III [silent T→ C mutation at nucleotide (nt) 1674] and *Bgl*II restriction sites are introduced at the 3′ end of VC1a, using PCR methodology; the 5′ primer contains a *Bam*HI site. This PCR fragment is then cloned into *Bam*HI–*Bgl*II-treated plasmid pTrcHisB, which contains cDNA encoding the C1 and C2 regions of ACV joined by an artificial linker (Scholich *et al.*[13], see Section I). The resulting construct containing ACV C1a–C2 with the engineered, unique *Eco*47III site is digested with *Eco*47III and the PCR-generated IC1b cDNA (encoding aa 472–607 of ACI), engineered to contain *Eco*RV sites at the 5′ and 3′ ends, is then inserted into this new restriction site. This generates VC1aIC1b. VC2. Using PCR methodology, this latter construct is then used as a template to generate the individual, chimeric C1 region in which the C1a portion is derived from ACV and the C1b domain is from ACI; the 5′ and 3′ primers are designed to include unique *Bam*HI and *Hin*dIII sites, respectively; a stop codon is placed after the *Hin*dIII site. The cDNA encoding the chimeric C1 regions is cloned into the *Bam*HI–*Hin*dIII sites in plasmid pTrcHisB.

C. Expression of Recombinant Proteins

The AC subunits IC1, IC2, VC1, VC2, and VC1aIC1b are expressed in the TP2000 strain of *E. coli*. Expression of proteins is induced with IPTG (100 μM) and is performed at 23° for 21 hr as described above under Expression of Soluble Adenylyl Cyclases. $G_i\alpha_1$ protein is expressed in *E. coli* JM109(DE3) that has been cotransformed with pBB 131[29] to express *N*-myristoyltransferase and ensure synthesis of myristoylated G protein.[30] The expression of both *N*-myristoyltransferase and $G_i\alpha_1$ is induced by addition of 100 μM IPTG and incubation at 30° for 16 hr. The bacterial pellets are harvested as described above and $G_i\alpha_1$ protein is purified by the method of Mumby and Linder.[31] Bovine brain $\beta\gamma$ subunits of heterotrimeric G proteins are purified to homogeneity as described by Mumby *et al.*[32] with the modifications of Neer *et al.*[33]

Reconstitution of Enzyme Activity from Type V and Type I Adenylyl Cyclase Subdomains. One of the unique features of using unlinked AC domains to reconstitute enzyme activity is that the two regions (C1 and C2) must be titrated against

[29] R. J. Duronio, D. A. Rudnick, S. P. Adams, D. A. Towler, and J. I. Gordon, *J. Biol. Chem.* **266,** 10498 (1991).

[30] M. E. Linder, I.-H. Pang, R. J. Duronio, J. I. Gordon, P. C. Sternweis, and A. G. Gilman, *J. Biol. Chem.* **266,** 4654 (1991).

[31] S. M. Mumby and M. E. Linder, *Methods Enzymol.* **237,** 254 (1994).

[32] S. Mumby, I.-H. Pang, A. G. Gilman, and P. C. Sternweis, *J. Biol. Chem.* **263,** 2020 (1988).

[33] E. J. Neer, J. M. Lok, and L. G. Wolf, *J. Biol. Chem.* **259,** 14222 (1984).

each other. In the example shown in Fig. 5B, supernatant of lysates from bacteria expressing one of the ACV domains is kept constant (10 μg of total protein) and the amount of supernatant from bacterial lysates containing the complementary domain of ACV is varied. The amount of C1 or C2 proteins in the lysates is estimated by densitometry of Western blots performed with anti-Xpress (Invitrogen) antibody, using anti-Xpress epitope-tagged pure $G_s\alpha$ as standard. Moreover, to facilitate the optimization of conditions, enzyme activity is measured in the presence of both $G_s\alpha$ (100 nM) and forskolin (100 μM). As shown in Fig. 5B, the ratio of the two domains of ACV increases the activity of AC until a peak is observed when similar amounts of each subunit are present (ratio of 1.0 in Fig. 5B). Thereafter, increasing concentrations of C2 decrease enzyme activity. Therefore, it would appear that for optimal activity the C1 and C2 regions must be in a 1 : 1 ratio and at higher concentrations of either domain, homodimeric forms interfere with the formation of a heterodimeric catalytically active enzyme.

After optimizing the amounts of the various subunits[||] that need to be mixed to reconstitute activity in the presence of both $G_s\alpha$ and forskolin, the activity[#] of various combinations of subunits is determined in the absence and presence of $G_s\alpha$ or forskolin (100 μM). With the exception of IC1.IC2, all other forms of soluble AC as well as full-length ACI expressed in Sf9 cells are stimulated by $G_s\alpha$ (Fig. 6A). On the other hand, forskolin is able to stimulate the activity only of mixtures containing either C1 or C1a and C2 domains of ACV (Fig. 6A). Therefore, forskolin is not used to stimulate enzyme activity in our experiments to elucidate the actions of $G_i\alpha$ and $G\beta\gamma$ subunits.

Because the C1b region of ACI binds Ca^{2+}/CaM[23–25] it would follow that C1 domains of AC containing the C1b region from ACV should be stimulated by Ca^{2+}/CaM. Except for IC1.IC2 all other forms of soluble ACs and the full-length ACI that contains the C1b region of ACI are stimulated by Ca^{2+}/CaM (Fig. 6B). Note that the substitution of the C1b region of ACV with the C1b region of ACI permitts stimulation of enzyme activity by Ca^{2+}/CaM. The inability of either $G_s\alpha$ or Ca^{2+}/CaM to stimulate the IC1.IC2 form of soluble AC cannot be attributed to the lack of interaction between the two domains of the enzyme because the basal activity of this mixture is measurable and higher than that of the other combinations of C1 and C2 domains. More likely, the complex formed by IC1 and IC2 does not allow access to the $G_s\alpha$- and Ca^{2+}/CaM-binding sites. It should be noted that the optimal concentration of $G_s\alpha$ required to stimulate enzyme activity in Fig. 6A is different and determined empirically

---

[||] For optimization, the amount of AC subdomain proteins in the supernatant of bacterial lysates is measured by densitometry of Western blots performed with the anti-Xpress (Invitrogen) epitope antibody. For standards, the anti-Xpress epitope-tagged pure $G_s\alpha$ is used.

[#] In this section, for expression of specific activity of enzyme, the amount of protein corresponding to the two AC domains is measured by Western analyses using the anti-Xpress epitope antibody (Invitrogen) as described in footnote [||].

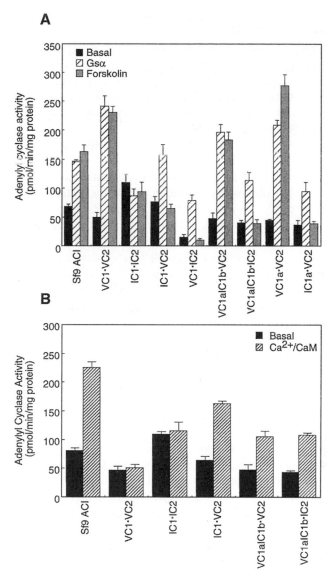

FIG. 6. Stimulation of adenylyl cyclase activity reconstituted by mixing the cytosolic C1 and C2 regions of ACI and/or ACV. (A) Adenylyl cyclase activity was measured under basal conditions or in the presence of $G_s\alpha^*$ or forskolin. Either membranes of Sf9 cells expressing full-length ACI or bacterial lysates expressing the C1 or C1a or chimeric C1 region (VC1aIC1b) were mixed with lysates expressing C2 domains of ACI or ACV to reconstitute enzyme activity. For all forms of AC shown, either 100 μM forskolin or the amount of $G_s\alpha^*$ that was required to stimulate maximal activity (see text) was used. Means $\pm$ SEM are shown ($n = 3$ experiments). (B) Membranes of Sf9 cells expressing the full-length ACI and various mixtures of C1 or chimeric C1 and C2 domains of ACI or ACV were assayed for adenylyl cyclase activity in the presence and absence of CaM (500 nM). Ca^{2+} was added to these assays at a final concentration of 30 μM and was present under basal conditions in the absence of CaM. Means $\pm$ SEM ($n = 3$ experiments) are shown. [These data have been published elsewhere; see Ref. 26.]

as described for the soluble forms of ACV in Fig. 2. Thus, VC1.IC2 requires 120 nM $G_s\alpha$; all other combinations require a 80 nM concentration of the G protein.

Using $G_s\alpha$-stimulated activity of the various mixtures of AC domains and the full-length ACI, the effects of purified bovine brain $G\beta\gamma$ subunits are monitored. As demonstrated previously by others,[6,34,35] $\beta\gamma$ subunits, in a concentration-dependent manner, inhibit the activity of full-length ACI (Fig. 7A). Likewise, consistent with previous findings that ACV is not inhibited by $\beta\gamma$ subunits (reviewed in Sunahara et al.[1] and Iyengar[2]), the soluble form of ACV (VC1.VC2) is not inhibited by $\beta\gamma$ subunits (Fig. 7A). To determine which portions of ACI is (are) necessary to observe the inhibition of activity mediated by $\beta\gamma$ subunits, experiments are performed with the C1 or C1a and C2 regions of ACI mixed with the complementary domains of ACV. When AC activity is reconstituted with the C1 domain or its N-terminal C1a region from ACI and the C2 region of ACV (IC1.VC2 and IC1a.VC2, respectively), the $G_s\alpha^*$-stimulated activity is inhibited by $G\beta\gamma$ subunits in a manner similar to that observed with the full-length ACI (Fig. 7A). On the other hand, when the C2 region of ACI is reconstituted with the C1 domain of ACV (VC1.IC2), $G_s\alpha^*$-stimulated activity is not altered by $G\beta\gamma$ subunits (Fig. 7A). These data demonstrate that the C1a region of ACI is sufficient to observe inhibition of enzyme activity by $G\beta\gamma$ subunits of G proteins. To determine whether the C1b region of ACI also contributes to inhibition of activity by $G\beta\gamma$ subunits of G proteins, the chimeric C1 region comprising C1a of ACV and C1b of ACI (VC1aIC1b) is reconstituted with the C2 region of ACV (VC1aIC1b.VC2). The $G_s\alpha^*$-stimulated AC activity of this enzyme is not altered by $G\beta\gamma$ subunits, indicating that the C1b region alone is not sufficient to observe $\beta\gamma$-mediated inhibition. Interestingly, however, when the C2 region of type I AC is used to reconstitute AC activity with VC1aIC1b, $G\beta\gamma$ subunits inhibit the $G_s\alpha^*$-stimulated enzyme as effectively as that observed with the full-length ACI and IC1a.VC2 (Fig. 7A). These findings, coupled with the observations that neither the C1b nor C2 region of ACI by itself is sufficient to observe $G\beta\gamma$-mediated inhibition of enzyme activity, suggest that the C1b and C2 domains of ACI interact with each other to form a $G\beta\gamma$-interacting site. This contention is also supported by the data in Fig. 7B. Hence, when the activity of VC1aIC1b.IC2 is elevated by Ca^{2+}/CaM, $\beta\gamma$ subunits of heterotrimeric G proteins inhibit activity. However, Ca^{2+}/CaM-stimulated activity of the VC1aIC1b.VC2 enzyme is not altered by $G\beta\gamma$ subunits (Fig. 7B). As expected, Ca^{2+}/CaM stimulates the activity of IC1.VC2 AC and this activity is also inhibited by $G\beta\gamma$ subunits in a concentration-dependent manner (Fig. 7B). Taken together, the data in Fig. 7 demonstrate that the C1a region of ACI is sufficient to observe $G\beta\gamma$-mediated inhibition of enzyme activity and that the C1b and C2 domains of ACI cooperate to form a $G\beta\gamma$-interacting site that also permits

[34] R. Taussig, L. M. Quarmby, and A. G. Gilman, J. Biol. Chem. **268,** 9 (1993).
[35] W.-J. Tang and A. G. Gilman, J. Biol. Chem. **266,** 8595 (1991).

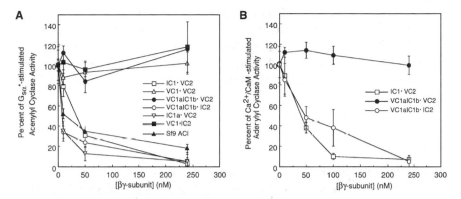

FIG. 7. Inhibition of activity of different AC forms by $\beta\gamma$ subunits of heterotrimeric G proteins. (A) AC activity in either membranes of Sf9 cells expressing the full-length ACI or in mixtures of subdomains from ACI and ACV were stimulated with $G_s\alpha^*$. The effect of various concentrations of $\beta\gamma$ subunits to modulate AC activity was monitored. Data are presented as percent inhibition of $G_s\alpha^*$-stimulated activity and represent means $\pm$ SEM ($n = 3$ experiments). (B) Same as (A), except that the AC activity was stimulated by addition of CaM (500 nM). Ca^{2+} was present at a final concentration of 30 μM. Percent inhibition of Ca^{2+}/CaM-stimulated activity is shown as means $\pm$ SEM ($n = 3$ experiments). [These data have been published elsewhere; see Ref. 26.]

these latter G protein subunits to inhibit enzyme activity. The requirement for both C1b and C2 regions of ACI to observe $G\beta\gamma$ effects is reminiscent of our previous findings that the C1b region of ACV interacts with a 10-amino acid region on its C2 domain and that this intramolecular interaction modulates the ability of $G_s\alpha$ to stimulate enzyme activity.[17]

Using the subdomains and chimeric domains described in Fig. 7, we have also been able to show that the $G_i\alpha$ interaction sites on ACV and ACI are located in their C1a and C1 regions, respectively. Furthermore, these regions from the two AC isoforms are sufficient to observe inhibition of enzyme activity. These data are described in detail in Wittpoth *et al.*[26] and, for the ACV, are similar to those of Dessauer *et al.*[36] Moreover, we have also observed that the presence of the C1b region of either ACI or ACV in conjunction with the C1a region of ACV enhances the sensitivity of AC to inhibition by $G_i\alpha$ (see Wittpoth *et al.*[26] for details). Overall, therefore, the examples of data described above show the utility of the experimental methodology involving AC domains and their chimeras.

IV. GTPase-Activating Protein Activity of Type V Adenylyl Cyclase

A. Introduction

As reviewed in several articles in volume 344 of this series, the GTPase activity of several α subunits of heterotrimeric G proteins is regulated by regulators of G protein signaling (RGS proteins). To date, however, no RGS protein that modulates $G_s\alpha$ function has been identified. Therefore, we propose the hypothesis that AC, the effector of $G_s\alpha$, may serve as a GAP for the G protein. Essentially our data (see Scholich et al.[37]) have demonstrated that the C1–C2 ACV, its C2 domain, and the C2I and C2II subdomains act as GAPs for $G_s\alpha$. In this section, we describe the methods that we have used to identify the GAP function of ACV and its subdomains.

B. Expression and Purification of C1, C2, C2I, and C2II Domains of Type V Adenylyl Cyclase

The cDNAs encoding the C1 and C2 domains of ACV are generated by PCR, using restriction enzyme site-tagged primers, and cloned into plasmid pTrcHisB. These proteins are expressed and purified exactly as described above for C1–C2 and C1a–C2 ACV (Section I). During the concentration step when buffer is exchanged, Centricon 10 (Amicon) is used for the C2 protein and Centricon 30 (Amicon) is used for the C1 protein.

C2I and C2II regions of ACV are also expressed as His$_6$-tagged proteins. The cDNAs encoding the C2I and C2II regions (nt 2983–3174 and 3271–3453, respectively, of canine ACV) are generated by PCR, using the appropriate restriction enzyme site-tagged oligonucleotide primers, and ligated into the plasmid pTrcHisB. The C2I and C2II proteins are expressed as described for soluble AC forms in Section I, except that the C2I protein is expressed at 14°. The C2I protein is purified exactly as described above for the larger domains of ACV. For the purification of the C2II subdomain the TALON column is washed with 20 column volumes of 100 mM imidazole, pH 8.0, and then eluted with 5 ml of 250 mM imidazole, pH 8.0. Significant loss of C2I and C2II proteins occurs during the concentration of samples with the Centricon spin columns. To minimize this loss, a 1-mg/ml solution of BSA is placed in the Centricon 10 (Amicon) columns for 5–10 min at room temperature. Unbound BSA is removed by washing the columns five times with 50 mM Tris (pH 8.0), 1 mM DTT, and 1 mM EDTA before use. By this method, the C2II subdomain is purified without any contamination.

C. GTPase-Activating Protein Assay for $G_s\alpha$

Monomeric $G_s\alpha$ (37 nM) or $G_i\alpha_1$ (54 nM) is incubated for 10 min at room temperature with $[\gamma\text{-}^{32}P]GTP$ (100 nM) in medium containing 25 mM HEPES–NaOH (pH 8.0), 0.1 mM EDTA, and 1 mM DTT. The final incubation volume is 1.5 ml. The temperature is then lowered in a refrigerated water bath to 14° and the incubation proceeds for another 5 min. At this point (time 0), GTP (0.1 mM

[36] C.W. Dessauer, J. J. G. Tesmer, S. R. Sprang, and A. G. Gilman, J. Biol. Chem. **273**, 25831 (1998).
[37] K. Scholich, J. B. Mullenix, C. Wittpoth, H. M. Poppleton, S. C. Pierre, M. A. Lindorfer, J. C. Garrison, and T. B. Patel, Science **283**, 1328 (1999).

final concentration) in the presence or absence of the indicated protein (1.1 μM) is added. These concentrations of C1–C2 ACV and its subdomains are similar to the concentration of RGS4 required to stimulate GTPase activity of $G_i\alpha_1$.[38] Aliquots (100 μl) are withdrawn at various times and reactions are terminated by mixing with 750 μl of an ice-cold 5% suspension of activated charcoal in 50 mM NaH_2PO_4, pH 7.4. The charcoal is pelleted in a tabletop centrifuge at 14,000g for 10 min at 4°. Aliquots (650 μl) of the supernatant are pipetted in fresh Eppendorf tubes and the spin is repeated at least once to remove residual charcoal. After the final spin, 625 μl of the supernatant is removed and the free ^{32}P is determined in a scintillation counter. Because this assay essentially measures the GTP hydrolysis from a single cycle of GTPase activity, below we refer to this assay as single cycle GTPase activity.

Using the aforementioned assay, the effect of C1–C2 ACV and the proteins corresponding to its C1 and C2 domains on the single-cycle GTPase activity of $G_s\alpha$ is monitored. As demonstrated in Fig. 8A, the addition of C1–C2 ACV to $G_s\alpha$ that has been loaded with $[\gamma$-^{32}P]GTP increases the rate of single-cycle GTP hydrolysis such that the $t_{1/2}$ for complete hydrolysis of $[\gamma$-^{32}P]GTP bound to $G_s\alpha$ is decreased from 55.1 ± 3.08 to 11.1 ± 2.52 sec. Similarly, protein corresponding to the C2 domain of ACV also increases the GTPase activity of $G_s\alpha$ and decreases the $t_{1/2}$ of single-cycle GTP hydrolysis by the G protein (Fig. 8A). In contrast, the C1 domain of ACV, at 50-fold molar excess over $G_s\alpha$, does not alter the rate of single-cycle GTP hydrolysis by $G_s\alpha$ (Fig. 8A). These data demonstrate that the C1–C2 ACV and its C2 domain act as GAPs for $G_s\alpha$. To identify the minimal region(s) on the C2 domain that can interact with $G_s\alpha$ and retain the $G_s\alpha$ GAP activity, the following approach is taken.

D. Yeast Two-Hybrid Assay to Determine Minimal Regions of C2 Domain of Type V Adenylyl Cyclase That Interact with $G_s\alpha$

The structural data from cocrystallization of $G_s\alpha$ with the C1a domain of ACV and the C2 region of ACII demonstrate that the G protein interacts with amino acids in AC that reside in two highly conserved regions within the C2 domain of all isoforms of the enzyme[19]; in ACV, these highly conserved regions in the C2 domain comprise aa 995–1058 and 1091–1151. These two regions are referred to here as C2I and C2II, respectively. To determine whether these regions, when taken out of the context of the ACV molecule or the complete C2 region, can interact with $G_s\alpha$, the yeast two-hybrid assay is employed. However, before using cDNAs encoding small subdomains of the C2 region of ACV, we have investigated whether the proteins corresponding to larger domains in ACV interact with $G_s\alpha$. Essentially, the same system (Matchmaker; Clontech) as described above in

---

[38] D. M. Berman, T. M. Wilkie, and A. G. Gilman, *Cell* **86**, 445 (1996).

A   **B**

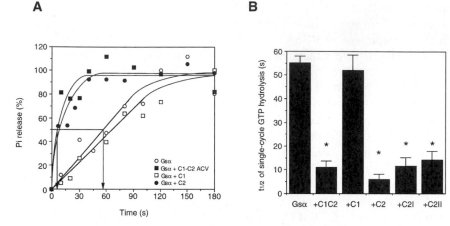

FIG. 8. C1–C2 ACV and its C2 domain act as GAPs for $G_s\alpha$. (A) $G_s\alpha$ (37 nM) was incubated for 10 min at room temperature with [γ-^{32}P]GTP (100 nM) and then at 14° as described in text. At time 0, 0.1 mM GTP and the indicated protein (1.1 μM) were added. Aliquots were withdrawn at the indicated times and free PO$_4$ was measured as described in text. The open circles represent $G_s\alpha$ alone, closed squares indicate $G_s\alpha$ in the presence of C1–C2 ACV, closed circles represent $G_s\alpha$ in the presence of C2, and open squares indicate $G_s\alpha$ in the presence of C1 ACV. Arrows depict the half-time ($t_{1/2}$) for complete hydrolysis of GTP bound to $G_s\alpha$ under control conditions and in the presence of C1–C2 ACV or its C2 domain. Values are presented as a percentage of total ^{32}P$_i$ released. Each experiment was repeated at least three times. Total P$_i$ released under the different conditions was as follows: $G_s\alpha$ alone, 50 fmol; $G_s\alpha$ + C1–C2 ACV, 48 fmol; $G_s\alpha$ + C2, 48 fmol; $G_s\alpha$ + C1, 42 fmol. (B) $t_{1/2}$ of complete hydrolysis of GTP bound to $G_s\alpha$ in the absence and presence of the indicated protein (each at 1.1 μM). Experimental conditions were the same as in (A). Each value is the mean $\pm$ SEM of at least four determinations. *$p < 0.001$ as compared with $G_s\alpha$ alone (Student unpaired t test). [These data have been published elsewhere; see Ref. 37.]

Section II is utilized. Our data demonstrate that as monitored by both *HIS3* and *lacZ* reporter genes, the C2 domain of ACV interacts better with $G_s\alpha^*$ as compared with wild-type $G_s\alpha$ (not shown). Notably, in the same two-hybrid system, using the epidermal growth factor receptor and either wild-type or constitutively active $G_s\alpha$, we have previously shown that the wild-type $G_s\alpha$ interacts better with the receptor as compared with $G_s\alpha^*$.[21] Thus, in the two-hybrid assay, the interactions of $G_s\alpha$ and $G_s\alpha^*$ are consistent with the notion that the active form of $G_s\alpha$ interacts better with receptor and that $G_s\alpha^*$ interacts better with the effector (AC). Most importantly, the findings described above demonstrate that the yeast two-hybrid method is useful to monitor interactions between C2 domain of ACV and $G_s\alpha^*$.

E. Interactions of C2 Subdomains with $G_s\alpha$

Further experiments are performed to define the subdomain(s) within this region that are involved in the interactions with $G_s\alpha^*$. Therefore, by employing the plasmids pGBT9 and pGAD424 containing the C2I (aa 995–1058) and C2II (aa 1091–1151) subdomains of ACV (Fig. 1) and $G_s\alpha^*$ or $G_s\alpha$, experiments similar to those described above are performed. The constructs used for C2I and C2II are described above under Section II. The $G_s\alpha$ constructs have been described previously.[21] The C2I and C2II subdomains of ACV are highly homologous in all forms of mammalian AC cloned and characterized. Although amino acids in this region of ACII have been shown to interact with $G_s\alpha$,[19] in yeast cells transformed with the C2II region and $G_s\alpha^*$ or $G_s\alpha$ no interaction is evident (not shown). However, in identical experiments with HF7c cells transformed with the C2I subdomain of ACV and $G_s\alpha$ or $G_s\alpha^*$, growth of cells is observed in Trp$^-$/Leu$^-$/His$^-$ medium; controls transformed with C2I in pGBT9 and pGAD424 without cDNA insert or vice versa do not grow on the latter medium, although transformation has occurred as demonstrated by growth of all cells in the Trp$^-$/Leu$^-$ medium (not shown). To monitor expression of the *HIS3* gene more quantitatively, six colonies each of HF7c cells transformed with the various constructs in pGBT9 and pGAD424 are grown in liquid Trp$^-$/Leu$^-$/His$^-$ medium. After 24 hr of growth in this medium, the $OD_{600\,nm}$ is determined for each culture (Fig. 9A). Similarly, β-galactosidase activity is measured in an equal number of cells (Fig. 9B). As shown in Fig. 9, by both criteria, that is, HIS3 and LACZ expression, the C2I region interacts better with $G_s\alpha^*$ than with $G_s\alpha$. As described in Section III, to facilitate the comparisons, all transformations and analyses are performed simultaneously on three separate occasions. Moreover, by both criteria, growth on Trp$^-$/Leu$^-$/His$^-$ medium and β-galactosidase activity, the interaction between the C2I region of ACV and $G_s\alpha$ is best observed when the former protein is expressed as a chimera with the GAL4-binding domain and $G_s\alpha$ as chimera with the GAL4-activating domain (Fig. 9). As mentioned previously, this is not unique to our studies, and a similar phenomenon has previously been observed with other proteins (reviewed in Fields and Sternglanz[22]).

The data obtained with the two-hybrid assay demonstrate that C2 as well as the 64-amino acid-long region of C2 (C2I) can interact with active $G_s\alpha$ and provide the rationale for expression of the smaller protein fragments for use in the GAP assay.

F. GTPase-Activating Protein Activity of C2I and C2II Subdomains of Type V Adenylyl Cyclase

To determine whether the C2I subdomain that interacts with $G_s\alpha^*$ in the two-hybrid assay also exhibits GAP activity against the G protein, the C2I region purified as described above is used. Although the C2II region of ACV does not interact with $G_s\alpha^*$ in the two-hybrid assay, because amino acids in this region interact with $G_s\alpha$,[19] the purified C2II region is also employed. Using the C2I and C2II proteins, the $G_s\alpha$ GAP assay described above is performed. In these experiments, the half-time for complete GTP hyrolysis from the single-cycle GTPase

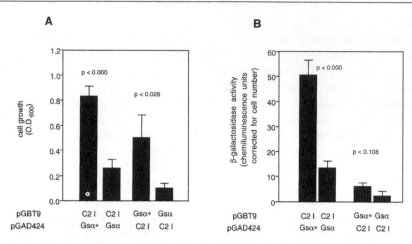

FIG. 9. Interactions between the $G_s\alpha$ or $G_s\alpha^*$ and C2I domain of ACV. (A) After growth of the various transformants in medium devoid of L-His for 24 hr the cell growth was determined by measuring the OD at 600 nm for six colonies each of the transformants. (B) β-Galactosidase activity in six colonies each of the transformants was measured after growing cells for 24 hr in medium devoid of L-His followed by incubation for 3 hr in complete medium. Equal numbers of cells as monitored by $OD_{600\ nm}$ were used to assay β-galactosidase activity. All transformations and assays were performed simultaneously. Student unpaired t test analyses was employed to assess significance of differences shown.

assay is measured in several experiments. As shown in Fig. 8B, both the C2I and C2II regions of ACV, like the larger C2 domain, exert $G_s\alpha$ GAP activity. Thus, the small domains in ACV, which include the amino acids that interact with $G_s\alpha$, are by themselves sufficient to act as $G_s\alpha$ GAPs. Notably, although the C2II subdomain of ACV does not interact with $G_s\alpha^*$ in the two-hybrid assay, it is capable of interacting with $G_s\alpha$ in the GAP assay. This is a classic example of why the negative data from the two-hybrid system should be cautiously interpreted.

Overall, the approaches described above have permitted us to identify the $G_s\alpha$ GAP activity of ACV and its subdomains.

V. Guanine Nucleotide Exchange Factor-Enhancing Activity of Type V Adenylyl Cyclase

Because the C2 region and its subdomains in ACV can serve as $G_s\alpha$ GAPs and thereby expedite the termination of signal, we have examined whether signal onset can also be modulated by ACV and its subdomains. For adenylyl cyclase activation via $G_s\alpha$, the G protein in its heterotrimeric form must be activated by a receptor (see Birnbaumer and Birnbaumer[39] for review). We and others[16,40,41] have previously

[39] L. Birnbaumer and M. Birnbaumer, *J. Receptor Signal Transduction Res.* **15,** 213 (1995).
[40] T. Okamoto, Y. Murayama, Y. Hayashi, M. Inagaki, E. Ogata, and I. Nishimoto, *Cell* **67,** 723 (1991).
[41] T. Okamoto and I. Nishimoto, *J. Biol. Chem.* **267,** 8342 (1992).

shown that peptides corresponding to short regions in G_s-coupled receptors can activate the heterotrimeric G protein *in vitro*. Therefore, these peptides can be regarded as constitutively active receptor.[40,41] In our assays we have employed the peptide βIII-2 corresponding to amino acids 259–273 in the β_2-adrenergic receptor.[40] The purified $G_s\alpha$ is mixed with purified bovine brain $G\beta\gamma$ subunits in a 1 : 5 molar ratio and incubated on ice for 60 min. Note that to achieve proper reconstitution of recombinant $G_s\alpha$ with bovine brain $G\beta\gamma$ subunits, GDP (1 μM) should be included. The purified soluble form of ACV, C1–C2 ACV, is used as the active enzyme.

Receptor-mediated activation of heterotrimeric G proteins results in an augmentation of GTP–GDP exchange on the α subunit.[39] This increase in GTP–GDP exchange can be conveniently measured by monitoring the steady state GTPase activity of G proteins.[16,40,42,43] Therefore, in our studies we have monitored the steady state GTPase activity of G_s. To measure receptor mimetic peptide-mediated increase in steady state GTPase activity, the following experimental conditions are used. The reconstituted heterotrimeric G_s (27.7 nM) is incubated with or without the peptide and/or proteins of interest at 25° in medium containing 100 nM $[\gamma$-^{32}P]GTP, 25 mM HEPES (pH 8.0), 100 μM EDTA, 120 μM MgCl$_2$, and 1 mM DTT. Aliquots (50 μl) are withdrawn and hydrolysis of $[\gamma$-^{32}P]GTP (100 nM), which is linear for more than 30 min, is monitored in duplicate at the 20-min time point as described above for the $G_s\alpha$ GAP assay.

As demonstrated previously[16,40] the peptide βIII-2, by enhancing the rate of GTP–GDP exchange, increases activation of the heterotrimeric G_s in a concentration-dependent manner (Fig. 10A). In the presence of either C1–C2 ACV (1.1 μM) or the C2 domain of ACV (1.1 μM; see Scholich *et al.*[37]), the concentration–response curve for βIII-2-mediated activation of G_s is shifted to the left such that in the presence of C1–C2 ACV or its C2 domain, 100-fold lower concentrations of the peptide are required to activate G_s to the same extent as that observed without AC or its C2 domain (Fig. 10A). Notably, in control experiments performed in the absence of peptide βIII-2, neither C1–C2 ACV nor the C2 domain of the enzyme alters steady state GTPase activity of the heterotrimeric G_s (see, e.g., Fig. 10A). Moreover, in the presence of peptide βIII-2 at a concentration (100 nM) that by itself does not increase steady state GTPase activity of heterotrimeric G_s, the effect of C1–C2 ACV on βIII-2-mediated activation of G_s is concentration dependent (see Scholich *et al.*[37] for details). Interestingly, like the C2 domain, the C2I and C2II regions retain the capacity to enhance the βIII-2-mediated increase in the steady state GTPase activity of G_s.[37] However, the protein corresponding to the C1 region of ACV does not alter the ability of βIII-2 to increase steady state GTPase activity of G_s.[37]

[42] D. M. Berman, T. M. Wilkie, and A. G. Gilman, *Cell* **86**, 445 (1996).

[43] T. Higashijima, K. M. Ferguson, M. D. Smigel, and A. G. Gilman, *J. Biol. Chem.* **262**, 757 (1987).

A **B**

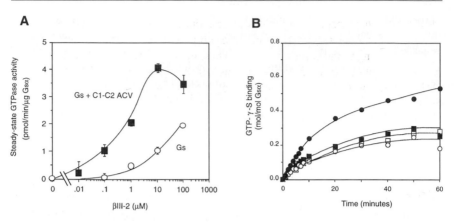

βIII-2 (μM)   Time (minutes)

FIG. 10. C1–C2 ACV enhances the β_2-adrenergic receptor peptide (βIII-2)-mediated activation of heterotrimeric G_s. (A) C1–C2 ACV augments the ability of different concentrations of βIII-2 to increase steady state GTPase activity of G_s. Steady state GTPase activity of reconstituted G_s (27.7 nM) was monitored in the presence of different concentrations of peptide βIII-2 with (filled squares) or without C1–C2 ACV (1.4 μM) as described in text. The rates of P_i release per microgram of $G_s\alpha$ in the heterotrimer are presented and represent means $\pm$ SEM ($n = 3$). (B) C1–C2 ACV increases the rate and extent of [^{35}S]GTPγS binding to reconstituted G_s. [^{35}S]GTPγS binding to G_s (10 nM) alone (open square), in the presence of 200 nM C1–C2 ACV (open circle), in the presence of 0.1 μM βIII-2 peptide (closed square), or in the presence of both C1-C2-ACV and βIII-2 (closed circle) was monitored in buffer containing 100 nM GTPγS, 25 mM Hepes (pH 8.0), 100 μM EDTA, 1 mM DTT, and 150 μM MgCl$_2$. Data are presented as moles of [^{35}S]GTPγS bound per mole of $G_s\alpha$ in the heterotrimer and are representative of two experiments. [These data have been published elsewhere; see Ref. 37.]

As a second method to evaluate the ability of C1–C2 ACV to augment guanine nucleotide exchange activity of βIII-2, the binding of GTPγS to G_s is monitored. These experiments are performed under conditions that do not favor rapid and maximal GTPγS binding to the G protein. Essentially, heterotrimeric G_s is incubated in buffer containing 100 nM [^{35}S]GTPγS, 25 mM HEPES (pH 8.0), 100 μM EDTA, 1 mM DTT, and 150 μM MgCl$_2$. For the time course of [^{35}S]GTPγS binding, aliquots (100 μl) are withdrawn at the times indicated and binding is terminated by mixing with 2.0 ml of ice-cold stop buffer containing 20 mM Tris-HCl (pH 7.4), 10 mM NaCl, and 25 mM MgCl$_2$. This mixture is then filtered through BA82 nitrocellulose disks (25-mm diameter; Schleicher & Schuell Keene, NH) under vacuum. The nitrocellulose disks are washed twice with 2.0 ml of stop buffer while still in the vacuum manifold. The BA82 filters are dissolved in 2 ml of ethylene glycol monomethyl ether and then counted for [^{35}S]GTPγS bound to the $G_s\alpha$. Nonspecific binding is determined in parallel incubations that are identical except that excess unlabled GTPγS (100 μM) is present.** Under these conditions, which are

---

** The background or nonspecific binding of [^{35}S]GTPγS may be high if fresh DTT is not used in the assay.

ideal to monitor receptor–mimetic peptide-induced activation of GTPγS binding to the heterotrimeric G_s, receptor-mimetic peptides alter both the rate and extent of GTPγS binding to the G protein.[16,37,40] Thus, in the presence of C1–C2 ACV, the rate and extent of threshold concentrations (100 nM; Fig. 10A) of βIII-2-mediated GTPγS binding to G_s is augmented (Fig. 10B). C1–C2 ACV by itself does not bind any GTPγS and does not by itself modulate GTPγS binding to G_s (Fig. 10B). Similar results are obtained with the C2 domain of ACV (see Scholich *et al.*[37] for details). These data along with the steady state GTPase assays demonstrate that the C1–C2 ACV and its C2 domain can augment the ability of the β-adrenergic receptor mimetic peptide to augment guanine nucleotide exchange on G_s.[††]

Using a peptide corresponding to amino acids 382–400 in the M_4 muscarinic receptor (MIII-4), we have investigated whether the C1 domain of ACV, which interacts with G_i, alters the ability of the muscarinic receptor mimetic peptide to activate G_i. Essentially, using approaches similar to those described above for G_s, we have shown that the C1–C2 ACV and the C1 domain of ACV augments MIII-4-mediated activation of G_i.[44] Thus, while the C2 region of ACV increases receptor-mediated activation of G_s by receptors coupled to this G protein, the C1 region of the enzyme enhances the activation of G_i-coupled receptors to activate the inhibitory G protein. These actions of ACV and its subdomains would ensure that the respective G proteins that modulate its activity are activated in the presence of low amounts of active receptors or in the presence of small amounts of receptor ligands. This mechanism would serve to amplify the signal initiated by activated receptors and at the level of AC ensure the rapid onset of signal. Thus, besides catalyzing the synthesis of cAMP, AC also has two other activities, namely $G_s\alpha$ GAP and enhancer of receptor function. Because of our findings with G_i,[44] this paradigm is applicable to both the stimulatory and inhibitory GTP-binding proteins of AC.

[††] For both C1–C2 ACV and the C2 domain to increase the guanine nucleotide exchange activity of βIII-2, the proteins have to be active. Thus, if C1–C2 ACV has lost enzymatic activity or if the C2 domain has lost the ability to reconstitute enzyme activity when reconstituted with the C1 region of ACV, then these proteins cannot enhance βIII-2-mediated GTP-GDP exchange on G_s.

[44] C. Wittpoth, K. Scholich, J. D. Bilyeu, and T. B. Patel, *J. Biol. Chem.* **275,** 25915 (2000).

[14] Photoaffinity Labeling of Adenylyl Cyclase

By MICHAEL K. SIEVERT, GÜLHAN PILLI, YU LIU,
ELIZABETH M. SUTKOWSKI, KENNETH B. SEAMON, and ARNOLD E. RUOHO

Introduction

Photoaffinity is an effective tool to covalently tag amino acid residues that are in or near ligand-binding sites.[1-3] The polypeptide chains that contribute to the binding site of a given receptor can be identified by sodium dodecyl sulfate–polyacrylamide gel electrophoresis (SDS–PAGE) if they are specifically labeled with a radioactive marker. Purification and subsequent N-terminal sequencing of these labeled peptides identify their location in the primary sequence of the receptor.

The specificity of photoaffinity labeling arises from the following characteristics: the products of photolysis of phenyl azides react in the binding site before dissociation of the ligand occurs; the lifetimes of phenyl nitrenes or azepines, which are derived from phenylazide-containing compounds, are extremely short (nanoseconds to milliseconds), so even ligands with equilibrium dissociation constants between 10^{-6} and 10^{-4} M can react before dissociation can occur from the binding site. Higher affinity photolabels ($K_D \leq 10^{-10} M$), however, increase the ease of the photolabeling experiment because excess photolabel can be removed from the receptor preparation before photolysis either by centrifugation or dilution. This fact enhances the signal-to-noise ratio and has been the reason that this technique is so effective for identification of most ligand-binding sites, where it is not uncommon to have equilibrium dissociation constants in the subnanomolar range.

Forskolin, a naturally occurring diterpene, has direct stimulatory effects on adenylyl cyclase.[4] It also has actions at other membrane-bound proteins such as inhibition of the glucose transporter,[5] modulation of voltage-dependent K^+ channels[6,7] and γ-aminobutyric acid type A (GABA$_A$)-gated chloride channels,[8] inhibition of the P-glycoprotein,[9] and modulation of gating of the nicotinic

[1] Y. Rong, M. Arbabian, D. S. Thiriot, A. Seibold, R. B. Clark, and A. E. Ruoho, *Biochemistry* **38**, 11278 (1999).

[2] M. K. Sievert and A. E. Ruoho, *J. Biol. Chem.* **272**, 26049 (1997).

[3] Z. Wu, D. S. Thiriot, and A. E. Ruoho, *Biochem. J.* **354**, 485 (2001).

[4] K. B. Seamon and J. W. Daly, *J. Cyclic Nucleotide Res.* **7**, 201 (1981).

[5] S. Sergeant and H. D. Kim, *J. Biol. Chem.* **260**, 14677 (1985).

[6] B. McNamara, D. C. Winter, J. E. Cuffe, G. C. O'Sullivan, and B. J. Harvey, *J. Physiol.* (*Lond*) **519**, 251 (1999).

[7] T. Hoshi, S. S. Garber, and R. W. Aldrich, *Science* **240**, 1652 (1988).

[8] G. Heuschneider and R. D. Schwartz, *Proc. Natl. Acad. Sci. U.S.A.* **86**, 2938 (1989).

[9] D. I. Morris, L. A. Speicher, A. E. Ruoho, K. D. Tew, and K. B. Seamon, *Biochemistry* **30**, 8371 (1991).

acetylcholine receptor.[10–12] The specificity of forskolin (Fsk) for different proteins can be modulated by altering the structure.[13] This article discusses the synthesis and use of two forskolin photoaffinity labels, N-[3-(4-azido-3-[^{125}I]iodophenyl) propionyl]forskolin (^{125}I-6-AIPP-Fsk) and ^{125}I-7-AIPP-Fsk. Although both compounds bind to adenylyl cyclase, ^{125}I-6-AIPP-Fsk is a more effective photolabel for derivatization of adenylyl cyclase.

Methods

Materials

Lubrol-PX is from Pierce (Rochford, IL); GTPγS, pepstatin A, and leupeptin are from Boehringer Mannheim (Indianapolis, IN); cytochalasin E, D-glucose, L-glucose, benzamidine, 2-mercaptoethanol, and Tween 60 are from Sigma (St. Louis, MO); cytochalasin B is from Aldrich (Milwanku, WI); [^{32}P]ATP and [^{3}H]cAMP are from DuPont-New England Nuclear (Boston, MA); forskolin and 1,9-dideoxyforskolin are the generous gift of R. Allen (Hoechst-Celanese, Kronberg, Germany); 7-bromoacetyl-7-deacetylforskolin is synthesized as described by Laurenza et al.[14]; 6-aminoethylcarbamylforskolin (6-AEC-Fsk) and 7-aminoethylcarbamylforskolin (7-AEC-Fsk) are synthesized as described by Robbins et al.[15] N-[3-(4-Azido-3-[^{125}I]iodophenyl)propionyl] succinimide (^{125}I-AIPPS) is synthesized carrier free with a specific activity of 2200 Ci/mmol as previously described.[16] The forskolin affinity column is synthesized by reacting Affi-Gel 15 with 7-AEC-Fsk.

Isolation of Brain Membranes

Crude washed membranes are isolated from bovine brains and frozen in liquid nitrogen.[9]

Synthesis of Photoaffinity Labels

(See Fig. 1.) ^{125}I-6-AIPP-Fsk and ^{125}I-7-AIPP-Fsk are synthesized by reacting 6-AEC-Fsk and 7-AEC-Fsk, respectively, with ^{125}I-AIPPS. ^{125}I-AIPPS (1 mCi, 2200 Ci/mmol) in ethyl acetate is evaporated to dryness and 20 μl of 6-AEC-Fsk (1 mg/ml in CH_2Cl_2) is added to the reaction vial and allowed to react at room temperature overnight. The reaction is monitored by thin-layer chromatography developed with ethyl acetate and the products are identified by autoradiography.

[10] M. L. Aylwin and M. M. White, Mol. Pharmacol. 41, 908 (1992).

[11] Z. Chen and M. M. White, Cell. Mol. Neurobiol. 20, 569 (2000).

[12] F. Grassi, L. Monaco, and F. Eusebi, Biochem. Biophys. Res. Commun. 147, 1000 (1987).

[13] K. B. Seamon, J. W. Daly, H. Metzger, N. J. de Souza, and J. Reden, J. Med. Chem. 26, 436 (1983).

[14] A. Laurenza, E. M. Sutkowski, and K. B. Seamon, Trends Pharmacol. Sci. 10, 442 (1989).

[15] J. D. Robbins, D. L. Boring, W. J. Tang, R. Shank, and K. B. Seamon, J. Med. Chem. 39, 2745–2752.

[16] J. M. Lowndes, M. Hokin-Neaverson, and A. E. Ruoho, Anal. Biochem. 168, 39 (1988).

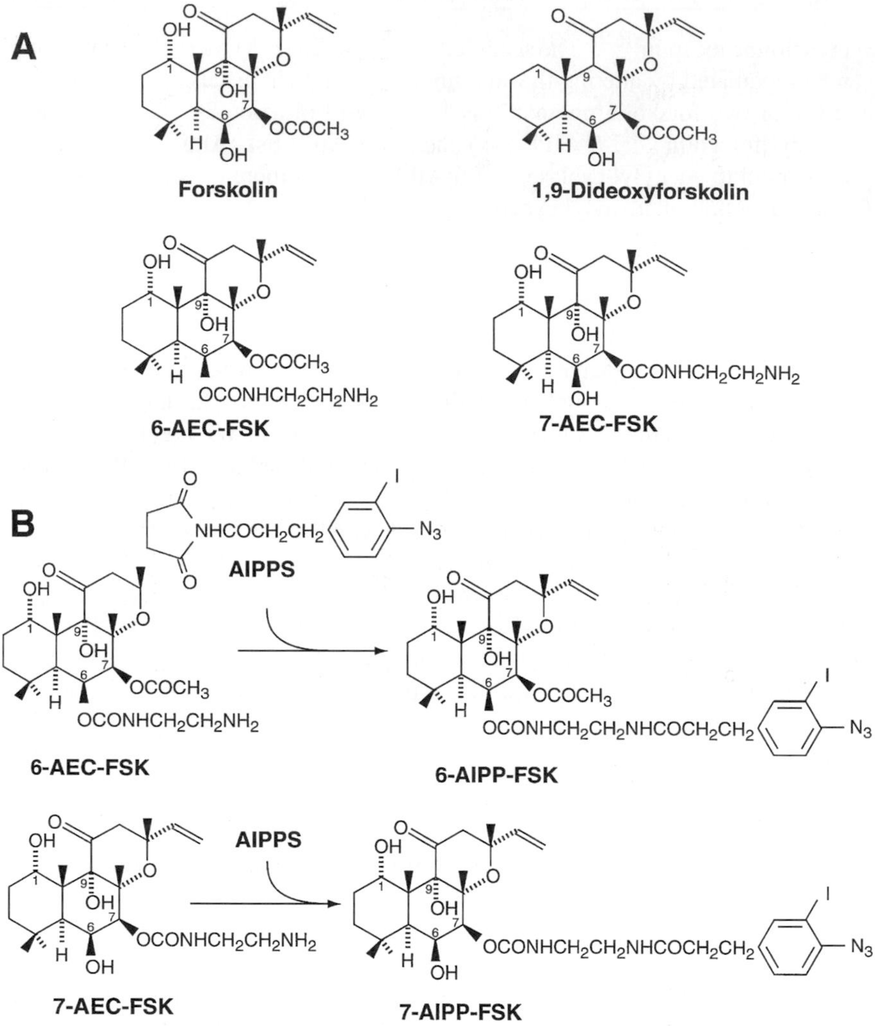

FIG. 1. (A) Structures of forskolin and derivatives. (B) Synthesis of ^{125}I-6-AIPP-Fsk and ^{125}I-7-AIPP-Fsk by reaction of ^{125}I-AIPPS with 6-aminoethylcarbamylforskolin and 7-aminoethylcarbamyl-forskolin, respectively.

^{125}I-AIPPS migrates with an R_f of 0.9 and ^{125}I-6-AIPP-Fsk and ^{125}I-7-AIPP-Fsk migrate with R_f values of 0.6. The starting materials, 6-AEC-Fsk and 7-AEC-Fsk, do not migrate from the origin under these conditions. Thus, an excess of 6-AEC-Fsk or 7-AEC-Fsk is used to improve the yield without decreasing the specific activity of the product. The reaction mixture is applied to a small silica column (0.9 × 2 cm) equilibrated with ethyl acetate. Fractions (0.3 ml) are collected, and each fraction is analyzed by thin-layer chromatography on silica plates followed by autoradiography. The ^{125}I-6-AIPP-Fsk and ^{125}I-7-AIPP-Fsk usually

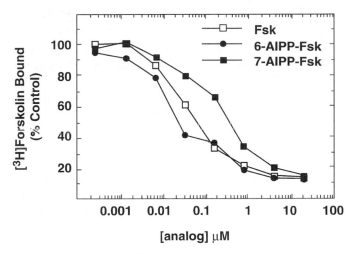

FIG. 2. Competitive displacement of [³H]forskolin binding to bovine brain membranes by 6-AIPP-Fsk and 7-AIPP-Fsk. Brain membranes were incubated with 30 nM [³H]forskolin and increasing concentrations of 6-AIPP-Fsk or 7-AIPP-Fsk. Data are plotted as a percentage of control (no drug). [Reprinted from D. I. Morris, J. D. Robbins, A. E. Ruoho, E. M. Sutkowski, and K. B. Seamon, *J. Biol. Chem.* **266**, 13377 (1991), with permission.]

elute in two or three fractions between the eighth and eleventh fractions. The fractions corresponding to ^{125}I-6-AIPP-Fsk or ^{125}I-7-AIPP-Fsk are pooled, the ethyl acetate is removed under N_2, and the material is redissolved in ethanol at a concentration of about 1 mCi/ml. ^{125}I-6-AIPP-Fsk and ^{125}I-7-AIPP-Fsk are completely separated from 6-AEC-Fsk and 7-AEC-Fsk under these chromatographic conditions and, therefore, the labeled compounds are assumed to be carrier free with the specific activity of the ^{125}I-AIPPS reagent, about 2200 Ci/mmol. The yields for the radioactive products are typically about 80% of the starting radioactivity. The purified ^{125}I-6-AIPP-Fsk and ^{125}I-7-AIPP-Fsk are stable for at least 3 months when stored in ethanol at room temperature.

[³H]Forskolin binds with high affinity to adenylyl cyclase, but not the glucose transporter, in bovine brain membranes.[17] The ability of 6-AIPP-Fsk and 7-AIPP-Fsk to bind to adenylyl cyclase is determined by their inhibition of [³H]forskolin binding to brain membranes (Fig. 2). 6-AIPP-Fsk inhibits [³H]forskolin binding with a 50% inhibitory concentration (IC_{50}) of 15 nM and is slightly more potent than forskolin, which has an IC_{50} of 30 nM. In contrast, 7-AIPP-Fsk is about 10-fold less potent than forskolin and 6-AIPP-Fsk (IC_{50} of 200 nM).

Purification of Adenylyl Cyclase

Adenylyl cyclase is purified with a forskolin affinity column. Bovine brain membranes are preactivated with 100 μM GTPγS with 10 mM $MgCl_2$ for 30 min

[17] A. Laurenza and K. B. Seamon, *Methods Enzymol.* **195,** 52 (1991).

at 30°, solubilized for 1 hr with a 5 mM Lubrol buffer containing 10 mM morpholine propane sulfonic acid (MOPS, pH 7.5), 5 mM EDTA, 1 mM dithiothreitol (DTT), and 1 mM MgCl$_2$, and ultracentrifuged at 35,000 rpm for 2 hr at 4° in a Ti 40 rotor. The solubilized proteins are diluted with an equal volume of column equilibration buffer [10 mM MOPS (pH 7.5), 1 mM Tween 60, 1 mM dithiothreitol, 1 mM EDTA, 1 mM MgCl$_2$, 0.1 M NaCl plus protease inhibitors (1 mM benzamidine, 1 μM leupeptin, 1 μM pepstatin)] and applied to a forskolin affinity column overnight. The column is washed with equilibration buffer containing 1 M NaCl, and then 0.5 M NaCl. Bound proteins are eluted with this buffer containing 100 μM forskolin. Column fractions are assayed for adenylyl cyclase activity with 100 μM forskolin and 5 mM MgCl$_2$ as previously described.[18] Peak fractions containing adenylyl cyclase are pooled, concentrated 5-fold, and frozen in aliquots in liquid nitrogen. This preparation is a complex of the cyclase with stimulatory G protein α subunit (G$_s\alpha$).

Expression and Purification of Adenylyl Cyclase Fragments IC1 and IIC2 in Escherichia coli

A functional adenylyl cyclase molecule can be formed from the combination of two intracellular domains of cyclase (C1 and C2).[19] The type I C1 domain (IC1) and the type II C2 domain (IIC2) are expressed and purified as described by Tang et al. The cocrystal of forskolin with the C2 homodimer[20,21] or with the C1/C2 heterodimer[22] indicates that forskolin binds at the interface between the monomers.

Photoaffinity Labeling Procedure

Brain Membranes. Brain membranes are incubated on ice in 15-ml Pyrex Sorvall tubes (2-mm-thick Pyrex, 18 × 100 mm; DuPont, Wilmington, DE) in the presence of buffer alone (10 mM Tris, pH 7.4) or buffer containing drug in a total volume of 100 μl for 30 min. ^{125}I-6-AIPP-Fsk or ^{125}I-7-AIPP-Fsk is added to the membranes to a final concentration of approximately 3 nM. The labels are added directly from the ethanol solutions, and the final concentration of ethanol in the assays does not exceed 2% (v/v). The membranes are incubated for an additional 30 min on ice in the dark. Immediately before photolysis, 0.9 ml of the corresponding buffer solution containing the drug used for competition is added, and the diluted samples are quickly photolyzed (5 sec) in an ice–water bath at a distance of 10 cm from a high-pressure 1-kW AH-6 mercury lamp. Alternatively,

[18] D. Morris, E. McHugh-Sutkowski, M. Moos, Jr., W. F. Simonds, A. M. Spiegel, and K. B. Seamon, *Biochemistry* **29,** 9079 (1990).

[19] W. J. Tang and A. G. Gilman, *Science* **268,** 1769 (1995).

[20] G. Zhang, Y. Liu, j. Qin, B. Vo, W. J. Tang, A. E. Ruoho, and J. H. Hurley, *Protein Sci,* **6,** 903 (1997).

[21] G. Zhang, Y. Liu, A. E. Ruoho, and J. H. Hurley, *Nature* (*London*) **386,** 247 (1997).

[22] J. J. Tesmer, R. K. Sunahara, A. G. Gilman, and S. R. Sprang, *Science* **278,** 1907 (1997).

samples can be photolyzed in microcentrifuge tubes in a Stratagene (La Jolla, CA) UV Stratalinker (energy, 200,000 μJ) for 1 min. Immediately after photolysis, 10 μl of 2-mercaptoethanol (final concentration of 1%, v/v) is added to each tube as a scavenger for long-lived reactive species. The membranes are centrifuged at 16,000g for 15 min at 4°, and the supernatants are removed. Membrane pellets are resuspended in SDS–PAGE sample buffer. After SDS–PAGE on an 10% (w/v) gel, the labeled proteins are detected by autoradiography.

As can be seen in Fig. 3, a number of protein bands are labeled by [125]I-6 AIPP-Fsk and ^{125}I-7-AIPP-Fsk in the absence of any competitors and can be observed after overnight exposure of the dried gels to X-ray film (Fig. 3, lanes marked *None*). The specificity of labeling is determined by preincubating the membranes with buffer, 100 μM forskolin, 100 μM 1,9-dideoxyforskolin, or 100 μM cytochalasin B and then with the photoaffinity labels. ^{125}I-6-AIPP-Fsk labels a 115-kDa protein that is inhibited by forskolin but not by 1,9-dideoxyforskolin or cytochalasin B (Fig. 3A, position of cyclase indicated by the arrow). ^{125}I-6-AIPP-Fsk also labeled three other proteins of 60, 33, and 18 kDa; however, the labeling of these proteins is unaffected by any of the drugs added for competition, and is therefore nonspecific.

The concentration dependence for forskolin inhibition of photolabeling of ^{125}I-6-AIPP-Fsk is determined in bovine brain membranes. Membranes are preincubated with increasing concentrations of forskolin and then with ^{125}I-6-AIPP-Fsk. Labeling of the 115-kDa protein is completely inhibited by forskolin at concentrations greater than 0.1 μM (Fig. 4). 1,9-Dideoxyforskolin does not inhibit the labeling of the 115-kDa protein when used at the same concentrations. Occasionally, the labeling of the 115-kDa protein can be inhibited by high concentrations (100 μM) of 1,9-dideoxyforskolin and cytochalasin B; however, this inhibition is slight and much less than that observed with forskolin (not shown).

[125]I-7-AIPP-Fsk labels proteins of 60, 33, and 18 kDa that are similar to those labeled by ^{125}I-6-AIPP-Fsk; however, a 115-kDa protein is not labeled (Fig. 3B). Two proteins of 94 and 43 kDa, also photolabeled by ^{125}I-7-AIPP-Fsk, are not detected with ^{125}I-6-AIPP-Fsk (Fig. 3A). The 43-kDa protein labeled with ^{125}I-7-AIPP-Fsk is the only protein whose labeling is inhibited by forskolin, 1,9-dideoxyforskolin, and cytochalasin B. The ability of these drugs to inhibit the labeling of the 43-kDa protein suggests that this protein may be the glucose transporter. Highly specific labeling of the glucose transporter (GLUT1) has been demonstrated with ^{125}I-7-AIPP-Fsk.[23,24]

Affinity-Purified Cyclase. For photolabeling of purified adenylyl cyclase, excess forskolin is removed before photolysis from the preparation by gel filtration on a 10-ml Ultra Gel 202 column equilibrated with 10 mM MOPS (pH 7.5), 1 mM Tween 60, 1 mM EDTA, 5 mM MgCl$_2$, and 0.1 M NaCl. Concentrated purified adenylyl cyclase (1 ml) is applied to the column eluted with equilibration buffer.

[23] B. E. Wadzinski, M. F. Shanahan, and A. E. Ruoho, *J. Biol. Chem.* **262,** 17683 (1987).
[24] B. E. Wadzinski, M. F. Shanahan, K. B. Seamon, and A. E. Ruoho, *Biochem. J.* **272,** 151 (1990).

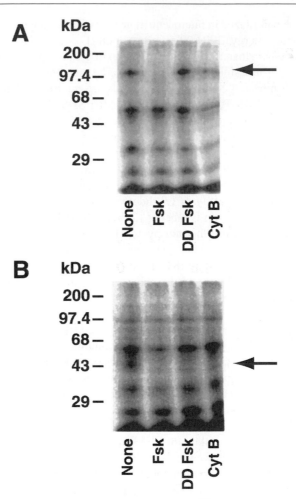

FIG. 3. Photolabeling of bovine brain membranes with (A) ^{125}I-6-AIPP-Fsk and (B) ^{125}I-7-AIPP-Fsk. Brain membranes were preincubated with either buffer alone (*None*), 100 μM forskolin *(Fsk)*, 100 μM 1,9-dideoxyforskolin *(DDFsk)*, or 100 μM cytochalasin B *(CytB)*. Shown is the overnight autoradiogram of the 10% (w/v) SDS–polyacrylamide gel. The positions of the cyclase (A) and glucose transporter (B) are indicated by arrows. [Reprinted from D. I. Morris, J. D. Robbins, A. E. Ruoho, E. M. Sutkowski, and K. B. Seamon, *J. Biol. Chem.* **266**, 13377 (1991), with permission.]

Fractions (1 ml) are assayed for adenylyl cyclase activity, and the two peak fractions are pooled. The photolabeling is performed according to the same procedure outlined for the brain membranes described above. Adenylyl cyclase (activity, ~660 pmol/min) is added per sample in a final incubation volume of 100 μl. After photolysis and addition of 10 μl of 2-mercaptoethanol, the proteins are precipitated by adding 1/10 volume of 0.15% (w/v) deoxycholate and incubated for 10 min, followed by the addition of 1/10 volume of 72% (w/v) trichloroacetic acid. The

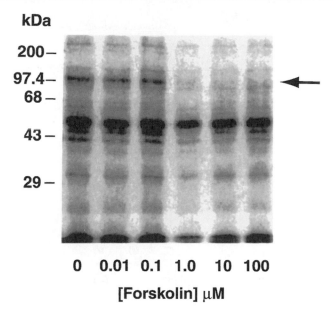

kDa

200—

97.4—
68—

43—

29—

0   0.01   0.1   1.0   10   100

[Forskolin] μM

FIG. 4. Forskolin concentration dependence for inhibition of ^{125}I-6-AIPP-Fsk photolabeling. Bovine brain membranes were preincubated with the indicated concentrations of forskolin and then photolabeled with ^{125}I-6-AIPP-Fsk. Shown is the overnight autoradiogram of the 10% (w/v) SDS–polyacrylamide gel. The position of the cyclase is indicated by the arrow. [Reprinted from D. I. Morris, J. D. Robbins, A. E. Ruoho, E. M. Sutkowski, and K. B. Seamon, *J. Biol. Chem.* **266,** 13377 (1991), with permission.]

proteins are collected by centrifugation at 16,000*g* for 15 min at 4°, washed with ice-cold acetone, and centrifuged as described previously before for 5 min. The pellets are resuspended in SDS–PAGE sample buffer and elecrophoresed on an 8% (w/v) polyacrylamide gel. Labeled proteins are detected by autoradiography of the dried gel.

As can be seen in Fig. 5, two protein bands of 102 and 25 kDA are labeled by ^{125}I-6-AIPP-Fsk. Labeling of the 102-kDa protein is inhibited by forskolin but not by 1,9-dideoxyforskolin. In contrast, the labeling of the 25-kDa protein is not affected by forskolin or 1,9-dideoxyforskolin. The inability of 1,9-dideoxyforskolin to inhibit the labeling of the 102-kDa protein is consistent with derivatization of adenylyl cyclase.

Soluble Fragments of Adenylyl Cyclase. For the soluble IC1 and IIC2 fragments of adenylyl cyclase, 1 μg of purified IIC2 and 0.1 μg of purified IC1 are combined in a Pyrex Sorvall tube in 100 μl of 10 m*M* Tris, pH 7.4. The C1 fragment is needed in small quantities in order to achieve high cyclase activity. The samples are preincubated for 30 min on ice in the presence of buffer or buffer containing drug. ^{125}I-6-AIPP-Fsk is added to a final concentration of approximately 1 n*M* and the samples are incubated for an additional 30 min on ice in the dark. The samples

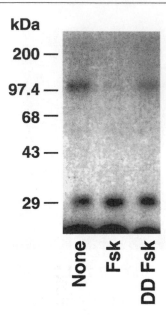

FIG. 5. Photoaffinity labeling of partially purified adenylyl cyclase. Adenylyl cyclase was purified with a forskolin affinity column. The purified sample was photolabeled with ^{125}I-6-AIPP-Fsk in the absence (*None*) or presence of 100 μM forskolin (*Fsk*) or 100 μM 1,9-dideoxyforskolin (*DDFsk*). Shown is the autoradiogram of an 8% (w/v) SDS–polyacrylamide gel. [Reprinted from D. I. Morris, J. D. Robbins, A. E. Ruoho, E. M. Sutkowski, and K. B. Seamon, *J. Biol. Chem.* **266,** 13377 (1991), with permission.]

are photolyzed as described above without dilution and 1 μl of 2-mercaptoethanol is added immediately after photolysis. The samples are not manipulated further before analysis by SDS–PAGE. After electrophoresis on a 10% (w/v) polyacrylamide gel, labeled proteins are detected with a Molecular Dynamics (Sunnyvale, CA) PhosphorImager.

Figure 6 is a PhosphorImager scan of the ^{125}I-6-AIPP-Fsk-labeled IIC2 fragment. A polypeptide with an approximate molecular mass of 27 kDa is labeled by ^{125}I-6-AIPP-Fsk. The IIC2 fragment can be seen to be specifically labeled as further suggested from the forskolin/IIC2 homodimer cocrystal structure.[21] IIC2 alone can also be specifically labeled with ^{125}I-6-AIPP-Fsk, although not as efficiently as in the presence of IC1 (data not shown). The labeling is protected by 100 μM forskolin (Fig. 6, lane *Fsk*) but not by 1,9-dideoxyforskolin (Fig. 6, lane *DDFsk*), consistent with binding to adenylyl cyclase.

Comments

^{125}I-6-AIPP-Fsk and ^{125}I-7-AIPP-Fsk are easily synthesized from the reaction of the *N*-hydroxysuccinimide-activated ester, ^{125}I-AIPPS, with the appropriate

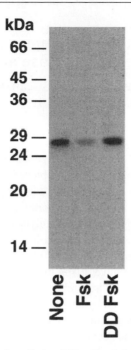

FIG. 6. Photoaffinity labeling of purified soluble adenylyl cyclase fragments IC1 and IIC2 by ^{125}I-6-AIPP-Fsk. The soluble cyclase fragments IC1 and IIC2 were expressed in *E. coli* and purified. IC1 and IIC2 were incubated at a ratio of 1 : 10 (w : w) in the absence (*None*) or presence of 100 μM forskolin (*Fsk*) or 100 μM 1,9-dideoxyforskolin (*DDFsk*). Shown is an overnight PhosphorImager scan of the dried 10% (w/v) SDS–polyacrylamide gel.

primary amine derivative of forskolin, 6-AEC-Fsk or 7-AEC-Fsk. This reaction is rapid and results in the incorporation of approximately 80% of the starting radioactivity into the photolabel. The purification is straightforward and completely separates the final products from the starting materials. While both photoaffinity labels have binding affinity for adenylyl cyclase, ^{125}I-6-AIPP-Fsk is a more effective photolabel for cyclase. This may be due to its higher affinity or, more likely, to a more favorable positioning of the photoreactive phenyl azide in the binding site as evidenced from the forskolin/IIC2 homodimer[21] and the VC1/IIC2 heterodimer cocrystal structures.[22]

[15] Crystallization of Complex between Soluble Domains of Adenylyl Cyclase and Activated $G_s\alpha$

By JOHN J. G. TESMER, ROGER K. SUNAHARA, DAVID A. FANCY,
ALFRED G. GILMAN, and STEPHEN R. SPRANG

X-ray crystallography has opened an exciting window into the physical chemistry of adenylyl cyclase. The three-dimensional structures that have been determined of individual domains[1] and complexes of the essential catalytic core with activators and substrate analogs[2–4] have provided important insights into the mechanism of catalysis and activation. Described here are the procedures used to generate crystals of the adenylyl cyclase catalytic core bound to its activators $G_s\alpha$ (stimulatory G protein α subunit) and forskolin.

Vertebrate G-protein-regulated isoforms of adenylyl cyclase are integral membrane proteins that appear to have arisen by gene duplication. Following a short amino-terminal segment is a hydrophobic region predicted to contain six transmembrane helices (M_1) followed by an ~40-kDa cytoplasmic domain (C_1). These two segments are approximately duplicated in the membrane-spanning (M_2) and cytosolic (C_2) domains of the second repeat.[5] Each of the two cytoplasmic segments contains a domain of 200–250 amino acid residues (C_{1a}, C_{2a}) that is roughly 30% identical in sequence to its partner and 60–80% identical to the corresponding domains of other adenylyl cyclase isoforms. Both conserved domains precede cytoplasmic segments (C_{1b}, C_{2b}) that are variable in sequence and length, and that may confer isoform-specific properties. It has been established that the catalytic apparatus,[6,7] as well as sites involved in the regulation of adenylyl cyclase by forskolin and heterotrimeric G protein α and $\beta\gamma$ subunits,[5,6,8,9] are contained within the cytoplasmic regions of the enzyme. The C_{1a} and C_{2a} domains, produced as soluble recombinant proteins in *Escherichia coli,* can, when combined, reconstitute forskolin and $G_s\alpha$-regulated adenylyl cyclase activity at levels comparable to that seen in membrane preparations of the holoenzyme, and to fusion proteins

[1] G. Zhang, Y. Liu, Y. A. E. Ruoho, and J. H. Hurley, *Nature (London)* **386,** 247 (1997).

[2] J. J. G. Tesmer, R. K. Sunahara, R. A. Johnson, A. G. Gilman, and S. R. Sprang, *Science* **285,** 756 (1999).

[3] J. J. G. Tesmer, R. K. Sunahara, A. G. Gilman, and S. R. Sprang, *Science* **278,** 1907 (1997).

[4] J. J. T. Tesmer, C. W. Dessauer, R. K. Sunahara, L. D. Murray, R. A. Johnson, A. G. Gilman, and S. R. Sprang, *Biochemistry* **39,** 14464 (2000).

[5] R. K. Sunahara, C. W. Dessauer, and A. G. Gilman, *Annu. Rev. Pharmacol. Toxicol.* **36,** 461 (1996).

[6] W.-J. Tang and A. G. Gilman, *Science* **268,** 1769 (1995).

[7] C. W. Dessauer, T. T. Scully, and A. G. Gilman, *J. Biol.Chem.* **272,** 22272 (1997).

[8] R. E. Whisnant, A. G. Gilman, and C. W. Dessauer, *Proc. Natl. Acad. Sci. U.S.A.* **93,** 6621 (1996).

[9] C. W. Dessauer, J. J. G. Tesmer, S. R. Sprang, and A. G. Gilman, *J. Biol.Chem.* **273,** 25831 (1998).

composed of linked C_{1a} and C_{2a} domains.[8] This fortunate circumstance has made structural studies of adenylyl cyclase feasible, because it is impracticable to produce the intact enzyme in sufficient quantities for crystallization. Even the soluble C_{1a}–C_{2a} fusion constructs are largely susceptible to degradation when expressed in heterologous systems.[6]

Here we describe the procedures used to obtain crystals of a complex comprising the C_{1a} and C_{2a} domains of adenylyl cyclase and activated, guanosine 5'-(γ-thio)triphosphate (GTPγS)-bound $G_s\alpha$. Formation of this complex is strongly promoted by forskolin, or its soluble analogs. Both the catalytic activity of C_{1a}:C_{2a}, and the affinity of the two domains for each other, are simulated synergistically by forskolin and $G_s\alpha \cdot$ GTPγS. Complexes formed in the absence of forskolin have been purified when all components are present in high concentration, and a complex of C_{1a} with C_{2a} can be purified without $G_s\alpha \cdot$GTPγS so long as a tight-binding inhibitor such as $2',5'$-dideoxyadenosine-$3'$-triphosphate ($2'5'$dd3'ATP$\cdot$Mg^{2+})[10] is present. However, such complexes have yet to be crystallized despite several attempts. Soluble forms of forskolin, however, can be backsoaked out of crystals of the complex, although the diffraction quality of the crystals ultimately suffers. In the absence of either activator, both domains form homodimers in solution.[11]

The C_{1a} and C_{2a} domains that we have used are derived from different adenylyl cyclase isoforms and mammalian species. Isoforms have been selected largely on the basis of ease of expression and availability of clones. We have grown crystals containing C_{1a} from isoform V of canine adenylyl cyclase (VC$_1$) and C_{2a} from isoform II of rat (IIC$_2$) together with bovine $G_s\alpha$.[3] Regulation by $G_s\alpha$ and forskolin is a common property of all isoforms of adenylyl cyclase (except isoform IX in the latter case[5]), and so it is not surprising that hybrid cyclase combinations retain the catalytic and regulatory properties of native enzyme.[11] On the other hand, regulation by inhibitory G protein α subunit (G$_i\alpha$), calmodulin, $G\beta\gamma$, Ca^{2+}, and phosphorylation is isoform specific. Hence, crystallization of complexes with other regulators is expected to require domain constructs that are derived from different isoforms and contain polypeptide segments outside of the C_{1a} and C_{2a} domains.

Reagents

Purification of Hexahistidyl-VC$_1$

The VC$_1$ domain used for these experiments is produced by expression of a cDNA encoding residues 364–580 of canine adenylyl cyclase V. The expression

[10] L. Desaubry, I. Shoshani, and R. A. Johnson, *J. Biol. Chem.* **271,** 2380 (1996).
[11] R. K. Sunahara, C. Dessauer, R. Whisnant, C. Kleuss, and A. G. Gilman, *J. Biol. Chem.* **272,** 22265 (1997).

construct is ligated into pQE60-H6 (Qiagen, Chatsworth, CA) at NcoI and HindIII sites. The expressed protein contains a hexahistidine tag at its amino terminus, followed by the residue corresponding to Met-564. *Escherichia coli* BL21(DE3) is cotransformed with pREP4[12] and pQE60-H6-VC$_1$ and grown overnight at 30° in 200 ml of Luria–Bertani (LB) medium containing 50 μM ampicillin and 50 μM kanamycin. This culture is used to inoculate 10 liters of T7 medium,[13] which is then incubated at 30°. Cells are then induced with 30 μM isopropyl-β-D-thiogalactopyranoside (IPTG) at an OD$_{600}$ of 1.2 and further incubated for 4 hr. Cells are harvested by centrifugation at 3000g for 15 min at 4°, frozen in liquid nitrogen, and stored at $-70°$. All subsequent protein extraction and purification steps are carried out at 4°. For lysis, cells are pulverized and suspended with a Polytron homogenizer (Brinkmann, Westbury, NY) in 1 liter of buffer A [50 mM Tris-HCl, (pH 8.0), 120 mM NaCl, 1 mM 2-mercaptoethanol (2-ME), and the following proteinase inhibitors at 22 mg/liter each: L-1-tosylamido-2-phenylethyl chloromethyl ketone, 1-chloro-3-tosylamido-7-amino-2-heptanone, and phenylmethylsulfonyl fluoride, plus 3.2 mg/liter each of leupeptin and lima bean trypsin inhibitor] for 30 min. Lysis is initiated by addition of lysozyme (0.25-μg/ml concentration) to the stirring cell suspension and incubated for 30 min on ice. The slurry is treated with DNase I (8 μg/ml final) and 1 mM MgCl$_2$ for an additional 30 min. Cellular debris is removed by centrifugation at 100,000g for 40 min. Lysates should be prepared for immediate purification and not frozen.

VC$_1$ is purified by a two-step procedure. Clarified cell lysate (1 liter) is applied to 7-ml TALON metal chelate column (Clontech, Palo Alto, CA) preequilibrated in buffer A, and subsequently washed with 20 column volumes of buffer A containing 500 mM NaCl, followed by 10 column volumes of buffer B [20 mM Tris-HCl (pH 8.0), 50 mM NaCl, and 1 mM 2-ME including protease inhibitors]. Hexahistidyl (His$_6$)–VC$_1$ is eluted from the column in 2-ml fractions with buffer B containing 100 mM imidazole. Peak fractions are pooled, and diluted with 2 volumes of buffer C [20 mM Na–HEPES (pH 8), 2 mM MgCl$_2$, 1 mM EDTA, 1 mM dithiothreitol (DTT) and protease inhibitors] and applied to a Mono S 5/5 column (Amersham Pharmacia Biotech, Piscataway, NJ). The column is washed with a 5-ml linear gradient of NaCl (20–400 mM NaCl in buffer C) and eluted with a 25-ml gradient of NaCl (400–1000 mM NaCl in buffer C) at a flow rate of 1 ml/min in 0.5-ml fractions. Peak fractions, eluting between 500 and 600 mM NaCl, are pooled and concentrated to 3 mg/ml, using a Centricon 30 (Millipore, Bedford, MA) concentration device. Typically, 3–5 mg of pure protein is obtained from 25 g of cells. The catalytic activity of hexahistidyl-VC$_1$ in the presence of IIC$_2$ and activators is measured by the procedure described in [10] this volume.[13a]

[12] D. Stuber, H. Matile, and G. Garotta, *Immunol. Methods* **4,** 121 (1990).

[13] E. Lee, M. Linder, and A.G. Gilman, *Methods Enzymol.* **237,** 146 (1994).

[13a] M.E. Hatley, A.G. Gilman, and R. K. Sunahara, *Methods Enzymol.* **345,** [10] 2001 (this volume).

Expression and Purification of IIC$_2$

The IIC$_2$ domain is derived from a cDNA clone encoding residues 874–1081 of rat ACII, in the expression vector pQE60. The amino and carboxyl termini of this protein correspond to a clostripain cleavage product that was found to fully retain the catalytic activity of larger IIC$_2$ constructs. The untagged protein is expressed in BL21 (DE3) cells cotransformed with pREP4. Ten liters of LB medium (and ampicillin and kanamycin at 50 μg/ml each) is inoculated with a 100-ml cell culture and allowed to grow at 30° to an OD$_{600}$ of 0.6. Cells are then induced with IPTG (30 μg/ml) and allowed to grow for an additional 12 hr at room temperature. Cells are harvested by centrifugation and stored as described for VC1.

For lysis, cells are pulverized and suspended with a Polytron homogenizer in 1 liter of buffer D [50 mM Tris-HCl (pH 8.0), 50 mM NaCl, 2 mM DTT, and proteinase inhibitors]. Lysis is initiated by addition of lysozyme (0.2 mg/ml final) to the stirring cell suspension and incubated for 30 min on ice. The slurry is treated with DNase I (8 μg/ml final) and 1 mM MgCl$_2$ for an additional 30 min. Cellular debris is removed by centrifugation at 100,000 g for 40 min at 4°. Clarified lysates are filtered through a 0.22-μm pore size filter and applied to tandem 75-ml Q-Sepharose (Amersham Pharmacia Biotech) fast protein liquid chromatography (FPLC) columns. The column is washed with 10 column volumes of buffer C (containing protease inhibitors). IIC2 is eluted with a linear NaCl gradient (50–300 mM NaCl in buffer D). Peak fractions are determined by adenylyl cyclase activity measurements[13a] and typically elute at 150–250 mM NaCl. Peak fractions are passed through a 50-ml ceramic hydroxyapatite (Bio-Rad, Hercules, CA) column to remove major contaminants. Hydroxyapatite flowthrough fractions (typically 3–5 g) are divided into four equal aliquots and separately diluted into ammonium sulfate (AS, 1.2 M final concentration in buffer C) and applied to a 100-ml phenyl Superose (Amersham Pharmacia Biotech) FPLC column. IIC2 is eluted from the column with a decreasing linear gradient of AS (1.2–0 M final in buffer C). IIC2-containing peaks (typically eluting between 0.4 and 0.2 M AS) are easily identified by sodium dodecyl sulfate–polyacrylamide gel electrophoresis (SDS–PAGE) and Coomassie staining. Protein-containing peaks are serially dialyzed (using membranes with a 10,000 molecular weight cutoff) against 1000 volumes of buffer C. Typical yields are greater than 1 g of purified IIC2 per 10-liter culture. Protein is concentrated to 10–20 mg/ml, snap frozen in liquid N$_2$, and stored at −70°.

Expression and Purification of Hexahistidyl-G$_s\alpha$

VC1 · IIC2 complexes are formed with the short-splice form of bovine G$_s\alpha$ in the activated, GTPγS-bound state. Amino-terminally hexahistidine-tagged G$_s\alpha$ from a pQE60 expression system is produced in BL21 (DE3) cells, and purified as described in this series.[13] The GDP present in recombinant $G_s\alpha$ must be replaced with $GTP\gamma S$. Nucleotide exchange is accomplished by incubating $G_s\alpha \cdot GDP$ (4 mg/ml) in 50 mM Na–HEPES, (pH 8.0), 10 mM MgSO$_4$, 1 mM EDTA, 2 mM DTT, and 800 μM $GTP\gamma S$ for 30 min at 30° (or overnight at 4°). $G_s\alpha$ is then subjected to limited proteolysis by incubating the $GTP\gamma S$-labeled $G_s\alpha$ with trypsin ($G_s\alpha$: trypsin molar ratio, 100 : 1) for 1 hr at room temperature. Trypsin cleaves $G_s\alpha \cdot GTP\gamma S$ after Arg-38, removing the N-terminal hexahistidine tag and amino-terminal helix, and after Arg-389, at the end of the α_5 helix. The $GTP\gamma S$-labeled and trypsinized $G_s\alpha$ is passed through a Ni^{2+}–nitrilotriacetic acid metal chelate column (Ni–NTA) to remove undigested His$_6$-tagged protein and peptide fragments. The flowthrough from the Ni–NTA column is applied to a Mono Q 10/10 FPLC column (Amersham Pharmacia Biotech) preequilibrated in buffer C. The $GTP\gamma S$-labeled and partially trypsinized $G_s\alpha$ is eluted with an NaCl gradient (0–300 mM NaCl in buffer C). Peak fractions are identified by SDS–PAGE and Coomassie blue staining.

Formation of VC1:IIC2:forskolin:$G_s\alpha$·$GTP\gamma S$ complex

VC1 (100 μM), IIC2 (200 μM), $G_s\alpha \cdot GTP\gamma S$ (100 μM), and forskolin (50 μM) are incubated for 30 min on ice in gel-filtration buffer containing 20 mM Na–HEPES (pH 8.0), 2 mM MgCl$_2$, 1 mM EDTA, 2 mM DTT, 100 mM NaCl, 25 μM forskolin, and 10 μM $GTP\gamma S$. The mixture is then applied to a Superdex 75 (HR 10/30) gel-filtration column in tandem with a Superdex 200 (HR 10/30) column (Amersham Pharmacia Biotech). The column is eluted with gel-filtration buffer and 0.5-ml fractions containing complex are pooled (Fig. 1); DTT, 7-deacetyl-7-(O-N-methylpiperazino)-γ-butyrylforskolin (MPFsk), a soluble forskolin derivative, and $GTP\gamma S$ are added to final concentrations of 5 mM, 200 μM, and 500 μM, respectively. The complex is concentrated to approximately 8 mg/ml with a Centricon 50 concentrator, aliquoted, and frozen in liquid nitrogen. About half of the preparations of this complex failed to produce useful crystals, either resulting in poorly formed specimens or no crystals at all. We have been unable to determine why certain preparations do not yield crystals, as all appear to be pure by polyacrylamide gel electrophoresis and contain 1 : 1 : 1 complexes of the three protein components.

Crystallization of VC1:IIC2:forskolin:$G_s\alpha$·$GTP\gamma S$ Complex

Crystals of the complex are obtained by vapor diffusion at 20°, either by a sitting drop or hanging drop configuration as described elsewhere in this series.[14] For a hanging drop experiment, a 2- to 3-μl drop containing concentrated solution of

[14] P. Weber, "Overview of Protein Crystallization Methods" (C. Carter, Jr. and R. Sweet, eds.). *Methods Enzymol.* **276,** 13 (1997).

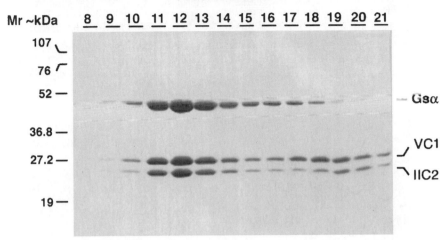

FIG. 1. Purification of VC_1 associated with IIC_2 and $GTP\gamma S \cdot G_s\alpha$. Trypinized $G_s\alpha \cdot GTP\gamma S$ (50 nmol), VC_1 (75 nmol), and IIC_2 (50 nmol) were incubated for 30 min on ice in gelfiltration buffer [20 mM Na–HEPES (pH 8.0), 50 mM, NaCl, 2 mM $MgCl_2$, 1 mM EDTA] containing 100 μM forskolin. The complex was concentrated to approximately 100 μM with a Centricon 50 and applied to tandemly arranged Superdex 75 and 200 columns. The complex was eluted in gel-filtration buffer containing 100 μM forskolin. Fractions (400 μl) were collected. Ten microliters of each fraction was loaded and resolved on a 15% SDS–polyacrylamide gel and stained with Coomassie blue. The positions to which molecular weight markers (not shown) migrate are indicated on the left margin of the gel.

the complex is deposited on a clean 22-mm siliconized glass coverslip (Hampton Research, Laguna Niguel, CA). Into this drop is injected an equal volume of well solution containing 7.2–7.5% (w/v) polyethylene glycol (PEG) 8000, 500 mM NaCl, and 100 mM 2-(N-morpholino)-ethanesufonic acid (MES, pH 5.4–5.6). The coverslip is then inverted over a well of either a Linbro or VDX 24-well plate (Hampton Research) containing 1 ml of the reservoir solution. To maintain an air-tight seal between the coverslip and the reservoir, a continuous bead of silicone grease is applied to the lip of the well before affixing the coverslip. In placing the inverted coverslip over the well, care is taken to ensure that a seal is made over the entire contact surface between the coverslip and the lip of the well. The crystallization plate is placed in a vibration-free, constant-temperature incubator maintained at either 15 or 20°. We have used a Precision Scientific (Winchester, VA) model 115 low-temperature incubator for this purpose. To minimize vibration, incubator shelves are covered with 1-inch-thick polyurethane mattress foam. Crystal plates are supported on a 3/8-inch-thick sheet of Plexiglas placed over the foam. Crystals typically appear as haystacks of thin plates within 2 days and grow to their maximum size over a period of 2 weeks (Fig. 2). It is advisable to set up a grid of experiments, varying PEG concentration and pH, in increments of 0.25% (w/v) PEG and 0.2 pH unit around the precipitant composition given above. Larger droplet volumes can be

FIG. 2. Crystals of the complex of VC_1 with IIC_2, forskolin, and GTPγS $\cdot$ G$_s\alpha$. Grown in a hanging drop experiment under conditions described in text, these crystals attain maximum dimension of approximately 200 μm and are 10–20 μm thick.

accommodated in a sitting drop apparatus. For these experiments, 5–10 μl of protein solution is mixed with an equal volume of precipitant solution in the central reservoir of one well of a Cryschem 24-well plate (22-mm well diameter) obtained from Hampton Research. The surrounding well is filled with 1 ml of precipitant solution; the lip of the well is greased and sealed with a glass coverslip as described above.

Single crystals of suitable size for data collection (Fig. 2; typically 200 × 100 × 10 μm) can be obtained by breaking up the stacks with a cat's whisker. Crystals are plucked out of solution with a 100- or 200-μm-diameter cryoloop mounted on a stainless steel pin (Hampton Research) and then transferred to a harvesting solution containing 100 mM MES (pH 5.4), 9% (w/v) PEG 8000, 500 mM NaCl, 5 mM MgCl$_2$, 20 mM Na–HEPES (pH 8.0), 1 mM EDTA, 2 mM DTT, 200 μM MPFsk, 100 μM GTPγS, and 30% (v/v) PEG 400 as a cryoprotectant. The well of a sitting-drop plate is convenient for this purpose. Inhibitors and substrate analogs, or other potential ligands, can be introduced at this stage. For example, P-site-inhibited complexes are obtained by soaking crystals in a harvesting solution that also includes either 3.5 mM 2'-deoxyadenosine 3'-monophosphate and 3.5 mM pyrophosphate or 100 μM 2'-cleoxy-3'-ATP and 2 mM pyrophosphate for times from 40 min to 2 hr. Crystals are extracted from harvesting solution in a 0.2-mm-diameter cryoloop, flash frozen, and stored in liquid nitrogen or liquid nitrogen-cooled propane as described elsewhere in this series.[15] Crystals are typically useful for 1 to 2 months after they stop growing, at which point they begin to turn yellow, become extremely difficult to separate from each other, and diffract poorly. Most of the crystals used for our experiments were harvested 2–4 weeks after the hanging drops were set up.

Crystals of the adenylyl cyclase:G$_s\alpha$ complex belong to the orthorhombic space group $P2_12_12$, with unit cell constants $a = 118.0$–119.1 Å, $b = 133.9$–134.9 Å, and $c = 70.7$–72.2 Å. Variations in unit cell dimensions can result from introduction of ligands, which can induce a conformational change in the structure of the catalytic domains, or from variations of other components, such as cryoprotectant, in the harvesting solution. Diffraction from complex crystals is anisotropic, extending to d-spacings of 2.3–2.8 Å in the $\mathbf{a}^*$ and $\mathbf{b}^*$ directions, and to 3.0–3.9 Å in the $\mathbf{c}^*$ direction. Anisotropy appears to be a consequence of sparse packing contacts between unit cells along the c axis, and this is reflected in the platelike morphology of the crystals themselves (the c axis corresponds to the thinnest dimension of the plate). Typically, introduction of substrate analogs and P-site inhibitors into the harvesting solution degrades crystal quality. This, however, is not always the case; crystals harvested in the presence of Mg^{2+} 2'-iodo-2'-deoxy-ATP, diffracted to higher resolution than native crystals, although only the density for the β and γ phosphates of this inhibitor is evident in subsequent electron density maps.

Limitations and Outlook

The methods described here provide a route to structural elucidation of G$_s\alpha$- and forskolin-coactivated catalytic domains of adenylyl cyclase. Crystals described here can be used to investigate the binding modes and conformational effects of active site-based ligands in the presence of Zn^{2+}, Mg^{2+}, and Mn^{2+}.[2,4] Although we have done little work in this area, it should also be possible to study a variety of

[15] D. W. Rodgers, *Methods Enzymol.* **276**, 183 (1997).

forskolin analogs and mutants of $G_s\alpha$. On the other hand, because VC1 and IIC2 homodimerize in the absence of forskolin and $G_s\alpha$, we have not yet achieved crystallization of a VC1:IIC2 complex in the absence of either activator. We anticipate that improved design of VC1:IIC2 fusion proteins or possibly advances in methods to express and purify active, intact adenylyl cyclase will provide opportunities to crystallize this enzyme in its native (or near native) state and with other activators and regulators not described here.

[16] Generation of Adenylyl Cyclase Knockout Mice

By SCOTT T. WONG and DANIEL R. STORM

Introduction

Adenylyl cyclases (ACs) are a family of structurally homologous enzymes that catalyze the formation of the second messenger cAMP. This activity was first characterized by E. Sutherland and colleagues in the early 1960s,[1] and decades later was found to represent a family of at least 10 individual enzymes, each possessing unique patterns of expression and regulation by cellular effector molecules.[2] Adenylyl cyclases are generally activated by external stimuli, such as hormones and neurotransmitters, through interaction with cell surface receptors, coupled to the activation of stimulatory or inhibitory G proteins, G_s or G_i. In turn, activation of G_s or G_i is thought to promote direct interaction with AC and regulation of its activity. However, it is now appreciated that regulation of these enzymes extends far beyond simple G protein–AC interaction. The discovery of isoform-dependent sensitivities to additional effector molecules, such as Ca^{2+}/calmodulin (CaM),[3,4] $G\beta\gamma$ subunits,[5] protein kinase C,[6–8] CaM kinase II,[9,10] and caveolin[11] has led

[1] E. W. Sutherland, T. W. Rall, and T. Menon, *J. Biol. Chem.* **237,** 1220 (1962).
[2] Z. Xia and D. R. Storm, "Regulatory Properties of the Mammalian Adenylyl Cyclases." R. G. Landes, New York, 1996.
[3] J. Krupinski, F. Coussen, H. A. Bakalyar, W. J. Tang, P. G. Feinstein, K. Orth, C. Slaughter, R. R. Reed, and A. G. Gilman, *Science* **244,** 1558 (1989).
[4] E. J. Choi, S. T. Wong, T. R. Hinds, and D. R. Storm, *J. Biol. Chem.* **267,** 12440 (1992).
[5] W. J. Tang and A. G. Gilman, *Science* **254,** 1500 (1991).
[6] E. J. Choi, S. T. Wong, A. H. Dittman, and D. R. Storm, *Biochemistry* **32,** 1891 (1993).
[7] O. Jacobowitz, J. Chen, R. T. Premont, and R. Iyengar, *J. Biol. Chem.* **268,** 3829 (1993).
[8] M. Yoshimura and D. M. F. Cooper, *J. Biol. Chem.* **268,** 4604 (1993).
[9] G. A. Wayman, S. Impey, and D. R. Storm, *J. Biol. Chem.* **270,** 21480 (1995).
[10] J. Wei, G. A. Wayman, and D. R. Storm, *J. Biol. Chem.* **271,** 24231 (1996).
[11] Y. Toya, C. Schwencke, J. Couet, M. P. Lisanti, and Y. Ishikawa, *Endocrinology* **139,** 2025 (1998).

to novel insights into cAMP-dependent processes including CRE-mediated gene transcription,[12] olfactory-dependent adaptation[13,14] and hippocampus-dependent learning and memory.[15]

Because each isoform of AC can confer to a cell unique specificity (i.e., ability to generate a cAMP signal) toward a given effector molecule, there is considerable interest in identifying the roles of specific isoforms of AC in a variety of phenomena, including neuroplasticity, vascular function, and obesity, because of their potential to be novel therapeutic targets for drug studies. In the absence of potent, isoform-specific antagonists of AC, it is difficult to evaluate the role of individual ACs in the intact animal because many ACs are often expressed in the same cell type.[16] Therefore, targeted gene disruption in embryonic stem (ES) cells was a particularly attractive methodology. An individual isoform of AC could be selectively ablated from the genome and, therefore, every cell in the mouse. This technology is an advance to the field of neurobiology because it provides investigators with the ability to create powerful tools to evaluate the role of any gene in an intact animal. We have previously used gene targeting to produce type1 AC (AC1)- and AC8-deficient animals (knockouts, KOs) for studying the role of Ca^{2+}-stimulated ACs in mechanisms underlying learning and memory.[15]

This review addresses the techniques used to generate AC-deficient mice. Our disruption studies using AC1, AC3, and AC8 mutant mice are used as examples to illustrate points made throughout this article. Because the technology involved in targeted disruption of specific alleles in ES cells in culture and incorporation of mutant ES cells into the mouse germ line have been the topic of many excellent texts and review articles,[17–19] this review limits its discussion to the methods and issues associated with the generation of AC KO mice.

Reagents and Materials

The materials and methods required for ES cell gene disruption and the production of gene-disrupted mice have been elaborately described in a variety of comprehensive research works. Noted examples are *Manipulating the Mouse Embryo: A Laboratory Manual* (1994)[20] by Hogan *et al.; Gene Targeting: A Practical*

[12] S. Impey, G. Wayman, Z. Wu, and D. R. Storm, *Mol. Cell. Biol.* **14,** 8272 (1994).
[13] J. Wei, S. Impey, A. Z. Zhao, G. C. K. Chan, L. P. Baker, J. A. Beavo, and D. R. Storm, *Neuron* **21,** 495 (1998).
[14] T. Leinders-Zufall, M. Ma, and F. Zufall, *J. Neurosci.* **19,** 1 (1999).
[15] S. T. Wong, J. Athos, X. A. Figueroa, V. Pineda, M. L. Shaefer, C. Chavkin, L. J. Muglia, and D. R. Storm, *Neuron* **23,** 787 (1999).
[16] R. T. Premont, *Methods Enzymol.* **238,** 116 (1994).
[17] M. R. Capecchi, *Science* **244,** 1288 (1989).
[18] A. Bradley, B. Zheng, and P. Liu, *Int. J. Dev. Biol.* **42,** 943 (1998).
[19] R. Ramírez-Solis, A. C. Davis, and A. Bradley, *Methods Enzymol.* **225,** 855 (1993).
[20] B. Hogan, F. Costantini, and E. Lacy, "Manipulating the Mouse Embryo: A Laboratory Manual," 2nd Ed. Cold Spring Harbor Laboratory Press, Cold Spring Harbor, New York, 1994.

Approach (1996),[21] edited by A. Joyner; and *Teratocarcinomas and Embryonic Stem Cells: A Practical Approach* (1987), [22] edited by E. J. Robertson.

The following are reagents for procedures described in this article.

Cell culture dishes: 10-cm and 96-well dishes are commercially available from Corning (Corning, NY)

Electroporator (Gene Pulser II): Commercially available from Bio-Rad (Hercules, CA)

Electroporation cuvettes: Bio-Rad Gene Pulser cuvettes

Fibroblast feeder cells and embryonic stem cells: Fibroblast feeder layers and ES cells are prepared according to previously published procedures[20–22]

Fetal calf serum (FCS): Each lot of FCS should be tested for plating efficiency and growth rate of ES cells.[22] Working ES medium contains 10% (v/v) FCS

Phosphate-buffered saline (PBS): Mix 8 g of NaCl, 0.2 g of KCl, 0.2 g of KH_2PO_4, and 1.072 g of $Na_2HPO_4 \cdot H_2O$ in water. Adjust to pH 7.4 and bring to a final volume of 1 liter with water

Dulbecco's modified Eagle's medium (DMEM): Commercially available from GIBCO-BRL (Gaithersburg, MD). Supplement with $NaHCO_3$ (2.2 g/liter) and adjust to pH 7.4 with HCl

L-Glutamine (200×): L-Glutamine is commercially available from GIBCO-BRL. It should be stored in 10-ml aliquots at $-20°$. Glutamine is heat sensitive and should be added to the medium at 2-week intervals at a final concentration of 1 mM

Penicillin–streptomycin (100×; Pen–Strep): Pen–Strep is commercially available from GIBCO-BRL. Pen–Strep should be added to ES medium to make the final solution 1% (v/v). Store at $-20°$

Murine leukemia inhibitory factor (ESGRO): Commercially available from GIBCO-BRL

Nonessential amino acids (100×): Commercially available from GIBCO-BRL

2-Mercaptoethanol (2-ME): To prepare diluted 2-ME, add 4.2 μl of 2-ME to 6 ml of DMEM and filter sterilize. Make up fresh

ES medium: Prepare 500 ml of DMEM supplemented with L-glutamine to a final concentration of 1 mM. Add 90 ml of FCS, 6 ml of Pen–Strep, 6 ml of nonessential amino acids, and 6 ml of diluted 2-ME

Ganciclovir (100×): Commercially available from Sigma-Aldrich (St. Louis, MO). Ganciclovir should be prepared as a 200 μM stock. The final concentration in ES medium during selection should be 2 μM

G418 (100×; Geneticin): To make a 100× stock solution, dissolve 180 mg of active ingredient (GIBCO-BRL) in 10 ml of PBS. Filter sterilize. Store at 4°

[21] A. Joyner (ed.), "Gene Targeting: A Practical Approach." IRL Press, Washington, D.C., 1996.

Trypsin–EDTA solution: Commercially available from GIBCO-BRL. Aliquot
and store at $-20°$

Aqueous gelatin: 0.15% (w/v) cell culture-grade gelatin mixed in water and
autoclaved. Store at room temperature

Mitomycin C: Commercially available from Sigma-Aldrich. Mitomycin C
should be prepared as a 1% (w/v) stock and stored at $-20°$

Embryonic Stem Cell Culture

The discovery that heterologous introduction of pluripotent ES cells to the
inner cell mass of a murine blastocyst leads to incorporation of the foreign ES
cells into the germ layer has accelerated our ability to design and create model
systems for studying important questions in neurobiology. The ES cells that we
use to generate AC KO mice are derived from 129 Sv mice. The quality and health
of the ES cells are critical parameters for the successful creation of genetically
altered mice. To ensure this, early-passage ES cells (i.e., three to six passages
from original isolation) should be used. Furthermore, ES cell preparations should
be well documented, such that if successful germ line transmission is achieved,
the investigator can have confidence that a particular batch of ES cells has a high
probability of achieving germ line transmission. It is critical to maintain ES cells
in a pluripotent and undifferentiated state. To do this, ES cells are grown in the
presence of leukemia inhibitory factor (LIF). LIF is presented to ES cells either
passively from primary mouse embryonic feeder (PMEF) cells[23] or directly by
exogenous LIF supplementation to the medium. However, we have found that a
combination of the two methods works best. We are able to maintain undifferentiated ES cells more consistently when the combination of PMEF feeder cells and
LIF supplementation is used.

ES cells in culture tend to aggregate into clumps of cells over time. Because
the availability of LIF to the cells in the interior of these clumps is limited, the cells
may differentiate. To avoid this problem, we passage the cells every 3 days and
replate them at 1/10 the original cell density. It should also be noted that care also
is taken not to plate ES cells too sparsely, for this too can lead to differentiation
via loss of contact inhibition with other ES cells.[19] The ES cells are fed with fresh
medium every 3 days or when the medium becomes overacidic.

The evening before AC DNA transfection, ES cells are expanded and then
replated at 30–50% of the original cell density. This is to ensure that we have a
population of actively dividing cells for the transfection. After the identification of
homologous recombinants, we karyotype each ES cell before blastocyst injection.

[22] E. J. Robertson (ed.), "Teratocarcinomas and Embryonic Stem Cells: A Practical Approach." IRL
Press, Washington, D.C., 1987.
[23] S. J. Abbondanzo, I. Gadi, and C. L. Sterwart, *Methods Enzymol.* **225,** 803 (1993).

Because many cells with abnormal karyotype can have a selective growth advantage, they can easily overrun the culture. In fact, 30–50% of newly derived ES cells can have an abnormal diploid karyotype.[23] Because these abnormal cells are unlikely to produce germ line transmittants, karyotyping is an extremely important step before blastocyst injection. Because of the great commitment of both time and expense paid to generate transgenic mice, it is imperative to use ES cells with the best possibility of transmitting the intended mutations into the germ layer.

Targeted Gene Disruption

General Strategy

The general strategy used to disrupt the *ac1, ac3,* and *ac8* alleles in transgenic mice is similar in all cases. Separate gene replacement vectors are constructed to inactivate each AC gene. They contain large contiguous regions of DNA having sequences identical (homologous) to that of the gene we wished to disrupt. These targeting vectors can specifically hybridize to the gene of interest, recombine with it, and afford an adduct that is structurally different from the original gene. The linearized vectors are then transfected into pluripotent ES cells, where they can incorporate themselves into the host genome either by homologous or nonhomologous recombination. If the vector integrates by homologous recombination (HR), it will specifically target and recombine with the gene of interest such that a region of this gene is deleted (Fig. 1a). However, nonhomologous recombination (non-HR) can arise at random locations within the mouse genome as well (Fig. 1b).

Therefore, a positive–negative drug selection is used to select homologous over nonhomologous recombinant ES cell clones.[24] This method takes advantage of the fact that during HR, the homologous flanking sequences on the targeting vector will replace only the sequences spanning the two cross-over points, but the external sequences will remain in the vector (Fig. 1a). In contrast, non-HR results in the incorporation of the entire vector (Fig. 1b). The positive–negative selection method utilizes two selection markers, neomycin (*neo*) and a herpes simplex virus thymidine kinase (HSV-*tk*) gene. The *neo* cassette is used as the replacement marker to identify cells that have stably integrated the AC targeting vector into their genomes. The HSV-*tk* cassette is placed outside of the regions of homology, such that it is incorporated into the genomes of nonhomologous recombinant clones and excised from homologous recombinant cell clones. Thus, cells harboring the *neo* gene will be resistant to G418, an aminoglycoside antibiotic. To selectively kill nonhomologous recombinants, cells are treated with ganciclovir. Ganciclovir is a prodrug that is activated by phosphorylation. On activation of ganciclovir by HSV-Tk, the nonlethal prodrug is converted to a phosphorylated active form that

[24] F. Köntgen and C. L. Steward, *Methods Enzymol.* **225,** 878 (1993).

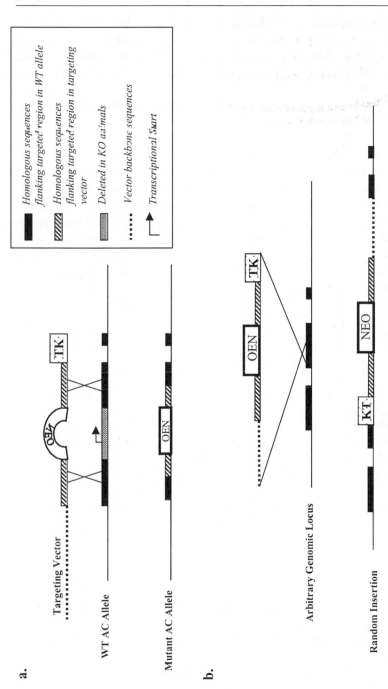

FIG. 1. Homologous and nonhomologous recombination in ES cells. (a) Schematic diagram of homologous incorporation of a targeting vector into a specific gene locus. The targeting vector contains flanking regions that are homologous to sequences that flank the desired disruption region. The neomycin cassette is oriented in the antisense orientation with respect to the original gene polarity to avoid misexpression products. Incorporation of the targeting vector sequences by homologous recombination leads to exclusion of the thymidine kinase cassette. (b) Schematic diagram of nonhomologous recombination. The targeting vector is fully inserted into a random gene locus as depicted. Note the incorporation of the thymidine kinase cassette by this mechanism.

incorporate into the DNA of replicating cells and cause death.[25] Because non-HRs will incorporate HSV-*tk* into their genomes, only those cell clones will be able to activate the ganciclovir and kill themselves. This double selection process will highly enrich the population of homologous recombinant clones on each plate.

After identification of both neomycin and ganciclovir resistant colonies, the ES cells are injected into blastocyst-stage C57BL/6 embryos and then transferred to surrogate mothers to bring to fruition the birth of chimeric mice. These mice contain a mosaic of 129 Sv cells from the mutant ES cell colony and the C57BL6 cells from the recipient blastocyst. If the 129 ES cells have migrated into the germ layer of the developing C57BL/6 blatstocyst, propagation of the mutated alleles to the subsequent generation can take place. By mating the chimeras to pure-bred C57BL/6, germ line transmission of the mutated allele can be monitored by observing the coat color of the offspring. Agouti-colored mice contain the mutant AC alleles, whereas black-colored mice do not.

Targeting Vector Construction

To create an AC knockout targeting vector, it is necessary to isolate 3–5 kb of genomic DNA that flanks each side of the desired gene locus to be deleted. The AC targeting constructs are designed to delete the promoter region and part of first exon of each gene. To accomplish this, the 5′ and 3′ genomic flanking sequences are ligated to either side of the *neo* resistance gene.[26] Expression of the *neo* resistance gene is governed by the RNA polymerase II (Pol II) promoter.[26] Because the known genomic structures of AC genes indicate they can be as large as 200 kb and have up to 30 or more exons,[27] we position the *neo* gene in the reverse orientation to the reading frame of each AC gene to avoid Pol II-dependent expression of AC splice variants or truncated forms of AC. The HSV-*tk* gene[28] is attached to either end to complete the construction of the targeting vector.

Generating and Isolating Homologous Recombinant Embryonic Stem Cells

Introduction of foreign DNA into ES cells that can homologously recombine with the host genome and isolation of monoclonal ES cell colonies containing these mutated genes are well-established techniques used for the successful disruption

[25] L. Z. Rubsam, P. D. Boucher, P. J. Murphy, M. KuKuruga, and D. S. Shewach, *Cancer Res.* **59,** 669 (1999).
[26] P. Soriano, C. Montgomery, R. Geske, and A. Bradley, *Cell* **64,** 693 (1991).
[27] L. M. Muglia, M. L. Schaefer, S. K. Vogt, G. Gurtner, A. Imamura, and L. J. Muglia, *J. Neurosci.* **19,** 2051 (1999).
[28] F. Colbere-Garapin, S. Chousterman, F. Horodniceanu, P. Kourilsky, and A. C. Garapin, *Proc. Natl. Acad. Sci. U.S.A.* **76,** 3757 (1979).

of several hundred genes in transgenic mice.[20–22] Although there are a variety of methods to generate and isolate homologous recombinant ES cells, we prefer the following.

Cell Culture. The isolation, preparation, and storage of ES cells and PMEF cells are performed according to previously published methods.[20–22] Additional manipulation of these cells to prepare them for electroporation is described.

1. In advance, prepare gelatinized cell culture dishes by adding 2 ml of aqueous gelatin for several hours. Then aspirate gelatin away and air to dryness.

2. Plate out 2×10^6 PMEF cells onto three gelatinized cell culture dishes and grow them until they form a confluent monolayer (3–5 days).

3. On the morning of electroporation, treat cells with mitomycin C at a final concentration of 0.1% (w/v) for 3 hr. Mitomycin C is an inhibitor of cell growth. This treatment is necessary because PMEF cells grow much more rapidly than ES cells and can overtake the plate at the expense of ES cell growth.

Preparation of Embryonic Stem Cells and Targeting Construct Transfection. ES cells previously plated out onto PMEF cells in the presence of added LIF are grown to approximately 80% confluency. Precautions as described above are taken to ensure that these cells are maintained in an undifferentiated state.

1. In advance, prepare 125 μg of $CsCl_2$-purified, targeting vector DNA per millilite of ES cells electroporated. This DNA is linearized at a unique restriction site at the very 5' end of the targeting vector, ethanol precipitated, and then resuspended in sterile PBS.

2. Wash the ES cells with sterile PBS. Then add 2 ml of trypsin–EDTA for 2–5 min, transfer the cells into a fresh 15-ml conical tube, and centrifuge (3000g). Aspirate away the supernatant and then wash the cell pellet (a mixture of ES cells with mitomycin C-treated PMEF cells) with PBS by resuspending and then respinning. Finally, resuspend the cells in PBS at 1.2×10^7 cells/ml. A typical ES cell preparation allows for three or four electroporations, that is, 3–4 ml of cells.

Embryonic Stem Cell Transfection and Antibiotic Selection. Methods to introduce DNA into ES cells has been extensively reviewed by Lovell-Badge in the Robertson text.[22] Our preferred method of transfecting ES cells is electroporation because it is a quick and highly efficient method by which to introduce linearized DNA into ES cells.

1. Place 1 ml of cell suspension in an electroporation cuvette, and then add 25 μg of linearized DNA (from step 1) and allow to stand at room temperature for 10 min.

2. Electroporate (100 μF, 250 V), using a full discharge. Split the cells from each electroporation cuvette between two (gelatinized only) cell culture dishes containing LIF and 10 ml of ES medium.

3. Approximately 24 hr later, change the medium, adding G418 (1× from stock). Set aside one dish with G418 only to calculate efficiency.

4. Six hours later, supplement G418-containing medium with ganciclovir (1× from stock).

5. Wash the cells with fresh antibiotic medium as needed until most dead cells have been removed; usually 2–3 days later and the 4 days after that.

6. Pick colonies as soon as the formation of foci is prominent. To facilitate visualization of foci, remove all medium except for 2 ml from the dish. Foci are removed with a P20 Pipetman (Rainin, Emeryville, CA) with sterile tips. Typically this is 11 days after electroporation or slightly earlier. In general, try to avoid picking large, rapidly growing foci.

7. Placed picked colonies into individual wells of a 96-well cell culture plate containing 20 μl of trypsin–EDTA. Allow colony to digest for 5 min, triturate, and then split the material onto two pregelatinized 96-well plates: one containing only ES medium and the other containing PMEF cells, ES medium, and LIF. The PMEF cell-containing plate will be used for colony expansion and the other plate will be used for Southern blotting. Generally, to curtail cost, it is not necessary to add LIF to the Southern blotting dish because the purpose of this dish is to grow as much material as possible to perform Southern blots. Therefore, it does not matter whether they are undifferentiated.

Freezing and Genotyping. Because it is important to maintain ES cells in an undifferentiated state, we preferred to freeze our ES cells after they have grown to 80% confluency to ensure low passage number. ES cells grown in 96-well plates are frozen and genotyped according to the procedure of Ramírez-Solis *et al.*[19] Clones positive for homologous recombination are then karyotyped and then expanded for blastocyst injection.

Characterization of Disrupted Gene Loci

Southern Blot Analysis. Southern blotting is the best method for screening ES cells for homologous recombinants and later for genotyping mice carrying the mutated genes. This method requires the identification of restriction enzyme sites within the wild-type AC locus that differ from the mutant gene locus. The blot is then probed with a DNA fragment that lies outside of the homologous regions contained within the targeting vector to ensure identification of the targeted locus over regions where nonhomologous recombination occurs. For example, *Hin*dIII is the restriction enzyme used to characterize the *ac3* gene locus for Southern blot analysis. The targeting vector of AC3 contains a novel *Hin*dIII site that is incorporated into the *ac3* locus after homologous recombination (Fig. 2, left). Probe

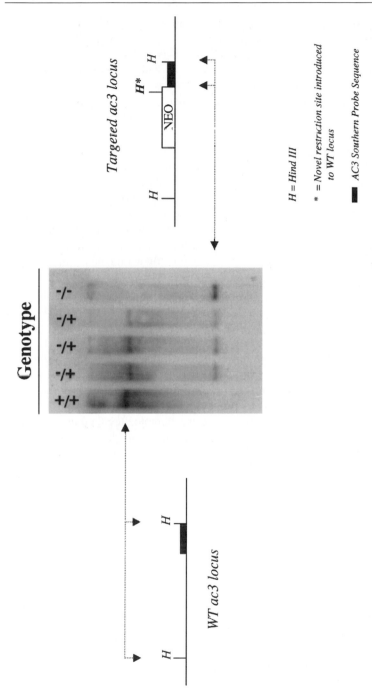

FIG. 2. Southern blotting strategy for AC3. Depicted is a representative Southern blot to analyze the *ac3* gene locus in wild type (WT, +/+), heterozygous (+/−), and homozygous (−/−) AC3 animals. Digestion of genomic DNA with *Hind*III leads to a 5.0-kb fragment for wild type, both a 5.0-kb and a 1.5-kb fragment for heterozygous, and a 1.5-kb fragment for homozygous AC3 animals. The schematic representations on either side of the blot depict how these bands are afforded from the wild-type and mutated allelic loci.

pND22 recognizes a 5.0-kb fragment in the wild-type allele and a 1.5-kb fragment in the mutated allele such that the presence of each allele is easily identified in wild-type, heterozygous, and AC3 mutant mice (Fig. 2, right).

Once a putative homozygote mouse has been identified, further characterization of the mutated allele and the mutant mouse genome can be achieved with Southern blot as well. To characterize the mutated allele, it is advisable to ensure that the desired allele is in fact deleted. To examine for this, the Southern blot can be stripped and then reprobed with a probe that would hybridize to sequences missing in the mutated allele. If the putative homozygotes identified with the first probe lack bands by Southern blot using the deleted region probe, then it is highly likely that the desired allele has been disrupted. Last, to ensure that both homologous and nonhomologous targeting events have not occurred in the same ES cell, the genomic DNA is digested with several different restriction enzymes that surround the mutant gene locus. Southern blot analysis is then performed with a *neo* probe. The Southern blot should reveal only one band for each enzyme, provided none of the restriction sites lie directly within the neomycin probe region.

Polymerase Chain Reaction. Because Southern blotting requires considerably more time to prepare and analyze, polymerase chain reaction (PCR) genotyping is an excellent alternative. PCR is advantageous because it is fast and does not require as much material as the Southern blot. There are a number of methods to screen for ES genotype; however, the simplest is to use a forward primer that hybridizes to the 3'-most edge (with respect to the targeted gene) of the neomycin cassette in the mutant allele, and a reverse primer that hybridizes just outside of the homologous flanking regions within the targeting vector (Fig. 3a).

Using a primer that hybridizes outside the homologous flanking regions is required to identify recombination events occurring at the desired gene locus. If the distance between the neomycin cassette and the "outside" primer is greater than 1.5 kb, the number of false negatives may increase under standard PCR conditions.[29] Therefore, if an investigator wishes to screen ES cell colonies by PCR, it is advisable to make one homologous arm of the targeting vector 1.5 kb or shorter. However, it is also advised not to make this arm too short for this may affect the rate of homologous recombination.[30]

PCR is also a cost-effective and rapid method to genotype mice for colony maintenance after germ line transmission of the deleted AC allele. The strategy for identification of the mutant allele is similar to that used for ES cell screening; however, the amplified region can be made arbitrarily small, provided the mutant gene locus has been previously characterized by Southern blot. To identify the presence or absence of the wild-type allele, two primers hybridizing to sequences within the deleted region are used. We try to design the PCRs such that the amplified wild-type and mutant PCR bands are less than 600 base pairs (bp) and are different sizes (Fig. 3b). Considering the possibility of false positive or negative

a.

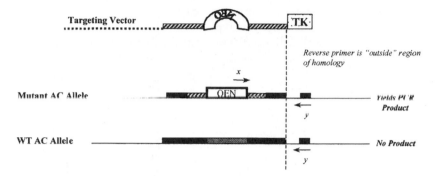

b.

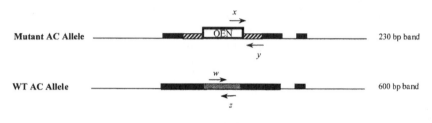

c.

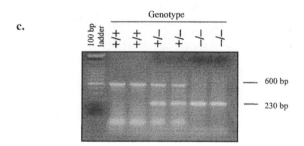

Fig. 3. Representative PCR genotyping strategy for AC3. (a) Schematic diagram of the PCR strategy used to screen for homologous recombinants in ES cells. A forward primer x that hybridizes to sequences within the neomycin casette and a reverse primer y that lies just outside the regions of homology are used to amplify DNA sequences between each other. The reverse primer is positioned outside the homology region to ensure that homologous recombinants will give rise to a band whereas nonhomologous recombinants will not. (b) Schematic diagram for PCR genotyping of transgenic animals. After confirmation of AC gene disruption, the reverse primer can be moved to an internal region to afford a smaller band on amplification. The PCR genotyping of the AC3 KO mouse will illustrate the basic strategy. The mutant allele is amplified with primers x and y to afford a 230-bp fragment. A separate set of primers, w and z, are used to amplify the wild-type (WT) allele. Primers w and z hybridize to sequences that are present in the wild-type allele but are absent in the mutant allele.

amplification of PCR bands, investigators should periodically verify their PCR results by Southern blot to ensure accurate genotyping.

Complications with PCR genotype analysis of AC knockouts can arise because of the considerably high GC content at the 5' ends of most ACs.[31] All of our KOs have deletions in this GC-rich region. For example, the sequences deleted in the AC1 KO is >75% GC. Thus, by the previously described genotyping strategy, amplification of this deleted region could be difficult because primer sets may not be specific enough to select for this particular region within the context of the entire mouse genome. In this situation, Southern blotting is the best alternative.

Mouse Breeding

Chimeras

After the mutant ES cells are injected into C57BL/6 blastocyst-stage embryos, the newly transplanted ES cells integrate into the inner cell mass of the host blastocyst to form a mosaic of 129Sv cells and C57BL6 cells. These blastocysts are then transferred into pseudopregnant host females to mature and ultimately become newborn mice. At gastrulation, the embryo becomes segmented into three cell regions, referred to as germ layers. For these mice to pass the 129Sv genetic material to the next generation, the heterologous ES cells must differentiate properly to become germ cells, which eventually migrate to the gonads and differentiate there into gametes. Ultimately, these mixed embryos become newborn mice that have a mottled coat color, because of the contributions of 129Sv and C57BL/6 coat color genes expressed in these mice. These mice are called chimeras.

F_1 Generation Mice

To determine whether the chimeras are able to transfer the mutant AC allele to the next generation of animals, the chimeras are mated to C57BL/6 mice. The gametes in chimeras, if the 129Sv cells had indeed differentiated into germ cells, should contain a mixture of either 129Sv or C57BL/6 gametes. For a male chimera, if the experiment was successful, a 129Sv sperm will fertilize the C57BL6 egg and the resultant offspring would be an F_1 hybrid—which is agouti colored. These pups are often referred to as germ line transmittants. However, if the 129Sv cells did not propagate and differentiate into germ cells, the chimera would be able to produce only C57BL/6 gametes. These chimeras will afford only black C57BL/6 pups when mated to purebred C57BL/6 mice.

(c) Two separate reactions for wild-type and mutant allele amplification are combined. The reactions can be then separated on a 1.4%, (w/v) agarose gel by electrophoresis to visualize the bands. Depicted is a representative PCR gel with amplicons from the wild-type allele (+/+) reactions running at 600 bp and the mutant allele (–/–) reactions running at 230 bp.

Production of Adenylyl Cyclase Knockout Mice

Germline transmittants are heterozygous for the mutant AC allele. Therefore, to generate AC KO mice an intercross between these F_1 hybrids is performed. By Mendelian genetics, the offspring from an F_1 cross of AC1 heterozygous animals would have the following genotypes: 25% *XX,* 50% *Xx,* and 25% *xx,* with *X* representing the wild-type allele of AC1 and *x* representing the mutant.

For studying the role of ACs in behavioral assays in the F_2 background, it is necessary to derive wild-type and KO animals from the F_1 cross and use litter mate controls to ensure proper control over genetic drift, which is an important concern when working with the transgenic mice.[32] Because the complement of 129Sv:C57BL/6 alleles in wild-type animals is not equivalent to the complement found in double knockout (DKO) animals, it is necessary to average out genetic background issues by using large numbers of animals to obtain a representative population from each genotype in the F_2 generation. Investigators, however, should not inbreed wild-type or KO mice from the F_2 generation to propagate their KO lines because genetic drift can be significant.[33] The KOs generated in the F_2 generation are produced at reasonable yield and the time and expense are not too great to necessitate other breeding strategies to control for genetic background, as one may have to do for doubly homozygous animals. These issues are discussed in greater detail in the next section.

If investigators wish to propagate inbred strains of mice, they should perform them on fully inbred backgrounds. The most efficient route toward obtaining KOs on a pure genetic background is to mate the chimera with a purely inbred 129Sv animal. Because the 129Sv gamete is genetically pure, this would result in 100% pure 129Sv heterozygotes, which give rise to KO animals in the F_2 generation. Alternatively, if mice on the C57BL/6 generation are desired, then an investigator must backcross F_1 hybrid mice at least five or six generations into C57BL/6.[32,33] The advantage to these strategies is that homozygote and wild-type breedings would give rise to progeny that are congenic with one another. Furthermore, less time will be required to genotype and maintain the mouse colonies.

Production of Adenylyl Cyclase Double Knockout Mice

One confounding problem to interpreting many transgenic mouse phenotypes is gene compensation. Because many gene products are functionally redundant

[29] H. Zhang and H. E. Henderson, *Biotechniques* **25,** 784 (1998).

[30] P. Hasty, J. Rivera-Perez, and A. Bradley, *Mol. Cell. Biol.* **11,** 5586 (1991).

[31] J. Krupinski and J. J. Cali, *Adv. Phosphoprotein Second Messenger Res.* **32,** 53 (1998).

[32] A. J. Silva, E. M. Simpson, and J. S. Takahashi *et al., Neuron* **19,** 755 (1997).

[33] L. Silver, "Mouse Genetics: Concepts and Applications." Oxford University Press, New York, 1995.

TABLE I
MOUSE CHROMOSOMAL MAP POSITIONS OF CLONED
ADENYLYL CYCLASES

AC type	Chromosome	Map position	Locus ID[a]	Ref.
1	11	1.25 cM	11507	b
2	13	41.0 cM	11508	b
3	12	A–B	11509	b
4	14	D3	11510	b
5	16	B-5	11511	b
6	15	F	11512	b
7	8	40.0 cM	11513	c
8	15	37.5 cM	11514	d
9	16	B1	11515	e

[a] As per GenBank and EMBL.
[b] S. Edelhoff, E. C. Villacres, D. R. Storm, and C. M. Disteche, *Mammalian Genome* **6,** 111 (1995).
[c] J. Doyle, K. Hellevuo, and L. Stubbs, *Mammalian Genome* **7,** 320 (1996).
[d] L. Z. Rubsam, P. D. Boucher, P. J. Murphy, M. KuKuruga, and D. S. Schewach, *Cancer Res.* **59,** 669 (1999).
[e] R. T. Premont, I. Matsuoka, M. G. Mettei, Y. Puille, N. Defer, and J. Hanoune, *J. Biol. Chem.* **271,** 13900 (1996).

among family members, it is often necessary to disrupt more than one gene to observe a desired phenotype. Disruption of either the *ac1* or *ac8* gene leads to insignificant differences in long-term potentiation (LTP) in hippocampal area CA1[34] and certain hippocampus-dependent memory tasks.[15] We hypothesized that because AC1 and AC8 are Ca^{2+}/CaM-stimulated ACs, the presence of either activity could be sufficient to sustain CA1 LTP and, therefore, we created doubly homozygous mice (DKO) for both AC1 and AC8.

To create this strain, a simple breeding strategy was used because the *ac1* and *ac8* alleles reside on different chromosomes (Table I[27,35-37]), which gives confidence that these AC alleles would segregate independently.

In contrast, some ACs, such as the *ac5* and *ac9* alleles, reside near each other on the same chromosomal map position. The creation of DKO AC5/AC9 animals,

[34] Z. L. Wu, S. A. Thomas, E. C. Villacres, Z. Xia, M. L. Simmons, C. Chavkin, R. D. Palmiter, and D. R. Storm, *Proc. Natl. Acad. Sci. U.S.A.* **92,** 220 (1995).

[35] S. Edelhoff, E. C. Villacres, D. R. Storm, and C. M. Disteche, *Mammalian Genome* **6,** 111 (1995).

[36] J. Doyle, K. Hellevuo, and L. Stubbs, *Mammalian Genome* **7,** 320 (1996).

[37] R. T. Premont, I. Matsuoka, M. G. Mettei, Y. Pouille, N. Defer, and J. Hanoune, *J. Biol. Chem.* **271,** 13900 (1996).

using a simple breeding method, may prove to be difficult and time intensive, especially if the alleles are genetically linked. For generation of these types of DKO, double disruption at the ES cell level may be the best option. A variety of breeding strategies can be used to create doubly homozygous AC animals. The benefits and limitations of each strategy depend primarily on the nature of the experiment(s) being conducted and the genetic background requirements for each. The following section focuses on three possible methods to create AC DKO animals and the potential problems that may be encountered. The generation of the AC1/AC8 DKO is used as an example.

Animals singly homozygous for either of the two AC alleles[33,38] can be described genotypically as *xxYY*, homozygous for AC1; and *XXyy*, homozygous for AC8.

In the first round of mating, these animals are crossed to afford offspring that are all dihybrid and can be described genotypically as 100% *XxYy*, doubly heterozygous for AC1 and AC8 (dihybrid). From this stage, an investigator has a number of options. One is to intercross the dihybrid offspring to one another. This cross in combination with the exclusion of brother–sister matings and constant mixing between sublines is preferred by many laboratories to control for genetic drift between wild-type and DKO animal backgrounds.[39–41] The major limitation to this breeding strategy is the time and expense required for generating enough animals to conduct experiments.

If an investigator chooses to employ this breeding strategy to make AC DKO mice, careful planning is required to minimize cost. A large source of waste in colony management expense is underestimation of how many mating pairs are required for generating the desired number of animals for a given experiment. Lack of planning can lead to expansion of a sizable colony that will yield too few animals of a given age group, sex, and genotype to conduct an experiment. It would take 3 months for another set of animals to be produced. This problem can be minimized by understanding more thoroughly the probabilities of generating either wild-type or DKO mice.

By classic Mendelian genetics, the probability that this cross will produce either wild-type or DKO progeny is 1 in 16. This means that the probability of obtaining either a wild-type or a DKO animal in a litter arbitrarily large will be 6.25%. However, because mice have finite litter sizes, the probability of obtaining a wild-type or a DKO should also be dependent on litter size.

[38] E. L. Watson, K. L. Jacobson, J. C. Singh, R. Idzerda, S. M. Ott, D. H. DiJulio, S. T. Wong, and D. R. Storm, *J. Biol. Chem.* **275**, 14691 (2000).
[39] R. Janz, T. C. Sudhof, R. E. Hammer, V. Unni, S. A. Siegelbaum, and V. Y. Bolshakov, *Neuron* **24**, 687 (1999).
[40] R. Janz, Y. Goda, M. Geppert, M. Missler, and T. C. Sudhof, *Neuron* **24**, 1003 (1999).
[41] K. A. Roth, C. Kuan, T. F. Haydar, C. D'Sa-Eipper, K. S. Shindler, T. S. Zheng, K. Kuida, R. A. Flavell, and P. Rakic, *Proc. Natl. Acad. Sci. U.S.A.* **97**, 466 (2000).

So, for a litter containing n pups, the probability of finding a wild-type pup can be described by the following expression:

$$P_{WT,n} + P_{X,n} = 1 \qquad (1)$$

or

$$P_{WT,n} = 1 - P_{X,n} \qquad (2)$$

where X is the probability that a pup is not wild type and n is the number of pups.

By Mendelian segregation, the probability that a single pup in a dihybrid cross will be wild type is one-sixteenth. Thus,

$$P_{WT,pup} = (1 - P_{X,pup}) \qquad (3)$$

Rearrangement and substitution yields

$$P_{X,pup} = \left(1 - \frac{1}{16}\right) = 0.9375 \qquad (4)$$

Therefore, the probability that a given litter will afford no wild-type animals as a function of pup number is

$$P_{X,n} = \prod_{0}^{n}(0.9375)_i = (0.9375)^n \qquad (5)$$

Substitution of the result from Eq. (5) back into Eq. (2) then yields the probability that a given litter of n pups will afford a wild-type pup, as given by Eq. (6):

$$P_{WT,n} = 1 - (0.9375)^n \qquad (6)$$

where n is the number of pups. Table II demonstrates the probability of obtaining a wild-type pup as a function of litter size. Clearly, as n becomes arbitrarily large, $Pwt_{,n}$ approaches unity.

In terms of actual litter size, this means that there is a 40.3% chance that in a litter of eight pups one will be a wild-type animal. This percentage also represents the probability of obtaining one DKO animal in an equivalent litter because the probability of a pup being DKO is equal to wild type. It should be noted here that the probability of finding two wild type, two DKO, or one of each genotype in the same litter (about eight pups) is much lower—roughly on the order of 16%. Therefore to have 80% confidence in obtaining at least one wild-type or one DKO pup, enough mating pairs need to be set up to afford 25 mice. Because the minimum number of mice to perform behavioral experiments is approximately 5 age-matched animals representing each genotype, this means that a minimum of 18–20 mating pairs, per genotype, would be required. Furthermore, if the experimental design stipulates using all male or female mice, then two to three times as many mating pairs per genotype might be needed. This could mean maintaining and genotyping up to 700 mice for a single experiment.

TABLE II
PREDICTED OCCURRENCE OF ONE WILD-TYPE OR ONE
DOUBLE KNOCKOUT ANIMAL IN A LITTER OF n PUPS
FROM A DIHYBRID CROSS

n	P	% Chance
0	0.0000	0
1	0.0625	6.2
2	0.1211	12.1
3	0.1760	17.6
4	0.2275	22.8
5	0.2758	27.6
6	0.3211	32.1
7	0.3635	36.4
8	0.4033	40.3
9	0.4406	44.1
10	0.4755	47.6
15	0.6202	62.0
20	0.7249	72.5
25	0.8008	80.1
30	0.8557	85.6
35	0.8955	89.6
40	0.9243	92.4
45	0.9452	94.5
50	0.9603	96.0
75	0.9920	99.2
100	0.9984	99.8
∞	1.0000	100

A possible compromise to balance the concern for genetic background and expense might be to intercross the wild-type offspring from the dihybrid crosses with one another and do the same with the DKOs. This scheme would alleviate the problem with having to mate such large numbers of animals and genotype all of them. The risk with interbreeding animals this way is that care must be taken to ensure that lines do not become inbred. This will require multiple lines of dihybrid founders, avoiding sister–brother matings, and extensive intercrossing of various wild-type and DKO sublines. The main limitation, however, is that the rate-limiting step is founder production because it still relies on a dihybrid cross.

A second alternative, which was previously reported by our laboratory,[15] allows us to generate DKO mice quickly with higher pup numbers. We simply mated an AC1/AC8 dihybrid mouse to an AC1KO mouse at the second step to afford an offspring that can be genotypically described as 25% xxYY, 25% XxYY, 25% xxYy, and 25% XxYy. From these offspring we took the xxYy mice and intercrossed them to generate DKO animals as follows: 25% xxYY, 50% xxYy, and 25% xxyy (DKO).

Our control strain was developed by using mice from the mice that were *XxYY* and mating them to each other. This yielded 25% *xxYY*, 50% *XxYY*, and 25% *XXYY* (wild type).

For our behavioral experiments, we use only age-matched mice derived from the same filial generation. We use mixtures of several different sublines of wild-type and DKO mice derived from different dihybrid/AC1KO crosses to ensure we had a representative population of animals from each genotype.

Genetic Background Issues

Differences in genetic background between wild-type and DKO animals can lead to the appearance of phenotypic differences that are unrelated to the gene disruption. Therefore, strict control over genetic background has been the topic of many research articles[42,43] and an excellent review by the Banbury Conference on Genetic Background.[35] However, in practice, precise control over genetic background in DKO mice is difficult to achieve because of the length of time required to derive mice in the best genetic backgrounds and expense involved. For our AC DKO studies, this would require breeding the mutant mice into a constant genetic background. Because generating DKO mice in the proper genetic backgrounds could take 3–5 years to generate, the cost of creating these strains without certainty of the phenotype was a tremendous concern.

Therefore, DKOs bred onto a mixed genetic background were used for our previously published studies[15] as a more affordable alternative. In fact, many laboratories that have generated DKO animals have studied their mutant animals on mixed genetic backgrounds.[39–41] These methods were not intended to supplant the need for rigorous control over genetic background, but rather to demonstrate alternative methods that can be used if a breeding scheme to maintain rigorous control over the DKO genetic background is financially prohibitive. Still, if an investigator chooses to evaluate a DKO on a hybrid 129Sv:C57BL/6 background, for instance, it is important to ensure that the observed phenotype is due to the disrupted AC allele and not due to nontargeted alleles.

Although many investigators may feel that the dihybrid cross provides a better control over genetic background drift, careful analysis of the genetic background suggests that this strategy is no more advantageous than the alternative strategies proposed above. This is because of the low probability for producing wild-type and DKO animals, as well as the generational distance away from the original chimeras. For example, if each single KO were the F_2 generation of a 129Sv:C57BL/6 F_1 dihybrid cross. The dihybrids, therefore, would be identical to the F_3 generation

[42] L. P. Baker, M. D. Nielson, S. Impey, B. M. Hacker, S. W. Poser, M. Chan, and D. R. Storm, *J. Neurosci.* **19**, 180 (1999).
[43] Y. Salomon, D. Londos, and M. Rodbell, *Anal. Biochem.* **58**, 541 (1974).

of this cross. It follows that the wild-type and DKO offspring of a dihybrid cross would arise in the F_4 generation. On the other hand, in the other mating strategies proposed in the previous section, the animals in the test groups would be in the F_5 generation. Because the F_1 generation is the only place that will provide rigorous control over the genetic background between wild-type and DKO animals,[32,33] wild-type pups afforded from neither the dihybrid cross nor the other proposed crosses will provide adequate control over genetic background in DKOs In this light, the presumption that wild-type and DKO animals must be generated from a dihybrid cross to properly control genetic drift between the two genotypes is not well founded. The more important issue in either strategy importance of proper mixing of sublines and a requirement to study a larger number of animals from each genotype to ensure a representative population. Because of the low frequency of obtaining wild-type and DKO animals, the assurance of a representative population may be particularly difficult.

Because these methods are intended only to minimize differences in genetic drift, it is advisable also to rescue the phenotype either genetically or pharmacologically. Genetic rescue would entail generating a transgenic mouse that overexpresses the disrupted AC. This transgenic mouse would subsequently be crossed with the AC-deficient animals to restore AC function. Alternatively, an investigator can take advantage of the fact that multiple ACs are often expressed in the same cell type.[16,42] Using forskolin, which is a general AC stimulator, it is possible to overcome deficits in cAMP production in each KO background by adding forksolin (Fig. 4).

If forskolin can rescue the deficits observed in the DKO, then this suggests a deficit in cAMP production, which is precisely what was disrupted.

Phenotypes of Adenylyl Cyclase Mutant Animals

After confirming the disruption of the AC allele by Southern and/or PCR analysis, there still remains the possibility that some truncated form of AC could have been made. For this reason, Northern and Western blotting should be performed to verify that both message for the AC of interest and the protein are not expressed. With the availability of AC-specific antibodies (Santa Cruz Biotechnology, Santa Cruz, CA), Western blotting for the existence of isoform-specific ACs is straightforward. Northern blots should be conducted with probes that hybridize near the 3' end of the AC message to ensure that truncated forms of AC are not produced.

Biochemical Characterization of Adenylyl Cyclase Mutant Animals

The presumption that AC mutant phenotypes arise from deficits in cAMP production should be tested biochemically. Because multiple isoforms of AC can exist in the same cell,[16] compensatory changes in AC activity can mask the

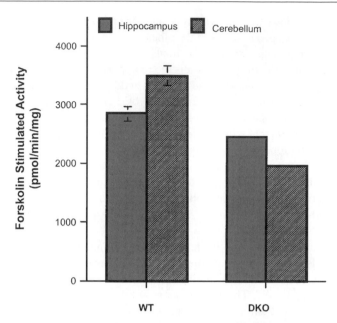

FIG. 4. Forskolin-stimulated activity in DKO hippocampus and cerebellum. In the hippocampus and cerebellum of AC1/AC8 double knockout mice (DKO), there remain additional AC isoforms that can be stimulated by forskolin ($10\ \mu M$). Although DKO mice are no longer stimulated by Ca^{2+}/CaM,[15] stimulation of these residual activities can be used to generate cAMP artificially with forskolin.

effect of AC disruption. There exist a variety of methods to monitor differences in AC activity between wild-type and mutant animals for a given tissue. The methods for these techniques have been well described in detail elsewhere.[43–45] Generally, AC assays can be divided into three main groups: (1) the radioimmunoassay (RIA), (2) the cell accumulation assay (CAA), and (3) the membrane assay. The RIA and CAA are similar in the sense that they monitor changes in cell accumulation of cAMP over time. This assay allows an investigator to study the regulation of ACs *in vivo,* in the context of the whole cell. The major differences between these two assays is that the RIA requires a cAMP-specific antibody and much less starting material is required. The membrane assay, on the other hand, monitors the specific activity of the enzyme in a broken cell preparation.

For example, the membrane assay has been helpful in the biochemical characterization of the Ca^{2+}/CaM-stimulated KOs.[15,38] We assayed for Ca^{2+}-stimulated

[44] S. R. Post, R. S. Ostrom, and P. A. Insel, *Methods Mol. Biol.* **126,** 363 (2000).

[45] Y. H. Wong, A. Federman, A. M. Pace, I. Zachary, T. Evans, J. Pouyssegur, and H. R. Bourne, *Nature (London)* **351,** 63 (1991).

activity in isolated hippocampi and cerebella from wild-type and mutant animals to determine whether disruption of either AC1, AC8, or both had any effect on cAMP production (Fig. 5).

These studies allowed us to determine that AC1 and AC8 were the only Ca^{2+}-stimulated isoforms of AC in mice. Furthermore, they demonstrated that there was no compensation for this activity by the other isoforms of AC expressed in these tissues. If biochemical characterization reveals a significant deficit between wild-type and mutant animals, then the investigator can be more confident that the KO will serve as a good model system to study effects at the tissue and whole animal level.

Biological Applications for Adenylyl Cyclase Knockout Mice

The creation of AC KO mouse strains has provided powerful and unique model systems to examine the role of specific AC isoforms in the control of cAMP expression and have furthered our understanding of many areas of neurobiology. We have created mice deficient in AC1, AC3, and AC8 to examine specific questions, which lent insight into the role of these ACs in specific forms of neuroplasticity in the hippocampus and olfactory epithelium. Table III[15,34,46,48–50] summarizes the findings that we have gathered to date on these AC-deficient systems.

Clearly, disruption of these AC isoforms has led to several interesting and revealing deficiencies in transgenic mice. Although we are far from a comprehensive understanding of the role of AC1 and AC8 in hippocampus-dependent plasticity or of AC3 in the olfactory system, the generation of these model systems has at least provided important information toward this end.

Perspectives

In summary, targeted disruption of specific isoforms of AC has provided a powerful model system to study their roles at many levels of study, including the cellular, tissue, and intact animal levels. The intent of this article has been to review the methods and potential caveats in making AC mutant mice. Great advances in the field of transgenic mouse research have provided even better research tools to study AC function *in vivo* because of the initial creation of our KO mice. These have overcome many potential problems inherent in germ line disruption of ACs

[46] R. M. Abdel-Majidl, W. L. Leong, L. C. Schalwyk, D. S. Smallman, S. T. Wong, D. R. Storm, A. Find, M. J. Dobson, D. L. Guernsey, and P. E. Neumann, *Nat. Genet.* **19**, 289 (1998).

[47] M. L. Schaefer, S. T. Wong, D. F. Wozniak, L. M. Muglia, A. Nardi, R. E. Hartman, G. Gurtner, S. K. Vogt, C. E. Luedke, D. R. Storm, and L. J. Muglia, *J. Neurosci.* **20**, 4809 (2000).

[48] E. C. Villacres, S. T. Wong, C. Chavkin, and D. R. Storm, *J. Neurosci.* **18**, 3186 (1998).

[49] D. R. Storm, C. Hansel, B. Hacker, A. Parent, and D. Linden, *Neuron* **20**, 1199 (1998).

[50] S. T. Wong and D. R. Storm, unpublished observations (2000).

a.

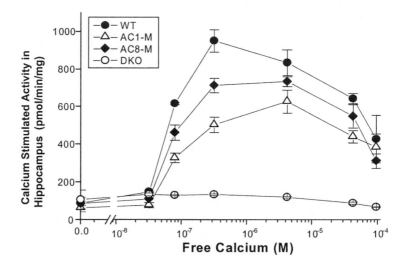

b.

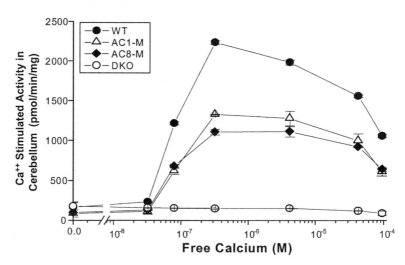

FIG. 5. Calcium-stimulated AC activity in AC1, AC8, and DKO hippocampus and cerebellum. These graphs demonstrate the disruption of Ca^{2+}-stimulated activities in hippocampi (a) and cerebella (b) of AC1, AC8, and DKO mice. Monitoring this AC activity is a useful measure to ensure that the disruption of the AC gene leads to a phenotypic change in biochemical activity.

TABLE III
PHENOTYPES FROM AC-DEFICIENT ANIMALS

Null allele	Phenotype			
	Anatomical	Biochemical	Electrophysiological	Behavioral
ac1	Loss of somatosensory barrel fibers[a]	Partial deficit in Ca^{2+}-stimulated activity[b] in hippocampus, cerebellum, and neocortex	CA3/mossy fiber LTP deficiency[c]	Deficit in probe trial of Morris water task[b] and rotorod test[d]
ac8	None	Partial deficit in Ca^{2+}-stimulated activity[b] in hippocampus, cerebellum, thalamus, and olfactory bulb. Full deficit in hypothalamus	Studies pending	Altered anxiety phenotype[b]
ac1/ac8	ND[e]	Complete ablation of Ca^{2+}-stimulated AC activity in all areas assayed	CA1/Schaffer collateral L-LTP deficiency[f]	Deficiency in passive avoidance and contextual fear conditioning[f]
ac3	ND	ND	Electroolfactogram is affected[g]	Olfactory-dependent behavior affected[g]

[a] R. M. Abdel-Majidl, W. L. Leong, L. C. Schalwyk, D. S. Smallman, S. T. Wong, D. R. Storm, A. Find, M. J. Dobson, D. L. Guernsey, and P. E. Neumann, *Nat. Genet.* **19**, 289 (1998).

[b] Z. L. Wu, S. A. Thomas, E. C. Villacres, Z. Xia, M. L. Simmons, C. Chavkin, R. D. Palmiter, and D. R. Storm, *Proc. Natl. Acad. Sci. U.S.A.* **92**, 220 (1995).

[c] E. C. Villacres, S. T. Wong, C. Chavkin, and D. R. Storm, *J. Neurosci.* **18**, 3186 (1998).

[d] D. R. Storm, C. Hansel, B. Hacker, A. Parent, and D. Linden, *Neuron* **20**, 1199 (1998).

[e] ND, Not determined.

[f] S. T. Wong, J. Athos, X. A. Figueroa, V. Pineda, M. L. Shaefer, C. Chavkin, L. J. Muglia, and D. R. Storm, *Neuron* **23**, 787 (1999).

[g] Preliminary phenotype.[50]

in ES cells, including embryonic lethality. Indeed, creation of mice by these newer methods can contribute a temporal and spatial aspect to AC KO research and thus can provide technology to address the relevance of individual ACs in subsets of tissues.

For example, region-specific disruption of genes, using the CRE–*lox* system,[51] could aid in the understanding of specific ACs in the CA1 region of the hippocampus. The *ac2* and *ac4* genes are also highly expressed in this area of the hippocampus; however, *ac2* and *ac4* are also expressed in many of the vital organs including the lung, spleen, and liver.[52,53] Therefore, to address tissue-specific actions of individual isoforms in the hippocampus, either the *ac2* or *ac4* gene would have to be "floxed" at the ES cell level. Floxing is the introduction of specific sequences, on either side of a targeted gene, that ultimately can lead to deletion of the intervening sequences in the presence of CRE recombinase. Thus an *ac2*-floxed mouse would then be mated to a mouse that expresses CRE recombinase in a CA1-specific manner. The resultant progeny that contain both CRE recombinase and the "floxed" allele would have that gene specifically disrupted in CA1; in our example this gene is *ac2*.

A second technique is inducible gene expression. Two successful approaches in transgenic mice rely on either the tetracycline (TET)[54] or ecdysone-inducible[55] gene expression system. Essentially, these systems are used to drive the expression of a given gene under the control of a heterologous promoter that cannot be expressed by the murine cell transcriptional machinery. For appropriate inducible expression, a tetracycline or ecdysone receptor is expressed that can drive the expression of the given gene in the presence of either TET or ecdysone. Usually this requires the creation of two independent strains, one harboring the gene in question and the other harboring the receptor to drive its expression. However, it is also possible to make the contruct bicistronic, such that both receptor and heterologous genes are expressed together. Nevertheless, an investigator would now have the ability to express a given protein at any time during development.

Conceivably, one could combine the CRE–*lox* system with an inducible system to create an inducible AC knockout system. Although potentially complicated to do in practice, in theory a null mutant of AC would be bred to a mouse harboring both the TET (or ecdysone) receptor and a floxed AC transgene, governed by the wild-type or region-specific promoter. This mouse would then be bred to a mouse

[51] J. Z. Tsien, D. F. Chen, D. Gerber, C. Tom, E. H. Mercer, D. J. Anderson, M. Mayford, E. R. Kandel, and S. Tonegawa, *Cell* **87,** 1317 (1996).

[52] P. G. Feinstein, K. A. Schrader, H. A. Bkalyar, W. J. Tang, J. Krupinski, A. G. Gilman, and R. R. Reed, *Proc. Natl. Acad. Sci. U.S.A.* **88,** 10173 (1991).

[53] B. N. Gao and A. G. Gilman, *Proc. Natl. Acad. Sci. U.S.A.* **88,** 10178 (1991).

[54] M. Mayford, M. E. Bach, Y. Y. Huang, L. Wang, R. D. Hawkins, and E. R. Kandel, *Science* **274,** 1678 (1996).

[55] D. No, T. P. Yao, and R. M. Evans, *Proc. Natl. Acad. Sci. U.S.A.* **93,** 3346 (1996).

that has a CRE recombinase gene, governed by the TET (or ecdysone) promoter. Thus, in the presence of drug, CRE would be expressed and disrupt the given AC at any time during development. A system such as this could be particularly useful for AC null mutants that are embryonic or perinatal lethal.

The combination of null mutations and these techniques is likely to be the future of transgenic work to study AC function. Undoubtedly, the information gained from the sum of null mutant, region-specific knockout, inducible expression, and other transgenic mouse systems will help in understanding the roles of individual AC isoforms in many areas of biology. Moreover, these studies pose tremendous potential to lay the groundwork for novel drug design and the treatment of human disease.

Acknowledgments

Work in the authors' laboratory was supported by National Institutes of Health Grants DC04156 and NS357056. We gratefully acknowledge the scientific help for this body of work from collaborators Zhengui Xia, Louis J. Muglia, Richard D. Palmitter, and Kien Trinh. The AC3 and AC8 mutant mice were prepared at the NIEHS Center for Ecogenetics and Environmental Health, Transgenic Animal Support Service, University of Washington.

[17] Construction of Soluble Adenylyl Cyclase from Human Membrane-Bound Type 7 Adenylyl Cyclase

By SHUI-ZHONG YAN and WEI-JEN TANG

Introduction

Adenylyl cyclase is the sole enzyme to synthesize cyclic AMP (cAMP), a key second messenger that regulates diverse physiological responses including sugar and lipid metabolism, olfaction, and cell growth and differentiation. Most adenylyl cyclase activity in mammalian tissues is found in plasma membrane preparations with the exception of the reproductive organs, which have both membrane-associated and soluble forms of adenylyl cyclase.[1-3] To date, nine membrane-bound and one soluble adenylyl cyclases from mammals have been cloned and characterized. Each isoform has its own pattern of tissue distribution and regulation.

[1] W.-J. Tang and J. H. Hurley, *Mol. Pharmacol.* **54,** 231 (1998).

[2] R. K. Sunahara, C. W. Dessauer, and A. G. Gilman, *Annu. Rev. Pharmacol. Toxicol.* **36,** 461 (1996).

[3] J. Buck, M. L. Sinclair, L. Schapal, M. J. Cann, and L. R. Levin, *Proc. Natl. Acad. Sci. U.S.A.* **96,** 79 (1999).

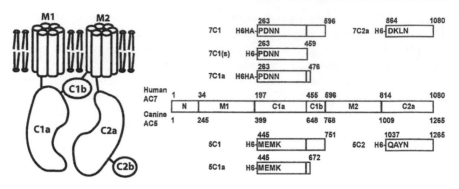

FIG. 1. A model of mammalian adenylyl cyclase and the schematic representation of the relevant soluble adenylyl cyclase constructs from human type 7 and canine type 5 adenylyl cyclases. The residue numbers for the boundaries of six domains of human AC7 (top) and canine AC5 (bottom) adenylyl cyclases are indicated. The four constructs of human AC7 adenylyl cyclase [7C1, 7C1(s), 7C1a, and 7C2a] described in this article are shown with their initiation and termination sites. The initial four residues of each construct are also shown. For comparison, three constructs of canine AC5 adenylyl cyclase (5C1, 5C1a, and 5C2) described by Gilman's group are listed.[16,23] On the basis of the sequence homology, 5C1a is 20 and 10 amino acids longer than 7C1a from N and C termini, respectively, and 5C2a is 23 amino acids longer than 7C2a at its N terminus.

Membrane-bound adenylyl cyclase is an integral membrane protein regulated by numerous extracellular signals such as hormones and neurotransmitters. All nine cloned isoforms of mammalian membrane-bound adenylyl cyclase share a common structure including two ~40-kDa cytoplasmic domains (C1 and C2), each following an ~20-kDa hydrophobic domain (M1 and M2)[1,2] (Fig. 1). The hydrophobic domains, the least conserved of these domains, made this enzyme difficult to express and purify. As a result, detailed biochemical and structural analyses of membrane-bound adenylyl cyclases have not been achieved.

Careful truncation analysis has permitted the construction of soluble adenylyl cyclases by using the conserved portions of the C1 and C2 domains, C1a and C2a.[4] C1a and C2a can be expressed and purified separately and then mixed together to reconstitute a functional adenylyl cyclase.[5] The soluble enzyme exhibits regulatory characteristics comparable to membrane-bound cyclases, including activation by the stimulatory G protein α subunit ($G_s\alpha$) and by the diterpene forskolin, and inhibition by the inhibitory G protein α subunit ($G_i\alpha$) and by adenosine analogs termed P-site inhibitors. Thus the soluble adenylyl cyclase system serves as a model system to study the regulation of mammalian adenylyl cyclase at both biochemical and structural levels.[5–22] Success in making soluble

[4] W. J. Tang and A. G. Gilman, Science 268, 1769 (1995).
[5] S. Z. Yan, D. Hahn, Z. H. Huang, and W. J. Tang, J. Biol. Chem. 271, 10941 (1996).

adenylyl cyclase is dependent on proper engineering of C1a and C2a. Here, we describe how to make a soluble adenylyl cyclase from the human type 7 isoform.

Expression of Functional 7C2a

The first functional soluble enzyme derived from membrane-bound adenylyl cyclase was constructed from the C1 and C2 domains of two different isoforms of adenylyl cyclase, type 1 and type 2, respectively.[4] This heterodimer has provided an effective model to study the catalytic mechanism and regulation of mammalian adenylyl cyclase.[5–22] However, the chimeric soluble models may have functional differences from the nonchimeric model. To date, only the type 5 and type 9 isoforms have yielded functional, nonchimeric soluble C1/C2 models.[8,17,23]

To extend this list, we made nonchimeric C1/C2 models from both type 7 and type 8 adenylyl cyclases. When *Escherichia coli* lysates containing 7C1/7C2 or 8C1/8C2 were mixed, a significantly increased $G_s\alpha$- and forskolin-sensitive enzyme activity was detected. Immunoblot revealed that the quantities of 7C1 and 7C2 domains were significantly higher than those of 8C1 and 8C2 (data not shown). Thus, our subsequent study focused on the expression of the C1 and C2 domains from type 7 adenylyl cyclase.

---

[6] S.-Z. Yan, Z.-H. Huang, R. S. Shaw, and W.-J. Tang, *J. Biol. Chem.* **272,** 12342 (1997).

[7] S. Z. Yan, Z. H. Huang, V. D. Rao, J. H. Hurley, and W. J. Tang, *J. Biol. Chem.* **272,** 18849 (1997).

[8] S. Z. Yan, Z. H. Huang, R. K. Andrews, and W. J. Tang, *Mol. Pharmacol.* **53,** 182 (1998).

[9] T. Mitterauer, M. Hohenegger, W. J. Tang, C. Nanoff, and M. Freissmuth, *Biochemistry* **37,** 16183 (1998).

[10] R. E. Whisnant, A. G. Gilman, and C. W. Dessauer, *Proc. Natl. Acad. Sci. U.S.A.* **93,** 6621 (1996).

[11] S. Doronin, C. Dessauer, and R. A. Johnson, *J. Biol. Chem.* **273,** 32416 (1998).

[12] C. W. Dessauer and A. G. Gilman, *J. Biol. Chem.* **272,** 27787 (1997).

[13] C. W. Dessauer and A. G. Gilman, *J. Biol. Chem.* **271,** 16967 (1996).

[14] C. W. Dessauer, T. T. Scully, and A. G. Gilman, *J. Biol. Chem.* **272,** 22272 (1997).

[15] C. W. Dessauer, J. J. Tesmer, S. R. Sprang, and A. G. Gilman, *J. Biol. Chem.* **273,** 25831 (1998).

[16] R. K. Sunahara, C. W. Dessauer, R. E. Whisnant, C. Kleuss, and A. G. Gilman, *J. Biol. Chem.* **272,** 22265 (1997).

[17] K. Scholich, A. J. Barbier, J. B. Mullenix, and T. B. Patel, *Proc. Natl. Acad. Sci. U.S.A.* **97,** 2915 (1997).

[18] K. Scholich, C. Wittpoth, A. J. Barbier, J. B. Mullenix, and T. B. Patel, *Proc. Natl. Acad. Sci. U.S.A.* **94,** 9602 (1997).

[19] K. Scholich, J. B. Mullenix, C. Wittpoth, H. M. Poppleton, S. C. Pierre, M. A. Lindorfer, J. C. Garrison, and T. B. Patel, *Science* **283,** 1328 (1999).

[20] G. Zhang, Y. Liu, A. E. Ruoho, and J. H. Hurley, *Nature* (*London*) **386,** 247 (1997).

[21] J. J. G. Tesmer, R. K. Sunahara, A. G. Gilman, and S. R. Sprang, *Science* **278,** 1907 (1997).

[22] J. J. Tesmer, R. K. Sunahara, R. A. Johnson, G. Gosselin, A. G. Gilman, and S. R. Sprang, *Science* **285,** 756 (1999).

[23] C. W. Dessauer, J. J. Tesmer, S. R. Sprang, and A. G. Gilman, *J. Biol. Chem.* **273,** 25831 (1998).

In previous work purifying 2C2a, we found that the C2 N terminus [amino acids (aa) 821–855] was subjected to proteolytic cleavage. As type 7 enzyme is highly homologous to type 2 enzyme, we construct the 7C2a with an N-terminal hexahistidine (His$_6$) tag based on our shorter version of 2C2a. The coding sequences of 7C2a are amplified by 18 cycles of polymerase chain reaction (PCR), using human AC7 cDNA as the template (kindly provided by M. Yoshimura, University of Colorado Health Sciences Center, Denver, CO), Vent DNA polymerase, and the following primers: 5'-CCACTGAATTCCTGACAAGTTAAACGAGGACTGGTACC-3' and 5'-GCAGAAGCTTTCAGTTCAGCCCCAGCCCCTGAAA-3'. The PCR product is digested with EcoRI and HindIII and ligated into pProExH6 digested with the same enzymes, resulting in pProExH6-7C2a. The coding region of 7C2a is confirmed by DNA sequencing analysis.

The optimal expression of 7C2a [E. coli BL21 (DE3) cells carrying pProExH6-7C2] is examined through small-scale induction (10 ml). The expected 25-kDa protein in the resulting lysate can be detected by immunoblot (Fig. 2A). This 7C2a is functional as judged by $G_s\alpha$-stimulated activity resulting from the mixture of E. coli lysate containing 7C2 and lysate containing the C1 domain of type 1, type 5, or type 7 adenylyl cyclase (data not shown). Optimal expression conditions for 7C2a are significantly different from those for 2C2a despite high sequence homology. Whereas 2C2a expression is greatest 4 hr after isopropyl-β-D-thiogalactopyranoside (IPTG) induction at 30°, 7C2a expression peaks after 19 hr at 25° (Fig. 2A).

Materials and Methods

The following are needed for the purification of 7C2a as well as 7C1.

Solutions for Escherichia coli Culture

T7 medium: (20 g/liter; Difco, Detroit,MI), Tryptone peptone yeast extract (10 g/liter; Difco), NaCl (5 g/liter), 50 mM potassium phosphate (pH 7.2)
Ampicillin (100 mg/ml)
Kanamycin (100 mg/ml)
Chloramphenicol (1 mg/ml)
IPTG (100 mM)

Solutions for protein Purification

Tris-HCl, 1 M (pH 7.7, pH 8.0 at 4°)
EDTA, 0.1 M (pH 7.0)
NaCl, 4 M
Imidazole, 1 M (pH 7.0)
Dithiothreitol (DTT, 1 M)
2-Mercaptoethanol, 14.3 M
Phenylmethylsulfonyl fluoride (PMSF, 100 mM)

A. 2C2/7C2   B. 7C1/7C1a

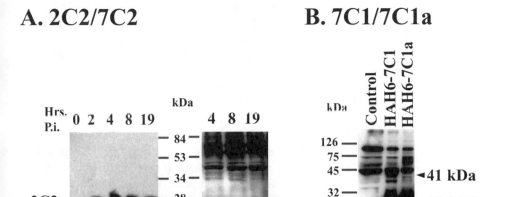

FIG. 2. The optimization for the expression of 7C2 and 7C1 proteins in *E. coli* BL21 (DE3). (A) The expression of 2C2 or 7C2 at the indicated times after IPTG induction. p.i., Post-IPTG induction. Bacterial lysate (10 μl) from BL21(DE3) cells containing 2C2 or 7C2, taken at different times after IPTG induction, were analyzed on a 13% (w/v) SDS–polyacrylamide gel and immunoblotted with 12CA5 (2C2a, for epitope tag of influenza virus hemagglutinin, HA) or anti-pentahistidine antisera (7C2a; Qiagen). (B) Bacterial lysates (10 μl) from BL21(DE3) cells containing 7C1 and 7C1a were analyzed on a 13% (w/v) SDS–polyacrylamide gel and immunoblotted with 12CA5.

Columns Needed for Purification

Ni^{2+}–nitriloacetic acid (Ni–NTA) column (5–10 ml; Qiagen, Valencia, CA)
Q-Sepharose column (30 ml; Pharmacia, Piscataway, NJ)

Both column media are regenerated as described on the basis of the manufacturer instructions.

We usually purify recombinant 7C2a from a 4-liter culture that yields about 50 mg of 7C2a that is more than 95% pure. *Escherichia coli* cells for the expression of recombinant protein need to be freshly grown. Less than 14 hr of incubation at 30° is allowed for the transformants to grow on a Luria–Bertani (LB) medium plate containing ampicillin at 50 μg/ml. The minimal time of incubation prevents the selection of genetic variants that can affect 7C2a production; similar precaution is also applied to the expression of 7C1 described below.[24] BL21 (DE3) cells (4 ml) that harbor pProExH6-7C2a from the freshly grown colony are grown in 4 liters of T7 medium containing ampicillin (50 μg/ml) at 25° in a shaker with agitation set at 250 rpm. IPTG to 100 μM is added when the culture reaches an A_{600} of 0.4. Cells are harvested 19 hr after induction, and spun at 6000g for 15 min at 4°. After removal of culture medium, the cell pellet is frozen at −80°.

[24] M. M. Zambrano, D. A. Siegele, M. Almiron, A. Tormo, and R. Kolter, *Science* **259,** 1757 (1993).

Frozen cells are thawed in 200 ml of solution A [20 mM Tris-HCl (pH 8.0), 5 mM 2-mercaptoethanol, 0.1 mM PMSF]. The cells are lysed by adding lysozyme at 0.1 mg/ml and applying 4 min of sonication on ice, with a cycle of 1 sec on and 3 sec off. The lysate is spun at 150,000g for 30 min at 4° and the supernatant is collected. NaCl is added to a final concentration of 100 mM.

All the following purification steps are performed in a 4° cold room. The supernatant is loaded onto the Ni–NTA column equilibrated with solution A containing 100 mM NaCl. The Ni–NTA column is washed with 10 column volumes of solution A containing 500 mM NaCl followed by 20 column volumes of solution A containing 100 mM NaCl and 20 mM imidazole (pH 7.0). The column is then eluted with 100 ml of solution A containing 100 mM NaCl and 150 mM imidazole (pH 7.0) at a 2-ml/min flow rate. The peak fraction (60 ml) is combined and then diluted with a 5-fold volume (about 300 ml) of solution B [20 mM Tris (pH 8.0), 1 mM DTT, 0.5 mM EDTA, 0.1 mM PMSF]. The diluted eluate is loaded onto the Q-Sepharose column, which is equilibrated with solution B. The absorbed proteins are eluted at 2 ml/min with a 240-ml linear gradient of NaCl (100–500 mM) in solution B; 6-ml fractions (2 ml/min) are collected. The peak fractions of 7C2a are then separated by electrophoresis on a 13% (w/v) sodium dodecyl sulfate (SDS)–polyacrylamide gel and analyzed for purity by Coomassie blue staining. The purified 7C2a is concentrated by ultrafiltration [Amicon (Danvers, MA) positive pressure ultrafiltration device with PM10 membrane] followed by use of a Centricon 10 microconcentrator (Amicon). The concentrated 7C2a is aliquoted (typically 100 μl) and stored at −80°, at a concentration greater than 2 mg/ml.

To determine whether the purified 7C2a is functional, its ability to enhance $G_s\alpha$ stimulation of 7C1a is assessed. $G_s\alpha$ is purified according to the protocol described.[25] To assay adenylyl cyclase activity, recombinant $G_s\alpha$ is activated by 50 μM AlCl$_3$ and 10 mM NaF or by 100 μM GTPγS. Excess GTPγS can be removed by gel-filtration chromatography. The assay can use adenylyl cyclase expressed in *E. coli* or in Sf9 (Spodoptera frugiporda ovary) cells.[26,27] Adenylyl cyclase assays are performed in the presence of 10 mM MgCl$_2$ and 0.5 mM ATP at 30° for 20 min as described.[28] Protein concentration is determined by using Bradford reagent with bovine serum albumin (BSA) as standard. After 45-fold purification, using both Ni–NTA and Q-Sepharose columns, greater than 95% pure 7C2a is obtained with a yield of 13 mg/liter of *E. coli* culture (Table I and Fig. 3).

[25] S.-Z. Yan and W.-J. Tang, *Methods Enzymol.* in press (2001).
[26] S. Z. Yan, D. Hahn, Z. H. Huang, and W. J. Tang, *J. Biol. Chem.* **271,** 10941 (1996).
[27] W.-J. Tang, J. Krupinski, and A. G. Gilman, *J. Biol. Chem.* **266,** 8595 (1991).
[28] Y. Salomon, C. Londos, and M. Rodbell, *Anal. Biochem.* **58,** 541 (1976).

TABLE I
PURIFICATION OF 7C2a[a]

Purification step	Quantity (mg)	Specific activity (nmol/min/mg)	Recovery (%)	Purification (-fold)
Lysate	5700	11	100	—
Ni–NTA	270	150	67	14
Q Sepharose	78	460	60	45

[a] Adenylyl cyclase activity assayed in the presence of 30 μM AlF$_3$, 10 mM MgCl$_2$, 10 mM NaF, 240 nM G$_s\alpha$, and 100 μM forskolin.

Expression of Functional 7C1a

To express a properly folded C1 protein, the initiation and termination sites of the C1 domain need to be determined empirically. We express 7C1(s) that contains only 7C1a [Fig. 1; 7C1(s) is the shorter form of 7C1a, described below]. All vectors for 7C1 expression are made similarly to that for 7C2a. 7C1(s) is expressed as a 23-kDa protein (data not shown). When *E. coli* lysate containing

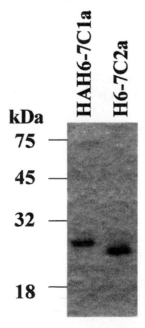

FIG. 3. Coomassie blue stain of purified 7C1a and 7C2a (2 μg each).

TABLE II
PURIFICATION OF 7C1a[a]

Purification step	Quantity (mg)	Specific activity (nmol/min/mg)	Recovery (%)	Purification (-fold)
Lysate	4000	4.5	100	—
Ni–NTA	40	290	64	65
Q-Sepharose	20	490	54	108

[a] Adenylyl cyclase activity assayed in the presence of 30 μM AlF$_3$, 10 mM MgCl$_2$, 10 mM NaF, 240 nM G$_s\alpha$, and 100 μM forskolin.

7C1(s) is mixed with *E. coli* lysate containing either 7C2a or 2C2a, increased G$_s\alpha$ and forskolin-stimulated activity is observed (not shown). 7C1(s) that is 90% pure can be obtained using only an Ni–NTA column. Highly enriched 7C1(s) does not migrate as a uniform peak in either an anion or cation exchanger and consistently precipitates during the concentration process. This suggests that 7C1(s) may be partially unfolded. To solve this problem, we express a full-length 7C1 containing both 7C1a and 7C1b.

Escherichia coli lysate containing full-length 7C1 appears to be enzymatically active when added to lysate containing either 7C2a or 2C2a. However, immunoblot analysis reveals only a small quantity of full-length 41-kDa protein accompanied by several proteolytic products (Fig. 2B). After purification by Ni–NTA and Q-Sepharose column chromatography, a catalytically active, ~26-kDa 7C1 protein is obtained that migrates as a distinct peak on the Q-Sepharose column. The molecular weight of this protein, determined by electrospray mass spectrometry, is 26,325 Da. Thus the proteolytic site should be at amino acid residue 476 (Fig. 1).

The shorter, 26-kDa version of 7C1 (hereafter referred to as 7C1a) is then constructed by introducing a termination codon at residue 477 of full-length 7C1, using Quick-Change site-directed mutagenesis (Stratagene, La Jolla, CA), with pProExHAH6-7C1 as a template. *Escherichia coli* BL21 (DE3) is transformed with the resulting vector and grown in T7 medium containing ampicillin (50 μg/ml) at 25°. When the A_{600} reaches 0.4, IPTG is added to a concentration of 30 μM. At 19 hr postinduction, the culture is harvested and lysed. Immunoblot analysis indicates that 7C1a migrates similarly to the 26-kDa proteolytic product of full-length 7C1 (Fig. 2B).

Recombinant 7C1a is purified with Ni–NTA and Q-Sepharose columns, similarly to 7C2a purification. The function of purified 7C1a is assessed by its ability to enhance G$_s\alpha$ stimulation of 7C2a. After 108-fold purification, 7C1a (>95% pure) is obtained with a yield of 5 mg/liter of *E. coli* culture (Table II and Fig. 3).

Comparison of 7C1a/7C2a with Membrane-Bound Adenylyl Cyclase

The baculovirus expression system is often used to analyze the biochemical properties of membrane-bound adenylyl cyclase.[29-35] To compare soluble adenylyl cyclase with its native membrane-bound form, recombinant baculovirus is constructed to express membrane-bound type 7 adenylyl cyclase. The coding sequence for human AC7 is excised from pCMV/SK-AC7 by *MluI* (filled in by DNA polymerase Klenow) and *XbaI* and inserted into pFastBac1 that has been digested by *NotI* (filled in by Klenow) and *XbaI*. The resulting plasmid is then used to construct recombinant baculovirus, using the GIBCO-BRL (Gaithersburg, MD) Bac-to-Bac system. Expression of recombinant adenylyl cyclases in insect Sf9 cells is performed as described.[27]

On the basis of the *in vitro* adenylyl cyclase assay, both membrane-bound AC7 and soluble 7C1a/7C2a exhibit substantial $G_s\alpha$ stimulation, consistent with the observation that a coexpressed G_s-coupled receptor could raise cAMP formation over 20-fold in mammalian cells overexpressing AC7 (Fig. 4A and C).[36,37] Both membrane-bound AC7 and soluble 7C1a/7C2a can be readily stimulated by forskolin (Fig. 4B and D). Interestingly, the magnitude of forskolin stimulation of 7C1a/7C2a is only 10- to 20-fold, significantly less than other soluble adenylyl cyclase systems.[4,26]

General Principles in Construction and Expression of Soluble Adenylyl Cyclase

The construction of a soluble adenylyl cyclase system is a process of protein engineering to identify the domains that can fold properly and resist proteolysis. The sequence homology across cyclases provides some guidance but exact domain boundaries for each cyclase need to be determined empirically. In this article we report success in making 7C2a; experience in constructing 2C2a helped to avoid proteolysis. We also show that the full-length 7C1 domain is highly sensitive to proteolysis while the core of 7C1, 7C1(s), is functional but only partially folded. Adding 17 amino acids to the C-terminal end of 7C1(s) eliminates the folding

[29] W. J. Tang, J. Krupinski, and A. G. Gilman, *J. Biol. Chem.* **266,** 8595 (1991).
[30] W.-J. Tang, M. Stanzel, and A. G. Gilman, *Biochemistry* **34,** 14563 (1995).
[31] R. Taussig, L. M. Quarmby, and A. G. Gilman, *J. Biol. Chem.* **268,** 9 (1993).
[32] R. Taussig, W.-J. Tang, J. R. Hepler, and A. G. Gilman, *J. Biol. Chem.* **269,** 6093 (1994).
[33] R. Taussig, J. A. Iniguez-Lluhi, and A. G. Gilman, *Science* **261,** 218 (1993).
[34] G. Zimmermann and R. Taussig, *J. Biol. Chem.* **271,** 27161 (1996).
[35] G. Zimmermann, D. Zhou, and R. Taussig, *J. Biol. Chem.* **273,** 6968 (1998).
[36] M. Yoshimura, H. Ikeda, and B. Tabakoff, *Mol. Pharmacol.* **50,** 43 (1996).
[37] K. Hellevuo, M. Yoshimura, N. Mons, P. L. Hoffman, D. M. Cooper, and B. Tabakoff, *J. Biol. Chem.* **270,** 11581 (1995).

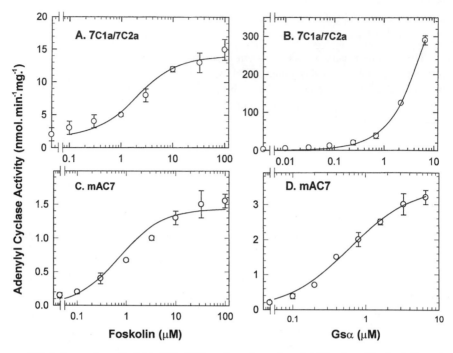

Fig. 4. Comparison of soluble 7C1a/7C2a and membrane-bound AC7 activated by forskolin and $G_s\alpha$. Adenylyl cyclase activities of soluble 7C1a/7C2a (0.4 μM each; A and B) and membrane-bound AC7 (40 μg; C and D) were assayed with the indicated concentration of forskolin (A and C) or $G_s\alpha$ (B and D).

problem. A similar approach has been used to construct soluble type 5 adenylyl cyclase (Fig. 1). On the basis of sequence homology, 7C1a and 7C2a are slightly shorter than 5C1a and 5C2a. Alignment of these two successful sets with other isoforms should provide guidance to identify initiation and termination points for other C1 and C2 domains.

Our results also emphasize the importance of optimal conditions for the expression of recombinant protein.[38] We show that altering induction conditions can significantly improve expression of 7C2a and 7C1a. Although the combinations of variables affecting expression are unlimited, researchers are advised to use a sparse-matrix approach to determine "optimal" conditions. We usually test expression by varying *E. coli* strains, inducer concentration, induction time, and temperature. It generally provides at least 3- to 4-fold enhancement in protein production. Such improvement has significantly reduced the cost and labor for purification of recombinant proteins in our laboratory.

[38] R. C. Stevens, *Structure* **8**, R177 (2000).

Acknowledgments

This research was supported by NIH Grant R01GM53459, an American Heart Association Established Investigator Award, a Brain Research Foundation grant to W.-J. Tang, a Fellowship from the University of Chicago Committee of Cancer Biology, and a Ralph S. Zitnik, M.D. Clinical Research Investigatorship from the American Heart Association of Metropolitan Chicago Affiliate to S.-Z. Yan.

[18] Genetic Selection of Regulatory Mutants of Mammalian Adenylyl Cyclases

By PETER CLAPP, AUSTIN B. CAPPER, and RONALD TAUSSIG

Regulation of intracellular cyclic AMP concentrations is principally controlled at the level of its synthesis, through the hormonal regulation of adenylyl cyclase, the enzyme responsible for the conversion of ATP into cyclic AMP. Currently, nine isoforms of membrane-bound adenylyl cyclases have been identified by molecular genetic approaches; and studies of these enzymes reveal both common and unique regulatory features.[1,2] All isoforms tested to date are activated by the GTP-bound form of the stimulatory G protein α subunit ($G_s\alpha$) and all but one is activated by forskolin; for some of the isoforms, such as type V, these stimulators synergistically activate the enzyme. All isoforms of adenylyl cyclase are further regulated by additional inputs in an isoform-specific pattern.

The yeast adenylyl cyclase, encoded by the *CYR1* gene,[3,4] is structurally distinct from and exhibits much different regulation than the mammalian adenylyl cyclases. Yeast strains containing a disruption of the *CYR1* gene are not viable but can be propagated in media containing cAMP, demonstrating the importance of cAMP for growth of the organism.[5] On the basis of this requirement of cAMP for growth, we have expressed mammalian adenylyl cyclases in a *CYR1*-deleted yeast strain to take advantage of the readily manipulatable genetic properties of the yeast for the isolation of mutant mammalian adenylyl cyclases defective in their regulatory responses.[6] We describe a genetic system utilizing the expression of mammalian type V adenylyl cyclase in yeast and its use in isolating mutant adenylyl cyclases defective in their regulation by $G_s\alpha$.

[1] R. Taussig and G. Zimmermann, *Adv. Second Messenger Phosphoprotein Res.* **32**, 81 (1998).
[2] D. M. F. Cooper, "Adenylyl Cyclases." Lippincott-Raven, New York, 1998.
[3] G. F. Casperson, N. Walker, and H. R. Bourne, *Proc. Natl. Acad. Sci. U.S.A.* **82**, 5060 (1985).
[4] T. Kataoka, D. Broek, and M. Wigler, *Cell* **43**, 493 (1985).
[5] K. Matsumoto, A. Toh-e, and Y. Oshima, *Mol. Cell. Biol.* **1**, 83 (1981).
[6] G. Zimmermann, D. Zhou, and R. Taussig, *J. Biol. Chem.* **273**, 6968 (1998).

Yeast Strains and Media

Yeast strain TC41-1[3] (*MATa leu2-3 leu2-112 ura3-52 his3 his4 cam1 cam2 cam3 cyr1 Δ::URA3*) is grown in YPAD medium supplemented with 2 mM cAMP (YPAD–cAMP). Yeast strain 12229 is isogenic to TC41-1 and contains a disruption of the *Trp1* gene with a construct that allows copper-dependent expression[7] of $G_s\alpha^6$; this strain is similarly grown in YPAD–cAMP medium.

SC medium contains the following components per liter: 1.2 g of yeast nitrogen base without amino acids and ammonium sulfate (Difco, Detroit, MI); 5 g of ammonium sulfate; 10 g of sodium hydroxide; 20 mg each of adenine, uracil, tryptophan, histidine, arginine, and methionine; 30 mg each of tyrosine and lysine; 50 mg of phenylalanine; 60 mg of leucine; and 200 mg of threonine. −Leu medium is SC medium without leucine added. YPAD medium contains 10 g of yeast extract, 20 g of Bacto-peptone, 0.1 g of adenine, and 2% (w/v) glucose per liter. Additions to the medium include cAMP (2 mM), forskolin (200 μM), and copper sulfate (100 μM). When used for plates, 20 g of Bacto-agar is added to 1 liter of medium.

Expression of Mammalian Adenylyl Cyclases in Yeast

Expression of mammalian adenylyl cyclase can be observed after introduction of the yeast expression vector encoding type V adenylyl cyclase (Fig. 1a[7a]) into cyclase-deficient yeast TC41-1 and 12229. Expression of the cyclase is driven by the yeast alcohol dehydrogenase (ADH) promoter; the plasmid[8] also contains sequences for high-copy replication in yeast (2μ origin) and the *Leu2* gene for selection in the Leu2⁻ strains TC41-1 and 12229. Plasmids are introduced into these strains by standard yeast transformation protocols,[9] and transformants are selected for on SC −Leu plates containing 2 mM cAMP.

Replica plating the TC41-1 and 12229 transformants on different growth media can be performed to assess expression and appropriate regulation of mammalian adenylyl cyclase. Growth is dependent on the regulatory state of the cyclase (Fig. 2). Basal and forskolin-stimulated conditions can be assessed with the TC41-1 transformants; this strain should grow only in the presence of 200 μM forskolin in the medium, or if cAMP is added (bypassing the need for a functional adenylyl cyclase). $G_s\alpha$-stimulated or synergistic stimulation by $G_s\alpha$ plus forskolin can be determined with the 12229 transformants. In the presence of 100 μM copper in the medium, growth is observed. Further addition of 200 μM forskolin results

---

[7] D. J. Thiele, C. F. Wright, M. J. Walling, and D. H. Hamer, *Experientia Suppl.* **52,** 423 (1987).

[7a] Y. Ishikawa, S. Katsushika, L. Chen, N. J. Halnon, J. Kawabe, and C. J. Homcy, *J. Biol. Chem.* **267,** 13553 (1992).

[8] T. Vernet, D. Dignard, and D. Y. Thomas, *Gene* **52,** 225 (1987).

[9] R. H. Schiestl and R. D. Gietz, *Curr. Genet.* **16,** 339 (1989).

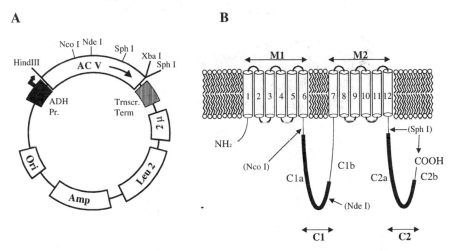

FIG. 1. Yeast expression plasmid encoding mammalian type V adenylyl cyclase. (A) Features of the plasmid used for the expression of mammalian cyclases in yeast are as follows: ADH Pr and Trnscr. Term, yeast alcohol dehydrogenase promoter and transcription terminator; ACV, dog type V adenylyl cyclase cDNA.[7a] Leu 2 and 2μ, and Ori and Amp, are sequences required for selection and high-copy propagation of the plasmids in yeast or bacteria, respectively. Restriction sites used for the introduction of the adenylyl cyclase cDNA into the parental vector or those used for the generation of error-prone PCR mutational libraries are indicated. (B) Predicted topology of membrane-bound adenylyl cyclases. Numbered cylinders represent membrane-spanning regions, and thick lines indicate regions of high amino acid conservation among all of the adenylyl cyclase family members. Nomenclature from amino terminus (NH$_2$) to carboxy terminus (COOH) is as follows: M$_1$, first set of membrane-spanning regions; C$_1$, the first large cytoplasmic domain, composed of conserved C$_{1a}$ and variable C$_{1b}$ subdomains; M$_2$, the second set of membrane-spanning regions; C$_2$, the second large cytoplasmic domain, composed of conserved C$_{2a}$ and variable C$_{2b}$ subdomains. Regions of the cyclase identified for targeted PCR mutagenesis reside between the indicated restriction sites (C$_{1a}$, NcoI and NdeI; C$_2$, SphI); the restriction sites correspond to those shown in (A).

in a no-growth phenotype, presumably because of excessively high intracellular cAMP concentrations. Under this condition, it is possible to isolate mutant adenylyl cyclases that are resistant to stimulation by G$_s\alpha$; such mutants should behave as if only forskolin is present as the sole activator, and would therefore permit growth. Procedures for isolating these G$_s\alpha$-insensitive mutants are outlined in the following sections.

Generation of Mutant Libraries

The first step in generating libraries encoding randomly mutated adenylyl cyclases is to choose the region of the molecule to be targeted for mutagenesis (Fig. 1). Alternatively, the entire adenylyl cyclase molecule can be targeted for random mutagenesis when an unbiased approach is favored, such as when the identification

A   B

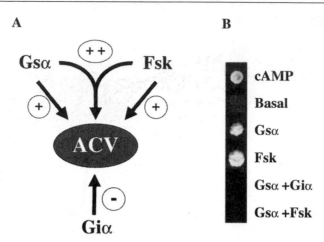

FIG. 2. Expression of mammalian adenylyl cyclase in yeast. Mammalian type V adenylyl cyclase was expressed in *CYR1*-disrupted strains of *Saccharomyces cerevisiae*. (A) Regulation of type V adenylyl cyclase (ACV) by $G_s\alpha$, forskolin (Fsk), and $G_i\alpha$. Stimulation, inhibition, and synergistic stimulation are indicated by (+), (−), and (++), respectively. (B) The growth characteristics of these strains were evaluated under the following conditions: addition of 2 mM cAMP to the medium (cAMP), no regulator (Basal), addition of 100 μM forskolin to the medium (Fsk), coexpression of $G_s\alpha$ ($G_s\alpha$), coexpression of $G_s\alpha$ and addition of 100 μM forskolin to the medium (Fsk + $G_s\alpha$), or coexpression of $G_s\alpha$ and $G_i\alpha$ ($G_s\alpha$ + $G_i\alpha$). [Adapted from G. Zimmermann, D. Zhou, and R. Taussig, *J. Biol. Chem.* **273,** 6968 (1998).]

of specific regions of the molecule may not be apparent. Both procedures are described below.

Plasmid libraries of randomly mutated adenylyl cyclases can be generated by passage in a mutator bacterial strain such as XL1 Red[10] (Stratagene, La Jolla, CA); this strain is deficient in three of the primary DNA repair pathways (*mutS, mutD,* and *mutT*), and possesses mutation rates ~5000-fold higher than that of wild type. Libraries can be prepared by the following protocol.

1. Introduce the pADHprACVLeu yeast expression plasmid containing the adenylyl cyclase sequences into the mutator bacterial strain, using a small-scale transformation protocol.[11]

2. Select >200 colonies at random from the transformation plate and inoculate 5–10 ml of Luria–Bertani (LB) broth containing ampicillin (50 μg/ml). Grow this culture overnight at 37°.

3. Isolate plasmid from 1.5 ml of culture, using standard minipreparation protocols.[11]

[10] A. Greener and M. Callahan, *Strategies* **7,** 32 (1994).
[11] J. Sambrook, E. F. Fritsch, and T. Maniatis, "Molecular Cloning." Cold Spring Harbor Laboratory Press, Cold Spring Harbor, New York, 1989.

4. Prepare libraries with higher extents of mutagenesis by inoculating another 5- to 10-ml culture with 10–50 μl of the overnight culture and, after a second overnight incubation at 37°, isolating plasmid. By propagating the plasmid in the mutator strain for 24, 48, and 72 hr, multiple libraries can be generated that differ in the extent of mutagenesis.

5. Amplify each of the libraries by introducing the plasmids into a bacterial strain such as DH5α. As the complexity of the libraries will be determined by the transformation efficiency of this step, protocols that yield a high efficiency of transformation should be used.[12] The transformants recovered on LB plates containing ampicillin are then collected and used to inoculate 500 ml of LB medium containing ampicillin. Plasmids are prepared from these cultures after incubation at 37° with shaking.

Plasmid libraries of adenylyl cyclases containing random mutations in a targeted region of the molecule (Fig. 1) can be generated by error-prone polymerase chain reaction (PCR) methods.[13,14] Libraries can be prepared by the following protocol.

1. For the C$_{1a}$ domain, sense (nucleotides 1125–1144) and antisense (nucleotides 1807–1826) primers are used. These primers were chosen because they flank restriction sites (*NcoI* and *NdeI* at positions 1235 and 1693, respectively) that are both present only once in the plasmid, and thus ensures the direct replacement of the PCR products in the correct location and orientation. Similarly, libraries targeting the C$_2$ region of the cyclase are generated with sense (nucleotides 2918–2937) and antisense (nucleotides 3754–3773) primers; this PCR product is introduced into the parental plasmid after digestion with *SphI*. PCR is performed under standard conditions for *Taq* polymerase except that MgCl$_2$ and deoxynucleotide triphosphate (dNTP) concentrations are altered depending on the degree of mutagenesis that is desired. Low ($\sim$0.1% error rate), medium ($\sim$0.5% error rate), and high ($\sim$2% error rate) levels of mutagenesis contain the following concentrations of these reagents, respectively: 1.5 mM MgCl$_2$, 0.2 mM dNTP; 7 mM MgCl$_2$, 1 mM dNTP; 7 mM MgCl$_2$, 1 mM dCTP, 1 mM dGTP, 0.2 mM dATP, and 0.2 mM dTTP. All PCRs are performed for 30 cycles (1 min at 94°, 1 min at 45°, 1 min at 72°).

2. PCR fragments are ligated to vector (pADHprACVLeu) after digestion with restriction enzymes and gel purification; for example, *NcoI* and *NdeI* digestion is used to generate C$_{1a}$ domain mutants.

3. Each of the libraries is generated by introducing the ligation reactions into a bacterial strain such as DH5α. Here again, the complexity of the libraries will be

[12] W. J. Dower, J. F. Miller, and C. W. Ragsdale, *Nucleic Acids Res.* **16**, 6127 (1988).

[13] D. W. Leung, E. Chen, and D. V. Goeddel, *Technique J. Methods Cell Mol. Biol.* **1**, 11 (1989).

[14] J. L. Lin-Goerke, D. J. Robbins, and J. D. Burczak, *BioTechniques* **23**, 409 (1997).

determined by the transformation efficiency of this step, and protocols that yield a high efficiency of transformation should be used. The transformants recovered on LB plates containing ampicillin are then collected and used to inoculate 500 ml of LB medium containing ampicillin. Plasmids are prepared from these cultures as outlined above.

Transformation of Yeast and Selection of Mutants

The following protocol is a modification of published methods[9] optimized for use with the TC41-1 and 12229 yeast strains.

1. Grow a 5-ml culture of the 12229 yeast strain in YPAD–cAMP medium at 30° (overnight).
2. Inoculate 200 ml of the same medium with an aliquot of the overnight culture. The goal is to find a dilution that places the 200-ml culture at early midlog phase the next day (OD_{600} between 0.5 and 1.0).
3. Pellet the cells by centrifugation ($850g$ for 5 min at room temperature).
4. Wash the cell pellet twice by resuspending in 100 ml of sterile, deionized water and centrifuging as before. Repeat this washing process twice more in 50 ml of TE–LiAc [10 mM Tris, 1 mM EDTA, 100 mM lithium acetate (pH 7.5)]. The final pellet is resuspended in 400 μl of TE–LiAc and incubated for 10 min at 30°.
5. Combine 500 μl of the yeast cell suspension with 50 μl of sheared salmon sperm DNA (10 mg/ml), 50 μg of library plasmid DNA, and 3 ml of 40% polyethylene glycol (w/v). Vortex the mixture for a few seconds and incubate at 30° for 20 min. Transfer the reactions to a 42° water bath, and incubate for 30 min.
6. Add the transformation reaction to 100 ml of YPAD–cAMP medium, and allow yeast to recover for 1 hr at 30° with shaking.
7. To determine the transformation efficiency, plate a range of volumes (i.e., 5, 20, and 50 μl) directly onto SC −Leu plates containing 2 mM cAMP. The total number of transformants obtained can be calculated from the number of colonies arising on these plates after several days of incubation at 30°.
8. Pellet the remainder of the transformation reaction by centrifugation at room temperature (5 min at 850 g). Resuspend the cells in 400 μl of SC −Leu medium. Plate all the cells (about 100 μl/plate) on SC −Leu plates containing 100 μM copper sulfate, 200 μM forskolin, and 2% (v/v) dimethyl sulfoxide. Colonies will appear after 1 week of incubation at 30°.

Secondary Genetic Screens

Mutant adenylyl cyclases exhibiting defects in their ability to bind $G_s\alpha$ would be expected to give rise to a subset of the mutants isolated. However, additional defects would also be expected to produce the same growth pheno-type; among these would be subclasses of adenylyl cyclase mutants unresponsive to forskolin or mutations that reduce the catalytic efficiency or stability of the enzyme. In addition, mutations in the yeast chromosome (independent of adeny-lyl cyclase mutation) may likely underlie the rescue of yeast growth for yet an-other subclass of mutants. The use of secondary genetic screens is therefore a necessary step in distinguishing the class of $G_s\alpha$-insensitive mutant adenylyl cy-clases from these "false positives." This is accomplished by recovering the plas-mids encoding the candidate adenylyl cyclase mutants, reintroducing them into yeast, and examining the growth of these transformed yeast strains under various conditions.

Plasmids are recovered from the selected positive yeast by the following mod-ification of published procedures.[15]

1. Inoculate 3 ml of SC −Leu medium containing 2 mM cAMP with colonies present on the primary yeast transformation plate. Grow the culture overnight at 30° with shaking.

2. Pellet the cells by centrifugation (5 min at 850g, room temperature).

3. Resuspend the cell pellet in 1 ml of autoclaved sterile water and transfer to a 1.5-ml microcentrifuge tube. Pellet the cells by centrifugation.

4. Discard supernatant and resuspend in 200 μl of yeast lysis buffer [10 mM Tris (pH 8), 1 mM EDTA, 100 mM NaCl, 2% (v/v) Triton, and 1% (w/v) sodium dodecyl sulfate (SDS)].

5. Add 0.2 g of glass beads (500-μm diameter, acid washed) and 200 μl of phenol–chloroform (1 : 1, v/v).

6. Vortex for 10 min and centrifuge at room temperature for 5 min at 13,000g.

7. Use 1–5 μl of the upper aqueous phase to transform DH5α bacteria. Trans-formants are recovered on LB plates containing ampicillin.

8. Plasmids are then isolated from 1.5 ml of culture inoculated with bacterial transformants, using standard minipreparation protocols.

Growth characteristics of yeast expressing each of the originally isolated adeny-lyl cyclase mutants can be determined by replica plating these resulting yeast trans-formants on various media as illustrated in Fig. 3. Basal and forskolin-stimulated activities of the mutants can be assessed by examining the growth characteristics of TC41-1 transformants plated on SC −Leu or SC −Leu plus forskolin medium, respectively. Defects in the $G_s\alpha$-stimulated activities of these mutants are revealed by the growth characteristics of the 12229 yeast transformants plated on SC −Leu

[15] F. M. Ausubel, R. Brent, R. E. Kingston, O. D. Moore, J. G. Seidman, J. A. Smith, and K. Struhl, (eds.),"Current Protocols in Molecular Biology." John Wiley & Sons, New York, 1995.

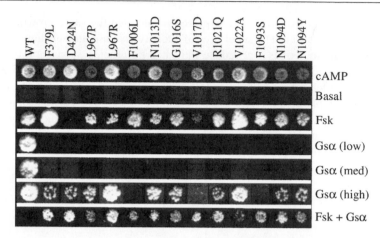

FIG. 3. Growth phenotypes of adenylyl cyclase mutants. Growth characteristics of *CYR1*-disrupted yeast expressing wild-type or mutant type V adenylyl cyclase constructs were evaluated in the absence of cyclase activators (Basal), addition of 2 mM cAMP (cAMP), addition of 100 μM forskolin (Fsk), coexpression of $G_s\alpha$ ($G_s\alpha$), and coexpression of $G_s\alpha$ and addition of 100 μM forskolin (Fsk + $G_s\alpha$). Intracellular $G_s\alpha$ concentrations (under the control of the copper-inducible *CUP1* promoter) were modulated by the omission of copper [$G_s\alpha$ (low)], addition of 10 μM CuSO$_4$ [$G_s\alpha$ (med)], or addition of 100 μM CuSO$_4$ [$G_s\alpha$ (high)] to the growth medium. [Reprinted from G. Zimmermann, D. Zhou, and R. Taussig, *J. Biol. Chem.* **273,** 6968 (1998).]

plus copper, either in the absence or presence of forskolin. As $G_s\alpha$ expression is under the control of a copper-inducible promoter, an indication of the severity of the $G_s\alpha$ insensitivity can be assessed by observing the growth characteristics at different copper concentrations. Mutants that display defects in growth in the presence of $G_s\alpha$ expression but maintain normal growth in the presence of forskolin are subjected to further analysis.

Further Characterization of Mutants

Once identified by genetic criteria, mutants are subjected to DNA sequence analysis to localize point mutations responsible for the observed phenotype. In the event that multiple point mutations are uncovered in a particular isolate, each point mutation must be examined in isolation. This can be accomplished by reintroducing each point mutation back into the cyclase expression plasmid (Fig. 1A) either on restriction fragments or by site-directed mutagenesis.[16]

---

[16] R. Higuchi, *in* "PCR Protocols: A Guide to Methods and Application" (M. A. Innis, D. H. Gelfand, J. J. Sninsky, and T. J. White, eds.), p. 177. Academic Press, San Diego, California, 1990.

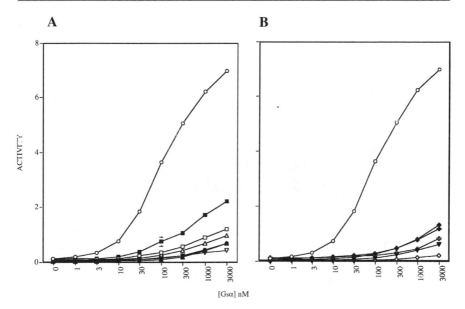

FIG. 4. (A and B) Regulation of C_2 adenylyl cyclase mutants by $G_s\alpha$. Membranes (20 μg) from Sf9 (*Spodoptera frugiperda* ovary) cells expressing either the wild-type recombinant type V adenylyl cyclase or the indicated C_2 mutants [wild type (O), L967P (□), L967R (■), F1006L (▼), N1013D (△), G1016S (▽), V1017D (◇), R1021Q (◆), V1022A (✛), F1093S (✚), N1094D (●), N1094Y (▲)] were assayed for adenylyl cyclase activity in the presence of varying amounts of $G_s\alpha$. Activities are expressed as nanomoles per minute per milligram, and have been adjusted to reflect the differences in mutant expression levels. Assays were performed in duplicate (bars, SD), and results are representative of at least three experiments. [Reprinted from G. Zimmermann, D. Zhou, and R. Taussig, *J. Biol. Chem.* **273**, 6968 (1998).]

The genetic approach outlined above provides a powerful tool in isolating mutations resulting in altered regulation by $G_s\alpha$. Biochemical approaches will be required to further characterize these defects. One approach is to utilize the baculovirus-mediated expression[17] of these cyclase mutants in insect cells; this provides a source of material for reconstitution experiments. These approaches have previously been described for the expression and purification of mammalian adenylyl cyclase isoforms.[18] As shown in Fig. 4, $G_s\alpha$-insensitive mutants isolated by yeast genetic approaches should display a reduced sensitivity toward stimulation by $G_s\alpha$ *in vitro,* requiring higher concentrations of $G_s\alpha$ to elicit an

[17] M. D. Summers and G. E. Smith, in Texas Agricultural Experiment Station Bulletin 1555, College Station, Texas (1987).

[18] R. Taussig, W. J. Tang, and A. G. Gilman, *Methods Enzymol.* **238**, 95 (1994).

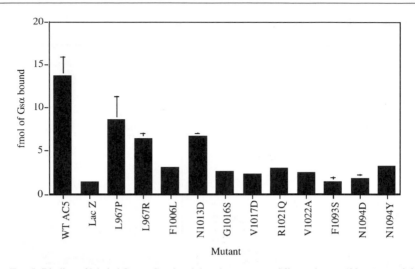

FIG. 5. Binding of labeled $G_s\alpha$ to C_2 adenylyl cyclase mutants. Sf9 membranes (20 μg) containing the wild-type adenylyl cyclase construct or the indicated C_2 mutants were incubated with labeled $G_s\alpha$ in the presence of forskolin. Specific binding to the wild-type, mutant, and control membranes was assessed by calculating the radioactivity competed with 1 μM unlabeled $G_s\alpha$. $G_s\alpha$ binding was adjusted to reflect the differences in mutant expression levels. Assays were performed in duplicate (bars, SD), and results are representative of three independent binding experiments. [Reprinted from G. Zimmermann, D. Zhou, and R. Taussig, *J. Biol. Chem.* **273,** 6968 (1998).]

activating response. In addition, mutations mapping to the $G_s\alpha$-binding site should display a decrease in the ability of the mutant to bind radiolabeled $G_s\alpha$, as shown in Fig. 5.

Summary

Initial steps in the identification of the $G_s\alpha$-binding site present in mammalian adenylyl cyclases can be achieved with the use of the yeast genetic system described. It must be stressed that this system serves as a means to identify mutants that are candidates; biochemical analysis of these mutants is a next and necessary step in the confirmation of these phenotypes. The system described can be readily adapted for the isolation of additional classes of mammalian adenylyl cyclase mutants including mutants with altered regulation toward forskolin, catalytic abnormalities, or enhanced sensitivities toward activators.[19,20] In addition, this system can be employed for the isolation of constitutively active adenylyl cyclase mutants,

---

[19] G. Zimmermann, D. Zhou, and R. Taussig, *J. Biol. Chem.* **273,** 19650 (1998).
[20] G. Zimmermann, D. Zhou, and R. Taussig, *Mol. Pharmacol.* **56,** 895 (1999).

or by coexpressing other adenylyl cyclase isoforms and their known regulators, mutations in the binding sites for these molecules can be elucidated.

Acknowledgments

Work from the author's laboratory was supported by the National Institutes of Health Grant GM53645, and the Burroughs Wellcome Foundation.

Section V

Phospholipases and Lipid-Derived Products

[19] Expression and Characterization of Rat Brain Phospholipase D

By ZHI XIE, HA KUN KIM, and JOHN H. EXTON

Introduction

Phospholipase D (PLD) is distributed widely in eukaryotes and prokaryotes, and is found in all mammalian cells. The mammalian enzyme is activated by a large variety of hormones, neurotransmitters, growth factors, and cytokines, and has been implicated in the control of cell growth, secretion, and the actin cytoskeleton.[1] Alternatively spliced forms of two mammalian PLD isozymes (PLD1 and PLD2) have been cloned.[2–10] The PLD1 isoforms are directly regulated by classic protein kinase C (PKC) isozymes and members of the families of the two small GTPases ARF and Rho.[1] The PLD2 isoforms show high basal activity and respond minimally, if at all, to PKC or the small GTPases. The first mammalian PLD to be cloned (PLD1b) was obtained from a human (HeLa) cDNA library, using polymerase chain reaction (PCR) primers based on an expressed sequence tag (EST) with homology to a yeast PLD gene.[2] Subsequently, an alternatively spliced PLD1 isozyme (PLD1a) was cloned from human promyelocytic (HL-60) cells,[3] and its properties were demonstrated to be almost identical to those of PLD1b.[3] A different gene product (PLD2) was cloned from a mouse embryonic cDNA library and shown to be less subject to regulation *in vitro*.[4] Since then, PLD isoforms have been cloned from other mammalian tissues.[5–10] In general, their properties are similar or identical to those of the initially cloned enzymes.

[1] J. H. Exton, *Biochim. Biophys. Acta* **1439**, 121 (1999).

[2] S. M. Hammond, Y. M. Altshuller, T.-C. Sung, S. A. Rudge, K. Rose, J. A. Engebrecht, A. J. Morris, and M. J. Frohman, *J. Biol. Chem.* **270**, 29640 (1995).

[3] S. M. Hammond, J. M. Jenco, S. Nakashima, K. Cadwallader, Q.-M. Gu, S. Cook, Y. Nozawa, G. D. Prestwich, M. A. Frohman, and A. J. Morris, *J. Biol. Chem.* **272**, 3860 (1997).

[4] W. C. Colley, T.-C. Sung, R. Roll, J. Jenco, S. M. Hammond, Y. Altshuller, D. Bar-Sagi, A. J. Morris, and M. A. Frohman, *Curr. Biol.* **7**, 191 (1997).

[5] W. C. Colley, Y. M. Altshuller, C. K. Sue-Ling, N. G. Copeland, D. J. Gilbert, N. A. Jenkins, K. D. Branch, S. E. Tsirka, R. J. Bollag, W. B. Bollag, and M. A. Frohman, *Biochem. J.* **326**, 745 (1997).

[6] T. Kodaki and S. Yamashita, *J. Biol. Chem.* **272**, 11408 (1997).

[7] S.-K. Park, J. J. Provost, C. D. Bae, W.-T. Ho, and J. H. Exton, *J. Biol. Chem.* **272**, 29263 (1997).

[8] K. Katayama, T. Kodaki, Y. Nagamachi, and S. Yamashita, *Biochem. J.* **329**, 647 (1998).

[9] I. Lopez, R. S. Arnold, and J. D. Lambeth, *J. Biol. Chem.* **273**, 12846 (1998).

[10] P. M. Steed, K. L. Clark, W. C. Boyar, and D. J. Lasala, *FASEB J.* **12**, 1309 (1998).

Studies of Rat Brain PLD1 in COS 7 Cells

Expression of Rat Brain PLD1 in COS 7 Cells

Rat brain PLD1 (rPLD1) can be expressed in COS 7 cells and in Sf9 (*Spodoptera frugiperda*) insect cells. Attempts to express the enzyme in an active form in prokaryotic cells have been unsuccessful to date. Stable expression of PLD1 in mammalian cells has also met with limited success. The procedures for expression of rPLD1 in COS 7 cells largely follow those described by Park *et al.*[7] rPLD1 cDNA obtained from a rat brain cDNA library is subcloned in pcDNA 3.1 vector (Invitrogen, Carlsbad, CA) as described previously.[7] LipofectAMINE (Life Technologies, Rockville, MD) and FuGENE 6 (Roche Molecular Biochemicals, Indianapolis, IN) have been found effective to transfect rPLD1 into COS 7 cells (American Type Culture Collection, Manassus, VA). The COS 7 cells are cultured at $37°$ in a humidified 5% CO_2 medium in high-glucose Dulbecco's modified Eagle's medium (DMEM; Life Technologies) supplemented with 10% fetal bovine serum (FBS; Sigma, St. Louis, MO). At 18–24 hr before transfection, COS 7 cells are seeded in six-well or 35-mm tissue culture plates at a density of 2×10^5 cells per well. When the cells reach 70–80% confluence, LipofectAMINE or FuGENE 6 can be used to transfect the COS 7 cells with comparable efficiency.[7,11]

Transfection with LipofectAMINE requires the cells to be in a serum-free medium. After the cell density reaches 70–80% confluence, DMEM containing 10% fetal bovine serum is replaced with 2 ml of Opti-MEM per well (Life Technology), and 2 μg of DNA (rPLD1 cDNA) cloned in pcDNA 3.1 vector (Invitrogen) with 6 μl of LipofectAMINE is added per well to achieve optimal transfection efficiency. After 5 hr of incubation, the transfection mixture is removed and replaced with DMEM supplemented with 10% FBS. After approximately 16 hr of growth and recovery in the serum-containing medium, the COS 7 cells can be serum starved overnight in 3 ml of DMEM containing 0.5% FBS and then the expression of rPLD1 can be readily detected by Western blotting using rabbit polyclonal antibodies raised against a C-terminal peptide of the enzyme, as described in detail below.

Transfection with FuGENE 6 can be carried out in the presence of serum. The cells are cultured in 2 ml of DMEM containing 10% FBS, and 0.6 μg of DNA and 2 μl of FuGENE are added for efficient transfection per well, when the cell density reaches 70 to 80% confluence. After 5 hr of transfection, the COS 7 cells can be immediately starved in DMEM containing 10% fetal bovine serum. After overnight starvation, the COS 7 cells are harvested in 200 μl of phosphate-buffered saline (PBS: 137 mM NaCl, 2.7 mM KCl, 10 mM Na_2HPO_4, 1.8 mM KH_2PO_4,

[11] Z. Xie, W.-T. Ho, and J. H. Exton, *J. Biol. Chem.* **273**, 34679 (1998).

pH 7.2) and the expression of rPLD1 is detected by Western blotting with rPLD1 antibodies as described below. Compared with the transfection procedure using LipofectAMINE, transfection of COS 7 cells with FuGENE 6 extensively shortens the PLD assay cycle (see below).

In Vivo Assay of Phospholipase D in COS 7 Cells

The assay is based on the transphosphatidylation reaction, which is unique to PLD. It involves measuring the transfer of the phosphatidyl moiety of phospholipids to a primary alcohol. In most cases, phosphatidylcholine ([³H]PtdCho) is utilized as the substrate either *in vitro* or in cells in which the lipids are labeled by incubation with [³H]myristic acid. Ethanol and 1-butanol are most commonly employed as nucleophiles (phosphatidyl acceptors).

COS 7 cells transfected with rPLD1 cDNA or empty vector as described above are incubated with a 3-ml/well concentration of DMEM (Life Technologies) containing 1 μCi/ml [9,10-³H]myristic acid (1 μCi/ml; NEN Life Science Products, Boston, MA) and 0.5% fetal bovine serum for 12–16 hr. Unincorporated label is removed by washing with PBS and the cells are incubated with a 3-ml/well concentration of DMEM containing 0.3% (w/v) bovine serum albumin (BSA) for 1 hr at 37°. 1-Butanol is then added to 40 mM and, after 10 min, agonists, for example, 4β-phorbol 12-myristate 13-acetate (PMA) or lysophosphatidic acid, are added and incubation is continued for up to 60 min. To terminate the reaction, the medium is aspirated and each well is washed twice with 3 ml of ice-cold PBS and then with 1.5 ml of ice-cold methanol is added. The cells are scraped into a mixture of 1.5 ml of chloroform and 1.5 ml of 0.1 M HCl. The mixture is then vortexed and allowed to stand for separation and extraction of lipids overnight. The lower organic phase is then transferred to a 1.5-ml Eppendorf tube, evaporated under N_2, and resuspended in 30 μl of chloroform–methanol (2 : 1, v/v), which is then applied to silica gel 6A thin-layer chromatography plates (Whatman, Clifton, NJ). The plates are developed with the upper phase of an ethyl acetate–isooctane–H_2O– acetic acid (55 : 25 : 50 : 10, v/v/v/v) solvent system. The plates are then exposed to iodine vapor to visualize the lipids and phosphatidylbutanol (PtdBut) is identified by comparison with a standard (Avanti Polar Lipids, Alabaster, AL). The labeled PtdBut is scraped off the plate and its radioactivity is measured by scintillation spectrometry.

To correct for possible variations in the incorporation of [³H]myristic acid into PtdCho and other phospholipids, the radioactivity in an aliquot of the lipid extract is also measured, and the results of the assay are usually expressed as the percentage of the total lipid radioactivity incorporated into PtdBut. Because COS 7 cells have significant, regulated endogenous PLD activity, the use of control (vector only) cells is essential for correct interpretation of data.

In Vitro Assay of Phospholipase D in COS 7 Cells

To study the regulation of rPLD1 *in vitro,* membranes can be prepared from COS 7 cells expressing rPLD1. The procedures described above for transfecting the cells with rPLD1 cDNA, using FuGENE 6, are employed. After transfection and expression, the cells are washed twice with lysis buffer composed of 20 mM HEPES (pH 7.2), 10% (v/v) glycerol, 1 mM EDTA, 1 mM EGTA, 1 mM dithiothreitol (DTT), 1 mM phenylmethylsulfonyl fluoride (PMSF), and a cocktail of protease inhibitors—1 tablet/50 ml (Roche Molecular Biochemicals). The cells are resuspended in this lysis buffer and disrupted by passage through a 27-gauge needle five times. The lysate is first centrifuged at 500g for 10 min at 4° to remove unbroken cells and then centrifuged at 120,000g for 45 min at 4° to separate cytosol and membrane fractions. The membrane fraction is washed with lysis buffer and resuspended by passage through a 27-gauge needle. The protein content of the cytosol and membranes is measured by the Bradford method.

For measurement of PLD activity, the procedures of Brown *et al.*[12] are followed. Membranes or cytosol (3 μg of protein) is incubated with phospholipid vesicles containing labeled PtdCho in a medium (60 μl) consisting of 50 mM HEPES (pH 7.5), 3 mM EGTA, 80 mM KCl, 1 mM DTT, 3 mM MgCl$_2$, and 2 mM CaCl$_2$. The phospholipids are added after mixing the membranes or cytosol with the incubation medium. They are prepared by dissolving phosphatidylethanolamine, dipalmitoyl-PtdCho (both from Avanti), and phosphatidylinositol 4,5-bisphosphate (Sigma) in chloroform and mixing them at a ratio of 16 : 1 : 1.4 (v/v/v). Dipalmitoyl[2-*palmitoyl*-9,10-^{3}H]PtdCho (NEN Life Sciences Products) (transphosphatidylation assay) or dipalmitoyl[*choline-methyl*-^{3}H]PtdCho (NEN Life Sciences Products) (choline release assay) is also included at 0.5 μCi/assay. The mixture is then dried under N$_2$ and sonicated for 1 min at 25° in sonication buffer [10 mM HEPES (pH 7.2), 0.6 mM EGTA, 16 mM KCl]. The phospholipid vesicles (10 μl) are added to the reaction mixture to give a final PtdCho concentration of 10 μM, and incubations are at 37° for 15–30 min. In the case of the transphosphatidylation assay, 1% (135 mM) 1-butanol is added to the reaction mixture and the reaction is terminated by the addition of 375 μl of chloroform–methanol–HCl (50 : 98 : 2, v/v/v). The lipids are then extracted and [^{3}H]PtdBut measured as described above for the *in vivo* PLD assay. In the case of the choline release assay, butanol is omitted and the reaction is terminated by addition of 200 μl of ice-cold 10% (w/v) trichloroacetic acid and 100 μl of ice-cold 1% (w/v) BSA. The mixture is centrifuged at 15,000 g for 10 min at 4° in a microcentrifuge and an aliquot (0.3 ml) of the supernatant is measured for radioactivity ([^{3}H]choline) by scintillation spectrometry.

[12] H. A. Brown, S. Gutowski, C. R. Moomaw, C. Slaughter, and P. C. Sternweis, *Cell* **75,** 1137 (1993).

Determination of Rat Brain PLD1 Expression in COS 7 Cells

Polyclonal antibodies[13] generated in rabbits against the C-terminal dodecapeptide sequence of rPLD1 (which is identical to that of human hPLD1) are commonly used for assaying the expression of rPLD1 in various cell lines. Whole cell lysate (2–5 μl) or membrane and cytosol fractions, prepared as described in the preceding section, are mixed with 2× sodium dodecyl sulfate–polyacrylamide gel electrophoresis (SDS–PAGE) sample buffer [4% (w/v) SDS, 200 mM dithiothreitol, 120 mM Tris-HCl (pH 6.8), 0.002% (w/v) bromphenol blue]. The samples (3–5 μg, 20 μl) are applied to 4–12% (w/v) Tris–glycine polyacrylamide gels (Novex, San Diego, CA) with high-range molecular weight standards (Amersham Pharmacia, Piscataway, NJ) and subjected to electrophoresis (125 V for 1.5 hr) in a minicell (Novex). The gels are transferred to polyvinylidene difluoride membranes (Millipore, Bedford, MA), using a blot module in the minicell (Novex). The membranes are then incubated with 5% (w/v) fat-free milk for 1 hr or overnight and washed with TTBS [100 mM Tris-HCl (pH 7.5), 154 mM NaCl, 0.1% (v/v) Tween 20]. They are then incubated with the anti-rPLD1 antibody (1:1000 dilution in TTBS) for 1 hr at room temperature, followed by washing three times in TTBS. The membranes are then incubated with horseradish peroxidase conjugated with anti-rabbit antibodies for Luminol-based enhanced chemiluminescence detection (ECL, Amersham Pharmacia) of rPLD1. The enzyme migrates on SDS–polyacrylamide gels as a 120-kDa protein.

Immunoprecipitation of Rat Brain PLD1 from COS 7 Cells

rPLD1 can be fused to a well-characterized epitope [e.g., Xpress, hemagglutinin (HA), Myc] and its expression can then be examined with antibodies against the corresponding epitope. To maintain PLD activity, the epitope tag should be fused to the N terminus of the enzyme, because addition of a tag at the C terminus abolishes the catalytic activity of rPLD1 (Z. Xie and J. H. Exton, unpublished observation, 2000) and also human PLDs.[14] rPLD1 tagged at the N terminus with Xpress epitope can be easily detected and immunoprecipitated from COS 7 cells as described by Xie *et al.*[11] To immunoprecipitate rPLD1, COS 7 cells cultured on 10-cm dishes are transfected and starved as described above. The cells are washed twice with ice-cold PBS and then resuspended in immunoprecipitation buffer containing 25 mM HEPES, 10% (v/v) glycerol, 1 mM EDTA, 1 mM EGTA, 50 mM KCl, 1% (v/v) Triton X-100, 0.1% (w/v) SDS, 10 mM NaF, 10 mM Na$_4$P$_2$O$_7$, 1.2 mM Na$_3$VO$_4$, and two tablets of protease cocktail (Roche Molecular Biochemicals). The cell suspension is then passed through a 27-gauge needle five times

[13] D. S. Min, E.-G. Kim, and J. H. Exton, *J. Biol. Chem.* **273,** 29986 (1998).
[14] T.-C. Sung, Y. M. Altshuller, A. J. Morris, and M. A. Frohman, *J. Biol. Chem.* **274,** 494 (1999).

and the resulting cell lysate is centrifuged at 15,000 rpm in a 1.5-ml Eppendorf tube for 10 min at 4° to pellet the unbroken cells. The supernatant is then precleared by mixing it with 1 μg of affinity-purified mouse IgG (Vector Laboratories, Burlingame, CA) and 20 μl of a 1:1 slurry of protein A beads for 1 hr at 4°. The mixture is centrifuged at 500 g for 5 min at 4° and the resulting supernatant is incubated overnight with 3 μl of Xpress mouse antibody (Invitrogen) and 20 μl of protein A–agarose beads (GIBCO, Grand Island, NY). The immunoprecipitates are washed four times with the immunoprecipitation buffer and then resuspended in SDS sample buffer and analyzed by Western blotting, using the procedures described above. The enzyme can be blotted with either Xpress antibody or rPLD1 antibody.

Properties of Rat Brain PLD1 Expressed in COS 7 Cells

COS 7 cells respond to PMA and lysophosphatidic acid (LPA), but not to other agonists because they lack their receptors. Cells transfected with rPLD1 cDNA or open vector are serum starved and labeled with [^{3}H]myristic acid in DMEM and 0.5% (v/v) FBS overnight as described above. Before starting the PLD assay, the cells are made quiescent by incubation in DMEM and 0.3% (w/v) BSA. Then 1-100 nM PMA or 1–100 μM LPA is added after a 10-min preincubation with 40 mM 1-butanol. The incorporation of label into PtdBut and total lipids is measured, as described above, at intervals during incubation for 60 min. Vector controls are necessary because the endogenous PLD of COS 7 cells shows large responses to PMA and LPA. The role of PKC in the response to LPA is usually evaluated by treating the cells with a PKC inhibitor such as 5 μM Ro 31-8220 or 5 μM bisindolylmaleimide (Sigma) 30 min before stimulation with agonist. Calphostin C should not be employed as a PKC inhibitor because it inhibits PLD activity.

The small GTPases ARF and Rho are important regulators of rPLD1.[3,15] The constitutively active mutant form of RhoA (V^{14}RhoA) can activate the enzyme when expressed in COS 7 cells[7] (Fig. 1). Cotransfection of COS7 cells with V^{14}RhoA and vector or rPLD1 is accomplished with the pcDNA 3.1 vector and FuGENE 6 as described above for rPLD1 transfection. The cells are assayed for PLD activity as described above and the expression of rPLD1 and V^{14}RhoA is assessed by Western blotting, using the rPLD1 C-terminal peptide antibody and monoclonal antibodies to RhoA (Santa Cruz Biotechnology, Santa Cruz, CA). Activation of rPLD1 by V^{14}RhoA is readily apparent because the endogenous PLD does not respond to this GTPase (Fig. 1).[7] However, constitutively active ARF3

[15] D. S. Min, S.-K. Park, and J. H. Exton, *J. Biol. Chem.* **273**, 7044 (1998).

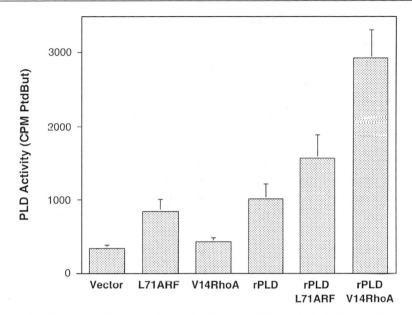

FIG. 1. Effects of constitutively active ARF and Rho on rPLD activity in COS cells. COS 7 cells were transfected with vector or rPLD1 either alone or in combination with L^{71}ARF3 or V^{14}RhoA, using the methods described in text. PLD activity was assayed by [^{3}H]PtdBut formation as described in text.

(L^{71} ARF3) stimulates the endogenous PLD of COS 7 cells, and its effect on rPLD1 is difficult to detect in this system (Fig. 1).[7]

Activation of rPLD1 by ADP ribosylation factor (ARF), Rho, and PKC can also be studied with membranes from COS 7 cells transfected with vector or rPLD1 cDNA.[7] The cells are transfected and their membranes are isolated as described above. The membranes are incubated with phospholipid vesicles containing labeled PtdCho, using the procedures[12] documented above. Additions are 30 μM guanosine 5′-O-(3-thiotriphosphate) (GTPγS) alone or in combination with 1 μM myristoylated ARF3 or 0.1 μM geranylgeranylated RhoA, prepared as described elsewhere.[16,17] In the case of stimulation by PKC, 50 nM recombinant PKCα (PanVera, Madison, WI) and 100 nM PMA are added. Because the activation of rPLD1 by PKC does not involve phosphorylation, ATP is not included in the incubation medium. In both studies, the formation of [^{3}H]PtdBut is measured over 15 or 30 min at 37°, as described above.

[16] O. Weiss, J. Holden, C. Rullea, and R. A. Kahn, *J. Biol. Chem.* **274,** 21066 (1989).
[17] C. D. Bae, D. S. Min, I. N. Fleming, and J. H. Exton, *J. Biol. Chem.* **273,** 11596 (1998).

Studies of Rat Brain PLD1 in Insect Cells

Expression of Rat Brain PLD1 in Sf9 Cells

The procedures largely follow those described by Min *et al.*[15] rPLD1 cDNA is inserted into the baculovirus transfer vector pBlueBacHis2B (Invitrogen) in frame with the sequence encoding the hexahistidine tag. This plasmid and linearized *Autographa calipornica* nuclear polyhedrosis virus (AcMNPV) viral DNA (Invitrogen) are cotransfected into *Spodoptera frugiperda* (Sf9) cells by lipofection, using insect cell-specific lipids (Invitrogen). The cells are grown at 27° in SF900II SFM medium (Life Technologies) supplemented with 5% (v/v) fetal bovine serum and gentamicin (100 μg/ml). Putative recombinant viral plaques are identified by color screening and their ability to express rPLD1 assessed by Western blotting of cell lysates with antibodies to the C terminus of rPLD1 (see above) or to the Xpress tag fused to the N terminus of rPLD1 (Invitrogen). Positive plaques are purified and the virus is amplified by infection of spinner cultures of 2×10^6 Sf9 cells/ml, stirred at 70 rpm and incubated at 27° in the medium described above. High-titer virus stocks are stored at 4° in the dark.

Monolayers of Sf9 cells (5×10^6 cells/100-mm dish) are infected with the recombinant baculovirus encoding (His$_6$)rPLD1 at a multiplicity of 10 and cultured at 27° for 12–48 hr. The cells are then washed in ice-cold PBS, scraped into lysis buffer [200 mM HEPES (pH 8.0), 50 mM NaCl, leupeptin (10 μg/ml), aprotinin (10 μg/ml), 1 mM phenylmethylsulfonyl fluoride, 5 mM MgCl$_2$, and 10% (v/v) glycerol], and lysed by passage through a 25-gauge needle five times. Disruption of cells, assessed by trypan blue staining, should be 95%. The lysate is centrifuged at 100,000g for 1 hr at 4° to prepare cytosol and membrane fractions. Membranes are washed once in lysis buffer. Expression of rPLD1, as assessed by Western blotting or PLD activity measurement (see above) is usually greater in the membrane fraction.

For purification of rPLD1, 10 dishes of Sf9 cells (3×10^7/150 mm dish) are used. They are infected at a multiplicity of 10 for 48 hr with the recombinant virus encoding (His$_6$)rPLD1. The medium is removed and the cells are lysed by incubation with 10 ml of lysis buffer containing detergent [50 mM HEPES (pH 8.0), 1% (v/v) β-octylglucopyranoside, 0.6 M NaCl, 5 mM MgCl$_2$, leupeptin (10 μg/ml), aprotinin (10 μg/ml), 1 mM phenylmethylsulfonyl fluoride, and 15% (v/v) glycerol] for 30 min at 4°. The suspension is centrifuged at 100,000g for 30 min at 4°. The supernant (10 ml) is incubated with 0.5 ml of Ni^{2+}–nitriloacetic acid (NTA)–agarose (Qiagen, Valencia, CA) at 4° overnight with constant agitation. The mixture is then placed in a chromatography column (Poly-Prep; Bio-Rad, Hercules, CA) and the resin is washed with 50 column volumes of buffer containing 20 mM HEPES (pH 8.0), 1% (v/v) β-octylglucopyranoside, 1 M NaCl, 20 mM imidazole, 5 mM MgCl$_2$, 10 mM 2-mercaptoethanol, and 20% (v/v) glycerol. Proteins are then eluted in four fractions with 80 mM imidazole, 0.6 M NaCl, 20 mM

HEPES (pH 8.0), 5 m*M* MgCl$_2$, and 20% (v/v) glycerol. The fractions are subjected to SDS–PAGE and Western blotting, using the rPLD1 antibody, as described above. The gels are also silver stained to determine enzyme purity, and those fractions showing the highest rPLD1 content and purity are pooled. The membrane fraction

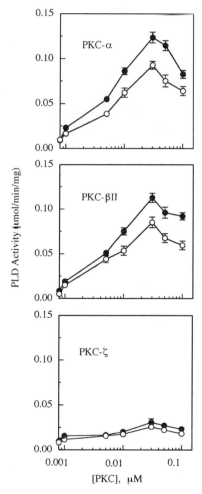

FIG. 2. Effects of PKC isozymes on rPLD activity. rPLD1 purified from Sf9 cells was incubated with phospholipid vesicles containing PtdInsP$_2$ and [^{3}H]PtdCho, and with increasing concentrations of recombinant PKC-α, PKC-β, and PKC-ζ with (●) or without (○) 100 n*M* PMA. The incubation medium and conditions and the PLD assay procedure were as described in text. Recombinant PKC isozymes are intrinsically active, hence their activation by PMA is relatively small. Other PKC isozymes were without effect (not shown).

can also be extracted with 0.6 M NaCl and subjected to adsorption chromatography as described for the soluble enzyme.

Properties of Rat Brain PLD1 Purified from Sf9 Cells

Recombinant rPLD1 is totally dependent on phosphatidylinositol 4,5-bisphosphate (PtdInsP$_2$) or phosphatidylinositol 3,4,5-trisphosphate (PtdInsP$_3$) for activity.[3,12,15] PtdInsP$_2$ is routinely used in *in vitro* assays of mammalian PLD isozymes. The composition and preparation of phospholipid vesicles containing PtdInsP$_2$ and labeled PtdCho are described in the preceding section describing the *in vitro* assay of PLD in COS 7 cells. When purified enzyme is employed the preferred assay is that of choline release.

The enzyme is specific for PtdCho. This is demonstrated by incubating the enzyme with phospholipid vesicles containing 1-palmitoyl-2-[1-^{14}C]linoleoyl-phosphatidylethanolamine (Amersham Pharmacia Biotech) or 1-stearoyl-2-[1-^{14}C] arachidonoylphosphatidylinositol (NEN Life Sciences Products) in place of the [^{3}H] PtdCho routinely used, and carrying out the transphosphatidylation reaction in the presence of 135 mM 1-butanol as described above. The incorporation of label into PtdBut is determined after incubation for 30 min at 37°.

The dependence of the enzyme on divalent cations is determined by incubating the enzyme with phospholipid substrate in the presence of 30 μM GTPγS plus 1 μM myristoylated ARF3, with Ca^{2+}/Mg^{2+}-EGTA buffers at pH 7.5 replacing the EGTA, CaCl$_2$, and MgCl$_2$ in the lipid reconstitution (reaction) buffer. The composition of the buffers is altered to give free cation concentrations from 0.1 μM to 1 mM according to the COMICS program (*http://www.pence.ualberta.ca/ftp*). The enzyme is stimulated by concentrations of Ca^{2+} and Mg^{2+} up to 1 mM.[15]

PKC plays an important role in the regulation of PLD *in vivo*. The enzyme also markedly stimulates rPLD1 and hPLD1 *in vitro*. However, the *in vitro* stimulation is limited to the α- and β-isozymes of PKC (Fig. 2) and occurs by a nonphosphorylating mechanism. To demonstrate this stimulation, recombinant rPLD1 is incubated with the standard incubation medium in the presence of recombinant PKC-α or PKC-β and 100 nM PMA. Stimulation of PLD activity is observed with 1 nM PKC isozyme and is maximal at 20 nM (Fig. 2). The addition of MgATP produces inhibition rather than stimulation.[15]

Small GTPases of the ARF and Rho families also markedly stimulate recombinant rPLD1.[3,13] This is illustrated by direct addition of recombinant forms of these GTPases to the purified enzyme in the presence of GTPγS.[3,15] Myristoylated ARF3 and geranylgeranylated RhoA, prepared as described elsewhere,[16,17] are added to the reaction mixture at concentrations from 0.01 to 5 μM together with 30 μM GTPγS. Stimulation of PLD activity is observed at 0.1 μM and is maximal at 2 μM for both GTPases. The maximal stimulation with ARF is greater than that of Rho. Rac and Cdc42Hs give less stimulation than Rho, and ARF1, -3, and -5 are more efficacious than ARF6.

[20] G-Protein-Coupled Receptor Regulation of Phospholipase D

By Guangwei Du, Andrew J. Morris, Vicki A. Sciorra, and Michael A. Frohman

Introduction

Phospholipase D (PLD) catalyzes the hydrolysis of phosphatidylcholine (PC) to generate phosphatidic acid (PA) and choline (reviewed in Frohman *et al.*[1]). PA has been implicated in signal transduction, membrane vesicular trafficking, cytoskeleton reorganization, and cell proliferation (reviewed in Jones *et al.*[2] and Liscovitch *et al.*[3]). PLD activity is present in many mammalian cells and tissues and is upregulated in response to a wide variety of agonists that signal through heterotrimeric G-protein-coupled or tyrosine kinase receptors. Receptor stimulation initiates multiple signal transduction cascades, ultimately including activation of protein kinase C (PKC), ADP-ribosylation factor (ARF), and Rho family members, which have been well characterized as activators of PLD in *in vitro* and *in vivo* assay systems. Because ARF, Rho, and PKC stimulate multiple downstream effector pathways that ultimately regulate cellular morphology, proliferation, and secretion, there has been intense interest in determining the relationship of PLD stimulation through each activator to these cell biological events.

The regulation of mammalian PLD by G-protein-coupled receptors is complex and not fully understood at present. The two isoforms, PLD1 and PLD2, are seemingly activated differently, although the details and degree of difference are not yet resolved. Despite the consensus that PKC, ARF, and Rho family members directly and potently stimulate PLD (specifically PLD1) *in vitro*,[4] the relative importance of each effector class in stimulation of PLD *in vivo*, and whether synergy between them is required, remain topics of current research interest in many laboratories.

Such studies can be approached through the manipulation of PKC, Rho, or ARF activity using overexpression of activated or inactive alleles, toxins, or pharmacological agents to activate or inhibit them (reviewed in Liscovitch *et al.*[3]). Although traditional and valuable, these approaches involve complicated interpretations, because manipulation of PKC, Rho, and ARF levels of activity affects

[1] M. A. Frohman, T.-C. Sung, and A. J. Morris, *Biochim. Biophys. Acta* **1439,** 175 (1999).

[2] D. Jones, C. Morgan, and S. Cockcroft, *Biochim. Biophys. Acta* **1439,** 229 (1999).

[3] M. Liscovitch, M. Czarny, G. Fiucci, and X. Tang, *Biochem. J.* **345,** 401 (2000).

[4] S. M. Hammond, J. M. Jenco, S. Nakashima, K. Cadwallader, S. Cook, Y. Nozawa, M. A. Frohman, and A. J. Morris, *J. Biol. Chem.* **272,** 3860 (1997).

many things aside from PLD regulation (discussed in more detail in Zhang *et al.*[5] and Du *et al.*[6]). As an alternative approach, we have generated alleles of PLD that exhibit altered regulation including mutants that are inactive,[7] selectively nonresponsive to PKC[5] and/or Rho,[6] or insensitive to phosphatidylinositol 4,5-bisphosphate [PI(4,5)P₂].[8] Using these molecular reagents, we have demonstrated that signal integration through both PKC and Rho via direct contact is required for significant physiological *in vivo* activation of PLD1 through G-protein-coupled receptors. The caveat to this approach is the one common to all overexpression studies, that the results obtained may not mirror with precision the regulation of the endogenous proteins.

Our laboratory uses two main experimental systems, *in vitro* and *in vivo* cell-based assays, to study PLD activation through G-protein-coupled receptors. The *in vitro* system is used to assess the interaction and direct activation of PLD by its regulators, all of which have been well characterized as important components in the signaling pathways mediated by G-protein-coupled receptors. On the basis of the molecular reagents we create and characterize in the *in vitro* system, the *in vivo* system is then used to explore more complicated questions regarding signaling in intact cells. This article discusses some of these assays. Standard protocols to measure PLD activity *in vitro* and *in vivo* have been described previously[9] and are not described here.

Protocols

Expression and Purification of Glu–Glu-Tagged Phospholipase D Isoform 1, Using Baculoviral System

In our previous experience, human PLD1 protein purified using the baculoviral system [either untagged or with a hexahistidine (His₆) tag] exhibited limited stability. Activity decreased noticeably each day and little was left after 3–4 days. To assist with purification, we explored expression of PLD1 with a Glu–Glu tag (MEYMPMEG) fused to the amino terminus. The purified protein was unexpectedly found to be relatively stable; no detectable loss of enzyme activity is observed over a period of storage at 4° for 10 days. One possibility for the improved protein

---

[5] Y. Zhang, Y. A. Altshuller, S. A. Hammond, F. Hayes, A. J. Morris, and M. A. Frohman, *EMBO J.* **18,** 6339 (1999).

[6] G. Du, Y. M. Altshuller, Y. Kim, J. M. Han, S. H. Ryu, A. J. Morris, and M. A. Frohman, *Mol. Biol. Cell* **11,** 4359 (2000).

[7] T. C. Sung, R. Roper, Y. Zhang, S. A. Rudge, R. Temel, S. M. Hammond, A. J. Morris, B. Moss, J. Engebrecht, and M. A. Frohman, *EMBO J.* **16,** 4519 (1997).

[8] V. A. Sciorra, S. A. Rudge, G. D. Prestwich, M. A. Frohman, J. Engebrecht, and A. J. Morris, *EMBO J.* **18,** 5911 (1999).

[9] M. A. Frohman, Y. Kanaho, Y. Zhang, and A. J. Morris, *Methods Enzymol.* **325,** 177 (2000).

stability may lie in the elution method rather than the tag. The current protocol uses peptide competition to recover the purified protein rather than changes in pH. The purification and concentration of the PLD stimulators PKC, Rho, and ARF have been described previously.[4]

Phospholipase D Isoform 1 Expression and Purification

1. The full-length PLD1 wild-type and mutant cDNAs are subcloned into a pFASTBAC vector (Life Technologies, Rockville, MD) modified to contain the sequence MEYMPMEG (Glu–Glu tag) at the amino terminus.[10] Recombinant bacmids are prepared by transfection of DH10Bac cells with the pFASTBAC plasmids. Recombinant baculoviruses are amplified and propagated by standard procedures as described in the manual about the Bac-to-Bac baculovirus expression system (Life Technologies). High-titer stocks of recombinant baculoviruses for expression of Glu–Glu-tagged wild-type and mutant PLD1 and PLD2 alleles are available on request.

2. To express PLD1, monolayers of exponentially growing *Spodoptera frugiperda* (Sf9) cells (3×10^7 cells/225-cm^2 flask, cultured at 27° in complete Grace's medium supplemented with glutamine, lactalbumin, Yeastolate (from Life Technologies), and 10% (v/v) fetal bovine serum containing penicillin (100 U/ml) and streptomycin (100 mg/ml); generally, two flasks of cells are used for each purification] are infected with recombinant baculoviruses at a multiplicity of 10 for 1 hr with gentle rocking. [If large amounts of protein are required (e.g., for concentration), 300–500 ml of the cells at a density of 1×10^6 cells/ml is seeded in a 500-ml spinner flask and is infected with recombinant baculoviruses at a multiplicity of 10. The infected cells are grown for 48 hr without removing the virus, and then the procedure described below is scaled up correspondingly.]

3. The virus-containing medium is then removed and replaced with fresh supplemented complete Grace's medium. The infected cells are grown for 48 hr, shaken off the flask, pelleted at 2000g for 5 min at 4°, and washed once with ice-cold PBS.

4. The cells are then lysed on ice by the addition of 5 ml of ice-cold lysis buffer per 225-cm^2 flask. After 30 min on ice, the cell lysate is centrifuged at 50,000g for 30 min at 4°.

5. While the centrifugation step is in progress, 0.3 ml of anti-Glu–Glu tag immunoaffinity resin (kindly provided by N. Pryor, Onyx Corp., Richmond, CA; the immunoaffinity resin is commercially available from BAbCo, Berkeley, CA) is washed with 5–10 ml of lysis buffer in a 15-ml tube several times (spin down the resin by gentle centrifugation to remove the lysis buffer, and then add fresh lysis buffer).

---

[10] L. R. Stephens, A. Eguinoa, H. Erdjument-Bromage, M. Lui, F. Cooke, J. Coadwell, A. S. Smrcka, M. Thelen, K. Cadwallader, P. Tempst, and P. T. Hawkins, *Cell* **89,** 105 (1997).

6. The lysate supernatant obtained (10 ml) is mixed with the immunoaffinity resin and rotated at 4° for at least 1 hr. The resin is then placed in a 10-ml Bio-Rad (Hercules, CA) disposable chromatography column. Remove the bottom cap and let the extract flow through. Use the flowthrough to wash the 15-ml tube and the sides of the column until all of the resin is packed (all steps at 4°).

7. Wash the column with 10 ml of lysis buffer, and then 10 ml of stock buffer, and finally 10 ml of salt wash buffer (all steps at 4°).

8. Move the column to room temperature. Wash the column with 5 ml of salt wash buffer. During the wash, prepare 0.9 ml of elution buffer.

9. Cap the column bottom. Resuspended the resin with 0.3 ml of elution buffer and let it stand for 2 min. Remove the bottom cap and recover the eluent. Elute twice more with 0.3 ml each time.

10. Check the purity of the protein by 8% (w/v) sodium dodecyl sulfate–polyacrylamide gel electrophoresis (SDS–PAGE). A 5- to 10-μl aliquot from each fraction is added to 2× SDS–PAGE sample loading buffer containing 8 M urea, denatured at room temperature, and then electrophoresed. The protein should be detectable by standard Coomassie blue staining. Alternatively, use ARF-stimulated PLD activity in each fraction to determine the respective amounts of active protein.

11. Measure the concentration of the protein by the Bradford method, using the Coomassie Plus-200 protein assay reagent (Pierce, Rockford, IL). Briefly, 20 μl of bovine serum albumin (BSA) standards or PLD samples is mixed with 200 μl of Bradford reagent in a 96-well plate and the absorbances at 595 nm are measured with a plate reader. Use the standard curve from BSA to determine the concentration of PLD.

12. Create a dose–response curve, using the *in vitro* PLD assay, to choose the suitable protein concentration for further experiments. Usually, 10–20 ng of purified protein per sample is appropriate for the *in vitro* PLD assay. In our hands, two 225-cm^2 flasks should yield approximately 20–70 μg of purified PLD1 protein.

Solutions

> Stock buffer: 20 mM Tris-HCl, (pH 7.8), 1 mM EDTA, 1 mM dithiothreitol (DTT), 20 μM leupeptin, 0.1 mM phenylmethylsulfonyl flouride (PMSF), 0.1 mM benzamidine
> Lysis buffer: Stock buffer plus 1% (v/v) Nonidet P-40 (NP-40) (final concentration)
> Salt wash buffer: Stock buffer plus 400 mM NaCl (final concentration)
> Elution buffer: Stock buffer plus 400 mM NaCl plus Glu–Glu peptide (100 μg/ml, final concentration; make fresh)
> Glu–Glu peptide: EYMPTD (the free amine of the E should be acetylated; the C terminus should be left as a carboxyl)

Notes. An example of the purification procedure is shown in Fig. 1A–C for a mutant PLD1 protein denoted "mini-PLD1," which lacks the amino terminus

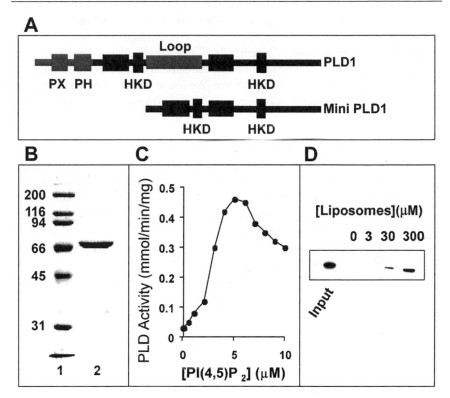

FIG. 1. Purification and characterization of mini-PLD. (A) Sequence alignment of PLD1 and mini-PLD1. (B) SDS–PAGE of purified mini-PLD1. Lane 1, molecular weight markers; lane 2, 10 μg of purified mini-PLD1. (C) Activation of mini-PLD1 by PI(4,5)P$_2$. Mini-PLD1 activity was determined with sonicated lipid vesicles containing 2.6 mM PC, the indicated concentrations of PI(4,5)P$_2$, and varying amounts of PE to give a final lipid concentration of 100 mM. Assays contained 3 μM GTPγS activated ARF1. (D) Binding of mini-PLD1 to PI(4,5)P$_2$ containing sucrose-loaded liposomes. Mini-PLD1 was incubated with the indicated concentrations of PC : PS : PE liposomes containing 5 mol% PI(4,5)P$_2$. The vesicles were sedimented by ultracentrifugation and vesicle-bound protein was detected by Western blotting.

and the central "loop" region. As reported previously,[11] the 70-kDa mini-PLD1 is catalytically active in a PI(4,5)P$_2$-dependent manner (see Fig. 1C), indicating that the amino-terminal PH domain does not mediate the activity-dependent PIP$_2$ interaction. Instead, the PI(4,5)P$_2$ interaction is mediated by a binding site located between the 2 HKD motifs (not shown; see also Sciorra et al.[8]). Figure 1B shows the recovery of Glu–Glu-tagged mini-PLD, using the protocol described above as visualized by Coomassie blue staining; virtually pure PLD protein can be generated

[11] T.-C. Sung, Y. Zhang, A. J. Morris, and M. A. Frohman, J. Biol. Chem. **274**, 3659 (1999).

in a one-step procedure by this approach. Figure 1C demonstrates that the protein is active in the presence of PLD stimulators [GTPγS-activated ARF1 and PI(4,5)P$_2$].

Concentration and Buffer Exchange of Phospholipase D Isoform 1 by Ultrafiltration. For concentration and buffer exchange of purified PLD1, we initially tried ultrafiltration powered by centrifugal force. We found that most of the protein was lost and little concentration occurred. More successful was the stirred-cell ultrafiltration system powered by nitrogen gas pressure forces described below. We have been able to concentrated full-length PLD1 from 20–50 to ~200 μg/ml successfully in 10 mM piperazine-N,N'-bis (2-ethanesulfonic acid) (PIPES), 50 mM potassium glutamate (pH 7.4), which is a buffer appropriate for microinjection.[12] The advantage of the stirred-cell ultrafiltration system over the centrifugal filter device is that it avoids concentration polarization, which may have caused protein precipitation or adherence onto the membrane surface. All of the procedures below are performed at 4° or in a cold room.

1. For buffer exchange and further concentration of PLD, the protein eluted from the column is diluted more than 10-fold into the desired buffer. To concentrate PLD without exchanging the buffer, the protein can be loaded directly into the stirred cell.

2. Load an Amicon PM10 membrane (Millipore, Bedford, MA) into the stirred cell (Millipore, model 8003) with the glossy side up. Handle the membrane carefully with powder-free gloves and hold by the edge to avoid scratching the smooth, glossy surface. Check carefully to make sure that the stirred cell is properly assembled and that no leakage is visible around the membrane when pressure is applied.

3. Pour the protein solution into the stirred cell (up to the maximum operating volume) and place the unit on a magnetic stirrer. Connect the cell to the nitrogen gas regulator and pressurize to 40–50 lb/in^2 (pressure should not exceed 70 lb/in^2).

4. Monitor the flow rate and ultrafiltrate volume until the desired rate of concentration has been achieved. The cell is refilled when the volume has decreased by one-third if the total sample volume initially exceeded the cell capacity.

5. Turn off the stirrer and the nitrogen flow to the cell. Slowly release the internal cell pressure. Disassemble the stirred cell and transfer the concentrated sample to a 1.5-ml centrifuge tube, using a Pipetman (Rainin, Woburn, MA).

Notes. This protocol is still in evolution. The recovery of PLD1 is variable, and the factor(s) influencing this are not certain. More difficulty seems to be encountered when the starting protein concentration is low or when low-salt buffers are

---

[12] N. Vitale, A.-S. Caumont-Primus, S. Chasserot-Golaz, G. Du, S. Wu, V. A. Sciorra, A. J. Morris, M. A. Frohman, and M.-F. Bader, *EMBO J.* **20,** 2424 (2001).

used for the exchange. In more recent preliminary experiments, Amicon YM10 membranes have exhibited better recoveries than the PM membranes described above.

Preparation of Tissue Culture Cells for Use in in Vivo Phospholipase D Assay for G-Protein-Coupled Receptor Signaling

Molecular Reagents. A wide variety of expression plasmids have been used for functional studies of PLD in mammalian cells. Our laboratories use the cytomegalovirus (CMV) promoter-driven plasmid pCGN, which appends an N-terminal hemagglutinin (HA) epitope to the PLD1 and PLD2 proteins. To dissect the signal pathways regulating PLD1 activation, we have generated a useful series of mutants, which includes inactive allele K898R,[7] an allele selectively nonresponsive to PKC PIM87,[5] an allele selectively nonresponsive to Rho 1870R,[6] and an allele selectively nonresponsive to both PKC and Rho PIM87/1870R.[6]

Using cultured mammalian cells with either endogenous, transiently expressed, or stably expressed G-protein-coupled receptors, PLD1 activation can be studied by using these alleles. The receptors thus far used in our studies are the m1, m2, and m3 muscarinic acetylcholine receptors (mAChRs), the Edg2 and Edg4 lysophosphatidic acid (LPA) receptors, and the angiotensin II (AT1a) receptor.

Notes. In our system, we have been able to examine PLD signaling in combination with G-protein-coupled receptors that signal through G_q/G_{11} and G_{12}/G_{13} (Edg4, AT1a, and m1 and m3 mAChRs), but not receptors that signal through G_i only (Edg2 and m2 mAChR), for which only small agonist responses were observed. This suggests that different types of G-protein-coupled receptors have different potentials to activate PLD1.[6]

Cell Maintenance. Our laboratory uses HEK-293 and COS-7 cells to study the regulation of PLD through G-protein-coupled receptors. The cells are maintained in complete Dulbecco's modified Eagle's medium (DMEM) supplemented with 10% fetal calf serum, penicillin (100 U/ml), and streptomycin (100 mg/ml). In some cases, we use HEK-293 cells stably expressing the m2 and m3 mAChRs to study the signaling. Those cells are maintained in DMEM containing G418 (0.5 mg/ml).

Transfection and Stimulation of Cells

1. On day 1 (morning), the cells are passed in 35-mm dishes ($3-4 \times 10^5$ cells/dish) and are grown for 20–24 hr. In some cases, especially when it is planned to transfect the HEK-293 cells with constitutively activated Rho, Rho kinase, or ARF, which can cause the cells to round up and detach from the tissue culture dishes, more reproducible assay data are achieved by using poly-L-lysine hydrobromide (Sigma, St. Louis, MO)-coated dishes instead of regular tissue culture (TC) dishes. For preparation of poly-L-lysine, dissolve it in sterile tissue culture-grade water at

0.1 mg/ml and store it at $-20°$; pipette 1 ml/dish and incubate overnight at room temperature; aspirate and rinse once with sterile tissue culture-grade water. Allow the plates to dry with the lid off (in a sterile laminar flow hood). The plates can be stored dry and sterile for months.

2. On day 2 (the cells should be $\sim$70–80% confluent), the cells are transfected in the morning with PLD1 alleles and/or G-protein-coupled receptors. Prepare the following solutions in 12×75 mm sterile tubes.

Solution A: For each transfection, dilute 1 μg of each DNA (plasmid) in 100 μl of Opti-MEM I containing 6–8 μl of PLUS reagent (Life Technologies) and incubate for 15 min at room temperature

Solution B: For each transfection, dilute 4–6 μl of Lipofectamine reagent in 100 μl of Opti-MEM I (the cocktails of Opti-MEM I-diluted PLUS and LipofectAMINE reagent can be made and added to the individual tubes in advance of the DNA)

Combine the DNA and lipid solutions, mix gently, and incubate at room temperature for 15 min. Wash the cells once with 1 ml of Opti-MEM I. For each transfection, add 0.8 ml of Opti-MEM I to each tube containing the lipid–DNA complexes. Mix gently and overlay the diluted complex solution onto the washed cells. Incubate the cells at $37°$ in a CO_2 incubator.

3. Four to 5 hr posttransfection, the medium is replaced with complete DMEM and the cells are incubated for an additional 22–24 hr. For *in vivo* PLD assays, the transfection mixtures are replaced with complete DMEM containing 2 μCi of [^{3}H]palmitic acid (from American Radiolabeled Chemicals, St. Louis, MO): To prepare [^{3}H]palmitic acid, 50 μl of a 10-mCi/ml stock is dried in a Speed-Vac concentrator. The [^{3}H]palmitic acid is then resuspended in 100 μl of sterile TE (5 mCi/ml) and sonicated for 30 sec at high power in a water sonicator before use. Use 0.4 μl of resuspended [^{3}H]palmitic acid per milliliter of medium.

4. On day 3, the cells are ready for PLD assays. Briefly, the cells are washed with 1 ml of warm, fresh Opti-MEM I once, and incubated in the same medium for 1 to 2 hr, after which the medium is replaced with fresh Opti-MEM I containing 0.3% (v/v) 1-butanol and optionally various receptor agonists (the preparation and working concentration of the agonists are listed below). After 30 min, the stimulating medium is replaced with 300 μl of ice-cold methanol to halt further synthesis of the labeled lipid end product. The dishes are placed on ice and processed for the *in vivo* PLD assay, which has been described in a previous volume of this series.[9]

Preparation of Agonists

Carbachol (carbamylcholine chloride; Sigma): Stock solution, 400 mM; store at room temperature. Working concentration, 1 mM

LPA (1-oleoyl-2-hydroxy-*sn*-glycero-3-phosphate, 18 : 1 lyso-PA; Avanti, Alabaster, AL): Store in chloroform at 25 mg/ml at $-20°$. May precipitate and need to be warmed for use. To prepare it for LPA receptors, dry down LPA under nitrogen gas, then resuspend in BSA (0.1 mg/ml in H_2O or medium) (e.g., Opti-MEM I), using sonication to make a 10 mM stock solution. Use a 100 μM final concentration to stimulate cells

Angiotensin II (Sigma): Stock solution, 10 μM (dissolve in sterile H_2O), store at $-20°$. Working concentration, 100 nM

Measurement of Phospholipase D Binding to Sucrose-Loaded Phospholipid Vesicles

PLD1 and PLD2 are extrinsic membrane proteins. Interaction of these enzymes with phospholipid bilayers is therefore important for their biological activity. The generation of PLD1 alleles for dissection of G protein receptor-coupled regulation raises the possibility that the activities could be altered if the mutations affect the interaction of PLD1 with the membrane itself.

Binding of the PLD enzymes to artificial membranes can be measured with sucrose loaded phospholipid vesicles. The PLD proteins bind to vesicles composed of a equal molar ratio of PC : PS : PE with an affinity of approximately 200 μM. This affinity is not increased by inclusion of 5 mol% PI in the vesicles. However, PLD1 and PLD2 bind to vesicles containing 5 mol% PI(4,5)P_2 with an affinity that is at least 10-fold higher. Structural integrity of a short region rich in basic amino acid residues is required for high-affinity binding of the enzymes to PI(4,5)P_2-containing lipid vesicles.

The basic procedure used is adapted from that described by Buser and McLaughlin,[13] and has been described previously for studies of PLD2.[8] In brief, large unilamellar vesicles are loaded with sucrose buffer and then incubated with the PLD proteins in an isotonic buffer solution. The higher density of the vesicle-encapsulated solution allows the vesicles to be sedimented by ultracentrifugation. The PLD binding is quantitated by either measurements of catalytic activity remaining in the supernatant or by SDS–PAGE of vesicle-bound proteins.

Generation of Sucrose-Loaded Phospholipid Vesicles

1. Lipids are combined from stock solutions in $CHCl_3$. In general, it is best to prepare about twice the amount of lipid required for the binding experiments because of losses during the procedure. The standard composition of the vesicles used in these experiments is 1 : 1 : 1 PC : PS : PE. A trace quantity of [³H]PC is included in the lipid mixture to monitor recovery.

---

[13] C. A. Buser and S. McLaughlin, *Methods Mol. Biol.* **84,** 267 (1998).

2. The lipid mixture is dried completely in a small glass flask, using a rotary evaporator. Complete drying of the lipids is essential for success of the procedure.

3. The lipids are resuspended by vortex mixing in 176 mM sucrose, 1 mM morpholinepropanesulfonic acid (MOPS, pH 7.6) (sucrose buffer). In general, the total lipid concentration at this stage will be about 1 mM. The lipid suspension is subjected to five freeze–thaw cycles in a liquid N_2 bath with thawing at 30°.

4. Vesicles are prepared by extruding the suspension through two 0.1-μm pore size polycarbonate filters, using either a hand-held extruder (Avanti Polar Lipids) or an N_2 driven extruder (Lipex, Vancouver, Canada) according to the manufacturer directions.

5. The extruded vesicles are diluted 5-fold into 100 mM KCl, 1 mM MOPS (pH 7.0) (salt buffer), sedimented by centrifugation (100,000g for 30 min at 4°), and resuspended in the same buffer. A sample is taken for scintillation counting to determine the concentration of vesicles.

Measuring Binding of Phospholipase D Proteins to Sucrose-Loaded Vesicles

1. PLD proteins are incubated with the vesicles on ice in siliconized 1.5-ml microcentrifuge tubes. The vesicles are diluted in salt buffer to give a range of final concentrations.

2. The vesicles are sedimented by centrifugation at 100,000 g for 30 min at 4°. PLD protein remaining in the supernatant is determined by measurement of enzymatic activity. Proteins associated with the sedimented vesicles are solublized in SDS–PAGE sample buffer and analyzed by SDS–PAGE and Western blotting.

Notes. An example of the PLD protein–vesicle interaction is visualized by Western blotting as a function of liposome concentration in Fig. 1D, using mini-PLD1. Despite loss of the PH domain, the mini-PLD1 still interacts with the vesicles in a concentration- and (not shown) PI(4,5)P$_2$-dependent manner.

Acknowledgments

This work was supported by grants from the NIH to M.A.F. (GM54813) and A.J.M. (GM50388). G.D. is a fellow of the American Heart Association.

[21] Analysis and Quantitation of Ceramide

By PAOLA SIGNORELLI and YUSUF A. HANNUN

Introduction: Ceramide Metabolism and Function

It is now well appreciated that many membrane lipids play important roles in the specific responses of cells to various stimuli. The sphingolipid ceramide has emerged as a relatively new addition to the group of bioactive lipid molecules, with suggested roles in the regulation of the stress response. Indeed, many inflammatory cytokines [such as tumor necrosis factor α (TNF-α), interleukin 1 (IL-1), and interferon γ], stresses (such as heat, ischemia/reperfusion, and oxidation), and cytotoxic agents (such as radiation and the chemotherapeutic agents Ara-C, vincristine, etoposide, and daunorubicin) affect ceramide metabolism and induce increases of its cellular levels.[1,2]

Data are beginning to implicate ceramide in cell responses to the above-described agents. For example, preventing the intracellular increase in free ceramide can protect from the effects of many of these agents on cell growth and apoptosis. On the other hand, treatment with exogenous ceramides or agents that induce accumulation of endogenous ceramide (e.g., by blocking its metabolism) can mimic many of the responses to a stress stimulus. These include effects on cell differentiation, apoptosis, and cell senescence.[3-5] Such effects are not induced by closely related compounds such as dihydroceramide and stereoisomers of the natural ceramides.

Intracellular levels of ceramide can be increased by its *de novo* synthesis (a multistep pathway involving several enzymes) or by the hydrolysis of more complex sphingolipids (such as cerebrosides and sphingomyelin via activation of cerebrosidases and sphingomyelinases, respectively). Reciprocally, ceramide levels can be decreased via metabolic and/or catabolic reactions, including incorporation of ceramide into other sphingolipids (via activation of enzymes such as sphingomyelin synthase or cerebroside synthase) or its hydrolysis into fatty acids and sphingosine (by the action of ceramidases).

Several stimuli are known to influence the activity of these enzymes, thus modulating intracellular levels of ceramide and its induced biologic effects. For example, activation of neutral sphingomyelinase occurs in response to treatment with TNF-α, Fas ligands, IL-1, and serum deprivation,[6,7] whereas activation of

[1] Y. A. Hannun, *Science* **274**, 1855 (1996).
[2] S. Mathias, L. A. Pena, and R. N. Kolesnick, *Biochem. J.* **335**, 465 (1998).
[3] M. Liscovitch, *Trends Biochem. Sci.* **17**, 393 (1992).
[4] L. M. Obeid, C. M. Linardic, L. A. Karolak, and Y. A. Hannun, *Science* **259**, 1769 (1993).
[5] Y. A. Hannun and C. Luberto, *Trends Cell Biol.* **10**, 73 (2000).

acidic sphingomyelinase has been shown to participate in the response to treatments with TNF, Fas ligands, as well as ionizing radiation, endotoxin, and anti-CD28.[2,7]

On the other hand, *de novo* synthesis of ceramide is induced by chemotherapeutic agents (such as etoposide, daunorubicin), angiotensin II, TNF-α, UV radiation, and cell loading with fatty acids.[2–4] Importantly, flux through this pathway can be inhibited by a number of specific enzyme inhibitors such as fumonisin B1, which inhibits dihydroceramide and ceramide synthase,[8] and myriocin or ISP-1, which inhibits serine palmitoyltransferase, the first and rate-limiting enzyme in sphingolipid biosynthesis[9] (see Fig. 1).

Many of the biological responses mediated by ceramide are due to its direct or indirect effects on specific cellular targets. *In vitro,* ceramide activates a ceramide-activated protein phosphatase[10,11] and a ceramide-activated kinase[12] and it inactivates mitochondrial respiratory complex III.[13] *In vivo,* the accumulation of ceramide leads to a number of specific biochemical responses including inhibition of phospholipase D and protein kinase Cα,[14,15] activation of the retinoblastoma gene product,[16,17] alteration of mitochondrial membrane potential and release of cytochrome C,[18,19] activation of death-promoting caspases,[20–24] activation of c-*jun* N-terminal kinase (jnk)/stress-activated protein kinases (SAPKs)[25–31] and regulation of expression of several genes such as c-*myc* and c-*jun*.[32,33]

Given the multiple connections of ceramide metabolism to cell agonists and to specific cell responses, it becomes important to evaluate ceramide formation and its mechanisms as a prelude to any study aimed at exploring the roles of ceramide in signal transduction and cell regulation.

In this article we describe methods to detect the changes in levels of intracellular ceramide, which can be measured using the diacylglycerol kinase (DGK) assay or by steady state labeling methods. Subsequently, we discuss methods to dissect the

[6] B. Liu, L. M. Obeid, and Y. A. Hannun, *Semin. Cell Dev. Biol.* **8,** 311 (1997).

[7] K. Kishikawa, C. E. Chalfant, D. K. Perry, A. Bielawska, and Y. A. Hannun, *J. Biol. Chem.* **274,** 21335 (1999).

[8] A. H. Merrill, Jr., E. M. Schmelz, D. L. Dillehay, S. Spiegel, J. A. Shayman, J. J. Schroeder, R. T. Riley, K. A. Voss, and E. Wang, *Toxicol. Appl. Pharmacol.* **142,** 208 (1997).

[9] S. Furuya, J. Mitoma, A. Makino, and Y. Hirabayashi, *J. Neurochem.* **71,** 366 (1998).

[10] J. T. Nickels and J. R. Broach, *Genes Dev.* **10,** 382 (1996).

[11] C. E. Chalfant, K. Kishikawa, M. C. Mumby, C. Kamibayashi, A. Bielawska, and Y. A. Hannun, *J. Biol. Chem.* **274,** 20313 (1999).

[12] Y. Zhang, B. Yao, S. Delikat, S. Bayoumy, X. H. Lin, S. Basu, M. McGinley, P. Y. Chan-Hui, H. Lichenstein, and R. Kolesnick, *Cell* **89,** 63 (1997).

[13] T. I. Gudz, K. Y. Tserng, and C. L. Hoppel, *J. Biol. Chem.* **272,** 24154 (1997).

[14] M. E. Venable, G. C. Blobe, and L. M. Obeid, *J. Biol. Chem.* **269,** 26040 (1994).

[15] J. Y. Lee, Y. A. Hannun, and L. M. Obeid, *J. Biol. Chem.* **271,** 13169 (1996).

[16] G. S. Dbaibo, M. Y. Pushkareva, S. Jayadev, J. K. Schwarz, J. M. Horowitz, L. M. Obeid, and Y. A. Hannun, *Proc. Natl. Acad. Sci. U.S.A.* **92,** 1347 (1995).

[17] S. Jayadev, B. Liu, A. E. Bielawska, J. Y. Lee, F. Nazaire, M. Pushkareva, L. M. Obeid, and Y. A. Hannun, *J. Biol. Chem.* **270,** 2047 (1995).

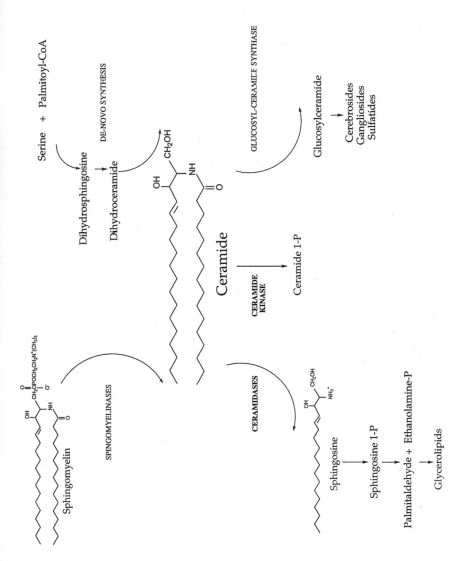

FIG. 1. Metabolism of sphingolipids.

two most common pathways of ceramide generation: the *de novo* pathway and the activation of sphingomyelin hydrolysis.

Measurement of Changes in Ceramide: Mass Assay by Diacylglycerol Kinase and Radiolabeling Procedures

Principle

Variations of cellular content of ceramide can be measured by labeling cells with radioactive precursors of lipids or by *in vitro* radioactive derivatization utilizing enzymatic reactions such as DGK. Fluorescence detection can be used as well, but this article focuses on the use of radioisotopes and the DGK assay.

Values derived from these methods are normalized to parameters related to cellular mass such as total phospholipids or protein content.

Concepts of in Vivo Labeling

When labeling cells with radioactive precursors it is important to differentiate between two approaches: a pulse label and a steady state label followed by a chase.

[18] P. Zhang, B. Liu, G. M. Jenkins, Y. A. Hannun, and L. M. Obeid, *J. Biol. Chem.* **272**, 9609 (1997).

[19] R. M. Kluck, E. Bossy-Wetzel, D. R. Green, and D. D. Newmeyer, *Science* **275**, 1132 (1997).

[20] T. Okazaki, U. Chung, T. Nishishita, S. Ebisu, S. Usuda, S. Mishiro, S. Xanthoudakis, T. Igarashi, and E. Ogata, *J. Biol. Chem.* **269**, 27855 (1994).

[21] G. S. Dbaibo, D. K. Perry, C. J. Gamard, R. Platt, G. G. Poirier, L. M. Obeid, and Y. A. Hannun, *J. Exp. Med.* **185**, 481 (1997).

[22] R. W. Keane, A. Srinivasan, L. M. Foster, M. P. Testa, T. Ord, D. Nonner, H. G. Wang, J. C. Reed, D. E. Bredesen, and C. Kayalar, *J. Neurosci. Res.* **48**, 168 (1997).

[23] N. Waterhouse, S. Kumar, Q. Song, P. Strike, L. Sparrow, G. Dreyfuss, E. S. Alnemri, G. Litwack, M. Lavin, and D. Watters, *J. Biol. Chem.* **271**, 29335 (1996).

[24] M. J. Smyth, D. K. Perry, J. Zhang, G. G. Poirier, Y. A. Hannun, and L. M. Obeid, *Biochem. J.* **316**, 25 (1996).

[25] J. K. Westwick, A. E. Bielawska, G. Dbaibo, Y. A. Hannun, and D. A. Brenner, *J. Biol. Chem.* **270**, 22689 (1995).

[26] P. Zhang, B. S. Miller, S. A. Rosenzweig, and N. R. Bhat, *J. Neurosci. Res.* **46**, 114 (1996).

[27] O. Cuvillier, G. Pirianov, B. Kleuser, P. G. Vanek, O. A. Coso, S. Gutkind, and S. Spiegel, *Nature (London)* **381**, 800 (1996).

[28] K. M. Latinis and G. A. Koretzky, *Blood* **87**, 871 (1996).

[29] M. Verheij, R. Bose, X. H. Lin, B. Yao, W. D. Jarvis, S. Grant, M. J. Birrer, E. Szabo, L. I. Zon, J. M. Kyriakis, A. Haimovitz-Friedman, Z. Fuks, and R. N. Kolesnick, *Nature (London)* **380**, 75 (1996).

[30] C. Huang, W. Ma, M. Ding, G. T. Bowden, and Z. Dong, *J. Biol. Chem.* **272**, 27753 (1997).

[31] B. Brenner, U. Koppenhoefer, C. Weinstock, O. Linderkamp, F. Lang, and E. Gulbins, *J. Biol. Chem.* **272**, 22173 (1997).

[32] R. A. Wolff, R. T. Dobrowsky, A. Bielawska, L. M. Obeid, and Y. A. Hannun, *J. Biol. Chem.* **269**, 19605 (1994).

[33] J. G. Reyes, I. G. Robayna, P. S. Delgado, I. H. Gonzalez, J. Q. Aguiar, F. E. Rosas, L. F. Fanjul, and C. M. R. Galarreta, *J. Biol. Chem.* **271**, 21375 (1996).

In the initial labeling phase (0–6 hr; this interval depends on several parameters such as metabolic rates and phase of the cell cycle) the radioactive precursor is taken up from the medium and used by the cell to form new metabolites. During this phase, precursors are mostly incorporated in newly synthesized molecules. Such a short labeling time ("pulse") can be used to follow preferentially specific metabolites and fluxes through pathways.

The "equilibrium" phase, or "steady state," is a phase that is reached when the precursor has been already metabolized and the amount of radioactivity per each lipid specie is more or less constant over the duration of the experiment, because of a balance of formation and loss. The time frame required to reach such a state is related to cell cycle and metabolism but, for most lipids, it lies approximately between 48 and 72 hr. This is also shown by the fact that the ratio between the amount of radioactivity taken up from the medium and released in the medium by proliferating cells becomes constant. After cells have been labeled, it is important to precede any treatment intended to affect labeled metabolites with a wash in phosphate-buffered saline (PBS) and period of "chase" in fresh medium not containing radioactive label (2 hr). During such a step, nonincorporated radioactivity will be released. Then cells can be changed to new medium and treated for the experiment.

It should be noted that with *in vivo* labeling it is possible to detect variations of ceramide content in response to different stimuli but such approaches will not give a good indication of changes of total mass of the metabolites.

The total intracellular amount of ceramide can be evaluated by a steady state labeling of cells with radioactive precursors (Method 1) or by utilizing the DG kinase *in vitro* assay (Method 2).

Method 1: Steady State Labeling

^{3}H- and ^{14}C-labeled serine, palmitate, or dihydrosphingosine can be delivered to cells, which incorporate them into more complex lipids. Subsequently, lipids can be extracted by total extraction (Folch *et al.* or Bligh–Dyer procedures[34,35]). Glycerolipids containing ester bonds can be hydrolyzed before ceramide separation by thin-layer chromatography in order to minimize the background.

Labeling in Vivo

1. Human (adherent) adenocarcinoma MCF7 cells (2×10^5) are seeded in 100-mm petri dishes and cultured for 36–48 hr.

2. The medium is replaced with new medium containing a 2-μCi/ml concentration of [^{3}H]palmitate or [^{3}H]serine (specific activity, 20 and 43 Ci/mmol,

[34] J. Folch, M. Lees, and G. H. S. Stanley, *J. Biol. Chem.* **226,** 497 (1957).
[35] E. G. Bligh and W. J. Dyer, *Can. J. Biochem. Physiol.* **37,** 911 (1959).

respectively, using a total volume of 6–8 ml in order to have initially 6–8 μCi/10^6 cells). Cells are cultured for an additional 48–72 hr (final number of cells will be about 2×10^6). If suspension cells are studied, $3–4 \times 10^5$ cells ml can be grown in medium containing radioactive precursors (2 μCi/ml) for 60–72 hr. The initial number of cells must be adjusted in order to have, after 3 days of labeling, at least $1–2 \times 10^6$ cells per sample. *Note:* Uptake of precursor can be monitored by measuring clearance of radioactivity in the medium and increases of counts in cell pellets (counts per minute per 10^6 cells): after 2 or 3 days, these values tend to become stable (steady state).

3. Cells are washed with PBS and allowed to rest for 2–4 hr in fresh medium (chase). Then the medium is changed again and the cells are restimulated.

4. At the time of determination, cells are washed with PBS, directly scraped into methanol (1 ml), and transferred to 13×100 mm glass screw-cap test tubes.

5. Cell pellets, resuspended in methanol, can be stored for up to 15 days at $-80°$. *Note:* From this step on, it is recommended that any step in which solvents are used be performed in capped tubes.

Extraction (Bligh–Dyer)

Lipids are extracted by the Bligh–Dyer procedure[35] by adding methanol, chloroform, and water. The procedure consists of two steps: first, lipids are extracted in nonpolar solvents. To have a single-phase mixture, it is important to maintain the following condition: methanol–chloroform–water (2 : 1 : 0.8, v/v/v). Then a double-phase system is obtained by adding chloroform and water in order to have a final methanol–chloroform–water ratio of 2 : 2 : 1.8 (v/v/v). The lower phase, containing less polar lipids, is separated from the upper phase, containing more polar lipids together with nonlipidic molecules.

1. About 2×10^6 MCF7 cells (one 10-cm petri dish, 80% confluent) or $1–2 \times 10^6$ suspension cells (MOLT4, Jurkat, HL60) can be processed using a volume of 2 ml of methanol, 1 ml of chloroform, and 0.8 ml of water. For higher mass (tissue extractions, e.g.) larger volumes of solvent are needed. Solvents are added to samples (dry pellets of cells, previously washed with PBS) in screw-cap glass test tubes (13×100 mm), and capped tubes are vortexed vigorously for at least 30 sec. *Note 1:* If the solvent mixture is not a single homogeneous phase, add a few drops of methanol. It is important to have a single phase during the extraction step. *Note 2:* Acid conditions (1% perchloric acid instead of neutral water) allow a better extraction of short-chain ceramides and anionic lipids.

2. After keeping the samples on ice for at least 15 min, vortex again for 30 sec.

3. Chloroform (1 ml) and water (1 ml) are then added and samples are vortexed again for 30 sec [final proportions are methanol–chloroform–water, 2 : 2 : 1.8 (v/v/v)]. This results in a biphasic mixture.

4. Centrifugation at 3000 rpm for 5 min at 4° breaks the two phases: the organic phase (lower one) can be collected in new glass tubes and lipids are dried down under a nitrogen stream or vacuum.

5. If needed, the lower phase can be aliquoted for ceramide measurement and total phospholipid measurement. *Note:* Starting with the above-suggested amounts of cells, aliquot 1 ml of the lower phase for ceramide measurement and 0.3 ml in duplicates for phospholipid measurement. This amount will provide values for phospholipids within the range of standard values

6. Dried lipids can be stored for up to 30 days at −80°.

Lipid Phosphate Measurement

The measurement of lipid phosphate offers an evaluation of the cell mass of the samples (especially when treating cells with agents affecting proliferation rate or inducing death) as well as a control for extraction efficiency.

The following method is a colorimetric assay for cellular phosphate, which can be extracted together with lipid and quantified by referring to a standard curve.

1. Standard values ranging from 5 to 80 nmol of phosphate are obtained by aliquoting into screw-cap glass test tubes 5 to 80 μl of a 1 mM NaH$_2$PO$_4$ aqueous solution.

2. A volume of 0.6 ml of ashing buffer [10 N H$_2$SO$_4$–70% HClO$_4$–H$_2$O, 9 : 1 : 40, (v/v/v) respectively] is added to dried lipid samples and standard tubes.

3. Open tubes are placed in a heating block at 160° for 8 to 12 hr. This will evaporate the aqueous phase, leaving in the tubes approximately 100 μl containing free phosphate (mineralization). *Note:* Without preheating the block, let the samples slowly reach this temperature.

4. Once samples are cooled, 0.9 ml of water is added and samples are vortexed for 30 sec to resuspend. *Note:* Pouring water around the tube walls will help in collecting residues of samples spread by vapors during heating.

5. A 0.5-ml volume of ammonium molybdate [0.9% (w/v) in water] is added and samples are vortexed. Phosphate and molybdate will combine as phosphomolybdic acid.

6. A 0.2-ml volume of L-ascorbic acid (9%, w/v) is added and samples are mixed gently. This step will reduce phosphomolybdic acid to molybdenum blue, which has two absorbance peaks at 600 at 820 nm.

7. After incubation at 45° (water bath) for 30 min, samples are cooled for 5–10 min and then absorbance is read at 820 nm (wavelength corresponding to the higher peak of absorbance). Standard absorbances are used to construct a standard curve. Quantitation of phosphate can be determined by referring to the standard curve and considering the appropriate dilution factors from the extraction (0.3 ml out of 2 ml). *Note:* Approximately, 1×10^6 MCF7 cells correspond to

100–150 nmol of lipid phosphate; 1×10^6 MOLT4 (or HL60, Jurkat) suspension cells correspond to 30–40 nmol of lipid phosphate.

Mild Base Hydrolysis

In the case of labeling with radioactive precursors such as palmitate, it may be helpful to hydrolyze ester bonds under mild basic conditions. Because labeled sphingolipids are resistant to such conditions, they can be separated from glycerolipids that are labeled as well but are subjected to the hydrolysis of their ester bonds.

1. Dry lipids are resuspended in 0.5 ml of chloroform and 0.5 ml of 0.2 N methanolic NaOH, vortexed for 30 sec and incubated for 30 min at room temperature.

2. After vortexing for 30 sec, the mixture is neutralized by addition of 0.5 ml of 0.2 N methanolic HCl. After adding 0.4 ml of water, lipids are extracted according to the Bligh–Dyer procedure (single-phase step). Samples are vortexed again for 30 sec.

3. A 0.5-ml volume of chloroform and 0.5 ml of water are added to break the mixture into two phases and samples are vortexed again for 30 sec.

4. After centrifugation at 3000 rpm for 5 min at 4°, the lower phase (containing sphingolipids, resistant to the hydrolysis) is collected in new tubes and the lipids are dried under a nitrogen stream or vacuum.

Separation of Ceramide by Thin-Layer Chromatography

Radioactive precursors are incorporated into a variety of cellular lipids. Most of them (except for the most polar such as sulfatides and gangliosides) are extracted by the Bligh–Dyer procedure. It is important to obtain a clear separation of ceramide from other labeled species in order to be able to quantify it. The use of thin-layer chromatography (TLC) and an appropriate developing solvent system allows the identification of ceramide according to the specific R_f in that system.

1. Dried lipids are resuspended in 60 μl of a mixture of chloroform–methanol (1 : 1, v/v) by vortexing capped tubes for 30–60 sec.

2. Immediately after, 20 μl is applied at the origin of a prescored silica gel 60 plate. *Note 1:* A sufficient volume of solvent is necessary to resuspend all the lipids contained in the tubes. In aliquoting 20 μl out of 60 μl, it is important not to let the solvents evaporate because this will alter the concentration of the applied lipids. The remaining lipids can be dried and saved for another run, if this is needed. *Note 2:* It is important to preclean the silica plate by placing it in acetone for 2 hr before use, making sure it is completely dry before applying samples.

3. Once the silica plate has dried, run in an appropriate solvent system until the solvent front reaches the top.

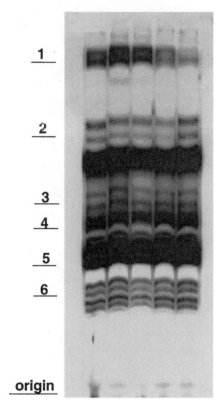

FIG. 2. Serine labeling (steady state, 72 hr). Lipids extracted according to Bligh–Dyer procedure and separated by TLC, using a solvent mixture of chloroform–methanol–calcium chloride (see text for references). (1) Ceramides R_f 0.9–0.85; (2) glucosylceramides R_f 0.73–0.68; (3) lactosylceramides R_f 0.54–0.5; (4) phosphatidylserine R_f 0.46; (5) sphingomyelin R_f 0.4; (6) gangliosides R_f 0.3–0.24.

4. To obtain a good separation of ceramide, the following solvent systems can be used:

Chloroform–methanol–calcium chloride, 15 mM (60 : 35 : 8, v/v/v)

In this system, ceramide is a double band with an R_f of 0.9–0.85 (Fig. 2).

Ethyl acetate–trimethylpentane–acetic acid (9 : 5 : 2, v/v/v)

In this system ceramide is a double band with an R_f of 0.55–0.5 (Fig. 3).

Chloroform–methanol–25% ammonium hydroxide (40 : 10 : 1, v/v/v)

In this system ceramide is a double band with an R_f of 0.75 (Fig. 4).

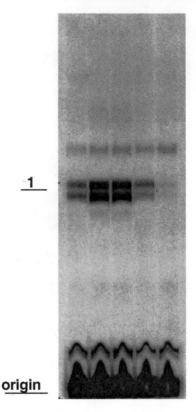

FIG. 3. Serine labeling (steady state, 72 hr). Lipids extracted according to Bligh-Dyer procedure and separated by TLC, using a solvent mixture of ethyl acetate–trimethylpentane–acetic acid (9 : 5 : 2, v/v/v). (1) Ceramides R_f 0.55–0.5.

The solvent mixture is poured into a well-sealed thin-layer chromatography chamber (use silicone and apply pressure in order to obtain good adherence of the lid). The chamber is saturated with vapors by using a sheet of filter paper as a wick (make a circular strip, approximately the same height as the tank). For a $26 \times 20 \times 7$ cm chamber, use 150 ml of solvent and allow saturation for 5–12 hr. The silica plate is placed in the chamber (inside the paper ring) and the solvent front is allowed to migrate to the top of the plate. *Note 1:* Prepare a well-sealed chamber at least 4–6 hr before use (depending also on the size of the tank) in order to reach an equilibrium between liquid–vapor phases of the solvent mixture. *Note 2:* Because saturating conditions are important in the run, it is important to leave the tank open as briefly as possible while inserting the plate.

5. After the solvent front runs to the top, the plate is dried in a chemical hood for 15–30 min before exposure to X-ray film. *Note:* Exposure time can be reduced by

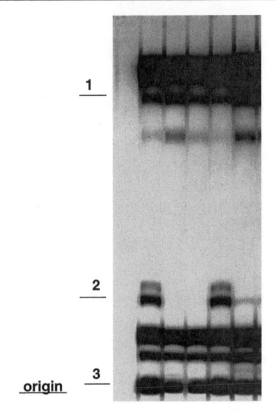

FIG. 4. Palmitic acid labeling (pulse, 6 hr). Lipids extracted according to Bligh–Dyer procedure and glycer a lipids hydrolyzed under basic conditions; sphingolipids separated by TLC, using a solvent mixture of chloroform–methanol–ammonium hydroxide (40 : 10 : 1, v/v/v). (1) Ceramides R_f 0.76; (2) glucosylceramides R_f 0.29–0.23; (3) sphingomyelin R_f 0.03.

coating plates evenly with a surface autoradiography enhancer (En[3]hance spray; New England Nuclear, Boston, MA). About 2–4 days of exposure is normally needed when using enhancer. The enhancer should be evaporated from plate before scraping the silica.

6. Radioactivity associated with ceramide is identified according to the R_f and by comparison with radioactive standards for long-chain ceramides. Spots of interest can be scraped from the plate and quantified by liquid scintillation counting. *Note:* To elute radioactive lipids and solubilize them, resuspend scraped silica in 400 μl of methanol–water (1 : 1, v/v) and let stand for 30–60 min before adding scintillation liquid.

7. The values of radioactive counts are normalized on the basis of the corresponding concentrations of total phospholipids in the samples.

Method 2: Diacylglycerol Kinase

Ceramides are first extracted by the Bligh–Dyer procedure and then labeled *in vitro* using DG kinase,[36,37] which phosphorylates diacylglycerols, yielding phosphatidic acid, and phosphorylates ceramides into ceramide phosphate. The kinase is commercially available or it can be overexpressed in bacteria, and their membranes used as a source for the enzymatic activity.[38] Extracted lipids can be phosphorylated by the kinase, using $[\gamma\text{-}^{32}P]ATP$. The resulting radioactive ceramide 1-phosphate can be separated by thin-layer chromatography and identified according to its R_f value.

This *in vitro* assay is accurate and allows a quantitative analysis of ceramide. It is essential to perform the assay under conditions in which the enzyme concentration is not a limiting factor and in which the substrate is totally converted to product.[38,39]

Culturing of Cells and Sample Collection

Human adenocarcinoma MCF7 cells (2×10^5) are seeded in 100-mm petri dishes and cultured for about 4 days (80% confluence, about 2×0^6 cells) or $2\text{--}4 \times 10^6$ suspension cells (MOLT4, Jurkat) are rested overnight before treatment.

Once treated, cells are washed twice in ice-cold PBS and either trypsinized, spun, and resuspended in 1 ml of methanol, or preferably scraped directly from the plate into 1 ml of methanol (suspension cells are resuspended directly in methanol after the PBS wash); samples are collected in screw-cap glass tubes and stored at $-80°$ (up to 2 weeks).

Extraction (Bligh–Dyer)

Lipids can be extracted according to the Bligh–Dyer procedure, aliquoting the extracted lipids for the DGK assay (1 ml of the lower phase) and for total phospholipid measurement (300 μl in duplicate samples). *Note:* Refer to Method 1 for details of lipid extraction.

Diacylglycerol Kinase Assay

1. A standard curve can be constructed with dioctanoylglycerol and a mixture of natural ceramides. Once resuspended in chloroform, 40 to 1280 pmol of these lipids is used to construct a standard curve by aliquoting in test tubes and drying down. Conversion of ceramide to ceramide 1-phosphate can be obtained and quantitated over a range from 25 pmol to 2 nmol.

[36] M. R. Hokin and L. E. Hokin, *Biochim. Biophys. Acta* **31**, 285 (1959).
[37] J. Priess, *J. Biol. Chem.* **261**, 8597 (1986).
[38] D. K. Perry, A. Bielawska, and Y. A. Hannun, *Methods Enzymol.* **312**, 22 (2000).
[39] P. P. Van Veldhoven, W. R. Bishop, D. A. Yurivich, and R. M. Bell, *Biochem. Mol. Biol. Int.* **36**, 21 (1995).

2. Extracted lipids as well as standards are resuspended in the assay buffer by using mixed micelles containing a nonionic detergent and phospholipids. A 20-μl volume of mixed micelles [octyl-β-D-glucopyranoside: dioleoylphosphatidylglycerol (βOG : DOPG, 7.5% (w/v) : 25 mM] (see below) is added to samples, and lipids are vortexed for 30 sec, rested for 5 min at room temperature, sonicated in a water bath for 30 sec, and immediately after that vortexed again for 30 sec.

OCTYL-β D GLUCOPYRANOSIDE: DIOLEOYLPHOSPHATIDYLGLYCEROL MIXED MICELLE PREPARATION

DIOLEOYLPHOSPHATIDYLGLYCEROL. L-α-Dioleoylphosphatidylglycerol sodium salt (20 mg/ml or 27 mM) is dissolved in chloroform, the solution is aliquoted (0.97 ml) in screw-cap glass test tubes, and the lipids are dried under a nitrogen stream or vacuum.

OCTYL-β-D-GLUCOPYRANOSIDE. To obtain pure crystals, recrystallize the compound before using it.

a. Octyl-β-D-glucopyranoside (5 g) is dissolved in 20 ml of acetone by heating at 40° until completely solubilized.

b. Preclean the solution by filtering it through a Pyrex scintered glass funnel. Add 100 ml of diethyl ether to the solution. If precipitation appears, the solution can be reheated.

c. Crystals are formed by maintaining the solution at −20° for not less than 12 hr.

After filtration through a scintered glass funnel, crystals can be recovered and washed with 500 ml of ice-cold diethylether.

Completely dried crystals are then scraped from the funnel, weighed, and stored at −20° for up to 12 months.

MIXED MICELLES. One milliliter of 7.5% (w/v) βOG in water is added to the tubes containing the above-mentioned aliquots of DOPG (final concentration will be 25 mM). Solubilization is obtained by vigorous and repeated vortexing before and after the solution is allowed to rest at 4° for 12 hr. Resulting micelles can be stored at −20°.

3. Samples are then resuspended gently (do not vortex) in 70 μl of buffer (assay mixture) containing the kinase.

Assay mixture
    50 μl of 2× buffer [0.1 M imidazole (pH 6.6, grade II), 0.1 M LiCl, 25 mM MgCl$_2$, 2 mM EGTA (pH 6.6)]: Store at 4° up to 6 months
    Dithiothreitol (DTT, 3 mM; 0.2 μl of a 1 M solution): add just before use
    5 μg of proteins from DG kinase-containing membranes: Add just before use

Up to 70 μl with dilution buffer [10 mM imidazole (pH 6.6), 1 mM Diethylenetriaminepentaacetic acid, pH 7]: Store at 4° up to 6 months

The reported amount of enzyme per sample can change according to the activity of the protein. A suitable way to determine such variations is to add to the samples known amounts of a long-chain ceramide (such as C_{16}-ceramide) and to compare their conversion into ceramide 1-phosphate by evaluating the mass of lipid phosphate of the radioactive product. It is necessary to start with at least 5 nmol of ceramide in order to have phosphate quantitation. After a 100% conversion of ceramide is established, it is convenient to increase the concentration of the enzyme used (e.g., to twice the amount used for a 100% conversion).

4. The reaction is begun by addition of 10 μl/sample of the following mixture.

ATP mixture:
ATP, 1 mM in water [γ-^{32}P]ATP (about 1.5–2 μCi/sample, from 3000 Ci/ mmol specific activity)

Samples are vortexed gently and briefly, and the reaction is allowed to proceed for 30 min at room temperature; the reaction is then stopped by the addition of chloroform (1 ml)–methanol (2 ml)–water (0.7 ml) and vortexing vigorously for 30 sec.

5. Phosphorylated lipids can be reextracted according to the Bligh–Dyer procedure by the addition of chloroform (1 ml)–water (1 ml) and vortexing. Separation of the two phases can be obtained by centrifugation at 3000 rpm for 5 min at 4°.

6. The upper phase, containing about 95% of the radioactivity, can be discarded. From the lower phase, containing phosphorylated lipids, 1.5 ml can be aliquoted into new tubes and dried under a nitrogen stream or vacuum.

Separation of Ceramide 1-Phosphate by Thin-Layer Chromatography

1. Lipids can be resuspended in 80 μl of chloroform–methanol (1 : 1, v/v) by vortexing for 30 sec, and 20 μl is applied at the origin of the prescored silica gel 60 plate. *Note:* Considerations similar to those discussed in Method 1 can be applied.

2. Lipids can be separated by migration on the silica, using an appropriate solvent system. A mixture of chloroform–acetone–methanol–acetic acid–water (10 : 4 : 3 : 2 : 1, v/v/v/v/v) is poured into a well-sealed thin-layer chromatography chamber. *Note:* For preparation of the chamber, refer to Method 1.

3. After marking the solvent front lane, the plate is dried in a chemical hood for 15–30 min before exposure to X-ray film (with the reported conditions, 10–14 hr of exposure at −80° is usually needed).

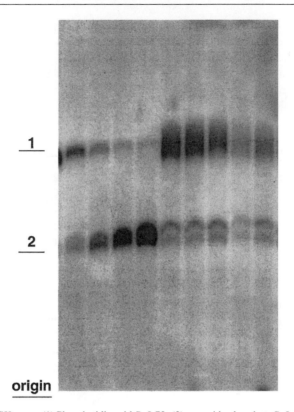

FIG. 5. DGK assay. (1) Phosphatidic acid R_f 0.73; (2) ceramide phosphate R_f 0.47–0.41.

4. Radioactivity associated with ceramide 1-[^{32}P]phosphate is represented by two spots of R_f 0.47–0.41: these are a mixture of long-chain ceramides and dihydroceramides (most of the natural ceramides; Fig. 5). The corresponding spots on the silica can be scraped from the plate resuspended in scintillation liquid (by vortexing vigorously and then allowed to stand for 1–2 hr to avoid quenching), and counted. Diacylglycerols and short-chain ceramides will also appear as their phosphorylated derivatives with R_f values of 0.62 and 0.25–0.3, respectively.

5. The counts in the standard samples are used to construct a standard curve. To obtain the real concentration in the samples, all the dilution factors used in the procedure must be included in the calculation (1 ml out of 2 ml in the Bligh–Dyer lipid extraction, 1.5 ml out of 2 ml in the extraction after the phosphorylation reaction, 20 μl out of 80 μl in the final aliquoting on the silica plate).

Alternatively and preferably, phosphorylated ceramide can be quantitated from the specific activity of [^{32}P]ATP. Because the reaction is equimolar and should go to completion, by measuring the emissions (cpm) from a known aliquot of the radioactive mixture and by measuring the nucleotide concentration (absorbance at 260 nm), it is possible to calculate the number of moles of ceramide from the counts of the samples.

6. Moreover, it is important to normalize these values to the total content of cellular phospholipids.

Evaluation of *de novo* Synthesis of Ceramide

In vivo labeling can be extremely useful in dissecting the involvement of various sphingolipid pathways in the response to a given stimulus. Ceramide can be labeled by cellular uptake of labeled serine, palmitate, or dihydrosphingosine. Because these lipids are first incorporated into ceramide by the enzymes that are involved in its *de novo* generation, a "pulse" (0–6 hr) of the label can be used to evaluate the rate of the activities of those enzymes. Another advantage is the ability to evaluate the biosynthesis of ceramide separately from the quantitation of total intracellular content (which includes ceramide derived from the catabolism of more complex sphingolipids such as sphingomyelin and cerebrosides).

Moreover, because dihydrosphingosine enters the *de novo* pathway downstream of serine and palmitate, it can be used to differentiate between the activity of the proximal and distal enzymes of the pathway. The use of specific enzyme inhibitors together with differential labeling can offer a broad spectrum of information on the regulation of this pathway.

Method

1. About 2×10^6 adherent cells (one 10-cm petri dish, 80% confluent) or $1–2 \times 10^6$ suspension cells are seeded in an appropriate volume of medium, rested for 2 hr, and then radioactive precursors are added (2–5 μCi/ml from specific activities ranging from 20 Ci/mmol for serine to 40 Ci/mmol for palmitate) followed by specific treatments of cells. *Note:* A good point of reference is to have about 4–5 μCi/10^6 cells.

Because the aim is to pulse-label cells, treatments must be considered in the time frame of the pulse. For some experiments, it may be necessary to stimulate the cells first, and then pulse with the label. This becomes important with delayed responses to stimuli.

2. At the appropriate time, based on the treatments and the pulse, cells are washed with PBS. Aliquots from medium and cell pellets can be checked for radioactivity that has been incorporated over time, comparing with the initial counts.

3. Pellets are resuspended in methanol and lipids are extracted according to the Bligh–Dyer procedure. *Note:* Refer to Method 1 for details.

4. Sphingolipids can be separated from glycerolipids by mild basic hydrolysis. *Note:* Refer to Method 1 for details.

5. Ceramides can be evaluated by thin-layer chromatography, as mentioned for the steady state labeling. *Note:* Refer to Method 1 for details.

Measurement of Sphingomyelin

Sphingomyelin Labeling

Sphingomyelin is an abundant sphingolipid in cellular membranes: its hydrolysis releases ceramide and phosphocholine and several stimuli are known to activate sphingomyelin hydrolysis. In evaluating cell responses, it is important to follow the variation in sphingomyelin mass and compare it with the course of ceramide increase. For this purpose, cells are labeled to a steady state with [^{3}H]- or [^{4}C]phosphocholine, which is incorporated into sphingomyelin. Then lipids are extracted and separated by thin-layer chromatography.

Method

1. Cells are seeded under the same conditions as for the steady state labeling: 2×10^5 human adenocarcinoma MCF7 cells are seeded in 100-mm petri dishes and cultured for 24 hr (cells will double, approximately).

2. Cells are then incubated in fresh medium containing a 1-μCi/ml concentration of [^{3}H]choline chloride (75 Ci/mmol; 5–6 ml of medium is an appropriate volume to use, in order to have initially about 5 μCi/10^6 cells). Cells are then cultured for an additional 48–72 hr (final number of cells will be about 2×10^6).

Alternatively, if working with cells in suspension, $3–4 \times 10^5$ cell/ml can be grown in radioactive medium (1 μCi/ml) for 3 days. The initial number of cells seeded must be enough to be aliquoted, after 3 days of labeling, into samples of at least $1–2 \times 10^6$ cells each. *Note:* For checking the uptake of the precursor until steady state is reached, see discussion above.

3. Cells are washed with PBS and allowed to rest for 2–4 hr in fresh medium (chase). Then the medium is changed again and the cells are stimulated.

4. Once collected, the cells are resuspended in 0.6 ml of water and lysed by ultrasonication (1% power, MCF7 adherent cells, three times for 15 sec, MOLT4 suspension cells, two times for 12–15 sec).

5. Five hundred microliters of the lysates can be used for sphingomyelin measurement by aliquoting in 13×100 mm screw-cap glass test tubes and by adding 2.5 ml of chloroform–methanol (2 : 1, v/v) final partition will be water–chloroform–methanol (0.3 : 0.5 : 0.25, v/v/v).

6. Capped tubes are vigorously vortexed for 30 sec, allowed to rest for a few minutes, vortexed again, and then centrifuged at 3000 rpm for 10 min at 4°.

7. The upper phase can be aspirated and discarded, whereas the lower phase is washed (using approximately the same volume and mixture of the upper phase) by addition of 860 μl of chloroform–methanol–water (1 : 16 : 15.6, v/v/v).

8. Samples are vortexed and spun again at 3000 rpm for 10 min at 4°.

9. The lower phase is collected into new tubes (aliquoting 1.5 ml for sphingomyelin determination and 0.2 ml in duplicate for phosphate measurement). Lipids are dried under a nitrogen stream or vacuum and stored at −80° for up to 30 days.

10. Labeled glycerolipids, such as phosphatidylcholine, can be hydrolyzed under basic conditions by adding 250 μl of 0.5 N methanolic NaOH and 250 μl of chloroform.

11. After vortexing for 30 sec, capped samples are incubated at 37° (water bath) for 2 hr.

12. Neutralization is obtained by addition of 250 μl of 0.5 N methanolic HCl and lipids are reextracted by addition of 430 μl of water, 500 μl of chloroform–methanol (2 : 1, v/v), and 850 μl of chloroform [final partition will be chloroform–methanol–water, 1.5 : 1 : 0.8 (v/v/v)]. Samples are vortexed vigorously for 30 sec and centrifuged at 3000 rpm for 10 min at 4°.

13. After discarding the upper phase, the lower phase is washed by addition of 860 μl of a mixture of chloroform–methanol–water (1 : 16 : 15.6, v/v/v), vortexed, and spun at 3000 rpm for 10 min at 4°.

14. The lower phase is collected in new tubes and lipids are dried under a nitrogen stream or vacuum.

Separation of Sphingomyelin by Thin-Layer Chromatography

1. Dried lipids can be resuspended in 80 μl of chloroform by vortexing for 30 sec and 40 μl is applied at the origin of prescored silica gel 60 plate. *Note:* Refer to Method 1.

2. When the plate is dried, lipids can be separated by migration on the silica plate, using an appropriate solvent system. A mixture of chloroform–methanol–acetic acid–water (50 : 30 : 8 : 5, v/v/v/v) is poured into a well-sealed thin-layer chromatography chamber. *Note:* For preparing the chamber, refer to Method 1.

3. After the solvent reaches the top, the plate is dried under a chemical hood for 15–30 min before exposure to X-ray film. *Note:* For usage of enhancer, refer to Method 1.

4. Radioactivity comigrating with sphingomyelin can be identified according to the R_f value, which in this solvent system is 0.44.

An amount of standard natural sphinaomyelin in the range of 50–100 nmol can be run as well and visually identified by exposure to iodine vapors for a few minutes.

5. Radioactive spots of silica can be scraped from the plate and quantified by liquid scintillation counting. *Note:* For elution of radioactivity from silica, refer to Method 1.

6. The values of radioactive counts are normalized on the basis of the corresponding concentrations of total phospholipids in the samples.

Analysis and Discussion

We have presented two complementary strategies to evaluate changes in cellular levels of ceramide. The first strategy is to measure ceramide levels by one of two methods, and the second strategy is to determine the source of ceramide, which not only informs us of the specific metabolic pathway that is activated in response to agonists and stimuli but also provides independent confirmation of the changes in ceramide.

We routinely use the DGK assay and labeling procedures to determine the cellular levels of ceramide. The DGK assay is sensitive (to 25 pmol of ceramide)[37,38] and highly reproducible. It has distinct advantages over mass spectroscopy, which is usually semiquantitative (but much more analytical in that it provides insight into the molecular species of ceramide). The DKG assay also provides a direct assessment of total mass of ceramide, which cannot be provided by radiolabeling methods.

The DGK method has provided a wealth of information about changes in ceramide levels, and has become the current standard for measuring ceramide. We have noticed, however, that one of the basic tenets of this procedure, that is, the requirement for total conversion of ceramide by DGK, has not been uniformly observed by some studies. This cannot be overemphasized as an essential criterion for the reliability of the assay. Refer to a previous discussion on the conceptual background for the assay.[38,40]

Radiolabeling procedures for the measurement of ceramide are more accessible for cell biologists. Although radiolabeling is not an ideal method for mass quantitation (unless the specific activity is verified for key substrates and steady state is truly established), it is convenient and it also stands to provide information about several metabolites simultaneously (depending on the precursor used). For example, labeling with palmitate can provide insight into changes in various phospholipids and glycerolipids whereas labeling with serine provides a relatively clean assessment of changes in various sphingolipids.

Once changes in ceramide are established in response to agonist, we suggest the pursuit of studies to determine the biochemical pathways of sphingolipid metabolism that are activated in response to stimuli. We have described methods to evaluate the *de novo* pathway and hydrolysis of sphingomyelin: the two most commonly regulated pathways that result in ceramide formation. This evaluation provides further evidence of the regulation of ceramide metabolism. Subsequently, investigators may chose to use inhibitors of the two pathways for further evaluation. Fumonisin B1 has remained a specific inhibitor of ceramide synthase and

[40] Y. A. Hannun and D. K. Perry, *Trends Biochem. Sci.* **24,** 226 (1999).

ISP1/myriocin is a potent inhibitor of serine palmitoyltransferase. Both are useful to evaluate the *de novo* pathway.[8,9,41] Scyphostatin is an inhibitor of neutral sphingomyelinase, but it has not been widely studied yet.[42] There are several nonspecific inhibitors of acid sphingomyelinase (such as desipramine[43]) but we do not recommend their use in cell biology studies because they inhibit many other enzymes. For the acid sphingomyelinase, knockout mice and cells from knockout mice or humans having a deficiency in the enzyme are more useful tools.[44–47]

[41] Y. Miyake, Y. Kosutsumi, S. Nakamura, T. Fujita, and T. Kawasaki, *Biochem. Biophys. Res. Commun.* **211,** 396 (1995).

[42] F. Nara, M. Tanaka, S. Masuda-Inoue, H. Doi-Yoshioka, K. Suzuki-Konagai, S. Kumakura, and T. Ogita, *J. Antibiot.* **52,** 531 (1999).

[43] S. Albouz, M. T. Vanier, J. J. Hauw, F. Le Saux, J. M. Boutry, and N. Baumann, *Neurosci. Lett.* **36,** 311 (1983).

[44] K. Horinouchi, S. Erlich, D. P. Perl, K. Ferlinz, C. L. Bisgaier, K. Sandhoff, R. J. Desnick, C. L. Stewart, and E. H. Schuchman, *Nat. Genet.* **10,** 288 (1995).

[45] B. Otterbach and W. Stoffel, *Cell* **81,** 1053 (1995).

[46] P. B. Schneider and E. P. Kennedy, *J. Lipid Res.* **8,** 202 (1967).

[47] J. T. Bernert, Jr. and M. D. Ullman, *Biochim. Biophys. Acta* **666,** 99 (1981).

[22] Assays for Phospholipase D Reaction Products

By YUHUAN XIE and KATHRYN E. MEIER

Introduction

The phospholipase D (PLD) reaction generates molecules with important biological activities. The direct product, phosphatidic acid (PA), has generally been considered as the major potential messenger molecule generated by PLD activation.[1] However, a potential downstream metabolite of PA, lysophosphatidic acid (LPA), has also attracted extensive attention because of its diverse biological activities.[2] LPA plays a potential role in some human cancers, on the basis of the clinical observation that LPA levels are elevated in serum and ascites fluid from human ovarian cancer patients.[3] Both PA and LPA are not only second-messenger molecules, but also precursors and/or intermediates for some lipids important in signaling, such as diacylglycerol (DG) and monoglyceride (MG). Figure 1 diagrams the metabolites potentially generated from PLD activation.

[1] D. English, Y. Cui, and R. A. Siddiqui, *Chem. Phys. Lipids* **80,** 117 (1996).

[2] W. H. Moolenaar, *Exp. Cell Res.* **253,** 230 (1999).

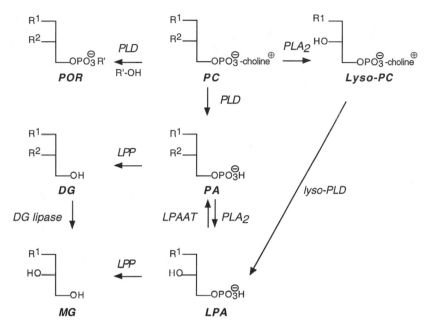

FIG. 1. Metabolic pathways for PLD reaction products. PLD cleaves PC to produce PA. PA can be further converted to DG or LPA by LPP or PLA$_2$, respectively. In the presence of an alcohol (R'-OH), PLD catalyzes a transphosphatidylation reaction to form phosphatidyl alcohol (POR), which is a unique product of the PLD reaction. LPA can be metabolized by LPP or LPAAT to give MG or PA, respectively. PLD, Phospholipase D; PC, phosphatidylcholine; PA, phosphatidic acid; DG, diacylglycerol; LPA, lysophosphatidic acid; LPAAT, LPA acyltransferase; LPP, lipid phosphate phosphatases; PLA$_2$, phospholipase A$_2$; MG, monoglyceride.

PLD was first cloned from plants, and subsequently from yeast and mammalian cells. Two mammalian isoforms, PLD1 and PLD2, have been characterized. All known PLD genes belong to an extended superfamily.[4] Family members share a relatively conserved structure that consists of domains I–IV. Domains II and IV have an HKD motif that is critical for enzymatic activity. Studies have established that the histidine residue in the HKD motif forms a transient phosphatidyl–enzyme intermediate, which is then attacked by an activated water molecule (or primary short-chain alcohol) to release PA (or a phosphatidyl alcohol). Plant PLD enzymes are soluble, whereas the yeast and mammalian enzymes are membrane localized. Most mammalian cells express both PLD1 and PLD2.[5] Mammalian PLDs are

[3] Y. Xu, D. C. Gaudette, J. D. Boynton, A. Frankel, X.-J. Fang, A. Sharma, J. Hurteau, G. Casey, A. Goodbody, A. Mellors, B. J. Holub, and G. B. Mills, *Clin. Cancer Res.* **1**, 1223 (1995).
[4] M. A. Frohman, T.-C. Sung, and A. J. Morris, *Biochim. Biophys. Acta* **1439**, 175 (1999).
[5] K. E. Meier, T. C. Gibbs, S. M. Knoepp, and K. M. Ella, *Biochim. Biophys. Acta* **1439**, 199 (1999).

activated in response to hormones, growth factors, cytokines, neurotransmitters, adhesion molecules, drugs, and physical stimuli. Many agonists that activate PLD bind to G-protein-coupled receptors. Studies of the mechanisms of receptor–PLD coupling have implicated multiple pathways including protein kinase C (PKC), small GTPases, and tyrosine kinases.[6]

Phospholipase D Assays

As shown in Fig. 1, PLD enzymes catalyze the hydrolysis of phosphatidyl-choline (PC) to generate choline and PA. In addition, a unique property of PLDs is that they can catalyze a transphosphatidylation reaction in which a relative stable product, phosphatidyl alcohol, is formed in the presence of a simple alcohol. A number of effective procedures have been used to measure PLD activity by detecting products such as choline, PA, and phosphatidyl alcohol.[7,8] These procedures can be divided into two categories: detection of PLD activity *in vivo,* using an endogenous phospholipid substrate, and measurement of PLD activity *in vitro,* using exogenous or endogenous substrates. *In vivo* or intact cell PLD assays require that endogenous substrates be prelabeled by suitable radioactive tracers. The radiolabeled PLD products (choline, PA, or phosphatidyl alcohol) are then analyzed and quantified. For *in vitro* assays, exogenous PCs labeled with either radioisotopes or fluorophores are commercially available. Alternatively, some investigators have used endogenous substrates for *in vitro* reaction by prelabeling cells and then preparing membranes that contain labeled phospholipids. The reaction products that are generated can be separated by a variety of chromatographic procedures. Product formation is quantitated by liquid scintillation spectrometry for radiolabeled products, or by digital imaging for fluorescent products. Herein, we briefly describe two general procedures for PLD activity. One is an assay that uses intact mammalian cells, and the other uses broken cell preparations.

Phospholipase D Assay in Intact Mammalian Cells

In this assay, PLD substrates are labeled with radioisotope in intact mammalian cells. Formation of a unique transphosphatidylation product of PLD, phosphatidylethanol (PEt), is measured to reflect PLD activity. The materials and procedures used for this assay have been detailed previously.[8] Briefly, cells growing in six-well plates are incubated with ^{3}H-labeled fatty acid (e.g., palmitic acid) for 10–24 hr. The labeled cells are then incubated with 0.5% (v/v) ethanol (substrate

[6] J. H. Exton, *Biochim. Biophys. Acta* **1439,** 121 (1999).

[7] A. J. Morris, A. A. Frohman, and J. Engebrecht, *Anal. Biochem.* **252,** 1 (1997).

[8] K. E. Meier and T. C. Gibbs, *in* "Signal Transduction: A Practical Approach," 2nd Ed., p. 301. IRL Press, Oxford, 1999.

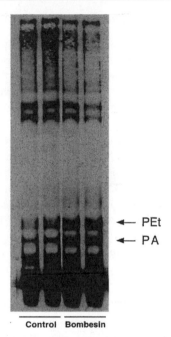

← PEt

← PA

Control   Bombesin

FIG. 2. Separation of products from an intact cell PLD assay. PC-3 cells were metabolically labeled with [³H]palmitic acid, and then treated with or without 100 nM bombesin in the presence of 0.5% (v/v) ethanol for 20 min. Cellular lipids were extracted and separated by the methods described in text. The plate was sprayed with En³Hance and exposed to film for 24 hr. The positions of PEt and PA standards are indicated. PEt, Phosphatidylethanol; PA, phosphatidic acid. Duplicate samples were run for each condition.

for the transphosphatidylation reaction) and the desired experimental agents in Dulbecco's modified Eagle's medium supplemented with 10 mM HEPES (pH 7.5) at 37° in a cell culture incubator for the desired time. Cellular lipids are extracted by adding 1 ml of methanol–HCl (50 : 2, v/v), followed by 0.5 ml of chloroform and 0.28 ml of 1 M NaCl. Dried lipid samples and PEt standard are loaded on thin-layer chromatography (TLC) plates (60 Å, 20 × 20 cm, 250-μm thickness, 19 channels, with preabsorbent strip, without fluorescent indicator; Whatman, Clifton, NJ), and then separated with a solvent consisting of the upper phase of ethyl acetate–trimethylpentane–acetic acid–water (90 : 50 : 20 : 100, v/v/v/v). Spots containing PEt, as well as the remainder of each sample lane, are scraped and then quantitated by liquid scintillation spectrometry. PEt formation is calculated as a percentage of total radioactivity recovered. The R_f for PEt in this system is 0.3, whereas that of PA is 0.2. Figure 2 shows the autoradiograph of lipid products separated by this method. The data demonstrate that bombesin increases PLD activity in PC-3, a human prostate cancer cell line.

Phospholipase D Assay with Membrane Preparations

The assay described here utilizes a fluorescent substrate, BODIPY-phosphatidylcholine (BPC), which is a substrate for plant, yeast, and mammalian PLD enzymes. BPC is an 1-alkyl-2-acyl-PC analog. When incubated with a source of PLD in the presence of alcohol (1–5%, v/v), BPC is converted to BODIPY-phosphatidyl alcohol. The detailed procedures for this assay have been described elsewhere.[8] A brief protocol for use with mammalian cell membranes is provided here. Cells grown in 60- or 100-mm dishes are incubated with or without agonist, washed with phosphate-buffered saline (PBS), resuspended in ice-cold lysis buffer [20 M HEPES (pH 7.5), 80 mM β-glycerophosphate, 10 mM EGTA, 2 mM EDTA, 5 mM dithiothreitol (DTT)], sonicated, and sedimented at 100,000g for 20 min at 4°. The pellet (membrane fraction) is resuspended in lysis buffer before use in phopholipase assays. An aliquot of 1 mM BPC (Molecular Probes, Eugene, OR) in ethanol is dried under nitrogen and briefly sonicated in 500 μM octylglucoside, 400 mM NaCl, 66 mM HEPES (pH 7.0) before use to make a 250 μM solution. The reaction mixture (12.5 μl) contains 6 μl of membrane suspension (1–10 μg of protein) in lysis buffer, 5 μl of 250 μM BPC, and 1.5 μl of

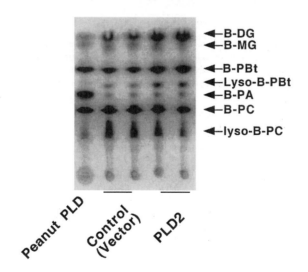

FIG. 3. Separation of products from an *in vitro* PLD assay. PC-3 cells were transfected with or without HA–hPLD2 for 48 hr. Cell membranes were prepared and then incubated with BPC substrate for 1 hr. Reaction products were separated by thin-layer chromatography. Peanut PLD was included as a positive control. The positions of the PLD products, B-PBt and PA, are indicated along with other products of the reaction. B-PC, BODIPY-phosphatidylcholine; B-PA, BODIPY-phosphatidic acid; Lyso-B-PBt, BODIPY-lysophosphatidylbutanol; B-PBt, BODIPY-phosphatidylbutanol; B-MG, BODIPY-monoglyceride; B-DG, BODIPY-diglyceride. PLD activity in control cells, transfected with empty vector, reflects endogenous PLD2 activity. Duplicate samples were run for each condition.

9% (v/v) butanol. The reaction, initiated by addition of membrane protein, is incubated for 60 min at 30°. A 5-μl aliquot of the reaction mixture is loaded on a TLC plate (60 Å, 20 × 20 cm, without fluorescent indicator; EM Science, Gibbstown, NJ) and developed with chloroform–methanol–water–acetic acid (45 : 45 : 10 : 2, v/v). Developed plates can be photographed on Polaroid-type film under UV light, imaged under UV light with a digital camera system, or imaged and quantified with a STORM system (Molecular Dynamics, Sunnyvale, CA). PLD activity will be represented by the formation of BODIPY-phosphatidylbutanol (B-PBt) and PA (if present). As an example, an experiment showing that PLD activity is increased in PC-3 cells transfected with human PLD2 (hPLD2) is presented in Fig. 3. The R_f values for B-PBt and PA are 0.8 and 0.65, respectively.

Assay for Lysophosphatidic Acid Production

Pathways for LPA production have not been fully delineated. Early studies indicated that activated platelets were the primary source of LPA in the plasma.[9] The roles of other cell types in LPA production are currently being evaluated.[10] A general concept is that LPA is generated by the sequential actions of PLD and phospholipase A_2 (PLA$_2$) (see Fig. 1). One possible pathway is that PA, formed from the PLD reaction, is hydrolyzed by PLA$_2$ to generate LPA. Another potential pathway is that a lyso-PLD utilizes lyso-PC, formed from the PLA$_2$ reaction, to produce LPA. Lyso-PLD activity has not yet been fully characterized, but does not appear to be conferred by PLD1 or PLD2. LPA can be further metabolized by lipid phosphate phosphatases (LPP) and acyltransferases. LPPs hydrolyze LPA to monoglyceride (MG), while acyltransferases convert LPA to PA by transferring an acyl chain to its sn-2 position. Therefore, the level of cellular LPA depends on the dynamic equilibrium of multiple enzymes.

LPA elicits a variety of effects in mammalian cells.[11] These effects can be classified into two groups. One group includes growth-related activities that include stimulation of proliferation, prolongation of survival, suppression of apoptosis, and alterations in differentiation. The other group includes cytoskeletal effects such as shape changes, aggregation, adhesion, chemotaxis, contraction, and secretion. These two groups of responses are interrelated in that cytoskeletal changes are involved in mitogenic signal transduction. Many effects of LPA appear to be mediated by G-protein-coupled receptors of the Edg family.[12] Edg receptors continue to be identified and characterized. Edg-2, Edg-4, and Edg-7 are considered to bind LPA preferentially, whereas Edg-1 and Edg-3 prefer a

[9] T. Eichholtz, K. Jalink, I. Fahrenfort, and W. H. Moolenaar, *Biochem. J.* **291**, 677 (1993).
[10] Z. Shen, J. Gelinson, R. E. Morton, Y. Xu, and Y. Xu, *Gynecol. Oncol.* **71**, 354 (1998).
[11] S. An, E. J. Goetzl, and H. Lee, *J. Cell Biochem. Suppl.* **30–31**, 147 (1998).
[12] J. Chun, J. J. Contos, and D. Munroe, *Cell Biochem. Biophys.* **30**, 213 (1999).

related lipid mediator, sphingosine 1-phosphate (SPP). The G_q, G_i, and $G_{12/13}$ isoforms of heterotrimetric G proteins have been implicated in initiating LPA signaling. Activation of G_q stimulates phospholipase C (PLC), whereas pertussis toxin (PTX)-sensitive G_i induces mitogen-activated protein kinase (MAPK) activity and inhibits adenylcyclase. Activation of $G_{12/13}$ activates the small GTPase Rho to evoke actin-based cytoskeletal rearrangement. Cross-talk among these signaling pathways further increases the diversity of LPA responses. Studies indicating that LPA participates in cellular secretion[13] demonstrate another interesting function related to the structure of LPA, rather than to its receptor-mediated effects.

Materials

Cells, growing either in six-well plates or in suspension
Cell culture medium, usually containing serum
Cell culture facility
[^{3}H]Lyso-platelet-activating factor (lyso-PAF, 1-*O*-alkyl-*sn*-glycerylphos-phorylcholine; Amersham Pharmacia Biotech, Piscataway, NJ)
Radioactive isotope usage and disposal facilities
Incubation medium, generally without serum
Stock solution of agonist/antagonist
Methanol–HCl (10 : 1, v/v)
Chloroform
NaCl, 1 M
Vacuum aspirator
Nitrogen gas with multiport evaporating system
Silica gel thin-layer chromatography (TLC) plates: Whatman, 60 Å, 20 × 20 cm, 250-μm thickness, 19 channels, with preabsorbent strip, without fluorescent indicator
Oxalic acid, 0.5 M in methanol
Oven for activating plates
Chromatography solvent: chloroform–methanol–HCl (87 : 13 : 0.5, v/v/v)
TLC tank (glass) containing solvent
Fume hood
Lysophosphatidic acid (LPA, 10 mg/ml in chloroform; Avanti Polar Lipids, Alabaster, GA)
TLC tank (glass) containing iodine crystals
En^{3}Hance spray (Dupont/NEN, Wilmington, DE)
Film (XAR-5; Eastman Kodak, Rochester, NY)
Scintillation vials
Liquid scintillation cocktail
Liquid scintillation counter
Calculator or computer

Procedure

Metabolic Labeling of Cells with [³H]Lyso-Platelet-Activating Factor. Seed cells in six-well tissue culture plates the day before the experiment. It is recommended that triplicate or quadruplicate wells be used for each experimental condition. Use 90% confluence for adherent cells, or a density of suspended cells that allows proliferation. One day after cell seeding, prepare the isotope ([³H]lyso-PAF) in medium [with or without serum, but containing 0.2% (w/v) fatty acid-free bovine serum albumin (BSA) as a carrier for lyso-PAF] at 5 mCi/ml. Either replace the cell culture medium with fresh isotope-containing medium, or directly add the isotope to the existing culture medium. Incubate for 24 hr to allow the isotope to be incorporated into the cellular phospholipid pool.

Incubation of Cells with Agonists. As incubation medium, prepare Dulbecco's modified Eagle's medium (or other compatible medium) supplemented with 10 mM HEPES, pH 7.5 (DME/H); warm the medium to 37°. For attached cells, use a disposable pipette to remove the isotope-containing culture medium. Rinse each dish twice by gently adding 2 ml of DME/H and agitating the dish manually for a few seconds, and then remove the medium with a pipettor. For suspended cells, go through the same procedure, collecting the cells by centrifugation between each wash. Transfer all media and washes to a radioactive waste container. Finally, add 2 ml of DME/H as incubation medium. Add the desired concentration of agonist, mixing it with the culture medium by a gentle swirl, and then incubate for the desired duration in a cell culture incubator.

Extraction of Lipid from Cells and Culture Medium. Working in a fume hood, transfer the incubation medium to a glass tube, using a disposable pipette. Add 1 ml of cold methanol–HCl (10 : 1, v/v) to each tube. Wash the cells twice with 4 ml of cold PBS. After washing, add 1 ml of cold methanol–HCl (10 : 1, v/v) to each well. Remove the cells with a disposable cell scraper. Transfer the resulting cell suspension to a glass tube. After this step, samples from media and cells will be handled by the same procedure. Add 0.5 ml of chloroform to each sample. Mix each sample well, using a vortex mixer. Leave the samples on ice for 40 min to extract lipid. Add 0.28 ml of 1 M NaCl to each sample; mix again. After mixing, the samples will become turbid. Centrifuge the samples at 700g for 10 min at room temperature. Carefully remove the upper phase, using a pipette attached to a vacuum aspirator, and retain the lower lipid phase. Dry the lower phase under nitrogen gas at room temperature. Samples can either be stored at −20° overnight, or run on TLC plates immediately.

Preparation of Thin-Layer Chromatography Plates. Prepare 1000 ml of 0.5 M oxalic acid in methanol. Working under a fume hood, transfer the oxalic acid

---

[13] A. Schmidt, M. Wolder, C. Thiele, W. Fest, H. Kratzin, A. V. Podtelejnikov, W. Witke, W. B. Huttner, and H.-D. Soling, *Nature (London)* **401,** 133 (1999).

solution to a container that can hold 20×20 cm TLC plates. Immerse two TLC plates (back to back) completely in the solution for 15 min. Remove the plates and dry them in the hood. Activate the plates by heating at $100°$ in an oven for 1 hr before use. A number of plates can be prepared at one time, but should be stored in a desiccator after activation.

Separation of Lysophosphatidic Acid by Thin-Layer Chromatography. Prepare the solvent mixture [chloroform–methanol–HCl (87 : 13 : 0.5, v/v/v), approximately 100 ml] in a TLC tank lined with chromatography paper. Cover the tank; seal tightly to prevent evaporation. If necessary, apply silicone vacuum grease to the upper edge of the TLC tank. Evaporation of solvent will cause uneven migration of the solvent front. Add 30 μl of chloroform–methanol (90 : 10, v/v) to each dried sample, along with 2 μl of LPA (10 mg/ml) as standard. Use a vortex mixer to dissolve the samples. Keep the sample tubes on ice to avoid evaporation of solvent. Mark the TLC plate with a pencil to indicate where the samples should be loaded (2.5 cm from the bottom of the plate, in the center of each lane). Use a pipettor with a disposable tip to apply the entire sample to the TLC plate. Apply each sample to a separate lane. On each plate, use one lane at each end to load LPA standard alone for purposes of calibration and background subtraction. Dry the applied sample spots by applying warm air to the plate, using a heat gun. Place the TLC plate in the solvent tank at a $30°$ angle. The sample origin should be above the level of the solvent. Cover the tank tightly. Develop the plate until the solvent has migrated to 3 cm from the top of the plate (30–60 min).

Remove the plate from the TLC tank. Dry in the fume hood. Stain the dried plate with iodine vapor. Remove the plate after the LPA bands (yellow–brown) are clearly visible, and mark the position of the bands immediately with a pencil. Leave the plate in the fume hood until the iodine staining disappears. Spray the plate with En³Hance according to the manufacturer recommendations. Expose the plate to Kodak XAR-5 film in a $-80°$ freezer. Develop the film after 48 hr or longer.

Quantitation of Lysophosphatidic Acid Production. Align the TLC plate with the autoradiogram to locate the radioactive LPA ([³H]LPA). In our experience, the [³H]LPA bands run slightly higher (within 4 mm) than the unlabeled oleoyl-LPA bands stained with iodine. This disparity may be due to heterogeneity between the LPA species in cells versus the oleoyl-LPA standard. Experiments in our laboratory have confirmed that acyl-LPA species with different fatty acid composition migrate with similar R_f values, and co-migrate with the [³H]-LPA bands. It is therefore important to localize the LPA bands by autoradiography. The R_f values for the major cellular lipids in this solvent system are as follows: PC, 0.23; phosphatidylinositol (PI), 0.23; LPA, 0.43; PA, 0.73; phosphatidylethanolamine (PE), 0.76. When [³H]lyso-PAF is used as the metabolic label, there are no other major radioactive species migrating near the LPA band. Figure 4 shows an autoradiograph of LPA separated by this method.

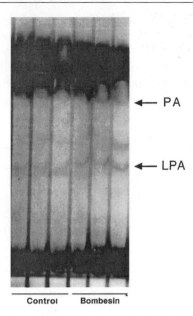

FIG. 4. Separation of LPA by thin-layer chromatography. PC-3 cells were metabolically labeled with [³H]lyso-PAF, and then treated with or without 100 n*M* bombesin for 20 min. Cellular lipids were extracted and separated by the methods described in text. The plate was sprayed with En³Hance and exposed to film for 48 hr. The positions of PA and LPA standards are indicated. Triplicate samples were run for each condition.

Prepare two labeled scintillation vials for each sample. One vial is for the LPA band, and the other is for the remainder of the lane. Scrape the LPA band and remainder from each sample into separate vials. Repeat the same procedure for all lanes. The unlabeled LPA lanes can serve as background for the TLC plate. The background can vary from plate to plate. Add scintillation fluid to each vial. Cap each vial and mix well. Quantitate the radioactivity, using a liquid scintillation counter. Calculate the results by subtracting the corresponding background values for LPA and remainder (as determined from lanes lacking radioactive sample) from the values obtained for the sample lanes. LPA production can be calculated as a percentage of the total radioactivity recovered. This corrects for differences in cell number, isotope uptake, and lipid recovery among samples. Perform further statistical analysis (e.g., mean ± error) of the replicate samples as desired.

Comments

Most of the reported methods for measuring LPA production involve labeling cellular phospholipids with an isotope precursor. LPA is then separated from other lipids by thin-layer chromatography. Several isotope precursors have been used.

Ortho[^{32}P]phosphoric acid is the major precursor used in earlier studies of LPA production.[14–16] Limitations of this isotope include the relatively large amount of radioactivity used (30–700 μCi/ml), requirement for low-phosphate medium, and complicated handling procedure. ^{3}H-Labeled fatty acids, commonly used as precursors in assays for PLD activity, can also be used for LPA assays. Results from our laboratory have indicated that either [^{3}H]palmitic acid or [^{3}H]lyso-PAF can be used. Palmitate labeling sometimes generates additional bands that migrate near the LPA bands in our TLC system, potentially interfering with quantitation of the results. Because lyso-PAF is an alkyl-linked precursor, only alkyl-linked LPA will be generated when this precursor is used. It should be noted that PLDs can utilize alkyl-linked PC as substrate, both *in vivo* and *in vitro,* and that both acyl- and alkyl-linked LPAs are biologically active. Although our data establish lyso-PAF as an appropriate precursor for LPA assays, this substrate is much more costly than ^{3}H-labeled fatty acids, and probably does not reflect the full spectrum of LPA species produced by cells.

Optimal LPA production will depend on optimal duration of isotope labeling, allowing the precursor to enter the appropriate metabolic pool. We have labeled the prostate cancer cell line Dul45 with [^{3}H]lyso-PAF for varying times (4–48 hr) before assaying LPA production. The results showed that LPA production in response to agonist was maximal by 24–48 hr (Y. Xie and K. E. Meier, unpublished data, 2000). Thus, shorter labeling times do not improve the results when [^{3}H]lyso-PAF is used as the precursor.

Efficient extraction of LPA from cells is important for detection, because LPA is a minor phospholipid. The classic Bligh–Dyer procedure can be modified to achieve better LPA extraction. Acidic extraction conditions will extract LPA into the chloroform phase by converting LPA salts to the free acids.[17] Although an acidic solvent improves LPA recovery, conversion of PA to LPA has been reported to occur in the presence of acid.[15] We therefore tested the effect of extraction with varied concentrations of acid (HCl) on LPA recovery. Concentrations as high as 1.2 N HCl in the final extraction solvent do not cause conversion of PA to LPA after 4 hr at 4°. HCl or water-saturated butanol has also been used to extract LPA by some investigators.[15,16,18]

Although it is not necessary to have an autoradiographic image to localize the position of labeled LPA on the TLC plate before scraping it for quantitation, an autoradiogram will establish whether the separation is adequate. Using the autoradiographic image to quantity the results (e.g., by densitometry) will be

[14] G. Mauco, H. Chap, M.-F. Simon, and L. Dousteblazy, *Biochimie* **60,** 6 (1978).
[15] J. M. Gerrard and P. Robinson, *Biochim. Biophys. Acta* **795,** 487 (1984).
[16] D. A. Kennerly, *J. Biol. Chem.* **262,** 16305 (1987).
[17] D. J. Hanahan, "A Guide to Phospholipid Chemistry," p. 177. Oxford University Press, New York, 1997.
[18] J. M. Gerrard, P. Robinson, and M. Narvey, *Biochem. Cell Biol.* **71,** 432 (1993).

problematic, because the intensity of image is critically dependent on the application of the enhancer. It is difficult to ensure that En³Hance has been sprayed evenly on the plate.

Only a few TLC procedures have been used to separate LPA from other lipids. Cohen and Derksen developed a solvent system to separate PA, phosphatidylglycerol, and cardiolipin from other phospholipids.[19] Mauco and co-workers proved that this method could separate LPA from PA and other phospholipids.[14] The system we describe here is modified on the basis of their work. A solvent system of benzene–pyridine–formic acid (50:40:11, v/v/v) has also been used to separate LPA from other lipids, using high-performance silica gel 60 plates.[20] Kennerly reported a double one-dimensional system that provided good separation of LPA.[16] In this method, the TLC plate was developed with a basic solvent of chloroform–methanol–ammonium hydroxide (65:35:7.5, v/v/v), followed by an acidic solvent of chloroform–acetone–methanol–acetic acid–water (50:20:10:15:5, v/v/v/v/v), the latter run in the reverse direction after removing all of the lipids that ran above PC. We have tested methods based on those described above, but found them to be less convenient and/or to provide suboptimal separation of LPA from other lipids than the method described herein. Two-dimensional TLC systems also can be used to detect LPA,[21] but are impractical for analysis of multiple samples. Future developments in LPA analysis will include the further development of improved analytic techniques (e.g., mass spectrometry) that will permit evaluation of bulk levels of LPA as well as specific LPA species.

[19] P. Cohen and A. Derksen, Br. J. Haematol. 17, 359 (1969).
[20] A. Tokumura, W. Kramp, and D. J. Hanahan, Arch. Bichem. Biophys. 247, 403 (1986).
[21] K. Liliom, Z.-W. Guan, J.-L. Tseng, D. M. Desiderio, G. Tigyi, and M. A. Watsky, Am. J. Physiol. 274, C1065 (1998).

[23] Determination of Strength and Specificity of Membrane-Bound G Protein–Phospholipase C Association Using Fluorescence Spectroscopy

By SUZANNE SCARLATA

As biochemical assays are identifying more and more interacting proteins, questions as to which interactions predominate under particular conditions need to be addressed because their relative strengths will determine the route a particular signal will be transduced. There are several techniques to measure affinities between proteins that freely diffuse in solution. However, when proteins are confined to membrane surfaces, the methods to measure their interactions are limited. Fluorescence spectroscopy is one of the few techniques that can monitor the association between membrane proteins in real time and allows the quantification of the energy of these interactions. It should also be considered that analysis of these quasi two-dimensional interactions poses particular problems. We present here fluorescence methods that we have been using to characterize the affinities between members of the G_q–phosphoinositol-specific phospholipase C (PLC) signaling system. Although measuring the methods described here entails purified fluorescence-labeled proteins, information can also be obtained from complex mixtures. Background information about fluorescence methods and theory can be found in most physical biochemistry textbooks (e.g., Cantor and Schimmel[1] and van Holde *et al.*[2]) and in a comprehensive text that focuses on biochemical systems.[3] Here, we focus on the use of fluorescence spectroscopy to measure membrane protein affinities.

Types of Methods

The two general fluorescence methods to measure the interactions between membrane-bound proteins are (1) changes in the emission spectrum/fluorescent lifetime of a protein, and (2) changes in fluorescence resonance energy transfer between two fluorophores. We note that fluorescence anisotropy, which is sensitive to the rotational volume, is often used in soluble systems to monitor protein associations. However, these changes are not applicable when the proteins

---

[1] C. Cantor and P. Schimmel, "Biophysical Chemistry." W. H. Freeman and Company, San Francisco, 1980.

[2] K. van Holde, W. C. Johnson, and P. S. Ho, "Principles of Physical Biochemistry." Prentice-Hall, Upper Saddle River, New Jersey, 1998.

[3] J. Lakowicz, "Principles of Fluorescence Spectroscopy," 2nd Ed. Plenum, New York, 1999.

are bound to membranes. Both emission and energy transfer methods involve labeling the protein of interest with a fluorescent probe (see below) and monitoring the selected fluorescence parameter as the protein partner is incrementally added. In these methods, the proteins are usually covalently labeled with a fluorescent probe because in most systems of interest the two proteins contain several intrinsic fluorophores (tyrosines and tryptophans), making it difficult to discern changes due to protein association. By plotting the change in the observed fluorescence parameter versus protein concentration, an apparent dissociation constant can be determined. If necessary, this apparent constant can then be recalculated to take into account that the proteins are associating on membrane surfaces. We note that the intrinsic fluorescence is usually a sensitive method to measure the binding constant of surface-associating proteins to lipid bilayers because these residues are quenched by the ionic charges of the head group. For example, the intrinsic fluorescence of $PLC\beta_{1-3}$ and $PLC\delta_1$ undergoes a 20–30% decrease on membrane binding.

Monitoring Protein–Protein Interactions by Changes in Emission

In some cases, the protein can be labeled at a site where the probe is located at or close to the protein interaction site or in a region that undergoes a change on binding. Then, the probe is sensitive to the polarity of its local environment, then a change in fluorescence will occur. The emission of several types of probes, most notably naphthalene derivatives such as acrylodan (6-acryloyl-2-dimethylaminonaphthalene), will increase in intensity and shift to higher energies (lower wavelengths) when their local environment is made less polar. A good source of data of different types of probes is the Molecular Probes (Eugene, OR) catalog. Usually, changes in the local environment also affect the time that the fluorophore spends in the excited state. Thus, protein–protein associations can also be monitored by time-resolved measurements, although the instrumentation for these types of studies is more complicated than for steady state (see Lakowicz[3]).

An example of how the emission of a probe can be used to monitor protein–protein associations is shown in Fig. 1a. In this study, we labeled purified $G\beta\gamma$ subunits with acrylodan and reconstituted the protein on model membranes as detailed below.[4] We then titrated an unlabeled partner into the solution under conditions where it would be membrane bound and substituted dialysis buffer as a control. In this example, the protein partner was the pleckstrin homology domain (PH) of $PLC\delta$, which comprises the first ~140 residues at the N terminus.[5] At

[4] L. W. Runnels, J. Jenco, A. Morris, and S. Scarlata, *Biochemistry* **35**, 16824 (1996).
[5] P. Garcia, R. Gupta, S. Shah, A. J. Morris, S. A. Rudge, S. Scarlata, V. Petrova, S. McLaughlin, and M. J. Rebecchi, *Biochemistry* **34**, 16228 (1995).

**a.   Binding of PH-PLCδ to 8 nM Acrylodan-Gβγ
on 455 μM POPC:POPS (2:1)**

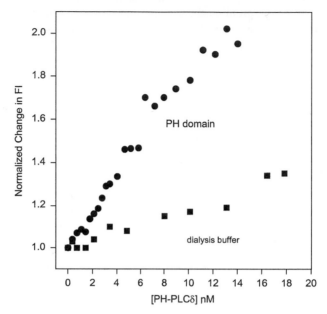

FIG. 1. (a) Increase in the fluorescence intensity of Gβγ labeled with an thiol-reactive probe, acrylodan, and reconstituted on lipid bilayers (see text) as an unlabeled preparation of the pleckstrin homology domain of PLCδ (PH–PLCδ), or dialysis buffer, is added. Data were taken by exciting the probe at 350 nm and scanning the emission from 400 to 600 nm, calculating the integrated area under the curve, and then normalizing to the first point. (b) The binding curve obtained when the buffer background has been subtracted from the sample data for the point in (a). PC, Phosphatidylcholine; PS, phosphatidylserine.

each point on the titration curve, the emission spectrum was recorded and the area under the curve was calculated (most commercial fluorometers come with software to calculate the integrated spectral area and center of spectral mass). In Fig. 1a, we find that the acrylodan intensity undergoes almost a 2-fold increase when PH–PLCδ is added. The data also show that the dialysis buffer itself increases acrylodan fluorescence, but not to such a large extent, and were corrected to give an association curve (Fig. 1b).

The emission spectrum of acrylodan will also shift to higher energies when the polarity of its local environment is reduced. Shown in Fig. 2 is a study similar to the one described in Fig. 1, where we recorded the emission spectrum at several points along the titration curve. From the emission spectrum, we obtained the center of

b.

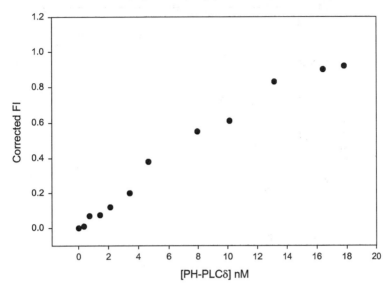

FIG. 1. (*Continued*)

spectral mass (CSM), which relates to the shift in emission wavelength:

$$\text{CSM} = \sum_{i=1} \text{EE}(\lambda_i) \times I(\lambda_i) \Big/ \sum_{i=1} I(\lambda_i) \tag{1}$$

where EE refers to the emission energy in units of energy [usually in kilokaisers (kk) or wavenumbers (cm^{-1}), where 1 kk = 1000 cm^{-1}] at each wavelength (λ_i) and I is the corresponding intensity. Note that accurate binding can be derived only from changes in energy rather than changes in wavelength. Also, the CSM is the preferred method to evaluate changes in emission energy because it takes into account changes in the skewness of the peak, which sometimes occur without a change in emission maximum.

Probes that are quenched by water or ionic groups, such as dansyl chloride, are usually sensitive to protein association.[3] In Fig. 3 we show the emission spectra of dansyl–G$\beta\gamma$ reconstituted on membrane surfaces in the absence and presence of excess PH–PLCδ. Whereas in most cases an increase in dansyl fluorescence on binding is seen, caused by a decrease in the local concentration of water, in this case the dansyl must be located on a site that moves it into a more quenching environment (i.e., water or an ionic charge). These data illustrate the point that

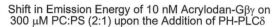

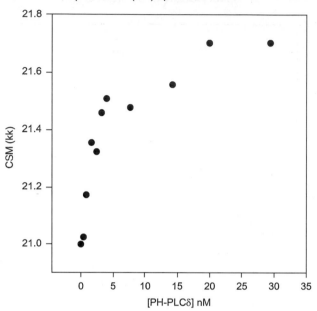

FIG. 2. A similar study to the one in Fig. 1, showing the increase in emission energy of acrylodan–G$\beta\gamma$ as PH–PLCδ is added. The emission energy is given in terms of kilokaisers (kk), or $10,000 \times$ [(1/CSM (nm)], where CSM refers to the center of spectral mass and is given in Eq. (1). Note that 1 kk is 1000 cm^{-1}.

the changes in fluorescence that are observed on protein association may vary depending on the local environment of the probe. Because the extent and sign of the change in fluorescence may vary with the labeling site (which is usually not known and could be variable for a large protein; see below), and with the nature of the membrane surface and/or the protein partner, it is important to conduct a battery of control studies, as discussed below, to verify that the observed changes are due to protein–membrane or protein–protein association and not to experimental artifacts, such as oxygen quenching, photobleaching, buffer components, adhesion to cuvette walls, etc.

Advantages and Disadvantages of Emission Studies. Using the changes in emission of a labeled protein to monitor protein associations has several advantages and disadvantages. On the positive side, this method requires the labeling of only a single species, unlike the transfer methods discussed below. On the negative side, if the probe is directly in the labeling site, it may affect the interaction energy. This possibility can be addressed by measuring the interaction energy when the protein is labeled with a probe that has a different reactive group and thus attaches

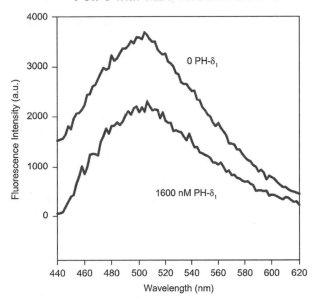

FIG. 3. Spectra of dansyl (DNS)–G$\beta\gamma$ in the absence and presence of PH–PLCδ taken at λ(ex) = 340 nm.

to a different residue, or by labeling the protein in the presence of the protein partner and repurifying the target protein. This latter method works well if one of the proteins can bind to nickel or another type of affinity resin.

Another disadvantage is that the probes are sensitive to local conditions and small variations in the labeling procedure may result in attachment of the probe to a site that is insensitive to binding of a partner. This is frequently the case, mainly because the external positions in which the probes usually attach are not affected by protein binding. For this reason, fluorescence energy transfer is often used because it is much less sensitive to local environment and is generally sensitive to all types of protein interactions.

Fluorescence Resonance Energy Transfer

In fluorescence resonance energy transfer (FRET), one protein is labeled with a probe that is capable of donating its excited state energy to a nearby acceptor. For energy transfer to occur, the absorption spectrum of the acceptor must have some overlap with the emission spectrum of the donor, and also the two probes must be within a certain distance from each other. Energy transfer distances are usually reported by the parameter R_0, which is the distance at which 50% of the donor emission is lost to transfer. Most probe pairs fall in the range of 20–50 Å,

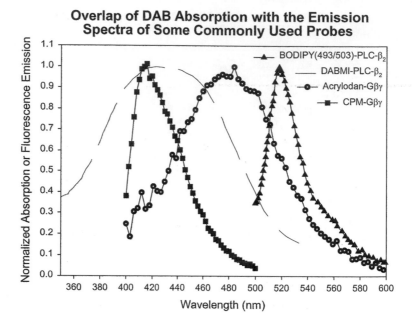

FIG. 4. Plots showing the absorption spectrum of DABMI–PLCβ_2 (dashed line) and the fluorescence spectra of BODIPY–PLCβ_2, acrylodan–G$\beta\gamma$, and coumarin (CPM)–G$\beta\gamma$ dispersed either in solution or in 0.7% CHAPS detergent. As can be seen, the absorption of DAB completely overlaps coumarin, making them an excellent energy transfer pair.

although pairs with smaller and larger values of R_0 exist. Some investigators have used this method as a ruler to measure intramolecular distances, using probe pairs with different energy transfer distances, and distances for various energy transfer pairs have been tabulated.[6]

Experimentally, energy transfer is detected by the loss of emission intensity of the donor or the gain in emission intensity of an acceptor. Control studies using unlabeled proteins are generally required to determine whether these parameters change in the absence of transfer, that is, simply by virtue of protein–protein association.

The acceptor that we prefer is 4-(4'-dimethylaminophenylazo)benzoic acid (DABCYL, as a thiol-reactive probe) or 4-dimethylaminophenylazophenyl-4'-maleimide (DABMI, as an amine-reactive probe). The latter fluorophore has a broad absorption band that has good overlap with the emission spectrum of many donors (Fig. 4). The main advantage of this acceptor is that it is nonfluorescent

[6] W. van der Meer, G. Coker, and S. S.-Y. Chen, "Resonance Energy Transfer, Theory and Data." VCH Publishers, New York, 1994.

a.

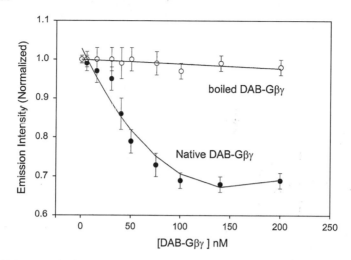

FIG. 5. (a) Decrease in the emission intensity of 28 nM coumarin–G$_q\alpha$(GDP) reconstituted on 75 μM PC : PS : PE (phosphatidylethanolamine) membranes on the addition of DAB–G$\beta\gamma$ subunits dispersed in 0.7% CHAPS. The decrease in emission intensity is due to energy transfer from the coumarin probe donors to the nonfluorescent DAB acceptors as the two proteins associate. As a control, boiled DAB–G$\beta\gamma$ was used. [Reprinted from L. W. Runnels and S. Scarlata, *Biochemistry* **38,** 1488 (1999), with permission from the American Chemical Society.] (b). A plot showing the sensitivity of coumarin to its local environment. In this study, G$_q\alpha$(GTPγS) was labeled with coumarin and the increase in integrated area on binding to lipid membranes was obtained. We have used this change in coumarin intensity to monitor the lipid binding of other proteins.[16] POPC, 1-Palmitoyl-2-oleoylphosphatidylcholine; POPS, 1-palmitoyl-2-oleoylphosphatidylserine.

and thus the decrease in donor fluorescence can be monitored directly without the need to correct for acceptor contribution. All of these probes can be commercially purchased (Molecular Probes).

Many probes are capable of donor emission in response to DABMI, and some of these are shown in Fig. 4. We have been using coumarin (CPM), because it is bright and allows us to detect nano- and picomolar quantities of some proteins. An example of a coumarin–DAB FRET study is shown in Fig. 5a, where G$_q\alpha$(GDP) was labeled with coumarin and reconstituted on model membranes (see below[7]). As DAB–G$\beta\gamma$ binds to membranes and encounters CM–G$_q\alpha$(GDP) subunits, the probes come into energy transfer distance, which is seen as a decrease in coumarin fluorescence. Although the plateau at a $\sim$30% loss in intensity is below the theoretical limit of transfer, it is typical of these proteins, which we find vary between 5 and 35% depending on the size of the protein and the particular labeling conditions.

[7] L. W. Runnels and S. Scarlata, *Biochemistry* **38,** 1488 (1999).

b.

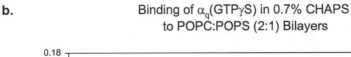

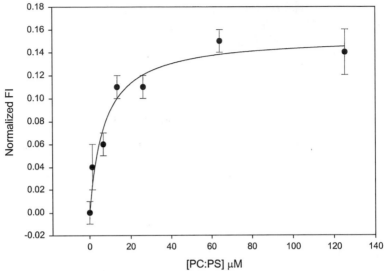

FIG. 5. (*Continued*)

It should be kept in mind that coumarin is also sensitive to polarity and undergoes large changes in intensity on membrane binding. An example is shown in Fig. 5b, where we have titrated large, unilamellar vesicles into a dilute solution of CM–$G_q\alpha$(GTPγS) dispersed in 0.01% Lubrol, and found that complete binding occurs at $\sim$100 μM lipid. This sensitivity of coumarin to local dielectric can be used as a guideline to verify reconstitution. These data also show the importance of verifying that the observed changes in coumarin fluorescence are not due to a local change in polarity.

Fluorescence Homotransfer. Fluorescence homotransfer refers to energy transfer between identical fluorophores.[6] Ideally, the fluorophore should have good overlap between its absorption and emission spectrum (small Stokes shift). Fluorescein and 8-bromomethyl-4,4-difluoro-1,3,5,7-tetramethyl-4-bora-3a,4a-diaza-s-indacene methyl bromide (BODIPY) are good candidates for homotransfer. The advantage of homotransfer as opposed to heterotransfer is that only one type of fluorescence label is required, making it well suited for monitoring the self-association of proteins. The most important advantage of homotransfer is that it potentially allows the determination of the number of subunits in a transferring complex, which can be useful in determining whether G proteins and their receptors and effectors form higher order complexes.[8]

[8] L. W. Runnels and S. F. Scarlata, *Biophys. J.* **69**, 1569 (1995).

Fluorescence homotransfer is detected by fluorescence anisotropy, which is a measure of the polarization of emitted light. When a sample is excited with polarized light, its fluorescence will be partially depolarized because of the position of its emission dipole and because of its rotational motion. If the sample can transfer its energy to a neighbor while it is in the excited state, then the emitted light becomes further depolarized. Each transfer results in more depolarization. We have developed theoretical methods to determine the number of subunits in an energy transfer complex from the loss in anisotropy and used these methods to determine that PLCβ_2 does not simultaneously bind G$_q\alpha$ and G$\beta\gamma$.

The disadvantage of this method is the loss of light through the polarizers (~70%), making it impractical to work with weakly fluorescing signals. Also, the total anisotropy is the sum of the anisotropies of the individual components multiplied by their fractional contribution. Therefore, it is critical to accurately determine these parameters when only one fluorescent species is present before assessing whether there is a loss in anisotropy of the mixture resulting from homotransfer.

Detection of Protein Associations in Complex Systems. Although quantitative information about the energetics of protein–protein interactions can be obtained only with purified proteins (unless it is certain that the contaminants will not interfere with the association, which is usually impossible), energy transfer studies can still be used to qualitatively detect protein associations. For example, experiments using energy transfer between chimeras with green fluorescent protein (GFP) analogs have detected protein–protein associations in living cells.

It is possible to obtain affinities for protein associations in impure systems by using fluorescently tagged antibodies.[9] The protocol involves labeling the commercial preparation of antibody with a fluorophore and adding the labeled antibody into the solution, optimally at a 1 : 1 concentration. The 1 : 1 stoichiometry is optimal because if too much free antibody is in solution, it will contribute to the background signal and mask the experimental signal. On the other hand, if not all of the proteins are labeled with antibody then energy transfer will be reduced, because there would be a decrease in the number of donor or acceptors available to the partners. The concentration of target protein can be obtained by estimating the amount of protein by electrophoresis. Another method to assess optimal binding of the antibody to the protein is by measuring the rotational volume by fluorescence anisotropy as labeled antibody is titrated into the protein solution. The binding of antibody to the protein will produce an elevated anisotropy, which will begin to decrease after all of the protein is tagged because free antibody will begin to contribute to the signal.

Unfortunately, it is possible only to estimate the affinities between the antibody-tagged proteins in a complex mixture because proteins or other factors may

---

[9] F. Bouamr, S. Scarlata, and C. Carter (submitted).

interfere with the interaction of interest, and also the presence of the antibody may affect the protein association site.

Factors Contributing to Extent of Energy Transfer. In these types of studies, the main factor that regulates the extent of transfer is the distance between the probes on the proteins. As mentioned, the extent of transfer depends on the protein pair, which in turn depends on the size of the protein and the position of the probe on the protein. The R_0 for most of the probe pairs we use ranges from 30 to 40 Å and energy transfer decreases with the sixth power of the distance (see van der Meer *et al.*[6]). Therefore, if the probes are attached to the far ends of the proteins, the energy transfer may not be observed.

A parameter often mentioned in energy transfer studies is the orientation factor or κ^2. This factor considers whether the dipoles of the donor and acceptor are positioned in such a way as to allow for transfer. If the probes can rotate freely, then the orientation of their dipoles is random and transfer can occur. Although knowledge of κ^2 is critical when determining the distance between a donor and acceptor, or when working in a system that does not allow probe rotation, it is not critical for protein association measurements because we are not interested in knowing the distance between the probes and because the most prominent attachment sites for the probe are external residues, which usually allow the probes rotational freedom. This rotational freedom can be readily verified by a low fluorescence anisotropy value.

Preparing Samples for Fluorescence Measurements

Unless the intrinsic fluorescence of the protein can be isolated from its partner, it must be modified with a fluorescent probe to isolate its signal. Unfortunately, protein labeling requires a fairly pure preparation because contaminants may interact with the target protein, causing unreliable results. Thus, at least one series of studies must be done with a highly pure preparation or with proteins tagged GFP analogs or tagged 1 : 1 with fluorescently labeled antibodies. Although labeling of G protein subunits is discussed elsewhere in this volume, we reiterate some important points here.

Probe Handling

Many of the reactive probes used in labeling are sensitive to water and light (see Molecular Probes literature). Thus, it is important to store the probes in brown bottles or bottles wrapped in aluminum foil, and to purge the material with an inert gas for storage. It should also be kept in mind that some solvents, such as chloroform, will quench the fluorescence of some probes. Our laboratory usually makes concentrated probe stock in dry dimethylformamide (DMF), which is then placed in dark bottles, purged with nitrogen, and stored in a desiccator at $-20°$.

Labeling

In our hands, low-scale labeling of PLCs with either amine-reactive or thiol-reactive probes works equally well. It should be kept in mind when modifying with a thiol-reactive probe that exposing PLCβ to an oxidizing environment decreases its enzymatic activity over a period of time, and so the protein should be transferred to a reducing environment immediately after the reaction. On the other hand, G protein subunits should be labeled with amine-reactive probes because some postsynthetic modifications such as palmitylation are sensitive to the strong reducing agents used to quench thiol reactions. Also, removal of reducing agents appears to induce aggregation during the time of the labeling procedure even in the presence of detergents.

Almost all of our protein solutions are in the 100–10,000 nM range, but presumably the same labeling procedure should work well at other concentrations. Labeling involves the addition of a 3- to 10-fold molar excess of the fluorophore on ice for 30–60 min, under optimal buffer conditions (i.e., above pH 8 for amine reactions and in the absence of reducing agents for thiol reactions). Although some investigators directly quench the reactions by adding reducing agents for thiol modification or a concentrated buffer at pH <7.5 for amine modification we usually combine this step with dialysis, using a quenching buffer to end the reaction. It is this dialysis step that we find to be the most problematic because the reaction volumes are usually small and ~20–50% of the material sticks to the dialysis tubing walls. Commercial microcaps appear to minimize the loss of protein.

Reconstitution

We usually reconstitute the G protein subunits onto membrane surfaces simply by adding the desired amount of G protein dispersed in a detergent-dialyzable solution into preformed membranes and dialyzing out the detergent. In most studies, we use large, unilamellar vesicles prepared by extrusion through a 0.1-μm filter. Because most of the fluorescence probes we use have limited membrane solubility, we can combine the dialysis step to remove unreacted probe and detergent at the same time.

We have not yet found any changes in protein–protein affinities with varying membrane composition and our studies imply that composition becomes an important parameter in terms of protein–membrane interactions rather than driving protein–protein interactions (S. Scarlata, L. Runnels, and M. BonHomme, unpublished results, 2000). However, certain compositions that affect the disposition of the protein on the surface may be expected to alter its affinity with a protein partner. We note that we have worked only with large, unilamellar vesicles and it is possible that protein interactions may change when they are bound to the highly stressed surface of small, unilamellar vesicles.

Membrane Crowding

Because we are viewing the interactions between membrane proteins, their concentration on the membrane surface must be considered. It is critically important to keep the protein concentration on the membrane surface low to avoid viewing nonspecific interactions due to surface crowding. Crowding effects can be estimated simply by calculating the surface area available to the proteins and the surface area that the proteins occupy.

For example, if the area of a lipid head group is $\sim 75 \text{ Å}^2$,[10] then for a 300 μM lipid solution the surface area of lipid will be

$$(0.5)(300 \times 10^{-6} \text{ mol/liter}) (75 \text{ Å}^2/\text{molecule}) (6.02 \times 10^{23} \text{ molecules/mol})$$

$$= 2.7 \times 10^{22} \text{Å}^2/\text{liter}$$

where the factor of 0.5 comes in because only lipids in the outer leaflet will be available for binding.

If we assume the dimensions of the G protein subunit to be $50 \times 50 = 2500 \text{ Å}^2$,[11,12] and those of PLC to be $150 \times 50 = 7500 \text{ Å}^2$,[13] as determined from their crystal structures, then 2.25 μM total protein or a 1.175 μM concentration of each must be used to fully cover the surface. This value can be used as a guideline to assess whether the observed affinities are specific or nonspecific.

Fluorescence Titrations

To minimize the amount of material used in these measurements, we conduct studies in a 3×3 mm cuvette that allows the viewing of samples at volumes as low as 100 μl. There are cuvettes available that hold smaller volumes, but mixing of the titration material is difficult. The protein partner is added incrementally and the spectrum is recorded until a clear plateau is obtained after the necessary corrections (e.g., dilution and background). The absence of a plateau would indicate nonspecific interaction or some experimental artifact.

Fluorescence titrations to determine the association between membrane-bound proteins are carried out by reconstituting one of the components on the membrane and titrating in the other either from solution if the protein is soluble, or in a concentrated detergent stock. As a general guideline, we have found that the addition

[10] R. B. Gennis, "Biomembranes: Molecular Structure and Function" (C. R. Cantor, ed.), p. 533. Springer Advanced Texts in Chemistry. Springer-Verlag, New York, 1989.

[11] M. A. Wall, D. E. Coleman, E. Lee, J. A. Iniguez-Lluhi, B. A. Posner, A. G. Gilman, and S. R. Sprang, *Cell* **83,** 1047 (1995).

[12] J. Sondek, A. Bohm, D. G. Lambright, H. E. Hamm, and P. B. Sigler, *Nature (London)* **379,** 369 (1996).

[13] L. O. Essen, O. Perisic, R. Cheung, M. Katan, and R. L. Williams, *Nature (London)* **380,** 595 (1996).

of ~20 mol% detergent will begin to cause leakage from the inner contents of large, unilamellar vesicles, and so the amount of added detergent should be kept below this value. It is important to work at lipid concentrations high enough to ensure complete membrane binding of the proteins. If not, then it is necessary to consider the membrane–protein along with the protein–protein equilibrium when analyzing data. Also, when adding a protein from a detergent stock, it is important to determine whether the addition of detergent affects the fluorescence signal in the absence of protein–protein interactions because the detergents will tend to partition into the bilayer and may alter the protein–lipid interactions. If it is not possible to have a partner protein in a detergent solution, then titrations can be done by reconstituting the protein pairs at different ratios on the membranes, which is somewhat labor intensive. Alternatively, if the off rate for membrane association is fast, so that the proteins can transfer to other vesicles in the system, then titrations with membrane-bound proteins can be conducted.

Detection Systems

Our fluorometer has photon-counting electronics, which allows us to detect nanomolar and sometimes picomolar amounts of labeled proteins. If photon counting is not available, then the amount of protein used must be increased ~10-fold. It is easy to compensate for these higher protein concentrations by reconstituting them with higher amounts of lipid so as to keep the proteins dilute on the lipid surface (see below). Because the presence of lipids will scatter light and perturb the emission spectrum, bandpass filters can be used to reduce the contribution of scattered light when this value becomes a significant part of the signal. If scattering is a problem, it is advisable to try to increase the separation between the exciting and emitting wavelengths. Alternatively, because scattered light is completely polarized, the polarizers can be "crossed" by setting the excitation polarizer at 90° and the emission at 0° or vice versa. Even though this latter method increases the signal-to-noise ratio, it greatly reduces the amount of emitted light that makes it way to the detector and is not recommended for weakly fluorescing samples.

Control Studies and Experimental Pitfalls

Many factors besides protein binding can alter fluorescence, and it is critical to carry out at least one and usually several types of control studies. These studies may include titrating in the buffer of the partner protein, or denaturing the partner protein, or titrating in a nonbinding protein and recording the spectra. These spectra are usually recorded and subtracted from the experimental spectrum, eliminating the need for dilution corrections.

The first point on the titration curve, which corresponds to the sample in the absence of added protein, is critical in terms of fitting to a dissociation constant. It is our experience that this point is the most variable, usually because the sample has not been thermally equilibrated with the chamber, or because settling of some of the contents in the cuvette has occurred. Solutions that have been kept on ice will sometimes form small, light-scattering bubbles in a cuvette after a few minutes at room temperature. Therefore, immediately before recording the spectrum, the cuvette should be mixed by inversion or gentle vortexing after the cuvette solution has equilibrated in the sample chamber for several minutes. This mixing will remove bubbles and suspend the solution contents. Precipitated material can be removed by centrifugation, and settling material can be made more buoyant by adding salt or sucrose. The titration should be started only after verifying that the initial spectrum is reproducible.

One factor that may interfere with reproducibility is photobleaching. This process occurs when oxygen reacts either reversibly or irreversibly with the excited state probe.[3,14] Photobleaching can be assessed by determining whether the emission intensity decreases with time, and can be avoided by flowing an inert gas (usually nitrogen or argon) into the sample chamber. Note that the emission of many probes will decrease when illuminated over time because of local heating, and so it is recommended that the sample be subjected to exciting light only when taking a spectrum.

Analysis of Experimental Data

Determination of the affinity from the experimental titration curves involves two key factors. First, as with all studies of this type, control studies must be done to verify that the binding is under equilibrium rather than stoichiometric conditions. Stoichiometric binding occurs when studies are conducted at concentrations far above the dissociation constant. In this case, the added protein completely binds to its partner and saturation is achieved when all of the partner is bound. Note that under these conditions, the stoichiometry of binding can be readily obtained simply by comparing the number of moles of added protein at saturation versus the moles of the protein partner. Background information about the thermodynamic treatment of protein–protein interactions can be found in biophysical chemistry textbooks (e.g., Cantor and Schimmel[1] and van Holde et al.[2]) or in the excellent comprehensive text of Weber.[15]

[14] G. Weber, "Protein Interactions." Chapman & Hall, New York, 1992.
[15] T. Wang, S. Pentyala, M. J. Rebecchi, and S. Scarlata, *Biochemistry* **38**, 1517 (1999).
[16] T. Wang, S. Pentyala, M. J. Rebecchi, and S. Scarlata, *Biochemistry* **38**, 1517 (1999).

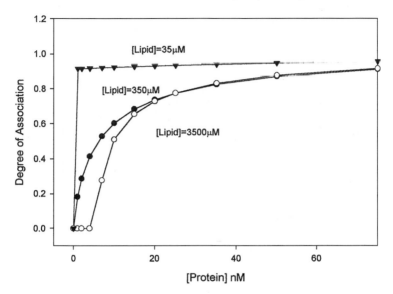

FIG. 6. Simulated association curves showing the change in the midpoint [i.e., K_d (app)] at varying lipid concentrations. As noted in text, soluble proteins will show a similar behavior, except the midpoints will be completely dependent on the bulk concentration of protein rather than on the total concentration.

When Membrane Concentrations Must Be Considered

In Fig. 6 we present idealized curves for equilibrium association between two membrane proteins. Although the midpoint shifts with membrane concentration of one species, the K_d is constant. It is this shift with concentration that should be used to verify that protein–protein associations are being viewed whether for a soluble or membrane-bound species. If the membrane concentration of one protein is much higher than the K_d, then a sharp break is seen on the association curve, indicating that all of the protein partner is bound. It is this break point that corresponds to the stoichiometry of the association.

From Fig. 6 we find that the apparent affinity between membrane-bound proteins depends on the amount of lipid used or on the surface area in which the proteins can associate. When both proteins are bound to membrane surfaces, then expressing interactions by solution association constants is not appropriate because the effective concentration of the proteins on the membrane surface is effectively

**Increase in Acrylodan-Gβγ on 355 μM POPC:POPS
upon binding of PLC-β₂**

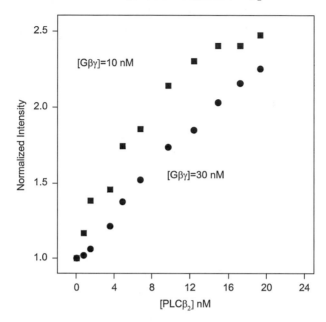

FIG. 7. An example of the shift in midpoint of two laterally associating proteins, acrylodan–G$\beta\gamma$ and unlabeled PLCβ_2 in the same type of study as that shown in Fig. 1. In this case, the midpoint of the shifts ~2- to 3-fold with a 3-fold increase in the amount of acrylodan–G$\beta\gamma$ used.

higher than in solution. This higher effective concentration greatly increases the probability of encounter. However, this higher effective concentration comes into play only when both proteins are membrane bound. If one of the proteins is freely diffusing in solvent, then traditional bimolecular association equations will hold. The simplest method to verify that the interaction being viewed occurs by two proteins interacting laterally on the membrane surfaces is by observing a shift in the titration curve to higher concentrations as the membrane concentration is raised. An example is shown in Fig. 7.

Determination of Bimolecular Constants

In general, we initially treat all protein associations as occurring in solution. Membrane-bound species are then further analyzed by assuming that all of the species are functional and can participate in binding. This assumption is usually tested by biochemical methods such as electrophoresis on native gels, capture on a resin having the bound protein partner, etc. If a significant population is not functional then it effectively reduces the concentration of the particular protein component and will cause the apparent K_d to appear weaker.

The degree of association (α) between PLCβ_2 and G$\beta\gamma$, or between two other proteins, can be obtained by the change in fluorescence along the titration curve by the general equation

$$\alpha(x) = \frac{F(x) - F(x = \text{initial})}{F(x = \text{final}) - F(x = \text{initial})} \tag{2}$$

where $F(x)$ is the fluorescence parameter that varies along the titration curve. $F(x = \text{initial})$ is the initial fluorescence parameter before the addition of species x. $F(x = \text{final})$ is the final fluorescence value once saturating concentrations of species x have been added.

We describe the methods used to analyze the association of PLCβ to G protein $\beta\gamma$ subunits reconstituted on model membranes, but this analysis will hold for any protein system. First, we assume that the proteins do not interact on the membrane surface of large unilamellar vesicles (LUVs), but in bulk solution phase:

$$[\text{PLC}\beta] + [\text{G}\beta\gamma] \underset{k_r}{\overset{k_f}{\rightleftharpoons}} [\text{PLC}\beta \cdot \text{G}\beta\gamma]$$

$$K_d = \frac{k_r}{k_f} = \frac{[\text{G}\beta\gamma][\text{PLC}\beta]}{[\text{G}\beta\gamma \cdot \text{PLC}\beta]} \tag{3}$$

letting $[\text{G}\beta\gamma] + [\text{G}\beta\gamma \cdot \text{PLC}\beta] = [\text{G}\beta\gamma]_0$; and $[\text{PLC}\beta] + [\text{G}\beta\gamma \cdot \text{PLC}\beta] = [\text{PLC}\beta]_0$. The degree of association (α) of PLCβ with G$\beta\gamma$ can be described analytically.

$$\alpha = \frac{1}{2}(x + b) - \frac{1}{2}\sqrt{x^2 + 2x(b - 2) + b^2} \tag{4}$$

$$\alpha \equiv \frac{[\text{G}\beta\gamma \cdot \text{PLC}\beta]}{[\text{G}\beta\gamma]_0} \qquad x = \frac{[\text{PLC}\beta]_0}{[\text{G}\beta\gamma]_0} \qquad b = 1 + \frac{K_d}{[\text{G}\beta\gamma]_0}$$

$\alpha(x)$ may then be fit to a simple bimolecular association curve for two proteins interacting in the bulk solution, rather than on the membrane surface, to give an "apparent" dissociation constant.

Determination of On and Off Rates by Fluorescence

A clear advantage of fluorescence spectroscopy is that it allows the viewing of interactions of membrane-bound proteins in real time, and measuring the kinetic parameters of these associations may not only give insight into signaling rates but also verify the observed apparent steady state dissociation constant.

The encounter rate of proteins in solution is rapid and is greatly enhanced when the proteins are bound to membranes. For these reasons, a stop-flow apparatus is usually required to measure association rates. The methodology we have used is analogous to steady state measurements, in which we measure the change in fluorescence of a labeled protein when the partner protein is injected. If the partner binds to membranes, then initial studies must be done to assess the rate of membrane binding, and to work at high enough membrane concentrations so that this rate can be neglected. In this way, the rate will be under second-order conditions (i.e., dependent on both species).

If measurements can be carried out at high concentrations so that the equilibrium is almost completely on the side of the complex, the forward rate (k_f) can be determined without the use of the dissociation rate (k_r):

$$[PLC\beta_2] + [G\beta\gamma] \underset{k_r \cong 0}{\overset{k_f}{\rightleftharpoons}} [PLC\beta_2 \cdot G\beta\gamma] \tag{5}$$

$$K_{eq} = \frac{k_f}{k_r}$$

The analytical solution describing the degree of association with time (t) for second-order conditions, where the degree of saturation (α) is defined as $\alpha = [PLC\beta_2 \cdot G\beta\gamma]/P_0$ and $[PLC\beta_2]_0 = [G\beta\gamma]_0 = P_0$, is

$$\alpha = \frac{k_f t P_0}{(k_f t P_0 + 1)} \tag{6}$$

For experiments in which $[PLC\beta_2] \neq [G\beta\gamma]$, we can replace P_0 by the average value of the two initial species:

$$P_0 = \frac{[PLC\beta_2]_0 + [G\beta\gamma]_0}{2} \tag{7}$$

$$\alpha = e^{-k_r t} \quad \text{where} \quad \alpha \equiv \frac{[G\beta\gamma \cdot PLC\beta]}{[G\beta\gamma]_0} \tag{8}$$

The rate of dissociation is sometimes more interesting than the association rate because this rate will estimate the time that the signaling complex will be active. Dissociation rate constants ("off rates") can be measured under first-order conditions by first forming the energy transfer complex and adding an unlabeled protein to displace one of the partners. If the steady state dissociation constant is strong and the association rate is diffusion controlled, then the time scale of dissociation can be measured by a conventional fluorometer and a stop-flow apparatus may not be necessary.

Protein–Protein Interactions on Membrane Surface

It is useful to convert the observed dissociation constants measured for membrane-bound proteins to parameters that are independent of lipid concentration and that can be compared with affinities between soluble proteins.

In our experiments, we studied G protein–PLC interactions when the proteins were bound to 50–500 μM large unilamellar vesicles (LUVs) prepared by extrusion through a 100-nm-diameter membrane. Both PLCβ_2 and G$\beta\gamma$ are confined to interact on the outer membrane surface of these vesicles and so the proteins can be imagined as interacting within a surface phase of volume v, equal to the surface area of the vesicle with outer radius r, which in this case is 100 nm/2, multiplied by the thickness from the membrane surface into the solvent in which the proteins interact (d). Because protein association occurs on the membrane surface, the value for d is approximately equal to the thickness of the proteins themselves (~10 nm for G$\beta\gamma$ or PLCβ_2) as estimated from crystallographic data. Thus,

$$v = 4\pi r^2 d \qquad (9)$$

The extent to which the concentration of the membrane-bound protein [P_m] in this membrane surface volume v is related to the concentration when it is not membrane bound [P_{bs}] but freely diffusing in the volume of the bulk solvent, V_{bs}, is

$$[P_m] = \frac{V_{bs}}{v}[P_{bs}] \qquad (10)$$

From the preceding relationship we can calculate how the measured or apparent dissociation constant for proteins interacting on the membrane surface ($K_{d\text{-}m}$) can be related to the dissociation constant that would be measured if the proteins were interacting in the bulk solution volume ($K_{d\text{-}bs}$).

Consider the following simple bimolecular reaction and its material balance analogous to the PLCβ_2–G$\beta\gamma$ association described above:

$$A + B \rightleftharpoons AB$$

$$K_{d\text{-}bs} = \frac{[A][B]}{[AB]}$$

$$[A_0] = [A] + [AB]$$
$$[B_0] = [B] + [AB] \qquad (11)$$

where [A_0] and [B_0] are the total concentration of species A and B, respectively.

$$K_{d\text{-}bs} = \frac{([A_0] - [AB])([B_0] - [AB])}{[AB]} \qquad (12)$$

The degree of association (α) of species A to B, and ratio (r) of species A and B, are defined as

$$\alpha = \frac{[AB]}{[B_0]} \qquad r = \frac{[A_0]}{[B_0]} \qquad (13)$$

$K_{d\text{-bs}}$ can be expressed in terms of α and r:

$$K_{d\text{-b}} = [B_0]\frac{(r - \alpha)(1 - \alpha)}{\alpha} \qquad (14)$$

Because the change in the degree of association with [A] (and thus r) is obtained by fluorescence measurements, the preceding equations can be used to fit these variables to a $K_{d\text{-b}}$ for the reaction.

Membrane-Bound Proteins. The association between two proteins interacting in the membrane surface phase volume can also be described by an identical bimolecular association reaction:

$$A + B \rightleftharpoons AB$$
$$K_{d\text{-m}} = \frac{[A][B]}{[AB]} \qquad (15)$$

The material balance on species A and B is

$$[A_0]_m = [A] + [AB]$$
$$[B_0]_m = [B] + [AB] \qquad (16)$$

Here, $[A_0]_m$ and $[B_0]_m$ are the total concentration of species A and B in the membrane surface phase volume, respectively. $[A_0]_m$ and $[B_0]_m$ can be related to the total concentration of species A and B in three dimensions, that is, $[A_0]$ and $[B_0]$:

$$[A_0]_m = \gamma[A_0]; \qquad [B_0]_m = \gamma[B_0]; \qquad \gamma = \frac{V_{bs}}{v} \qquad (17)$$

The dissociation constant for the membrane-bound species is

$$K_{d\text{-m}} = \frac{([A_0]_m - [AB])([B_0]_m - [AB])}{[AB]} \qquad (18)$$

Defining the degree of association (α) of species A to B, and ratio (r) of species A and B:

$$\alpha = \frac{[AB]}{[B_0]_m} \qquad r = \frac{[A_0]_m}{[B_0]_m} \qquad (19)$$

$K_{d\text{-m}}$ can be expressed in terms of α and r:

$$K_{d\text{-m}} = [B_0]_m\frac{(r - \alpha)(1 - \alpha)}{\alpha} \qquad (20)$$

Substituting for $[B_0]_m$ gives

$$K_{\text{d-m}} = \gamma [B_0] \frac{(r - \alpha)(1 - \alpha)}{\alpha} \tag{21}$$

Comparing this dissociation constant for membrane-bound species with one for the freely diffusing species yields

$$K_{\text{d-m}} = \gamma K_{\text{d-bs}} \tag{22}$$

Thus, the dissociation constant for two protein interacting in the membrane surface phase volume ($K_{\text{d-m}}$) is proportional to the dissociation constant for two proteins interacting in the bulk solvent volume ($K_{\text{d-bs}}$) by a factor γ, relating surface phase volume to the bulk volume.

For the experiments described in this article, the concentrating factor (γ) can be calculated from the concentration of lipid in the experiment:

$$\gamma = (1/2)[\text{lipid}]N_A d\rho(10^{-24}\,\text{liters/nm}^3) \tag{23}$$

where [lipid] is in moles per liter, N_A is Avogadro's number; d is the width of the surface phase (nm), and ρ is the average surface area of a lipid head group (nm^2). For a lipid concentration of 350×10^{-6} M, $d = 100$ nm, and $\rho = 0.7$ nm^2, $\gamma = 0.0074$. Thus for an apparent K_d of 10 nM, as measured for PLCβ_2 and G$\beta\gamma$ interacting on a 350 μM lipid surface phase volume, the dissociation constant that would be observed if the two were freely diffusing in bulk solvent would be 10 n$M \times 136 = 1.35$ μM.

Concluding Statements

The measurements we describe here may seem ominous at first, but they are actually fairly simple even to those without experience in spectroscopy. We describe how the problems of complex solutions, possible interference by the fluorescence label, and the assessment of the fraction of nonreactive material can be addressed. The information gained by these types of measurements can be powerful in understanding the varying routes of cellular signals.

Acknowledgments

The author thanks the students who were the driving forces in working out these methods: Loren Runnels, Marjorie BonHomme, Nathan Manes, Tieli Wang, Louisa Dowal, Vijaya Narayanan, and Paxton Provitera.

[24] Assays and Characterization of Mammalian Phosphatidylinositol 4,5-Bisphosphate-Sensitive Phospholipase D

By XUEJUN JIANG, STEPHEN GUTOWSKI, WILLIAM D. SINGER, and PAUL C. STERNWEIS

Introduction

Phospholipase D (PLD) hydrolyzes phospholipids into phosphatidic acid (PA) and the base head groups. In mammalian cells, this activity plays an active role in signal transduction by acting as a mediator for a variety of extracellular stimuli.[1,2] Two mammalian PLD genes have been identified and the encoded enzymes expressed and characterized. The activity of PLD1 is regulated by the ADP-ribosylation factor (ARF) and Rho families of monomeric GTPases, classic iso-forms of protein kinase C (PKC), and the signaling lipid phosphatidylinositol 4,5-bisphosphate (PIP_2). PLD2, although highly sensitive to PIP_2, has shown only modest response to ARF and appears unresponsive to Rho and PKC, *in vitro* (see Refs. 2–4 for reviews).

Characterization of PLD activity and identification of various regulatory mole-cules were greatly facilitated by the establishment of an assay for *in vitro* studies.[5] These assays utilized phospholipid vesicles as the medium for delivery of the sub-strate, [^{3}H-*choline*]dipalmitoylphosphatidylcholine. An essential element in this assay, which allowed consistent and reliable measurement of activities both in membranes and solubilized preparations, and the observation of regulation by activating molecules, was the inclusion of PIP_2 in substrate vesicles. Although fre-quently referred to as a cofactor for PLD, this lipid is not absolutely required for activity of the enzyme. Rather, the lipid acts as a bona fide activator that stimulates activity by itself but also facilitates measurement of activities stimulated by other regulatory molecules. Therefore assessment of basal PLD activity and almost all stimulated activities *in vitro* is measured in the presence of PIP_2. Alternative meth-ods for the assay of PLD described in this article facilitate measurement of basal activity and regulation in the absence of PIP_2.

Reagents and Solutions

Lipids and Detergent

Dipalmitoylphosphatidylcholine (DPPC), dipalmitoylphosphatidic acid (DPPA), bovine brain phosphatidylethanolamine (PE), phosphatidylinositol (PI),

---

[1] D. English, *Cell. Signal.* **8,** 341 (1996).

and phosphatidylserine (PS) are purchased from Avanti (Alabaster, AL). PIP_2 and n-octyl-β-D-glucopyranoside (OG) are purchased from Boehringer Mannheim (Indianapolis, IN) and Calbiochem (La Jolla, CA), respectively. Radioactive phosphatidylcholine   (L-α-[^{3}H-methylcholine]dipalmitoylphosphatidylcholine) ([^{3}H]PC) is obtained from DuPont (Wilmington, DE). Other agents are defined in procedures.

Solutions

Solution A: 50 mM Na–HEPES (pH 7.5), 80 mM KCl, 3 mM EGTA, 1 mM dithiothreitol

Solution B: 20 mM Na–HEPES (pH 7.5), 25 mM KCl, 1 mM EDTA, 2 mM dithiothreitol

Solution C: 20 mM Na–HEPES (pH 7.5), 1 mM EDTA, 2 mM dithiothreitol, 1% OG (w/v)

Solution D: 20 mM Na–HEPES (pH 7.5), 1 mM EDTA, 1 mM dithiothreitol

Protease Inhibitors

Where indicated, protease inhibitors (PIs) are added to solutions to achieve the following final concentrations: N^α-p-tosyl-L-lysine chloromethyl ketone, 21 μg/ml; tosylphenylalanyl chloromethyl ketone, 21 μg/ml; phenylmethylsulfonyl fluoride, 21 μg/ml; leupeptin, 2.5 μg/ml; pepstatin A, 1 μg/ml; N^α-p-tosyl-L-arginine methyl ester, 21 μg/ml. The inhibitors are made and stored as stock solutions at 1000× final concentrations.

Preparation of Substrate Vesicles

Method 1: Sonicated Vesicles

Mix DPPC (36 nmol), PE (360 nmol), PIP_2 (36 nmol), and [^{3}H]PC (about 4 μCi) in chloroform in a glass test tube. Evaporate the chloroform with a stream of nitrogen. Add 600 μl of solution A to the dried lipids, vortex to suspend the lipids, and sonicate for 2–10 min at room temperature to form unilamellar vesicles (particulate material is clarified such that the appearance of the lipid mixture changes during sonication to translucent or slightly cloudy). If flocculent material remains after sonication, one or several of the lipids have deteriorated. This mixture provides enough substrate (about 100 cpm/pmol PC) for at least 100 assays of 30 μl (see Assay of Phospholipase D Activity, below). Vesicles are used on the same day.

[2] J. H. Exton, *Biochim. Biophys. Acta* **1439**, 121 (1999).

[3] W. D. Singer, H. A. Brown, and P. C. Sternweis, *Annu. Rev. Biochem.* **66**, 475 (1997).

[4] M. A. Frohman, T. C. Sung, and A. J. Morris, *Biochim. Biophys. Acta* **1439**, 175 (1999).

[5] H. A. Brown, S. Gutowski, C. R. Moomaw, C. Slaughter, and P. C. Sternweis, *Cell* **75**, 1137 (1993).

Method 2: Gel-Filtered Vesicles

Mix DPPC (66 nmol), PE (660 nmol), PIP$_2$ (66 nmol), and [^{3}H]PC (about 20 μCi) in chloroform in a glass test tube. Evaporate the chloroform with a stream of nitrogen. Add 300 μl of solution A containing 1% (w/v) OG to the dried lipids, vortex, and sonicate for 1–2 min at room temperature to suspend the lipids in detergent micelles (solution should be clear).

Unilamellar vesicles are formed by gel filtration at 4°. Apply the micelle mixture to a column (0.7 × 30 cm) of AcA34 (Biosepra, Marlborough, MA), which has been equilibrated with solution A, and elute with the same solution at a rate of 5 ml/hr. Collect fractions of 300 μl, analyze aliquots by scintillation spectroscopy, and pool the four or five fractions containing the peak of radioactivity. The concentration of DPPC in the vesicles is determined from the specific activity measured in the original lipid mixture. When examined in detail, the recovery of all phospholipids in a single preparation was similar and ranged from 40 to 60%; thus the molar ratio of lipids in the vesicles was essentially the same as the starting mixtures. Aliquots of the vesicles can be used directly or stored at −80° after rapid freezing with liquid nitrogen.

Preparation of Recombinant Phospholipase D

Solubilized, Enriched Preparation of Recombinant PLD1

Infect cultures (1.6–2.4 liters) of *Spodoptera frugiperda* (Sf9) for 48 hr with baculovirus directing the expression of PLD1.[6] Detailed procedures can be found in [28] in this volume.[7]

Harvest the cells by centrifugation at 700g for 15 min at 4°, resuspend in 80 ml of solution B containing PIs, and lyse the cells by nitrogen cavitation in a Parr (Moline, IL) bomb. Remove unbroken cells and nuclei by centrifugation at 700g for 15 min at 4°. Adjust the remaining lysate to 500 mM KCl and 1% (w/v) OG and stir on ice for 60 min. Remove particulate material by centrifugation at 100,000g for 60 min at 4°. Dilute the supernatant with an equal volume of solution C and apply to a 100-ml column of SP-Sepharose (Pharmacia, Piscataway, NJ). Use a 300-ml linear gradient of 200–1000 mM KCl in solution C to elute bound protein. PLD activity elutes in a broad peak at approximately 700–1000 mM KCl. Pool the fractions containing the peak of activity, add PIs, and store in aliquots at −80°. The SP-Sepharose step enriches PLD about 20-fold and serves to separate the enzyme from endogenous regulatory proteins and lipids. Stimulated activities

---

[6] S. M. Hammond, Y. M. Altshuller, T. C. Sung, S. A. Rudge, K. Rose, J. Engebrecht, A. J. Morris, and M. A. Frohman, *J. Biol. Chem.* **272,** 3860 (1997).

[7] C. Wells, X. Jiang, S. Gutowski, and P. C. Sternweis, *Methods Enzymol.* **345,** [28] 2001 (this volume).

of 20 nmol/min/mg with ARF and PIP$_2$ are about 10% of those reported for preparations of purified recombinant rat or human PLD1.[6,8]

Enriched Membrane Preparation of PLD1

PLD enzymes are expressed for 48 hr after infection of Sf9 cells grown in monolayer (0.5 ml of viral stock per 150-mm culture plate containing about 1×10^7 cells). After 48 hr of expression, gently wash the cells with a solution containing 137 mM NaCl, 2.7 mM KCl, 1.5 mM KH$_2$PO$_4$, 8 mM Na$_2$HPO$_4$ (adjusted to pH 7.4). Harvest the cells from each dish with 5 ml of solution D containing PIs and collect by centrifugation at 400g for 5 min at 4°. Suspend the pelleted material with solution D containing PIs (0.4 ml/plate) and homogenize by repeated passage of the mixture through a 25-gauge needle. Mix the homogenate with 2 volumes of 65% (w/v) sucrose in the same buffer to achieve a final concentration of 47% (w/v). Place aliquots (1.2 ml, material from one plate) in centrifuge tubes for a TLS-55 swinging bucket rotor (Beckman, Fullerton, CA) and overlay with equal volumes of the same solution containing 40 and 20% sucrose, respectively. Centrifuge at 80,000g for 60 min at 4° and collect the membranes that migrate to the 20–40% sucrose interface. Utilize this fraction as the light membrane fraction. About 500 μg of protein (enough material for 100–200 assays) can be obtained from two plates of infected cells.

Preparation of Activators: ARF, Rho, PKC

ARF can be purified from porcine brain through five chromatographic steps ending with Mono Q chromatography as described previously.[5] Recombinant, partially myristoylated ARF1 is coexpressed with yeast N-myristoyltransferase in *Escherichia coli* and purified as described previously.[9] Recombinant PKCα is produced in Sf9 cells, using a baculovirus expression system, and purified.[10]

Recombinant prenylated RhoA is expressed in Sf9 cells as a complex with a recombinant fusion protein of glutathione S-transferase (GST) and RhoGDI (GDP dissociation inhibitor). The RhoA–GDI complex is isolated by affinity chromatography with glutathione resin and the prenylated RhoA is eluted by dissociation from the immobilized GDI. This procedure is described in detail in [28] in this volume.[7]

Assay of Phospholipase D Activity

Assays are usually performed in 12×75 mm polypropylene tubes and a volume of 30 μl, but can be scaled to larger reactions. To each sample, add successively

[8] D. S. Min, S. K. Park, and J. H. Exton, *J. Biol. Chem.* **273**, 7044 (1998).
[9] P. A. Randazzo, O. Weiss, and R. A. Kahn, *Methods Enzymol.* **257**, 128 (1995).

and mix 5 μl of 4× reaction solution [200 mM Na–HEPES (pH 7.5), 12 mM EGTA, 4 mM dithiothreitol, 320 mM KCl], 5 μl of water or activators (see below), 5 μl of PLD sample (about 5 ng of the enriched solubilized PLD or 1–2 μg of enriched membranes), and 5 μl of a solution containing 18 mM CaCl$_2$ and 18 mM MgCl$_2$. Start the assay by addition of 10 μl of gel-filtered vesicles or 5 μl of solution A and 5 μl of sonicated vesicles. Incubate at 37° for 20–60 min (lower temperature can be used for longer times). Stop reactions by addition of 200 μl of 10% (w/v) trichloroacetic acid and 100 μl of 10% (w/v) bovine serum albumin. Mix vigorously and move the reaction tubes to ice. Spin the mixtures for 10 min at 3600 rpm and 4° in a J-6 centrifuge (Beckman). Carefully remove 270 μl of the supernatant to assess the amount of free [³H]choline present by scintillation spectroscopy. Experimental duplication is usually within 5–10% error.

The trichloroacetic acid causes denaturation of protein and precipitation of both proteins and phospholipids; the bovine serum albumin is added as a carrier to help pellet the precipitated material. This procedure leaves the hydrolysis product, choline, in solution. In general, about 50% of the substrate is available for hydrolysis. Activities measured are linear during hydrolysis of up to 10–20% of the total substrate. Calcium is not required for PLD activity but can enhance activities when Mg^{2+} is lowered and acts as an activator of PKC.

Final concentrations of activators used for maximum stimulation are as follows: recombinant ARF (5 μM), RhoA (1 μM of the preactivated GTPase), and PKCα (50–100 nM). Activation of the monomeric GTPases is achieved by inclusion of 10 μM GTPγS in assays including ARF and by pretreatment of RhoA. For maximal activation, RhoA (6 μM) is incubated with 20 mM Na–HEPES, (pH 8.0), 1 mM EDTA, 1 mM dithiothreitol, 400 mM NaCl, and 60 μM GTPγS for 20 min at room temperature; 5 μl of the preactivated material is added to the assays. ARF and PKCα are diluted on ice with 20 mM Na–HEPES (pH 7.5), 1 mM EDTA, 1 mM dithiothreitol, and 50 mM NaCl as necessary just before use in assays. The amount of PKCα required for maximal stimulation varies with the preparation and assay conditions; because too much PKC can reduce the maximal signal, titration of this activator is recommended for new preparations and conditions.

Characterization: Discussion of Activities and Problems

Two examples are shown to demonstrate the scope of the PLD assays. Figure 1 compares activities obtained with substrate vesicles produced by the two alternative methods described. The original use of vesicles formed by direct sonication of phospholipids in an aqueous suspension provided the first assay for measurement of PLD activity that could be regulated by GTP-binding proteins. Measurement of activity shows a high degree of dependence on the presence of PIP$_2$ in the vesicle composition. Formation of vesicles by gel filtration (Fig. 1B) provides

---

[10] W. D. Singer, H. A. Brown, X. Jiang, and P. C. Sternweis, *J. Biol. Chem.* **271**, 4504 (1996).

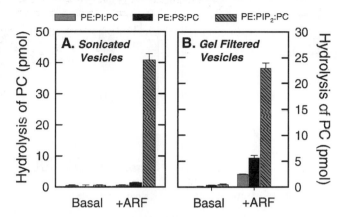

FIG. 1. Comparison of PLD activities obtained with different substrate vesicles. Solubilized recombinant PLD1 was expressed in Sf9 cells, extracted and enriched by chromatography through SP-Sepharose, and assayed as described. Substrate vesicles composed of PE:DPPC:acidic lipid (molar ratio, 9:1:1) were made by direct sonication of lipids or by gel filtration. Acidic lipids used were phosphatidylinositol (PI), phosphatidylserine (PS), or PIP_2 as indicated. (A) Sonicated vesicles; (B) gel-filtered vesicles.

a more facile assay and convincingly demonstrates that PIP_2 is not required for expression of hydrolytic activity by PLD. Thus activity is readily seen with ARF as an activator when either PI or PS is substituted for PIP_2. Although both molecules can act alone, the synergism between ARF and PIP_2 is readily observed with both preparations of vesicles. The ability of PIP_2 to enhance activations observed with the other regulators of PLD1 explains its value as a constant component of substrate vesicles. The potency of PIP_2 as an activator of PLD1 is demonstrated in Fig. 2, as well as the abilities of three other regulators to stimulate in the presence of the lipid.

The two substrate vesicle preparations differ in advantages and disadvantages. Direct sonication of the lipids is rapid and efficient and readily adaptable to different sizes of preparation. However, the preparations of lipids work best when used on the day of preparation (often not useful the next day), have more problems with lipid aging and contamination, and produce higher blanks (usually about 2–3% of total signal). The latter property makes measurement of basal activities and low-abundance preparations of enzyme difficult or inaccurate. Vesicles made by gel filtration are relatively stable and may be stored on ice overnight or flash frozen and stored for future use with little loss in signal output. In addition, these vesicle provide for more sensitive and accurate assessment of samples with lower PLD activities due to reduced background (<0.5%). Negative features include a longer preparation time and less efficient use of procured lipids.

Troubleshooting

There are several factors that can interfere with this assay. The three most common are the age of the lipids, the use of plastic pipette tips, and the handling of

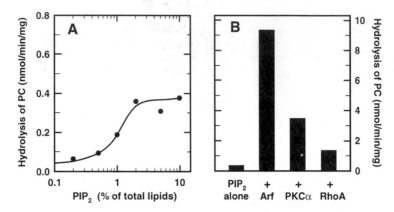

FIG. 2. Potency of PIP$_2$ for the activation of PLD. Solubilized recombinant PLD1 was expressed in Sf9 cells, extracted and enriched through one column, as described. (A) Assays were performed with gel-filtered vesicles composed of PE : PI : DPPC (molar ratio, 5 : 1 : 1) and the indicated molar percentage of PIP$_2$. In (B), assays utilized vesicles with 10% PIP$_2$ and also contained 1 μM ARF, 100 nM PKCα, or 670 nM RhoA as indicated.

enzymes. Aging of lipids is most common with the stock of PE. Frequently, use of a fresh preparation of this lipid will alleviate problems involving low activities or precipitation of sonicated lipid. The use of plastic tips to dispense stock solutions of lipids can cause deterioration of vesicle formation, presumably because of leaching of chemicals out of the plastic by the organic solvents in which the lipids are stored. Once the lipids are in aqueous medium, the use of plastic causes no problem.

Detergents markedly inhibit these assays, presumably by disruption of the vesicular substrate. Both PLD1 and some preparations of the monomeric GTPases are extracted and prepared with detergent. Final steps of preparation contain OG because the assay tolerates small amounts of this detergent better than others examined. Protein preparations containing detergent are diluted into assay buffers such that the final concentration of OG added in the assay is less than 0.033% (w/v).

For the most part, the enzymes utilized in these assays remain viable when stored frozen at $-80°$. Preparations of PLD do tend to deteriorate over several months of storage and their enzymatic activity is not retained well on ice over several days. Multiple cycles of freezing and thawing are particularly deleterious to PLD and PKC. Nevertheless, the most likely cause for low activities with freshly thawed preparations is the assay rather than the enzymes.

Acknowledgments

This work was supported by the National Institutes of Health (GM 31954), the Robert A. Welch Foundation, and the Alfred and Mabel Gilman Chair in Molecular Pharmacology.

[25] Characterization and Purification of Phosphatidylinositol Trisphosphate 5-Phosphatase from Rat Brain Tissues

By RUDIGER WOSCHOLSKI

Introduction

Phosphoinositides (PIs) are important regulators of many signal transduction processes that are important for cell growth, communication, and differentiation.[1,2] The generation or removal of these inositol lipids is performed by PI kinases and phosphatases, which in turn results, respectively, in a corresponding initiation or termination of PI-dependent signaling, such as cell growth or apoptosis.[3–5] Although different types of PI phosphatases have been described, PI 5-phosphatases in particular seem to play a prominent role in signal transduction mechanisms. Because removal of the 5-phosphate of the inositol head group, a process performed by all PI 5-phosphatases, is the preferred way to metabolize phosphatidylinositol trisphosphate (PtdInsP₃), the potential enzymological properties of the 5-phosphatases are relevant for their biological function.[6,7]

The 5-phosphatases consist of two different types: type I and type II phosphatases. The latter are involved in PI metabolism whereas the type I 5-phosphatases work only on soluble inositol polyphosphates. The type II phosphatases could be classified with respect to their domain organization as GTPase-activating protein (GAP) domain-containing inositol phosphatases (GIPs), Src homology 2 (SH2) domain-containing inositol phosphatases (SHIPs), and Sac1-containing inositol phosphatases (SCIPs).[3,8] As shown in Fig. 1, each of these groups consists of two mammalian members that possess a unique substrate specificity (see Fig. 1). However, the overall substrate recognition seems to be preserved within each class. Whereas SHIPs tend to work on PtdInsP₃ and inositol tetrakisphosphate (InsP₄) only, GIPs tend to have a preference for soluble inositol phosphates as well as PtdIns(4,5)P₂, and SCIPs prefer phosphatidylinositol bisphosphate (PtdInsP₂) and PtdInsP₃. All type II 5-phosphatases have the ability to hydrolyze PtdInsP₃, although not always as efficiently as their preferred substrates. Both possession of

[1] P. De Camilli, S. D. Emr, P. S. McPherson, and P. Novick, *Science* **271,** 1533 (1996).

[2] X. Zhang and P. W. Majerus, *Semin. Cell Dev. Biol.* **9,** 153 (1998).

[3] R. Woscholski and P. J. Parker, *Trends Biochem. Sci.* **22,** 427 (1997).

[4] B. Vanhaesebroeck and M. D. Waterfield, *Exp. Cell Res.* **253,** 239 (1999).

[5] C. L. Carpenter and L. C. Cantley, *Biochim. Biophys. Acta* **1288,** M11 (1996).

[6] L. R. Stephens, K. T. Hughes, and R. F. Irvine, *Nature (London)* **351,** 33 (1991).

[7] R. Woscholski, M. D. Waterfield, and P. J. Parker, *J. Biol. Chem.* **270,** 31001 (1995).

[8] C. Erneux, C. Govaerts, D. Communi, and X. Pesesse, *Biochim. Biophys. Acta* **1436,** 185 (1998).

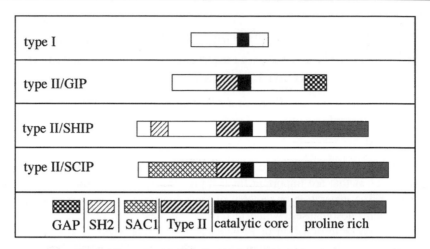

FIG. 1. Alignment of the type II phosphoinositide 5-phosphatases.

regulatory domains such as an SH2 domain and/or proline-rich domain (PRD) as well as substrate specificity will therefore define the functional role of the corresponding phosphatase.

The SCIPs represent the only class of type II 5-phosphatases that contain two independent, intrinsic phosphatase activities with distinct substrate specificity and catalytic mechanism defining it as a multienzyme.[9,10] Phosphatases of the synaptojanin subgroup (SCIPs) are characterized by the presence of an N-terminal Sac1 domain that is homologous to a *Saccharomyces cerevisiae SAC1* gene product that produces inositol auxotrophy.[11] This domain has subsequently been identified as a phosphoinositide phosphatase related to the phosphoinositide 3-phosphatase PTEN (phosphatase and tensin homolog) as judged by sequence comparison.[9] As the Sac1 domain is responsible for hydrolyzing PtdIns(4)P, which is the natural precursor for PtdIns(4,5)P_2, involvement of this particular activity in synaptojanin function must be considered. Indeed, there is some evidence that synaptojanin is influencing the cytoskeleton, probably via its action on PtdIns(4,5)P_2 bound to cytoskeletal proteins such as profilin or actin.[12] Furthermore, a knockout of synaptojanin 1 in mice revealed its essential role in postbirth life as well as its effect on PI 3-kinase signaling.[13] The latter observation is in agreement with an earlier

[9] S. Guo, L. E. Stolz, S. M. Lemrow, and J. D. York, *J. Biol. Chem.* **274,** 12990 (1999).

[10] W. E. Hughes, R. Woscholski, F. T. Cooke, R. S. Patrick, S. K. Dove, N. Q. McDonald, and P. J. Parker, *J. Biol. Chem.* **275,** 801 (2000).

[11] E. A. Whitters, A. E. Cleves, T. P. McGee, H. B. Skinner, and V. A. Bankaitis, *J. Cell Biol.* **122,** 79 (1993).

[12] T. Sakisaka, T. Itoh, K. Miura, and T. Takenawa, *Mol. Cell. Biol.* **17,** 3841 (1997).

[13] O. Cremona, G. Di Paolo, M. R. Wenk, A. Luthi, W. T. Kim, K. Takei, L. Daniell, Y. Nemoto, S. B. Shears, R. A. Flavell, D. A. McCormick, and P. De Camilli, *Cell* **99,** 179 (1999).

finding that synaptojanin is the major constitutively active PtdInsP$_3$ 5-phosphatase activity in mammalian brain tissue.[14]

This article focuses on the major PtdInsP$_3$ 5-phosphatase in rat brain tissue, synaptojanin (i.e., synaptojanin 1). As pointed out above, this enzyme contains two independent phosphoinositide phosphatase activities that have distinct substrate specificity. Because the 5-phosphatase activity is characterized by a stronger catalytic activity as compared with the phosphoinositide monophosphatase activity of the Sac1 domain, procedures and methods described concentrate on this 5-phosphatase activity. However, all described assays include comments on the determination of the Sac1 phosphatase activity as well.

Assay Methods

Creation of Radioactive Labeled Phosphoinositides

Phosphoinositide phosphatase assays have always relied on the availability of lipid kinases, which could be employed for [^{32}P]phosphate labeling of commercial phosphoinositide preparations from natural sources.[7,15,16] Several phosphoinositide kinases have been sufficiently characterized with respect to their assay conditions. Thus, even crude tissue preparations could provide all the necessary enzymes for [^{32}P]phosphate labeling of phosphoinositides.[7,10,17] Given that these crude preparations do contain a great deal of phosphatases and phospholipases it is not surprising that their use is less efficient than the use of partially purified phosphoinositide kinases.[7] Purification of these lipid kinases could be achieved either by classic chromatographic methods or by employment of commercial antibodies against these enzymes or known binding partners. For example, there are many antibodies available that would immunoprecipitate PI 3-kinase or their corresponding binding partners such as the platelet-derived growth factor (PDGF) receptor. To facilitate the latter, it is best to use cells that do possess the receptor and treat these with the growth factor and then immunoprecipitate the receptor together with the associated PI 3-kinase. This method, although sufficient for providing sustainable amounts of PI 3-kinase, is not practical if bulk amounts of lipid must be labeled. For these occasions it would be advisable to employ the crude tissue preparations as a source for further purification. For example, PI 3-kinase is abundant in brain

[14] R. Woscholski, P. M. Finan, E. Radley, N. F. Totty, A. E. Sterling, J. J. Hsuan, M. D. Waterfield, and P. J. Parker, *J. Biol. Chem.* **272,** 9625 (1997).

[15] S. Giuriato, B. Payrastre, A. L. Drayer, M. Plantavid, R. Woscholski, P. Parker, C. Erneux, and H. Chap, *J. Biol. Chem.* **272,** 26857 (1997).

[16] S. P. Jackson, S. M. Schoenwaelder, M. Matzaris, S. Brown, and C. A. Mitchell, *EMBO J.* **14,** 4490 (1995).

[17] F. T. Cooke, S. K. Dove, R. K. McEwen, G. Painter, A. B. Holmes, M. N. Hall, R. H. Michell, and P. J. Parker, *Curr. Biol.* **8,** 1219 (1998).

tissue,[18] which could be employed to affinity purify it.[18,19] Indeed, it is possible to obtain pure PI 3-kinase by using a short tyrosine-phosphorylated peptide resembling the PDGF receptor insert and to which the SH2 domain of PI 3-kinase will bind. However, the PI 3-kinase cannot be eluted without denaturing and thus destroying its lipid activity.[20–22] Therefore, like with the immunoprecipitated PI 3-kinase described above, phosphoinositide labeling must be done with the immobilized enzyme.

Preparation of [^{32}P]Phosphate-Labeled Phosphoinositides

All 3-phosphorylated phosphoinositides can be easily phosphorylated by using affinity- or immuno-purified PI 3-kinase preparations and the corresponding phosphoinositides as substrates. However, each of the substrates should be employed under conditions that favor a high turnover.[7] The radioactive label of the phosphoinositides facilitates the detection of the lipids after the assay. Phospholipids will be separated from the water phase by chloroform extraction. Acidification of the assay mix ensures that the negatively charged phosphoinositides will be neutralized and subsequently partition into the organic phase. The organic phase ($CHCl_3$) contains all the lipids while the water phase has all the released phosphate and inositol phosphates. Other phosphoinositides such as PtdIns(4)P and PtdIns(4,5)P$_2$ can be produced by employing PI 4-kinase or PIP kinase preparations, respectively, if [^{32}P]phosphate-labeled phosphoinositides are needed. However, these phosphoinositides are available as [^{3}H]inositol-labeled compounds as well, which have been employed in phosphatase and phospholipase assays.[7,23]

Procedure. The phosphoinositides (0.1–0.2 mg) are dried in a nitrogen stream or under vacuum and then sonicated in 0.2 M Tris-HCl, pH 7.4 (0.2 ml). After adding 0.1 ml of enzyme preparation (purified PI 3-kinase), 0.05 ml of 20 mM MgCl$_2$ and 0.05 ml of 40 mM EGTA-Na$_2$ are added. It is important to add the divalent cations after the lipid has been mixed with the enzymes, because the metal would otherwise induce precipitation, which would affect the performance of this assay. The inclusion of EGTA inhibits any contaminating phospholipase C

[18] F. Ruizlarrea, P. Vicendo, P. Yaish, P. End, G. Panayotou, M. J. Fry, S. J. Morgan, A. Thompson, P. J. Parker, and M. D. Waterfield, *Biochem. J.* **290,** 609 (1993).

[19] S. J. Morgan, A. D. Smith, and P. J. Parker, *Eur. J. Biochem.* **191,** 761 (1990).

[20] M. Otsu, I. Hiles, I. Gout, M. J. Fry, F. Ruiz-Larrea, G. Panayotou, A. Thompson, R. Dhand, J. Hsuan, N. Totty, A. D. Smith, S. J. Morgan, S. A. Courtneidge, P. J. Parker, and M. D. Waterfield, *Cell* **65,** 91 (1991).

[21] R. Woscholski, T. Kodaki, M. McKinnon, M. D. Waterfield, and P. J. Parker, *FEBS Lett.* **342,** 109 (1994).

[22] M. J. Fry, G. Panayotou, R. Dhand, F. Ruiz-Larrea, I. Gout, O. Nguyen, S. A. Courtneidge, and M. D. Waterfield, *Biochem. J.* **288,** 383 (1992).

[23] J. K. Chung, F. Sekiya, H. S. Kang, C. Lee, J. S. Han, S. R. Kim, Y. S. Bae, A. J. Morris, and S. G. Rhee, *J. Biol. Chem.* **272,** 15980 (1997).

activities. If this assay is employed to generate [^{32}P]PtdInsP$_3$ from PtdInsP$_2$, using PI 3-kinase, 0.05 ml of 5% (w/v) sodium cholate solution is added, otherwise (PtdIns and PtdInsP as substrate) 0.05 ml of water is used instead. Adding 0.05 ml of 10 mM [^{32}P]ATP solution (neutralized with Tris) starts the reaction. The incubation can be performed for up to 1 hr depending on the enzyme purity. If crude extracts are employed the incubation should not exceed 30 min. Best turnover is achieved by incubation at 37°; however, it is possible to choose lower temperatures such as 30° or room temperature as well. This protocol results in up to 50% turnover of the substrate if sufficient amounts of PI 3-kinase activity are employed. As turnover rates are linked with the amount of PI 3-kinase present in the assay, it is advisable to obtain and concentrate as much lipid kinase activity as possible.

The lipid incubation with the enzyme is stopped by adding 1 ml of CHCl$_3$–methanol (1 : 2, v/v), followed by the addition of 0.05 ml of concentrated HCl. After mixing thoroughly phase separation can be accelerated by centrifugation. The bottom phase (CHCl$_3$ phase) is moved into a new tube and washed several times with methanol–1 N HCl (1 : 1, v/v). The lipids so obtained can be dried under vacuum and resuspended in CHCl$_3$–methanol (1 : 1, v/v) for storage. It is noteworthy that this lipid preparation still contains all the lipids present in the kinase assay, that is, substrate and products of this reaction. If pure [^{32}P]phosphate-labeled phosphoinositides are needed, it is recommended that the lipids obtained be purified by preparative thin-layer chromatography (TLC). In this case the lipids are scraped from the TLC plates and eluted by the developing solution. Any carryover of silica material must be separated before the organic solution so cleared is dried. Purified [^{32}P]phosphate-labeled phosphoinositides can be supplemented with unlabeled lipids in order to compensate for any losses due to the procedure.

Thin-Layer Chromatography

A valuable and efficient way to analyze the quality of the produced [^{32}P]phosphate-labeled lipids is the TLC system. Although several TLC systems have been described that are able to separate phosphoinositides isomers [e.g., PtdIns(3)P and PtdIns(4)P],[24] most TLC systems are designed to separate the phosphoinositides according to their amount of phosphorylation. Thus, these TLC systems are suitable for the detection of phophoinositide phosphatases or kinases and have consequently been employed successfully for this purpose.[7,17,22] The inorganic phosphate of the water phase can be separated from the water-soluble inositol phosphates by molybdate treatment. The resulting phosphomolybdate complex can be easily extracted into the organic phase, leaving the inositol phosphates behind in the remaining water phase.

[24] T. Kodaki, R. Woscholski, S. Emr, M. D. Waterfield, P. Nurse, and P. J. Parker, *Eur. J. Biochem.* **219**, 775 (1994).

Procedure. Phosphoinositides present in the organic phase can be separated by TLC employing two distinct developing solutions, each designed to achieve maximum separation between substrate and product. Both systems rely on the use of silica 60 TLC plates (Merck, Rahway, NJ) and a propanol–acetic acid developing solution. If PtdInsP$_3$ is to be separated from PtdInsP$_2$ the TLC plate should be developed in *n*-propanol–2 *N* acetic acid (1 : 1, v/v), whereas PtdInsP$_2$ separation from PtdInsP can be achieved by development in *n*-propanol–2 *N* acitic acid (2 : 1, v/v). It is worth mentioning that the described system is based on the use of untreated TLC plates and will not work if the plates are treated or activated in any way.

Phosphoinositide Phosphatase Determination

Synaptojanin, the major PtdInsP$_3$5-phosphatase activity in rat brain tissue,[14,25] contains two distinct intrinsic phosphoinositide phosphatase activities.[9] The differential substrate specificity of these two phosphatases allows testing for them individually. For example, employment of PtdInsP$_3$ will result in the exclusive determination of the 5-phosphatase activity, because the Sac1 phosphatase activity is unable to hydrolyze either PtdInsP$_3$ or PtdIns(3,4)P$_2$, the product of the 5-phosphatase action. Alternatively, employment of PtdIns(4)P or PtdIns(3)P will allow the detection of the Sac1 phosphatase activity without interference from the 5-phosphatase activity. Another possibility to exclude one of the synaptojanin specific phosphatase activities is to employ inhibitors that will affect only one of the phosphatases (see Inhibitor Profile, below).

Procedure. For the determination of the phosphoinositide 5-phosphatase activity phosphoinositides are dried under vacuum and then dissolved in 2% (w/v) cholate in 0.2 M Tris pH 7.4 by sonication. The presence of the detergent cholate inhibits the Sac1 phosphatase activity, thus allowing exclusive detection of the 5-phosphatase activity of synaptojanin. The 5-phosphatase activity is determined in the presence of 5 m*M* MgCl$_2$, 2.5 m*M* EGTA, 50 m*M* Tris (pH 7.4), 0.5% (w/v) cholate and 3–100 μ*M* PtdInsP$_3$ [containing an excess of PtdIns(4,5)P$_2$ because of lipid production; the inclusion of this lipid blocks PtdIns(4,5)P$_2$ 5-phosphatases and works as a carrier lipid]. The incubation (0.05 ml) is stopped by adding 0.2 ml of CHCl$_3$–mathanol (1 : 2, v/v), followed by the addition of 0.2 ml of 1 *N* HCl. The extracted inositol lipids (bottom phase) are then separated by TLC (see above) and the corresponding spots of PtdInsP$_3$ and PtdInsP$_2$ are scraped out and counted. The amount of [^{32}P]PtdIns(3,4)P$_2$ present in the lipid preparation is determined and subtracted (blank incubation). The same procedure can be employed to test for PtdIns(4,5)P$_2$ 5-phosphatase activity, using [^{3}H]PtdIns(4,5)P$_2$. The assay should be performed under the same conditions as for the PtdIns(3,4,5)P$_3$ 5-phosphatase activity (see above).

[25] P. S. McPherson, E. P. Garcia, V. I. Slepnev, C. David, X. Zhang, D. Grabs, W. S. Sossin, R. Bauerfeind, Y. Nemoto, and C. P. De, *Nature (London)* **379,** 353 (1996).

The Sac1 phosphatase is tested in the same way as the 5-phosphatases except that either no detergent or 0.25% (w/v) octylglucoside is employed in the assay. Because this phosphatase works well on PtdIns(3)P, a lipid that can be readily labeled with [^{32}P]phosphate by PI 3-kinase, testing for Sac1 phosphatase activity is less laborious than the 5-phosphatase assay. Because the phosphatase results in the release of [^{32}P]phosphate from the lipid substrate, activity can be determined by Cerenkov counting of the aqueous phase after phase separation (see above).

Colorimetric Determination of Phosphoinositide Phosphatase Activity

The advent of commercially available phosphoinositides has made it possible to employ a colorimetric phosphate release assay. This assay is based on the detection of inorganic phosphate by molybdate, which in turn is visualized through a reaction with malachite green.[26] This assay has been successfully employed in many phosphatase assays, but lacks the sensitivity of the assay based on radioactively labeled substrates.[27–29] Thus, colorimetric determination of phosphoinositide phosphatases has been restricted to assays for abundant enzymes and/or enzymes with high catalytic activity.[10] Hence, this assay is not the first choice if crude cell or tissue extracts are to be tested. If substrates are employed that possess several phosphates, this method will be unable to distinguish between the possible phosphatase activities. This indiscriminate detection of phosphatase activities is less troublesome once the enzymes to be tested are free of any contaminating activity. Therefore, this assay is ideal for the use of recombinant enzymes as well as sufficiently characterized purified enzyme preparations. However, if synaptojanin is tested it is important to employ substrates that are unique for the corresponding phosphatase activity. Furthermore, the availability of all relevant phosphoinositides allows the investigation of enzymological properties that are normally more difficult to pursue if the relevant lipid kinases are not readily at hand. In particular, a comprehensive analysis of the substrate specificity is an obvious application of this type of assay.[10]

Procedure. The activity of the phosphoinositide phosphatases is determined in the presence of 0.125 mM phosphoinositide, 0.25% (w/v) octylglucoside, 4 mM MgCl$_2$, and 0.1 M Tris (pH 7.4). After incubation at 37° (5 to 30 min) released phosphate is assessed calorimetrically as described previously.[20] Briefly, the assay is stopped by adding a similar volume of malachite green/molybdate reagent, followed by 3 volumes of water. After 30 min measurements are taken at 610 nm. Because this assay is based on the visualization of the released phosphate, it is

[26] A. A. Baykov, O. A. Evtushenko, and S. M. Avaeva, *Anal. Biochem.* **171,** 266 (1988).

[27] T. P. Geladopoulos, T. G. Sotiroudis, and A. E. Evangelopoulos, *Anal. Biochem.* **192,** 112 (1991).

[28] E. B. Cogan, G. B. Birrell, and O. H. Griffith, *Anal. Biochem.* **271,** 29 (1999).

[29] D. H. Ng, K. W. Harder, I. Clark-Lewis, F. Jirik, and P. Johnson, *J. Immunol. Methods* **179,** 177 (1995).

important to avoid any contamination by phosphate through either water or equipment. For the same reason it is essential to employ controls and blanks in order to be able to correct for any carryover of phosphate through the ingredients of this assay (e.g., phosphoinositides or detergent). It is noteworthy that the commercial synthetic phosphoinositides sometimes vary in their quality. It is therefore advisable to establish the actual amount/concentration of phosphoinositides by ashing the lipids before colorimetric phosphate determination.[30]

Purification of Phosphatidylinositol Trisphosphate 5-Phosphatase (Synaptojanin) Activity from Rat Brain

Any attempt to purify an enzyme will likely be preceded by an evaluation of the tissue distribution of this enzyme. As pointed out above, determination of phosphoinositide phosphatases in crude extracts is best performed by employing radioactively labeled lipids. As shown in Fig. 2, PtdInsP$_3$ 5-phosphatase activity is abundant in the cytosolic fraction derived from rat brain tissue. Thus, purification of this particular activity from the soluble fraction of rat brain tissue is likely to produce sufficient amounts of activity. Indeed, the PtdInsP$_3$ 5-phosphatase can be obtained in reasonable purity by employing several ion-exchange and size-exclusion chromatography procedures. The latter is facilitated by the high molecular mass (145 kDa) of the PtdInsP$_3$ 5-phosphatase.[7]

Procedure. Rat brain tissue (200 g) is homogenized in a Dounce homogenizer in 600 ml of buffer A [40 mM Tris-HCl (pH 7.4), 1 mM EDTA, 10 mM benzamidine, 1 mM phenylmethylsulfonyl fluoride (PMSF), soybean trypsin inhibitor (0.1 mg/ml)]. The cytosol (400 ml) obtained by unltracentrifugation (100,000g, 1 hr, 4°) is loaded onto a 50-ml HiLoad Q column and washed with 300 ml of buffer A followed by 300 ml of 0.1 M NaCl in buffer A. The phosphatase activity is eluted from the column with buffer A containing 0.3 M NaCl (300 ml). This step separates the 5-phosphatase activity from most of the contaminating phosphoinositide activities. The eluate is then diluted 3-fold in buffer A and loaded onto a 15-ml HiTrapS column (made from three 5-ml HiTrapS columns that have been connected). The column is washed with buffer A containing 0.1 M NaCl (100 ml) and step eluted with buffer A containing 0.2 M NaCl (100 ml). The eluate is diluted 2.5-fold with buffer A and loaded onto a 5-ml phosphocellulose column. After washing with buffer A most of the phosphoinositide phosphatase activity is eluted with buffer A containing 0.1 M NaCl. The eluate is diluted 2.5-fold with buffer A and loaded onto a 2-ml HiTrap heparin column (made from two connected 1-ml columns). The heparin column is washed with buffer A containing 0.2 M NaCl and bound phosphatase activity is eluted with buffer A containing 0.3 M NaCl. The heparin eluate is already a sufficiently pure preparation, but can be further purified by

---

[30] M. Petitou, F. Tuy, and C. Rosenfeld, *Anal. Biochem.* **91**, 350 (1978).

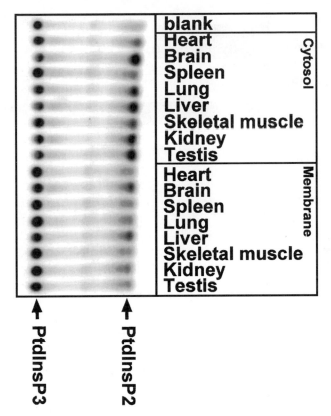

FIG. 2. Tissue distribution of PtdInsP$_3$ 5-phosphatases. Crude extracts of different tissues were separated into cytosolic and particulate fraction by centrifugation and subsequently tested for [^{32}P]PtdInsP$_3$ 5-phosphatase activity. Equal relative amounts of these fractions were employed, resulting in a comparable presentation of the data. The lipids were separated by TLC and visualized by autoradiography. Positions of the corresponding lipids are indicated.

size-exclusion chromatography (S-300 column). Active fractions are pooled and concentrated by loading onto a 1-ml HiTrapQ column. After washing with 0.1 M NaCl in buffer A bound activity is eluted by a linear 0.1–0.3 M NaCl gradient in buffer A. The active fractions at about 0.2 M NaCl are combined and concentrated by ultrafiltration with a 100-kDa cutoff membrane. The enzyme can be stored frozen at $-20°$ (several months) or for longer storage at $-80°$ (several years).

Characterization of Phosphoinositide Phosphatases

Substrate Specificity

The substrate specificity is best established with recombinant enzyme preparations that allow the employment of the colorimetric phosphatase assay.[10] As most

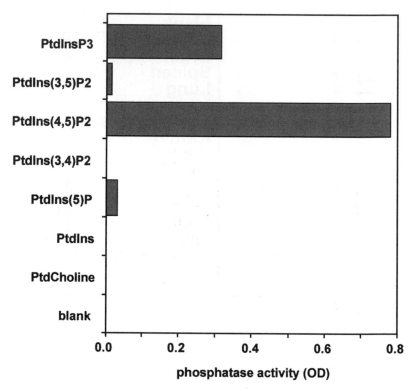

FIG. 3. Substrate specificity of recombinant synaptojanin. Synaptojanin 1 from rat was expressed as a GST fusion protein without the proline-rich domain (C-terminal deletion) in bacteria. The affinity-purified phosphatase was then tested for the indicated substrates as described in text, using the colorimetric assay. The activity is shown as the optical density (OD) after 30 min of incubation with the corresponding lipids.

relevant inositol phosphates and phosphoinositides are available as synthetic compounds, testing the substrate specificity can be performed in a comparable fashion. As shown in Fig. 3, recombinant synaptojanin (expressed without its proline-rich domain) is able to hydrolyze all inositol lipids that contain a 5-phosphoester bond on the inositol head group. However, more highly phosphorylated lipids such as PtdIns(4,5)P$_2$ and PtdIns(3,4,5)P$_3$ are by far the best substrates. As this assay is performed under conditions that allow the detection of both intrinsic phosphatase activities, it cannot be ruled out that the observed turnover of PtdIns(3,5)P$_2$ and PtdIns(5)P is due to the Sac1 phosphatase activity. To distinguish the two phosphatase activities it is necessary to employ conditions that favor or inhibit one of these activities, a task facilitated by the knowledge of the corresponding inhibitor profile.

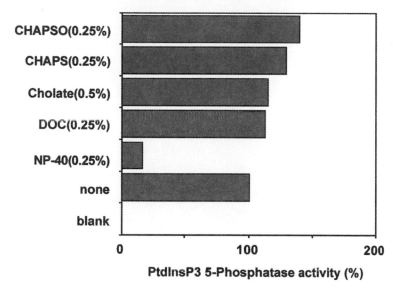

FIG. 4. Detergent sensitivity of the purified PtdInsP$_3$ 5-phosphatase. The purified PtdInsP$_3$ 5-phosphatase (synaptojanin) was tested in the presence of the indicated detergents, using [^{32}P]PtdInsP$_3$ as described in text. The relative activity as compared with an assay without any detergent is shown.

Inhibitor Profile

The abundance of inositol and phosphoinositide phosphatases has led to a classification that is based on chromatographic properties as well as the inhibitor profile of these enzymes.[3] Apart from testing known inhibitors of similar phosphatase activities, it is useful to determine effects of different chemicals that might be able to influence the enzyme activity. For example, detergents, divalent metals, and phosphate-containing lipids are likely to interfere with lipid presentation and thus should affect phosphatase activities. Indeed, while the 5-phosphatase activity of synaptojanin is more robust in its detergent tolerance (see Fig. 4), the Sac1 phosphatase is inhibited by charged (e.g., cholate and CHAPS) and nonionic detergents (e.g., Triton X-100).[9] The only detergent that can be employed for this phosphatase activity is octylglucoside (see colorimetric assay above). All type II 5-phosphatases, including synaptojanin, rely on the presence of magnesium for catalytic activity,[7] whereas the Sac1 phosphatase activity is fully active in the presence of chelators.[9] The 5-phosphatase activity of synaptojanin is resistant to sulfhydryl-modifying agents such as N-ethylmaleimide and iodoacetamide,[7] a feature not shared with the Sac1 phosphatase activity, which relies on the presence of reducing agents for full activity.[9]

Section VI

Small GTP-Binding Proteins

[26] Assay of Cdc42, Rac, and Rho GTPase Activation by Affinity Methods

By VALERIE BENARD and GARY M. BOKOCH

Introduction

The Rho GTPases are members of the Ras superfamily and include several isoforms of Cdc42, Rac, and Rho. The Rho subfamily regulates a variety of signal transduction pathways in eukaryotic cells, including cell adhesion and migration, by modulating cytoskeletal dynamics. Cdc42 induces formation of filopodia, whereas Rac regulates actin polymerization at the plasma membrane where ruffles are formed. Both induce the formation of cytoskeletal/signaling aggregates known as focal complexes. Rho controls the assembly of focal adhesions and the reorganization of actin into stress fibers. More recently, it has been recognized that Rho GTPases also initiate signaling pathways that impact on gene expression and cell growth regulation, and that each of these GTPases is essential for Ras to induce malignant transformation. A specific role for Rac has been identified in phagocytic cells, such as neutrophils or monocytes, where it activates the NADPH oxidase enzymatic complex.

GTPases cycle from inactive (GDP-bound) to active (GTP-bound) forms that interact with and regulate components of intracellular signaling pathways. This cycle is under the control of several classes of regulatory protein. Under basal conditions, the Rho GTPases are complexed with their cytosolic GDP-dissociation inhibitor (GDI), which stabilizes the inactive GDP-bound form. On cell stimulation, the GTPase dissociates from the GDI and translocates to the plasma membrane. Subsequently the GTPase releases GDP and binds GTP, a reaction promoted by guanine nucleotide exchange factors (GEFs), which leads to the interaction with specific molecular targets. Inactivation of the GTPase results from hydrolysis of GTP into GDP, and the intrinsic rate of hydrolysis by the GTPase is enhanced by the action of GTPase-activating proteins (GAPs). The active form of the GTPase appears to be transient and labile, which makes its measurement difficult. Indirect methods based on monitoring association with or activation of regulatory proteins or on relocalization within the cell have been used to assess Rho GTPase activation during the dynamic cell response. These methods have their problems and limitations, and to characterize Rho GTPase activation a direct measurement of the formation of the active form within the cell is necessary. A classic method to analyze direct GTPase activation consists in measuring the GTP : GDP ratio after ^{32}P labeling of cells and immunoprecipitation of the GTPase. In addition to the need for high levels of radioactivity, this approach is limited for Rho GTPases by the lack of efficient immunoprecipitating antibodies and the rapid turnover of the

METHODS IN ENZYMOLOGY, VOL. 345

GTP-bound state. More recently a new assay measuring Rho GTPase activation has been developed on the basis of the fact that only the active form of the GTPase interacts with downstream effectors. Identification of the GTPase-binding region on target proteins, and the expression of these binding domains in active soluble forms, allows them to be used as specific probes to detect the active form of the GTPase. This article describes an assay developed for Rac and Cdc42 activation using the GTPase/p21-binding region of p21-activated kinase 1 (PAK1). Similar strategies developed as assays for Rho GTP are discussed.

Principle of Assay

The Rac and Cdc42 (p21)-activated kinase PAK1 contains in the N-terminal regulatory region a specific site for interaction with the active form of these two GTPases. This region, referred to as the CRIB domain (Cdc42/Rac interactive binding domain) or p21-binding domain (PBD), was first described by Burbelo et al.[1] and the minimal consensus sequence identified for specific GTPase binding corresponds to amino acids 74 to 89 in PAK1. A homologous, but Cdc42-selective, CRIB domain is also found in the Wiskott Aldrich syndrome protein (WASP), amino acids 235 to 268, which shares 50% homology with the PAK CRIB domain. Binding affinities of the PAK PBD range from $\sim$20 nM to $\sim$1 μM, depending on the length of the peptide encompassing the CRIB domain used.[2] More recently, distinct interactive sites for GTP–RhoA, RhoB, and RhoC were identified on the Rho effectors Rhotekin (amino acids 7–89)[3] and Rho kinase (amino acids 934–1015).[4]

The affinity precipitation assay detecting active Rac and Cdc42 uses the p21-binding domain (PBD) of the target PAK1 fused to glutathione S-transferase (GST) to isolate (precipitate) a complex containing the GST–PBD bound to the active GTPase (Rac–GTP or Cdc42–GTP). The precipitated GTPase is then quantified by Western blotting, using specific antibodies (Fig. 1).

Method for GST–PBD Preparation

Construction of GST–PAK1 PBD Fusion Protein

The cDNA of the coding sequence for PAK1 PBD corresponding to amino acids 67–150 is amplified by polymerase chain reaction (PCR), cloned into a pGEX2T vector at the *Bam*HI–*Eco*RI sites, and used to transform the DH-10B strain of *Escherichia coli*. The bacteria then produces a fusion protein with glutathione

[1] P. D. Burbelo, D. Drechsel, and A. Hall, *J. Biol. Chem.* **270**, 29071 (1995).

[2] G. Thompson, D. Owen, P. Chalk, and P. N. Lowe, *Biochemistry* **37**, 7885 (1998).

[3] T. Reid, T. Furuyashiki, T. Ishisaki, G. Watanabe, N. Watanabe, K. Fujisawa, N. Morii, P. Madaule, and S. Narumiya, *J. Biol. Chem.* **271**, 13556 (1996).

[4] K. Fujisawa, A. Fujita, T. Ishizaki, Y. Saito, and S. Narumiya, *J. Biol. Chem.* **271**, 23022 (1996).

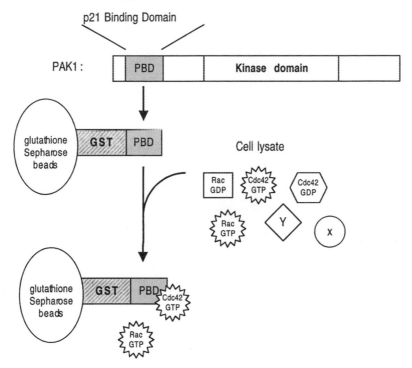

FIG. 1. Principle of the affinity precipitation assay to detect active Rac and Cdc42, using the GST–PAK1 PBD. *Note:* x and Y represent nonrelevant proteins in the lysate.

S-transferase (GST) upstream of the PBD. Transformed bacteria are stored at −80° in 20% (v/v) glycerol L-broth medium.

Expression and Purification of Fusion Protein

L-broth medium (30 ml) containing ampicillin (100 μg/ml) is inoculated with the glycerol stock and grown overnight at 37°. The overnight culture is used to inoculate 1 liter of L-broth medium–ampicillin (100 μg/ml) and is grown for 3 hr at 37°. Transcription is then induced by adding 0.8 mM isopropyl-β-D-thiogalactopyranoside (IPTG) for 3 hr at 30°. The bacterial suspension is centrifuged for 10 min at 5000 rpm at 4° and resuspended in 10 ml of lysis buffer [50 mM Tris-HCl (pH 7.5), 150 mM NaCl, 5 mM MgCl$_2$, 1 mM dithiothreito (DTT), 1 mM EDTA, 1 mM phenylmethylsulfonyl fluoride (PMSF), lysozyme (1 mg/ml), DNase I (20 μg/ml), and aprotinin (1 μg/ml)]. The *E. coli* lysate is incubated for 30 min on ice, and then sonicated and centrifuged for 10 min at 10,000 rpm at 4°.

The fusion protein is purified from the supernatant with glutathione–Sepharose 4B beads (Pharmacia, Piscataway, NJ) previously washed five times in lysis buffer.

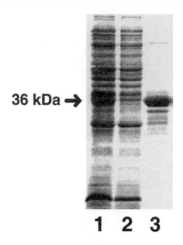

36 kDa →

1 2 3

1 *E.coli* supernatant after induction
2 unbound proteins in *E.coli* supernatant
3 GST-PBD bound to glutathione Sepharose beads

FIG. 2. Analysis of GST–PAK1 PBD preparation on a 12% (w/v) SDS–polyacrylamide gel. Lane 1, *E. coli* lysate; lane 2, unbound proteins in *E. coli* lysate; lane 3, Sepharose bead-isolated GST–PBD.

Glutathione–Sepharose beads are incubated with the supernatant for 2 hr at 4°, and then the beads are centrifuged for 5 min at 2000 rpm at 4° and washed five times with washing buffer [50 mM Tris-HCl (pH 8), 150 mM NaCl, 5 mM MgCl$_2$, 1 mM dithiothreitol, 1 mM phenylmethylsulfonyl fluoride, aprotinin (1 μg/ml)]. Beads can be aliquoted at this point in washing buffer with 10% (v/v) glycerol and stored at −80°. The integrity of the GST–PBD (∼36 kDa) is checked by sodium dodecyl sulfate–polyacrylamide gel electrophoresis (SDS–PAGE) (Fig. 2) and the final protein concentration is determined. The presence of a minor fraction of PBD breakdown products is typical in these preparations, and these do not affect PBD activity.

The GST–PAK1 PBD specifically interacts with the active conformation of Rac or Cdc42, but not Rho. This should be verified after each GST–PBD purification by using recombinant GTPases loaded with GDP as a negative control or with GTPγS as a positive control (Fig. 3A). Accordingly, recombinant GTPases (50–100 ng) are diluted in TEDM buffer [25 mM Tris-HCl (pH 7.5), 1 mM EDTA, 1 mM DTT, and 5 mM MgCl$_2$] and 26.4 μl of the GTPase is mixed with 15 μl of 0.1 M EDTA, 15 μl of 1 mM GTPγS, or 15 μl of 10 mM GDP and incubated for 10 min at 30°. The exchange reaction is stopped by the addition of 4 μl of 1 M MgCl$_2$. Another control is to use expressed mutants of Rac, Cdc42, or Rho: dominant active mutants such as G12V/G14V or Q61L/Q63L maintain bound GTP and should bind to the GST–PBD, whereas the GDP-associated dominant negative T17N/T19N mutant does not (Fig. 3B).

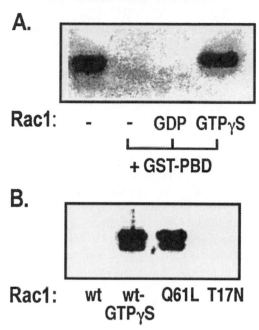

A.

Rac1:   -   -   GDP GTPγS
              └─────┴─────┘
                 + GST-PBD

B.

Rac1:   wt   wt-   Q61L T17N
              GTPγS

FIG. 3. GST–PBD affinity precipitation assay performed with *in vitro*-loaded GTPase. (A) Precipitation with GST–PBD, using recombinant Rac1 either unloaded (–), loaded with GDP, or loaded with GTPγS. The first lane represents the total Rac1 in the sample before PBD precipitation. (B) Precipitation with GST–PBD, using BHK cell lysates expressing various Rac1 mutants and blotted with an anti-Myc antibody to detect transfected GTPases only. Rac1 wild type (wt) was precipitated either as is (GDP form) or after loading with GTPγS. Rac1 Q61L is constitutively GTP bound, whereas Rac1 T17N is in the GDP-bound state.

Simple control experiments can be performed by generating GDP-bound GTPase and GTP-bound GTPase *in vitro* from the cell lysates to be used for study. Indeed, this is an effective means to assess the total amount of activatable GTPase present in the sample. Either expressed recombinant GTPases or endogenous GTPases can be loaded with nucleotides in this manner. Briefly, the lysate should be incubated with nucleotides in excess, $(100-200 \ \mu M)$ GTPγS, a nonhydrolyzable GTP analog, or 1 mM GDP, in a high concentration of EDTA $(10-20 \ \text{m}M)$ and a low concentration of Mg^{2+} (5 mM maximum) at 30° to induce the binding of the nucleotide.[5] Subsequently increasing the Mg^{2+} concentration (to $50-70$ mM) will stop the nucleotide exchange. Specific conditions are as follows: 100 μl of cell lysate containing the GTPase is mixed with 12 μl of 0.2 M EDTA and 12 μl of 1 mM GTPγS or 12 μl of 10 mM GDP and incubated for

[5] A. J. Self and A. Hall, *Methods Enzymol.* **256,** 67 (1995).

10 min at 30°. The exchange reaction is stopped by the addition of 8 μl of 1 M MgCl$_2$. Loaded GTPases should be used immediately in the affinity precipitation assay.

Preparation of Cell Extract to Measure GTP-Bound GTPase

Preparation of cell extract represents a critical point in the assay and determination of the most appropriate lysis buffer must be established before any affinity precipitation. Cells are pretreated or stimulated to generate the active form of the GTPase. For cells in suspension the stimulation is stopped by addition of an equal volume of cold 2× lysis buffer. For adherent cells, the stimulation medium is removed, cold 1× lysis buffer is added, and the cells are scraped from the plates while on ice. The resulting cell lysates are kept on ice for several minutes, and then clarified by centrifugation at 4° for 10 min at 10,000 rpm in a table-top microcentrifuge. Cell lysate can be used directly for the affinity precipitation assay, or can also be frozen in liquid nitrogen and stored at −80° until use. For each affinity precipitation sample, an equal number of cells or the same amount of total cell protein should be mixed with the GST–PBD.

Detergents are necessarily used in the lysis buffer for cell extracts in order to disrupt cells and solubilize membrane proteins. Each cell type has a different detergent requirement to obtain a good protein solubilization without disruption of the nucleus. In neutrophils and breast cancer cell lines, we have used the following lysis buffer: 50 mM Tris-HCl (pH 7.5), 10 mM MgCl$_2$, 200 mM NaCl, 1% (v/v) Nonidet P-40, 5% (v/v) glycerol, 1 mM phenylmethylsulfonyl fluoride, leupeptin (1 μg/ml), and aprotinin (1 μg/ml).[6,7]

Detergents are also present in the binding buffer for the affinity precipitation. In general, detergents help to prevent nonspecific binding to the GST–PBD; however, some detergents can also disrupt specific binding. We have observed that the specific binding of GST–PAK1 PBD to active Rac and Cdc42 is decreased in the presence of 1% (v/v) 3-[(3-cholamidopropyl)-dimethyl-ammonio]-1-propane sulfonate (CHAPS), 1% (v/v) sodium deoxycholate, or 0.1% (w/v) SDS, but is not affected by 1% (v/v) Triton X-100 or 1% (v/v) Nonidet P-40. Analysis of the effect of detergents and detergent concentration on specific versus nonspecific binding is recommended when setting up the PBD assay.

Affinity Precipitation Assay for Rac and Cdc42

Cell extract or recombinant GTPases are incubated with GST–PBD in a maximal final volume of 500 μl for 1 hr at 4°. If necessary the solution containing

[6] V. Benard, B. P. Bohl, and G. M. Bokoch, *J. Biol. Chem.* **274,** 13198 (1999).
[7] J.-P. Mira, V. Benard, J. Groffen, L. C. Sanders, and U. G. Knaus, *Proc. Nat. Acad. Sci. U.S.A.* **97,** 185 (2000).

the GTPase is diluted in binding buffer [25 mM Tris-HCl (pH 7.5), 1 mM DTT, 30 mM MgCl$_2$, 40 mM NaCl, 0.5% (v/v) Nonidet P-40, leupeptin (1 μg/ml), aprotinin (1 μg/ml), and 1 mM PMSF] to have an equal volume in each sample. The bead pellet is then centrifuged for 2 min at 2000 rpm, 4° and washed three times with washing buffer [25 mM Tris-HCl (pH 7.6), 1 mM DTT, 30 mM MgCl$_2$, 40 mM NaCl, 1% (v/v) Nonidet P-40, plus antiproteases] and twice with the same buffer without Nonidet P-40. The bead pellet is finally suspended in 20 μl of Laemmli sample buffer. Proteins bound to the beads are separated on a 12% (w/v) SDS–polyacrylamide gel, transferred onto a nitrocellulose membrane, and blotted for the appropriate GTPase using specific antibodies. The exact blotting conditions should be optimized depending on the antibody used for detection; blots can be quantified by PhosphorImager or densitometry. Blots can be stripped and reprobed for Rac or Cdc42. Typical results obtained with the PAK1 PBD assay can be found in Refs. 6–9.

Ratio of GST–PBD and GTPase

To recover 80% of recombinant Rac1 loaded with GTPγS, we determined that the amount of GST–PBD added needs to be 20 times the amount of the GTPase (Fig. 4A). When starting with cell extract expressing endogenous levels of GTP-bound GTPase, 10^6 to 10^7 cells or 300 to 800 μg of total protein was necessary to observe a signal in an affinity precipitation assay performed with 10 to 15 μg of GST–PBD. Variations are probable depending on the cell type or the agonist used to stimulate the GTPase pathway. Because the amount of activated (GTP-bound) GTPase usually represents only a small fraction of the total available GTPase, we recommend that Western blots initially be performed on the total cell lysates and PBD-precipitated GTPγS-loaded lysates in order to assess the level of signal detectable with the antibodies to be used for detection of the bound (active) GTPase.

Time Course of Interaction

The binding of GTPase to GST–PBD already coupled to Sepharose beads is rapid and reaches about 75% of maximum after 30 min, and is maximal after 1 hr at 4° (Fig. 4B). We typically fix the incubation time at 1 hr for enhanced sensitivity; however, this should be balanced with the rates of GTP hydrolysis occurring in the samples (see below).

Stability of Interaction

When bound to GST–PBD, active GTPase stays tightly bound and is not readily exchanged for another GTPase. Figure 5A shows that the binding occurring

[8] N. Geijsen, S. van Delf, J. A. M. Raaijmakers, J.-W. J. Lammers, J. G. Collard, L. Koenderman, and P. J. Coffer, *Blood* **94,** 1121 (1999).

[9] T. Akasaki, H. Koga, and H. Sumimoto, *J. Biol. Chem.* **274,** 18055 (1999).

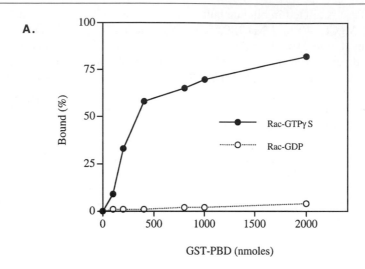

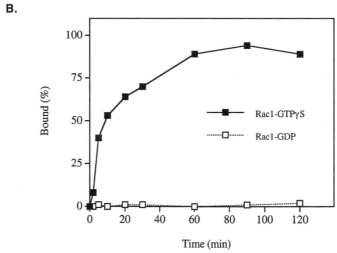

FIG. 4. (A) Binding of increasing amounts of GST–PBD to 20 pmol of recombinant Rac1 loaded with [^{35}S]GTPγS (solid line) or [^{3}H]GDP (dotted line). (B) Time course of the binding between 500 pmol of GST–PBD and recombinant Rac (50 pmol) loaded with [^{35}S]GTPγS (solid line) or [^{3}H]GDP (dotted line).

between Rac1–[^{35}S]GTPγS and GST–PBD is not reversed by addition of increasing amounts of unlabeled Rac1–GTPγS, unless it is added at the same time as Rac1–[^{35}S]GTPγS.

We determined whether the binding of GST–PAK1 PBD to active GTPase is sufficient to block GTP hydrolysis. Recombinant Rac1 or Cdc42 was loaded with [γ-^{32}P]–GTP and GTPase activity was measured at 20° for 20 min.[5] The binding of

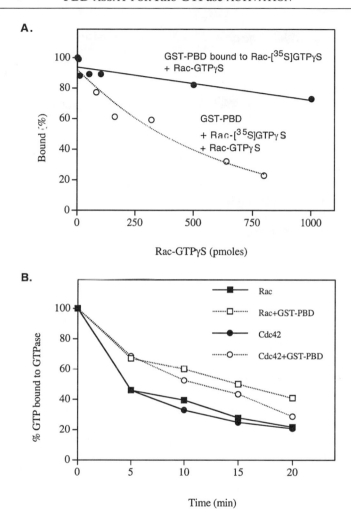

FIG. 5. (A) Stability of the PBD–GTPase interaction: 20 pmol of Rac1-[^{35}S]GTPγS was incubated for 1 hr with (solid lines) or without (dashed lines) 4 μg of GST–PBD before the addition of increasing amount of unlabeled Rac1-GTPγS and incubation for another 1 hr. The radioactivity bound to GST–PBD was then quantified. (B) GTPase activity: 50 ng of Rac1 was loaded with [γ-^{32}P]GTP, and then GTP hydrolysis was initiated at 20° in the presence (solid lines) or the absence (dashed lines) of 4 μg of GST–PBD.

active Rac and Cdc42 to GST–PBD under these conditions does not totally inhibit GTP hydrolysis, although the rate of hydrolysis was significantly decreased by ~30–50% (Fig. 5B). GTP hydrolysis is largely decreased at 4°, so we recommend that cell lysate preparation and affinity precipitation be performed at 4° or on ice, and all buffers used should be kept ice-cold.

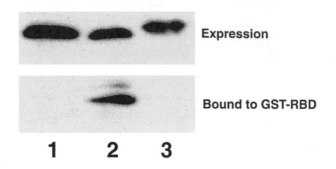

1 Rho A wild-type
2 Rho A Q63L
3 Rho A T19N

FIG. 6. Affinity precipitation assay of various Rho GTPases overexpressed in COS-7 cells, using GST–RBD from Rhotekin. RhoA wild type and RhoA T19N are GDP bound, while RhoA Q63L is constitutively GTP bound.

Comparison with Affinity Precipitation Assay for Rho

The N-terminal region of Rhotekin (amino acids 7 to 89) produced as a fusion protein with GST interacts with GTP–Rho and has been used as the basis for an affinity isolation assay specific for Rho.[10] The principle of the assay using the Rho-binding domain (RBD) of Rhotekin is similar to the assay using the PAK1 PBD. However, major differences are noted in the optimum buffer compositions. The Rhotekin/GST–RBD assay was described with Swiss 3T3 and COS-7 cells lysed in a classic radioimmunoprecipitation assay (RIPA) buffer containing 50 mM Tris-HCl (pH 7.2), 10 mM $MgCl_2$, 500 mM NaCl, 1% (v/v) Triton X-100, 0.5% (v/v) sodium deoxycholate, 0.1% (w/v) SDS, 1 mM phenylmethylsulfonyl fluoride, leupeptin (10 μg/ml), and aprotinin (10 μg/ml).[10] Under these conditions, specific binding to the RBD is maximized while nonspecific binding is minimized (Fig. 6).

Conclusion

We have described a nonradioactive affinity-based method for analysis of Rac, Cdc42, or Rho activation in cell lysates. We have tried to point out areas in which problems might occur, and simple procedures to check whether the PBD/RBD probe is functioning correctly. The assay is, of course, limited by the levels of

[10] X.-D. Ren, W. B. Kiosses, and M. A. Schwartz, *EMBO J.* **18,** 578 (1999).

GTPases present in the samples analyzed, as well as by the quality and sensitivity of the antibodies used for detection. The use of these affinity-based assays provides a simple and effective means to directly assess the biochemical pathways regulating Rho GTPase activation.

[27] Assays of ADP-Ribosylation Factor Function

By JUN KUAI and RICHARD A. KAHN

ADP-ribosylation factors (ARFs) comprise a family of 20-kDa GTPases first identified as the protein cofactor required for the ADP ribosylation of the $G_s\alpha$ protein, catalyzed by the A_1 subunits of the bacterial toxins cholera toxin (CT) and *Escherichia coli* heat-labile toxin (LT).[1,2] The high degree of structural and functional conservation of ARFs cloned from a wide variety of eukaryotic organisms has allowed a diversity of experimental approaches addressing the mechanism(s) and role(s) of ARFs in cell regulation. In mammalian cells, the role of ARFs in the regulation of intracellular membrane traffic is well studied but the mechanism of this regulation is poorly understood. In the yeast *Saccharomyces cerevisiae,* ARFs are required for protein secretion,[3] sporulation,[4] mitotic growth,[5] and respiration (Zhang et al.[5]; M. Cavenagh and R. A. Kahn, unpublished observation, 2000). Thus, ARFs clearly play regulatory roles in a number of processes in addition to that in vesicular traffic. A variety of approaches have been used to identify downstream effectors. With a continuously expanding list of identified effectors for ARFs,[5-7] the complexity of ARF signaling is clearly more extensive than previously thought and so new techniques are required to study specific aspects of the cellular roles of ARF.

The two best characterized effectors are the heptameric coat complex, termed coatomer or COP-1, and phospholipase D1 (PLD1). The presence of ARF on *in vitro*-generated coated vesicles[8] and the recruitment to membranes of COP-I by ARF have been proposed as the nucleating step in budding and subsequent

[1] R. A. Kahn and A. G. Gilman, *J. Biol. Chem.* **261,** 17 (1986).

[2] J. Moss and M. Vaughan, *Adv. Enzymol. Relat. Areas Mol. Biol.* **61,** 303 (1988).

[3] T. Stearns, R. A. Kahn, D. Botstein, and M. A. Hoyt, *Mol. Cell. Biol.* **10,** 12 (1990).

[4] S. A. Rudge, M. M. Cavenagh, R. Kamath, V. A. Sciorra, A. J. Morris, R. A. Kahn, and J. Engebrecht, *Mol. Biol. Cell* **9,** 8 (1998).

[5] C. J. Zhang, M. M. Cavenagh, and R. A. Kahn, *J. Biol. Chem.* **273,** 31 (1998).

[6] A. L. Boman, J. Kuai, X. Zhu, J. Chen, R. Kuriyama, and R. A. Kahn, *Cell Motil. Cytoskeleton* **44,** 2 (1999).

[7] A. L. Boman, C. J. Zhang, X. J. Zhu, and R. A. Kahn, *Mol. Biol. Cell.* **11,** 1241 (2000).

[8] T. Serafini, L. Orci, M. Amherdt, M. Brunner, R. A. Kahn, and J. E. Rothman, *Cell* **67,** 2 (1991).

formation of coated vesicles in transport from Golgi stacks.[9,10] The finding that ARF is a potent activator of PLD1 leads to an alternative proposal in which activation of PLD1 is the initiating event preceding the protein coat recruitment in ARF-regulated vesicle formation.[11]

Membrane traffic is mediated by transport vesicles. This vesicular transport is important not only in cargos that are selectively shuttled between compartments, but also in the maintenance of gradients or differences in lumenal contents or lipid composition of donor and acceptor membranes, in other words in the maintenance of organelle integrity. A number of different protein complexes are reversibly recruited to membranes during vesicle formation. There are currently six different, multimeric, protein complexes designated as coats: COP-I, COP-II, AP-1, AP-2, AP-3, and AP-4. Each of these has been reported to be recruited to membranes by a member of the ARF family; COP-I,[8] AP-1,[12] AP-2,[13] AP-3,[14] and AP-4[15] by an ARF itself and COP-II by the more divergent SAR1 proteins.[16] The activation of ARF is proposed as the initiating event, leading directly to recruitment of coats and later steps that result in the formation of coated vesicles.

With the discovery that ARF proteins are direct activators of PLD1 comes the realization that another model for ARF action in the regulation of coat protein recruitment is also plausible. It is that the activation of PLD1 is the initiating step that leads to the recruitment of coat proteins and later budding and releasing of vesicles from the donor membranes. PLD catalyzes the conversion of phosphatidylcholine (PC) to phosphatidic acid (PA). This model was tested and evidence to support it was provided by Ktistakis et al.[11] The localization of PLD1 to Golgi membranes[17] further strengthens this model.

The functions of ARF proteins have been studied in intact cells by expressing the dominant, activating mutant Q71L, as this renders the protein refractory to the hydrolysis of bound GTP that normally accompanies interaction with ARF GAPs (GTPase-activating proteins). Induction of the expression of [Q71L]ARF1,[18,19]

[9] L. Orci, D. J. Palmer, M. Ravazzola, A. Perrelet, M. Amherdt, and J. E. Rothman, *Nature (London)* **362,** 6421 (1993).

[10] L. Orci, D. J. Palmer, M. Amherdt, and J. E. Rothman, *Nature (London)* **364,** 6439 (1993).

[11] N. T. Ktistakis, H. A. Brown, M. G. Waters, P. C. Sternweis, and M. G. Roth, *J. Cell Biol.* **134,** 2 (1996).

[12] M. A. Stamnes and J. E. Rothman, *Cell* **73,** 5 (1993).

[13] M. A. West, N. A. Bright, and M. S. Robinson, *J. Cell Biol.* **138,** 6 (1997).

[14] C. E. Ooi, E. C. Dell'Angelica, and J. S. Bonifacino, *J. Cell Biol.* **142,** 2 (1998).

[15] J. Hirst, N. A. Bright, B. Rous, and M. S. Robinson, *Mol. Biol. Cell* **10,** 8 (1999).

[16] A. Nakano and M. Muramatsu, *J. Cell Biol.* **109,** 2677 (1989).

[17] N. T. Ktistakis, H. A. Brown, P. C. Sternweis, and M. G. Roth, *Proc. Natl. Acad. Sci. U.S.A.* **92,** 11 (1995).

[18] C. J. Zhang, A. G. Rosenwald, M. C. Willingham, S. Skuntz, J. Clark, and R. A. Kahn, *J. Cell Biol.* **124,** 3 (1994).

[19] C. Dascher and W. E. Balch, *J. Biol. Chem.* **269,** 2 (1994).

[Q71L]ARF3, or [Q71L]ARF4 (C. Zhang and R. A. Kahn, unpublished observations, 2000) in stably transfected NRK cells causes the vesiculation of Golgi stacks and expansion of the endoplasmic reticulum (ER) lumen.[18] The effects on Golgi morphology are so dramatic as to be readily seen by light microscopy. These effects on Golgi and ER structures are assumed to be mediated by the increased stabilization of COP-I binding at the target membranes, but could result from sustained activation of PLD1. Results[20] indicate that neither of these models can explain the effects of activated ARFs on Golgi structure. We have developed a more rapid and quantifiable assay for this effect of activated ARFs on Golgi morphology in intact cells.[20]

The following three assays were developed to allow quantitative comparisons between point mutants in ARF proteins but should be readily adaptable to the study of related proteins or processes.

Methods

Preparation of Purified, Recombinant ADP-Ribosylation Factor

ARF proteins are expressed in bacteria and purified as previously described.[21,22] Typical yields are 5–50 mg of purified ARF per liter of bacterial culture, depending on the isoform and conditions used. Partially N-myristoylated ARF is produced in bacteria by coexpression of ARF with a second plasmid encoding human or yeast N-myristoyltransferase (NMT) as described elsewhere in this volume.[23] The extent of protein myristoylation is about 30%, as determined by reversed-phase high-performance liquid chromatography (HPLC) analysis.[24]

Assay of ADP-Ribosylation Factor-Dependent Phospholipase D Activity

Preparation of Human PLD1. Hexahistidine (His$_6$) tagged, human PLD1 is expressed in Hi5 cells, using a baculovirus construct (the generous gift of S. H. Ryu, Postech University, Pohang, Korea). To eliminate interference by PLD inhibitors and enrich the PLD activity, the PLD1 is extracted from membranes with detergent and enriched on a HiTrap SP column. Fractions containing the PLD activity are pooled and used in the PLD assay. This partially purified PLD1 is stable at $-80°$ for more than 1 year without detectable loss of activity. Activity is linear over a range of $0–3$ μg of protein and typically 0.5 μg is used in each reaction.

[20] J. Kuai, A. L. Boman, B. S. Arnold, X. J. Zhu, and R. A. Kahn, *J. Biol. Chem.* **275,** 6 (2000).

[21] P. A. Randazzo, O. Weiss, and R. A. Kahn, *Methods Enzymol.* **257,** 128 (1995).

[22] P. A. Randazzo and R. A. Kahn, *Methods Enzymol.* **250,** 394 (1995).

[23] H. VanValkenburgh and R. A. Kahn, *Methods Enzymol.* (2000).

[24] P. A. Randazzo, T. Terui, S. Sturch, H. M. Fales, A. G. Ferrige, and R. A. Kahn, *J. Biol. Chem.* **270,** 24 (1995).

1. Baculovirus is amplified in Spodoptera frugiperda (Sf9) cells. Virus from plaques (2 ml) is inoculated into 100 ml of Sf9 cells in midlog phase and grown in Sf-900II SFM (GIBCO-BRL, Gaithersburg, MD). The virus is harvested after 5 days of infection by centrifugation at $1000g$ for 10 min at room temperature to pellet the cell debris.

2. End-point titration of the above-described virus supernatant is performed in a 12-well plate. Each well contains 1 ml of Hi5 cells at a cell density of 5×10^5 cells/ml. Cells are inoculated with 100, 10, 1, and 0 μl of virus supernatant. Three days after infection, cells are collected and PLD activity is measured (see below) in membranes (see steps 4 and 5 for preparation). PLD activity typically is maximal with 10 μl of virus. This titer (10 μl of virus per 5×10^5 cells) can then be scaled up linearly for optimal multiplicity of infection (MOI) for large-scale infection.

3. PLD is expressed in Hi5 cells, grown in EX-CELL 401 (JRH Biosciences, Lenexa, KS) medium. (*Note:* Expression in Hi5 cells gives higher yields of PLD activity than expression in Sf9 cells.) Grow 1 liter of Hi5 cells to a density of 1.0×10^6 cells/ml and add 20 ml (or the appropriate volume) of the above-described viral stock for infection.

4. After 3 days, harvest the infected cells by centrifugation at $3000g$ for 30 min at $4°$. Resuspend the cells in 50 ml of buffer A [20 mM HEPES (pH 7.5), 1 mM dithiothreitol (DTT), 1 mM EDTA] and protease inhibitor cocktails [aprotinin (2 μg/ml), leupeptin (1 μg/ml), pepstatin A (1 μg/ml), 1 mM phenylmethylsulfonyl fluoride (PMSF)]. Cells can be frozen at this point in aliquots and stored at $-80°$ for at least 1 year.

5. An aliquot (5 ml) of cells is broken by nitrogen cavitation in a Parr (Moline, IL) bomb. Pressure (500 lb/in^2) is applied for 25 min with shaking every 5 min. The cells are slowly released from the chamber to avoid bubbles. Centrifuge the cell lysate at $3000g$ for 10 min at $4°$. The membranes are then collected by centrifugation at $100,000g$ for 90 min at $4°$.

6. The PLD1 activity is solubilized by resuspension of the membrane pellet in 4 ml of buffer C [500 mM KCl, 1% (w/v) n-octyl-β-D-galactopyranoside in buffer A] and incubated with shaking at $4°$ for 1 hr. Centrifuge the extract at $100,000g$ for 1 hr at $4°$. The supernatant is then diluted with buffer D [1% (w/v) n-octyl-β-D-galactopyranoside in buffer A] to a final concentration of 100 mM KCl before applying to a 5-ml HiTrap SP column (Amersham Pharmacia Biotech, Piscataway, NJ) previously equilibrated with 100 mM KCl in buffer D. Proteins are eluted with a linear (100–500 mM) gradient of KCl in buffer D over a 25-min period at 1 ml/min and collected in 0.5-ml fractions. The fractions containing PLD activity (fractions 35–45) are pooled to give a final protein concentration of 50 μg/ml. The PLD activity in this preparation is linear over a range of 0–3 μg of total protein and is stable at $-80°$ for more than 1 year without loss of activity after snap freezing in liquid nitrogen.

Phospholipase D Assay. The PLD assay is performed according to a modification of the original method of Brown *et al.*[25,26] using phosphatidylcholine (PC), radiolabeled in the head group, as substrate in synthetic lipid vesicles and monitoring the release of the water-soluble, labeled choline. The ARF is preincubated with GTPγS to equilibrium in the presence of lipid vesicles to give more consistency in quantitation of ARF-dependent PLD activity.

1. Lipid vesicles are prepared by mixing 10 mol% [^{3}H-*methylcholine*]dipalmitoylphosphatidylcholine (DPPC), 86 mol% dioleoylphosphatidylethanolamine (DOPE), and 4 mol% phosphatidylinositol 4,5-bisphosphate (PIP$_2$). The lipid mixture is dried under nitrogen before resuspension in 50 mM HEPES (pH 7.5), 3 mM EGTA, and 80 mM KCl at a final concentration of 690 μM lipid. The lipid vesicles are prepared by sonication for 5 min with a 30-sec pause every 1 min.

2. In a glass test tube, mix myristoylated ARF (myrARF), GTPγS, and lipid vesicles in reaction buffer containing 2.5 mM MgCl$_2$, 1.7 mM CaCl$_2$, 1 mM DTT, 3.5 mM EGTA, 80 mM KCl, 1.2 mM NaCl, and 20 mM HEPES, pH 7.5. The final concentration of GTPγS in the reaction is 30 μM and that of phospholipids is 115 μM. The mixture is incubated at 37° for 40 min before the addition of PLD1.

3. The final reaction volume in the PLD1 assay is 150 μl. Typically, 4 μM myrARF and 0.5 μg of partially purified PLD1 are used in each reaction. Product formation is linear under these conditions for at least 60 min. We typically stop the reaction after 20 min by the addition of 1 ml of chloroform–methanol–HCl (50 : 50 : 0.3, v/v/v), followed by 350 μl of 1 M HCl with 5 mM EGTA. After a brief centrifugation at 3000g for 5 min at 4°, a portion of the aqueous phase (500 μl) is analyzed for the release of [^{3}H]choline by scintillation counting. PLD activity is expressed as the amount of released choline per minute per milligram of partially purified PLD1 protein. Figure 1 shows the ARF and nucleotide dependence of the PLD activity. The half-maximal concentration of ARF3 (EC$_{50}$) is 50–100 nM in this assay, which is similar to that reported for PLD1 purified from bovine brain (20–30 nM).[25]

Coat Protein Recruitment Assay

ARF is found to copurify with coatomer-coated vesicles that are generated *in vitro* by the incubation of Golgi-enriched membranes with cytosol and GTPγS.[8] The ARF dependence of coatomer recruitment can be assayed *in vitro* with purified myrARF, ARF-free cytosol, and Golgi-enriched membranes. The binding of

[25] H. A. Brown, S. Gutowski, C. R. Moomaw, C. Slaughter, and P. C. Sternweis, *Cell* **75**, 6 (1993).
[26] H. A. Brown, S. Gutowski, R. A. Kahn, and P. C. Sternweis, *J. Biol. Chem.* **270**, 25 (1995).

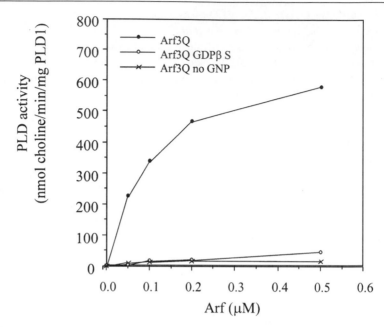

FIG. 1. PLD1 activity is highly dependent on added ARF proteins and GTP. The PLD assay is performed with different amounts of myristoylated [Q71L]ARF3 in the presence of 30 μM GTPγS (filled circles), GDPβS (open circles), or no nucleotides (crosses). Data shown are the average of duplicate samples with a difference of less than 5%.

coatomer to the Golgi membrane is detected by immunoblot using β-COP anti-body (EAGE[27]). Golgi membranes from Chinese hamster ovary (CHO) cells gives a lower background than membranes from other sources, for example, rat liver. The most variable aspect of this assay is the enriched Golgi membranes, as some preparations are less active than others. Controls for electrophoretic transfer to nitrocellulose and an internal control are required for quantification or comparison between experiments.

Preparation of ADP-Ribosylation Factor-Free Bovine Brain Cytosol. Bovine brain cytosol is prepared according to Serafini and Rothman.[28] The cytosol is resolved on a gel-filtration column and fractions containing coatomer proteins are pooled. The presence of either β-COP or ARF proteins is detected by im-munoblotting with β-COP (EAGE) or ARF (1D9) antibodies, respectively, and can be obtained from Affinity BioReagents (Golden, CO) or other commercial sources. Those fractions containing COP subunits are free of ARF proteins. From

[27] R. Pepperkok, J. Scheel, H. Horstmann, H. P. Hauri, G. Griffiths, and T. E. Kreis, *Cell* **74,** 1 (1993).
[28] T. Serafini and J. E. Rothman, *Methods Enzymol.* **219,** 286 (1992).

350 g of frozen bovine brains (Pel-Freez, Rogers, AR) we obtain 140 ml of cytosol with protein at 16.0 mg/ml. After resolution by gel filtration of only 15 ml of this cytosol we obtain 25 mg of ARF-free cytosol. This material is stable for long periods of time after quick freezing and storage at $-80°$.

1. Thaw a frozen bovine brain (350 g; Pel-Freez) in 250 ml of breaking buffer [500 mM KCl, 250 mM sucrose, 25 mM Tris-HCl (pH 8.0), 2 mM EGTA, 1 mM DTT, and protease inhibitor cocktail including aprotinin (2 μg/ml), leupeptin (1 μg/ml), pepstatin A (1 μg/ml), and 1 mM PMSF]. After the tissue is completely thawed, mince it coarsely with scissors and then bring the total volume to 700 ml with breaking buffer. Homogenize in a Polytron blender for 30-sec bursts alternating with 30 sec on ice between bursts.

2. Cell and tissue debris is removed by centrifugation at 9000g for 60 min at 4°. Carefully transfer the supernatant to clean tubes and clarify by centrifugation at 150,000g for 90 min at 4°.

3. Dialyze the supernatant in a Spectra/Por4 (12,000–14,000 cutoff) dialysis tubing four times for a minimum of 2 hr each against 5 liters of dialysis buffer [50 mM KCl, 25 mM Tris-HCl (pH 8.0), 1 mM DTT]. Clarify the dialyzed material by centrifugation and recover the supernatant, being careful not to dislodge the loosely pelleted material.

4. Add ammonium sulfate to 60% saturation slowly, with stirring, and continue stirring for an additional 30 min. Recover the precipitate by centrifugation at 9000g for 30 min at 4°. Resuspend the pellet in dialysis buffer to one-seventh of the original volume ($\sim$100 ml). Remove the remaining ammonium sulfate by dialyzing against dialysis buffer at least twice for >2 hr, the final time in dialysis buffer without DTT. Clarify again by centrifugation at 150,000 g for 90 min at 4°. This material may be directly applied to a gel-filtration column or frozen in liquid nitrogen and stored at $-80°$.

5. Apply 15 ml cytosol to a Sephacryl S-300, HiPrep 26/60 column (Amersham Pharmacia Biotech), previously equilibrated with 50 mM HEPES–KOH (pH 7.4), 200 mM KCl, and 10% (v/v) glycerol. The flow rate is constant at 0.3 ml/min and 4-ml fractions are collected. The presence of COP-I coatomer or ARF proteins is determined by immunoblotting with the EAGE anti-β-COP[27] or monoclonal ARF antibody, 1D9.[29] Fractions 40–52, which are free of ARF and enriched in coatomer, are pooled and snap frozen in aliquots for use in the coatomer recruitment assay. Fractions containing the partially purified ARF may be collected, concentrated, and used as a source of ARF, for example, in a positive control.

[29] M. M. Cavenagh, M. Breiner, A. Schurmann, A. G. Rosenwald, T. Terui, C. Zhang, P. A. Randazzo, M. Adams, H. G. Joost, and R. A. Kahn, *J. Biol. Chem.* **269,** 29 (1994).

Preparation of Golgi-Enriched Membranes. Golgi-enriched membranes are prepared from CHO cells as described in Beckers and Rothman,[30] with the modification that a hypertonic treatment of membranes is added to further reduce the amount of bound ARF and coatomer before use in the recruitment assay.

1. CHO cells at 90% confluence from 20 dishes (150 mm) are suspended with trypsin–EDTA and collected by centrifugation at 800g for 10 min at 4°. Cells are washed once with phosphate-buffered saline (PBS) and once with homogenization buffer [250 mM sucrose, 10 mM HEPES (pH 7.4)]. The final pellet (1.5–2 ml) is resuspended in 10 ml of homogenization buffer.
2. Cells are lysed by one cycle of freezing and thawing and then by five or six passes through a ball-bearing homogenizer [Industrial Tectonics (Ann Arbor, MI), with a 0.1562-in. bore in the stainless steel block and 0.1552-in. stainless steel ball]. The extent of cell breakage can be monitored by examining the cell lysate under a light microscope.
3. Load 6 ml of cell lysate at the bottom of a polycarbonate SW28 tube, and adjust the sucrose concentration to 1.4 M by adding 2 M sucrose and add EDTA to 10 mM. Overlay 15 ml of 1.2 M sucrose in 10 mM HEPES, pH 7.4, and 9 ml of 0.9 M sucrose in 10 mM HEPES, pH 7.4. Mark the interface with a marker pen. Centrifuge the gradients for 2.5 hr at 25,000 rpm at 4° in an SW28 rotor.
4. Collect the membranes at the 0.9–1.2 M interface by puncturing the side wall of the tube with a syringe. From 20 dishes of CHO cells, we typically obtain 3–4 ml of Golgi-enriched membranes with a protein concentration of 0.2–0.4 mg/ml. Membranes are frozen in aliquots. Immediately before use in the recruitment assay the membranes are thawed. An equal volume of 1 M KCl is added to thawed membranes and incubated on ice for 5 min. Then membranes are pelleted by centrifugation at 14,000g for 20 min at 4° and resuspended in the original volume of 0.4 M sucrose in 10 mM HEPES, pH 7.4. The salt-washed membranes contain no detectable endogenous ARF or coatomer proteins (see Fig. 2, lanes 3 and 4) and are used directly in the recruitment assay.

Coatomer Recruitment Assay. Myristoylated ARF proteins and the ARF-free, coatomer-enriched, bovine brain cytosol are cleared by centrifugation at 100,000g for 30 min at 4° in a TLA 100.1 rotor (Beckman, Fullerton, CA) before use in the assay. All reactions are performed in siliconized microcentrifuge tubes. Golgi membranes (10–12 μg of protein per reaction) are incubated with 4 μM myrARF and ARF-free cytosol (0.6 mg/ml) at 37° for 20 min in a final volume of 100 μl in the presence of 25 μM of GTPγS or GDPβS, in a buffer consisting of 2.5 mM MgCl$_2$, 1 mM DTT, 0.2 M sucrose, bovine serum albumin (BSA, 15 μg/ml),

[30] J. M. Beckers and J. E. Rothman, *Methods Enzymol.* **219**, 5 (1992).

myr-Arf3	+	+	-	+	+	+
GTPγS	+	-	+	+	-	+
GDPβS	-	-	-	-	+	-
Golgi membrane	-	+	+	+	+	+
coatomer	+	+	+	-	+	+

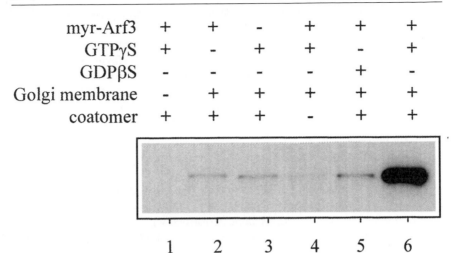

1   2   3   4   5   6

FIG. 2. Recruitment of β-COP to Golgi membrane is dependent on the addition of myristoylated ARF and GTPγS. Additions to the binding reaction are shown at the top, and the presence of coatomer in the membrane pellet is indicated by the immunoreactivity with the EAGE antibody on immunoblots, shown below. Note the requirement for ARF (compare lanes 3 and 6), coatomer (lane 4 vs lane 6), and for GTPγS (lane 5 vs lane 6).

1 mM ATP, 2 mM creatine phosphate, creatine phosphokinase (8 IU/ml), and 20 mM HEPES, pH 7.4. The reaction is terminated by centrifugation of the membranes through a 300-μl cushion of 20 mM HEPES (pH 7.4), 0.5 M sucrose, and 20 mM KCl. The pellet is washed with 20 mM HEPES, pH 7.4, and dissolved in 15 μl of sodium dodecyl sulfate (SDS) sample buffer. The proteins are resolved in gradient (8–16%) polyacrylamide gels and transferred to nitrocellulose membranes to allow immunoblotting, using the rabbit polyclonal β-COP antibody. Immunoreactivity is visualized with horseradish peroxidase-conjugated anti-rabbit IgG antibody (Amersham) and the enhanced chemiluminescence substrate from Amersham.

The ARF and GTPγS dependence of coatomer binding to the Golgi membranes is shown in Fig. 2.

Golgi Vesiculation by Transient Expression of Activated ADP-Ribosylation Factor 3 in Normal Rat Kidney Cells. Dramatic changes in Golgi morphology are originally observed in stably transfected NRK cells capable of inducible expression of the dominant activated mutant [Q71L]ARF1.[18] Alterations in the structure of the Golgi are so dramatic as to be readily observed at the light level, after staining cells with antisera directed against lumenal Golgi proteins, for example, mannosidase II. This observation has been modified for use in characterizing second site mutations in the background of the activated mutant. To screen a relatively large number of second site ARF mutants, in the activated mutant background, we have developed a transient expression system by microinjection into NRK cells of plasmids expressing ARF proteins.

Preparation of Plasmids for Microinjection. The coding region of ARF3 is subcloned into the mammalian expression vector pcDNA3 (Invitrogen, Carlsbad, CA) at the *Nde*I and *Xba*I sites to allow constitutive expression under control of the cytomegalovirus (CMV) promoter. Plasmids are purified with a Qiagen (Valencia, CA) DNA purification kit. Purified plasmids (120 ng/ml), lysine-fixable fluorescein isothiocyanate (FITC)–dextran (1.2 mg/ml; Molecular Probes, Eugene, OR), and 100 mM K$_3$PO$_4$ are mixed in a microcentrifuge tube and clarified at 14,000g for 10 min at 4° right before use.

Preparation of Normal Rat Kidney Cells. NRK cells are grown on coverslips in phenol red-free medium (RPMI medium 1640; GIBCO-BRL). Cells are used after 2–3 days, when the cells reach about 60% confluence.

Microinjection. The injection can be performed with an Eppendorf automatic microinjector system. The needles are prepared with a PC-10 micropipette puller

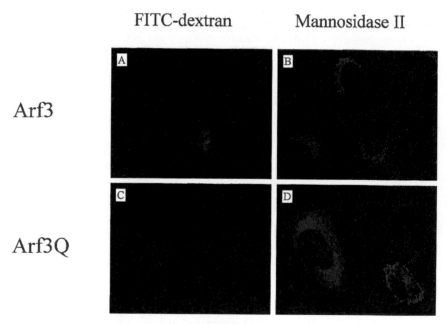

FITC-dextran   Mannosidase II

Arf3

Arf3Q

FIG. 3. Assay of Golgi vesiculation after expression of ARF3 and mutants by microinjection of plasmids into NRK cells. The pcDNA3-derived plasmids are coinjected with FITC–dextran into NRK cells grown on coverslips. (A and C) Injected cells, visualized by green fluorescence of the FITC–dextran. (B and D) Immunostaining of Man II, visualized with Texas Red-conjugated secondary antibody. (D) Expanded, perinuclear staining with Man II antibody resulting from the expression of [Q71L]ARF3. As control, (B) shows tight perinuclear staining of Man II that is not affected by overexpression of wild-type ARF3.

(Narishige International USA, New York, NY). Samples (3 μl) are back-loaded into the needle (outlet diameter is 0.5 ± 0.2 μm). The injection time is 0.5 sec. Injections are directed into the nucleus with the aid of a phase microscope. The injected volume is estimated to be about 50 nl.

After injection the cells are incubated at 37° for 7 hr before indirect immunofluorescence staining is performed with a rabbit polyclonal antibody to mannosidase II (Max II) as primary antibody and Texas Red-conjugated anti-rabbit IgG antibody as secondary antibody. Examples of the control and vesiculated Golgi morphology are shown in Fig. 3.

The vesiculation of Golgi stacks caused by the overexpression of the dominant, activating mutant ([Q71L]ARF3) must be scored within a limited time frame as long exposure to activated ARFs can be lethal to cells. For comparison of mutants it is necessary to optimize for the amount of DNA injected and time course of the effects. As seen in Fig. 4A, increasing the amount of DNA leads to increased effects until a plateau is reached between 80 and 120 μg of DNA. The time dependence for Golgi vesiculation caused by activated ARF3 is shown in Fig. 4B, with a peak at 5–7 hr. At times longer than 10 hr of incubation the number of cells that have been injected with activated ARF3 is decreased. This is likely due to cell death and not diffusion of the indicator as the FITC–dextran is still found in control cells at later times. The optimal response is found with DNA at 120 μg/ml and

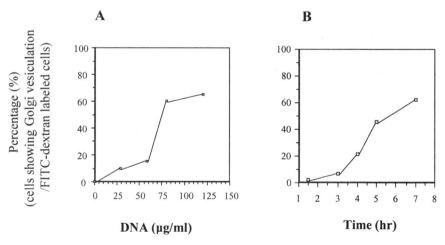

FIG. 4. DNA concentration and time dependence of Golgi vesiculation in NRK cells. The percentage represents cells with enlarged Golgi determined by mannosidase II immunostaining divided by the number of cells labeled with FITC–dextran at different DNA concentrations and different times after injection. (A) Increasing amounts of the pcDNA3-Arf3Q plasmid are coinjected with FITC–dextran in a constant volume and the percentage of injected cells with enlarged Golgi is scored after 7 hr. (B) Cells are fixed at different time points for immunostaining after coinjection of pcDNA3-Arf3Q plasmid (120 μg/ml) with FITC–dextran.

a 7-hr incubation. Under these conditions, the percentage of cells injected with [Q71L]ARF3 showing Golgi vesiculation varies between 40 and 70% in different experiments but can be as low as 20%. This incomplete response is likely due to damage caused by microinjection, missing the nucleus, or other problems. Similar results are found in our laboratory with unrelated proteins with no evident toxicities. The use of earlier passage cells correlates with higher percentages of injected cells expressing ARF3.

Comments

1. The percentage of activated GTPγS-bound ARF is lower under the PLD assay conditions than under conditions optimal for nucleotide exchange because the high concentration of Mg^{2+} dramatically slows GDP dissociation. Under our standard nucleotide exchange assay conditions, GTPγS binding reaches half-maximal levels within 5 min whereas under the PLD assay conditions, the rate is linear with time for at least 1 hr. Because we want to assay the ARF-dependent PLD1 activity, we perform this preloading step to generate activity that is more reflective of the degree of activation of the ARF in the assay.

2. The preparation of lipid vesicles is the most variable aspect of the PLD assay. It is important to use the same batch of lipids and to be uniform in formulating and sonicating the lipid mixture to produce consistent results.

3. While the maximum PLD activity is increased about 3- to 5-fold for myristoylated ARF3, and about 10-fold for myristoylated ARF3[Q71L], nonmyristoylated ARF proteins can also be used to activate PLD. In contrast, there is no detectable recruitment of coatomer with nonmyristoylated ARF proteins up to 4 μM ARF. The recruitment of coatomer to the Golgi membrane is detected with 0.5 μM myristoylated ARF proteins. The amount of recruited coatomer is increased with increased amounts of myristoylated ARF proteins in the assay and saturates above 5 μM ARF.

4. Because the addition of epitope tags at either end of the ARF protein can alter ARF function, untagged forms of ARF proteins are expressed in NRK cells in our assay system. The expression of ARFs from injected plasmids cannot be unambiguously confirmed because of a high level of endogenous ARF proteins, making microinjecting necessary. Transient transfection may be an attractive alternative when assaying for expression of proteins that can be monitored independently of endogenous protein.

[28] Functional Characterization of p115 RhoGEF

By CLARK WELLS, XUEJUN JIANG, STEPHEN GUTOWSKI,
and PAUL C. STERNWEIS

Introduction

The Rho family of monomeric GTPases plays a prominent role in the regulation of cell shape and movement.[1,2] The activity of these proteins is dependent on a nucleotide cycle that is regulated by a combination of guanine nucleotide exchange factors (GEFs) that facilitate exchange of GTP for GDP on the Rho proteins, and GTPase-activating proteins (GAPs) that stimulate hydrolysis of GTP bound to Rho. The GEFs promote activation, that is, formation of the GTP-bound state, whereas proteins with GAP function serve to inactivate the GTPases. The heterotrimeric G proteins that mediate regulation by a variety of extracellular stimuli are also controlled by a GDP/GTP cycle.[3,4] Integral membrane receptors for detection of hormones and other stimuli act as the GEFs to provide stimulatory inputs and a family of RGS (regulator of G protein signaling) proteins[5,6] act as GAPs that can effect either inhibitory or downstream regulation (see Fig. 1).

The GEFs for the Rho family of GTPases compose a large and growing family of proteins characterized by conserved, tandem DH (Dbl homology) and PH (pleckstrin homology) domains.[7,8] These exchange factors stimulate Rho proteins with various selectivities for three defined functional groups represented by RhoA, Rac1, and Cdc42.[8] p115 RhoGEF was purified and cloned as a GEF that was selective for Rho.[9] The subsequent identification of an RGS domain in the N terminus of p115, which had specificity for the G_{12} family of heterotrimeric G proteins, and the regulation of GEF activity by the activated $G\alpha_{13}$ subunit, defined this protein as a direct link for coupling regulation between these two G protein pathways[10,11] (Fig. 1).

[1] A. L. Bishop and A. Hall, *Biochem. J.* **348,** 241 (2000).

[2] K. Kaibuchi, S. Kuroda, and M. Amano, *Annu. Rev. Biochem.* **68,** 459 (1999).

[3] P. C. Sternweis, *in* "Signal Transduction" (C.-H. Heldin and M. Purton, eds.), Chap. 19. Chapman & Hall, London, 1996.

[4] J. R. Hepler and A. G. Gilman, *Trends Biochem. Sci.* **17,** 383 (1992).

[5] J. R. Hepler, *Trends Pharmacol. Sci.* **20,** 376 (1999).

[6] L. De Vries, B. Zheng, T. Fischer, E. Elenko, and M. G. Farquhar, *Annu. Rev. Pharmacol. Toxicol.* **40,** 235 (2000).

[7] R. A. Cerione and Y. Zheng, *Curr. Opin. Cell Biol.* **8,** 216 (1996).

[8] L. Kjoller and A. Hall, *Exp. Cell Res.* **253,** 166 (1999).

[9] M. J. Hart, S. Sharma, N. El-Masry, R. G. Qiu, P. McCabe, P. Polakis, and G. Bollag, *J. Biol. Chem.* **271,** 25452 (1996).

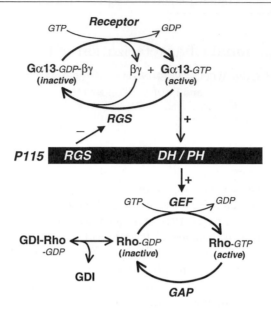

FIG. 1. Both heterotrimeric G proteins (top) and monomeric GTPases (bottom) are regulated by a guanine nucleotide cycle. p115 RhoGEF acts as both a downstream effector and an inhibitory regulator of the heterotrimeric G_{13} protein. See text for further details.

This article describes the preparation of two substrates ($G\alpha_{13}$ and RhoA) and four assays for *in vitro* assessment of the various functions of p115 RhoGEF and similar molecules.

Reagents and Solutions

Protease Inhibitors

Where indicated, protease inhibitors (PIs) are added to solutions to achieve the following final concentrations (N^{α}-p-tosyl-L-lysine chloromethyl ketone, 21 μg/ml; tosylphenylalanyl chloromethyl ketone, 21 μg/ml; phenylmethylsulfonyl fluoride, 21 μg/ml; leupeptin, 12.5 μg/ml; pepstatin A, 1 μg/ml; N^{α}-p-tosyl-L-arginine methyl ester, 21 μg/ml). The inhibitors are made and stored as stock solutions at 1000× final concentrations.

Detergents

Detergents used are as follows: n-octyl-β-D-glucopyranoside (OG), polyoxyethylene 10 lauryl ether ($C_{12}E_{10}$; Sigma, St. Louis, MO), sodium cholate

[10] T. Kozasa, X. Jiang, M. J. Hart, P. M. Sternweis, W. D. Singer, A. G. Gilman, G. Bollag, and P. C. Sternweis, *Science* **280,** 2109 (1998).
[11] M. J. Hart, X. Jiang, T. Kozasa, W. Roscoe, W. D. Singer, A. G. Gilman, P. C. Sternweis, and G. Bollag, *Science* **280,** 2112 (1998).

(repurified cholic acid that is neutralized to pH 6 with NaOH as a stock solution), Triton X-100.

Solutions

Solution A: 50 mM Na–HEPES (pH 7.5), 1 mM EDTA, 1 mM dithiothreitol, 100 mM NaCl

Solution B: 20 mM Na–HEPES (pH 7.5), 1 mM EDTA, 1 mM dithiothreitol, 100 mM NaCl

Solution C: 20 mM Na–HEPES (pH 7.5), 1 mM EDTA, 1 mM dithiothreitol, 300 mM NaCl, peptide EYMPME, synthesized locally (100 μg/ml)

Solution D: 20 mM Na–HEPES (pH 8.0), 10 mM 2-mercaptoethanol, 100 mM NaCl, 0.1 mM GDP

Solution E: 20 mM Na–HEPES (pH 8.0), 10 mM 2-mercaptoethanol, 300 mM NaCl, 0.1 mM GDP

Solution F: 20 mM Na–HEPES (pH 8.0), 10 mM 2-mercaptoethanol, 200 mM NaCl, 0.1 mM GDP

Solution G: 20 mM Na–HEPES (pH 8.0), 10 mM 2-mercaptoethanol, 100 mM NaCl, 0.1 mM GDP, 10 mM NaF, 10 mM MgCl$_2$, 0.1 mM AlCl$_3$

Solution H: 20 mM Na–HEPES (pH 8.0), 1 mM dithiothreitol, 200 mM NaCl, 0.8% (w/v) OG, 10 mM NaF, 5 mM MgCl$_2$, 0.1 mM AlCl$_3$

Solution I: 50 mM Na–HEPES (pH 7.5), 1 mM EDTA, 1 mM dithiothreitol, 50 mM NaCl, 5 mM MgCl$_2$, 10 $\mu$$M$ GDP

Solution J: 50 mM Na–HEPES (pH 7.5), 2 mM EDTA, 1 mM dithiothreitol, 250 mM NaCl, and 1% (w/v) sodium cholate

Solution K: 50 mM Na–HEPES, 1 mM EDTA, 1 mM dithiothreitol, 200 mM NaCl, and 5 mM MgCl$_2$

Solution L: 50 mM Na–HEPES (pH 7.5), 8 mM EDTA, 2 mM dithiothreitol, 100 mM NaCl, 1% (w/v) OG

Solution M: 50 mM Na–HEPES (pH 7.5), 1 mM EDTA, 50 mM NaCl, 5 mM MgCl$_2$, and 0.8% (w/v) OG

Solution N: 20 mM Tris-HCl (pH 8.0), 25 mM MgCl$_2$, 100 mM NaCl

Solution O: 160 mM Na–HEPES (pH 8.0), 40 mM EDTA, 8 mM dithiothreitol, 0.4% (v/v) C$_{12}$E$_{10}$

Solution P: 25 mM Na–HEPES (pH 8.0), 5 mM EDTA, 1 mM dithiothreitol, 0.05% (v/v) C$_{12}$E$_{10}$

Preparation of Protein

Expression of Proteins in Spodoptera frugiperda (Sf9) Cells

Most of the proteins utilized in this article require expression in eukaryotic systems. A general protocol for their expression using a baculovirus system follows.

Baculovirus Stocks

Baculovirus stocks are routinely prepared by infection of 50 ml of Sf9 cells in suspension culture (1×10^6 cells/ml) with 0.5 ml of a previous stock. The cells are incubated in suspension culture for 6 days. Debris is removed by centrifugation and the supernatant is retained as the viral stock.

Growth and Infection of Sf9 Cells

Growth of Sf9 cells is done at 28° in suspension culture with complete IPL-41 medium [450 ml of IPL-41, 50 ml of heat-inactivated fetal bovine serum, 5 ml of 10% (v/v) Pluronic F68, 0.5 ml of gentamicin (50 μg/ml)]. Prepare a 500-ml seed culture by growing Sf9 cells for 2–3 days from a starting density of $0.5–1 \times 10^6$ cells/ml. On day 1, expand the seed culture into several 800-ml cultures (4-liter flasks) with medium to a density of about 0.8×10^6 cells/ml. Grow cells in a rotating shaker (about 150 rpm) for 24 hr (cell density doubles). Infect each culture flask (800-ml culture) with viral stocks as indicated (usually 1 ml of stock per 100 ml of culture) and grow the infected cells for 48 hr. Harvest the cells (day 4) by centrifugation at 2000 rpm and 4° for 20 min in a JA-10 rotor (Beckman, Fullerton, CA). The supernatant is discarded and the pelleted cells are washed by suspension with a solution containing 137 m*M* NaCl, 2.7 m*M* KCl, 1.5 m*M* KH$_2$PO$_4$, and 8 m*M* Na$_2$HPO$_4$ (adjusted to pH 7.4) and collected by centrifugation as described.

p115 RhoGEF

The baculovirus construct for p115 RhoGEF contains an EE tag (EYMPME) at the N terminus of the protein for purification via an antibody affinity column.[9] The baculovirus can be obtained from Onyx Pharmaceuticals (Richmond, CA) and the anti-EE resin from BAbCo (Berkeley, CA) (affinity matrix, AFC-115P; Covance, Princeton, NJ). Express p115 RhoGEF with 1.6 liters of Sf9 cells as described above. Collect cells after expression and use directly or freeze as a pellet in liquid nitrogen for storage.

Disperse pellets of cells in 40 ml of solution A containing PIs. Lyse the cells with three cycles of rapid freeze–thawing [alternatively, cells can be lysed by nitrogen cavitation with a Parr (Moline, IL) bomb]. Remove particulate material from the lysed mixture by centrifugation at 100,000*g* for 45 min at 4°. Collect the supernatant (cytosol) and supplement with Triton X-100 to a final concentration of 0.75% (v/v); the detergent facilitates further isolation of protein.

Mix the cytosol (approximately 40 ml containing p115 RhoGEF) with 1 ml of anti-EE resin and incubate with mixing (a rocking platform is convenient) for 2 hr at 4°. After incubation, pour the mixture into a 0.6×10 cm column (Econo column; Bio-Rad, Hercules, CA) to collect the resin with bound p115 RhoGEF. Wash the retained resin at 4° with 10 column volumes of solution B containing PIs.

For elution, add 1 ml of solution C containing PIs to the resin and allow the flow to continue until about 100 μl remains above the resin. Stop the flow and incubate the resin (column) for 10 min at room temperature to allow exchange of peptide for protein on the immobilized antibodies. Restore the flow and collect six 500-μl fractions by passing an additional 2–3 ml of solution C slowly through the resin at room temperature. Place the collected fractions on ice. The bulk of the p115 RhoGEF is usually contained in fractions 2 and 3, but this will vary with dead space in the column and tubing. Fractions containing p115 RhoGEF can be pooled, aliquoted, and frozen for storage or further purified.

Approximately 70–80% of the p115 RhoGEF is recovered as full-length protein (FL-p115); about 20–30% of the protein is cleaved by proteolysis into two pieces. The N-terminal fragment containing about the first 250 amino acids of p115 is retained and eluted from the anti-EE matrix most efficiently and is the dominant contaminant. Isolate FL-p115 from fragments and other minor impurities by passage of 1 ml of the eluate from the anti-EE resin through a 75-ml column of AcA44 (BioSepra, Marlborough, MA) that is equilibrated and eluted with solution B. Collect fractions of 2 ml; FL-p115 elutes in fractions 10–15 (essentially excluded from the resin). Pool the fractions containing the desired protein, concentrate to 0.5–1 ml by pressure filtration on a PM30 membrane (Amicon, Danvers, MA), and supplement with PIs. Aliquot the concentrated FL-p115, freeze in liquid nitrogen, and store at $-80°$. A typical yield of purified protein from 1 liter of Sf9 cells is about 1 mg. The final step of gel filtration also removes EE–peptide from the preparation.

$G\alpha_{13}$

Routine preparations utilize coexpression of $G\alpha_{13}$ with tagged $\beta\gamma$ subunits as described previously for G_{12}.[12] Express the G protein in 3.2 liters of Sf9 cells as described above. Coinfect each 800-ml flask of cells with stock cultures of baculoviruses encoding the $G\alpha_{13}$ (12 ml/flask), $G\beta_1$ (8 ml/flask), and $G\gamma_2$ N-terminal hexahistidine (His_6) tag (6 ml/flask) subunits.

Suspend the cells collected from 3.2 liters of Sf9 expression cultures in 600 ml of ice-cold solution D (lysis buffer) containing fresh PIs. Unless indicated, subsequent procedures are performed at $4°$.

Lyse the cells by nitrogen cavitation (expose the cells in a Parr bomb to 500 lb/in^2 for 45 min; then release the cells through the outlet slowly over several minutes). Remove intact cells and nuclei from the lysates by centrifugation at 1400 rpm for 10 min in a JA6 rotor (Beckman). Separate the resulting supernatant into cytoplasmic and particulate fractions by centrifugation at 35,000 rpm for 30 min in a Ti45 rotor (Beckman). Discard cytoplasmic proteins (supernatant) and suspend the membranes (pellet) with 100 ml of ice-cold solution D containing PIs by repeated passage through a 20-gauge needle with syringe. Membranes are usually frozen in liquid nitrogen and stored at $-80°$.

---

[12] T. Kozasa and A. G. Gilman, *J. Biol. Chem.* **270,** 1734 (1995).

Thaw the membranes (1–1.5 g of protein) and dilute to 5 mg of protein/ml with solution D containing PIs. Add 0.05 volume of 20% (w/v) sodium cholate (to give ~ 1%) and stir on ice for 1 hr. Remove the membrane remnants by centrifugation at 35,000 rpm for 30 min. Discard the pellet and dilute the supernatant with 4 volumes of solution D containing 0.5% (v/v) $C_{12}E_{10}$ and PIs. If cloudiness develops on dilution, clarify the extract by repeating the centrifugation.

Apply the extracted proteins to a 5-ml column of Ni^{2+}-chelated resin (Ni–NTA–agarose; Qiagen, Valencia, CA) to collect the G protein heterotrimer via the His_6 tag on the γ subunit. Wash the resin sequentially with 30 ml of solution E containing 0.5% (v/v) $C_{12}E_{10}$ and 5 mM imidazole and 30 ml of solution F containing 1% (w/v) OG and 15 mM imidazole.

Warm the resin to room temperature for 15 min and continue elutions with solutions at room temperature but collect fractions of 3 ml on ice. Wash the column sequentially with 15 ml of solution F containing 1% (w/v) OG and 15 mM imidazole; 18 ml of solution G containing 1% (w/v) OG and 10 mM imidazole; 9 ml of solution G containing 0.9% (w/v) OG, 0.1% (w/v) sodium cholate, and 10 mM imidazole; 9 ml of solution G containing 0.7% (w/v) OG, 0.3% (w/v) sodium cholate, and 10 mM imidazole; 9 ml of solution G containing 1% (w/v) sodium cholate and 10 mM imidazole; and 9 ml of solution G containing 1% (w/v) sodium cholate and 150 mM imidazole. The addition of aluminum, fluoride, and magnesium initiates dissociation of $G\alpha_{13}$ from the $\beta\gamma$ subunits, increasing concentrations of cholate enhance this dissociation, and the final high concentration of imidazole effects elution of the bound $\beta\gamma$ subunits. Adjustments in imidazole, NaCl, and detergents are optimized to remove impurities.

Analyze fractions by separation on sodium dodecyl sulfate (SDS)–polyacrylamide gels (about 10 μl of fractions) and staining with silver. If $G\alpha_{13}$ is pure, prepare for storage. Otherwise, proceed with further enrichment on hydroxylapatite. Prepare a 1-ml column of hydroxylapatite (BioGel HTP; Bio-Rad) and equilibrate with solution H. Pool the fractions containing $G\alpha_{13}$ from the Ni–NTA column and slowly apply to the hydroxylapatite at 4°. Wash the resin with 5 ml of solution H and then elute with ten 300-μl aliquots of solution H containing 100 mM KPO_4 (pH 8.0). Add GDP to each fraction to achieve a concentration of 10 μM. Analyze the fractions for the presence of $G\alpha_{13}$, pool, and concentrate the appropriate fractions with a Centricon PM30. Freeze the purified protein in liquid nitrogen for storage at −80°.

Prenylated RhoA

Prenylated RhoA is prepared after coexpression of the GTPase in Sf9 cells with glutathione S-transferase (GST)–RhoGDI. RhoGDI (Rho guanine nucleotide dissociation inhibitor) is a protein that binds selectively to the prenylated form of Rho and maintains the GTPase as a soluble complex in cytosol.[13] The fusion protein, in which GST is fused to the N terminus of RhoGDI, then allows selective formation of complexes with the modified forms of the Rho proteins and their isolation by affinity techniques.

Infect 1.6 liters of Sf9 cells with 16 ml each of amplified baculoviruses encoding RhoA and GST–GDI; express the proteins for 48 hr and collect cells as described. Suspend the cell pellet in 35 ml of solution I containing 10 μM GDP and PIs. Cells are lysed by either three cycles of rapid freezing and thawing or by nitrogen cavitation with a Parr bomb. Clear cell debris by centrifugation at 100,000g for 45 min at 4°.

Pass the supernatant (clarified cytosol containing the RhoA/GST–GDI complex) through a 1-ml column of glutathione–Sepharose at 4° at a rate of <1 ml/min to allow time for binding. Wash the resin with 20 ml of solution I containing 10 μM GDP and elute the RhoA with ten 1-ml aliquots of solution J. The detergent lowers the affinity of Rho for GDI.

Pool fractions containing RhoA, add 50 ml of buffer I containing 10 μM GDP and 0.8% (w/v) OG, and concentrate the mixture to 1–2 ml by pressure filtration with a PM10 membrane (Amicon). Dilute the concentrated proteins with 50 ml of solution I containing 10 μM GDP and 0.8% (w/v) OG and repeat the concentration. This step serves to replace the cholate with OG as well as to concentrate the protein.

Aliquots of isolated RhoA are frozen in liquid nitrogen and can be stored at −80° for at least 2 years. The yield from 1.6 liters of cultured Sf9 cells is typically 2–3 mg of purified RhoA. The advantage of this procedure is the procurement of native (untagged) prenylated proteins (preparations of prenylated Rac 1 and Cdc42 can also be obtained by this procedure, although with lower yields). A disadvantage for some experiments is a minor but significant (2–10%) contamination of the preparations with endogenous Sf9 Rho GTPases. This can lead to misinterpretation of experiments that test for specificity within the Rho GTPase family, or can potentially cause variance in measurements of kinetic parameters.

Nonprenylated RhoA

The nonprenylated form of RhoA is made as a fusion protein with glutathione *S*-transferase via the pGEX-KG vector and expression in BL21 (DE3) cells (Novagen, Madison, WI). Use a fresh overnight culture of cells containing the RhoA plasmid to inoculate 1 liter of LB medium containing ampicillin (50 $\mu g/ml$). Grow the cells at 37° until the OD$_{600}$ reaches 0.5. Add isopropyl-β-D-thiogalactopyranoside (IPTG) to a final concentration of 200 μM to induce expression and grow the bacteria for an additional 6 hr at 30°. Collect the cells by centrifugation at 20,000g for 20 min. Pelleted cells are frozen in liquid nitrogen and stored at −80°.

Suspend cells from 1 liter of culture in 20 ml of solution K. Add lysozyme to a final concentration of 1 mg/ml and incubate at 4° for 10 min with intermittent

---

[13] B. Olofsson, *Cell. Signal.* **11**, 545 (1999).

mixing. The lysed cells should become viscous. Sonicate the lysate [e.g., 2 min with a Branson (Danbury, CT) probe sonifier 450 with a duty cycle of 50% and an output setting of 8] to ensure complete breakage of cells and shearing of DNA or add DNase to 20 μg/ml and incubate for an additional 10 min. Fractionate the lysate by centrifugation at 100,000g for 30 min at 4° and collect the supernatant containing GST–RhoA.

Apply the supernatant to a 1-ml column of glutathione–Sepharose and wash with 20 ml of solution K. Elute the resin with seven 1-ml aliquots of solution K containing 10 mM reduced glutathione. RhoA is usually eluted in fractions 2–5, which are pooled and diluted with 50 ml of solution K and concentrated to 1 ml. GDP is added to a final concentration of 10 μM to stabilize the protein.

The yield of GST–RhoA from 1 liter of cells is typically about 2 mg. Concentrated GST–RhoA is aliquoted, frozen in liquid nitrogen, and stored at −80°.

Assays of Guanine Nucleotide Exchange

Principle

The presumed primary activity of p115 RhoGEF is to facilitate exchange of nucleotide onto the guanine nucleotide-binding site of Rho. This is accomplished by stimulating the rate of dissociation of GDP from Rho under physiological conditions and thus allow binding of the activating ligand, GTP. Two strategies for measuring this exchange are to directly measure the dissociation of prebound [^{3}H]GDP or to assess this rate indirectly by measuring binding of [^{35}S]GTPγS.

Loading of RhoA with [^{3}H]GDP

Place 100 μCi of [^{3}H]GDP (New England Nuclear, Boston, MA) in a 12×75 glass test tube and dry under a stream of nitrogen on ice (about 1 hr). Dissolve the [^{3}H]GDP in 150 μl of solution L. Adjust the volume to 300 μl by adding unlabeled GDP to a final concentration of 25 μM, RhoA to a final concentration of 2.5 μM (750 pmol), and water. Mix well and incubate for 1 min at 30°. Add MgCl$_2$ to a final concentration of 10 mM and incubate for an additional 10 min. Use 5-μl aliquots to measure total counts per minute and for filter binding to measure bound nucleotide.

Separate free nucleotide from bound nucleotide by passage through two spin columns. Prepare two 2-ml volumes of Sephadex G-50 in 0.8×4 cm columns (PolyPrep; Bio-Rad) and equilibrate with buffer M. After equilibration, place 1.5-ml Eppendorf tubes in the bottom of 17×82 mm test tubes (Sarstadt, Newton, NC), which fit in the buckets of a TJ-6 rotor (Beckman). Place the columns containing Sephadex in the tubes such that they empty into the Eppendorf tubes and spin the assemblies for 4 min at 1000 rpm. Change Eppendorf tubes to collect

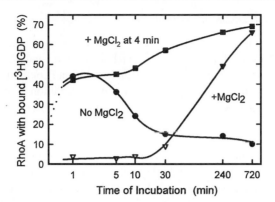

FIG. 2. Loading RhoA with [³H]GDP. The conditions used are those described in text; the amount of [³H]GDP bound is expressed as a percentage of the total binding sites present. In the absence of divalent metal (circles), the exchange rate (limited by dissociation of GDP) is rapid. Presumably because of this low affinity, however, the protein is unstable and sites are lost with extended incubation. In the presence of Mg^{2+} (triangles), exchange is slow but maximal exchange can be achieved. Sites that appear to be lost in the absence of Mg^{2+} can be slowly recovered in the presence of the metal. Rapid exchange that is achieved in the absence of Mg^{2+} can be stabilized by subsequent addition of the metal (squares).

samples. Split the loaded Rho sample into two equal aliquots and apply to the tops of the two columns. Repeat centrifugation and collect the filtrate. Add 180 μl of solution M to the columns, spin, and combine the filtrates containing RhoA. Free GDP remains in the resin. This step serves to reduce the background by removing free GDP and exchanges the RhoA into a solution desired for assay. This solution can be changed as desired. The presence of detergent is primarily for working with prenylated RhoA.

The dissociation of GDP from RhoA is rapid in the absence of Mg^{2+} but slow in its presence.[14] An example of loading is shown in Fig. 2. In the absence of Mg^{2+}, exchange is rapid but unstable, presumably because of partial denaturation of Rho under these conditions. If Mg^{2+} is present, exchange is slow but complete and stable. For convenience, loading is usually done by a short incubation (1–4 min) in the absence of divalent metal, followed by addition of Mg^{2+} to stabilize the protein.

The [³H]GDP is initially dried to concentrate the nucleotide and to remove [³H]H$_2$O and other volatile components. The mass of the [³H]GDP is significant (20–30 μM) and needs to be taken into account for calculating specific binding; additional but less significant amounts of unlabeled GDP in the reaction come from endogenous nucleotide bound to RhoA and any free GDP present in the preparation of Rho.

[14] A. J. Self and A. Hall, *Methods Enzymol.* **256**, 67 (1995).

Regulation of GDP Dissociation from RhoA

The dissociation assay is usually done in polypropylene tubes (12×75 mm) with $0.1 \ \mu M$ RhoA in a final volume of 20 μl of solution I and 100 μM cold GDP. Dilute activators or inhibitors of the reaction in solution I containing 10 μM GDP, mix in assay tubes on ice, and adjust the volume with solution I containing GDP to 18 μl. Add 2 μl of RhoA (1 μM, 5000–7000 cpm of bound [^{3}H]GDP) and incubate at 30°. Stop assays at the desired time by addition of 2 ml of ice-cold solution N. Separate the bound GDP from free GDP by vacuum filtration through a BA-85 nitrocellulose filter (Intermountain Scientific, Kaysville, UT). After application of sample, rinse the assay tube with 2 ml of solution N and add to the filter. Wash the filter four times with additional 2-ml aliquots of solution N. Dry the filters under a heat lamp for approximately 10 min and assess radioactivity by placing the filter in 5 ml of scintillation fluid and counting.

Regulation of GTPγS Binding to RhoA

Binding of [^{35}S]GTPγS to RhoA is done in polypropylene tubes (12×75 mm) and in 20 μl of solution I. Dilute activators or inhibitors in solution I and mix on ice. Add the reaction components in the following order: the amount of solution I needed to adjust the reaction to 20 μl; GTPγS (including about 200,000 cpm of [^{35}S]GTPγS) in solution I to achieve a final concentration of 10 μM; diluted activators and inhibitors. Start the binding reaction with the addition of RhoA to achieve a final concentration of 1 μM. Incubate at 30° for the desired time. Stop the reactions by the addition of 2 ml of ice-cold solution N. Rapidly filter the samples as described in the previous section and analyze dried filters for [^{35}S]GTPγS bound to RhoA.

Discussion

All preparations of G proteins should be assessed for function, usually their ability to bind GTPγS. The exchange assays described above are optimized to measure stimulation of guanine nucleotide exchange on Rho by GEFs. Although they can be used to measure maximal binding sites in preparations of Rho, especially in the presence of a potent GEF, more optimal conditions are described elsewhere.[14,15] Measurement of nucleotide binding to Gα_{13} is problematic and is discussed in the literature.[16]

P115 RhoGEF is an effective stimulator of guanine nucleotide exchange on Rho. However, its potency varies with the prenylation state of the GTPase. Whereas 0.3 nM p115 RhoGEF will cause stimulation of 50% of prenylated Rho within

[15] T. Mizuno, H. Nakanishi, and Y. Takai, *Methods Enzymol.* **256**, 15 (1995).
[16] W. D. Singer, R. T. Miller, and P. C. Sternweis, *J. Biol. Chem.* **269**, 19796 (1994).

10 min at 30°, 10–100 times as much p115 RhoGEF is required to obtain a similar stimulation of RhoA expressed in bacteria. This appears to be selective for the p115 interaction as the basal rates of exchange for Rho from both sources are similar.

$G\alpha_{13}$ has been shown to stimulate the exchange activity of p115 RhoGEF.[11] The easiest way to handle this reaction is to stimulate the $G\alpha_{13}$ with aluminum and fluoride. This is rapid and stabilizes binding of GDP on the G protein; the latter effect prevents binding of labeled nucleotide on $G\alpha_{13}$ and thus a potential correction when examining binding to RhoA. $G\alpha_{13}$ activated by preincubation with GTPγS can be assayed for stimulation of p115 by measuring dissociation of [^{3}H]GDP from RhoA.

Assay of GTPase Activity

Loading of $G\alpha_{13}$ with [γ-^{32}P]GTP

Association of $G\alpha_{13}$ with [γ-^{32}P]GTP is typically done in 150 μl and is prepared immediately before running GTPase assays. Because the intrinsic rate of $G\alpha_{13}$ GTPase activity causes rapid hydrolysis of bound GTP relative to rates of exchange, this procedure should be done rapidly and GTPase assays run within 30 min of the loading reaction. To a reaction tube on ice, add 18 μl of solution O, 15 μl of 50 μM GTP, 7.5×10^7 cpm of [γ-^{32}P]GTP (to give $\sim$100 cpm/fmol), and $G\alpha_{13}$ for a final concentration of 2 μM. Adjust the volume to 150 μl with water. Incubate the mixture for 15 min at 30°. Have prepared a 2-ml column of Sephadex

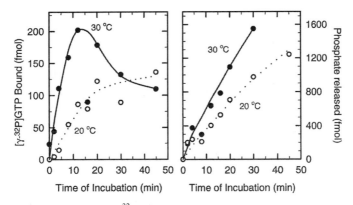

FIG. 3. Loading of $G\alpha_{13}$ with [γ-^{32}P]GTP. A time course for binding of the nucleotide was done as described; each time point contained 10 pmol of $G\alpha_{13}$. Duplicate samples were also analyzed for the presence of inorganic phosphate. The loading is a contest between slow rates of exchange and substantial intrinsic GTPase activity. While loading is done in the absence of Mg^{2+} the ability of the GTP-bound form of $G\alpha_{13}$ to scavenge the metal and effect hydrolysis of the nucleotide is exceptional. Therefore, the amount of bound GTP that can be obtained at equilibrium is small.

G-50 in a 0.8 × 4 cm column (PolyPrep; Bio-Rad) and equilibrated with solution P. When 6 min remains in the incubation of $G\alpha_{13}$, spin the column at 4° and 1000 rpm for 4 min in a TJ6 centrifuge to remove free liquid as described above for loading Rho with [^{3}H]GDP. After incubation, apply the $G\alpha_{13}$ to the Sephadex G-50 column, repeat the spin, and collect the eluate containing $G\alpha_{13}$ with bound GTP in a 1.5-ml Eppendorf tube; keep on ice.

Figure 3 shows a time course for loading GTP to $G\alpha_{13}$. While the rate of exchange is reasonably fast, the intrinsic rate of hydrolysis is faster, even in the apparent absence of Mg^{2+}. The left- and right-hand panels of Fig. 3 show GTP bound to $G\alpha_{13}$ and released inorganic phosphate, respectively. At maximal loading, about 0.5–1% of the $G\alpha_{13}$ contains bound [γ-^{32}P]GTP.

Assay for GAP Activity on $G\alpha_{13}$

These assays should be planned and set up before loading of $G\alpha_{13}$ with [γ-^{32}P]GTP. Prepare stop tubes (12 × 75 tubes) containing 750 μl of 50 mM NaH_2PO_4 (pH 7) containing 5% (w/v) activated charcoal (100-400 mesh; Sigma) as a suspension. Assays are usually done at 4° and in larger volumes (330 μl) for the removal of several aliquots over a time course. In 12 × 75 mm polypropylene tubes, add 3 μl of 1 M $MgCl_2$, 33 μl of 10 mM GTP, 10–20 μl of an RGS (regulator of G protein signaling) preparation or other protein (diluted with solution P), and solution L to a final volume of 297 μl. Start the reaction by the addition of 33 μl of $G\alpha_{13}$ loaded with [γ-^{32}P]GTP and incubation at 4°. At the desired times, transfer 50-μl aliquots to tubes containing activated charcoal and vortex vigorously to stop the reaction. Leave these tubes at 4°. Mix a 5-μl aliquot of the loaded $G\alpha_{13}$ preparation with charcoal to determine free phosphate at time 0. After completion of assays, centrifuge the stopped reactions at 3000 rpm for 10 min in a J6 centrifuge to remove charcoal. Withdraw 0.5 ml of the supernatant and mix with 5 ml of scintillation fluid to analyze for ^{32}P.

The charcoal absorbs GTP and protein to prevent further hydrolysis; inorganic phosphate released during the assay remains in solution. A low temperature is used to slow the intrinsic GTPase activity of $G\alpha_{13}$ and allow better observation of stimulated activities. Examples of stimulation by the RGS domain from p115 RhoGEF can be found in the literature.[10]

Acknowledgments

This work was supported by the National Institutes of Health (GM 31954), the Robert A. Welch Foundation, and the Alfred and Mabel Gilman Chair in Molecular Pharmacology.

[29] Nonisotopic Methods for Detecting Activation of Small G Proteins

By KENDALL D. CAREY and PHILIP J. S. STORK

Introduction

Rap1 proteins are members of the Ras superfamily of small G proteins and cycle between the GDP- and GTP-bound state.[1] In the GTP-bound state, Rap1 is biologically active and is capable of binding with high affinity to several downstream signaling molecules, including Raf-1, B-Raf, and RalGDS.[2–4] Activated Rap1 regulates several signaling pathways, including the mitogen-activated protein kinase cascade,[4–7] and plays a role in regulating cellular adhesion.[8–10] A number of stimuli that may regulate cell growth or differentiation utilize Rap1 signaling to achieve specific physiological effects, including T cell receptor activation,[11] B cell receptor activation,[6] growth factor action,[12–14] and hormonal stimulation.[4,7,15,16] All these stimuli increase levels of GTP-bound Rap1 through the activation of Rap1-specific guanine nucleotide exchange factors (GEFs), many of which are themselves regulated by multiple intracellular signals.[17,18] Therefore, the ability to measure Rap1 activation may provide a useful tool in the examination of a number of biological systems.

[1] F. J. Zwartkruis and J. L. Bos, *Exp. Cell Res.* **253,** 157 (1999).

[2] J. L. Bos, *EMBO J.* **17,** 6776 (1998).

[3] A. Wittinghofer and C. Herrmann, *FEBS Lett.* **369,** 52 (1995).

[4] M. Vossler, H. Yao, R. York, C. Rim, M.-G. Pan, and P. J. S. Stork, *Cell* **89,** 73 (1997).

[5] H. Kitayama, T. Matsuzaki, Y. Ikawa, and M. Noda, *Proc. Natl. Acad. Sci. U.S.A.* **87,** 4284 (1990).

[6] S. J. McLeod, R. J. Ingham, J. L. Bos, T. Kurosaki, and M. R. Gold, *J. Biol. Chem.* **273,** 29218 (1998).

[7] N. Mochizuki, Y. Ohba, E. Kiyokawa, T. Kurata, T. Murakami, T. Ozaki, A. Kitabatake, K. Nagashima, and M. Matsuda, *Nature (London)* **400,** 891 (1999).

[8] K. Katagiri, M. Hattori, N. Minato, S. Irie, K. Takatsu, and T. Kinashi, *Mol. Cell. Biol.* **20,** 1956 (2000).

[9] K. A. Reedquist, E. Ross, E. A. Koop, R. M. Wolthuis, F. J. Zwartkruis, Y. van Kooyk, M. Salmon, C. D. Buckley, and J. L. Bos, *J. Cell Biol.* **148,** 1151 (2000).

[10] C. S. Buensuceso and T. E. O'Toole, *J. Biol. Chem.* **275,** 13118 (2000).

[11] K. D. Carey, J. M. Schmitt, A. M. Baird, T. J. Dillon, A. D. Holdorf, A. S. Shaw, and P. J. S. Stork, *Mol. Cell. Biol.* **20,** 8409 (2000).

[12] R. D. York, H. Yao, T. Dillon, C. L. Ellig, S. P. Eckert, E. W. McCleskey, and P. J. S. Stork, *Nature (London)* **392,** 622 (1998).

[13] H. Yao, R. D. York, A. Misra-Press, D. W. Carr, and P. J. S. Stork, *J. Biol. Chem.* **273,** 8240 (1998).

[14] L. Xing, C. Ge, R. Zeltser, G. Maskevitch, B. J. Mayer, and K. Alexandropoulos, *Mol. Cell. Biol.* **20,** 7363 (2000).

[15] J. D. Jordan, K. D. Carey, P. J. S. Stork, and R. Iyengar, *J. Biol. Chem.* **274,** 21507 (1999).

[16] J. M. Schmitt and P. J. S. Stork, *J. Biol. Chem.* **275,** 25342 (2000).

Several methods for studying the activation of small G proteins have been described.[19-23] These methods rely on the use of radiolabeled purified proteins to measure exchange activity *in vitro* or radiolabeling cells with orthophosphate and immunoprecipitating the G protein of interest to measure the GTP/GDP-bound ratio. However, for some G proteins, such as Rap1, good immunoprecipitating antibodies are unavailable, requiring transfection of epitope-tagged Rap1 or the use of purified proteins to study Rap1 activation[4,22,24] (Fig. 1). Furthermore, these methods use large amounts of radioactivity to metabolically label cells with orthophosphate, or require the use of purified Rap1. Nonisotopic methods have been devised that enable the detection of endogenous Ras and Rap1 activation in treated cell lysates. The methods presented here take advantage of the ability of proteins to bind selectively to the GTP-bound state of the small G proteins Ras and Rap1. One of these methods uses epitope-tagged versions of Ras- or Rap1-binding proteins to selectively pull down GTP-bound Ras or Rap1 from treated cellular lysates. In this manner, activation of endogenous Rap1 or Ras can be detected directly through its ability to interact with downstream binding partners *in vitro,* such as the amino terminus of Raf-1, or the Ras-binding domain of RalGDS, respectively.[25,26] In a second method, transfected epitope-tagged Ras or Rap can be used to pull down endogenous downstream effectors such as Raf-1 or B-Raf *in vivo.*[11,16] Examples of both of these methods are illustrated in Fig. 2. In Fig. 2A, Rap1 activation is assayed by measuring its binding to RalGDS *in vitro*. Purified glutathione *S*-transferase (GST)–RalGDS is added to stimulated cellular lysates to pull down endogenous GTP-bound Rap1, which is then detected by Western blotting with Rap1 antiserum. In Fig. 2B, Ras activation is measured by examining the binding of Raf-1 *in vivo*. In this example, an epitope-tagged version of Ras (polyhistidine-tagged Ras, or His–Ras) is transfected into mammalian cells. On stimulation, His–Ras becomes GTP loaded via the activation of Ras exchange

[17] S. S. Grewal, R. D. York, and P. J. Stork, *Curr. Opin. Neurobiol.* **9,** 544 (1999).

[18] J. de Rooij, F. J. Zwartkruis, M. H. Verheijen, R. H. Cool, S. M. Nijman, A. Wittinghofer, and J. L. Bos, *Nature (London)* **396,** 474 (1998).

[19] J. Downward, *Methods Enzymol.* **255,** 110 (1985).

[20] J. E. DeClue, A. G. Papageorge, J. A. Fletcher, S. R. Diehl, N. Ratner, W. C. Vass, and D. R. Lowy, *Cell* **69,** 265 (1992).

[21] T. Satoh and Y. Kaziro, *Methods Enzymol.* **255,** 149 (1995).

[22] T. Gotoh, S. Hattori, S. Nakamura, H. Kitayama, M. Noda, Y. Takai, K. Katbuchi, H. Matsui, O. Hatase, H. Takahashi, T. Kurata, and M. Matsuda, *Mol. Cell. Biol.* **15,** 6746 (1995).

[23] J. B. Gibbs, M. D. Schaber, M. S. Marshall, E. M. Scolnick, and I. S. Sigal, *J. Biol. Chem.* **262,** 10426 (1987).

[24] D. L. Altschuler, S. N. Peterson, M. C. Ostrowski, and E. G. Lapetina, *J. Biol. Chem.* **270,** 10373 (1995).

[25] X.-f. Zhang, J. Settleman, J. M. Kyriakis, E. Takeuchi-Suzuki, S. J. Elledge, M. S. Marshall, J. T. Bruder, U. R. Rapp, and J. Avruch, *Nature (London)* **364,** 308 (1993).

[26] B. Franke, J.-W. Akkerman, and J. L. Bos, *EMBO J.* **16,** 252 (1997).

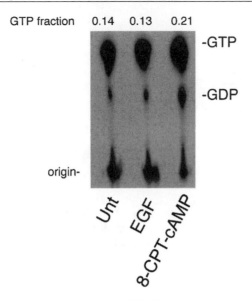

GTP fraction   0.14   0.13   0.21

-GTP

-GDP

origin-

Unt

EGF

8-CPT-cAMP

FIG. 1. GTP loading Rap; Rap1 activation by 8-CPT. GTP loading was assayed in PC12 cells after transfection of His–Rap (30 μg). PC12 cells were left untreated (Unt), or were treated with EGF or 8-CPT-cAMPN for 5 min as indicated. His–Rap was precipitated with Ni–NTA–agarose and eluates were analyzed for GTP and GDP content by thin-layer chromatography. The GTP fraction of total guanine nucleotide is given above each lane.

factors and binds with high affinity of Raf-1. Nickel–agarose beads are then used to selectively pull down both activated and inactive His–Ras and levels of Raf-1 are evaluated by Western blotting with Raf-1 antiserum. Because only activated Ras associates with Raf-1, the amount of Raf-1 detected by Western blotting reflects Ras activation. This method has also been used to detect other Ras and Rap1 effectors in treated cellular lysates, such as B-Raf.[4,12,27–29]

As mentioned above, the physiological actions of both Ras and Rap1 require activation, or GTP loading of the respective G proteins. Likewise, the association of RalGDS with Rap1 is dependent on the activation state of Rap1 itself. This is shown in Fig. 3A and B. The GTP dependence of Rap1 association with the Ras-binding domain of RalGDS is demonstrated in Fig. 3A.[33] Here PC12 cells were transfected with cDNAs of an N-terminally FLAG-tagged version of Rap1 with or without a cDNA for Rap1 GAP, a Rap1-specific GTPase-activating

[27] R. D. York, D. C. Molliver, S. S. Grewal, P. E. Stenberg, E. W. McCleskey, and P. J. S. Stork, *Mol. Cell. Biol.* **21,** 18069 (2000).
[28] M. G. Seidel, M. Klinger, M. Freissmuth, and C. Holler, *J. Biol. Chem.* **274,** 25833 (1999).
[29] S. Grewal, A. M. Horgan, R. D. York, G. S. Withers, G. A. Banker, and P. J. S. Stork, *J. Biol. Chem.* **274,** 3722 (2000).

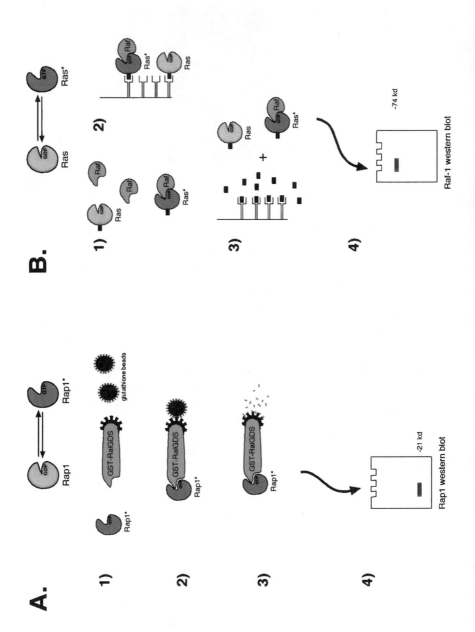

protein.[34,35] Rap1 GAP1 can selectively reduce the levels of GTP-bound Rap1 when expressed ectopically within cells.[9,27,36] In this example, cAMP is used as a Rap1 activator.[24] cAMP can activate Rap1 through specific cAMP-regulated exchangers or by protein kinase A (PKA) itself.[4,18] The actions of both cAMP and PKA can be mimicked by forskolin, a potent activator of adenylyl cyclase. As shown in Fig. 3A, this activation can be measured by GTP loading assays[4] and by RalGDS pull-down assays. In this example FLAG–Rap1 is transfected into cells

---

FIG. 2. Schematic of nonisotopic assays. (A) *In vitro* RalGDS assay. Rap1 exists in an equilibrium between GTP-bound (active) and GDP-bound (inactive) states. Effectors such as B-Raf and RalGDS bind with much higher affinity to the GTP-bound (active) state of Rap1. This can be monitored *in vitro* with chimeric proteins that contain the Rap1-binding domain of such effectors. One useful tool is GST–RalGDS, a chimeric protein combining the glutathione *S*-transferase domain with the RalGDS domain. Ral is a small G protein that can be activated by a guanine dissociation stimulator called RalGDS. This protein has been shown to bind to activated Rap1 with high affinity and can be used *in vitro* to select GTP-bound Rap1 molecules. (1) Cells are treated with potential Rap1 activators and lysed. Lysates are mixed with bacterially purified GST–RalGDS, at high concentrations to displace any bound endogenous effectors. Subsequently, glutathione-coupled beads are added to the mix. (2) Glutathione forms a tight association with the GST–RalGDS and bound active Rap1. Excess GST–RalGDS is free to bind to other molecules including GTP-bound Ras. (3) The addition of free glutathione displaces the beads from the GST–RalGDS complex. (4) SDS–PAGE separates out proteins within the GST–RalGDS complex and specific antibodies can be used in Western blotting to identify the relative amount of bound Rap1 (or Ras). Because little GDP-bound Rap1 can bind GST–RalGDS, the level of Rap1 recovered provides a useful index of Rap1 activation. This same assay can be used simultaneously to examine multiple proteins that bind RalGDS including Ras. Other GST fusions have been designed that display high affinities for selected G proteins. These fusion proteins can be utilized in a similar manner to examine the activation state of other G proteins including Ras, Rac, Ral, and Rho (see Table 1). (B) *In vivo* pull-down assay. In this example, we measure the activation of Ras. Ras exists is an equilibrium between GTP-bound (active) and GDP-bound (inactive) states. Effectors such as Raf-1 and RalGDS (see A) bind with much higher affinity to the GTP-bound (active) state of Ras [or Rap1; see (A)]. This can be monitored with epitope-tagged (black rectangle) Ras. Polyhistidine-tagged Ras (His–Ras) is used in this example, although other epitopes can be used as well. Cells are transfected with His–Ras and lysates prepared after stimulation, for example, with growth factors, to activate Ras. (1) Active GTP-bound Ras will remain associated with effectors, whereas inactive GDP-bound Ras will not. (2) Both active and inactive Ras and bound effectors can be immobilized by using a nickel affinity column, and (4) both active and inactive Ras can be eluted with imidazole (black rectangles), which competes with histidine for the nickel-binding sites. (4) SDS–PAGE separates out active and inactive Ras from any bound effectors, including Raf-1, as shown in this example. These associated proteins can be detected by Western blot. In this example, Raf-1 is examined with Raf-1 antiserum. Other Ras effectors can be examined from the same eluates, including B-Raf, RalGDS, PI3-K, and others, using the appropriate antibodies. In this way, the levels of effectors detected by Western blot provide an index of the activation state of Ras. This is largely because transfected His–Ras is in excess. Note that this assay measures the Ras activation state rather than the Raf-1 activation state. For Raf more signals may be needed in addition to recruitment of Raf-1 to Ras to achieve full activation.[30,31] This is particularly relevant for Rap1 pull-down assays. Rap1/Raf-1 pull downs are effective for monitoring Rap1 activation but are not appropriate to measure Raf-1 activation, because Raf-1 binding to Rap1 does not result in Raf-1 activation.[4,32]

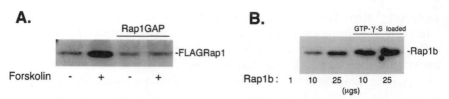

FIG. 3. GST–Ral GDS binds GTP-bound Rapl. (A) GST–RalGDS as a measure of Rap1 activation. Untransfected PC12 cells or PC12 cells transfected with a cDNA encoding Rap1GAP1 were stimulated with forskolin, as indicated. It has been previously demonstrated[4,24,33] that forskolin, through its stimulation of adenylyl cyclase, is a potent activator of Rap1. This assay shows that the activation of Rap1 can be monitored by the GST–RalGDS assay. Moreover, the fold increase in Rap1 activation is blocked in cells expressing Rap1GAP1. This demonstrates that GST–RalGDS is recognizing GTP-bound Rap1 The level of Rap1 detected in unstimulated cells is not a reflection of active Rap1 in the resting cells because this level is not further reduced by Rap1Gap1. Rather, it reflects the low level of GDP-bound Rap1 that may be detected by GST–RalGDS. (B) The GST–RalGDS assay is sensitive to GTP loading. Baculovirally expressed Rap1b was purified and left unloaded, or loaded with GTPγS *in vitro*. The resultant protein was mixed with purified GST–RalGDS protein and the levels of Rap1 protein measured as described in Fig. 2. The data demonstrate that the loading of Rap1 with GTP increases the level of Rap1 recovered after the GST–RalGDS assay. Note that unloaded Rap1 does bind to a limited degree to GST–RalGDS. This may be because of the high levels of protein used in this assay, and may reflect the residual GTP-loading of Rap1b recovered from baculovirus.

and the association of FLAG–Rap1 is monitored by Western blot, using FLAG antibodies. This use of epitope tagging to examine transfected Rap1 allows the simultaneous examination of cotransfected cDNAs. In this example, this approach is used to examine the action of Rap1 GAP1 on Rap1 activation. When transfected cells are stimulated with forskolin, intracellular cAMP levels are raised, leading to FLAG–Rap1 activation and increased association with GST–RalGDS (Fig. 3A). However, Rap1GAP1 blocks this association by stimulating the intrinsic GTPase activity of Rap1, thereby reducing the level of Rap–GTP within the cell and decreasing the amount of Rap1 pulled down by GST–RalGDS. The GTP-dependent association with RalGDS can also be shown directly with purified proteins *in vitro* (Fig. 3B). Purified His–Rap1 associates with GST–RalGDS, but the amount of HisRap pulled down is enhanced when it is loaded with GTPγS, a nonhydrolyzable analog of GTP. Because little GDP-bound Rap1 binds GST–RalGDS, the level

[30] R. Marais, Y. Light, H. F. Paterson, and C. J. Marshall, *EMBO J.* **14,** 3136 (1995).

[31] D. K. Morrison and J. R. E. Cutler, *Curr. Opin. Cell Biol.* **9,** 174 (1997).

[32] T. Okada, C. D. Hu, T. G. Jin, K. Kariya, Y. Yamawaki-Kataoka, and T. Kataoka, *Mol. Cell. Biol.* **19,** 6057 (1999).

[33] Y. Wan and X. Y. Huang, *J. Biol. Chem.* **273,** 14533 (1998).

[34] P. G. Polakis, B. Rubinfeld, T. Evans, and F. McCormick, *Proc. Natl. Acad. Sci. U.S.A.* **88,** 239 (1991).

[35] L. A. Quilliam, C. J. Der, R. Clark, E. C. O'Rourke, K. Zhang, F. McCormick, and G. M. Bokoch, *Mol. Cell. Biol.* **10,** 2901 (1990).

[36] C. Anneren, K. A. Reedquist, J. L. Bos, and M. Welsh, *J. Biol. Chem.* **275,** 29153 (2000).

of Rap1 protein binding to RalGDS that is detected by Western blotting provides a useful index of Rap1 activation, and provides a convenient nonisotopic method to examine stimuli that activate Rap1 *in vivo*. The purpose of this article is to provide protocols and points of discussion regarding the two nonisotopic methods outlined in Fig. 2 that are designed to detect Ras and Rap1 activation in mammalian cells.

Method 1: *In vitro* Assay of Small G Proteins: Precipitating Activated Rap by GST–RalGDS

Introduction

The method examined in this section is an adaptation of the method for examining Rap1 activation derived from the laboratory of J. Bos.[26] Because this method relies on purified proteins that are not commercially available, we will discuss the bacterial purification of GST–RalGDS as well as its utilization in mammalian cell lysates. For examining Ras activation, another GST fusion protein, GST–Raf1RBD [derived from the Ras-binding domain (RBD) of Raf-1, as described in Table I], can be used (Fig. 4, bottom[37–40]).

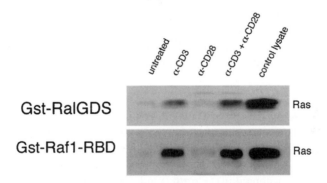

FIG. 4. Nonisotopic examination of Ras activation, using GST fusion proteins *in vitro*. *Top:* GST–RalGDS as a measure of Ras activation. Jurkat cells were incubated with anti-human CD3 (α-CD3) and/or anti-human CD28 (α-CD28) for 5 min or left untreated as indicated. Lysates were prepared and assayed for Ras activation, using GST–RalGDS and Western blots performed with Ras antiserum, and the position of Ras is shown. *Bottom:* Raf-1 RBD assay as a measure of Ras activation. Jurkat cells were incubated with anti-human CD3 (α-CD3), and/or anti-human CD28 (α-CD28), for 5 min or left untreated as indicated. Lysates were prepared and assayed for Ras activation, using GST–Raf1 RBD and Western blots performed with Ras antiserum. The position of Ras in control lysates and after isolation of glutathione-bound proteins is shown. The data show that Ras is activated by cross-linking the T cell receptor (α-CD3) in the presence and absence of simultaneous CD28 activation, and compares favorably with other determinations of Ras activation by these cross-linking antibodies.[37,38] Equal amounts of control lysates serve to normalize the levels of Ras pulled down in this assay as compared with the GST–RalGDS assay shown (*top*). Note that the results using GST–Raf1 RBD and GST–RalGDS are qualitatively similar; however, relatively less Ras is pulled down with GST–RalGDS, possibly reflecting a lower affinity of Ras for RalGDS.[39,40]

Preparation of Cell Lysate

Subconfluent cultured cells that have been maintained in appropriate medium are serum starved in low serum containing medium [0.2–1% (v/v) human serum (HS) or fetal calf serum (FCS), depending on cell type] for up to 24 hr before stimulation. It is important to maintain cells at subconfluency, because Rap1 activation may be regulated by cell density.[41] Cells are stimulated with the appropriate concentration of agonist for the desired time interval, rinsed twice in ice-cold (4°) phosphate-buffered saline (PBS) on ice, and then lysed in ice-cold (4°) RalGDS lysis buffer [50 mM Tris (pH 8.0), 150 mM NaCl, 1% (v/v) Nonidet P-40 (NP-40), 10% (v/v) glycerol, and 5 mM MgCl$_2$ with freshly added protease inhibitors phenylmethylsulfonyl fluoride (PMSF, 1 mM), aprotinin (1 μM), leupeptin 1 μM), sodium vanadate (Na$_3$VO$_4$, 1 mM), NaF (10 mM), and soybean trypsin inhibitor (10 μg/ml)] (0.5 ml/2 × 10^6 cells) for 5 min. Cell lysates are centrifuged briefly for 2 min at 2000 rpm to remove insoluble debris. A small (1- to 10-μl) aliquot is removed and protein levels for each condition are determined by the Bradford assay.[42] The cell lysates should be tested immediately. Also, the amount of lysate needed to detect Rap1–GTP varies depending on the strength of the signal induced by the stimuli being examined as well as the cell line used. This may not be due only to Rap1 protein levels within the cell, but may also be due to the particular Rap1 exchanger activated or the levels of endogenous Rap1GAP activity within the cell line examined. Because of these considerations, typically experiments should begin with as much as 1 mg of total cell lysate.

Purification of RalGDS Fusion Protein

The Ras-binding domain of RalGDS, consisting of amino acids 726–852, is expressed in DH5α bacteria as an N-terminally tagged GST fusion protein in pGEX-4T3. Ten milliliters of an overnight culture of pGEX-4T3-GST-RalGDS is added to 1 liter of Luria broth (10 g of tryptone, 5 g of yeast extract, and 10 g of NaCl per liter) with ampicillin (50 μg/ml) and grown to an OD$_{600}$ of approximately 0.6 at 37°. Expression of GST–RalGDS is induced by adding 1 mM isopropyl-β-D-thiogalactopyranoside, and the culture is incubated at 37° for an additional 6 hr. The GST–RalGDS protein is stable under these conditions, although slightly higher yields are obtained by growing the cells at 30° for 8–12 hr.

The culture is harvested by centrifugation, and the bacterial pellets can be stored at −80°. The bacterial paste is resuspended in ice-cold (4°) phosphate-buffered

[37] W. Li, C. D. Whaley, A. Mondino, and D. L. Mueller, *Science* **271,** 1272 (1996).

[38] P. E. Fields, T. F. Gajewski, and F. W. Fitch, *Science* **271,** 1276 (1996).

[39] M. Spaargaren and J. R. Bischoff, *Proc. Natl. Acad. Sci. U.S.A.* **91,** 12609 (1994).

[40] C. Herrmann, G. Martin, and Q. Wittinghofer, *J. Biol. Chem.* **270,** 2901 (1995).

[41] G. Posern, C. K. Weber, U. R. Rapp, and S. M. Feller, *J. Biol. Chem.* **273,** 24297 (1998).

[42] M. M. Bradford, *Anal. Biochem.* **72,** 248 (1976).

saline at approximately 0.1 g/ml containing freshly added protease inhibitors [1 mM PMSF, 1 μM aprotinin, 5 mM benzamidine, 1 μM leupeptin, and soybean trypsin inhibitor (10 μg/ml)]. Bacterial cell lysis is achieved by three passages through a French press, or by sonication (microtip, 10 times, 30 sec each). Triton X-100 is added to a final concentration of 1% (v/v) and the bacterial lysate is rocked at 4° for 30 min. Bacterial debris is pelleted at 12,000g for 10 min at 4°, and the supernatant is transferred to a fresh tube. At this point, the GST–RalGDS can be separated from bacterial proteins by either affinity chromatography or batch binding. For batch binding, 1 ml/50 ml of supernatant of a 1 : 1 slurry of glutathione–agarose is added and rocked at 4° for 1 hr. The GST–RalGDS/bead complex is pelleted at 2000 rpm for 2 min. The beads are washed three times with ice-cold PBS with protease inhibitors, and the GST–RalGDS is eluted with 5 mM reduced glutathione, 10 mM Tris (pH 8.0), with protease inhibitors. The eluted GST–RalGDS is then dialyzed against PBS. The purity of the GST–RalGDS protein is estimated by Coomassie blue staining and should typically be between 85 and 95%. Typical yields range from 0.3 to 1 mg of GST–RalGDS per liter of bacteria. Some researchers have found that the GST–RalGDS protein stability can be increased if the aliquots of GST–RalGDS are maintained at high protein concentrations. Alternatively, aliquots of the unpurified bacterial extracts can be maintained in a frozen state. In this case, the bacterial extracts containing GST–RalGDS can be loaded onto glutathione beads and the complex mixed directly with lysates from mammalian cells if necessary.[6]

RalGDS Assay

For each sample, add 40 μg of purified GST–RalGDS protein to 0.5–1 ml of cell lysate having 0.5–1 mg of total cellular protein. We have found that using 40–60 μg of purified GST–RalGDS protein works best for detecting Rap1 activation. This may be because the Ras-binding domain of RalGDS will also efficiently pull down activated forms of other small G proteins that compete with Rap1 for binding to RalGDS, including Ras and Rap2[43,44] (Fig. 4, top). Samples are rocked with the GST–RalGDS protein at 4° for 30 min. Add 30 μl of a 1 : 1 (w/v) (beads : water) slurry of glutathione–agarose beads (Sigma St. Louis, MO) and continue to rock the samples for an additional 30 min. Centrifuge the samples for 2 min at 14,000g to pellet the complexes. Rinse the samples twice with ice-cold (4°) RalGDS lysis buffer containing protease inhibitors as previously described. Proteins are eluted from the beads with 2× Laemmli buffer and applied to a 12% (w/v) sodium dodecyl sulfate (SDS)–polyacrylamide gel. Proteins are transferred to a polyvinylidene difluoride (PVDF) membrane, blocked for 1 hr, and

[43] Y. Ohba, N. Mochizuki, K. Matsuo, S. Yamashita, M. Nakaya, Y. Hashimoto, M. Hamaguchi, T. Kurata, K. Nagashima, and M. Matsuda, *Mol. Cell. Biol.* **20,** 6074 (2000).

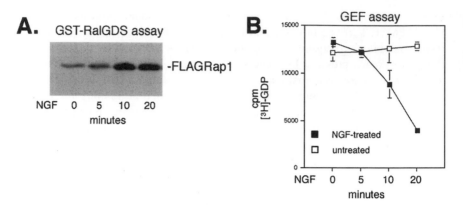

FIG. 5. Comparison of RalGDS assay and GEF assay. (A) PC12 cells transfected with FLAG–Rap1, were treated with NGF for the indicated times. RalGDS assays were performed on lysates and Western blots were probed for FLAG–Rap1, using anti-FLAG antibody. The position of FLAG–Rap1 is shown. (B) Untransfected PC12 cells were treated with NGF for the indicated times, lysates were prepared, and exchange assays were performed on lysates as described in Carey *et al.*[11] Open squares represent unstimulated cells, and solid squares represent NGF-treated cells.

probed with a 1 : 1000 dilution of an anti-Rap1 (Krev) antibody (Santa Cruz Biotechnology, Santa Cruz, CA) overnight at 4°, followed by a horseradish per-oxidase (HRP)-conjugated anti-rabbit secondary antibody. Rap1 is detected by enhanced chemiluminescence (DuPont-NEN, Boston, MA). We find that it helps to run a nonstimulated total lysate control lane as a positive control for the detection of Rap1 protein on Western blots.

Notes

1. An important variable to consider when designing experiments examining Rap1 activation is that the time course of Rap1 activation can vary depending both on the cell type and the stimulus. In PC12 cells, GTP loading of Rap1 is detectable when stimulated with nerve growth factor (NGF) after 5 min, Rap1 activation peaks at 10–15 min, and slowly decreases thereafter (Fig. 5A). However, in Jurkat cells, Rap1 activation by stimulation with anti-CD3 antibodies is detectable after 30 sec, is maximal at 2–5 min, and returns to basal levels after 10 min.[44] Both the amount of lysate to use and the time course of activation will depend on the cell type and the stimulus.

2. This method can also be used to detect activation of epitope-tagged G pro-teins, such as FLAG–Rap1, as shown in Fig. 3A. Technically, after transfection of FLAG–Rap1, cells are treated with specific stimuli and the activation state of FLAG–Rap1 is determined as described previously. It is important to note that GST–RalGDS will bind both endogenous GTP-bound Rap1 as well as GTP-bound

---

[44] K. A. Reedquist and J. L. Bos, *J. Biol. Chem.* **273,** 4944 (1998).

FLAG–Rap1. To identify transfected FLAG–Rap1 and endogenous Rap1, Western blotting is performed with an anti-FLAG antibody. Total Rap1 activation can also be monitored with antiserum to Rap1, although the separation of endogenous Rap1 and FLAG–Rap1 on the Western blot may be technically demanding. As is shown in Fig. 5A, the time course of activation of Rap1 by NGF stimulation in FLAG–Rap1-transfected PC12 cells reflects the kinetics seen in endogenously stimulated cells.[36] These nonisotopic methods give results similar to those of exchange assays (Fig. 5B). However, it should be noted that although the RalGDS assay (Fig. 5A) and the GEF assay (Fig. 5B) show similar kinetics, it is not always the case that exchange assays reflect the GTP-bound state of Rap1, because the GTP-bound state represents an equilibrium between exchange and GTPase action by Rap1 GAPs and GAP activity is not measured in the GEF assay. For example, it has been shown that Rap1 GAPs can be regulated independently by certain stimuli.[7,11,15] Loading purified Rap1 with [^{3}H]GDP will ensure that endogenous GAP activity will not influence the GEF assay because [^{3}H]GDP is not a substrate for GAPs. In this case, the contribution of GAP activity to the activation state of Rap1 *in vivo* will not be measured.

3. Although RalGDS binds tightly to Rap1, it has been suggested that it may be a physiological effector of Ras *in vivo*.[1,45] Indeed, RalGDS binds to Ras, but with a lower affinity that it does to Rap1.[39,46] Because RalGDS can bind to multiple G proteins, these assays can be used simultaneously to examine the activation of other small G proteins that recognize the same effector proteins, such as Ras (Fig. 4, top).

4. The peptide fragment of Raf-1 comprising the Ras-binding domain of Raf-1 (Raf-1 RBD) is now commercially available (Upstate Biotech, Lake Placid, NY) and provides a convenient nonisotopic method analogous to RalGDS for examining Ras activation (Fig. 4, bottom).

Method 2: *In Vivo* Pull Down of Ras/Rap1 Effectors

Introduction

Generally, the activation of small G proteins can be monitored by the recruitment of effectors to the small G protein itself. This recruitment is essential for activation of the recruited molecule but is not necessarily sufficient,[30,47] and subsequent activation steps may be required.[48] Indeed, not all molecules that are recruited to Rap1 are activated. For example, Raf-1 binds with high affinity to Rap1 but is not activated by Rap1 *in vivo*.[32] In contrast, B-Raf binds with lower

[45] J. L. Bos, B. Franke, L. M'Rabet, K. Reedquist, and F. Zwartkruis, *FEBS Lett.* **410,** 59 (1997).

[46] C. Herrmann, G. Horn, M. Spaargaren, and A. Wittinghofer, *J. Biol. Chem.* **271,** 6794 (1996).

[47] J. Avruch, X.-F. Zhang, and J. M. Kyriakis, *Trends Biochem. Sci.* **19,** 279 (1994).

[48] R. Marais, Y. Light, C. Mason, H. Paterson, M. F. Olson, and C. J. Marshall, *Science* **280,** 109 (1998).

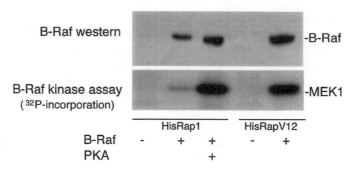

FIG. 6. Rap1 pull-down assay examining the association and activation of B-Raf with Rap1b. Using a pull-down assay with histidine-tagged Rap1 proteins, both the association of B-Raf as well as the kinase activity of associated B-Raf can be monitored. COS-7 cells were transfected with 10 μg of His–Rap, 12 μg of B-Raf, 5 μg of PKA, or 10 μg of His–RapV12 as indicated. His–Rap proteins were precipitated with Ni–NTA–agarose and associated B-Raf was detected by Western blotting. The position of B-Raf is shown (*top*). In addition, associated B-Raf kinase activity was immunoprecipitated with B-Raf antiserum and kinase assays were performed with MEK1 as a substrate *in vitro* as previously described[4] (*bottom*). The position of MEK-1 is shown. Note that in this example, the detection of B-Raf association with Rap1 is dependent on transfection of B-Raf, presumably because of the low levels of endogenous B-Raf in COS-7 cells.[4] The data show the effect of transfection of the catalytic subunit of PKA on Rap1 activation. In addition, the GTP dependence of the association of B-Raf with Rap1 is examined by comparing the action of His–Rap1b (wild type) with His–RapV12 (a constitutively activated mutant form of Rap1).

affinity to Rap1 than does Raf-1,[32] but is activated by Rap1.[4,49] With this caveat, the recruitment of endogenous molecules such as Raf-1 and B-Raf to Ras and Rap1 are powerful indicators of GTP loading of Ras and Rap1, which provide an index for the activation state of Ras and Rap1 *in vivo*. In this section, the methods for examining Ras- or Rap1-associated proteins are discussed.

Transfection and Lysis

Little GDP-bound Rap1 associates with GST–RalGDS or Raf-1, even when Rap1 is overexpressed after transfection of His–Rap1 (Fig. 3). This ability of transfected Rap1 proteins to retain the GTP dependence of action forms the basis for another nonisotopic method used to detect Rap1 activation and Rap1 effectors.

For this assay, cells are transfected with epitope-tagged Rap1, using methods that have been optimized for that cell type. The choice of epitope depends on the method of recovery of tagged Rap1. Polyhistidine-tagging methods utilize metal-chelate resins, such as nickel–nitrilotriacetic acid (Ni–NTA), from which associated proteins can be easily eluted for further biochemical characterization (see Fig. 6). FLAG and Myc epitopes have also been successfully used for

[49] T. Ohtsuka, K. Shimizu, B. Yamamori, S. Kuroda, and Y. Takai, *J. Biol. Chem.* **271**, 1258 (1996).

immunoprecipitation. In both cases, the use of multiple (tandem) epitope tags increases the efficiency of FLAG immunoprecipitation. Subsequent to transfection, the cells are allowed to recover for 24 hr. It is important to have at least 30–50% transfection efficiency in order for this method to work well. Cells are rinsed twice with ice-cold PBS, and then lysed in ice-cold buffer containing 1% (v/v) NP-40, 10 mM Tris (pH 8.0), 20 mM NaCl, 30 mM MgCl$_2$, 1 mM PMSF, aprotinin (0.5 mg/ml), leupeptin (0.1 mg/ml), 50 mM NaF, 1 mM Na$_3$VO$_4$, and supernatant is prepared by low-speed centrifugation. Lysates should be used immediately.

Elution of Proteins Associated with His–Rap1

For this example, we utilize polyhistidine-tagged Rap1 (His–Rap1) and recover His–Rap1 by nickel affinity chromatography. Transfected histidine-tagged proteins are separated from supernatants containing equal amounts of protein, using Ni–NTA–agarose (Qiagen, Chatsworth, CA) and washed with 20 mM imidazole in lysis buffer to remove nonspecific binding interactions. Proteins are eluted with 500 mM imidazole and 5 mM EDTA in PBS. The eluates containing histidine-tagged proteins and effectors are separated by SDS–PAGE and endogenous Raf-1 proteins are detected by Western blotting. This protocol is designed for the detection of proteins that associate with histidine-tagged GTP-bound Rap1 or Ras, but it can be adapted for FLAG epitope-tagged proteins as well, using the RalGDS buffer and lysis conditions as described in the first section.[11] Figure 6 shows an application of this method for the detection of B-Raf association with histidine-tagged Rap1 in PC12 cells stimulated with agents that elevate intracellular cAMP.[4]

Notes

1. Endogenous G proteins can also be examined with the method if available immunoprecipitating antibodies exist. Whereas good antibodies are available for Ras (Y13-238),[50] currently there are no immunopreciptating antibodies for Rap1. Therefore, for Rap1, epitope tagging provides the surest way to immunoprecipitate proteins that bind GTP-bound Rap1.

2. The source of proteins that bind epitope-tagged Ras or Rap1 can be endogenous or transfected. Transfection allows the examination of association with proteins that may not be expressed to high levels in the cell employed. This is the case in Fig. 6, where the association of B-Raf with Rap1 in COS-7 cells is dependent on transfection of B-Raf.[4]

3. One advantage of the pull-down assays for examining Ras and Rap1 function is that these methods examine the association of Ras or Rap1 with endogenous proteins in vivo. In addition, further biochemical studies on the eluates recovered

---

[50] M. E. Furth, L. J. Davis, B. Fleurdelys, and E. M. Scolnick, *J. Virol.* **43**, 294 (1982).

from these pull-downs can be performed. For example, the kinase activity associated with Rap1 can be examined. In Fig. 6, we show the examination of B-Raf kinase activity within the eluates from a His–Rap1 pull-down.

4. Because this method examines the association of endogenous proteins with transfected small G proteins, additional cDNAs can be transfected to examine the consequence of selected genetic manipulations on the recruitment of proteins to Ras or Rap1. The ability to cotransfect additional activators or interfering mutants can be informative. Of the two techniques presented here, the RalGDS assay is primarily a method to examine endogenous Rap1 activation, although it can be modified to examine transfected Rap1 (see above). In the second pull-down method, the examination of proteins that associate *in vivo* with transfected Rap1 works consistently in a variety of cells, and can be used to examine the association of transfected G proteins with endogenous or cotransfected proteins, as shown in Fig. 6.

Conclusions

We have described two simple nonisotopic methods for the detection of endogenous or transfected activated Rap1, which give results similar to those of isotopic methods. One method, the RalGDS assay allows examination of the activation state of endogenous Rap1, whereas a second method, the *in vitro* pull-down method, is better suited for examination of epitope-tagged Rap1. These methods should be broadly applicable for the detection of the activation of a number of small G proteins.

Methods similar to the RalGDS method described are currently being utilized for other small G proteins, taking advantage of the selective binding of specific proteins to GTP-bound active forms of the G proteins. Using various epitope-tagged fusion proteins that display high affinities for selected G proteins, researchers have

TABLE I
DETECTION OF SMALL G PROTEINS USING EFFECTOR DOMAINS

G protein	Recognition domain	Refs.
Rap1a, Rap1b	Ras-binding domain of RalGDS (aa 726–828)[a]	26, 39
Rap2	Ras-binding domain of RalGDS (aa 726–828)	43, 44
Harvey-Ras	N terminus of Raf-1 (aa 1–147)	25
Ral	Ral-binding domain of RalBP1 (aa 397–518)	51
Rac1	p21-binding domain of PAK1 (aa 67–150)	52, 53
Cdc42	p21-binding domain of PAK1 (aa 67–150)	52, 53
RhoA	Rho-binding domain of Rhotein (aa 7–89)	54

[a] aa, Amino acid.

developed tools to examine the activation state of other G proteins including Ras, Rac, Ral, and Rho (Table I [51–54]). Like all Western blotting techniques, these methods are not rigorously quantitative. However, one report suggests an adaptation that can render the RalGDS method semiquantitative.[55]

[51] T. Goi, G. Rusanescu, T. Urano, and L. A. Feig, *Mol. Cell. Biol.* **19,** 1731 (1999).
[52] V. Benard, B. P. Bohl, and G. M. Bokoch, *J. Biol. Chem.* **274,** 13198 (1999).
[53] S. Bagrodia, S. J. Taylor, K. A. Jordon, L. V. Aclst, and R. A. Cerione, *J. Biol. Chem.* **273,** 23633 (1998).
[54] X.-D. Ren, W. B. Kiosses, and M. A. Schwartz, *EMBO J.* **18,** 578 (1999).
[55] F. C. von Lintig, R. B. Pilz, and G. R. Boss, *Oncogene* **19,** 4029 (2000).

[30] BIG1 and BIG2: Brefeldin A-Inhibited Guanine Nucleotide-Exchange Proteins for ADP-Ribosylation Factors

By Gustavo Pacheco-Rodriguez, Joel Moss, and Martha Vaughan

Introduction

Recognition of the inhibitory effects of brefeldin A (BFA), a fungal macrocyclic lactone, on protein secretion in cultured rat hepatocytes[1] was followed by its widespread use in the experimental investigation of molecular mechanisms of vesicular transport. Although multiple intracellular structures are affected by BFA,[2] the drug has been, thus far, especially useful in elucidating processes that involve the Golgi apparatus. In mammalian cells, protein targets of BFA, all associated with Golgi membranes, include a mono-ADP-ribosyltransferase that is activated by BFA[3] and two BFA-inhibited guanine nucleotide-exchange proteins, BIG1 and BIG2.[4–7] These proteins accelerate GTP binding by, and thus activation of, ADP-ribosylation factors or ARFs.

[1] Y. Misumi, Y. Misumi, K. Miki, A. Takatsuki, G. Tamura, and Y. Ikehara, *J. Biol. Chem.* **261,** 11398 (1986).
[2] R. D. Klausner, J. G. Donaldson, and J. Lippincott-Schwartz, *J. Cell. Biol.* **116,** 1071 (1992).
[3] R. Weigert, A. Colanzi, A. Mironov, R. Buccione, C. Cericola, M. G. Sciulli, G. Santini, S. Flati, A. Fusella, J. G. Donaldson, M. DiGirolamo, D. Corda, M. A. DeMatteis, and A. Luini, *J. Biol. Chem.* **272,** 14200 (1997).
[4] N. Morinaga, S.-C. Tsai, J. Moss, and M. Vaughan, *Proc. Natl. Acad. Sci. U.S.A.* **93,** 12856 (1996).
[5] N. Morinaga, J. Moss, and M. Vaughan, *Proc. Natl. Acad. Sci. U.S.A.* **94,** 1226 (1997).
[6] N. Morinaga, R. Adamik, J. Moss, and M. Vaughan, *J. Biol. Chem.* **274,** 17417 (1999).
[7] A. Togawa, N. Morinaga, M. Ogasawara, J. Moss, and M. Vaughan, *J. Biol. Chem.* **274,** 12308 (1999).

ARF was first identified as an $\sim$20-kDa protein that activated cholera toxin-catalyzed ADP ribosylation of $G_s\alpha$ in a GTP-dependent fashion (reviewed in Moss and Vaughan[8]). Although understanding of their physiological functions is far from complete,[9] it appears that individual ARFs participate in different pathways of intracellular vesicular transport. All are capable of activating a specific phospholipase D (PLD) *in vitro*.[10] ARF activation of phosphatidylinositol 4-phosphate 5-kinase has also been reported.[11] All of these actions are properties of ARF with GTP bound. After hydrolysis of bound GTP, which requires the assistance of a GTPase-activating protein or GAP, inactive ARF–GDP dissociates from the membrane location where ARF–GTP is usually found. Activation of ARF requires the replacement of bound GDP with GTP (Fig. 1). Under physiological conditions, this appears to be slow in the absence of a guanine nucleotide-exchange protein or GEP.[9] GEPs, therefore, have critical roles in the regulation of ARF activity (as do GAPs).

A relationship between ARF activity and vesicular transport was unknown when BFA was first used in studies of protein secretion. As *in vitro* model systems began to yield information regarding mechanisms of vesicular transport, a role for ARF in vesicle formation from Golgi and endoplasmic reticulum (ER) membranes was defined (reviewed in Rothman and Wieland[12]). In 1992, two groups described BFA-inhibited enzymatic activity in Golgi membranes that catalyzed the release of ARF-bound GDP and binding of GTPγS.[13,14] Almost 5 years later, purification of a $\sim$200-kDa BFA-inhibited ARF GEP from bovine brain was reported.[4] Peptide sequences revealed the likely presence of a so-called Sec7 domain that has been found in each of the subsequently recognized ARF GEPs (reviewed in Jackson and Casanova[15]). This BFA-inhibited GEP and a similar $\sim$190-kDa protein purified at the same time were termed BIG1 and BIG2, and further characterized as recombinant proteins after preparation of both bovine and human cDNA clones.[5–7]

We describe here procedures for assay of ARF GEP activity in the early stages of purification or in the purified state and for preparation of native or recombinant BIG1 and BIG2.

[8] J. Moss and M. Vaughan, *J. Biol. Chem.* **270,** 12327 (1995).

[9] J. Moss and M. Vaughan, *J. Biol. Chem.* **273,** 21431 (1998).

[10] D. Massenburg, J.-S. Han, M. Liyanage, W. A. Patton, S. G. Rhee, J. Moss, and M. Vaughan, *Proc. Natl. Acad. Sci.U.S.A.* **91,** 11718 (1994).

[11] A. Honda, M. Nogami, T. Yokozeki, M. Yamazaki, H. Nakamura, H. Watanabe, K. Kawamoto, K. Nakayama, A. J. Morris, M. A. Frohman, and Y. Kanaho, *Cell* **99,** 521 (1999).

[12] J. E. Rothman and F. T. Wieland, *Science* **272,** 227 (1996).

[13] J. G. Donaldson, D. Finazzi, and R. D. Klausner, *Nature (London)* **360,** 350 (1992).

[14] J. B. Helms and J. E. Rothman, *Nature (London)* **360,** 352 (1992).

[15] C. L. Jackson and J. E. Casanova, *Trends Cell Biol.* **10,** 60 (2000).

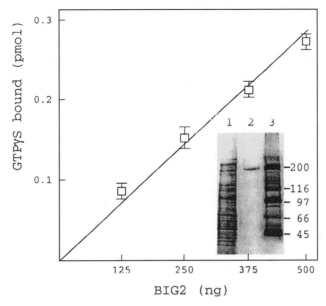

FIG. 1. Effect of His$_6$–BIG2 on [^{35}S]GTPγS binding by rARF1. rARF1 (0.5 μg, 25 pmol) was incubated with the indicated amount of BIG2 and 2 μM [^{35}S]GTPγS for 20 min at 25°. [^{35}S]GTPγS bound in the absence of BIG2 has been subtracted. Data represent means $\pm$ SE of values from triplicate assays. *Inset:* Silver-stained 8%, (w/v) gel after SDS–PAGE. Lane 1, total soluble protein (10 μg) from Sf9 cells containing His$_6$–BIG2; lane 2, His$_6$–BIG2 (2 μg) after purification with Ni–NTA agarose; lane 3, standard proteins with size (kDa) on the right. [Reprinted by permission from A. Togawa, N. Morinaga, M. Ogasawara, J. Moss, and M. Vaughan, *J. Biol. Chem.* **274,** 12308 (1999).]

Materials

Procedures for preparation of native ARF1/3[16] and recombinant myristoylated[17] and nonmyristoylated ARFs are published.[18] L-α-Phosphatidyl-L-serine (PS) is prepared by sonication as described by Tsai *et al.*[19] TENDS buffer contains 20 mM Tris-HCl (pH 8.0), 1 mM EDTA, 1 mM NaN$_3$, 10 mM dithiothreitol (DTT), and 250 mM sucrose. Protease inhibitors present throughout purification and assay are leupeptin, aprotinin, and soybean and lima bean trypsin inhibitors (each 1 μg/ml) and 0.5 mM AEBSF [4-(2-aminoethyl)benzenesulfonyl fluoride

[16] S.-C. Tsai, M. Noda, R. Adamik, P. P. Chang, H.-C. Chen, J. Moss, and M. Vaughan, *J. Biol. Chem.* **263,** 1768 (1988).

[17] R. S. Haun, S.-C. Tsai, R. Adamik, J. Moss, and M. Vaughan, *J. Biol. Chem.* **268,** 7064 (1993).

[18] G. Pacheco-Rodriguez, E. Meacci, N. Vitale, J. Moss, and M. Vaughan, *J. Biol. Chem.* **273,** 26543 (1998).

[19] S.-C. Tsai, R. Adamik, J. Moss, and M. Vaughan, *Proc. Natl. Acad. Sci. U.S.A.* **91,** 3063 (1994).

hydrochloride]. Solutions of BFA (10 μg/μl) in methanol or ethanol are freshly prepared and diluted in TENDS for addition to assays.

Two Assays for Guanine Nucleotide-Exchange Protein Activity

ADP-Ribosylation Factor Activation of Cholera Toxin A Subunit-Catalyzed ADP-Ribosylagmatine Synthesis

With enzyme preparations that contain other GTP-binding proteins, GTP binding by ARF can be assessed by quantifying activated ARF as reflected by its stimulation of cholera toxin A (CTA)-catalyzed ADP–ribosylagmatine formation in two consecutive incubations.[19,20]

Incubation 1: Guanine Nucleotide-Exchange Protein Activation of GTPγS Binding by ADP-Ribosylation Factor. Samples of GEP preparation with or without 1 μM ARF (native or recombinant) with or without 200 μM BFA in 50 μl of TENDS containing 4–20 μM GTPγS, 30–60 μg of bovine serum albumin (BSA), and 20 μg of PS are incubated for 40 min at 37°. Assay tubes are placed on ice while four additions are made for the second reaction.

TENDS buffer	100 μl
Solution containing 30 μg of ovalbumin, 20 μg of PS, and 120 μM Cibachrome blue	50 μl
Phosphate buffer (300 mM, pH 7.5), containing 60 mM agmatine, 1.2 mM [^{14}C]NAD (10^5 cpm), 30 mM MgCl$_2$, 3 mM ATP, and 120 mM DTT	50 μl
TENDS buffer containing 2 μg of CTA	50 μl

Incubation 2: Stimulation of Cholera Toxin A Subunit-Catalyzed ADP–Ribosylagmatine Synthesis by Active ADP-Ribosylation Factor. After incubation for 1 hr at 30°, 150-μl samples of reaction mixture are applied to 1 ml of anion-exchange resin AG 1-X2 (Bio-Rad, Hercules, CA) previously equilibrated with deionized water followed by five washes with 1 ml of water. ^{14}C-Labeled ADP–ribosylagmatine in the effluent is quantified by liquid scintillation counting. After subtraction of ADP–ribosylagmatine synthesized in the absence of ARF, with or without GEP as appropriate, the difference between synthesis in the presence of ARF with and without GEP represents GEP activity, which is assayed without and with BFA to calculate inhibition.

[^{35}S]GTPγS Binding by ADP-Ribosylation Factor

With GEP preparations that do not contain other GTP-binding proteins, GTPγS binding to native ARF1/3 or recombinant ARF can be measured directly.[18] Assays,

[20] S.-C. Tsai, R. Adamik, J. Moss, and M. Vaughan, *Proc. Natl. Acad. Sci. U.S.A.* **93**, 1941 (1996).

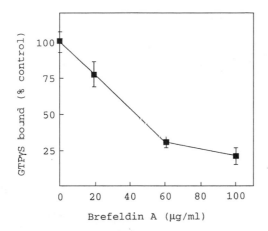

FIG. 2. Effect of BFA on BIG2 stimulation of GTPγS binding by rARF1. rARF1 (0.5 μg), 2 μM [^{35}S]GTPγS, and 250 μg of BIG2 were incubated with the indicated concentrations of BFA in 50 μl at 25° for 20 min. GTPγS bound is shown as a percentage of that bound in the absence of BFA. Data represent means ± SE of values from triplicate assays. [Reprinted by permission from A. Togawa, N. Morinaga, M. Ogasawara, J. Moss, and M. Vaughan, *J. Biol. Chem.* **274**, 12308 (1999).]

in 100 μl of TENDS, contain 30 μl of GEP preparation or vehicle, protease inhibitors, 30–60 μg of BSA, 20 μg of PS, 4–10 μM [^{35}S]GTPγS, 3–5 mM MgCl$_2$, about 200 nM native bovine ARF1/3[16] or 200–500 nM recombinant ARF or vehicle, without or with 200 μM BFA.

After incubation at 37° for 40 min (or at lower temperature and/or shorter time to ensure that the rate of binding is constant throughout incubation and binding is proportional to amount of GEP), samples are cooled on ice and transferred to nitrocellulose filters followed by an 8-ml wash of the incubation tube and 4-ml wash of the filter with 25 mM Tris-HCl (pH 8.0), containing 5 mM MgCl$_2$, 100 mM NaCl, and 1 mM DTT. Filters are dried before radioassay of protein-bound [^{35}S]GTPγS. After subtraction of GTPγS bound in the absence of ARF from that in its presence, without or with GEP, the increase in GTPγS binding by ARF due to GEP (Fig. 1) and the effect of BFA on it (Fig. 2) is calculated.

It is preferable to use myristoylated ARF in these assays because nonmyristoylated ARFs are relatively poor substrates for BIG1 and BIG2, as they are for other GEPs.[6] BIG1 and BIG2 are most active with myristoylated bovine ARF1 and ARF3. Activity toward myristoylated bovine ARF5 (class II) has been demonstrated,[6] but its physiological significance is unknown and no activity with ARF6 was found.[6] Caution is advised in drawing conclusions about intracellular functions from these *in vitro* assays. It is important to remember that guanyl nucleotide binding by different ARFs is affected differently by many variables, such as Mg^{2+} concentration and the presence of specific phospholipids.[9]

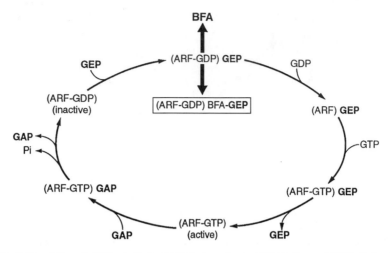

FIG. 3. Regulation of ARF activity by GEPs that accelerate GTP binding and GAPs that activate GTP hydrolysis by ARF. BFA blocks guanine nucleotide exchange (i.e., ARF activation), by forming a complex with ARF–GDP and GEP that prevents release of GDP and thus GTP binding.[21,22] BFA inhibition is reversed when it dissociates from the complex and ARF–GDP–GEP returns to the activation–inactivation cycle.

BFA acts as an uncompetitive inhibitor by interacting with and stabilizing the complex of ARF–GDP and the Sec7 domain of a BFA-sensitive ARF GEP.[21,22] This precedes formation of the normally more stable complex of the Sec7 domain with ARF lacking bound nucleotide, which is followed by GTP binding[21,22] This mechanism means that the inhibitory concentration of BFA varies with the concentration of ARF–GDP–GEP, that is, with the concentrations of ARF–GDP and GEP (Fig. 3). The elegant demonstration by two groups of this seemingly uncommon type of inhibition explains nicely earlier observations regarding BFA inhibition. It also suggests the value of thinking in a new way about molecular mechanisms of nucleotide exchange on ARF and related GTPases.

Preparation of Brefeldin A-Inhibited Guanine Nucleotide-Exchange Proteins 1 and 2

Purification of Native Proteins

BIG1 and BIG2 were first isolated from bovine brain cytosol as components of a large multiprotein complex.[4] After $(NH_4)_2SO_4$ precipitation followed by sequential

[21] A. Peyroche, B. Antonny, S. Robineau, J. Acker, J. Cherfils, and C. L. Jackson, *Mol. Cell* **3**, 275 (1999).

[22] S. J. Mansour, J. Skaug, X.-H. Zhao, J. Giordano, S. W. Scherer, and P. Melançon, *Proc. Natl. Acad. Sci. U.S.A.* **96**, 7968 (1999).

chromatography on DEAE Sephacel, hydroxylapatite, and Mono Q, BFA-sensitive GEP activity precipitated during dialysis at pH 5.8. Activity in this fraction was eluted from a column of Superose 6 at the position of thyroglobulin (669 kDa). Among the proteins in column fractions that were separated by sodium dode-cyl sulfate–polyacrylamide gel electrophoresis (SDS–PAGE), two silver-stained bands of approximately 200 and 190 kDa appeared to correlate best with the BFA-sensitive GEP activity.[4] Each of these proteins was found to be active af-ter electroelution and renaturation, and they were later named BIG1 and BIG2.[7] Because recovery of activity from Superose 6 was low (∼10%), most enzymatic characterization was carried out with the pH 5.8 precipitate.[4]

The composition (qualitative and quantitative) of the macromolecular com-plexes with which BIG1 and BIG2 are associated remains to be determined. Im-munoprecipitation with anti-peptide antibodies specific for each protein estab-lished that 70–75% of BIG1 and BIG2 in the cytosol of cultured HepG2 cells are apparently present together in the protein complexes.[23] The two proteins were also coimmunoprecipitated from microsomal fractions, consistent with preservation of the multimeric structure in translocation between Golgi and cytosol.[23]

By double-immunofluorescence microscopy, endogenous BIG1 and overex-pressed BIG2 were, to a large extent, colocalized in Golgi membranes.[23] Golgi localization of overexpressed BIG1 was demonstrated in other studies, which also showed that this distribution was determined by structure in the N-terminal part of the p200 (BIG1) molecule.[22]

Synthesis of Recombinant Brefeldin A-Inhibited Guanine Nucleotide-Exchange Proteins 1 and 2

Detailed methods for construction of bovine BIG1 and human BIG2 cDNA clones in the baculovirus transfer vector pAcHLT-C for synthesis of proteins with N-terminal hexahistidine (His$_6$) sequences in Sf9 (*Spodoptera frugiperda*) cells are published.[5–7] Baculogold DNA (PharMingen, Carlsbad, CA) and 2 μg of cDNA construct are mixed with 2×10^9 Sf9 cells in 3 ml of TNM-FH in-sect medium (PharMingen), followed by incubation for 5 days at 27°. A sample (10 μl) of cell supernatant is added to 1×10^5 Sf9 cells in 3 ml of the same medium and incubated for 5 days at 27° before collection of cell supernatant, 1 ml of which is added to 2×10^7 Sf9 cells in 30 ml of TNM-FH medium. Cells are harvested by centrifugation after 5 days at 27°, dispersed in 1 ml of ice-cold 10 mM sodium phosphate, pH 8.0, with 100 mM NaCl, 8 μg of benza-midine hydrochloride, 0.5 mM AEBSF, and 5 μg each of phenanthroline, apro-tinin, and leupeptin. After 45 min on ice, cells are lysed by freezing and thawing twice.

---

[23] R. Yamaji, R. Adamik, K. Takeda, A. Togawa, G. Pacheco-Rodriguez, V. J. Ferrans, J. Moss, and M. Vaughan, *Proc. Natl. Acad. Sci. U.S.A.* **97**, 2567 (2000).

The lysate is cleared by centrifugation (20,000g, 15 min, 4°) and 1 ml is mixed with 0.3 ml of nickel–nitrilotriacetic acid agarose (Qiagen, Valencia, CA). After 1 hr at 4°, the matrix is washed with five 1-ml portions of 5 mM imidazole and 50 mM sodium phosphate, pH 8.0, in 300 mM NaCl, 10% (v/v) glycerol, and 0.5 mM, AEBSF before elution of bound protein with 100 mM imidazole and 50 mM sodium phosphate, pH 6.0, in the same solution of NaCl, glycerol, and AEBSF. The eluted protein is dialyzed overnight at 4° against 20 mM Tris-HCl (pH 8.0), 1 mM EDTA, 1 mM NaN$_3$, 1 mM DTT, 0.5 mM AEBSF, 5 mM MgCl$_2$, 30 mM NaCl, and 0.25 M sucrose. Protein, in the same medium, is stored in small portions (to avoid repeated freezing and thawing). Analogous procedures are used for preparation of truncated and otherwise mutant proteins for structure–function studies.[6]

[31] Functional Interaction of Gα$_{13}$ with p115RhoGEF Determined with Transcriptional Reporter System

By JUNHAO MAO and DIANQING WU

Introduction

Four classes of heterotrimeric GTP-binding proteins—G$_s$, G$_i$, G$_q$, and G$_{12}$—are involved in transducing signals for many hormones, neurotransmitters, and other biologically active molecules.[1–3] The G$_{12}$ family of G proteins, consisting of Gα$_{12}$ and Gα$_{13}$,[4] regulates a number of cellular responses via the monomeric GTP-binding protein RhoA. These responses include cytoskeleton rearrangement, serum response factor (SRF)-mediated gene transcription, cell transformation, apoptosis, Na$^+$/H$^+$ exchanger regulation, and phospholipase D activation.[5–12] Moreover, a

[1] A. G. Gilman, *Annu. Rev. Biochem.* **56,** 615 (1987).

[2] L. Birnbaumer and M. Birnbaumer, *J. Recept. Signal Transduct. Res.* **15,** 213 (1995).

[3] E. J. Neer, *Cell* **80,** 249 (1995).

[4] M. I. Simon, M. P. Strathman, and M. Gautam, *Science* **252,** 802 (1991).

[5] A. M. Buhl, N. L. Johnson, N. Dhanasekaran, and G. L. Johnson, *J. Biol. Chem.* **270,** 24631 (1995).

[6] C. Fromm, O. A. Coso, S. Montaner, N. Xu, and J. S. Gutkind, *Proc. Natl. Acad. Sci. U.S.A.* **94,** 10098 (1997).

[7] H. Jiang, D. Wu, and M. I. Simon, *FEBS Lett.* **330,** 319 (1993).

[8] H. Althoefer, P. Eversole-Cire, and M. I. Simon, *J. Biol. Chem.* **272,** 24380 (1997).

[9] T. A. Voyno-Yasenetskaya, A. M. Pace, and H. R. Bourne, *Oncogene* **9,** 2559 (1994).

[10] T. A. Voyno-Yasenetskaya, B. R. Conklin, R. L. Gilbert, R. Hooley, H. Bourne, and D. Barbar, *J. Biol. Chem.* **269,** 4721 (1994).

[11] Z. S. Vexler, M. Symons, and D. L. Barber, *J. Biol. Chem.* **271,** 22281 (1996).

[12] S. G. Plonk, S. K. Park, and J. H. Exton, *J. Biol. Chem.* **273,** 4823 (1998).

study using mice lacking Gα_{13} indicates that Gα_{13} is involved in thrombin-induced fibroblast migration, which may also involve the Rho family small G proteins.[13]

The possible involvement of Rho-specific guanine nucleotide exchange factors (GEFs) in Gα_{13}-mediated Rho activation has attracted great attention. Among these potential candidate GEFs, p115RhoGEF and its mouse homolog Lsc have a unique structure—the RGS (regulators of G protein signaling)-like domains at the N termini of these proteins. RGS proteins are a group of proteins that share homologous sequence motifs, the RGS domains, and inhibit G protein function by stimulating the GTPase activity of the Gα subunits. The interaction of Gα_{13} with p115RhoGEF has been revealed with both *in vitro* reconstitution and cell transfection systems.[14–16] Although the RGS-like domain of p115RhoGEF inhibits both Gα_{12} and Gα_{13}, only Gα_{13} appears to use p115RhoGEF to activate RhoA. This article describes methods that utilize a transcriptional reporter gene assay system to study the functional interaction of Gα_{13} with p115RhoGEF and characterize RhoA-mediated signaling pathways.

Characterization of RhoA-Mediated Pathways Using SRE.L Luciferase Reporter Gene Assay

RhoA activates SRF in a *Clostridium batulinum* C3 toxin-sensitive manner. C3 toxin specifically ADP-ribosylates RhoA and inhibits RhoA, but not other members of the Rho small G protein family including Cdc42 and Rac proteins.[17] SRF activation can be evaluated by examining SRF-dependent transcriptional activation of a luciferase reporter gene, the transcription of which is controlled by a transcription regulatory element, called SRE.L. SRE.L is a derivative of c-*fos* serum response element (SRE), to which SRF, but not tertiary complex factors (TCFs), binds.[17] Thus, SRE.L-mediated production of luciferase mainly depends on the activity of SRF. To assess the regulation of RhoA by a signaling molecule, for example, Gα_{13}, Gα_{13}-mediated activation of SRE.L-mediated production of luciferase in the presence and absence of C3 toxin is compared. If C3 toxin suppresses Gα_{13}-induced activation of SRF, it suggests that Gα_{13} activates RhoA. As shown in Fig. 1, Gα_{13}-mediated SRF activation is indeed inhibited by C3 toxin. Because of the high cytotoxicity of C3 toxin, it is important to run appropriate controls for the effect of C3 toxin in addition to the normalization of transfection

[13] S. Offermanns, V. Mancino, J. P. Revel, and M. I. Simon, *Science* **275**, 533 (1997).
[14] T. Kozasa, X. J. Jiang, M. J. Hart, P. M. Sternweis, W. D. Singer, A. G. Gilman, G. Bollag, and P. C. Sternweis, *Science* **280**, 2109 (1998).
[15] M. J. Hart, X. J. Jiang, T. Kozasa, W. Roscoe, W. D. Singer, A. G. Gilman, P. C. Sternweis, and G. Bollag, *Science* **280**, 2112 (1998).
[16] J. Mao, W. Xie, H. Yuan, M. I. Simon, H. Mano, and D. Wu, *Proc. Natl. Acad. Sci. U.S.A.* **95**, 12973 (1998).
[17] C. S. Hill, J. Wynne, and R. Treisman, *Cell* **81**, 1159 (1995).

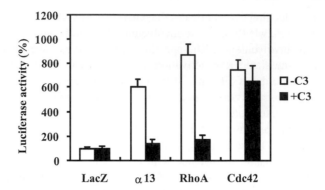

FIG. 1. Regulation of SRF-mediated gene transcription. NIH 3T3 cells were cotransfected with 0.1 μg of SRE.L-luciferase reporter plasmid, 0.1 μg of GFP expression construct, and 0.05 μg of activated Cdc42V12 or RhoV14, or 0.01 μg of Gα_{13}QL (α_{13}) in the presence (solid columns) or absence (open columns) of 0.02 μg of C3 toxin expression plasmid. LacZ expression plasmid was used to make the total amount of DNA equal (0.5 μg/well) in all transfection. One day later, cells were lysed, and GFP levels and luciferase activity were determined with a Wallac Victor^2 multilabel counter. The luciferase activities presented are normalized against the levels of GFP expression. Data show similar tendencies with or without normalization. Error bars represent standard deviations.

(see below). One of the controls we frequently use is to examine the effect of C3 toxin on Cdc42-mediated activation of SRF. C3 toxin does not modify Cdc42; thus, it should not inhibit Cdc42-induced SRF activation. This experiment, as shown in Fig. 1, can serve as an excellent control for the specific effect of C3 toxin on RhoA.

RhoA activates SRE.L-mediated transcription in a number of cell lines, including NIH 3T3 mouse fibroblast cells, human embryonic kidney (A293) cells, primary mouse embryonic fibroblast cells, Cos-7 monkey kidney cells, cardiac myocytes, and lymphocytes. It is preferable that transient transfection be used, because the chromosome-integrated reporter gene appears to be less efficiently regulated by the RhoA pathway.[18]

Functional Interaction between Gα_{13} and p115RhoGEF

Many lines of evidence have indicated that Gα_{13} activates RhoA. It is a reasonable hypothesis that RhoA-specific GEF may mediate Gα_{13}-induced RhoA activation. To prove the hypothesis with a transfection assay system, three criteria need to be met: (1) The RhoA-specific GEF activates SRF in a C3 toxin-sensitive manner, and it acts synergistically with Gα_{13} in activation of SRF; (2) a mutant of this RhoA-specific GEF inhibits Gα_{13}-mediated SRF activation; and (3) Gα_{13} physically interacts with the RhoA-specific GEF. p115RhoGEF meets all

[18] A. S. Alberts, O. Geneste, and R. Treisman, *Cell* **92,** 475 (1998).

three criteria. p115RhoGEF activates SRF, and the activation is sensitive to C3 toxin.[16] In addition, p115RhoGEF acts synergistically with $G\alpha_{13}$, but not $G\alpha_{12}$ or G_q, in activation of SRF.[16] p115RhoGEF, like other GEFs for the Rho family of small G proteins, contains a DH (Dbl homology) domain, which is flanked by a PH (pleckstrin homology) domain. The DH domain is required for the exchange activity,[19] and thus is responsible for regulation of downstream effectors. The DH deletion mutant is therefore a putative dominant negative mutant if the deletion does not affect the interaction with its upstream regulator. Interestingly, GEF115 also contains an RGS-like domain at its N-terminal region that is apart from the DH domain. Because the RGS domains in RGS proteins interact directly with G protein α subunits,[20] it is possible that the RGS-like domain of p115RhoGEF is responsible for the interaction with a $G\alpha$ protein, which may be the upstream regulator for the GEF. Thus, the DH-deletion mutant of GEF115 is likely to be a potent dominant negative mutant. In fact, the DH-deletion mutant of p115RhoGEF turned out to inhibit both $G\alpha_{12}$- and $G\alpha_{13}$-mediated activation of SRF.[16] However, because there is no synergistic effect between $G\alpha_{12}$ and GEF115, $G\alpha_{12}$ may not activate GEF115. This conclusion is consistent with the findings made *in vitro* with recombinant proteins.[14,15] This is a good example of why both the first and second criteria must be satisfied before a conclusion may be drawn that one protein activates the other in a signaling pathway, using the transfection assay. Establishment of the third criterion, if possible, will further cement the conclusion. Immunoprecipitation has been used to detect the physical interaction between $G\alpha_{13}$ and p115RhoGEF.[15]

Normalization of Reporter Gene Assay

Overexpression of proteins in a cell often leads to cell death. Both C3 toxin and the activated $G\alpha_{13}$ as well as other activated $G\alpha$ subunits are highly cytotoxic when overexpressed in cells. Many other factors, including the cell condition, the length and quality of cDNAs, and the number of cotransfected plasmids, could also affect the transfection efficiency. Thus, normalization of the transfection is important, especially when inhibition of SRF activation is observed. In addition, the normalization may allow the results from different experiments to be compared. Coexpression of β-galactosidase (LacZ) has been widely used to normalize the transfection. This method, however, suffers from labor intensiveness and a narrow linear dynamic range. We have been using coexpression of green fluorescence proteins (GFP) to normalize the transfection.[16,21-24] This method significantly

[19] M. J. Hart, S. Sharma, N. Masry, R. G. Qiu, P. McCabe, P. Polakis, and G. Bollag, *J. Biol. Chem.* **271,** 25452 (1996).

[20] J. J. Tesmer, D. M. Berman, A. G. Gilman, and S. R. Sprang, *Cell* **89,** 251 (1997).

[21] J. Mao, W. Xie, H. Yuan, M. I. Simon, H. Mano, and D. Wu, *EMBO J.* **17,** 5638 (1998).

Seed NIH3T3 cells (5×10^3/well) in 24-well plates

↓ 24h

Transfect the cells using lipofectamine

↓ 3h

Stop transfection by switching to 300 ul 0.5% Serum DMEM

↓ 24hr

Wash the cells once with PBS and lyse them with 200ul lysis buffer

↓ 5 mins on a shaker

Transfer the cell lysate (40 ul) into a well of a 96-well white fluorescence plate

↓

Measure the fluorescence

↓

Add luminance substrate (40 ul) and measure luminescence in the same plate

↓

Normalize luciferase activity using the following formula:

$$\text{Normalized luciferase activity} = \frac{\text{Sample Lumin.}}{\text{Control Lumin.}} \times \frac{\text{Control GFP}}{\text{Sample GFP}} \times 100$$

FIG. 2. Outline of the experimental procedures. Lumin., Luminescence reading; GFP, GFP fluorescence reading. We usually take samples transfected with GFP, *lacZ,* and reporter gene as the control. The normalized activity is the percent increase over the control.

streamlines the normalization process by taking advantage of multiwell plate readers that are capable of reading both fluorescence and luminescence. In addition, this method exhibits a wide linear dynamic range (see Fig. 2).

If significant differences between normalized and nonnormalized data occur, the experiment is suggested to be repeated with modification of DNA concentrations or shortening expression time. The amount of plasmids encoding proteins that are suspected to cause cell death should be reduced. Even though we favor an expression duration of 24 hr before luciferase assays, which allows moderate expression of exogenous proteins, the expression of GFP can be detected 14 hr after transfection.

[22] J. Mao, H. Yuan, W. Xie, and D. Wu, *J. Biol. Chem.* **273,** 27118 (1998).

[23] L. Li, H. Yuan, C. Weaver, J. Mao, G. H. Farr III, D. J. Sussman, J. Jonkers, D. Kimelman, and D. Wu, *EMBO J.* **18,** 4233 (1999).

[24] L. Li, H. Yuan, W. Xie, J. Mao, E. McMahon, D. Sussman, and D. Wu, *J. Biol. Chem.* **274,** 129 (1999).

Exemplary Experimental Procedures

For transfection of NIH 3T3 cells, which are maintained in Dulbecco's modified Eagle's medium (DMEM) containing 10% (v/v) fetal calf serum at 37° under 5% CO_2, 0.5 ml of NIH 3T3 cells (5×10^4 cells/well) is seeded into one well of a 24-well plate the day before transfection. Cells are transfected with 0.5 μg of DNA per well, using LipofectAMINE Plus (Life Technologies, Rockville, MD) in 250 μl of serum-free DMEM as suggested by the manufacturer. To avoid significant cytotoxicity, we usually use 0.01–0.025 μg per well of Gα13QL, an activated form of Gα13 and C3 toxin. In addition, we also use GFP expression plasmid (0.1 μg/well; Clontech, Palo Alto, CA) for normalization. The reporter gene is usually used at 0.1 μg/well. *lacZ* expression plasmid is used to make up the total DNA concentration to 0.5 μg/well. For NIH 3T3 cells, we found that 1 μl each of Plus and LipofectAMINE reagents gives optimal transfection efficiency. The amounts of plasmid DNA, Plus reagent, and LipofectAMINE need to be optimized for different cell lines. Transfection is usually stopped after 3 hr by switching to 300 μl of culture medium containing 0.5% (v/v) fetal bovine serum. Cells are subjected to analysis 24 hr after transfection. Although we generally use 24 hr of expression time before the luciferase assay, the expression duration should be determined on the basis of cell types and the strength of the promoters used in the experiments. We found that GFP can be detected 14 hr after transfection, using the cytomegalovirus (CMV) promoter, and that the level peaks 48 hr after transfection.

Luciferase assays are performed with a Roche Biochemical (Indianapolis, IN) luciferase assay kit as instructed by the manufacturer. Cells are washed with phosphate-buffered saline (PBS) and lysed in 200 μl of lysis buffer provided by the kit. Cells are incubated at room temperature for 5 min on a shaker. Cell lysate (40 μl) is transferred into a well of a 96-well white fluorescence plate (VWR, West Chester, PA) and the plate is read by a multilabel counter (Victor[2]; Wallac, Gaithersburg, MD) under the fluorescence mode. The luminance substrate (provided by the kit) is then added, and the same plate is subjected to luminescence reading.

Concluding Remarks

This article describes the use of a reporter gene assay system to investigate the functional interaction between Gα13, an α subunit of the G_{12} family of G proteins, and p115RhoGEF, a RhoA-specific guanine exchange factor. The Gα13–RhoA signaling pathway plays a critical role in a variety of cellular functions, including regulation of cell mobility and cell growth. Using this cellular system, we found that p115RhoGEF is involved in Gα13-mediated activation of the small G protein RhoA and that the functional interaction of Gα13 and p115RhoGEF requires the N-terminal RGS-like domain of p115RhoGEF. These findings are consistent with the data obtained from an *in vitro* reconstitution study showing that recombinant p115RhoGEF acts as a GAP for both $G\alpha_{12}$ and $G\alpha_{13}$, but only $G\alpha_{13}$ stimulates the guanine nucleotide exchange activity of p115RhoGEF.[14,15]

To obtain a fast, reliable, and reproducible measurement of reporter gene activity in different transfections, we have developed a novel, easy-to-use procedure to normalize the transfection between different samples. Comparing with other commonly used normalization methods, including those using β-galactosidase or alkaline phosphatase, the GFP-based normalization method has its obvious advantages: time and cost saving, high sensitivity, and measurement of fluorescence and luminescence of the same sample, thus avoiding the variation caused by sample handling.

p115RhoGEF may not be the sole intermediary mediator that links $G\alpha_{13}$ to RhoA. In an *in vitro* reconstitution study, $G\alpha_{13}$ only moderately increases the guanine nucleotide exchange activity of p115RhoGEF, and in our cellular reporter system, the dominant negative form of p115RhoGEF partially blocks $G\alpha_{13}$ or $G\alpha_{13}$-coupled receptor-induced activation of SRF. By using similar transfection and reporter gene assay systems we have found alternative signaling pathways that may lead to regulation of RhoA by $G\alpha_{12}$ and $G\alpha_{13}$. We found that the Tec/Btk family of protein tyrosine kinases are also involved in $G\alpha_{12/13}$-mediated regulation of RhoA and SRF.[21] This finding is supported by the observation that purified $G\alpha_{12}$ activates Btk *in vitro*.[25] However, little is known about the functional relationship between p115RhoGEF and the Tec/Btk family protein kinases in signal transduction for $G\alpha_{13}$. Is regulation of p115RhoGEF and the Tec/Btk family tyrosine kinases by $G\alpha_{13}$ cell-type specific? Can the Tec/Btk kinases phosphorylate and regulate p115RhoGEF or other Rho-specific guanine nucleotide exchange factors to activate RhoA? The approaches described in this article are suitable for addressing these questions.

Acknowledgments

We thank Alan Hall, J. Silvio Gutkind, and Matthew Hart for providing cDNAs. This work is supported by grants to D.W. from the NIH (GM53162 and GM54167) and from the National Heart Association.

[25] Y. Jiang, W. Ma, and X. Huang, *Nature (London)* **395,** 808 (1998).

Section VII

Protein Kinases and Phosphatases

[32] Analysis of c-Jun N-Terminal Kinase Regulation and Function

By J. PERRY HALL and ROGER J. DAVIS

Introduction

Mitogen-activated protein (MAP) kinase signal transduction pathways coordinate appropriate cellular responses to signals from the extracellular environment. These pathways play significant roles in a diverse set of physiological processes including cellular proliferation, cellular differentiation, oncogenic transformation, programmed cell death, and the development of acquired immunity. MAP kinases have also been demonstrated to participate in a variety of G-protein-coupled effector mechanisms.[1–3]

The c-Jun NH_2-terminal kinase (JNK) and p38 groups of MAP kinases are activated during the cellular responses to proinflammatory cytokines and various forms of environmental stress, and this distinguishes them from the extracellular signal-regulated kinase (ERK) group of MAP kinases, which are activated primarily in response to mitogens and growth factors. Protocols have been developed for analyzing the regulation and function of JNK activity. These methods allow the researcher to examine JNK activity, the subcellular localization of activated JNK, and the phosphorylation status of JNK substrates *in vivo*. In addition, methods have been developed to increase or diminish JNK activity in a controlled manner. In this article we discuss these methods and provide an indication of their applicability to different experimental systems.

Measurement of c-Jun N-Terminal Kinase Protein Kinase Activity

The immune complex protein kinase assay is now a standard method for measuring JNK activity, and it remains the method of choice for measuring the activity of exogenously expressed JNK in tissue culture cell lines.[4,5] In this method, JNK is immunoprecipitated from cell lysates with antibodies that are bound to either protein A– or protein G–Sepharose beads (depending on the antibody subclass and the animal species from which it is derived), and the immunoprecipitates are then incubated with an appropriate JNK substrate protein and $[\gamma\text{-}^{32}P]ATP$. The various proteins in the kinase assay mixture are then resolved by sodium dodecyl

[1] Y. T. Ip and R. J. Davis, *Curr. Opin. Cell Biol.* **10**, 205 (1998).

[2] L. A. Tibbles and J. R. Woodgett, *Cell. Mol. Life Sci.* **55**, 1230 (1999).

[3] T. S. Lewis, P. S. Shapiro, and N. G. Ahn, *Adv. Cancer Res.* **74**, 49 (1998).

[4] B. Derijard, *Cell* **76**, 1025 (1994).

[5] J. Raingeaud, *J. Biol. Chem.* **270**, 7420 (1995).

sulfate–polyacrylamide gel electrophoresis (SDS–PAGE), and the incorporation of [^{32}P]phosphate into the substrate protein is quantitated. To measure endogenous JNK activity by this method we now routinely use a goat antibody raised against the carboxy terminus of the JNK1α1 isoform (C17; Santa Cruz Biotechnology, Santa Cruz, CA). This antibody primarily immunoprecipitates the JNK1α1 isoform, and it is not suitable for determining the activity of other JNK isoforms. For experiments in which JNK is exogenously expressed in cells, JNK expression plasmids designed to incorporate an epitope tag at the JNK amino terminus can be employed. The tagged JNK proteins can then be immunoprecipitated with an antibody that recognizes this tag. Two widely used epitope/antibody systems are the M2 antibody from Sigma (St. Louis, MO), which recognizes the FLAG tag, and the 12CA5 antibody from Roche (Indianapolis, IN), which recognizes the hemagglutinin (HA) tag. Similar methods can be used to measure the activity of JNK purified by other methods.

An alternative to the anti-JNK immune complex kinase assay is the solid-phase kinase assay.[6] This method requires that a given kinase can bind tightly to a unique substrate before the phosphorylation event. All JNK isoforms form a stable association with the amino-terminal domain of the transcription factor c-Jun, and it was in fact this property of JNK proteins that facilitated the initial characterizations of their activities.[6] Therefore, by using the c-Jun amino terminus as the substrate in a solid-phase kinase assay one can measure the sum of all JNK activities in a given cell type. For this assay a glutathione S-transferase (GST)–c-Jun fusion protein containing the amino-terminal 79 amino acids of human c-Jun is first immobilized on glutathione–Sepharose beads, and the beads are then incubated with cell lysates to promote the GST–c-Jun/JNK association. After this, the beads are incubated with [γ-^{32}P]ATP. The proteins in the kinase assay are then resolved by SDS–PAGE, and the incorporation of [^{32}P]phosphate into the GST–c-Jun substrate is quantitated.

In certain cases it may be necessary to determine the relative contributions of JNK1, JNK2, and JNK3 isoforms to total cellular JNK activity. This can be done, in principle, by immunocomplex protein kinase assays using isoform-specific antibodies. However, commercially available antibodies that distinguish each JNK isoform are not available. JNK3 is expressed in only a limited number of cell types, including neurons.[7,8] JNK3 activity can be measured by immune depletion of the JNK1 and JNK2 protein kinases from cell extracts.[8] Measurement of the separate activity of JNK1 and JNK2 is difficult. One experimental strategy is to take advantage of the different masses of these JNK isoforms. While both JNK1 and JNK2 are alternatively spliced to form a group of 46- and 55-kDa protein

[6] M. Hibi, A. Lin, T. Smeal, A. Minden, and M. Karin, *Genes Dev.* **7,** 2135 (1993).

[7] S. Gupta, *EMBO J.* **15,** 2760 (1996).

[8] D. D. Yang, *Nature (London)* **389,** 865 (1997).

kinases,[7] we find that 46-kDa JNK1 and 55-kDa JNK2 are selectively expressed. Furthermore, the 46-kDa JNK1 and JNK2 proteins (and the 55-kDa JNK1 and JNK2 proteins) can be resolved by high-resolution SDS–PAGE.[9-11] Two methods can be used to examine JNK activation after SDS–PAGE. First, phosphorylation of JNK on threonine and tyrosine (the activating sites) can be monitored by Western blot analysis using anti-phospho-JNK antibodies (see below). The *in situ* (or in gel) protein kinase assay is an alternative method that can be used to detect JNK activity after SDS–PAGE.[4,6,12,13] In this procedure, the substrate protein [e.g., GST–c-Jun(1–79)] is uniformly polymerized into an SDS–polyacrylamide gel, cell lysates are electrophoresed through the gel, and the resolved proteins are then subjected to a denaturation/renaturation step. The gel is then incubated with $[\gamma\text{-}^{32}\text{P}]$ATP to allow refolded protein kinases to phosphorylate the polymerized substrate at their specific migrational position in the gel, and the incorporation of $[^{32}\text{P}]$phosphate into the polymerized substrate is quantitated. Thus, the in gel kinase assay can be used to determine the relative contributions of JNK1 and JNK2 isoforms to the overall JNK activity in a sample. It should be noted that autophosphorylating kinases will yield a signal in this assay; therefore, to distinguish bona fide substrate phosphorylation it is important to run a control gel in parallel that does not contain the polymerized substrate (e.g., GST alone).

Reagents

Triton lysis buffer (TLB): 20 mM Tris-HCl (pH 7.4), 137 mM sodium chloride, 25 mM sodium β-glycerophosphate, 2 mM sodium pyrophosphate, 2 mM EDTA, 1% (v/v) Triton X-100, and 10% (v/v) glycerol. Prepare fresh from a 2× stock (store at 4°), at which time phenylmethylsulfonyl fluoride (PMSF) and Na$_3$VO$_4$ (sodium orthovanadate) are each added to final concentrations of 1 mM, and leupeptin is added to a final concentration of 10 μg/ml

Kinase assay buffer (KB): 25 mM HEPES (pH 7.4), 25 mM sodium β-glycerophosphate, and 25 mM MgCl$_2$. Prepare fresh from a 2× stock (store at 4°), at which time Na$_3$VO$_4$ is added to a final concentration of 0.1 mM, and dithiothreitol (DTT) is added to a final concentration of 0.5 mM

Buffer A: 50 mM Tris-HCl (pH 8.0) and 5 mM 2-mercaptoethanol

Buffer B: 40 mM HEPES (pH 7.4), 5 mM MgCl$_2$, 0.1 mM EGTA, and 2 mM DTT

Phosphate-buffered saline (PBS): 20 mM sodium phosphate (pH 7.4) and 150 mM NaCl

[9] C. Dong, *Science* **282,** 2092 (1998).
[10] D. D. Yang, *Immunity* **9,** 575 (1998).
[11] C. Tournier, P. Hess, D. D. Yang, J. Xu, T. K. Turner, A. Nimnual, D. Bar-Sagi, S. N. Jones, R. A. Flavell, and R. J. Davis, *Science* **288,** 870 (2000).
[12] I. Kameshita and H. Fujisawa, *Anal. Biochem.* **183,** 139 (1989).
[13] Y. Gotoh, *Eur. J. Biochem.* **193,** 661 (1990).

2-Propanol (20%, v/v)–50 mM Tris-HCl (pH 8.0)

Trichloroacetic acid (TCA; 5%, w/v)–1% (w/v) sodium pyrophosphate

SDS–PAGE sample buffer (6×): 21 ml of 500 mM Tris-HCl (pH 6.8)–0.4% (w/v) SDS, 10.8 ml of glycerol, 3 g of SDS, 2.79 g of DTT, and 3.6 mg of bromphenol blue (aliquoted and stored frozen at −20°)

Protein G–Sepharose (Pharmacia, Piscataway, NJ): Swell 50:50 in 10 mM HEPES (pH 8.0)–0.025% (v/v) Triton X-100 and store at 4°

Glutathione–Sepharose 4B (Pharmacia)

Recombinant GST and GST–c-Jun (amino acids 1 to 79) proteins: Express in *Escherichia coli* and purify by affinity chromatography as described[14]

Goat anti-JNK1 antibody (C17; Santa Cruz Biotechnology), or mouse monoclonal anti-FLAG antibody (M2; Sigma), or mouse monoclonal anti-HA antibody (12CA5; Roche)

ATP (Sigma) and [γ-^{32}P]ATP (New England Nuclear, Boston, MA)

SDS–10% (w/v) polyacrylamide gels: Running gel: 10% (w/v) acrylamide (from 30:0.8 acrylamide:bisacrylamide mix; National Diagnostics, Atlanta, GA) and 0.375 M Tris-HCl (pH 8.8); stacking gel: 4.5% (w/v) acrylamide and 0.125 M Tris-HCl (pH 6.8)

Guanidine hydrochloride (Sigma)

Tween 40 (United States Biochemical, Cleveland, OH)

Immune Complex Protein Kinase Assay

The cells to be examined are first washed in PBS, and then scraped and/or resuspended in ice-cold TLB and transferred to a standard 1.7-ml microcentrifuge tube on ice. It is important to use enough TLB to ensure complete and efficient lysis. After 5 to 10 min on ice the lysis suspensions are pelleted at 15,000 rpm for 15 min at 4° to remove insoluble material. This procedure is sufficient for lysis of most, if not all, adherent and suspension cell types; however, tissue samples should be homogenized in TLB with a Dounce homogenizer and allowed to lyse on ice for ~30 min. Clarified lysates are then transferred to a fresh tube and kept at −70° until needed. It is often worthwhile to freeze several aliquots of the clarified lysates if multiple kinase assays are to be performed on the same sample. This can avoid some of the protein degradation problems caused by multiple freeze–thaws. Lysate protein concentrations are determined by the Bradford assay.

We routinely initiate our immune complex protein kinase assays by prebinding the antibody to the protein G–Sepharose. This tends to avoid potential problems associated with the presence of contaminants and/or non-IgG antibody classes found in many antibody preparations. For each assay approximately 1 μg of anti-JNK1, anti-FLAG, or anti-HA antibody is added to a microcentrifuge tube containing

[14] D. B. Smith and K. S. Johnson, *Gene* **67**, 31 (1988).

500 μl of TLB and 60 μl of a 50% slurry of protein G–Sepharose (i.e., prewashed several times with an equal volume of TLB just before use). (The choice of antibody will depend on whether endogenous JNK1α1 or exogenously expressed, epitope-tagged JNK proteins are being analyzed.) The prebinding mixtures are incubated for 45 to 60 min at 4° on a rotating platform, and the protein G–Sepharose/antibody aggregates are then pelleted briefly in a microcentrifuge and washed twice with 1 ml of TLB. After the last wash, 500 μl of TLB is added to each pellet. An equal amount of protein lysate is then added to the respective tubes, and the tubes are placed on a 4° rotating platform for 3 hr to overnight. The protein G–Sepharose/antibody/JNK aggregates are then washed three times with 1 ml of TLB and two times with 1 ml of KB. After the last wash all of the remaining KB is removed from the top of the bead pellets, using a small pipette tip. To initiate the kinase assay 50 μM [γ-^{32}P]ATP (10 Ci/mmol) and $\geq$2 μg of GST–c-Jun(1–79) in 25 μl of KB are added to each pellet, and the reactions are allowed to proceed at 30° for 30 min. The optimal reaction time must be determined empirically, as it should fall within the linear range of activity for all samples being analyzed. The reactions are terminated by the addition of 10 μl of 6× sample buffer. Just before electrophoresis, the samples are heated to $\geq$100° for 5 min and pelleted. A fraction of each supernatant is then examined on a 10% (w/v) SDS–polyacrylamide gel. The dye fronts are cut away from the running gel, or allowed to migrate off the gel, in order to remove all unincorporated radioactivity. The gel is fixed, stained with Coomassie blue, and subsequently destained enough to visualize equal loading of substrate protein and antibody heavy chain in each sample. After drying the gel at 80° it can be visualized by autoradiography, and the incorporation of [^{32}P]phosphate into GST–c-Jun(1–79) can be quantitated on a PhosphorImager (Molecular Dynamics, Sunnyvale, CA).

Solid-Phase Protein Kinase Assay

The protocol is adapted from Hibi *et al.*,[6] and it is similar in many respects to the immune complex kinase assay described above. For each assay, 10 μg of GST–c-Jun(1–79) is added to a microcentrifuge tube containing 500 μl of TLB and 60 μl of a 50% slurry of glutathione–Sepharose 4B (i.e., prewashed several times with an equal volume of TLB just before use). The tubes are incubated for 45 to 60 min at 4° on a rotating platform, and the glutathione–Sepharose/GST–c-Jun(1–79) aggregates are then washed twice with 1 ml of TLB. After the last wash 500 μl of TLB is added to each pellet. An equal amount of protein lysate is then added to the respective tubes, and the tubes are placed on a 4° rotating platform for 3 hr to overnight. The glutathione–Sepharose/GST–c-Jun(1–79)/JNK aggregates are then washed three times with 1 ml of TLB and two times with 1 ml of KB. After the last wash, all of the remaining KB is removed from the top of the bead pellets, using a small pipette tip. To initiate the kinase assay 50 μM [γ-^{32}P]ATP (10 Ci/mmol) in 10 μl of KB are added to each pellet, and the reactions are allowed to proceed at 30° for ~30 min. The reactions are terminated and analyzed by SDS–PAGE as described above.

In Situ Detection of Protein Kinase Activity after Sodium Dodecyl Sulfate–Polyacrylamide Gel Electrophoresis

The protocol is adapted from Gotoh.[13] To begin, 10% (w/v) SDS–polyacrylamide gels are prepared in which GST–c-Jun(1–79) (0.25 mg/ml) is included in the running gel and allowed to polymerize into the gel matrix. Control gels are also prepared in which GST (0.25 mg/ml) is used. For each cell lysate to be examined, 20 to 50 μg of total protein is resuspended in 6× SDS–PAGE sample buffer, heated to ≥100° for 5 min, and then loaded onto the GST–c-Jun(1–79) gel and, in parallel, the GST-alone control gel. Once electrophoresis is completed SDS is removed from the gels by washing twice in 100 ml of 20% (v/v) 2-propanol–50 mM Tris-HCl (pH 8.0) for 30 min at room temperature with gentle shaking. The gels are then washed twice in 100 ml of buffer A. The resolved proteins in the gels are fully denatured by washing the gels twice in 100 ml of 6 M guanidine hydrochloride–buffer A (filtered before use). The proteins are then renatured by shaking the gels overnight at 4° with several changes of 200 ml of buffer A–0.04% (v/v) Tween 40. The gels are then washed for 30 min at room temperature in 25 ml of buffer B, and then incubated for 60 min with gentle shaking in 25 ml of buffer B containing 0.25 μM [γ-^{32}P]ATP (10 Ci/mmol). The gels are then washed with several changes of 200 ml of 5% (w/v) TCA–1% (w/v) sodium pyrophosphate until the radioactivity at the edges of the gels reaches background levels. After drying the gels at 80° the incorporation of [^{32}P]phosphate into the polymerized GST–c-Jun(1–79) can be quantitated on a PhosphorImager. The GST-alone control gels allow for a determination of which signals arise from bona fide c-Jun(1–79) substrate phosphorylation, and which are the products of renatured, autophosphorylating kinases or GST-directed kinase activities.

Analysis of c-Jun N-Terminal Kinase Activation and Subcellular Localization, Using Phosphorylation-Specific Antibodies

MAP kinases require the phosphorylation of a conserved threonine and a conserved tyrosine residue in their kinase subdomain VIII in order to become fully activated.[1–3] In all JNK isoforms this Thr-X-Tyr motif is Thr-Pro-Tyr (TPY). Phospho-JNK antibodies are now commercially available from several suppliers [e.g., New England Biolabs (Beverly, MA) and Promega (Madison, WI)]. These antibodies are prepared with a phosphorylated peptide corresponding to the region surrounding the TPY motif. It is expected (but not experimentally demonstrated) that these phospho-JNK antibodies recognize all JNK isoforms equally because the sequence surrounding this TPY motif is conserved in all JNK isoforms.

Immunoblot analysis performed by probing with these anti-phospho-JNK antibodies provides only an estimation of JNK activities in cell lysates, because the phosphorylation state of any MAP kinase is not necessarily a direct measure of its activity. Other factors may be required for achieving a full complement of kinase activity in response to a given stimulus. For this reason, it is better to measure JNK activity directly in a protein kinase assay. However, this immunoblot analysis is simple to perform and represents a useful initial experimental approach. We perform immunoblot analysis with phospho-JNK antibodies by following the manufacturer recommendations for the probing conditions.

Immunocytochemistry experiments with these antibodies now allow researchers to visualize the subcellular localization of activated JNK molecules. This facilitates a qualitative assessment of JNK activities at the single-cell level, and this is particularly important for cell and tissue types that cannot be harvested in sufficient quantity for routine kinase assays and/or immunoblot analyses. However, as with any antibody, in certain cell types and under certain detection conditions these antibodies may not be completely specific for phosphorylated JNK proteins. Therefore, control experiments with blocking peptides are warranted in order to gain the necessary confidence that any signals observed with these antibodies derive from the presence of phosphorylated JNKs.

Analysis of c-Jun N-Terminal Kinase Signaling Activity

Activation of the c-Jun transcription factor—and the sequence-specific transcriptional activator AP-1, of which c-Jun forms a part—requires the phosphorylation of Ser-63 and Ser-73 in the c-Jun amino-terminal *trans*-activation domain by JNK.[1] For this reason, the function of JNK pathway activity can be assessed by measuring changes in the activation and/or phosphorylation of c-Jun.

Changes in c-Jun transcriptional activity can be conveniently measured with reporter gene assays in transfection experiments. AP-1 reporter plasmids can be used because c-Jun activated by JNK does increase AP-1 transcription activity. However, care needs to be taken because the JNK pathway represents only one of several pathways that can regulate AP-1 activity. Thus, cells derived from knockout mice demonstrate that Ras can activate AP-1 transcription activity independently of the JNK signaling pathway.[15] For this reason, we prefer to employ a more specific reporter, the activation domain of c-Jun fused to a heterologous DNA-binding domain. In this assay, cells are transfected with (1) a plasmid that expresses the c-Jun *trans*-activation domain fused in frame with the DNA-binding domain of the yeast transcription factor GAL4, and (2) a "reporter" plasmid that contains the firefly luciferase gene driven by a minimal promoter containing GAL4 DNA-binding sites. On phosphorylation the *trans*-activation potential of the fusion protein is

---

[15] D. Yang, *Proc. Natl. Acad. Sci. U.S.A.* **94**, 3004 (1997).

increased, and this produces a concomitant increase in expression of the luciferase reporter gene. Luciferase protein is then quantitated by measuring its enzymatic conversion of a luminescent substrate (see Reporter Gene Assays below). This approach allows the researcher to address changes in JNK pathway activity that occur in response to extracellular stimuli (e.g., UV light, bacterial endotoxin, proinflammatory cytokines) or ectopically expressed signaling proteins. Indeed, it remains a standard method for assessing the contribution(s) of exogenously expressed, putative JNK pathway components to overall pathway throughput. Unfortunately, it is not readily adaptable to analyses of JNK pathway activity in nontransfectable cell types such as primary cells or tissues; however, this problem can be circumvented by constructing transgenic animals that harbor a pathway reporter transgene.[8,10]

The JNK-dependent phosphorylation and activation of transcription factors other than c-Jun can be adapted to the reporter gene assay approach, but problems of specificity arise given that JNK is the only c-Jun amino-terminal kinase. For example, ATF2 is phosphorylated and activated by both JNK and p38; therefore, if a GAL4–ATF2 fusion is used in these assays then a JNK-specific activator or inhibitor must also be utilized in order to remain confident that changes in reporter gene expression are in fact JNK dependent.[16,17] This is a problem, because many of the stress stimuli and ectopically expressed activators which activate JNK also activate p38 (e.g., UV light, osmotic stress, bacterial endotoxin).

JNK can be activated in transfection assays by using upstream components of the JNK pathway. Unfortunately, MAPKKKs (MAP kinase kinase kinases) are not specific for JNK and can cause activation of the ERK and p38 MAPK. Similarly, the JNK activator MKK4 (MAPK kinase 4) can activate JNK, but also activates p38 MAPK.[18] However, JNK activation can be specifically induced by overexpression of MKK7, although large increases in JNK activity require that MKK7 be coexpressed with JNK.[19,20]

Inhibition of the JNK pathway can be achieved with dominant negative JNK. Examples include JNK with mutations in the ATP-binding site and mutations in the activating sites of phosphorylation.[4,21–23] Dominant negative MKK4 can also be used to inhibit JNK, although under some conditions inhibition of p38

[16] S. Gupta, D. Campbell, B. Derijard, and R. J. Davis, *Science* **267,** 389 (1995).

[17] J. Raingeaud, A. J. Whitmarsh, T. Barrett, B. Derijard, and R. J. Davis, *Mol. Cell. Biol.* **16,** 1247 (1996).

[18] B. Derijard, *Science* **267,** 682 (1995).

[19] C. Tournier, A. J. Whitmarsh, J. Cavanagh, T. Barrett, and R. J. Davis, *Proc. Natl. Acad. Sci. U.S.A.* **94,** 7337 (1997).

[20] C. Tournier, A. J. Whitmarsh, J. Cavanagh, T. Barrett, and R. J. Davis, *Mol. Cell. Biol.* **19,** 1569 (1999).

[21] M. Rincon, *J. Exp. Med.* **188,** 1817 (1998).

[22] A. J. Whitmarsh, S. H. Yang, M. S. Su, A. D. Sharrocks, and R. J. Davis, *Mol. Cell. Biol.* **17,** 2360 (1997).

[23] A. J. Whitmarsh, P. Shore, A. D. Sharrocks, and R. J. Davis, *Science* **269,** 403 (1995).

MAPK is also observed.[22,24] Dominant-negative MKK7 can also be used to inhibit JNK.[19,20,25] The most specific inhibitor for JNK that has been described is JNK-interacting protein 1 (JIP-1), which is a scaffold protein that binds components of the JNK pathway signaling module and functions as an enhancer of JNK pathway signaling.[24,26] Overexpression of JIP-1 (or the JNK-binding domain of JIP-1) causes marked inhibition of JNK by sequestration of JNK in the cytoplasm and inhibition of catalytic activity.

Monitoring the phosphorylation status of Jun on Ser-63 and/or Ser-73 is now a standard method for monitoring changes in JNK pathway signaling. Anti-phospho-c-Jun(Ser-63) and anti-phospho-c-Jun(Ser-73) rabbit polyclonal anti-peptide antibodies are commercially available [New England Biolabs; Upstate Biotechnology, Lake Placid, NY], and although they exhibit some cross-reactivity with JunD (a second Jun family member) these antibodies can be used for anti-phospho-c-Jun immunoblotting and immunocytochemistry experiments (see Anti-Phospho-c-Jun Immunoblotting, below). It is important that control experiments be performed to monitor the level of c-Jun expression, because changes in phospho-Jun immunoreactivity could be mediated by either increased Jun phosphorylation or increased Jun expression. The phosphorylation state of c-Jun is certainly not a direct measure of JNK activity, but it is clearly an approximation of the amount of signal transduction through multiple JNK pathways. This makes these antibodies extremely valuable for addressing the amount of JNK-dependent signaling in a biological system. For example, we have used antibodies to Jun and phospho-Jun to examine JNK activity in the hippocampus of wild-type and JNK knockout mice by immunocytochemistry.[8]

Reagents

Phosphate-buffered saline (PBS): 20 mM sodium phosphate (pH 7.4) and 150 mM NaCl

Potassium phosphate buffer, 0.1 M (pH 7.8)

Luciferase assay buffer: 15 mM potassium phosphate (pH 7.8), 25 mM glycylglycine (pH 7.8), 15 mM Mg$_2$SO$_4$, 4 mM EGTA, 2 mM ATP (100 mM stock, pH 7.8), and 1 mM DTT

Z buffer: 100 mM sodium phosphate (pH 7.0), 10 mM KCl, and 1 mM Mg$_2$SO$_4$

Na$_2$CO$_3$ (sodium carbonate), 1 M

D-Luciferin (PharMingen, Carlsbad, CA): Prepare 1 mM stock by dissolving 10 mg of D-luciferin in 31.5 ml of distilled H$_2$O and adjusting to pH 6.0–6.3 by dropwise addition of 100 mM NaOH; aliquot and store at $-20°$

[24] A. J. Whitmarsh, J. Cavanagh, C. Tournier, J. Yasuda, and R. J. Davis, *Science* **281**, 1671 (1998).
[25] T. Moriguchi, *EMBO J.* **16**, 7045 (1997).
[26] M. Dickens, *Science* **277**, 693 (1997).

O-Nitrophenyl-β-D-galactopyranoside (ONPG) (United States Biochemical): Prepare 4-mg/ml stock in 100 mM potassium phosphate (pH 7.0); sterilize by filtration, aliquot, and store at $-20°$

Plasmids: pSG424-c-Jun (encodes fusion of residues 1 to 79 of c-Jun with the GAL4 DNA-binding domain),[6] pG5E1bLuc (five GAL4-binding sites in minimal E1b promoter driving the luciferase reporter gene),[27] pCH110 (β-galactosidase reporter plasmid to control for transfection efficiency) (Pharmacia)

Triton lysis buffer (TLB): 20 mM Tris-HCl (pH 7.4), 137 mM sodium chloride, 25 mM sodium β-glycerophosphate, 2 mM sodium pyrophosphate, 2 mM EDTA, 1% (v/v) Triton X-100, and 10% (v/v) glycerol. Prepare fresh from a 2× stock (store at 4°), at which time PMSF and Na$_3$VO$_4$ (sodium orthovanadate) are each added to final concentrations of 1 mM, and leupeptin is added to a final concentration of 10 μg/ml

SDS–PAGE sample buffer (6×): 21 ml of 500 mM Tris-HCl (pH 6.8)– 0.4% (w/v) SDS, 10.8 ml of glycerol, 3 g of SDS, 2.79 g of DTT, and 3.6 mg of bromphenol blue (aliquoted and stored frozen at $-20°$)

Rabbit anti-phospho-c-Jun(Ser-73) antibody (Upstate Biotechnology)

Donkey anti-rabbit immunoglobulin–horseradish peroxidase (HRP) conjugate (Amersham, Arlington Heights, IL)

Transfer buffer (TB): 25 mM Tris-HCl (pH 7.4), 200 mM glycine, 20% (v/v) methanol, and 0.1% (w/v) SDS

Tris-buffered saline (TBS): 15 mM Tris-HCl (pH 7.4) and 150 mM NaCl

SDS–10% (w/v) polyacrylamide gels: Running gel: 10% (w/v) acrylamide (from 30 : 0.8 acrylamide:bisacrylamide mix; National Diagnostics) and 0.375 M Tris-HCl, pH 8.8); stacking gel: 4.5% (w/v) acrylamide and 0.125 M Tris-HCl (pH 6.8)

Tween 20 (United States Biochemical)

Nonfat dry milk

Western blot chemiluminescence reagent Plus kit (New England Nuclear Life Science Products)

Reporter Gene Assays

We use cells growing in 35-mm multiwell dishes for transfection assays using the LipofectAMINE reagent (Life Technologies). The cells are transfected with a GAL4-c-Jun expression vector, a GAL4-luciferase reporter plasmid, and a β-galactosidase expression vector to normalize transfection efficiency.[22] Control experiments are performed with the GAL4 DNA-binding domain and with a mutated GAL4–cJun protein in which the Ser-63 and Ser-73 phosphorylation sites

[27] S. Gupta, A. Seth, and R. J. Davis, *Proc. Natl. Acad. Sci. U.S.A.* **90,** 3216 (1993).

are replaced with alanine residues. After 48 hr, the cells are washed three times with ice-cold PBS, and then scraped into 200 μl of ice-cold potassium phosphate buffer (pH 7.8) and transferred to a standard 1.7-ml microcentrifuge tube. Samples are then frozen on dry ice for 5 min, and then thawed at 37° for 5 min, and this freeze–thaw cycle is repeated two more times. The samples are then clarified by centrifugation at 15,000 rpm for 15 min at 4°. The luminometer should be set to inject 100 μl of D-luciferin stock solution per sample, and 300 μl of luciferase assay buffer is added to a luminometer cuvette for each sample to be analyzed. The clarified samples are then added to the cuvettes and the luminescence values are read 10 sec after injection of the D-luciferin stock. To control for transfection efficiency, β-galactosidase expression in each sample is measured by adding 15 μl of the clarified sample to 200 μl of Z buffer and 100 μl of ONPG in a microcentrifuge tube and incubating at 37° until the mixture takes on a pale yellow color. The reactions are terminated by the addition of 500 μl of 1 M Na$_2$CO$_3$ and then clarified by centrifugation. The optical densities of these supernatants are then measured at 420 nm.

Anti-Phospho-c-Jun Immunoblotting

The method is adapted from the manufacturer protocol for detecting phosphorylated c-Jun in cell lysates, using the anti-phospho-c-Jun(Ser-73) antibody (Upstate Biotechnology). First, cell lysates are prepared and protein concentrations are determined according to the method outlined above (see Immune Complex Protein Kinase Assay). Equal amounts of cell lysates are then resuspended in an appropriate volume of 6× sample buffer, heated to ≥100° for 5 min, and resolved by electrophoresis on a 10% (w/v) SDS–polyacrylamide gel. The gel is then washed in TB for 5 min, and an appropriately sized piece of Immobilon-P nylon transfer membrane (Millipore, Bedford, MA) is wetted in 100% methanol and then soaked in TB for ~5 min. The blotting/transfer is constructed by sandwiching the gel and Immobilon-P between standard gel blot paper (Schleicher & Schuell Keene, NH) within the electrode plates of a semidry transfer apparatus (Hoefer Scientific, San Francisco, CA). Transfer is performed for 2–3 hr at ≤10 V. The blotted membrane is then blocked for 1 hr at 4° with constant agitation in TBS containing 5% (w/v) nonfat dry milk and 0.05% (v/v) Tween 20. The blot is then moved into fresh TBS–5% (w/v) milk–0.05% (v/v) Tween 20 containing a 1 : 1000 to 1 : 5000 dilution of rabbit anti-phospho-c-Jun(Ser-73), and agitation is continued overnight at 4°. The membrane is then washed three times (10 min each time) at room temperature with TBS containing 0.05% (v/v) Tween 20. It is then incubated in TBS–5% (w/v) milk–0.05% (v/v) Tween 20 containing a 1 : 3000 to 1 : 10,000 dilution of donkey anti-rabbit–HRP for 1 hr. The membrane is then washed three times (10 min each time) with TBS containing 0.05% (v/v) Tween 20 and subsequently developed with enhanced chemiluminescence reagents according to the manufacturer directions (New England Nuclear). Signals are detected by exposure to standard X-ray film.

Strategies for Inhibiting c-Jun N-Terminal Kinase Expression

Ten different mammalian JNK isoforms, encoded by the genes *Jnk1, Jnk2,* and *Jnk3,* have been identified.[7] *Jnk1* and *Jnk2* gene transcripts are alternatively spliced to yield four isoforms each, and *Jnk3* transcripts are processed to yield two isoforms.[7] JNK1 and JNK2 proteins are widely expressed in adult mouse tissues, but JNK3 expression is restricted to brain, heart, and testis. Deciphering the precise functions of these isoforms remains a difficult task given the paucity of reagents that can reliably distinguish one isoform from another, and this problem is only exacerbated by the daunting variety of cell types throughout the body and the complexity of their intracellular signaling mechanisms. Therefore, a clear understanding of the functions of these different isoforms requires that their expression be inhibited in a controlled manner from a given model system.

Mouse gene knockout experiments have provided many important insights into the functions of JNK1, JNK2, and JNK3 isoforms. It has been demonstrated that hippocampal neurons in $Jnk3^{-/-}$ mice show increased resistance to kainic acid-induced seizures and programmed cell death (i.e., apoptosis), and these observations support the notion that JNK3 isoforms may be instrumental in the development of certain neurodegenerative diseases.[8] Both JNK1 and JNK2 isoforms are required during early development because the $Jnk1^{-/-}$ $Jnk2^{-/-}$ compound mutation causes embryonic lethality and a dysregulation of programmed cell death in the developing brain.[28] Furthermore, it has been shown that the $Jnk1^{-/-}$ and $Jnk2^{-/-}$ mutations impair T cell development and function.[9,10] These studies indicate that JNK isoforms play important roles throughout mammalian growth and development. Further analysis of Jnk^- mice will undoubtedly facilitate our understanding of specific JNK isoforms in the broader context of JNK signaling physiology.

Unfortunately, mouse gene knockout strategies are not always appropriate, for example, when there is a need to delete a specific JNK isoform in a unique model system. This is particularly true with human cells. Although several studies have now been done in which both copies of a gene of interest have been deleted by homologous recombination in immortalized, diploid human cell lines, this approach cannot be used with primary human cell types.[29,30] These considerations warrant the need for a rapid, facile strategy that can be used with many different cell types. Several groups have demonstrated that JNK1 and JNK2 isoforms can be selectively downregulated from cultured human cells by using antisense oligonucleotides. The use of antisense JNK oligonucleotides has been reviewed in

---

[28] C. Y. Kuan, *Neuron* **22,** 667 (1999).

[29] F. Bunz, *Science* **282,** 1497 (1998).

[30] T. A. Chan, H. Hermeking, C. Lengauer, K. W. Kinzler, and B. Vogelstein, *Nature (London)* **401,** 616 (1999).

Methods in Enzymology.[31,32] Antisense oligonucleotides are thought to hybridize with complementary sequences in the targeted mRNA, causing degradation of the RNA:DNA hybrid and an inhibition of expression of the protein product. By this approach, it was demonstrated that JNK2 is primarily required for the proliferation of MCF human breast carcinoma[33] and A549 lung carcinoma cell lines[34]; that JNK1 is required for reoxygenation-induced apoptosis in primary human kidney epithelial cells[32]; that JNK2 is required for tumor necrosis factor α (TNF-α)-induced E-selectin expression in primary human dermal endothelial cells[32]; and that JNK2 is required for the proliferation of $p53^-$ RKO and HCT116 human colorectal cancer cell lines.[35] JNK antisense studies such as these provide an excellent opportunity for delineating the roles of JNK isoforms in different cell types, and the experimental use of JNK antisense should be adaptable to many model systems. However, it is important that the conclusions drawn from these studies be carefully considered because it is possible that the effects of the antisense oligonucleotides are mediated by mechanisms that are independent of JNK inhibition.

Concluding Remarks

In this article we discuss current and available technologies for studying the regulation and function of JNK pathways in mammalian cell types. In future these methods will be improved as new components of these pathways are cloned and characterized, and as JNK isoform-specific reagents become available. Together with ongoing JNK knockout and antisense experiments, these improvements will begin to unravel the complicated physiologies surrounding JNK signaling mechanisms. We expect to see the development of specific drugs that can be used to modify the function of the JNK pathway *in vivo*. These approaches represent powerful methods that can be applied to the study of the JNK signaling pathway.

---

[31] F. Bost, R. McKay, N. M. Dean, O. Potapova, and D. Mercola, *Methods Enzymol.* **314,** 342 (2000).
[32] C. L. Cioffi and B. P. Monia, *Methods Enzymol.* **314,** 363 (2000).
[33] O. Potapova, H. Fakhrai, S. Baird, and D. Mercola, *Cancer Res.* **56,** 280 (1996).
[34] F. Bost, R. McKay, N. Dean, and D. Mercola, *J. Biol. Chem.* **272,** 33422 (1997).
[35] O. Potapova, *Mol. Cell. Biol.* **20,** 1713 (2000).

[33] Double-Label Confocal Microscopy of Phosphorylated Protein Kinases Involved in Long-Term Potentiation

By Maria Grazia Giovannini

Introduction

There are a number of phenomenological features of memory that must have parallels at the synaptic/cellular level. Among the most important of these attributes is the ability to establish a learned modification relatively quickly and to have it persist over a relatively long period. The best candidate for a cellular mechanism by which synaptic strength is increased as a consequence of experience is long-term potentiation (LTP).[1] Much of the effort devoted to the cellular study of memory is devoted to the understanding of LTP and most of this work is conducted in hippocampal slices. Most plastic synapses in the central nervous system employ glutamate as their neurotransmitter and current models of LTP propose that the activation of N-methyl-D-aspartate (NMDA) channels is critical for the influx of Ca^{2+} into the postsynaptic cell. The increased postsynaptic concentration of Ca^{2+} activates a variety of calcium-binding proteins, such as adenylyl cyclase, protein kinase C (PKC), calmodulin kinase II (CaMKII), and others.[2,3] These in turn initiate a cascade of events leading to alterations in receptor and channel phosphorylation[4] and conductance as well as to changes in gene expression.[5] These postsynaptic biochemical cascades are capable of being influenced by neurotransmitters. Most notably, norepinephrine, serotonin, acetylcholine, and dopamine can modulate these signal transduction pathways and the resultant magnitude and duration of LTP in the postsynaptic cell.[6–9] This effect might be the correlate of *in vivo* responses to reinforcing environmental stimuli that result in the activation of one or more modulatory systems, which would then facilitate the retention of the experience that preceded the reinforcing event itself. Therefore, an interesting method for induction of hippocampal LTP is the pairing of theta pulse stimulation with

[1] T. V. Bliss and G. L. Collingridge, *Nature* (*London*) **361,** 31 (1993).

[2] T. R. Soderling, *Adv. Second Messenger Phosphoprotein Res.* **30,** 175 (1995).

[3] U. S. Bhalla and R. Iyengar, *Science* **283,** 381 (1999).

[4] A. Barria, D. Muller, V. Derkach, L. C. Griffith, and T. R. Soderling, *Science* **276,** 2042 (1997).

[5] P. V. Nguyen, T. Abel, and E. R. Kandel, *Science* **265,** 1104 (1994).

[6] C. W. Berridge and S. L. Foote, *J. Neurosci.* **11,** 3135 (1991).

[7] N. A. Otmakhova and J. E. Lisman, *J. Neurosci.* **16,** 7478 (1996).

[8] M. J. Thomas, T. D. Moody, M. Makhinson, and T. J. O'Dell, *Neuron* **17,** 475 (1996).

[9] D. G. Winder, K. C. Martin, I. A. Muzzio, D. Rohrer, A. Chruscinski, B. Kobilka, and E. R. Kandel, *Neuron* **24,** 715 (1999).

β-receptor stimulation. Hippocampal CA1 neurons burst at theta rhythm (3–12 Hz) when animals move through space[10] and the neurons of the locus ceruleus, the major noradrenergic input to the hippocampus, increase their firing during the states of increased attention and arousal that accompany learning behaviors.[11] The combination of theta rhythm and β-adrenergic stimulation may thus be of particular importance for learning. Coupling of β-adrenergic stimulation to theta pulse stimulation is particularly effective for induction of LTP, and may be more physiologically relevant than the high-frequency stimulation often used to induce LTP. LTP induced by both forms of stimulation is mediated by the influx of Ca^{2+} ions through NMDA receptors. However, much work remains to be done to understand the nature of these biochemical cascades, their interactions, and their regulation by both intrinsic and extrinsic factors.

It has previously been demonstrated[12,13] by Western immunoblot that the activation of the ERK (extracellular signal-regulated kinase) pathway is necessary for the induction and maintenance of LTP induced by high-frequency stimulation in the rat. Furthermore, Winder and co-workers[9] have demonstrated that ERK plays a regulatory role in LTP induced by theta frequency stimulation and β-adrenergic receptor activation in the mouse. Martin and co-workers[14] showed that activation of ERK by forskolin is followed by translocation of the phosphorylated form of the enzyme to the nucleus in mouse hippocampal slices.

Another kinase that has been demonstrated to play a pivotal role in the various phases of LTP is CaMKII.[15–17] Lisman has proposed that autophosphorylation of this enzyme after Ca^{2+} entry and activation of calmodulin enables it to maintain a biochemical "memory" at the synapse.[15,18] Furthermore, it has been shown that LTP is blocked in mice expressing a mutant form of CaMKII that cannot undergo autophosphorylation.[19] It has also been shown that CaMKII and ERK are both present in the postsynaptic density (PSD) fraction,[20,21] giving support to the possibility that these two kinases might interact in LTP induction and maintenance.

[10] A. A. Fenton and R. U. Muller, *Proc. Natl. Acad. Sci. U.S.A.* **95**, 3182 (1998).

[11] G. Aston-Jones, J. Rajkowski, P. Kubiak, and T. Alexinsky, *J. Neurosci.* **14**, 4467 (1994).

[12] J. D. English and J. D. Sweatt, *J. Biol. Chem.* **271**, 24329 (1996).

[13] J. D. English and J. D. Sweatt, *J. Biol. Chem.* **272**, 19103 (1997).

[14] K. C. Martin, D. Michael, J. C. Rose, M. Barad, A. Casadio, H. Zhu, and E. R. Kandel, *Neuron* **18**, 899 (1997).

[15] J. Lisman, *Trends Neurosci.* **17**, 406 (1994).

[16] J. E. Lisman and M. A. Goldring, *Proc. Natl. Acad. Sci. U.S.A.* **85**, 5320 (1988).

[17] Y. Ouyang, A. Rosenstein, G. Kreiman, E. M. Schuman, and M. B. Kennedy, *J. Neurosci.* **19**, 7823 (1999).

[18] J. E. Lisman and J. R. Fallon, *Science* **283**, 339 (1999).

[19] A. J. Silva, Y. Wang, R. Paylor, J. M. Wehner, C. F. Stevens, and S. Tonegawa, *Cold Spring Harb. Symp. Quant. Biol.* **57**, 527 (1992).

[20] H. Husi, M. A. Ward, J. S. Choudhary, W. P. Blackstock, and S. G. Grant, *Nat. Neurosci.* **3**, 661 (2000).

This article describes some of the methods and experimental paradigms we have used with the rat hippocampal slice preparation to elucidate the signal transduction mechanisms underlying the induction and maintenance of LTP. Double-label confocal microscopy is used to study the anatomical and cellular localization of the phosphorylated form of two kinases (CaMKII and ERK) that have been demonstrated to be important in LTP, and to verify their possible colocalization in neurons of area CA1 of the hippocampus.

Solutions

Electrophysiology

> Artificial cerebrospinal fluid (ACSF): $10\times$ stock solution containing 12.5 mM NaH$_2$PO$_4$, 35 mM KCl, 240 mM NaHCO$_3$, 1.18 M NaCl
> MgSO$_4$: 1 M stock solution
> CaCl$_2$: 1 M stock solution

On the day of the experiment, make up fresh $1\times$ ACSF from the above-described stock solutions. For 2 liters:

Deionized H$_2$O	1600 ml
MgSO$_4$ stock solution, 1 M	2.6 ml
CaCl$_2$ stock solution	5 ml
Glucose (15 mM final concentration)	5.41 g

Last, add 200 ml of $10\times$ ACSF stock solution, top off to 2 liters with deionized H$_2$O, and bubble with 95% O$_2$/5% CO$_2$. Final osmolarity of the solution is 290–300 mOsm.

> Isoproterenol (Sigma, St. Louis, MO): Dissolve in H$_2$O at a concentration of 10 mM, aliquot and store at $-20°$, and dilute in ACSF at the final concentration just before use

Immunohistochemistry

> Phosphate-buffered saline (PBS, 0.1 M): Prepare by diluting 1 : 10 the $10\times$ stock solution (pH 7.0) with distilled H$_2$O. The pH of the working solution should be 7.4. Concentration of the stock solution:
> KH$_2$PO$_4$, 10 mM
> NaH$_2$PO$_4$, 100 mM

---

[21] R. S. Walikonis, O. N. Jensen, M. Mann, D. W. Provance, Jr., J. A. Mercer, and M. B. Kennedy, *J. Neurosci.* **20**, 4069 (2000).

NaCl, 1.37 M

KCl, 27 mM

PBS, 0.2 M (2× PBS): Prepare by diluting 1 : 5 the 10× PBS solution with deionized H_2O

Paraformaldehyde (4%, w/v)–glutaraldehyde (0.1%, v/v) in PBS (pH 7.4): Make up fresh before the experiment. For 50 ml of the solution weigh 2 g of paraformaldehyde (Sigma) and dissolve in 25 ml of distilled H_2O by heating (maximum, 60°) under agitation. If by 30 min the solution is still cloudy, add 10–20 μl of 1 M NaOH, which helps the aldehyde go into solution. Add 25 ml of 2× PBS, cool the solution, and check the pH. It should be 7.4. Add 200 μl of glutaraldehyde [50% (v/v) in H_2O; Fisher, Pittsburgh, PA) and chill on ice

PBS–TX: 0.1 M PBS containing 0.3% (v/v) Triton X-100

PBS–TX containing 0.75% (v/v) H_2O_2: Prepare just before use by diluting 30% (v/v) H_2O_2 (Sigma)

Blocking solution: 10% (v/v) Normal goat serum (Vector Laboratories, Burlingame, CA) plus 10% (v/v) normal horse serum (Vector) in PBS–TX. This solution is used also to dissolve the primary and secondary antibodies.

Primary antibodies: Anti-phospho(Thr-202/Tyr-204)-ERK (rabbit polyclonal; New England Biolabs, Beverly, MA) is used at a 1 : 750 dilution. Anti-phospho(Thr-286)-CaMKII (mouse monoclonal; Upstate Biotechnology, Lake Placid, NY) is used at a 1 : 1000 dilution. The two antibodies are dissolved together in blocking solution containing 0.1% (w/v) NaN$_3$. Under these conditions the antibodies can be stored for up to 1 month at 4°, and can be used more than one time

Secondary antibody for phospho-ERK: Fluorescein isothiocyanate (FITC)-conjugated goat anti-rabbit IgG (Vectastain; Vector Laboratories), diluted 1 : 200 in blocking solution

Secondary antibody for phospho-ERK plus secondary antibody for phospho-CaMKII : FITC-conjugated goat anti-rabbit IgG (as above) plus Texas Red-conjugated horse anti-mouse IgG (Vectastain; Vector Laboratories), both diluted 1 : 200 in blocking solution

Methods

Electrophysiology

Hippocampal slices from area CA1 are valuable for studying the cellular processes that underlie changes occurring in neurons and at individual synapses. Many of the studies that have improved our understanding of the mechanisms involved in LTP have used technical procedures that are difficult *in vivo*, such as the ability to stimulate specific pathways and record from specific postsynaptic cells under controlled environmental conditions. Brain slices have been used to work out the second-messenger cascades and the array of enzymes involved in phosphorylation and dephosphorylation events during the induction and maintenance of LTP.[8,12,13,22]

A typical study of LTP involves the activation of afferents leading to the postsynaptic cells of interest. Monosynaptic excitatory postsynaptic potentials (EPSPs) are recorded either from individual neurons, using intracellular or whole-cell recording techniques, or from populations of neurons, using field recording techniques. The most common method for inducing LTP has been the delivery of a high-frequency tetanus to the presynaptic fibers (typically 100 Hz for 1 sec or less), and is revealed as a posttetanic increase in the magnitude or slope of the EPSP that can be relatively long-lasting (little decay up to 3 hr). For the induction of LTP we use the coupling of β-adrenergic stimulation to theta pulse stimulation.

Dissection and recording methods are similar to those previously described.[22] Young (100–125 g) male Sprague-Dawley rats are maintained and used according to the guidelines of the Institutional Animal Care and Use Committee, under the care of a fully trained veterinarian. The animals are fed and watered *ad libitum* and housed in a 12-hr light/dark environment. Rats, one per experiment, are anesthetized with halothane, and the brain is rapidly removed and rinsed with ice-cold ACSF. The hippocampus is quickly dissected out in a cold petri dish, and positioned on the chilled stage of a tissue chopper. Hippocampal slices (400 μm thick) are prepared with a McIlwain tissue chopper and placed in an interface chamber at room temperature. Slices are oxygenated by bubbling a gas mixture of 95% O_2/5% CO_2 through ACSF in the maintenance chamber, ensuring that the slices are properly oxygenated without drying them. Slices can be kept healthy in this manner for up to 12 hr.

For recording, the slices are transferred to a submersion chamber, where they are kept in place between two meshes, superfused with ACSF and bubbled with 95% O_2/5% CO_2.

LTP is induced in area CA1 by applying 1 μM isoproterenol (ISO) in the bath for 10 min (Fig. 1), followed by theta-pulse stimulation (TPS, 150 pulses at 10 Hz) of the Schaffer collaterals (stratum radiatum in area CA3). ISO is washed out immediately after TPS. Monophasic, constant-current stimuli (100 μsec) are delivered with a bipolar electrode (F. Haer & Company, Bowdoinham, ME), and LTP is monitored by field recording in the stratum radiatum of area CA1 (Fig.1). The stimulus intensity used for TPS is adjusted to initially produce a field EPSP of 1.0 to 1.5 mV. Before and after TPS, single pulses are delivered every 30 sec. The recordings are low-pass filtered at 3 kHz and digitized at 20 kHz. EPSP amplitude and slope (measured as the maximum negative slope in any 1-msec window during the EPSP) are calculated online, using the Axobasic program (Axon Instruments, Foster City, CA).

[22] R. D. Blitzer, T. Wong, R. Nouranifar, R. Iyengar, and E. M. Landau, *Neuron* **15,** 1403 (1995).

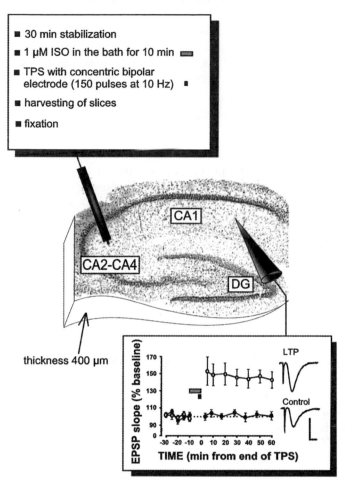

FIG. 1. A Nissl-stained 40-μm section obtained from a control hippocampal slice. The approximate location of the stimulating electrode (in area CA3) and the recording electrode (in the stratum radiatum of CA1) are shown (*middle*). *Top:* Schematic representation of the stimulation method used. *Bottom:* Traces of the EPSP in control slices (filled circles) and in slices stimulated with TPS–ISO (open circles). The traces show representative field EPSPs before and 60 min after stimulation.

All drugs are dissolved in ACSF and added to the maintenance chamber (for preincubation) or to the superfusate in the recording.

Each experiment includes two control groups: "controls" (harvested directly from the interface chamber) and "shams" (placed in the recording chamber and subjected to test stimuli only). Care is taken to allow recovery of the slices for at least 2 hr before harvesting, which is well beyond the time when any artifactual increase in phospho-ERK after slicing has faded.[14]

This stimulation paradigm induces a stable LTP (Fig. 1) for up to 1 hr. This form of LTP is dependent on phosphorylation of ERK1/ERK2, being inhibited by pretreatment with the MEK [MAPK (mitogen-activated protein kinase)/ERK kinase] inhibitor PD98059 (not shown).

Preparation of Slices for Immunohistochemistry

At different time points after the end of stimulation (typically 2, 15, and 60 min), TPS–ISO-treated slices and sham and control slices are quickly transferred into 30 × 10 mm petri dishes containing 2 ml of ice-cold 4% (w/v) paraformaldehyde–0.1% (v/v) glutaraldehyde in PBS (pH 7.4) and fixed overnight. The following day the slices are washed for 2–3 hr in PBS and sliced into 40-μm slices (thin slices), using a Vibratome (series 1000; Lancer, Bridgeton, MO). The slicing chamber of the Vibratome is filled with ice-cold PBS and one hippocampal 400-μm slice is glued to the block of the Vibratome, using instant glue. The amount of glue used is crucial, because too much glue will cover and damage the slice, whereas too little will not allow the slice to stick properly to the block. In the latter case, the slice will eventually be lifted during slicing by the vibration and friction of the blade, causing damage to the whole slice. Care must be taken to position the slice on the block "face up" (i.e., the upper side of the slice in the recording chamber is the upper side on the block). Orienting the slice in a consistent way will permit ordering the newly cut slices with respect to their location (top to bottom) in the thick slice. The immunohistochemistry can then be performed at the same depth of the thick slice, using the same-numbered thin slices. The slicing is performed with the blade set at the 4 speed and 8 amplitude. Under optimal conditions one can obtain four or five good thin slices from a thick one.

Thin slices are then transferred to 24-well plates (1 per well) containing 1 ml of PBS–0.01% (w/v) NaN$_3$ and kept at 4° for up to 1 month.

Immunohistochemistry

The anatomical and cellular localization of activated kinases involved in signal transduction pathways has been made possible by the development of antibodies raised against the phosphorylated form of the proteins. In particular, we use a rabbit polyclonal antibody raised against phospho-ERK and a mouse monoclonal antibody raised against phospho-CaMKII. The commercial availability of the two primary antibodies raised in different species makes it straightforward to perform the double labeling to study the colocalization of the activated forms of the enzymes.

Day 1

Immunohistochemistry is performed by the free-floating method, using 500-μl/well of the appropriate solution for each step. When the slices are incubated with the antibodies, we use a volume of 250 μl/well, enough to cover the slice and at the same time avoid wasting the antibodies. All the following steps are performed under light agitation, using an orbital shaker, at room temperature and in the light, unless otherwise stated. Addition and withdrawal of the solutions are done slowly, using a P1000 Gilson Pipetman (Rainin, Woburn, MA) being careful not to damage the slice.

Experimental Procedure

1. Transfer slices into 24-well plates (1 per well), each containing PBS–0.3% (v/v) Triton X-100 (PBS–TX).
2. Rinse for 10 min in PBS–TX.
3. Incubate for 15 min in PBS–TX containing 0.75% (v/v) H_2O_2.
4. Rinse three times with PBS–TX (10 min each).
5. Block with 10% (v/v) normal goat serum plus 10% (v/v) normal horse serum in PBS–TX (blocking solution) for 40 min.
6. Dissolve the two primary antibodies in blocking solution: anti-phospho-ERK (rabbit polyclonal, diluted 1 : 750) plus anti phospho-CaMKII (mouse monoclonal, diluted 1 : 1000). Primary antibodies can be reused for up to 1 month if dissolved in blocking solution containing 0.1% (w/v) NaN_3. Under these conditions they can be stored at 4°.
7. Incubate the slice with the primary antibodies for 72 hr at 4°.

Day 2

1. After incubation with the primary antibodies, wash the slices in PBS–TX (three times, 10 min each).
2. Incubate for 2 hr at room temperature in the dark with FITC-conjugated goat anti-rabbit IgG, diluted 1 : 200 in blocking solution.
3. Wash the sections (three times, 10 min each) with PBS–TX.
4. Incubate the slices for an additional 2 hr at room temperature in the dark with FITC-conjugated goat anti-rabbit IgG plus Texas Red-conjugated horse anti-mouse IgG, both diluted 1 : 200 in blocking solution.
5. Wash the slices extensively.
6. Following a final 30-min wash, the sections are mounted on gelatin-coated slides, dried overnight, and coverslipped with Vectashield (Vector Laboratories, Burlingame, CA) as mounting medium.

Aspecific labeling is visualized in slices incubated in PBS–TX without primary antibodies; these should show a complete absence of immunostaining.

Double-Label Laser Confocal Microscopy Immunohistochemistry

FITC/Texas Red double-labeled tissue sections are analyzed and imaged with a Zeiss (Oberkochen, Germany) LSM 410 inverted confocal microscope with Zeiss Plan-Neofluar objectives.

For visualization of FITC, an Ar/Kr 488/568 laser is used with 515- to 540-nm bandpass emission, and for visualization of Texas Red the same laser is used with a 590-nm longpass emission filter. The slices are imaged with a Zeiss Plan-Apochromat ×63/1.25 NA oil immersion objective. Switching of illumination is computerized, and dual images are collected sequentially. A confocal aperture is set digitally and this setting is maintained throughout the study (pinhole size of 15;

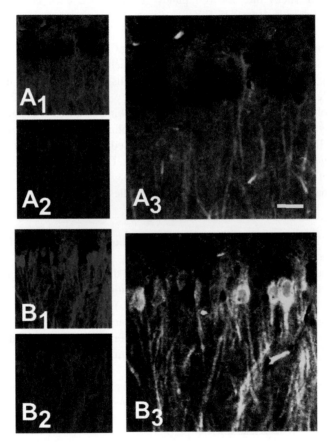

FIG. 2. A1–A3 represent sham slices, and B1–B3 represent TPS–ISO-stimulated slices 15 min after stimulation. The green label (FITC) represents phospho-ERK, and the red label (Texas Red) represents phospho-CaMKII. Sham-stimulated slices show some phospho-ERK and phospho-CaMKII immunoreactivity and some colocalization in apical dendrites and in the perinuclear region (A1–A3) of CA1 neurons. However, in slices stimulated with TPS–ISO and harvested 15 min later (B1–B3), phospho-ERK labeling increase above sham controls in the apical dendrites of stratum radiatum, in the cell bodies of stratum pyramidale and in the nuclei. Phospho-CaMKII immunoreactivity also increases both in the cell bodies and in the dendrites of CA1 pyramidal neurons. The yellow–orange color in the digital superimposition of the images (A3 and B3) is indicative of colocalization. Bars: 50 μm (A1, A2, B1, and B2) and 20 μm (A3 and B3).

Zeiss Imaging Software). All imaging parameters (contrast and brightness; Zeiss Imaging Software) are set in order to yield high-resolution images for both bright and dim sections. All settings must be kept constant for recording all data within an experiment so that images from control and treated slices can be compared. By using the above-described immunohistochemistry protocol and scanning the whole depth of the slices we find that the antibodies fully penetrate the slice from the surface. Therefore, scans are performed at a depth of 20 μm, corresponding to the brightest plane in the slice. Images are saved as TIFF files.

The resulting images demonstrate colocalization of phospho-ERK and phospho-CaMKII in neurons but represent only semiquantitative immunofluorescence intensity. To obtain images such as those shown in Figs. 2 and 3, two sets of two overlapping scans of sections of area CA1 are recorded from each slice, one set of phospho-MAPK images (digitally converted to green), and a second set of phospho-CaMKII (digitally converted to red). Scans are then digitally combined to obtain double-labeled FITC/Texas Red images. The images are then opened in Adobe Photoshop (Adobe Systems, Mountain View, CA) and assembled into montages.

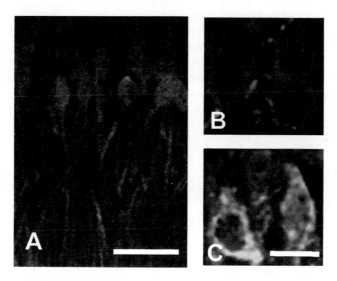

FIG. 3. Digitally combined signals of phospho-ERK immunoreactivity (FITC-labeled, green) and phospho-CaMKII immunoreactivity (Texas Red-labeled, red) in a sham slice (B) and in a TPS–ISO-stimulated slice (A and C) harvested 2 min after stimulation. The yellow–orange color in the digital superimposition of the images demonstrates that phospho-ERK and phospho-CaMKII colocalize in the apical dendrites and the perinuclear somata of CA1 pyramidal neurons of TPS–ISO-stimulated slices. Furthermore, in the TPS–ISO-stimulated slice, neurons where phospho-ERK and phospho-CaMKII are colocalized show strong phospho-ERK immunoreactivity in the nucleus (green label, C), indicating that in these neurons only phospho-ERK translocated into the nucleus. Bars: 50 μm (A); 10 μm (B and C).

Results

Our results demonstrate that stimulation of the Schaffer collaterals in CA3 using as stimulation paradigm the theta frequency stimulation (10 Hz) coupled to β-adrenergic receptor activation induces a long-lasting LTP in area CA1 of the hippocampus (Fig. 1).

The phosphorylation of both ERK and CaMKII increase 15 min after TPS–ISO stimulation (Fig. 3), and the activated forms of the enzymes colocalize both in dendrites and cell bodies of CA1 pyramidal cells. At this longer time point, the presence of the green label within the cell body is indicative of the presence of the activated form of ERK in the nucleus.

In slices stimulated by TPS–ISO, double-labeling immunohistochemistry shows that, as early as 2 min after the end of stimulation, phospho-ERK and phospho-CaMKII show some colocalization in dendrites and cell bodies of hippocampal CA1 pyramidal cells (Fig. 2A and C). The presence of the green label in the nuclei of stimulated (Fig. 2C) but not of sham (Fig. 2B) slices indicates that at this relatively early time point after the end of stimulation, the activated form of ERK has already translocated into the nucleus.

The presence of the phosphorylated form of ERK both in the nucleus and in the cytoplasm of CA1 neurons at the two time points tested suggests the involvement of this kinase in the activation of transcription factors [such as ELK1 and RSK2 (ribosomal S6 kinase 2)] and possibly cytoplasmic proteins (such as phospholipase A_2). Furthermore, the activation of ERK and CaMKII in TPS–ISO-induced LTP, and their colocalization in stimulated neurons of area CA1 of the hippocampus, suggest a possible interaction of these two signaling pathways in LTP.

Acknowledgments

This article is based in part on experiments supported in part by NIH Grants GM 5408 and NS 3346. Work by M.G.G. was partly supported by a CNR-NATO fellowship (N. 217.30). I thank Dr. R. Iyengar and Dr. E. M. Landau for the opportunity to work with them, Dr. R. G. Blitzer for invaluable advice and encouragement, Mr. T. Wong for performing the electrophysiology, and Mr. W. G. M. Jannsen for generously sharing technical knowledge of LCM.

[34] Regulation of Mitogen-Activated Protein Kinases by G-Protein-Coupled Receptors

By MARIO CHIARIELLO and J. SILVIO GUTKIND

Introduction

A remarkably diverse array of stimuli, including growth factors, vasoactive polypeptides, chemoattractants, neurotransmitters, hormones, phospholipids, photons, odorants, and taste ligands, can elicit biological responses by stimulating receptors that transmit signals through the activation of heterotrimeric G proteins. In fact, the family of G-protein-coupled receptors (GPCRs) represents the largest group of cell surface receptors involved in signal transmission, and accounts for more than 1% of the total proteins encoded by the human genome. This large number of GPCRs, together with their ability to stimulate distinct $G\alpha$ and $\beta\gamma$ subunits and intracellular effector molecules, explains the remarkable complexity of the GPCR–G protein signal-transducing system. Indeed, on GPCR activation, G protein α subunits and $\beta\gamma$ complexes can regulate the activity of a wide range of downstream effector molecules, including adenylyl cyclases, phosphodiesterases, phospholipases, ion channels, phosphatidylinositol 3-kinases (PI3 kinases), serine-threonine protein kinases and phosphatases, receptor and nonreceptor tyrosine kinases, and guanine nucleotide exchange factors for small GTP-binding proteins.[1]

Interestingly, stimulation of GPCRs at the level of the cell membrane can also provoke nuclear transcriptional events, and this is often mediated by the activation of one or more members of the mitogen-activated protein kinase (MAPK) super-family of proline-targeted serine-threonine kinases. This family includes extracellular regulated kinase 1 (ERK1) and ERK2, also known as p44^{MAPK} and p42^{MAPK}, respectively; c-Jun N-terminal kinase (JNK), also known as stress-activated protein kinase (SAPK1); members of the p38 family of MAP kinases, including p38α, p38β, p38γ (or SAPK3 or ERK6) and p38δ (or SAPK4); and ERK5 (or BMK1).[2] Besides a high level of protein sequence conservation, these kinases also share a common mechanism of regulation by their upstream regulators, which involves phosphorylation on tyrosine and threonine residues within a conserved activation loop by their corresponding dual-specificity kinase, known as MAP kinase kinases (or MAP/ERK kinases, MEKs).[3] These residues are dephosphorylated by a set of MAP kinase-specific phosphatases, which provide the basis for signal termination.[4,5]

[1] J. S. Gutkind, *Oncogene* **17,** 1331 (1998).

[2] M. J. Robinson and M. H. Cobb, *Curr. Opin. Cell Biol.* **9,** 180 (1997).

[3] C. Widmann, S. Gibson, M. B. Jarpe, and G. L. Johnson, *Physiol. Rev.* **79,** 143 (1999).

Many GPCR agonists are able to activate ERKs through the accumulation of membrane-bound forms of $\beta\gamma$ subunits, which, in turn, can stimulate Ras function resulting in ERK activation.[1] In more restricted cell types, both $\beta\gamma$- and Ras-independent pathways are also likely to participate in GPCR-induced ERK stimulation.[1] The existence of pathways connecting GPCRs to other newly discovered members of the MAP kinase family, such as JNK, p38α, p38γ, p38δ, and ERK5, have been described.[6] While accumulating evidence indicates that two small GTPases of the Rho family, Rac1 and Cdc42, are an integral part of the pathway connecting GPCRs to JNK,[7,8] the nature of the molecules connecting these receptors to other MAP kinase family members is much less understood.[6] Nevertheless, available information is sufficient to outline a model in which GPCRs communicate to the nucleus through several pathways involving the sequential activation of small GTP-binding proteins of the Ras superfamily, which, in turn, control the activity of divergent kinase cascades that culminate in the phosphorylation and stimulation of different MAP kinase family members. On activation, MAP kinases phosphorylate cytoplasmic targets and translocate to the nucleus, where they regulate the phosphorylation state, and thus the activity, of different nuclear proteins, including transcription factors. The final biological outcome most likely results from a complex network of interactions among these biochemical routes at the level of the membrane, cytoplasm, and nucleus, rather than from a single series of sequential events.

Efforts have helped identify numerous transcription factors whose activity can be regulated on phosphorylation by one or more members of the MAP kinase superfamily.[3] Thus, the ability to assess the activation of these MAP kinases by GPCRs has allowed important predictions to be made about the identity of nuclear targets for GPCRs in each cell type, as well as about the potential participation of these receptors in biological responses such as cell growth, differentiation, survival, or apoptosis. Furthermore, it also provides unique opportunities to unravel the basic mechanisms of signal transmission by GPCRs. The study of the regulation of MAP kinases by GPCRs also represents a relatively simple readout system to evaluate the effectiveness of drug candidates affecting GPCRs, which represent nearly 60% of all currently available therapeutic agents.

This article describes detailed protocols for the measurement of the ability of GPCRs to stimulate the ERK, JNK, p38, and ERK5 members of the MAP kinase

[4] K. Kelly and Y. Chu, in "Signaling Networks and Cell Cycle Control: The Molecular Basis of Cancer and Other Diseases" (J. S. Gutkind, ed.), p. 165. Humana Press, Bethesda, Maryland, 2000.

[5] M. Haneda, T. Sugimoto, and R. Kikkawa, Eur. J. Pharmacol. **365,** 1 (1999).

[6] M. J. Marinissen, M. Chiariello, M. Pallante, and J. S. Gutkind, Mol. Cell. Biol. **19,** 4289 (1999).

[7] O. A. Coso, H. Teramoto, W. F. Simonds, and J. S. Gutkind, J. Biol. Chem. **271,** 3963 (1996).

[8] M. V. Prasad, J. M. Dermott, L. E. Heasley, G. L. Johnson, and N. Dhanasekaran, J. Biol. Chem. **270,** 18655 (1995).

superfamily of proteins, with emphasis on the use of *in vitro* kinase assays. In addition, the development of antibodies highly selective for the dual phosphory-lated, active forms of MAP kinases has now afforded the possibility of examining the state of activation of these MAP kinases by conventional techniques such immunocytochemistry or Western blot analysis, and is also described below.

In Vitro Mitogen-Activated Protein Kinase Assays

Overview of Assay

In vitro MAP kinase assays are based on the ability of these proteins to act as phosphotransferase enzymes between a labeled "donor substrate," $[\gamma\text{-}^{32}\text{P}]\text{ATP}$, and a protein acting as "acceptor substrate" (the kinase-specific substrate). Al-though the source of enzyme can be as different as crude cell lysate, cellular frac-tions, immunoprecipitates, partially purified proteins, or purified enzymes, this review specifically deals with *in vitro* kinase assays performed on endogenous or transfected, epitope-tagged MAP kinases, isolated by immunoprecipitation with specific antibodies.

MAP kinases are relatively stable enzymes in the presence of protease in-hibitors such as phenylmethylsulfonyl fluoride (PMSF), leupeptin, and aprotinin. Calcium chelators (EDTA, EGTA) can similarly help to reduce the activity of calcium-activated proteases. Equally important, their activating tyrosine/threonine phosphorylation can be easily preserved for the duration of the assay by keep-ing them constantly at $4°$, in buffers containing common phosphatase inhibitors (see procedures).

The measurement of the kinase activity of immunoprecipitated endogenous MAP kinases often represents the most specific and sensitive approach to evaluate the effect of GPCR agonists on the *in vivo* activity of each MAP kinase pathway. The main limitation of this experimental approach is the availability of antibodies capable of immunoprecipitating active MAP kinase enzymes. Many commercial antibodies against ERKs, JNK, and p38α have demonstrated such an ability [exam-ples are anti-ERK2 (C-14; Santa Cruz Biotechnology, Santa Cruz, CA); anti-JNK1 (PharMingen, Carlsbad, CA); and anti-p38 (C-20 and N-20; Santa Cruz Biotech-nology)]. Commercial antibodies for the most recently described MAP kinases such as p38γ (ERK6), p38δ (SAPK4), and ERK5 still need further development.

To investigate the mechanism by which GPCRs or their downstream effectors such as Gα or $\beta\gamma$ subunits activate MAP kinases, several laboratories have used the transient expression of epitope-tagged forms of different MAP kinases, in readily transfectable cell lines. The advantage of this system is that several proteins can be transiently coexpressed at high levels, avoiding the influence of biological changes that might be manifested during prolonged culturing of stably expressing cells. On transfection, epitope-tagged forms of each MAP kinase can be immu-noprecipitated by specific anti-epitope monoclonal antibodies, and assayed by

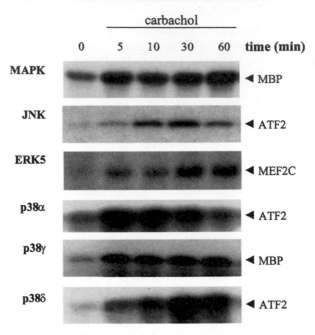

FIG. 1. Activation of different MAP kinase family members by M_1 G-protein-coupled receptors. NIH 3T3 cells expressing the M_1 muscarinic receptor (NIH 3T3-M1) were stably transfected with expression vectors containing HA-tagged MAPK (ERK), JNK, ERK-5, p38α, p38γ, and p38δ. After serum starvation, cell lines were treated with 1 mM carbachol for 5 to 60 min. Unstimulated cells were used as controls. After stimulation, lysates were immunoprecipitated with anti-HA antibodies and used for kinase reactions, as described in text. [Adapted with permission from M. J. Marinissen, M. Chiariello, M. Pallante, and J. S. Gutkind, *Mol. Cell. Biol.* **19,** 4298 (1999).]

standard *in vitro* kinase assays.[9] Among the different available epitopes, HA (a tag of nine amino acids derived from the influenza hemagglutinin HA1 protein[10]), has been proved to allow the efficient immunoprecipitation of active MAP kinases,[6] although other epitopes, such as Myc, AU5, and AU1 can be equally used. In addition, the use of stable or transiently expressed epitope-tagged MAP kinases allows the evaluation of the regulation of each MAP kinase by GPCRs by *in vitro* kinase assays, circumventing the need for specific anti-MAP kinase antibodies[6] (Fig. 1).

The following protocol is based on our experience with difference cell lines. It can be applied to evaluate the activity of endogenous as well as stably or transiently

[9] P. Crespo, N. Xu, W. F. Simonds, and J. S. Gutkind, *Nature (London)* **369,** 418 (1994).
[10] I. A. Wilson, H. L. Niman, R. A. Houghten, A. R. Cherenson, M. L. Connolly, and R. A. Lerner, *Cell* **37,** 767 (1984).

transfected MAP kinases. Furthermore, it can be successfully used for *in vitro* kinase assays for the measurement of the activity of a number of MAP kinases, including ERKs, JNK, p38α, p38γ (ERK6), p38δ (SAPK4) and ERK5, on their immunoprecipitation by the use of specific anti-epitope antibodies. The requirement for specific substrates for each kinase is discussed in the technical procedures. Note that many of these kinases are activated by different serum components. To avoid high background signals provoked by the culturing medium, cells are usually starved before appropriate stimulation. The length of the starvation period depends on the MAP kinase considered and can be empirically determined. In our experience, the best results are usually obtained by overnight (12-hr) serum starvation for ERKs and ERK5 kinase assays, and 2–3 hr of starvation for assays measuring the activity of JNK, p38α, p38γ, and p38δ. It is also important to remember that most of these kinases are sensitive to stress conditions such as those caused by changes in temperature and/or pH, or even prolonged serum starvation. Finally, to control the proper technical execution of the kinase assays, it is important to include an internal positive control for the experiment, treating the cells with a stimulus that is expected to increase the activity of the considered kinase. Examples of commonly used positive controls are as follows: lysophosphatidic acid (LPA, 5–10 μM) or serum (10%, v/v) for ERKs; anysomicin (10 μg/ml) or NaCl (300 mM) for JNK, p38α, p38γ, and p38δ; and H$_2$O$_2$ (200 μM) for ERK5.

Lysis of Mammalian Cells

Lysis buffer (store at 4°)

HEPES, pH 7.5	20 mM
EGTA, pH 8	10 mM
Nonidet P-40 (NP-40)	1% (v/v)
MgCl$_2$	2.5 mM
Dithiothreitol (DTT): Add immediately before use	1 mM
Na$_3$VO$_4$: Prepare fresh and add immediately before use	2 mM
β-Glycerophosphate: Prepare fresh and add immediately before use	20 mM
Aprotinin–leupeptin: Add immediately before use	20 μg/ml
Phenylmethylsulfonyl fluoride (PMSF); Add immediately before use from a 100 mM stock solution in ethanol	1 mM

1. After appropriate cellular stimulation, wash cells once with cold phosphate buffered saline (PBS). Aspirate the PBS and then immediately add 700–900 μl of cold lysis buffer per 100-mm plate (400–600 μl/60-mm plate; 200–300 μl/35-mm plate). Place the plates on ice for 20 min with occasional shaking. Scrape the plates with a cell lifter to collect the cellular lysates and transfer them to 1.5-ml microcentrifuge tubes.

2. Clarify lysates in a refrigerated centrifuge at ~10,000g for 10 min, to remove insoluble debris. Transfer the clarified lysates into new microcentrifuge tubes. Determine the protein concentration of the lysates, for example, by bicinchoninic acid (BCA) protein assay (Pierce Rochford, IL). The lysates are now ready for immunoprecipitation.

Immunoprecipitation

> PBS/NP-40 wash buffer (store at 4°)
> > PBS   1×
> > NP-40 (Triton X-100; or equivalent)   1% (v/v)
> > Na₃VO₄   2 mM
>
> Lithium/Tris wash buffer (store at 4°)
> > LiCl   0.5 M
> > Tris, pH 7.5   100 mM
>
> Kinase reaction buffer (store at 4°)

Morpholinepropanesulfonic acid (MOPS, pH 7.5)	12.5 mM
MgCl₂	7.5 mM
EGTA	0.5 mM
Na₃VO₄	0.5 mM
β-Glycerophosphate	12.5 mM
NaF	0.5 mM

1. Aliquot the desired amount of lysates into new microcentrifuge tubes, based on the protein concentration of the samples. (*Note:* Save at least 20–50 μl of the lysate for a Western blot analysis of the kinase of interest, which can be used to normalize the enzymatic activity of each MAP kinase by its expression level in the corresponding samples.) The amount of lysate needed to detect kinase activity is dependent on several factors (amount of the MAP kinases in each cell type, affinity of the antibody used for immunoprecipitation, efficiency of transfection for assays involving transiently transfected cells) and therefore should be empirically determined.

2. Add the specific anti-MAPK (or anti-epitope for epitope-tagged transfected kinases) antibodies to the samples and rotate at 4° for 2 hr to overnight. The amount of antibodies needed to immunoprecipitate each MAP kinase activity is antibody dependent and therefore the reader should consult the protocol recommended by the manufacturer. In general, 1 μg of antibodies or antiserum is sufficient to immunoprecipitate the majority of the MAP kinases from lysates containing 1 mg of total cellular proteins.

3. Add 20 μl of prewashed GammaBind G Sepharose (Pharmacia Biotech, Piscataway, NJ) or equivalent to the samples. Rotate at 4° for at least 1 hr.

4. Wash three times with cold PBS/NP-40 wash buffer. For each wash, centrifuge the samples in a refrigerated microcentrifuge at 10,000g for 25 sec. Aspirate

the supernatant with a flat gel-loading tip connected to a vacuum, being careful not to aspirate the beads. Add 400 μl of wash solution, mix and recentrifuge the samples, repeating the washing procedure for as many times as needed (usually three times).

5. Wash once with cold lithium/Tris wash buffer.

6. Wash once with cold kinase reaction buffer. Proceed to the *in vitro* kinase reaction.

In Vitro Kinase Reaction and Acquisition of Data

Kinase reaction mix (per sample)

Kinase reaction buffer	30 μl
[γ-^{32}P]ATP	1 μCi
unlabeled ATP (stock, 1 mM; store at $-20°$)	20 μM
DTT (stock, 100 mM; store at $-20°$)	3 mM
Specific substrate [myelin basic protein, 1.5 mg/ml; 1 μg of purified, bacterially expressed glutathione S-transferase (GST)–ATF2 or GST–MEF2C]	4–6 μl

1. Prepare the kinase reaction mix (always prepare for one extra sample). Add 35 μl of the kinase reaction mix to each sample and incubate the microcentrifuge tubes for 30 min at 30° in a water bath designated for radioactive samples.

2. Stop the reactions by adding 10 μl of 5× protein loading buffer to each sample. Boil the proteins for 5 min.

3. Load half of the samples onto a 10–15% (according to the molecular weight of the substrate) denaturing polyacrylamide gel. The remaining material can be stored at $-20°$ should it be needed. Run the gel until the dye front has reached the bottom. Dry the gel for 2 hr at 80° and then expose it either on a phosphoimager plate or on an X-ray film, with an intensifier screen. The length of the exposure depends on many different factors and thus needs to be empirically determined.

Note. Selecting a substrate normally depends on the specific kinase under evaluation. Even if each MAP kinase has been usually described to phosphorylate a number of specific proteins, only a few of them are of practical use as substrates in *in vitro* kinase assays. For example, myelin basic protein (MBP) serves as a good substrate for different MAPKs such as ERKs, p38α (HOG1), and p38γ (ERK6). Bacterially purified GST-tagged ATF2 is the substrate of choice for JNK, even if GST–c-Jun is an equally good alternative for this kinase. GST–ATF2 also works as a good substrate to assay p38α (HOG1) kinase activity. Bacterially purified GST–MEF2C is the best substrate to assay ERK5 kinase activity. Nevertheless, if not available, MBP has also been successfully used in *in vitro* ERK5 kinase assays.[11]

---

[11] J. Abe, M. Kusuhara, R. J. Ulevitch, B. C. Berk, and J. D. Lee, *J. Biol. Chem.* **271**, 16586 (1996).

Although abundant amounts of MBP can usually be purchased from different companies (e.g., Sigma, St. Louis, MO), bacterially purified GST–ATF2, GST–c-Jun, and GST–MEF2C can be easily prepared by the researcher (see following protocol), if expression vectors for the different fusion proteins are available. Otherwise, ready-to-use stocks of some of these proteins (GST–ATF2 and GST–c-Jun) can be purchased from a number of companies (e.g., Santa Cruz Biotechnology and New England Biolabs, Beverly, MA).

Expression and Purification of Glutathione S-Transferase Fusion Proteins

Bacterial lysis solution	
PBS	$1\times$
Triton X-100	1% (v/v)
EDTA	1 mM
Aprotinin–leupeptin: Add immediately before use	2 μg/ml
PMSF: Add immediately before use	1 mM
Elution buffer	
Tris, pH 7.5: Readjust the pH after the addition of glutathione	50 mM
Glutathione	10 mM
Aprotinin–leupeptin: Add immediately before use	2 μg/ml
PMSF: Add immediately before use	1 mM

1. Transform the BL21 Lys strain of *Escherichia coli* with pGEX vectors (or equivalent) encoding the fusion protein GST–ATF2, GST–c-Jun, or GST-MEF2C.

2. With a single colony inoculate a 20-ml overnight culture in LB broth containing ampicillin (100 μg/ml). Dilute the overnight culture in 500 ml of LB broth containing ampicillin and let it grow until the OD_{600} is 0.6–1.0 (approximately 1–2 hr). At this time add isopropyl-β-D-thiogalactopyranoside (IPTG; 0.5–1 mM final concentration) and continue to grow the bacteria for an additional 3 hr.

3. Collect the cells by centrifugation at 3000g for 30 min at 4° and resuspend them in 8 ml of bacterial lysis solution. Perform two freeze-and-thaw cycles and then sonicate the cell suspension (three times, 10 sec each). To remove cellular debris centrifuge at 10,000g for 15 min at 4°.

4. Mix the supernatant with 300 μl of glutathione–agarose beads (Pharmacia Biotech, Piscataway, NJ) and incubate for 1–2 hr at 4°, with shaking. Centrifuge at 3000g for 5 min at 4°. Wash the pellet three times with bacterial lysis solution, and then twice with PBS containing aprotinin (2 μg/ml), leupeptin (2 μg/ml), and 1 mM PMSF. Finally, elute purified fusion proteins in 300 μl of elution buffer, shaking the sample on a rocker for 5–10 min at 4°. Store the eluted protein at −80°

[in 10% (v/v) glycerol], saving a small aliquot to check its concentration on a denaturing polyacrylamide gel, by comparison with a protein of known concentration (usually bovine serum albumin, BSA).

Analysis by Anti-Phospho-Specific Antibodies

Overview of Assays

Although grouped in at least four different subfamilies (ERKs, JNKs, p38s, and ERK5), all MAP kinases require the contemporary phosphorylation of a tyrosine and threonine residue within their activation loop to be enzymatically active.[12] Specific phosphorylation of these residues is usually performed by MAP kinase kinases, also known as MEKs.[12] To date, phosphorylation of the Thr-X-Tyr motif is the best known mechanism of MAP kinase regulation, and the amount of the dual phosphorylated form of these proteins is usually considered a good estimate of their enzymatic activity. The use of anti-phospho-MAP kinase-specific antibodies is therefore a valuable tool to monitor the activity of endogenous MAP kinases in response to different GPCR ligands (Fig. 2), especially under conditions in which the amount of cellular lysate is not enough to perform immunoprecipitations. In addition, these reagents eliminate the need for ^{32}P labeling. The success in the use of such reagents is dependent on the quality of the phospho-specific antisera or monoclonal antibodies, which, because of the efforts of a number of companies, has improved dramatically. Examples of phospho-specific antibodies that, in our experience, have proved high levels of sensitivity and reproducibility are: phospho-p44/42 MAP kinase (Thr-202/Tyr-204) monoclonal antibody, phospho-p38 MAP kinase (Thr-180/Tyr-182), and phospho-SAPK/JNK (Thr-183/Tyr-185) polyclonal antibodies from New England Biolabs; phospho-JNK (pTPpY) polyclonal antibody from Promega; phospho-p38 and phospho-ERK5 from QCB-Biosource International (Hopkinton, MA). Ideally, parallel membranes should be probed with antisera or monoclonal antibodies reacting with both the phosphorylated and unphosphorylated species of these MAP kinases, thus facilitating the evaluation of the fractional increase in the levels of the activated forms of these kinases.

Analysis by Western Blotting

1. Prepare cellular lysates as described in Lysis of Mammalian Cells (above).

2. Aliquot the desired amount of lysates in a new microcentrifuge tube, on the basis of their protein concentration. The amount of lysates needed for the detection of phosphorylated kinases is dependent on different factors (expression levels of each MAP kinase in each cell type, affinity of the antibody used) and therefore should be empirically determined. Add required 5× protein loading buffer to the

---

[12] R. Seger and E. G. Krebs, *FASEB J.* **9**, 726 (1995).

A

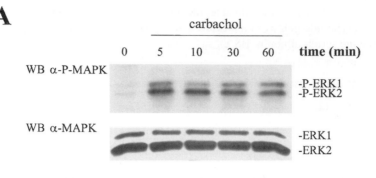

WB α-P-MAPK

WB α-MAPK

B

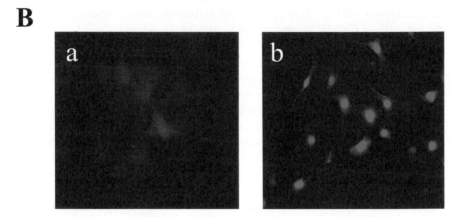

FIG. 2. Stimulation of MAPK (ERK1/ERK2) by M_1 G-protein-coupled receptors. (A) After serum starvation, NIH 3T3-M1 cells were treated with 1 mM carbachol for 5 to 60 min. Unstimulated cells were used as controls. Cell lysates were analyzed by Western blot with a specific anti-phospho-MAPK antibody (upper gel). The presence of equal amounts of MAPK in the cell lysates was assessed by Western blot analysis with anti-MAPK antibodies (lower gel). (B) After serum starvation, NIH 3T3-M1 cells were treated with 1 mM carbachol for 5 min (b). Unstimulated cells were used as controls (a). Cells were then fixed and analyzed by double immunofluorescence with specific anti-phospho-ERK antibodies.

different samples and boil them for 5 min. Load the samples onto a denaturing polyacrylamide gel. The remaining part of the cellular lysates may be stored at −80°, should they be needed. Run the gel until the dye front has reached the bottom.

3. Transfer the proteins to a nitrocellulose or nylon membrane for 5–6 hr at 150 mA (or overnight at 90 mA), in a cold room. If needed, please refer to specific laboratory manuals for a detailed description of protein transfer techniques.[13]

4. Block and incubate the membrane with the specific anti-phospho-MAPK antibodies or antisera, according to the manufacturer instructions. Wash the membrane and incubate with horseradish peroxidase (HRP)-conjugated anti-mouse/rabbit secondary antibody, accordingly.

5. Detect immunocomplexes by chemiluminescence by using ECL Western blotting detection reagents (Amersham, Arlington Heights, IL) or equivalent product, and exposing the membrane to an autoradiography film. The optimal time of exposure is dependent on many different factors and therefore should be determined empirically.

Analysis by Immunofluorescence

Note: Remember to seed the cells on glass coverslips.

1. After appropriate cellular stimulation, wash the cells three times with cold PBS.

2. Fix the cells with 4% (v/v) formaldehyde and 5% (w/v) sucrose in PBS for 10 min, and permeabilize them with 0.5% (v/v) Triton X-100 in PBS for 10 min.

3. Wash the cells three times with PBS (at this point the cells can be stored at 4°).

4. Block the cells in 1% (w/v) BSA in PBS for 1 hr at room temperature.

5. Incubate the cells with the specific anti-phospho-MAP kinase antibody or antiserum, according to the manufacturer instructions.

6. Wash three times with 1% (w/v) BSA in PBS, and then incubate the cells with a fluorescein- or tetramethylrhodamine isothiocyanate (TRITC)-conjugated goat F(ab')$_2$ IgG anti-mouse or anti-rabbit antibody (e.g., from Jackson Immuno-Research Laboratories, West Grove, PA).

7. Wash once with 1% (w/v) BSA in PBS and three times with PBS.

8. Wash the coverslips once in water and then mount them in mounting medium containing *p*-phenylenediamine (ICN, Costa Mesa, CA) at 1 mg/ml to inhibit photobleaching.

9. Analyze the samples, using a photomicroscope equipped for epifluorescence.

Acknowledgments

We thank Maria Julia Marinissen, Salvatore Pece, Roberta Visconti, and Ricardo Sanchez-Prieto for critical reading of the manuscript and many helpful discussions.

---

[13] J. Sambrook, E. F. Fritsch, and T. Maniatis, "Molecular Cloning: A Laboratory Manual" (C. Nolan, N. Ford, and M. Ferguson, eds.). Cold Spring Harbor Laboratory Press, Cold Spring Harbor, New York, 1989.

[35] Analysis of Protein Kinase B/Akt

By MICHELLE M. HILL and BRIAN A. HEMMINGS

Introduction

Protein kinase B (PKB or Akt) was initially identified as a serine/threonine kinase with homology to protein kinases A and C, and as an oncoprotein.[1-3] Three isoforms of PKB have subsequently been identified, termed PKBα, PKBβ, and PKBγ, which are 80% identical in amino acid sequence, and all contain an N-terminal pleckstrin homology (PH) domain and a C-terminal regulatory domain.[4-6] Activation of PKB by growth factors is mediated by phosphatidylinositol 3,4,5-trisphosphate (PI3,4,5P$_3$), generated by phosphoinositide 3-kinase (PI3K) at the inner surface of the plasma membrane. PKB is recruited to the plasma membrane through an interaction of its PH domain with PI3,4,5P$_3$, resulting in a conformational change of PKB, allowing it to be phosphorylated. Maximal activation of PKB requires the phosphorylation of two residues conserved in all PKB isoforms, namely threonine at position 308 (Thr-308) in the activation loop and serine at position 473 (Ser-473) in the C-terminal regulatory domain (numbering according to PKBα).[7] The kinase responsible for phosphorylating Thr-308 has been cloned and named 3-phosphoinositide-dependent kinase 1 (PDK1),[8] while the authentic upstream kinase for Ser-473 is yet to be identified.

Many substrates have now been identified that are phosphorylated by activated PKB.[4-6] Overall these data demonstrate the importance of PKB in insulin signaling and cell survival; however, the precise substrates responsible for PKB-mediated cell transformation have yet to be identified. The consensus substrate sequence determined for PKB is Arg-X-Arg-X-X-Ser/Thr-Hyd, where Hyd is a hydrophobic amino acid.[9] However, it is important to bear in mind that phosphorylation of a

---

[1] P. F. Jones, T. Jakubowicz, F. J. Pitossi, F. Maurer, and B. A. Hemmings, *Proc. Natl. Acad. Sci. U.S.A.* **88,** 4171 (1991).

[2] P. J. Coffer and J. R. Woodgett, *Eur. J. Biochem.* **201,** 475 (1991).

[3] A. Bellacosa, J. R. Testa, S. P. Staal, and P. N. Tsichlis, *Science* **254,** 274 (1991).

[4] I. Galetic, M. Andjelkovic, R. Meier, D. Brodbeck, J. Park, and B. A. Hemmings, *Pharmacol. Ther.* **82,** 409 (1999).

[5] T. O. Chan, S. E. Rittenhouse, and P. N. Tsichlis, *Annu. Rev. Biochem.* **68,** 965 (1999).

[6] S. R. Datta, A. Brunet, and M. E. Greenberg, *Genes Dev.* **13,** 2905 (1999).

[7] D. R. Alessi, M. Andjelkovic, B. Caudwell, P. Cron, N. Morrice, P. Cohen, and B. A. Hemmings, *EMBO J.* **15,** 6541 (1996).

[8] D. R. Alessi, S. R. James, C. P. Downes, A. B. Holmes, P. R. Gaffney, C. B. Reese, and P. Cohen, *Curr. Biol.* **7,** 261 (1997).

protein by PKB *in vitro* does not necessarily imply that it is a bona fide *in vivo* substrate.

This article outlines several methods for studying endogenous PKB and transiently transfected PKB. Transient expression of PKB in mammalian cells is useful in the analysis of PKB regulation and function, as it allows the analysis of different PKB mutants, such as constitutively active, kinase inactive, and phosphorylation site mutants. Some of the main PKB and PDK1 constructs used in numerous studies are listed in Table I.

Transient Transfection

Human embryonic kidney (HEK)-293 cells and COS-1 cells are maintained in Dulbecco's modified Eagle's medium (DMEM) supplemented with 10% (v/v) fetal calf serum (FCS) at 37° in a 5% CO_2 atmosphere. Cells are transfected at 40–60% confluence, using a modified calcium phosphate method.[10] Generally, a confluent 75-cm^2 flask of 293 cells is seeded into twenty 10-cm dishes 24 hr before transfection. For a 10-cm dish containing 10 ml of DMEM–FCS, 1 ml of transfection mix is prepared by sequential addition of 450 μl of sterile water, 10 μg of plasmid DNA, 50 μl of sterile 2.5 M $CaCl_2$, and 500 μl of sterile 2× BBS buffer [50 mM *N,N*-bis-(2-hydroxyethyl)-2-aminoethanesulfonic acid (BES)–NaOH (pH 6.95), 0.28 M NaCl, and 1.5 mM Na_2HPO_4]. The mix is vortexed, incubated at room temperature for 15 min, and then added dropwise to the cell medium. The cells are incubated in a 3% CO_2 incubator for 8–24 hr, after which they are washed with DMEM and then incubated in DMEM without serum at 5% CO_2 for an additional 16–24 hr.

Figure 1A shows an experiment in which HA–PKBα, HA–PKBβ, and HA–PKBγ were expressed in HEK-293 cells, and the expression was monitored by Western blotting with anti-HA (12CA5; Roche Biochemicals, Indianapolis, IN) antibody. The expression of Myc-tagged constructs is routinely monitored with anti-Myc antibody (9E10; Upstate Biotechnology, Lake Placid, NY).

Trouble Shooting

1. The pH of 2× BBS buffer is crucial to the efficiency of transfection. Buffers of different pH ranging from 6.9 to 7.0 should first be tested for transfection efficiency before performing experiments.

2. While it is not necessary to change cell culture medium for transfecting HEK-293 cells, for some cell types replacing the medium with fresh DMEM–10% (v/v) FCS before transfection may improve the efficiency of transfection.

---

[9] D. R. Alessi, F. B. Caudwell, M. Andjelkovic, B. A. Hemmings, and P. Cohen, *FEBS Lett.* **399**, 333 (1996).

[10] C. A. Chen and H. Okayama, *Biotechniques* **6**, 632 (1988).

TABLE I
PKB AND PDK1 CONSTRUCTS

Construct	Properties	Ref.
HA–PKBα	Hemagglutinin (HA) epitope-tagged wild-type human PKBα	a
HA–K179A-PKBα	Kinase-deficient PKBα mutant	b
HA–m/p-PKBα	Membrane-targeted PKBα containing Lck myristoylation/palmitylation signal at the N terminus, following HA epitope	c
HA–ΔPH-PKBα	PKBα lacking PH domain	d
HA–C1-PKBα-ΔPH	PKBα with PH domain replaced by C1 (diacylglycerol binding) domain from PKCα	d
HA–T308A/D-PKBα, HA–S473A/D-PKBα	Phosphorylation site mutants of PKBα	e
PKBα–CAAX	PKBα containing the CAAX box of Ki-Ras at the C terminus, acts as dominant negative when overexpressed	f
AAA–PKBα	Dominant negative PKBα, contains theree mutations: K179A, T308A, and S473A	g
HA–PKBβ	HA epitope-tagged wild-type human PKBβ	h
HA–m/p-PKBβ	Membrane-targeted PKBβ	h
HA–T309A-PKBβ, HA–S474A-PKBβ	Phosphorylation site mutants of PKBβ	h
HA–PKBγ	HA epitope-tagged wild-type human PKBγ	i
HA–ΔPH-PKBγ	PKBγ lacking PH domain	i
HA–T305A/D-PKBγ, HA–S472A/D-PKBγ	Phosphorylation site mutants of PKBγ	i
Myc–PDK1	Myc-tagged PDK1	j
Myc–PDK1-KD	Kinase-deficient PDK1 (K110Q)	j

[a] M. Andjelkovic, T. Jakubowicz, P. Cron, X. F. Ming, J. W. Han, and B. A. Hemmings, *Proc. Natl. Acad. Sci. U.S.A.* **93,** 5699 (1996).

[b] D. A. Cross, D. R. Alessi, P. Cohen, M. Andjelkovich, and B. A. Hemmings, *Nature (London)* **378,** 785 (1995).

[c] M. Andjelkovic, D. R. Alessi, R. Meier, A. Fernandez, N. J. Lamb, M. Frech, P. Cron, P. Cohen, J. M. Lucocq, and B. A. Hemmings, *J. Biol. Chem.* **272,** 31515 (1997).

[d] M. Andjelkovic, S. M. Maira, P. Cron, P. J. Parker, and B. A. Hemmings, *Mol. Cell. Biol.* **19,** 5061 (1999).

[e] D. R. Alessi, M. Andjelkovic, B. Caudwell, P. Cron, N. Morrice, P. Cohen, and B. A. Hemmings, *EMBO J.* **15,** 6541 (1996).

[f] P. C. van Weeren, K. M. de Bruyn, A. M. de Vries-Smits, J. van Lint, and B. M. Burgering, *J. Biol. Chem.* **273,** 13150 (1998).

[g] Q. Wang, R. Somwar, P. J. Bilan, Z. Liu, J. Jin, J. R. Woodgett, and A. Klip, *Mol. Cell. Biol.* **19,** 4008 (1999).

[h] R. Meier, D. R. Alessi, P. Cron, M. Andjelkovic, and B. A. Hemmings, *J. Biol. Chem.* **272,** 30491 (1997).

[i] D. Brodbeck, P. Cron, and B. A. Hemmings, *J. Biol. Chem.* **274,** 9133 (1999).

[j] N. Pullen, P. B. Dennis, M. Andjelkovic, A. Dufner, S. C. Kozma, B. A. Hemmings, and G. Thomas, *Science* **279,** 707 (1998).

A.

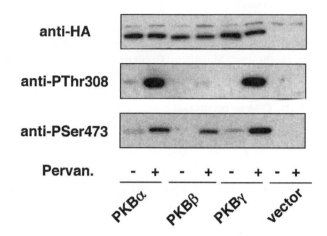

B.

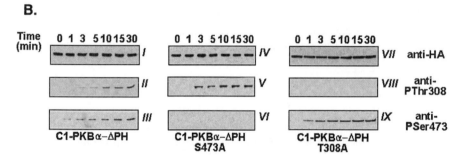

FIG. 1. (A) Expression of HA-tagged PKB constructs in HEK-293 cells, and analysis of their phosphorylation status on pervanadate stimulation. HEK-293 cells were transfected with 10 μg of plasmid DNA encoding HA-tagged PKB isoforms, or empty pCMV5 plasmid. After transfection, cells were serum starved for 16 hr, and then stimulated with 0.1 m*M* pervanadate for 10 min at 37°, or left unstimulated. Cell lysates were prepared as described, and 20 μg was analyzed for expression of HA-tagged PKB by Western blotting with 12CA5 anti-HA antibody (*top* gel). HA-tagged PKB isoforms were immunoprecipitated from 200 μg of lysate, and then analyzed by blotting with phospho-Thr-308 (*middle* gel) or phospho-Ser-473 antibodies (1 *bottom* gel.). (B) Use of phospho-specific antibodies to monitor phorbol ester-stimulated phosphorylation of HA–C1-PKB-ΔPH. HEK-293 cells transfected with HA–C1-PKB-ΔPH or phosphorylation site mutants of HA–C1-PKB-ΔPH were stimulated with TPA for 0 to 30 min as indicated. The phosphorylation of HA–C1-PKB-ΔPH on Thr-308 and Ser-473 was assessed by blotting 20 μg of cell lysate with phospho-specific antibodies, or anti-HA antibody as a control. [Reproduced from M. Andjelkovic, S. M. Maira, P. Cron, P. J. Parker, and B. A. Hemmings, *Mol. Cell. Biol.* **19,** 5061 (1999).]

Cell Treatment

Growth Factors and Inhibitors

Cells are washed twice in serum-free DMEM and then incubated in serum-free DMEM for 16–24 hr before experiment. Treatment with growth factors such as insulin (0.1 μM; Boehringer, Mannhein, Germany) or insulin-like growth factor type I (IGF-I; 50 ng/ml; Life Technologies, Rockville, MD) or inhibitors including LY294002 (50 μM; Alexis, San Diego, CA) and wortmannin (100 nM; Calbiochem, La Jolla, CA) is performed by direct addition to the cell culture medium (serum-free DMEM). Typically inhibitor treatments are initiated 30 min before growth factor stimulation (5–15 min).

Pervanadate

The insulin mimetic pervanadate[11] is prepared as follows. To prepare a 50-ml stock solution of 0.1 M sodium orthovanadate, add 0.92 g of powder (Sigma, St. Louis, MO) to 30 ml of H_2O, and adjust to pH 10 with HCl. The solution should turn yellow. Boil the solution until it is colorless, and recheck the pH. Filter-sterilized stocks of orthovanadate can be kept at 4° for at least 2 years. Alternatively, aliquots may be stored at −20°.

Pervanadate treatment is performed in a buffer containing 20 mM HEPES–NaOH (pH 7.4) and 120 mM NaCl (buffer A). Prepare 10 ml of 0.1 mM pervanadate solution as follows. Add 10 μl of 0.1 M sodium orthovanadate to 10 μl of 0.2 M hydrogen peroxide (diluted 50-fold from a 30% stock from Merck, Rahway, NJ). Incubate the solution (yellow color) at room temperature for 15 min. Inactivate hydrogen peroxide by adding 28 μl of buffer A and 2 μl of catalase [10 mg/ml, prepared by diluting a catalase solution (20 mg/ml) from Boehringer with phosphate-buffered saline (PBS), and centrifuging to remove insoluble material]. Incubate at room temperature for a further 5 min, then add the solution to 10 ml of buffer A. Replace cell culture media with pervanadate solution, and incubate at 37° for 5–15 min.

Immunocomplex Protein Kinase B Assay

Buffers

NP-40 lysis buffer: 50 mM Tris-HCl (pH 7.5), 120 mM NaCl, and 1% (v/v) nonidet P-40 (NP-40), supplemented with phosphatase inhibitors [25 mM sodium fluoride, 1 mM sodium pyrophosphate, 0.1 mM sodium orthovanadate, 2 μM microcystin LR (Alexis)], and protease inhibitors [1 mM phenylmethylsulfonyl fluoride (PMSF) and 1 mM benzamidine]

---

[11] B. I. Posner, R. Faure, J. W. Burgess, A. P. Bevan, D. Lachance, G. Zhang-Sun, I. G. Fantus, J. B. Ng, D. A. Hall, and B. S. Lum, *J. Biol. Chem.* **269,** 4596 (1994).

Kinase assay buffer (prepare fresh from stock solutions): 50 mM Tris-HCl (pH 7.5), 10 mM MgCl$_2$, 1 mM dithiothreitol (DTT), 1 μM protein kinase A inhibitor peptide (TTYADFIASGRTGRRNAIHD; Bachem, Bubendorf, Switzerland), 30 μM substrate peptide (Crosstide, GRPRTSSAEG; or PKBtide, RPRAATF), 1 mM PMSF, 1 mM benzamidine, and 2 μM microcystin LR (Alexis)

Preparation of Cell Lysate

Different types of cells grown in tissue culture have been used to study endogenous and/or transiently expressed PKB, including HEK-293 cells, COS-1 cells, Swiss 3T3 cells, and PC12 rat pheochromocytoma cells.

Serum-starved cells are treated as desired, washed with ice-cold phosphate-buffered saline [PBS, 10 mM Na$_2$HPO$_4$-HCl (pH 7.3), 1.5 mM KH$_2$PO$_4$, 3 mM KCl, 137 mM NaCl], and lysed in ice-cold NP-40 lysis buffer (400 μl for a 10-cm dish). All subsequent steps are performed at 4°. Lysates are cleared by centrifugation at 12,000g for 10 min, and the protein concentration of the supernatant is determined with the Bradford reagent (Bio-Rad, Hercules, CA) with fatty acid-free bovine serum albumin (BSA) as standard. Lysates may be snap-frozen in liquid nitrogen at this stage, and stored at −80°. Expression of transiently expressed PKB is monitored by Western blotting of 20 μg of cell lysate (Fig. 1A).

Immunoprecipitation

PKB may be immunoprecipitated with antibodies specific to the epitope tag (for transiently expressed protein), pan-specific PKB antibodies, or isoform-specific antibodies; some of these are listed in Table II. The amount of antibody required for each immunoprecipitation should be determined empirically for each antibody.

1. Wash protein A–Sepharose CL-4B (Amersham-Pharmacia, Piscataway, NJ) (15- to 20-μl packed bead volume per immunoprecipitation) twice with PBS, pelleting the beads by centrifugation at 8000 rpm for 1 min in an Eppendorf centrifuge. Nonspecific binding sites on the beads may be blocked with 1% (w/v) BSA in PBS for 30 min at room temperature, and washed as described above. This step is not necessary if a "blank" assay is to be performed to determine the background produced by beads alone.

2. Incubate beads with appropriate antibody in PBS at 4° with mixing for 2 hr to overnight.

3. Wash the antibody-coupled beads twice in lysis buffer, and split into the required number of Eppendorf tubes in the final wash.

4. Incubate the beads with 0.2 mg of lysate (per reaction) at 4° with mixing for 2 hr. For duplicate assays, the immunoprecipitation can be performed in one tube

TABLE II
PKB ANTIBODIES

Antibody	Antigen
PKBα(466–480)	Peptide corresponding to residues 466–480 of human PKBα (RPHFPQFSYSASSTA)
PH domain	Recombinant protein comprising the PH domain of human PKBα
PKBα specific	Peptide corresponding to residues 453–465 of murine PKBα (DQDDSMECVDSER)
PKBβ specific	Peptide corresponding to residues 454–466 of murine PKBβ (DRYDSLDPLELDQ)
PKBγ specific	Peptide corresponding to residues 450–463 of murine PKBγ (EKYDDDGMDGMDNE)
Phospho-Ser-473[a]	Peptide corresponding to residues 466–477 of human PKBα with phosphorylation at Ser-473 (RPHFPQFpSYSAS)
Phospho-Thr-308[a]	Peptide corresponding to residues 299–310 of human PKBγ with phosphorylation at Thr-305 (DAATMKpTFCGTP)
Phospho-Ser-124	Peptide corresponding to residues 116–129 of human PKBα with phosphorylation at Ser-124 (EEMDFRSGpSPSDNS)
Phospho-Thr-450	Peptide corresponding to residues 442–455 of human PKBα with phosphorylation at Thr-450 (FTAQMITIpTPPDQD)

[a] Also commercially available.

(i.e., ∼40 μl of beads plus 0.4 mg of lysate). The immunoprecipitation volume should be about 0.5 ml (make up with lysis buffer if protein concentration is high).

In Vitro Kinase Assay

1. Wash PKB immunoprecipitates with 1 ml of the following buffers kept at 4°: (a) lysis buffer supplemented with 0.5 M NaCl, (b) lysis buffer, and (c) 50 mM Tris-HCl (pH 7.4) supplemented with 1 mM DTT, 1 mM PMSF, and 1 mM benzamidine. For duplicate assays, divide beads in the last wash into two fresh tubes. Centrifuge to pellet the beads and remove the wash buffer without disturbing the beads.

2. Add 40 μl of cold kinase assay buffer to each tube.

3. Initiate the kinase reaction by adding 5 μl of ATP mix containing 50 μM ATP and 1.5–2 μCi of [γ-^{32}P]ATP per reaction. Incubate at 30° for 30 min with shaking.

4. Stop the kinase reaction by adding 5 μl of 0.5 M EDTA to each tube. Reaction tubes may be frozen at −20° at this stage.

5. Pellet the beads by centrifugation. Spot 30–35 μl of the supernatants onto squares (2 × 2 cm) of P81 phosphocellulose paper[12] (Whatman, Clifton, NJ;

[12] R. Roskoski, Jr., *Methods Enzymol.* **99,** 3 (1983).

numbered with a lead pencil). Allow to dry for a few seconds, and then drop the P81 paper into a beaker containing 1% (w/v) phosphoric acid.

6. Wash the P81 papers five times in 500 ml of 1% (w/v) phosphoric acid (5–10 min per wash). The setup used for washing filter papers includes a 1-liter glass beaker, a magnetic stirrer and stirring bar, and a plastic container with holes that is suspended in the glass beaker by a bulldog clip.

7. After a final rinse in acetone, filter papers are air dried and transferred to scintillation vials containing 2 ml of scintillant. Radioactivity incorporated into the substrate peptide is determined by scintillation counting (1 min).

Notes

1. It is imperative that phosphatase inhibitors be included in the lysis buffer used for kinase activity assays as PKB can be rapidly inactivated by dephosphorylation. Omission of phosphatase inhibitors can result in reduced or no kinase activity in subsequent assays.

2. Reaction blanks using immunoprecipitates from mock-transfected cells are routinely performed in experiments on transiently transfected PKB. For assaying endogenous PKB, a "blank" immunoprecipitation using protein A–Sepharose beads without antibody should be carried out.

3. Commercial preparations of $[\gamma-^{32}P]ATP$ may contain contaminants that bind to P81 paper. This background can be monitored by spotting an aliquot of a "reagent blank" reaction (containing 5 μl of H_2O instead of beads) onto a piece of P81 paper, and processing it in parallel with samples.

4. Washing of more than ~30 filter papers in one beaker can result in elevated background radioactivity. Thus, in experiments involving a larger number of assays, more than one beaker should be used for this step.

5. Crosstide and PKBtide produce similar results in immunocomplex PKBα assays. The sequence of Crosstide is based on the PKB phosphorylation site in glycogen synthase kinase 3, and it is also phosphorylated by mitogen-activated protein (MAP) kinase-activated protein kinase 1 (MAPKAPK-1) and p70 S6 kinase. PKBtide is a more specific substrate for PKBα.[9]

Analysis of Protein Kinase B Phosphorylation with Phospho-Specific Antibodies

Identification of the two critical regulatory phosphorylation sites on PKBα has enabled the development of phospho-specific antibodies, allowing a relatively rapid and easy method of determining the phosphorylation status of each of these sites. However, it is important to bear in mind that full activation of PKB requires phosphorylation at both sites. Thus, whereas phosphorylation of Thr-308 alone achieves only one-third of maximal activity, phosphorylation of Ser-473 alone has no significant effect on PKB activity.[7] This observation underscores the importance of analyzing the phosphorylation status of both sites, as a minimum, in an experimental paradigm.

Because of the sequence conservation surrounding these two phosphorylation sites, antibodies specific for phospho-Thr-308 or phospho-Ser-473 appear to recognize all three PKB isoforms (Fig. 1A). Thus, these antibodies are useful to monitor the phosphorylation status of PKB, either by analyzing PKB immunoprecipitates (Fig. 1A) or directly on cell lysates (Fig. 1B).

Production of Phospho-Specific Antibodies

Phospho-peptides corresponding to the sequences surrounding Thr-308 or Ser-473 are synthesized and coupled to keyhole limpet hemocyanin.[13] After immunization of rabbits, several bleeds of serum are obtained, and IgG is purified with a protein A–Sepharose column. Phospho-specific antibodies are affinity purified by a two-step procedure involving negative purification through a dephospho-peptide column, followed by specific binding to a phospho-peptide column. Affi-Gel (Bio-Rad) is the matrix used for peptide columns. The pI of the peptide to be coupled determines whether Affi-Gel 10 or Affi-Gel 15 should be used (consult the manufacturer instructions). All procedures are performed at 4°.

Protein A–Sepharose Column

1. Pack 1 ml of protein A–Sepharose CL-4B (Amersham-Pharmacia) onto a column and wash with 10 ml of PBS.

2. Add protease inhibitors (1 mM PMSF and 1 mM benzamidine) to 10 ml of serum and filter through a 0.22-μm pore size filter.

3. Pass serum through the column five times (∼40 min per pass).

4. Wash with 10 ml of 100 mM Tris-HCl (pH 8), and then with 10 ml of 10 mM Tris-HCl (pH 8).

5. Elute with 4.5 ml of 100 mM glycine-HCl (pH 3) into a tube containing 0.5 ml of 1 M Tris-HCl (pH 8).

Dephospho-Peptide Column

1. Wash 3 ml of Affi-Gel (Bio-Rad) twice with cold H_2O (30 ml each) and then incubate with 5 mg of dephospho-peptide in 2 ml of 0.1 M morpholine propane-sulfonic acid (MOPS)–NaOH (pH 7.5) overnight at 4°. (Read the OD_{280} of peptide solution before coupling.)

2. Load Affi-Gel onto a column and collect solution. (Add 100 μl of 0.1 N HCl and read the OD_{280} to check the efficiency of coupling.) Wash the column with 1 ml of H_2O.

3. Wash sequentially with the following:

[13] R. F. Doolittle, "Of URFs and ORFs: A Primer on How to Analyze Derived Amino Acid Sequences." Oxford University Press, London, 1986.

Tris-HCl (pH 8), 100 mM	5 ml
Glycine-HCl (pH 3), 100 mM	5 ml
Tris-HCl (pH 8), 100 mM	5 ml
Triethylamine (TEA), 100 mM	5 ml
Tris-HCl (pH 8), 100 mM	10 ml

4. Pass the eluate from the Protein A column through the peptide column 10 times (10–20 min per pass).

5. Wash column with the following:

| Tris-HCl (pH 8), 10 mM | 10 ml |
| Tris-HCl (pH 8), 10 mM, containing 0.5 M NaCl | 10 ml |

6. Elute with 4.5 ml of 100 mM glycine-HCl (pH 3) into a tube containing 0.5 ml of 1 M Tris-HCl (pH 8).

7. Wash the column with 10 ml of 100 mM Tris-HCl (pH 8).

8. Elute with 4.5 ml of 100 mM TEA into a tube containing 0.5 ml of 1 M Tris-HCl (pH 8).

9. Wash the column with 10 ml of 100 mM Tris-HCl (pH 8).

10. Pool the eluates (10 ml total), which represent antibodies that recognize unphosphorylated PKB.

Phospho-Peptide Column

1. Prepare the column by coupling 5 mg of phospho-peptide used for antibody production to Affi-Gel as described above.

2. Pass the flowthrough from the dephospho-peptide column through the phosphopeptide column 10 times (~10 min per pass).

3. Elute as described above to obtain phospho-specific antibodies.

Eluates are dialyzed against PBS overnight, and the antibody concentration is measured by OD_{280}, using the extinction coefficient for IgG (1.35), that is, IgG at 1 mg/ml has an OD_{280} of 1.35. The specificity of purified phospho-specific antibodies can be tested on PKB dephosphorylated *in vitro* as described below and shown in Fig. 1B.

Dephosphorylation of Protein Kinase B with λ-Phosphatase

Hemagglutinin (HA)-tagged PKBα is expressed in HEK-293 cells and then immunoprecipitated with anti-HA antibodies (12CA5; Roche Biochemicals) as described for the immunocomplex PKB kinase assay. After washing, immunoprecipitated PKBα is treated with or without 2000 units of λ phosphatase (New England BioLabs, Beverly, MA) for 2 hr at 30° according to the manufacturer instructions. After dephosphorylation, PKBα (still attached to the beads) is washed twice more in ice-cold lysis buffer and then analyzed by sodium dodecyl sulfate–polyacrylamide gel electrophoresis (SDS–PAGE) and immunoblotting with phospho-specific antibodies (see below).

Western Blotting with Phospho-Specific Antibodies

1. Perform SDS–PAGE according to the method of Laemmli,[14] using immunoprecipitated PKB or cell lysates. Transfer proteins onto polyvinylidene difluoride (PVDF) membranes (Immobilon-P; Millipore, Bedford, MA) according to the method of Towbin *et al.*[15]

2. Block the PVDF membranes in TBST [50 mM Tris-HCl (pH 7.4), 150 mM NaCl, 0.1% (v/v) Tween 20] containing 5% (w/v) skim milk powder (Fluka, Ronkonkoma, NY) for 1 hr at room temperature.

3. Rinse the membranes twice in TBST and then incubate with antibodies diluted in TBST at 4° overnight. Dilution of antibodies should be experimentally determined. In our experience, a dilution of 1 in 1000 is appropriate for most anti-phospho-Ser-473 antibodies, whereas anti-phospho-Thr-308 antibodies need to be used at 1 in 250 to obtain a good signal for direct analysis of endogenous PKB from cell lysates.

4. Wash the membranes in TBST (three times, 10 min each).

5. Incubate the membranes with alkaline phosphatase- or horseradish peroxidase-linked secondary antibodies (Amersham-Pharmacia) (diluted 1 in 2000 to 5000 in TBST) for 1 hr at room temperature.

6. Wash the membranes in TBST (three times, 10 min each) and detect immunoreactive signals with alkaline phosphatase substrate (Bio-Rad) or enhanced chemiluminescence (ECL) substrate (Amersham-Pharmacia) for alkaline phosphatase- or horseradish peroxidase-linked secondary antibodies, respectively.

Note

Antibodies against unphosphorylated PKB are often produced along with phospho-specific antibodies, thus it is important to affinity purify phospho-specific antibodies. Several phospho-specific PKB antibodies are now commercially available; however, batches of phospho-Ser-473 antibodies appear to be contaminated with different amounts of antibodies that recognize unphosphorylated PKB. As illustrated in Fig. 2, the addition of a negative purification step allowed the separation of antibodies that recognize unphosphorylated PKB (Fig. 2, top) from phospho-Ser-473-specific antibodies (Fig. 2, middle).

Analysis of Protein Kinase B Subcellular Localization

Membrane translocation is an important event in the stimulation of PKB activity,[16,17] thus, in addition to monitoring the phosphorylation status and kinase activity, it is also desirable to determine the subcellular localization of PKB in

---

[14] U. K. Laemmli, *Nature (London)* **227,** 680 (1970).
[15] H. Towbin, T. Staehelin, and J. Gordon, *Proc. Natl. Acad. Sci. U.S.A.* **76,** 4350 (1979).

**Eluate from
Dephospho-PKB
peptide column**

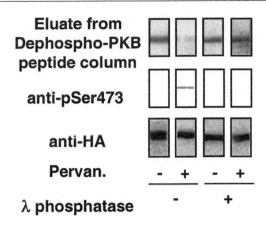

anti-pSer473

anti-HA

Pervan.   -   +   -   +

λ phosphatase   -   +

FIG. 2. Evaluation of the specificity of purified phospho-Ser-473 antibody. HEK-293 cells were transfected with HA–PKBα, serum starved for 16 hr, and then stimulated with 0.1 mM pervanadate for 10 min at 37°, or left unstimulated. HA–PKBα was immunoprecipitated from 100 μg of lysate, using a monoclonal antibody to HA (12CA5), treated with or without λ phosphatase for 2 hr at 30°, and then analyzed by SDS–PAGE. Immunoblotting was performed with antibodies eluted from the dephospho-peptide column, or from the phospho-peptide column (anti-pSer-473). Blotting with anti-HA antibody was performed as a control for the amount of HA–PKBα immunoprecipitated.

response to different stimuli. The subcellular location of PKB may be determined by immunofluorescence microscopy or subcellular fractionation techniques, with either endogenous or transiently expressed PKB.

Immunofluorescence

Cells are plated (and transfected) on glass coverslips. After treatment as desired, cells are washed with PBS and then fixed in 4% (w/v) paraformaldehyde in PBS for 10 min at room temperature. Fixed cells are permeabilized by incubating in PBS containing 0.2% (v/v) Triton X-100 for 10 min, followed by a 10-min PBS wash. Coverslips are incubated with the desired antibody (for PKB or for the appropriate epitope tag) diluted in PBS (generally diluted 1 : 10 to 1 : 50) at 37° for 1 hr. After three 10-min PBS washes, coverslips are incubated with the appropriate secondary antibody [e.g., fluorescein isothiocyanate (FITC)-conjugated anti-rabbit antibody from Sigma, diluted 1 in 100 in PBS] for 30 min, and then washed two more times in PBS. Coverslips are mounted onto glass slides with Fluoromount (Southern Biotechnology, Birmingham, AL), and viewed under a fluorescence microscope

[16] M. Andjelkovic, D. R. Alessi, R. Meier, A. Fernandez, N. J. Lamb, M. Frech, P. Cron, P. Cohen, J. M. Lucocq, and B. A. Hemmings, *J. Biol. Chem.* **272**, 31515 (1997).

[17] R. Meier, D. R. Alessi, P. Cron, M. Andjelkovic, and B. A. Hemmings, *J. Biol. Chem.* **272**, 30491 (1997).

(Leitz DRMBE; Leica, Bensheim, Germany). Images are captured with a digital camera (Leitz DMRD; Leica), and SPOT RT software (Diagnostic Instruments, Sterling Heights, MI).

Subcellular Fractionation

Preparation of subcellular fractions enriched in the plasma membrane allows the comparison of the relative proportion of membrane-bound and cytosolic PKB before and after stimulation. In addition, measurement of PKB activity in these fractions can be performed by immunocomplex PKB kinase assay. Two methods are described below, for preparation of crude total membrane fraction and crude plasma membrane fraction (modified from a protocol for preparation of plasma membrane from adipocytes[18]).

Buffers

Fractionation buffer: 20 mM HEPES–NaOH (pH 7.4), 250 mM sucrose, phosphatase inhibitors [25 mM sodium fluoride, 1 mM sodium pyrophosphate, 0.1 mM sodium orthovanadate, 2 μM microcystin LR (Alexis)], and protease inhibitors (1 mM PMSF and 1 mM benzamidine)
Sucrose cushion: 20 mM HEPES–NaOH (pH 7.4), 1.15 M sucrose

Cell Homogenization

HEK-293 cells (grown in 10-cm dishes) are treated as desired, and then placed on ice. After washing once in ice-cold PBS, cells are scraped in 500 μl of ice-cold fractionation buffer, and then homogenized by passing through a 26-gauge needle 10 times.

All centrifugation steps are performed at 4°.

PREPARATION OF CRUDE TOTAL MEMBRANE FRACTION

Centrifuge the cell homogenate at 1000g for 10 min in an Eppendorf microcentrifuge to remove the nuclei. Centrifugation at higher speeds is not recommended as the plasma membrane may pellet under these conditions (see Fig. 3). The supernatant is then centrifuged at 100,000g for 30 min, using a Beckman (Fullerton, CA) TLA100.3 rotor at 50,000 rpm. The resulting pellet containing cellular membranes, vesicles, and cytoskeletal elements is generally termed the P100 fraction and the supernatant is termed the S100 fraction. Figure 4A shows the analysis of membrane translocation and phosphorylation of HA–C1-PKBα-ΔPH induced by TPA (12-O-tetradecanoylphorbol 13-acetate), using this procedure.[19]

---

[18] M. M. Hill, S. F. Clark, and D. E. James, *Electrophoresis* **18,** 2629 (1997).
[19] M. Andjelkovic, S. M. Maira, P. Cron, P. J. Parker, and B. A. Hemmings, *Mol. Cell. Biol.* **19,** 5061 (1999).

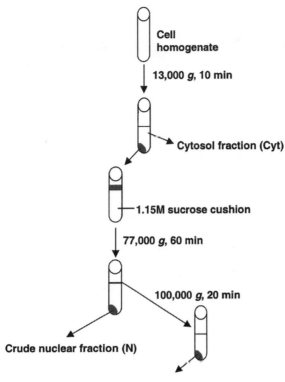

FIG. 3. Centrifugation steps used to prepare crude plasma membrane fraction from HEK-293 cells.

PREPARATION OF CRUDE PLASMA MEMBRANE FRACTION

1. Centrifuge cell homogenate at $14,000g$ (10,000 rpm in a Beckman JA-17 rotor) for 10 min to pellet the plasma membrane, along with dense organelles such as nuclei and mitochondria. Decant the supernatant into fresh tubes and label as the cytosol fraction.

2. Resuspend the pellet from the centrifugation step in 1 ml of fractionation buffer and carefully layer onto 10 ml of sucrose cushion placed in a 12.5-ml Beckman Ultraclear centrifuge tube. Fill to 2 mm from the top with fractionation buffer.

3. Centrifuge in a Beckman SW41 rotor at $77,000g$ (25,000 rpm) for 60 min.

4. The plasma membrane fraction appears as a diffuse band at the interface of the 0.25 M (fractionation buffer) and 1.15 M sucrose solutions. Collect 1 ml around this band and add to 2 ml of fractionation buffer in an ultracentrifuge tube. Centrifuge at $100,000g$ (35,000 rpm in a Beckman Ti50 rotor) for 20 min to

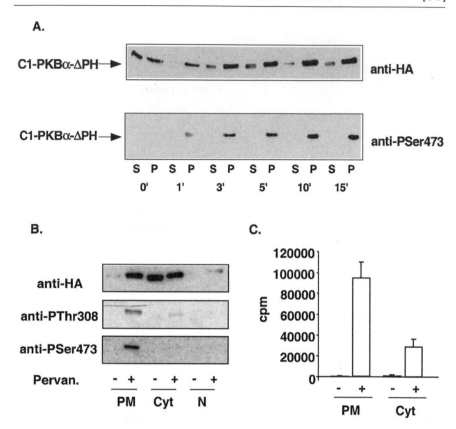

FIG. 4. Analysis of PKB in subcellular fractions. (A) HEK-293 cells were transfected with HA-tagged C1-PKBα-ΔPH, stimulated with TPA for 0 to 15 min, and then homogenized in hypotonic buffer.[19] S100 (S) and P100 (P) fractions were prepared and analyzed by immunoblotting with antibodies specific for the HA epitope tag or phospho-Ser-473. [Reproduced from M. Andjelkovic, S. M. Maira, P. Cron, P. J. Parker, and B. A. Hemmings, *Mol. Cell. Biol.* **19,** 5061 (1999).] (B and C) HEK-293 cells transfected with HA-tagged PKBα were serum starved and then stimulated with 0.1 m*M* pervanadate for 10 min at 37°, or left untreated. Subcellular fractionation was performed as depicted in Fig. 3. (B) Ten micrograms of each fraction was analyzed by SDS–PAGE and immunoblotting with antibodies specific for phospho-Thr-308, phospho-Ser-473, or the HA tag. (C) PKB was immunoprecipitated from 100 μg of plasma membrane and cytosol fractions, and then assayed for activity with Crosstide as substrate.

pellet the crude plasma membrane fraction. Figure 3 provides a summary of the centrifugation steps.

Analysis of Subcellular Fractions

P100 or crude plasma membrane pellets are solubilized in NP-40 lysis buffer and centrifuged at 12,000*g* for 10 min to remove insoluble material. Subcellular fractions are analyzed for the amount of PKB and phospho-PKB by Western blotting (Fig. 4A and B). In addition, PKB can be immunoprecipitated from these fractions and assayed for activity (Fig. 4C).

Notes

1. At least five 10-cm dishes of HEK-293 cells are required to prepare crude plasma membrane fractions. This is limited by the visual ability to detect a band at the interface of the sucrose solutions.

2. The amount of protein in the crude plasma membrane fraction prepared using this protocol represents approximately 1% of total protein in HEK-293 cells, whereas the P100 fraction contains approximately 30% of total cellular protein.

Large-Scale Immunoaffinity Purification of HA–PKB and Derivatives

To isolate modest amounts (500–1000 μg) of protein for *in vitro* studies, we routinely transfect about forty 10-cm dishes of HEK-293 or COS-1 cells with HA-tagged forms of PKB and isolate the expressed protein by immunoaffinity chromatography.[20]

1. Transfect and prepare cell lysate as described for the immunocomplex PKB assay. The following steps are performed at 4°.

2. Preabsorb the extract with Sepharose CL-6B (preequilibrated in lysis buffer). This step reduces nonspecific binding to Sepharose in the subsequent steps.

3. Prepare a column consisting of 12CA5 monoclonal antibody covalently linked to protein A–Sepharose CL-4B, using dimethyl pimelimidate (DMP), following instructions for the IgG orientation kit (Pierce, Rockville, IL).

4. Pass precleared extract through the column three times at a slow rate.

5. Wash five times with 10 ml of lysis buffer.

6. Elute bound proteins by first incubating in lysis buffer containing a 1-mg/ml concentration of HA peptide (YPYDVPDYA) for 1 hr at 4°.

We have also adapted this procedure to isolate endogenous PKB from HeLa cell extracts. For this purpose, we affinity purified peptide-specific antisera, which are subsequently covalently linked to protein A–Sepharose. The bound PKB can be eluted with the antigenic peptide.

Acknowledgments

We thank members of the Hemmings laboratory for their contributions to the development of these methods. These studies were supported by the Novartis Research Foundation and in part by the Swiss Cancer League. We thank the American Society for Microbiology for permission to reproduce Figs. 1B and 4A.

[20] M. Frech, M. Andjelkovic, E. Ingley, K. K. Reddy, J. R. Falck, and B. A. Hemmings, *J. Biol. Chem.* **272**, 8474 (1997).

[36] Direct Stimulation of Bruton's Tyrosine Kinase by G Protein α Subunits

By WILLIAM E. LOWRY, YONG-CHAO MA, SVETLANA CVEJIC, and XIN-YUN HUANG

Introduction

Heterotrimeric G proteins transduce signals from cell surface receptors to downstream effectors to control a wide variety of cellular responses.[1,2] One class of effectors for G proteins is tyrosine kinases.[3,4] The first tyrosine kinase effector discovered for $G\alpha_q$ protein is Bruton's tyrosine kinase (Btk).[3] Btk family tyrosine kinases include Btk, Tec, Itk/Tsk, Bmx/Etk, and Txk/Rlk.[5] These tyrosine kinases are characterized by an amino-terminal pleckstrin homology (PH) domain followed by a Tec homology (TH) domain [composed of a Btk motif (BM) and a proline-rich (PR) region], a Src homology 3 (SH3) domain, an SH2 domain, and a COOH-terminal kinase domain (Fig. 1). They are expressed in a variety of cells and tissues including hematopoietic, epithelial, and endothelial cells and they are involved in cell growth, differentiation, apoptosis, and other cellular signaling processes.[6] Defects in Btk are responsible for X chromosome-linked agammaglobulinemia (XLA) in humans and X chromosome-linked immunodeficiency (XID) in mice.[6] Combined deletion of Itk and Rlk in mice caused marked defects in T cell receptor-mediated responses including proliferation, cytokine production, apoptosis, and adaptive immune responses.[7] The *Drosophila* Btk homolog is required for adult survival and male genital formation.[8] We have shown that the kinase activity of Btk can be directly stimulated by $G\alpha_q$ and $G\alpha_{12}$.[3,4] Here we describe the *in vitro* assay for monitoring the regulation of Btk by G protein α subunits.

Purification of Recombinant Btk Protein from *Escherichia coli*

For purification, we generated a hexahistidine (His$_6$)-tagged Btk by subcloning a human Btk cDNA into pET21a plasmid vector. A plasmid DNA harboring

[1] A. G. Gilman, *Annu. Rev. Biochem.* **56**, 615 (1987).

[2] A. J. Morris and C. C. Malbon, *Physiol. Rev.* **79**, 1373 (1999).

[3] K. Bence, W. Ma, T. Kozasa, and X. Y. Huang, *Nature (London)* **389**, 296 (1997).

[4] Y. Jiang, W. Ma, Y. Wan, T. Kozasa, S. Hattori, and X. Y. Huang, *Nature (London)* **395**, 808 (1998).

[5] J. B. Bolen and J. S. Brugge, *Annu. Rev. Immunol.* **15**, 371 (1997).

[6] A. Satterthwaite and O. Witte, *Annu. Rev. Immunol.* **14**, 131 (1996).

[7] E. M. Schaeffer, J. Debnath, G. Yap, D. McVicar, X. C. Liao, D. R. Littman, A. Sher, H. E. Varmess, M. J. Lenardo, and P. L. Schwartzberg, *Science* **284**, 638 (1999).

[8] K. Baba, A. Takeshita, K. Majima, R. Veda, S. Kondo, N. Juni, and D. Yamamoto, *Mol. Cell. Biol.* **19**, 4405 (1999).

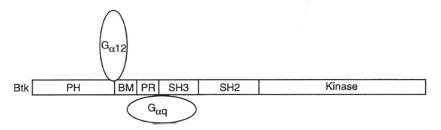

FIG. 1. Diagram of the domain structure of Btk and the interaction sites of $G\alpha_q$ and $G\alpha_{12}$ on Btk.

human Btk cDNA, pApuro-hBtk,[9] was used as template for polymerase chain reaction (PCR) with the following oligonucleotide primers: 5′ oligonucleotide primer ATACGGATCCATGGCCGCAGTGATTCTG and 3′ oligonucleotide primer CTA-GCTCGAGGGATTCTTCATCCATGAC. The PCR product (~1.9 kb) was subcloned into the BamHI and XhoI sites of pET21a vector (Novagen, Madison, WI). This construct is cotransformed into the BL21 bacterial strain with plasmid pREP4groESL carrying genes encoding GroES and GroEL chaperones.[10,11] Cultures are grown at 37° until the OD_{600} reaches 0.5. The expression of histidine-tagged Btk is induced by 1 mM isopropyl-β-D-thiogalactopyranoside (IPTG) for 8 hr at room temperature. After harvesting, the pellet is resuspended in lysis buffer: 25 mM Tris-HCl (pH 8), 100 mM NaCl, 50 mM sodium phosphate (pH 8.8), 10 mM 2-mercaptoethanol, 1% (v/v) Triton X-100, 5% (v/v) glycerol, 3 μM leupeptin, 3 μM pepstatin, aprotinin [0.15 TIU (trypsin inhibitor unit (TIU)/ml], 0.25 mM phenylmethylsulfonyl fluoride (PMSF), 1 mM benzadimine. For a 1-liter culture, 25 ml of lysis buffer is used. Lysozyme (0.1–0.5 mg/ml) is added, and the sample is put on ice for 30 min. After cells are lysed by nitrogen cavitation [Parr bomb (Moline, IL)] or sonication, the lysate is spun down at 100,000g for 1 hr at 4°. One milliliter of Ni^{2+}–NTA–agarose resin (50% slurry) (Qiagen, Valencia, CA), preequilibrated with 25 mM Tris-HCl (pH 8.8), 300 mM NaCl, is added to the supernatant. The mixture is incubated with gentle agitation for 1 hr at 4°. The Btk-bound beads are then poured onto a column (C10/20; Pharmacia, Piscataway, NJ). The column is washed with wash buffer 1 [25 mM Tris-HCl (pH 8.8), 250 mM NaCl, 5 mM imidazole, 5% (w/v) glycerol] until OD_{280} < 0.01, and then washed with 10 ml of wash buffer II [25 mM Tris-HCl (pH 8.8), 250 mM NaCl, 25 mM imidazole, 5% (v/v) glycerol]. Protein can be eluted with 5 ml of elution buffer (50 mM Tris, 300 mM NaCl, 250 mM imidazole, pH 6.0). To the eluate, dithiothreitol (DTT) is added to 5 mM. The eluted material is then dialyzed

---

[9] M. Takata and T. Kurosaki, *J. Exp. Med.* **184,** 31 (1995).

[10] K. E. Amrein, B. Takacs, M. Stieger, J. Molnos, N. A. Flint, and P. Burn, *Proc. Natl. Acad. Sci. U.S.A.* **92,** 1048 (1995).

[11] N. A. Flint, K. E. Amrein, T. Jascur, and P. Burn, *J. Cell. Biochem.* **55,** 389 (1994).

against 25 mM Tris (pH 8.0), 5 mM DTT, 100 mM NaCl, 5% (v/v) glycerol. Btk–His$_6$ is further purified with a Q anion-exchange column for fast protein liquid chromatography (FPLC; Pharmacia) with a linear gradient of NaCl (from 100 to 500 mM) in 25 mM Tris (pH 8.0), 5 mM DTT, 5% (v/v) glycerol. Protein fractions are then analyzed by silver stain, Western blot, and kinase assay. After dialysis against 50 mM Tris (pH 7.4), 10 mM MnCl$_2$, 100 mM NaCl, 5% (v/v) glycerol, samples are concentrated to ~1 mg/ml, frozen in liquid nitrogen, and stored at −80°. Purified Btk runs as an 80- to 85-kDa protein on sodium dodecyl sulfate–polyacrylamide gel electrophoresis (SDS–PAGE) based on prestained protein molecular mass markers (Bio-Rad, Hercules, CA).[3]

Btk Kinase Buffer

Btk kinase is sensitive to its surroundings and thus requires a specialized buffer to achieve maximum activity. We use a kinase buffer consisting of 50 mM Tris (pH 7.4), 10 mM MnCl$_2$. Tris or HEPES buffers between pH 7 and 8 are suitable for Btk. Kinases in general are known to require Mg^{2+} to perform phosphorylation. We found that Btk activity requires the presence of Mn^{2+}. For these reactions, 10 mM Mn^{2+} is sufficient. Bovine serum albumin (BSA, as high as 200 μg/ml) should be included in the kinase reaction to stabilize the enzyme. A final ATP concentration from 200 nM to 100 μM is sufficient. A 200 nM concentration of [γ-^{32}P] ATP (3000 Ci/mmol) is included in the reaction to visualize the phosphorylation of the peptide substrate. Nonradioactive ATP can be added to the reaction up to 100 μM, and will significantly promote substrate phosphorylation. However, the more nonradioactive ATP added, the less γ-^{32}P incorporated and the longer the exposure required.

Btk Kinase Substrate

To visualize the kinase activity of Btk, radiolabeled ATP must be incorporated into a substrate. We synthesized peptide substrate (Genemed, South San Francisco, CA) corresponding to the peptide sequence of the Btk autophosphorylation site (KKVVALYDYMPMN) and found this peptide to be a functional substrate recognized by Btk.[3,12] The peptide (~35 mg) is dissolved into 50 μl of dimethyl sulfoxide (DMSO) and then diluted into Btk kinase buffer to a stock concentration of 2 mM. The final concentration of the peptide substrate in each kinase reaction is 100 μM. A glutathione S-transferase (GST) fusion protein of the cytoplasmic domain of Band 3 (CDB3) protein can also be used as an alternative substrate for the Btk kinase assay.[13]

[12] H. Park, M. I. Wahl, D. E. Afar, C. W. Turck, D. J. Rawlings, C. Tam, A. M. Scharenberg, J. P. Kinet, and O. N. Witte, *Immunity* **4,** 515 (1996).
[13] Y. Wan, K. Bence, A. Hata, T. Kurosaki, A. Veillete, and X. Y. Huang, *J. Biol. Chem.* **272,** 17209 (1997).

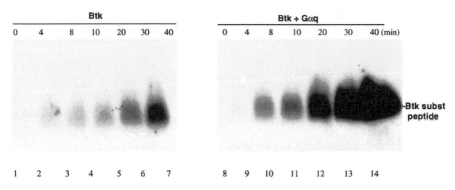

FIG. 2. Gα_q stimulation of Btk.

Reaction Conditions

The final concentration of Btk kinase used in the reaction is 10–50 nM. As little as 10 nM Btk can be used to produce sufficient signal, which can further be stimulated by G protein α subunits. When G protein α subunits (Gα_q or Gα_{12}) are included in the kinase reaction, the G protein concentration used can range from 1 to 300 nM. Boiled controls of these G protein preparations should be used alongside to monitor the effect of solvents.

Once [γ-^{32}P]ATP is added, the reaction mixture is placed at 30°. A time course analysis shows that within 40 min, the kinase activity is within the linear range (Fig. 2). The reactions can then be placed on ice and stopped by adding SDS–PAGE sample buffer.

Sodium Dodecyl Sulfate-Polyacrylamide Gel Electrophoresis Separation

The kinase reactions are heat inactivated for 5 min at 95°, and then loaded onto an SDS–polyacrylamide gel. Because the substrate is only 13 amino acids long, a high-percentage gel must be used. A 20% (w/v) gel (20 × 20 cm) has been found to provide adequate separation of peptide and free [γ-^{32}P]ATP. The gel is run until the sample dye is within 1 inch of the bottom. After cutting the gel right above the sample dye, the bottom gel piece that contains free [γ-^{32}P]ATP is properly discarded. The remaining gel piece, which contains phosphorylated peptides, is dried in a vacuum gel dryer. Radiolabel incorporation into substrate is visualized by autoradiography or PhosphorImager (Molecular Dynamics, Sunnyvale, CA) (Fig. 2).

Signal Quantification

To quantify the ^{32}P incorporation, the dried SDS–polyacrylamide gel is aligned with the autoradiogram. The gel pieces containing the phosphorylated substrate

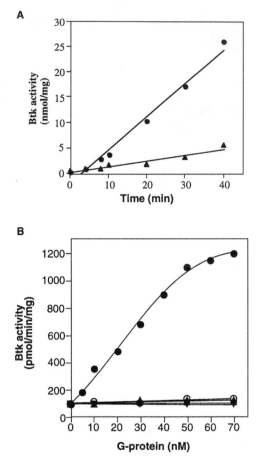

FIG. 3. Quantification of the stimulation of Btk by $G\alpha_q$. (A) Time course of Btk kinase activity. Triangles: Btk alone. Filled circles: Btk plus $G\alpha_q$. (B) Dosage curve of Btk stimulation by G proteins. Filled circles: $G\alpha_q$–GTPγS. Open triangles: $G\alpha_{i1}$–GTPγS. Filled triangles: $G\alpha_q$–GDP. Filled inverted triangles: heat-inactivated $G\alpha_q$–GTPγS.

are cut out of the gel, added to scintillation fluid, and counted. The activity is calculated and expressed as specific activity or activity per mass of protein over time (Fig. 3).

G-Protein Regulation of Tyrosine Kinases

Although the detailed activation mechanisms of Btk by $G\alpha_q$ and $G\alpha_{12}$ are not known, possible activation models have been proposed. $G\alpha_q$ binds directly to Btk to a region composed of a Tec homology (TH) domain and a Src homology 3

(SH3) domain (Fig. 1).[14] Mutations of Btk that disrupt its ability to bind $G\alpha_q$ also eliminated Btk stimulation by $G\alpha_q$, suggesting that this interaction is important for Btk activation. Remarkably, the structure of this TH–SH3 fragment of the Btk family of tyrosine kinases shows an intramolecular interaction.[15] Interestingly, the crystal structure of the Src family of tyrosine kinases reveals that the intramolecular interaction involving SH3 is the major determining factor keeping the kinase inactive.[16,17] Thus, an activation model has been proposed that entails binding of $G\alpha_q$ to the TH–SH3 region of Btk, thereby disrupting the TH–SH3 intramolecular interaction and activating Btk.[14]

$G\alpha_{12}$ interacts with a domain composed of the PH domain and the adjacent Btk motif (BM).[4] Because the crystal structure of this PH–BM module of Btk also shows an intramolecular interaction, an activation model has been proposed similar to the activation of Btk by $G\alpha_q$.[4]

Some cellular signaling pathways have been described to involve these G protein–Btk interactions. In Btk-deficient DT40 lymphoma cells, stimulation of p38 MAPK (mitogen-activated protein kinase) by G_q-coupled receptors is blocked; thus, stimulation of Btk by $G\alpha_q$ may be essential *in vivo* for activation of the p38 MAPK pathway.[3] In transfected cells, $G\alpha_{12}$ can also activate Tec and Bmx, two other members of the Btk family of tyrosine kinases.[18] It has also been suggested that Btk family tyrosine kinases can link $G\alpha_{12}$ and $G\alpha_{13}$ to a Rho-dependent pathway, leading to the activation of serum response factor.[18]

Acknowledgments

Our research is supported by grants from the National Institutes of Health, the American Cancer Society, the American Heart Association, and the Irma Hirschl Trust.

[14] Y. C. Ma and X. Y. Huang, *Proc. Natl. Acad. Sci. U.S.A.* **95,** 12197 (1998).
[15] A. H. Andreotti, S. C. Bunnell, S. Feng, L. J. Berg, and S. L. Schreiber, *Nature (London)* **385,** 93 (1997).
[16] W. Xu, S. C. Harrison, and M. J. Eck, *Nature (London)* **385,** 595 (1997).
[17] F. Sicheri, I. Moarefi, and J. Kuriyan, *Nature (London)* **385,** 602 (1997).
[18] J. Mao, W. Xie, H. Yuan, M. I. Simon, H. Mano, and D. Wu, *EMBO J.* **17,** 5638 (1998).

[37] Isozyme-Specific Inhibitors and Activators of Protein Kinase C

By DEBORAH SCHECHTMAN and DARIA MOCHLY-ROSEN

Introduction

In cells, signal transduction often occurs within "signaling complexes" where physical associations among several enzymes are coordinated by scaffold proteins.[1] The scaffold is thought to provide a means for subcellular localization of select signaling enzymes, resulting in coordinated responses to outside stimuli. Because transient translocation and binding of signaling enzymes to their anchoring proteins (termed also translocation) is a common theme in intracellular signal transduction,[2] interference in protein–protein interactions is expected to be useful in dissecting the molecular events in signaling. Such modulation of protein–protein interactions can be achieved with peptides derived from the interacting sites of each signaling enzyme and its respective anchoring protein. Peptide modulators for protein kinases have been discovered by systematic searches as well as by screening both random and biased peptide libraries (for review, see Souroujon and Mochly-Rosen[3]). Several rational approaches to identify such selective peptide modulators have also been used successfully.[3] Here we describe the methods that have been successfully used to identify short peptide modulators of protein kinase C (PKC)-mediated signal transduction.

PKC represents a family of isozymes, each presumably with different functions (Fig. 1[4–9]; see Nishizuka[10]). Multiple PKC isozymes may contribute in a pathway-specific and cell type-specific manner to signaling events, and sometimes even exhibit opposing effects. For example, we found that εPKC activation suppresses the contraction rate of cardiac myocytes, whereas δPKC activation enhances it.[11] Peptide-specific inhibitors and activators of the different PKC isozymes were

---

[1] T. Pawson and J. D. Scott, *Science* **278**, 2075 (1997).

[2] D. Mochly-Rosen, *Science* **268**, 247 (1995).

[3] M. C. Souroujon and D. Mochly-Rosen, *Nat. Biotechnol.* **16**, 919 (1998).

[4] Y. Nishizuka, *Science* **258**, 607 (1992).

[5] P. J. Parker, L. Coussens, N. Totty, L. Rhee, S. Young, E. Chen, S. Stabel, M. D. Waterfield, and A. Ullrich, *Science* **233**, 853 (1986).

[6] A. C. Newton, *Curr. Biol.* **5**, 973 (1995).

[7] W. S. Sossin and J. H. Schwartz, *Trends Biochem. Sci.* **18**, 207 (1993).

[8] R. B. Sutton and S. R. Sprang, *Structure* **6**, 1395 (1998).

[9] D. Mochly-Rosen, K. G. Miller, R. H. Scheller, H. Khaner, J. Lopez, and B. L. Smith, *Biochemistry* **31**, 8120 (1992).

[10] Y. Nishizuka, *FASEB J.* **9**, 484 (1995).

[11] J. A. Johnson, M. O. Gray, C. H. Chen, and D. Mochly-Rosen, *J. Biol. Chem.* **271**, 24962 (1996).

Protein kinase C isozymes

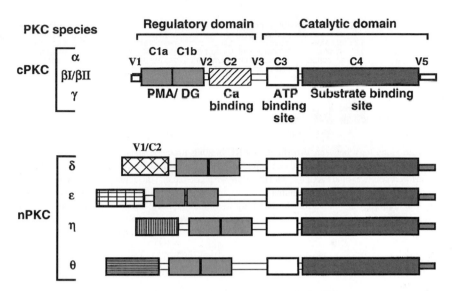

FIG. 1. Schematic depiction of functional domains in members of the PKC family. There are three or four subfamilies of PKCs, based on homology within the regulatory domain, and their sensitivity to activation by second messengers.[4] We focused our work on classic (cPKC) and novel (nPKC) PKCs only. The more homologous domains among the isozymes were termed common or C domains and the unique domains were called variable or V domains.[5] They are numbered from 1 to 5, starting from the N terminus. The C3/C4 domains constitute the catalytic domain. The C1 domain was found to be the diacylglycerol- and phorbol ester-binding domain (reviewed in Newton[6]). The C2 domain was initially identified only in cPKCs and was shown to bind phosphatidylserine and calcium. Further sequence analysis[7] and subsequent structural analysis[8] demonstrated that the first variable domain, V1, in the novel nPKCs is a C2-like domain. Our studies show that this domain contains at least part of the RACK-binding site on PKC (see text).[9]

identified to elucidate the varying roles played by PKC in normal and disease states. These peptides interfere with the interaction of individual PKC isozymes with their anchoring proteins, RACKs (receptor for activated C-kinases), and thereby specifically affect intracellular signaling events mediated by each isozyme in intact cells.

When we began searching for pharmacological regulators of PKC functions, we did not expect that short peptides from PKC or RACKs would properly mimic the binding sites between these proteins. However, after the serendipitous finding of peptide I in 1991 (see below and Fig. 2 [12–16]), we were encouraged to continue

---

[12] A. Aitken, C. A. Ellis, A. Harris, L. A. Sellers, and A. Toker, *Nature (London)* **344,** 594 (1990).

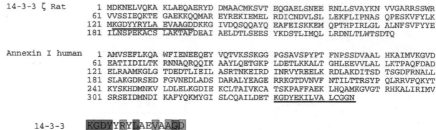

```
14-3-3 ζ Rat        1  MDKNELVQKA KLAEQAERYD DMAACMKSVT EQGAELSNEE RNLLSVAYKN VVGARRSSWR
                   61  VVSSIEQKTE GAEKKQQMAR EYREKIEMEL RDICNDVLSL LEKFLIPNAS QPESKVFYLK
                  121  MKGDYYRYLA EVAAGDDKKG IVDQSQQAYQ EAFEISKKEM QPTHPIRLGL ALNFSVFYYE
                  181  ILNSPEKACS LAKTAFDEAI AELDTLSEES YKDSTLIMQL LRDNLTLWTSDTQ

Annexin I human     1  AMVSEFLKQA WFIENEEQEY VQTVKSSKGG PGSAVSPYPT FNPSSDVAAL HKAIMVKGVD
                   61  EATIIDILTK RNNAQRQQIK AAYLQETGKP LDETLKKALT GHLEEVVLAL LKTPAQFDAD
                  121  ELRAAMKGLG TDEDTLIEIL ASRTNKEIRD INRVYREELK RDLAKDITSD TSGDFRNALL
                  181  SLAKGDRSED FGVNEDLADS DARALYEAGE RRKGTDVNVF NTILTTRSYP QLRRVFQKYT
                  241  KYSKHDMNKV LDLELKGDIE KCLTAIVKCA TSKPAFFAEK LHQAMKGVGT RHKALIRIMV
                  301  SRSEIDMNDI KAFYQKMYGI SLCQAILDET KGDYEKILVA LCGGN
```

FIG. 2. Homology between two different PKC-binding proteins, annexin I and 14-3-3 zeta. A homologous sequence between these two unrelated proteins identified by Aitkin[12] is underlined. Both proteins were found to bind PKC,[13–15] and therefore we predicted that the common sequence may serve the same function, namely a PKC-binding site. This prediction was confirmed (see text).[16] *Bottom:* The regions of homology are indicated. The dark gray areas mark identical amino acids and similar amino acids are marked by light gray and by dashes. Peptide I corresponds to the sequence in annexin I.

searching for other short peptides with similar inhibitory activities. This approach led to the identification of isozyme-selective inhibitor and activator peptides for all the classic and novel PKC isozymes (Table I[17–24]).[3] It is likely that such an approach will be useful for other protein kinases.

The following is a description of the methods used in our studies. We first describe how to identify a binding protein for a signaling enzyme, how to confirm that the interaction occurs in cells, and how to identify short peptides that modulate these protein–protein interactions *in vitro* and in cells.

[13] D. Mochly-Rosen, H. Khaner, and J. Lopez, *Proc. Natl. Acad. Sci. U.S.A.* **88,** 3997 (1991).

[14] A. Toker, C. A. Ellis, L. A. Selleres, and A. Aitken, *Eur. J. Biochem.* **191,** 421 (1990).

[15] A. Aitken, B. Amess, S. Howell, D. Jones, H. Martin, Y. Patel, K. Robinson, and A. Toker, *Biochem. Soc. Trans.* **20,** 607 (1992).

[16] D. Mochly-Rosen, H. Khaner, J. Lopez, and B. L. Smith, *J. Biol. Chem.* **266,** 14866 (1991).

[17] D. Ron, J. Luo, and D. Mochly-Rosen, *J. Biol. Chem.* **270,** 24180 (1995).

[18] Z.-H. Zhang, J. A. Johnson, L. Chen, N. El-Sherif, D. Mochly-Rosen, and M. Boutjdir, *Circ. Res.* **80,** 720 (1997).

[19] M. Yedovitzky, D. Mochly-Rosen, J. A. Johnson, M. O. Gray, D. Ron, E. Abramovitch, E. Cerasi, and R. Nesher, *J. Biol. Chem.* **272,** 1417 (1997).

[20] M. O. Gray, J. S. Karliner, and D. Mochly-Rosen, *J. Biol. Chem.* **272,** 30945 (1997).

[21] D. Mochly-Rosen, G. Wu, H. Hahn, H. Osinska, T. Liron, J. N. Lorenz, A. Yatani, J. Robbins, and G. W. Dorn II, *Circ. Res.* **86,** 1173 (2000).

[22] B. Hundle, T. McMahon, J. Dadgar, C. H. Chen, D. Mochly-Rosen, and R. O. Messing, *J. Biol. Chem.* **272,** 15028 (1997).

[23] G. C. Mayne and A. W. Myrray, *J. Biol. Chem.* **273,** 24115 (1998).

[24] G. W. Dorn II, M. C. Souroujon, T. Liron, C. H. Chen, M. O. Gray, H. Z. Zhou, M. Csukai, G. Wu, J. N. Lorenz, and D. Mochly-Rosen, *Proc. Natl. Acad. Sci. U.S.A.* **96,** 12798 (1999).

TABLE I

PEPTIDE MODULATORS OF PKC: SPECIFICITY AND BIOLOGICAL ACTIVITY

Peptide or fragment	Specificity	Effect on cellular functions
βC2-1, βC2-2, and βC2-4[a]	cPKC isozymes α, βI, βII, and γ	Inhibits PMA-induced modulation of L-type calcium channels.[b] Inhibits insulin secretion from pancreatic cells[c]
εPKC V1 fragment[d] and εV1-2[d,e]	εPKC	Inhibits negative chronotropic effect in cardiac myocytes.[d] Inhibits normal cardiac muscle development leading to dilated cardiomyopathy in transgenic mice expressing εV1 fragment in their hearts.[f] Prevents enhancement of nerve growth factor responses by ethanol and phorbol esters.[g] Inhibits cell death induced by ischemia.[e] Blocks inhibition of cell death induced by tumor necrosis factor α (TNF-α)[h]
$\psi\varepsilon$RACK[i]	εPKC	Increases resistance of isolated cardiac myocytes, and heart of transgenic mice expressing this peptide from prolonged ischemia.[i] Stimulates physiological cardiomyocyte growth and increase in heart weight/body weight ratio without altering normal cardiac functions[f]
δPKC V1 fragment[d] δV1-1[j]	δPKC	Stimulates negative chronotropic effect.[d] Decreases damage to cardiac myocytes induced by ischemia[j]
$\psi\delta$RACK[j]	δPKC	Increases damage to cardiac myocytes induced by ischemia[j]
ηV1-2[d]	ηPKC	No effect on the above negative chronotropic effect[d]

[a] D. Ron et al.[17]
[b] Z.-H. Zhang et al.[18]
[c] M. Yedovitzky et al.[19]
[d] J. A. Johnson et al.[11]
[e] M. O. Gray et al.[20]
[f] D. Mochly-Rosen et al.[21]
[g] B. Hundle et al.[22]
[h] G. C. Mayne and A. W. Myrray.[23]
[i] G. W. Dorn et al.[24]
[j] L. Chan et al., Submitted (2001a).

Methods

Step I. Determine That a Kinase Interacts with a Protein via a Site Other Than Its Catalytic Site: Overlay (Far Western), Interactive Cloning in Bacteria, and Coimmunoprecipitation Experiments

Immunofluorescence, genetic, and biochemical studies may suggest that the signaling enzyme anchors to another protein and that this interaction is important for the signaling event. The first step is to identify that protein partner. To this end, several methods and approaches have been successfully used. We limit the discussion here to those applied for the study of PKCs and RACKs.

PKC binding to anchoring proteins can be determined after sodium dodecyl sulfate–polyacrylamide gel electrophoresis (SDS–PAGE) of the preparation containing the anchoring proteins. This binding assay, termed "overlay" or "far Western," can be used to characterize the anchoring protein. A negative result is not informative, as the properties of the binding protein may be sensitive to denaturation. However, this assay is simple and fast, and therefore it is a good first approach. The steps outlined below are modified from the assay developed by Wolf and Sahyoun.[25]

A. Overlay (Far Western) Method

1. PREPARATION AND FRACTIONATION OF TISSUE OR CELL LYSATES

a. Reagents

Homogenization buffer (final concentrations): 20 mM Tris-HCl (pH 7.5), 2 mM EDTA, 10 mM EGTA, 0.25 M sucrose, and protease inhibitors diluted 1 : 1000 (added just before use). Stock solutions of protease inhibitors are prepared as follows:

1. Dissolve phenylmethysulfonyl fluoride (PMSF) in 2-propanol (17 mg/ml).
2. Dissolve soybean trypsin inhibitor (SBTI, 20 mg/ml), leupeptin (25 mg/ml), and aprotinin (25 mg/ml) in 20 mM Tris-HCl, pH 7.5.

DE buffer (final concentrations): 20 mM Tris-HCl (pH 7.5), 1 mM EDTA, 1 mM EGTA, 12 mM 2-mercaptoethanol, 10% (v/v) glycerol, 1% (v/v) Triton-X 100 and freshly added protease inhibitors (same concentrations as above).

b. Procedure

1. Immediately after harvesting, prepare tissue lysates in ice-cold homogenization buffer, using a cooled Dounce homogenizer.

2. For lysates prepared from tissue cultures, remove the cell culture medium and wash the cells once with phosphate-buffered saline (PBS). Scrape the cells in homogenization buffer with a rubber policeman, and pass several times through a 25- or 27-gauge needle. It is important to avoid foaming throughout the lysate preparation to minimize protein denaturation.

3. If separation of soluble and Triton-soluble fractions is desired, homogenates should be ultracentrifuged at 100,000g for 40 min at 4°.

4. After decanting the supernatant (soluble fraction), solubilize the pellet in DE buffer (15–30 min at 4°). Solubilization may be aided by passing the material through a 25-gauge needle.

[25] M. Wolf and N. Sahyoun, *J. Biol. Chem.* **261,** 13327 (1986).

5. Ultracentrifuge the solubilized material as described above, or maintain at −20° in the presence of 50% (v/v) glycerol to prevent freezing. The supernatant after centrifugation contains the Triton-soluble material, and the pellet contains the Triton-insoluble material.

6. Store samples at −20°.

7. Determine the protein concentration by the Bradford assay,[26] and run an SDS–polyacrylamide gel using at least 20 μg of lysate per lane (amount depends on the abundance of the binding protein in the lysate). Transfer the gel to nitrocellulose, using the Towbin method.[27] Immediately after transfer, overlay as follows.[13,16]

2. OVERLAY (FAR WESTERN)

a. Reagents

Overlay blocking buffer (final concentrations): 200 mM NaCl, 50 mM Tris-HCl (pH 7.5), 3% (w/v) bovine serum albumin (BSA), 0.1% (w/v) polyethylene glycol (PEG) 8000

Overlay buffer (final concentrations): 200 mM NaCl, 50 mM Tris-HCl (pH 7.5), 12 mM 2-mercaptoethanol, 1% (w/v) PEG 8000, protease inhibitors (same concentrations as above)

Overlay wash buffer (final concentrations): 200 mM NaCl, 50 mM Tris-HCl (pH 7.5), 0.1% (w/v) PEG 8000, 12 mM 2-mercaptoethanol

TBS–Tween: 20 mM Tris-HCl (pH 7.5), 100 mM NaCl, 0.05% (v/v) Tween 20

b. Procedure

1. Block the nitrocellulose for 30 min to 1 hr in overlay blocking buffer.

2. Overlay the nitrocellulose with purified PKC or PKC fragments expressed recombinantly in bacteria (10–50 μg/ml). If desired, the fragments may be epitope tagged to facilitate detection. Incubate at room temperature for 30 min to 1 hr in overlay buffer. To reduce the volume required, the nitrocellulose may be incubated in sealed plastic bags.

3. Wash the nitrocellulose three times in overlay wash buffer, for 15 min each.

4. Incubate the nitrocellulose for 1.5 hr in primary antibody (e.g., anti-PKC) diluted in overlay blocking buffer.

5. Wash the nitrocellulose in TBS–Tween, three times for 15 min.

6. Incubate in secondary antibody diluted in overlay blocking buffer for 2 hr. The secondary antibody may be coupled to ^{125}I, alkaline phosphatase, or peroxidase.

7. Wash in TBS–Tween three times for 15 min and detect binding according to label of the secondary antibody.

[26] M. M. Bradford, *Anal. Biochem.* **72**, 248 (1976).
[27] H. Towbin, T. Staehelin, and J. Gordon, *Proc. Natl. Acad. Sci. U.S.A.* **76**, 4350 (1979).

c. Observations

1. If activation of the kinase is required for binding to the anchoring protein, the kinase activators must be included in the above-described assay. For example, to demonstrate PKC binding to RACK, calcium and phospholipids need to be present.[13]

2. The proteins identified in this method bind PKC with a K_D of approximately 5 nM.[28,29]

3. Jaken and collaborators identified PKC-binding proteins by a similar method that included a glutaraldehyde cross-linking step to fix the protein complex to the nitrocellulose membrane.[30] This method has been useful in identifying proteins that are involved in PKC-mediated signaling but that bind PKC with lower affinity as compared with the above-described method.[31,32]

B. Expression–Interaction Cloning

Expression–interaction cloning is a powerful way to identify proteins that interact with each other. We have used expression libraries such as λGT11, to clone RACK1 (specific for βII PKC)[28] and RACK2 (specific for εPKC).[29] The screening is performed by binding the phage-infected bacterial colonies to nitrocellulose membranes, and overlaying with the protein of interest (PKC or PKC fragments), using the method described above. Positive clones are rescreened until they are purified and then sequenced. This method and the criteria for identifying positive clones have been described in detail elsewhere.[33] The method of Chapline et al.,[30] which includes a cross-linking step, was also successful in cloning PKC-binding proteins that are substrates of PKC rather than RACKs.[31,32]

C. In Vitro Pulldown Assay

The in vitro pulldown assay was first used in identifying signaling complexes in the tyrosine kinase cascade. For this assay, the protein or fragment of interest is expressed in bacteria or insect cells as a fusion protein with tags such as glutathione S-transferase (GST),[34] maltose-binding protein (MBP)[35] or polyhistidine.[36] Affinity columns for these tags are commercially available (e.g., GST–Sepharose,

[28] D. Ron, C. H. Chen, J. Caldwell, L. Jamieson, E. Orr, and D. Mochly-Rosen, Proc. Natl. Acad. Sci. U.S.A. 91, 839 (1994).

[29] M. Csukai, C. H. Chen, M. A. De Matteis, and D. Mochly-Rosen, J. Biol. Chem. 272, 29200 (1997).

[30] C. Chapline, K. Ramsay, T. Klauck, and S. Jaken, J. Biol. Chem. 268, 6858 (1993).

[31] C. Chapline, B. Mousseau, K. Ramsay, S. Duddy, Y. Li, S. Kiley, and S. Jaken, J. Biol. Chem. 271, 6417 (1996).

[32] C. Chapline, J. Cottom, H. Tobin, J. Hulmes, J. Crabb, and S. Jaken, J. Biol. Chem. 273, 19482 (1998).

[33] M. Csukai and D. Mochly-Rosen, Methods Mol. Biol. 88, 133 (1998).

[34] D.B. Smith and K.S. Johnson, Gene 67, 31 (1988).

[35] C. V. Maina, P. D. Riggs, A. G. Grandea III, B. E. Slatko, L. S. Moran, J. A. Tagliamonte, L. A. McReynolds, and C. D. Guan, Gene 74, 365 (1988).

amylose–Sepharose or a nickel column, respectively). Lysates are obtained in the presence of protease inhibitors and stored at −80° in small aliquots or at −20° in the presence of 50% (v/v) glycerol.

a. REAGENTS

Overlay wash buffer (final concentrations): 200 mM NaCl, 50 mM Tris–HCl (pH 7.5), 12 mM 2-mercaptoethanol, 0.1% (w/v) PEG 8000.
Overlay buffer (final concentrations): 200 mM NaCl, 50 mM Tris–HCl (pH 7.5), 12 mM 2-mercaptoethanol, 1% (w/v) PEG 8000, 0.1% (w/v) BSA, and protease inhibitors (same concentrations as in Step I).

b. PROCEDURE

1. Wash the affinity column beads three times in 1 ml of the same buffer used to purify the recombinant protein. Precipitate the beads with low-speed centrifugation (approximately 1000g for 1 min at room temperature) after each wash. Usually 25 μl of packed beads is sufficient for each binding reaction.

2. Add lysate containing the fusion protein to the beads (final volume of 200 μl) and incubate the mixture at room temperature for 15 min with continuous rocking.

3. Wash the beads three times in overlay wash buffer as described above.

Note that before the next step, studies should be performed to confirm that the amount of fusion protein added does not exceed the amount that can bind to the beads. In this example, a constant amount of fusion protein is bound to increasing amounts of control beads or beads bound to the binding protein. Nonspecific binding is expected to increase linearly with increasing amounts of added fusion protein, whereas specific binding should be saturable. (The optimal ratio of fusion protein to beads should result in no or only limited nonspecific binding and should be the minimum amount required to saturate the specific binding.)

4. Incubate the beads with lysate containing the diluted partner protein in overlay buffer for 15 min at room temperature with continuous rocking. Again, protein concentrations depend on the affinity of the proteins.

5. Wash the beads as in step 3, using overlay wash buffer with 1% (w/v) nonionic detergent (e.g., Igepal; Sigma, St. Louis, MO). Before the last wash, transfer the bead suspensions to a clean Eppendorf tube and pellet as described previously. (This step is included to reduce nonspecific binding to the test tube walls.)

6. After the last wash, resuspend the beads in 2.5× Laemmli sample buffer[37] and boil before running on an SDS–polyacrylamide gel.

7. Detect binding proteins by Western blot.[27]

[36] H. Berthold, M. Scanarini, C. C. Abney, B. Frorath, and W. Northemann, *Protein Expr. Purif.* **3,** 50 (1992).
[37] D. W. Cleveland, S. G. Fischer, M. W. Kirschner, and U. K. Laemmli, *J. Biol. Chem.* **252,** 1102 (1977).

c. OBSERVATIONS

1. As described above, in the case of PKC, the enzyme binds the anchoring protein after activation. Therefore, the kinase activators should be required for the binding.[28]

2. There may be nonspecific binding of the enzyme to the beads or to the fusion protein. One way to avoid this is to include an additional blocking step after binding of the first protein to the column. For example, when using GST–Sepharose bound with RACK1–GST and overlaid with an MBP–cPKC fusion protein, the beads were first blocked with MBP to eliminate nonspecific binding of MBP–cPKC to the affinity column.

3. If possible, reciprocal pulldown assays should also be carried out, for example, using an affinity column with PKC fragment and determining RACK binding with cell lysates.

Step II. Confirm That the Interaction Identified in Vitro Also Occurs in Vivo

Several criteria must be met in order to demonstrate that an interaction identified *in vitro* also occurs in cells:

The proteins should colocalize in the cell in an inducible manner (e.g., we showed that εRACK colocalizes with εPKC only after PKC activation).[29]

The endogenous proteins coimmunoprecipitate (εRACK coimmunoprecipitates with εPKC).[29]

Blocking the interaction between the kinase and its binding protein interferes with the function of the tested kinase [e.g., we showed that inhibition of εPKC binding to εRACK inhibited phorbol myristate acetate (PMA)-induced negative chronotropy in cardiac myocytes].[11]

A. Colocalization

Colocalization of proteins as evidenced by immunofluorescence is insufficient on its own to prove an association between two proteins. However, this study can still be informative, because a lack of colocalization strongly suggests that these two proteins do not interact.

a. PROCEDURE

i. Culture and fixation of cells

Culture the cells on eight-well plastic or glass chambers or on glass coverslips to approximately 30% confluence. After the treatment of choice (e.g., phorbol ester stimulation), rinse the cells with cold PBS and quickly fix with either methanol–acetone (1 : 1, v/v) for 3 min on ice or with 4% (v/v) formaldehyde at room temperature, diluted in PBS [the stock solution available by Polysciences (Warrington, PA) is 16% (w/v)]. After fixation, rinse the cells twice in PBS and keep in distilled water at 4° until staining.

ii. Staining of cells

1. Block in PBS containing 1% (v/v) normal goat serum and 0.1% (v/v) Triton X-100 (to permeabilize the cells). Incubate in a humid chamber at room temperature for 1 to 2 hr or overnight at 4°.

2. To colocalize two different proteins, it is necessary to use primary antibodies raised in two different species (e.g., one from mouse the other from rabbit). Aspirate the blocking solution and incubate in primary antibodies. The antibodies should be approximately five times more concentrated than the concentration required for Western blot analysis. Incubate at room temperature for at least 2 hr at room temperature or at 4° overnight (in a humid chamber).

3. Aspirate the primary antibodies and wash three times for 5 min each with PBS containing 0.1% (v/v) Triton X-100.

4. Add blocking buffer containing the secondary antibodies conjugated to two different fluorochromes [e.g., Texas Red anti-mouse antibodies and fluorescein isothiocyanate (FITC) anti-rabbit antibodies]. Incubate in the dark for 45 min to 2 hr.

5. Wash as in step 3.

6. If using eight-well chamber slides, remove the plastic wells and gaskets. Wash three times as described previously, followed by two washes in water (to decrease background fluorescence from buffer salts). Allow them to dry. Mount glass coverslips, using aqueous media. Avoid trapping air bubbles by tilting each coverslip over. A Slow Phade kit (Molecular Probes, Eugene, OR) may be used as recommended by the manufacturer to reduce bleaching of the fluorochrome.

b. OBSERVATIONS

1. Diluted antibodies should be prefiltered (e.g., using a 0.45-μm pore size Centrex filter; Schleicher & Schuell, Keene, NH) to remove aggregates.

2. During the procedure, buffer evaporation should be avoided, as salts increase background fluorescence.

3. For cells in suspension, labeling may be done in an Eppendorf tube, pelleting the cells at the lowest speed necessary (approximately 1000g for 1 min at room temperature). As a final step, wash the cells in deionized water and resuspend them in 30–50 μl of deionized water. Add an equal volume of the mounting reagent and analyze as before.

4. Coimmunolocalization by electron microscopy is superior to that observed in immunofluorescence study. However, it is not available in most laboratories.

B. Coimmunoprecipitation

Coimmunoprecipitation from cell lysates can provide further evidence that two proteins interact *in vivo*. In this assay, immunoprecipitation is carried out with an antibody against one of the proteins and the immunoprecipitated material is analyzed by Western blot with an antibody against the other protein partner.[29] However, not all antibodies are effective in immunoprecipitation, either because their affinity is too low or because the partner protein covers the antigenic site, and it may be necessary to test several antibodies to the partner proteins. In addition, some proteins may interact in cells but do not coimmunoprecipitate because their affinity for each other is too low. Alternatively, it is possible that two proteins coimmunoprecipitate but do not interact directly with each other; rather, they may each bind to a common third protein. Finally, some common proteins may coimmunoprecipitate with the protein of interest in a nonspecific manner. For these reasons, neither a negative nor a positive result is informative alone; this assay requires corroboration by other assays. It is common practice to try to demonstrate reciprocal coimmunoprecipitation. If the proteins interact *in vivo,* it is expected that antibodies to protein A will bring down protein B and antibodies to protein B will bring down protein A. For immunoprecipitation studies, lysates may be prepared and fractionated into soluble and Triton-soluble fractions as described above.

a. PROCEDURE

1. For each immunoprecipitation reaction, add 2–3 μg of primary antibody to 50 μl of recombinant protein G–agarose (Life Technologies, Rockville, MD) and incubate for 2 hr at 4°. After incubation, remove the supernatant and wash the agarose three times with PBS, 0.1% (v/v) Triton X-100 by centrifugation (approximately 1000g). For 1 min at 4°.

2. Add the Triton-soluble cell fraction to the protein G–agarose and incubate for 4 hr at 4°.

3. Remove the supernatant and wash the protein G–agarose again three times. Add Laemmli sample buffer to the beads. Boil the beads and run the supernatant on an SDS–polyacrylamide gel, followed by Western blot analysis with an antibody specific to the protein partner.[27]

b. OBSERVATIONS

1. A blocking step with 3% (w/v) BSA in wash solution may be added for 1 hr after incubation of the agarose with the antibodies to inhibit nonspecific binding.

2. Protein A–agarose may also be used to immobilize the antibodies.

C. Interfering with Function of Kinase by Inhibiting Interaction of Kinase with its Anchoring Protein

To inhibit the interaction between the kinase and its anchoring protein, it is first necessary to identify the interaction sites between the two proteins and to generate short peptides corresponding to these binding sites. In Step III, we describe sequence search methods used to identify the binding sites of PKCs and their

RACKs. Typically one to three peptides were made. With one exception (a peptide that was insoluble), all peptides that we made had biological activities. These sequence searches use a variety of simple rationales. However, one can always resort to systematic search by synthesizing peptides that span the whole protein of interest. The following is a description of several approaches we used to identify peptides that selectively interfere with the binding of PKC isozymes to their specific RACKs. We also describe a simple method to identify peptides that induce this binding site, and therefore serve as isozyme-selective agonists.

Step III. Identify Sequences That Are Involved in Protein–Protein Interactions

A. Common Sequences in Nonrelated Proteins That Bind to the Same Signaling Enzyme. Grouping proteins that have a common binding protein and the identification of homologous sequences between them may lead to mapping of the binding site for that binding protein. That is, if both protein A and B bind to protein C, sequences homologous between proteins A and B may contain the binding site for protein C. An example of a program that helps in finding similarities between two different proteins may be found at *http://www.expasy.ch/tools/sim-prot.html*. It may be necessary to increase or decrease the gap open penalty and gap extension penalty to identify the sequence similarities. The goal is to identify a short sequence of homology of approximately 10 amino acids or less. [It should be noted that more recent structural information indicated that, with one exception, all the biologically active peptides that we identified lie within a β strand or a loop. Should the interaction site constitute an α helix, it is expected that a larger peptide should be made to have proper folding (possibly 12–16 amino acids).]

Using this rationale, we identified a peptide that inhibited βPKC binding to its RACK. Both annexin I[13] and the 14-3-3 protein[14] have been described as PKC-binding proteins. A short sequence of homology between these nonrelated proteins was identified by Aitken *et al.*[12] Because annexin I can bind PKC,[13] we predicted that a peptide (peptide I) corresponding to the homologous sequence between the two nonrelated PKC-binding proteins, annexin I and 14-3-3, contained the binding site for PKC in these proteins as well as in its RACK (which had not yet been identified at that time). This peptide would thus inhibit PKC function if introduced into cells (Fig. 2). These predictions were confirmed. We showed that in addition to inhibiting βPKC binding to its RACK, when injected into oocytes, peptide I inhibits insulin-induced translocation of βPKC and oocyte maturation.[38] Interestingly, the annexin-like sequence in rat 14-3-3ζ is a conserved region between all 14-3-3 proteins from different species (Fig. 2), further suggesting its importance for protein function (see below). In addition, crystal structure analysis demonstrated subsequently that peptide I corresponds to an exposed α-helix

[38] D. Ron and D. Mochly-Rosen, *J. Biol. Chem.* **269**, 21395 (1994).

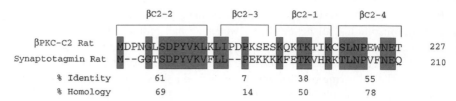

FIG. 3. Sequence homology in the C2 domain of synaptotagmin and PKC identifies critical interaction sites for PKC anchoring to RACK. The C2 domain of βPKC is present in other proteins involved in translocation from one cell compartment to another. We compared rat synaptotagmin (also called p65) with βPKC and found three areas of greater conserved sequences between these two homologous regions (shaded). Peptides βC2-1, βC2-2, and βC2-4 were found to inhibit βPKC translocation and function. βC2-3 peptide was used as a control peptide, because it is derived from a nonconserved region, and as expected had no activity.[17]

structure present in the groove formed between 14-3-3 dimers.[39] Finally, when RACK1 was cloned, a peptide I-like sequence was found in it and a peptide corresponding to this sequence modulated βPKC binding to RACK1.[38]

B. Conserved Sequences in Homologous Domains. Our studies suggested that at least part of the RACK-binding site on βPKC is located in the C2 region of the enzyme (Fig. 1).[9,17] As discussed above, the cPKCs are a subset of PKC isozymes (consisting of α, β, and γ) containing a homologous C2 domain. There are other proteins containing a C2 domain such as synaptotagmin and phospholipase Cγ. Because many of these proteins bind lipids in a calcium-dependent manner, it was proposed that the C2 domain mediates binding of these proteins to lipids and plasma membrane.[6,40–44] However, we showed that both synaptotagmin and phospholipase C also bind to RACK1 albeit with a lower affinity than βPKC,[9,45] suggesting that a binding site on PKC for RACK1 resides within the C2 domain. Three peptides derived from the homologous sequences within the C2 domains, between βPKC and synaptotagmin (βC2-1, βC2-2, and βC2-4) (Fig. 3), were made and found to inhibit PKC translocation and function.[17] A third peptide synthesized as a control because it corresponds to a nonconserved sequence in the C2 domain (βC2-3) had no biological activity, as predicted. Of interest, a more recently obtained crystal

[39] B. Xiao, S. J. Smerdon, D. H. Jones, G. G. Dodson, Y. Soneji, A. Aitken, and S. J. Gamblin, *Nature* (*London*) **376**, 189 (1995).

[40] E. Jorgensen, E. Hartwieg, K. Schuske, M. Nonet, Y. Jin, and H. Horvitz, *Nature* (*London*) **378**, 196 (1995).

[41] E. A. Nalefski, M. M. Slazas, and J. J. Falke, *Biochemistry* **36**, 12011 (1997).

[42] M. Medkova and C. Wonhwa, *J. Biol. Chem.* **273**, 17544 (1998).

[43] P. J. Plant, H. Yeger, O. Staub, P. Howard, and D. Rotin, *J. Biol. Chem.* **272**, 32329 (1997).

[44] N. Brose, A. G. Petrenko, T. C. Sudhof, and R. Jahn, *Science* **256**, 1021 (1992).

[45] M. H. Disatnik, S. M. Hernandez-Sotomayor, G. Jones, G. Carpenter, and D. Mochly-Rosen, *Proc. Natl. Acad. Sci. U.S.A.* **91**, 559 (1994).

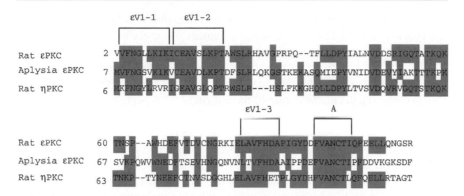

FIG. 4. Homology between rat εPKC, *Aplysia* εPKC, and rat ηPKC within the V1 region. The identical amino acids in the three proteins are shaded. Peptides corresponding to the conserved sequences εV1-1, εV1-2, and εV1-3 were synthesized. Other regions that were also conserved between *Aplysia californica* calcium-independent PKC and rat εPKC were not chosen because they were also homologous to sequences in rat ηPKC and would expect to be not specific for εPKC (e.g., peptide A). εV1-1 was not soluble, εV1-2 specifically inhibited εPKC translocation and function,[20] and εV1-3 partially overlapped with the ψRACK sequence[24] and had some εPKC agonist activity (see Fig. 5).

structure of the C2 domain of βPKC reveals that all the peptides with biological activities are on exposed surfaces in this domain.[8]

C. Importance of Evolutionary Conserved Sequences. When sequences of the same signaling protein from different evolutionary remote species are available, a homology search may identify functionally important sequences, such as these involved in anchoring. We found that the εV1 domain (a homologous region of the C2 domain in nPKCs) (Fig. 1) contains the εRACK-binding site; the εV1 fragment specifically inhibits εPKC translocation and function in cardiac cells.[11] Comparison of this region in rat εPKC with *Aplysia californica* calcium-independent PKC demonstrated an overall 62% sequence identity between these two εPKC sequences.[7] We identified clusters of the most conserved regions within the V1 domain of these two proteins and made three peptides (Fig. 4). εV1-1 peptide was not soluble. εV1-2 peptide specifically inhibits translocation of εPKC but not of other PKCs, when introduced into cells.[11,20] This peptide also inhibits the interaction of RACK2 (εPKC-RACK) with εPKC, and inhibits εPKC function in cells at an intracellular concentration below 50 nM.[11,20] The εV1-3 peptide overlaps in part with $\psi\varepsilon$RACK (pseudo-εRACK) described below, and has partial agonist characteristics. Other regions that were also conserved between *Aplysia californica* calcium-independent PKC and rat εPKC were not chosen because they were also homologous to ηPKC and would not be specific for εPKC (see Fig. 4; e.g., peptide A). Note that ηV1-2 peptide, which is derived from ηPKC and corresponds to the position of ηV1-2, did not have any effect on εPKC.[11]

D. Identification of Sequences Involved in Intramolecular Interactions. Kemp and Pearson identified pseudosubstrate sequences within the regulatory domain of several protein kinases.[46] These sequences, which resemble a substrate phosphorylation site but lack the phospho-acceptor amino acid serine or threonine, were found to mediate intramolecular interactions with the kinase active site, presumably maintaining the inactive state of the enzyme.[47,48] A peptide corresponding to the pseudosubstrate sequence is an inhibitor of the catalytic activity.[48] However, antibodies to that sequence activate PKC in the absence of activators, presumably by interfering with intramolecular interactions between the pseudosubstrate site and the active site in PKC.[49] Using the same rationale, we searched for a pseudo-RACK sequence in PKC that resembles a RACK sequence but lacks an important interaction site. Similar to the action of the antibodies to the pseudosubstrate site,[49] we predicted that a peptide corresponding to the pseudo-RACK site will act as a selective activator of PKC. We found a pseudo-RACK1 in βPKC[28] and a pseudo-εRACK in εPKC[24] that were homologous to the sequence in their corresponding RACKs but had a charge change from lysine (K) in RACK1 to glutamic acid (E) in pseudo-RACK1[28] and from asparagine (N) in RACK2 to aspartic acid (D) in pseudo-RACK2[24] (Fig. 5). Peptides corresponding to these sequences in β and εPKC, $\psi\beta$RACK and $\psi\varepsilon$RACK, respectively, were found to be selective agonist of the corresponding isozymes (Table I).[24,50] For example, when pseudo-εRACK was introduced into cardiac myocytes at an intracellular concentration of approximately 10 nM, it caused translocation and activation of the enzyme, resulting in inhibition of ischemia-induced cell death.[24]

The same homology search program described above (*http://www.expasy. ch/tools/sim-prot.html*) can be used to identify short homologies between protein kinase and its anchoring protein.

Step IV. Introduce the Peptides into Cells

As described above, peptides and protein fragments identified were tested in cells for their isozyme specificity in altering PKC interactions with RACKs and consequently inhibiting the function of the corresponding PKC isozyme. Peptides are not usually cell permeable. However, a variety of methods are now available to introduce peptides of interest into living cells. These include microinjection,[18] transient permeabilization,[11,20,51] use of cell-permeable carrier peptides[24] and transfer

[46] B. E. Kemp and R. B. Pearson, *Trends Biochem Sci.* **15,** 342 (1990).
[47] C. House and B. E. Kemp, *Science* **238,** 1726 (1987).
[48] C. House and B. E. Kemp, *Cell. Signal.* **2,** 187 (1990).
[49] M. Makowske and O. M. Rosen, *J. Biol. Chem.* **264,** 16155 (1989).
[50] D. Ron and D. Mochly-Rosen, *Proc. Natl. Acad. Sci. U.S.A.* **92,** 492 (1995).
[51] J. A. Johnson, M. O. Gray, J. S. Karliner, C. H. Chen, and D. Mochly-Rosen, *Circ. Res.* **79,** 1086 (1996).

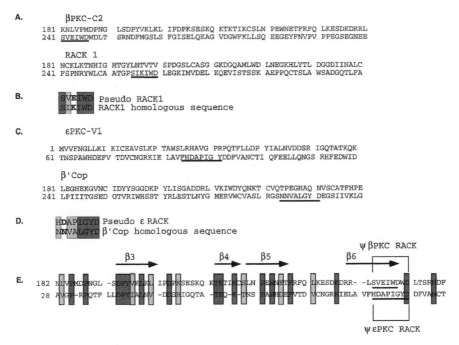

FIG. 5. Sequence homology of βPKC with RACK1, and of εPKC with εRACK. Underlined or shaded sequences indicate homology between βRACK and βPKC (A and B) and εRACK and εPKC (C and D). These sequences are homologous with one charge difference between them as depicted in boldface in (B) and (D) for βPKC and εPKC, respectively. (E) although the homology between εPKC and εRACK is lower than that between βPKC and its RACK, the pseudo-RACK sequence in the two enzymes fall at the same position within C2/V1 domain. β strands 3–6 in the structure are indicated with an arrow over the sequence. [After Sutton and Sprang.[8]]

of genes, expressing the peptides in transgenic mice[21,24] or by transient or stable transfection of such genes in cells.[22] These methods are briefly described below.

A. Microinjection

For studies in which single-cell analysis is possible, for example, by electrophysiology, microinjection of the peptide of interest is an effective method to introduce the peptide to test its biological activities. Usually the number of cells is too small to determine biochemically the effect of the peptide on PKC translocation. In this case, immunofluorescence analysis can be carried out provided that the injected cells can be identified.

a. REAGENTS

Peptides are prepared in an appropriate "intracellular" buffer (e.g., 139.8 mM CsCl, 0.1 or 10 mM K_2EGTA, 4 mM $MgCl_2$, 0.062 mM $CaCl_2$, 5 mM creatine phosphate, 10 mM HEPES, 3.1 mM Na_2ATP, and 0.42 mM Na_2GTP; adjust to pH 7.1 with KOH).[51]

b. OBSERVATIONS

1. Peptide aggregates (which can clog the pipette) are removed by high-speed centrifugation (e.g., 15,000g for 15 min at room temperature).

2. Peptides are either injected as in the study of *Xenopus* oocyte maturation,[52] or allowed to diffuse into the cells, as in the study of cardiac myocytes.[21]

B. Transient Permeabilization

Transient permeabilization was a successful approach in the introduction of peptides and fragments as big as 25,000 Da into cells.[11,20, 51] The amounts of peptide introduced into the cells can be determined with labeled peptides. The experimental conditions for each cell type may vary. Special attention needs to be taken to ensure that cells do not leak much of their content and that they reseal and fully recover before the study begins. We have used this method successfully to introduce peptides and fragments into neonatal cardiac myocytes.[11] Peptide entrance was confirmed by Western blot analysis, cell recovery was confirmed by resumption of synchronous and fast contraction rate (approximately more than 300 beats per minute).[11] The effect of the peptide could then be determined.

a. REAGENTS

Cell culture PBS: 1 mM $CaCl_2$, 0.4 mM $MgCl_2 \cdot 6H_2O$, 2.7 mM KCl, 0.15 mM KH_2PO_4, 8 mM $NaH_2PO_4 \cdot 7H_2O$, 136 mM NaCl

Double-strength (2×) skinning buffer: 20 mM EGTA, 280 mM KCl, 40 mM HEPES buffer (pH 7.4), saponin (100 μg/ml), 10 mM sodium azide, 10 mM potassium oxalate. This buffer should be made fresh and, if filtered, may be stored for 1 week at 4°

ATP, 200 mM in 20 mM HEPES buffer (pH 7.4)

b. PROTOCOL

1. Combine the protein or peptide of interest with double-strength permeabilization buffer and ATP stock, and adjust with deionized water to a final concentration of 1× skinning buffer, 6 mM ATP, and the desired protein concentration.

2. Keep the permeabilization buffer on ice. Set up one container of cell culture PBS (CC-PBS) at room temperature, one on ice, and one at 37°.

3. Remove conditioned medium from cells and save it at 37°.

4. Rinse the cells gently with CC-PBS at room temperature and incubate for 2 min.

5. Rinse the cells gently with cold CC-PBS and incubate for 2 min on ice.

6. Aspirate CC-PBS and gently add just enough chilled permeabilization buffer to cover the cell monolayer (for adherent cells, 7 ml for a 100-mm tissue culture dish). Incubate the cells on ice for 10 min.

---

[52] B. Smith, S. Garber, and D. Mochly-Rosen, *Biochem. Biophys. Res. Commun.* **188**, 1235 (1992).

7. Carefully aspirate all traces of skinning buffer and gently wash the plates four times with cold CC-PBS,

8. Incubate the cells on ice with cold CC-PBS for 20 min.

9. Incubate with CC-PBS for 2 min at room temperature.

10. Incubate with CC-PBS for 2 min at 37°, and replace the CC-PBS with conditioned medium reserved from step 4.

11. Allow the cells to recover in an incubator for 15–30 min and begin the experiment with the desired treatment.

c. OBSERVATIONS

Slow changes in temperature and gentle handling of the cells are essential to ensure the viability and recovery of the cells from transient permeabilization.

C. Carrier Peptide

Several laboratories have discovered short peptides that are capable of entering into cells and can serve as carrier for other peptides. These "carrier peptides" were reviewed by Lindgren et al.[53] The carrier peptide system we have used successfully is the Antennapedia carrier peptide.[54,55] This peptide is 16 amino acids long and corresponds to the third helix domain of Antennapedia, a Drosophila transcription factor. This peptide is internalized into cells when added to cell cultures and is not cell type dependent, because its internalization is receptor independent.[54,55] In our studies, we have cross-linked our peptides via an N-terminal cysteine–cysteine bond to the Drosophila Antennapedia homeodomain-derived carrier peptide (C-RQIKIWFQNRRMKWKK).[24] For example, introduction of $\psi\varepsilon$RACK octapeptide into cardiac myocytes either by transient permeabilization or by cross-linking to the Antennapedia peptide induced an increase in εPKC translocation and function.[24] More recently, we have used the Tat-derived peptide[56] as well as other newly identified peptides with equal or better success. These will be described elsewhere [L. chen et al., Submitted (2001b)].

D. Gene Transfer

a. TRANSFECTION OF CELLS. Cells may be transfected either transiently or by generating stably transfected cell lines. There are many systems for expression of fragments and peptides in eukaryotic cells. The role of εPKC in nerve growth factor-induced differentiation of PC12 cells was successfully demonstrated by specifically inhibiting εPKC translocation.[22] PC12 cell lines expressing εV1 were generated by using a plasmid containing εV1 tagged with a FLAG epitope, and cell lines were selected by drug resistance.[22]

[53] M. Lindgren, M. Hallbrink, A. Prochiantz, and U. Langel, Trends Pharmacol. Sci. 21, 99 (2000).

[54] D. Derossi, S. Calvet, A. Tembleau, A. Brunissen, G. Chassaing, and A. Prochiantz, J. Biol. Chem. 271, 18188 (1996).

[55] D. Derossi, A. H. Joliot, G. Chassaing, and A. Prochiantz, J. Biol. Chem. 269, 10444 (1994).

[56] S. R. Schwarze, A. Ho, A. Vocero-Akbani, and S. F. Dowdy, Science 285, 1569 (1999).

We have also transiently transfected cells with a fusion protein of the green fluorescent protein (GFP).[57] Expression of fragments can be obtained as fusion proteins to GFP, using the cytomegalovirus (CaMV) promoter to obtain high levels of expression of the fragments. An advantage of this fusion protein is that the internal fluorescence of GFP enables ready identification of cells containing the fragments. This is important in the cell-based assays to assess functional consequences of the peptides and fragments. Localization of RACK1 and endogenous RACK1 ligand protein can be determined after fixation with two other fluorescent dyes (e.g., Texas Red and Cascade Blue). This can also be compared with the GFP fusion protein fragment as has been demonstrated for transfection of different PKC and PKC fragments.[58-63] When fixing cells expressing GFP, one should use either 4% (w/v) formaldehyde or 4% (w/v) paraformaldehyde for 10 min at room temperature; under these conditions GFP fluorescence is maintained whereas fixation with methanol acetone is not appropriate.

b. TRANSGENIC MICE. The advantage of using transgenic mice is that the continuous expression of the peptide could mimic a chronic exposure of the animal to the activator or inhibitor isozyme-specific peptides. We find that expressing peptides and fragments is more effective than overexpression or knockout of whole PKC isozymes since compensation for a partial gain or loss of function by one PKC isozyme-modulating peptide was not observed (for further discussion see below and Mochly-Rosen et al.[21]). To this end, in collaboration with G. Dorn in Cincinnati, we have engineered mice to express postnatally one of the biologically active PKC-derived peptides. For example, $\psi\varepsilon$RACK peptide was expressed under regulation of the promoter for α-myosin heavy chain to obtain postnatal expression, and the transgenic mice demonstrated a selective increase in εPKC translocation of 20% in their hearts and were better protected from cardiac damage induced by ischemia as compared with the nontransgenic littermates.[24] In addition, the mice had an increase in cardiac muscle wall thickness, an increase in cardiac myocyte number, a 10% decrease in myocyte size, and no deleterious effects on cardiac functions.[24] We also engineered transgenic mice expressing the εV1 fragment.[21] These mice had characteristics opposing those of the $\psi\varepsilon$RACK peptide transgenic mice. These latter mice had a selectively decreased εPKC translocation of 30%, a decrease in muscle wall thickness, an increase in cardiomyocyte cell size of 20%,

[57] S. R. Kain, M. Adams, A. Kondepudi, T. T. Yang, W. W. Ward, and P. Kitts, *Biotechniques* **19,** 650 (1995).

[58] Y. Shirai, N. Sakai, and N. Saito, *Jpn. J. Pharmacol.* **78,** 411 (1998).

[59] Y. Shirai, K. Kashiwagi, K. Yagi, N. Sakai, and N. Saito, *J. Cell. Biol.* **143,** 511 (1998).

[60] N. Sakai, K. Sasaki, N. Ikegaki, Y. Shirai, Y. Ono, and N. Saito, *J. Cell. Biol.* **139,** 1465 (1997).

[61] S. Ohmori, Y. Shirai, N. Sakai, M. Fujii, H. Konishi, U. Kikkawa, and N. Saito, *Mol. Cell. Biol.* **18,** 5263 (1998).

[62] E. Oancea, M. N. Teruel, A. F. Quest, and T. Meyer, *J. Cell. Biol.* **140,** 485 (1998).

[63] E. Oancea and T. Meyer, *Cell* **95,** 307 (1998).

and impaired left ventricular fractional shortening.[21] In neither mouse line were there changes in the amount of any PKC isozymes or RACKs. Depending on the integration of the construct, mouse lines expressing different levels of the gene of interest can be identified. Analysis of mice with increased doses of the transgene demonstrated, for example, that εPKC is essential for postnatal cardiac development and that its inhibition after birth leads to a lethal dilated cardiomyopathy.[21]

c. METHOD. To obtain postnatal organ-specific expression the transgenes were engineered as follows.

1. The cDNA for each peptide, preceded by an eight-amino acid peptide encoding a FLAG epitope, was directionally cloned into exon 3 of the full-length mouse α-myosin heavy chain promoter.

2. The transgene constructs were separated from the vector backbone and were injected into male pronuclei of fertilized FVB/N mouse oocytes.

3. Southern genomic analysis of tail clip DNA was performed in order to identify transgenic mice.

Summary

We describe here the methods we have used to generate selective peptide inhibitors and activators of PKC-mediated signaling. These approaches should be applicable to any signaling event that is dependent on protein–protein interaction. Furthermore, targeting downstream enzymes in signal transduction has been notoriously difficult as there are often families of related enzymes in each cell. The approaches we have used overcame this difficulty and may prove useful not only in basic research, but also in drug discovery.

Acknowledgments

This work was supported by Grant HL52141 from the National Institutes of Health (to D.M.R.) and a fellowship from the Ares-Serono Foundation (to D.S.).

[38] Assay of Raf-1 Activity

By Jürgen Müller and Deborah K. Morrison

Introduction

The Raf-1 serine/threonine kinase is a key effector of Ras-mediated signal transduction. In response to a diverse array of extracellular signals, the evolutionarily conserved Ras GTPase is converted from its inactive GDP-bound form to its active GTP-bound form.[1] Activated Ras then interacts directly with Raf-1 and recruits Raf-1 from the cytosol to the plasma membrane.[2,3] At the membrane, Raf-1 becomes activated and phosphorylates MEK1 [mitogen-activated protein kinase (MAPK)/extracellular signal-regulated kinase (ERK) kinase] at serine residues 217 and 221.[4,5] These phosphorylation events result in the activation of MEK1, which in turn phosphorylates and activates MAPK (ERK1 and ERK2).[6–8] The cascade culminates in the translocation of activated MAPK to the nucleus, where it phosphorylates the targets needed for the subsequent transcription of genes involved in cell growth and differentiation.[9,10]

The role of Ras in the Raf-1 activation process was first indicated when Raf-1 was shown to specifically interact with GTP-bound Ras both *in vitro* and *in vivo*.[11–13] Although the binding event between Raf-1 and Ras was not found to stimulate the enzymatic activity of Raf-1, it did serve to localize Raf-1 to the plasma membrane. The importance of targeting Raf-1 to the membrane was further revealed by the finding that membrane-localized farnesylated Raf-1 is constitutively activated in a Ras-independent manner.[14,15] From these and other studies, the following model for Raf-1 activation has evolved. In a quiescent cell,

[1] A. Wittinghofer, *J. Biol. Chem.* **379,** 933 (1998).

[2] S. A. Moodie and A. Wolfman, *Trends Genet.* **10,** 44 (1994).

[3] D. K. Morrison and R. E. Cutler, *Curr. Opin. Cell. Biol.* **9,** 174 (1997).

[4] W. Huang, A. Alessandrini, C. M. Crews, and R. L. Erikson, *Proc. Natl. Acad. Sci. U.S.A.* **90,** 10947 (1993).

[5] D. R. Alessi, Y. Saito, D. G. Campbell, P. Cohen, G. Sithanandam, U. Rapp, A. Ashworth, C. J. Marshall, and S. Cowley, *EMBO J.* **13,** 1610 (1994).

[6] C. M. Crews, A. Alessandrini, and R. L. Erikson, *Science* **258,** 478 (1992).

[7] T. A. Haystead, P. Dent, J. Wu, C. M. Haystead, and T. W. Sturgill, *FEBS Lett.* **306,** 17 (1992).

[8] D. M. Payne, A. J. Rossomando, P. Martino, A. K. Erickson, J. H. Her, J. Shabanowitz, D. F. Hunt, M. J. Weber, and T. W. Sturgill, *EMBO J.* **10,** 885 (1991).

[9] R. Seger and E. G. Krebs, *FASEB J.* **9,** 726 (1995).

[10] J. Zuber, O. I. Tchernitsa, B. Hinzmann, A. C. Schmitz, M. Grips, M. Hellriegel, C. Sers, A. Rosenthal, and R. Schäfer, *Nat. Genet.* **24,** 144 (2000).

[11] A. B. Vojtek, S. M. Hollenberg, and J. A. Cooper, *Cell* **74,** 205 (1993).

[12] B. Hallberg, S. I. Rayter, and J. Downward, *J. Biol. Chem.* **269,** 3913 (1994).

[13] S. A. Moodie, B. M. Willumsen, M. J. Weber, and A. Wolfman, *Science* **260,** 1658 (1993).

Raf-1 exists in an inactive state in the cytosol. The inactive conformation of Raf-1 appears to be maintained by autoinhibitory interactions occurring between the N-terminal regulatory and the C-terminal catalytic regions[16,17] and by the binding of 14-3-3 to the N-terminal Ser-259 phosphorylation site of Raf-1.[18,19] In response to signaling events, Ras becomes activated and GTP loaded. The direct binding of Raf-1 to Ras at the plasma membrane relieves the autoinhibitory interactions,[16,17] disrupts 14-3-3 binding to Ser-259,[20] and allows Raf-1 to contact membrane phospholipids.[21–23] These events and other modifications occurring at the membrane, including phosphorylation of Ser-338 and Tyr-340/341,[24–26] result in Raf-1 activation. Once activated, Raf-1 then propagates the signal through MEK1 and MAPK.

In this article, we describe methods for determining the activation state of Raf-1. Because Raf-1 is a key effector of Ras signaling, the quantitation of Raf-1 activity can be a valuable assay for many scientists studying signal transduction. The protocols presented focus on the measurement of endogenous Raf-1 activity in mitogen- or growth factor-treated mammalian cells. However, the methods can easily be adapted for evaluating Raf-1 activity in tissue samples and in cells expressing exogenous wild-type or mutant Raf-1 proteins.

Procedures for Assaying Raf-1 Kinase Activity

Materials

Dulbecco's modified Eagle's medium (DMEM), glutamine, and penicillin–streptomycin are from Life Technologies (Gaithersburg, MD). Platelet-derived

[14] D. Stokoe, S. G. Macdonald, K. Cadwallader, M. Symons, and J. F. Hancock, *Science* **264,** 1463 (1994).

[15] S. J. Leevers, H. F. Paterson, and C. J. Marshall, *Nature (London)* **369,** 411 (1994).

[16] R. E. Cutler, Jr., R. M. Stephens, M. R. Saracino, and D. K. Morrison, *Proc. Natl. Acad. Sci. U.S.A.* **95,** 9214 (1998).

[17] D. G. Winkler, R. E. Cutler, Jr., J. K. Drugan, S. Campbell, D. K. Morrison, and J. A. Cooper, *J. Biol. Chem.* **273,** 21578 (1998).

[18] N. R. Michaud, J. R. Fabian, K. D. Mathes, and D. K. Morrison, *Mol. Cell. Biol.* **15,** 3390 (1995).

[19] G. Tzivion, Z. Luo, and J. Avruch, *Nature (London)* **394,** 88 (1998).

[20] C. Rommel, G. Radziwill, J. Lovric, J. Noeldeke, T. Heinicke, D. Jones, A. Aitken, and K. Moelling, *Oncogene* **12,** 609 (1996).

[21] S. Ghosh, J. C. Strum, V. A. Sciorra, L. Daniel, and R. M. Bell, *J. Biol. Chem.* **271,** 8472 (1996).

[22] R. A. McPherson, A. Harding, S. Roy, A. Lane, and J. F. Hancock, *Oncogene* **18,** 3862 (1999).

[23] M. A. Rizzo, K. Shome, C. Vasudevan, D. B. Stolz, T. C. Sung, M. A. Frohman, S. C. Watkins, and G. Romero, *J. Biol. Chem.* **274,** 1131 (1999).

[24] A. J. King, H. Sun, B. Diaz, D. Barnard, W. Miao, S. Bagrodia, and M. S. Marshall, *Nature (London)* **396,** 180 (1998).

[25] J. R. Fabian, I. O. Daar, and D. K. Morrison, *Mol. Cell. Biol.* **13,** 7170 (1993).

[26] C. S. Mason, C. J. Springer, R. G. Cooper, G. Superti-Furga, C. J. Marshall, and R. Marais, *EMBO J.* **18,** 2137 (1999).

growth factor (PDGF)-BB, inactive MEK1, kinase-defective MEK1 (K97A mutant), inactive MAPK, and kinase-defective MAPK (K71A mutant) are purchased from Upstate Biotechnology (Lake Placid, NY). Enhanced chemiluminescence (ECL) reagents, horseradish peroxidase-conjugated anti-mouse and anti-rabbit antibodies, protein G–Sepharose 4 Fast Flow, and $[\gamma\text{-}^{32}\text{P}]\text{ATP}$ (3000 Ci/mmol) are from Amersham Pharmacia (Arlington Heights, IL). Bovine serum albumin (BSA) (fraction V) is purchased from Sigma (St. Louis, MO) and nitrocellulose membranes are from Schleicher & Schuell (Keene, NH). The Raf-1 antibodies used in this analysis are from Calbiochem (San Diego, CA), PharMingen (San Diego, CA), Santa Cruz Biotechnology (Santa Cruz, CA), Sigma, Transduction Laboratories (Lexington, KY), and Upstate Biotechnology. The phospho-specific Ser-338 antibody is purchased from Upstate Biotechnology and the phospho-specific Tyr-340/341 antibody is from Biosource International (Camarillo, CA).

Buffers

Phosphate-buffered saline (PBS): 10 mM sodium phosphate buffer (pH 7.4), 137 mM NaCl, 1.5 mM KCl

Radioimmunoprecipitation assay (RIPA) lysis buffer: 20 mM Tris (pH 8.0), 137 mM NaCl, 10% (v/v) glycerol, 1% (v/v) Nonidet P-40 (NP-40), 0.1% (w/v) sodium dodecyl sulfate (SDS), 0.5% (w/v) sodium deoxycholate, 2 mM EDTA, 1 mM phenylmethylsulfonyl fluoride (PMSF), 1 mM aprotinin, 20 μM leupeptin, and 5 mM sodium vanadate

NP-40 lysis buffer: 20 mM Tris (pH 8.0), 137 mM NaCl, 10% (v/v) glycerol, 1% (v/v) NP-40, 2 mM EDTA, 1 mM PMSF, 1 mM aprotinin, 20 μM leupeptin, and 5 mM sodium vanadate

Tris-buffered saline (TBS): 10 mM Tris (pH 8.0), 150 mM NaCl

TBST: TBS containing 0.2% (v/v) Tween 20

Kinase buffer: 30 mM N-2-Hydroxyethylpiperazine-N'-2-ethanesulfonic acid (HEPES, pH 7.4), 7 mM MnCl$_2$ (made fresh), 5 mM MgCl$_2$, 1 mM dithiothreitol (DTT), 15 μM ATP

Sample buffer(4×): 33% (v/v) glycerol, 0.3 M DTT, 6.7% (w/v) SDS, 0.1% (w/v) bromphenol blue

Immunoblot stripping buffer: 62.5 mM Tris (pH 6.7), 2% (w/v) SDS, 100 mM 2-mercaptoethanol

All buffers are made with deionized water and reagents of the highest purity available. PBS, as well as RIPA and NP-40 lysis buffers, are stored at 4°. Protease and phosphatase inhibitors are added from concentrated stocks immediately before use. TBS, TBST, and immunoblot stripping buffer are stored at room temperature, whereas 4× sample buffer is stored at −20°. Kinase buffer is made fresh for each assay.

Preparation of Cell Extracts

We describe procedures for preparing cell extracts from PDGF-treated fibro-blasts. These protocols can easily be adapted for the analysis of other cell types treated with other mitogens and growth factors. Parameters to take into considera-tion when preparing cell extracts are the following. First, the basal activity of Raf-1 must be low in order to detect the activation of Raf-1 induced by mitogen or growth factor treatment. Therefore, the cells must be well quiesced before stimulation. Sec-ond, the choice of lysis buffer is critical. To prevent the coimmunoprecipitation of associating kinases and to efficiently solubilize activated Raf-1 from the plasma membrane, the cells must be lysed in a stringent lysis buffer containing SDS, NP-40, and deoxycholate. Incomplete solubilization and contaminating kinase ac-tivities have been observed when cells are lysed less stringently, using buffers that contain only 1% (v/v) NP-40 or Triton X-100 (see Fig. 1). The presence of these contaminating kinase activities can obscure the determination of authentic Raf-1 activity.

BALB/3T3 cells are grown to confluency at 37° on 150-mm tissue culture plates in DMEM supplemented with 2 mM glutamine, penicillin (50 U/ml), streptomycin (50 μg/ml), and 10% (v/v) fetal calf serum. Eighteen to 24 hr before stimulation, the cells are washed twice with serum-free DMEM and are incubated at 37° in DMEM supplemented with 2 mM glutamine, penicillin (50 U/ml), streptomycin (50 μg/ml), and 0.2% (v/v) fetal calf serum. After serum starvation, PDGF-BB

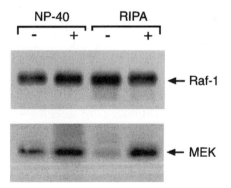

FIG. 1. Comparison of different cell lysis conditions. Untreated (–) or PDGF-treated (+) BALB/3T3 cells were lysed in NP-40 or RIPA lysis buffer as indicated and Raf-1 proteins were immunoprecipitated with 2 μg of Raf-1 antibody (Transduction Laboratories R19120). Direct kinase assays were performed on the Raf-1 immunoprecipitates and the reactions were terminated by the addition of 4× sample buffer. The samples were resolved on an 8% (w/v) SDS–polyacrylamide gel and blotted onto a 0.2-μm pore size nitrocellulose membrane. The phosphoproteins were visualized by autoradiography (MEK1; *bottom*) and the Raf-1 protein levels were determined by immunoblot analysis with the Transduction Laboratories R19120 antibody (Raf-1; *top*).

(20 ng/ml) is added directly to the medium and the cells are incubated at 37° for various time periods. Maximal activation of Raf-1 is achieved after 5 min of PDGF stimulation. Full activation of Raf-1 may vary in different cell types and in response to other mitogens; however, peak activity is usually observed between 1 and 15 min after stimulation at 37°. In addition, factors that stimulate cell proliferation typically induce a transient activation of Raf-1, whereas factors that promote differentiation often induce a more sustained Raf-1 activation.

Lysis of the cells is performed directly on the tissue culture dishes. The cells are washed twice with cold (4°) PBS, being careful not to disrupt the cell monolayer. Residual PBS is aspirated from the plates and ice-cold RIPA lysis buffer is added directly to the cells. Routinely, we use 1 ml of lysis buffer per 150-mm plate, 750 μl/100-mm plate, and 500 μl/60-mm plate. The dishes are then incubated at 4° on a rocking platform for 15 min. The lysed cells are scraped off the dish with a rubber policeman and transferred to 1.5-ml microcentrifuge tubes. The lysates are centrifuged at 16,000g for 10 min at 4° to remove insoluble cellular debris. The supernatants are transferred to new 1.5-ml microcentrifuge tubes and the clarified lysates are either used immediately for immunoprecipitation assays or are quick frozen in a dry ice–ethanol bath and stored at $-80°$. The enzymatic activity of Raf-1 is stable. Activated Raf-1 protein can be freeze–thawed, incubated for several days at 4°, and heated to 37° for 30 min without significant loss of activity.

Immunoprecipitation

Raf-1 activity cannot be assayed directly in cell extracts because of the presence of other kinases that can phosphorylate MEK1. Therefore, Raf-1 must be isolated from the complex mixture of proteins found in the lysates. By using a Raf-1-specific antibody, complicated purification protocols are not needed and Raf-1 can be immunoprecipitated from crude extracts without compromising sensitivity or specificity. Numerous Raf-1 antibodies are commercially available; however, for use in Raf-1 kinase assays, the antibody must meet the following criteria. First, it must efficiently and specifically immunoprecipitate the Raf-1 protein. Second, it must bind Raf-1 such that it does not inhibit the catalytic activity of the protein. We have evaluated several commercially available Raf-1 antibodies for their use in Raf-1 activity assays. As shown in Fig. 2, all of the antibodies directed against the C-terminus of Raf-1 (Santa Cruz Biotechnology 227, 133, and 7267; Sigma R5773 and R7773; and Upstate Biotechnology 06-393) immunoprecipitated equivalent amounts of Raf-1 activity and Raf-1 protein (a 9- to 14-fold activation was observed in BALB/3T3 cells treated for 5 min with PDGF). In contrast to the C-terminal antibodies, an antibody recognizing sequences surrounding the serine/threonine-rich CR2 domain (Calbiochem 543501) immunoprecipitated increased amounts of Raf-1 activity but much less Raf-1 protein. Thus, the specific activity of the protein immunoprecipitated by the CR2 domain antibody

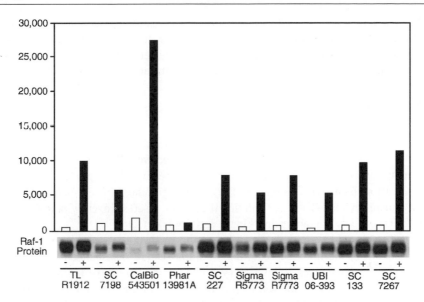

FIG. 2. Comparison of commercially available Raf-1 antibodies for use in Raf-1 kinase assays. Untreated (−) or PDGF-treated (+) BALB/3T3 cells were lysed in RIPA buffer and Raf-1 proteins were immunoprecipitated with 2 μg of the various Raf-1 antibodies. Direct kinase assays were performed and analyzed as described in Fig. 1. Relative Raf-1 activity was determined with a PhosphorImager and ImageQuant software to quantitate the amount of phosphorylated MEK1. The antibodies examined include Transduction Laboratories (TL) R19120; Santa Cruz Biotechnology (SC) 7198, 227, 133, and 7267; Sigma R5773 and R7773; Calbiochem (CalBio) 543501; PharMingen (Phar) 13981A; and Upstate Biotechnology (UBI) 06-393.

is significantly greater. The CR2 domain antibody may have a higher affinity for activated Raf-1 molecules or the binding of this antibody may help stabilize the active Raf-1 conformation. Alternatively, antibodies binding to the Raf-1 C-terminus may have some inhibitory activity. Finally, an antibody binding to residues surrounding the conserved APE sequence in the catalytic subdomain VIII (PharMingen 13981A) was found to immunoprecipitate the Raf-1 protein; however, little Raf-1 activity was detected. Subdomain VIII has been shown to be critical for substrate interaction[27] and it is likely that this antibody either inhibits the catalytic activity of Raf-1 or interferes with MEK1 binding.

For immunoprecipitation reactions, 1–4 μg of Raf-1-specific antibody and 25 μl of protein G–Sepharose beads (50 : 50% slurry in RIPA lysis buffer) are added to clarified cell lysates prepared as described above. The samples are then incubated on a rocking platform for 2–4 hr at 4°. The immunoprecipitated complexes are pelleted by centrifugation in a microcentrifuge at 700g for 1 min at 4°.

[27] S. K. Hanks and T. Hunter, *FASEB J.* **9,** 576 (1995).

The supernatant is discarded (being careful not to disturb the pelleted beads), and the beads are washed once with RIPA lysis buffer, three times with NP-40 lysis buffer, and once with 30 mM HEPES (pH 7.4) buffer. For each wash, the beads are pelleted at 700g for 1 min at 4° and gently resuspended in 1 ml of wash buffer.

Direct Kinase Assay

The direct kinase assay measures the ability of Raf-1 to phosphorylate its physiological substrate MEK1. Thus, the read out for this assay is the incorporation of ^{32}P into kinase-defective MEK1. Kinase-defective MEK1 can be generated by mutation (K97A) or by treating kinase-competent MEK1 protein with 5'-p-fluorosulfonylbenzoyladenosine (FSBA). Although other kinases such as MAPK will phosphorylate MEK1, we find that if cells are lysed stringently in RIPA lysis buffer little if any contaminating kinase activity is detected.

Raf-1 proteins are immunoprecipitated and washed as described above. After the final wash, the immunoprecipitated complexes are resuspended in 40 μl of kinase buffer containing 20 μCi of [γ-^{32}P]ATP and 0.1 μg of kinase-defective MEK1. The reaction is allowed to proceed for 30 min at room temperature. To terminate the assay, 15 μl of 4× sample buffer is added directly to the samples. The terminated kinase reactions are heated for 5 min in a boiling water bath and the proteins resolved by sodium dodecyl sulfate–polyacrylamide gel electrophoresis (SDS–PAGE). The proteins are then transferred electrophoretically onto 0.2-μm pore size nitrocellulose membranes and analyzed by autoradiography or by the use of a PhosphorImager.

Coupled Kinase Assay

Raf-1 activity can also be measured by examining the ability of Raf-1 to phosphorylate and activate MEK1 in a coupled kinase assay. The activation of MEK1 can either be assayed directly, in which case the readout of the assay is the incorporation of ^{32}P into kinase-defective MAPK, or indirectly, in which case the readout of the assay is the phosphorylation of myelin basic protein (MBP) by MAPK. A point of consideration for the coupled assay is that MEK1 is an excellent substrate for MAPK, and unlike the activating Raf-1-dependent phosphorylation (at Ser-217 and Ser-221), the MAPK-dependent phosphorylation of MEK1 (at Thr-291 and Thr-385) is inhibitory.[28] Therefore, as MAPK becomes activated in the indirect coupled assay, it may also phosphorylate and inactivate MEK1.

For the direct coupled kinase assay, 40 μl of kinase buffer containing 20 μCi of [γ-^{32}P]ATP, 0.1 μg of inactive MEK1, and 0.5 μg of kinase-defective (ERK1-K71A) MAPK is added to the washed Raf-1 immunoprecipitates and the samples

---

[28] Y. Saito, N. Gomez, D. G. Campbell, A. Ashworth, C. J. Marshall, and P. Cohen, *FEBS Lett.* **341**, 119 (1994).

are incubated at room temperature for 30 min. Fifteen microliters of 4× sample buffer is then added directly to the reactions and the samples are processed as described above. For the indirect coupled kinase assay, 40 μl of kinase buffer containing 20 μCi of $[\gamma\text{-}^{32}P]$ATP, 0.1 μg of inactive MEK1, and 0.5 μg of inactive MAPK is added to the Raf-1 immunoprecipitates and the samples are incubated at room temperature for 30 min. The samples are then spun at 700g for 1 min at 4° to pellet the beads. A 20-μl aliquot of the reaction mix is transferred to a new tube and 20 μl of kinase buffer containing 20 μCi of $[\gamma\text{-}^{32}P]$ATP and 5 μg of MBP is added. The kinase reactions are allowed to proceed for 20 min at room temperature. The reactions are terminated by the addition of 15 μl of 4× sample buffer and processed as described above.

Immunoblot Analysis

The Raf-1 protein levels are determined by immunoblot analysis. This control is important for demonstrating that equivalent amounts of Raf-1 protein were present in the various samples. This step is also required for comparing Raf-1 activity from sources that contain different concentrations of Raf-1 protein.

After the Raf-1 kinase reactions are transferred to nitrocellulose, residual binding sites on the membrane are blocked by incubating the filter with 2% (w/v) bovine serum albumin (BSA) in TBS for 1 hr at room temperature. The membrane is then washed three times at room temperature with TBST (5 min/wash) and incubated overnight at 4° with primary antibody diluted in TBST. After incubation with the primary antibody, the membrane is washed three times at room temperature with TBST (5 min/wash) and incubated for 1 hr at room temperature with a horseradish peroxidase-conjugated secondary antibody diluted 1 : 10,000 in TBST. The membrane is again washed three times at room temperature with TBST (5 min/wash) and immune reactions are detected by enhanced chemiluminescence, using ECL reagents. The amount of Raf-1 protein can then be quantitated by densitometric measurement.

Alternative Methods to Measure Raf-1 Activation

Phospho-specific antibodies allow for the phosphorylation state of individual protein residues to be determined by standard, immunoblotting techniques.[26] Several phospho-specific antibodies have been developed for Raf-1, including antibodies that recognize phospho-Ser-338 and phospho-Tyr-340/341. Because the activation of Raf-1 is often coincident with the phosphorylation of Ser-338 and/or Tyr-340/341,[24–26] these antibodies may be used to evaluate Raf-1 activation.

To use the phospho-specific antibodies as an assay for Raf-1 activation, the Raf-1 protein is first immunoprecipitated from cell lysates with an antibody that efficiently immunoprecipitates Raf-1. The immunoprecipitates are then washed, separated by SDS–PAGE, and transferred onto a 0.2-μm pore size nitrocellulose

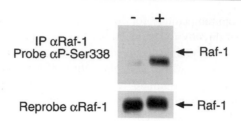

FIG. 3. Detection of activated Raf-1, using phospho-Ser-338 antibodies. Untreated (−) or PDGF-treated (+) BALB/3T3 cells were lysed in RIPA lysis buffer and Raf-1 proteins were immunopre-cipitated with 2 μg of Transduction Laboratories R19120 antibody. The proteins were resolved on a 7% (w/v) SDS–polyacrylamide gel and transferred to a 0.2-μm pore size nitrocellulose membrane. The membrane was then examined by immunoblot analysis with the phospho-Ser-338 antibody (*top*). After detection of the phosphorylated Raf-1 protein, the membrane was stripped and reprobed with the Transduction Laboratories R19120 antibody to visualize the total Raf-1 protein levels (*bottom*).

membrane as described above. After blocking in BSA, the membrane is incubated with the phospho-specific antibody (diluting the antibody in TBST to a concentration recommended by the manufacturer). The blot is then processed and the immune reactions are detected as described above. After the blot is probed with the phospho-specific antibody, the bound antibodies are stripped from the membrane by incubating the blot in stripping buffer for 30 min at 55°. After a brief wash in TBST, the membrane is blocked and reprobed with a Raf-1 antibody to normalize for the amount of total Raf-1 protein immunoprecipitated. Figure 3 demonstrates the use of the phospho-Ser-338 antibody to detect activated Raf-1 in PDGF-treated BALB/3T3 cells.

Concluding Remarks

In the assays described here, we use MEK1 as the Raf-1 substrate. While other substrates have been used to quantitate Raf-1 activity, MEK1 is the only physiological substrate of Raf-1 that is phosphorylated at significant rates. Other substrates previously used to measure Raf-1 activity were often assayed under less stringent cell lysis conditions, and most likely represent phosphorylation by contaminating kinase activities. In addition, Raf-1 will not phosphorylate denatured or improperly folded MEK1 proteins or peptides containing the Ser-217 and Ser-221 sites of phosphorylation, suggesting that tertiary structure plays an important role in the ability of Raf-1 to recognize and phosphorylate its substrate. Finally, the protocols described here can also be used to measure the activity of other Raf family members, including A-Raf and B-Raf. Antibodies that recognize these Raf proteins are commercially available, but as with Raf-1, the antibodies should be evaluated for their ability to immunoprecipitate active Raf protein.

[39] Analyzing Protein Kinase C Activation

By ALEXANDRA C. NEWTON

Introduction

Protein kinase C isozymes play a central role in mediating signals from lipid second messengers.[1,2] It has become apparent that they are activated not only by the canonical diacylglycerol (DG) pathway, but also through the phosphatidyl-inositol 3-kinase (PI3 kinase) signaling pathway.[3–5] Specifically, protein kinase C family members are activated by two sequential mechanisms: phosphorylation, which is required for catalytic competence, followed by allosteric regulation by membrane-bound cofactors. For conventional (α, βI, βII, γ) and novel (δ, ε, η, θ) isozymes, phosphorylation appears to be part of the maturation of the enzyme, but for atypical protein kinase Cs (ζ, ι), phosphorylation is the direct on/off switch for the enzyme. This article discusses how to measure the cofactor-stimulated activity of protein kinase C *in vitro* as well as how to assess its activation state in cells.

Phosphorylation

The phosphorylation state of protein kinase C is regulated by its upstream kinase, the phosphoinositide-dependent kinase PDK-1.[3–5] This kinase phosphory-lates the activation loop of conventional, novel, and atypical protein kinase Cs, an event that triggers the rapid autophosphorylation of two (conventional and novel) or one (atypical) carboxyl-terminal sites. Phosphorylation at the carboxyl-terminal sites (but not the activation loop) results in a pronounced decrease in the elec-trophoretic mobility of protein kinase C that is readily observed on 7.5% (w/v) poly-acrylamide gels. Phosphorylation on the activation loop as well as the hydrophobic site can be detected with commercially available phospho-specific antibodies to these sites.

Phosphorylation of conventional protein kinase C isozymes appears to be con-stitutive and part of the maturation process of these family members. That is, the conventional protein kinase C isozymes in cells are typically >80% phosphory-lated under unstimulated conditions. Although this phosphorylation is required for

[1] A. C. Newton, *Curr. Opin. Cell Biol.* **9**, 161 (1997).

[2] Y. Nishizuka, *FASEB J.* **9**, 484 (1995).

[3] M. M. Chou, W. Hou, J. Johnson, L. K. Graham, M. H. Lee, C. S. Chen, A. C. Newton, B. S. Schaffhausen, and A. Toker, *Curr. Biol.* **8**, 1069 (1998).

[4] J. A. Le Good, W. H. Ziegler, D. B. Parekh, D. R. Alessi, P. Cohen, and P. J. Parker, *Science* **281**, 2042 (1998).

[5] E. M. Dutil, A. Toker, and A. C. Newton, *Curr. Biol.* **8**, 1366 (1998).

TABLE I
COFACTOR REQUIREMENTS OF PROTEIN KINASE C FAMILY MEMBERS

Kinase	Ca^{2+}	DG	PS
Conventional	+	+	+
Novel	−	+	+
Atypical	−	−	−

Abbreviations: DG, Diacylglycerol; PS, phosphatidylserine.

catalytic competence, it does not, in itself, activate the enzyme. Rather, activation is achieved by allosteric activators, which remove the autoinhibitory pseudosubstrate from the substrate-binding cavity. In marked contrast, phosphorylation is sufficient to directly activate the atypical protein kinase Cs. Thus, probing the phosphorylation state of atypical protein kinase Cs is an effective measure of their activation state in the cell. This is not the case for conventional and novel protein kinase Cs, where phosphorylation is not sufficient for activation.

Cofactors

The activity of all isozymes of protein kinase C, including atypical protein kinase Cs, is stimulated by phosphatidylserine (PS) (Table I). In addition, DG specifically stimulates the activity of conventional and novel protein kinase Cs and Ca^{2+} stimulates the activity of conventional protein kinase Cs. Extensive biochemical analyses have unveiled the molecular mechanisms for this cofactor stimulation. In its unstimulated state, an autoinhibitory module, the pseudosubstrate sequence, blocks the substrate-binding cavity of protein kinase C. Binding of cofactors to the membrane-targeting modules of the enzyme recruits protein kinase C to the membrane, an interaction that provides the energy to release the pseudosubstrate from the substrate-binding cavity. Activity is directly related to the strength of the membrane interaction. As a result, sufficiently high levels of one cofactor can allow maximal activity in the absence of another cofactor. Thus, members of the Ca^{2+}-regulated conventional protein kinase Cs can be maximally activated by saturating levels of PS and DG, in the absence of Ca^{2+}. Maximal cofactor-stimulated activity does, however, require PS.

Reagents

Lipid

Lipid is a key component in protein kinase C assays. Because the enzyme specifically recognizes lipid structure, rather than membrane structure, lipid can be presented to the enzyme in a variety of ways, all of which support maximal

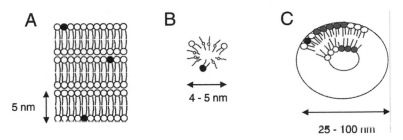

FIG. 1. Lipid structures supporting protein kinase C activity. (A) The standard assay uses multilamellar structures composed of PS (open circles) with trace DG (solid circles). Less than 5% of the lipid is surface exposed. (B) The mixed micelle assay uses Triton X-100 : lipid micelles. (C) The bilayer surface of small sonicated vesicles (diameter, 25 nm; 67% of lipid is surface exposed) or large unilamellar vesicles (e.g., diameter 100 nm; 50% of lipid is surface exposed) best approximates a biological membrane. A composition of 25 mol% PS in PC (gray circles) is optimal for measuring Ca^{2+} or DG dependencies.

activity of the enzyme. In the standard assay, the enzyme binds and is activated by multilamellar structures of PS containing trace DG (Fig. 1A). The high concentration of PS can result in maximal activation of conventional protein kinase Cs in the absence of Ca^{2+} or DG. Thus, this assay system is not appropriate for rigorous determination of cofactor dependencies of protein kinase C, but is excellent for routine assays of activity. An additional concern is that the high concentration of anionic lipid results in significant membrane aggregation in the presence of divalent cations.

The mixed micellar assay pioneered by Bell and co-workers[6] disperses PS and DG in Triton X-100 micelles, with maximal activity being supported with ≥ 15 mol% PS and ≥ 5 mol% DG (Fig. 1B). Such micelles form defined structures, with one micelle binding one molecule of protein kinase C, and are relatively stable to aggregation under normal assay conditions. This assay has proved valuable for systematic dissection of the cofactor requirements for protein kinase C activity.

The membrane bilayer of vesicles containing phosphatidylcholine as the bulk lipid with defined amounts of PS best approximates the structure and composition of a biological membrane (Fig. 1C). These can be small unilamellar vesicles formed by sonication or large unilamellar vesicles formed by extrusion. Protein kinase C effectively binds to, and is activated by, phosphatidylcholine vesicles containing 5 mol% DG and 25 mol% PS (e.g., see Mosior and Newton[7]).

It should be noted that the interaction of protein kinase C with membranes obeys the law of mass action so that membrane binding and activity depend not only on the composition of membranes, but on the total lipid concentration.[7]

---

[6] R. M. Bell, Y. Hannun, and C. Loomis, *Methods Enzymol.* **124,** 353 (1986).
[7] M. Mosior and A. C. Newton, *Biochemistry* **35,** 1612 (1996).

Substrates

Protein kinase C phosphorylates serine and threonine residues, typically surrounded by basic residues. Although basic domains surround phosphorylation sites of most known substrates, the requirement for basic residues is not absolute and the enzyme does not have a particularly strict consensus phosphorylation site sequence.[8] For *in vitro* assays, basic peptides are preferred substrates not only because they bind the anionic substrate binding cavity of protein kinase C, but because they bind the anionic membranes in the assay and thus their local concentration near protein kinase C is increased. *In vitro* assays commonly employ histone, myelin basic protein, or synthetic peptides. The latter has the advantage that background phosphorylation is considerably lower. A heptapeptide based on the phosphorylation sequence in the MARCKS protein, Ac-FKKSFKL-amide, is referred to as the protein kinase C selective peptide and is an excellent *in vitro* substrate.[9] The K_m for this peptide is on the order of 50 μM.[10] Other commonly used peptide substrates are based on the pseudosubstrate sequence with a serine at the position of the phospho-acceptor site. The pseudosubstrate sequence of protein kinase C ε is particularly enriched in arginine and thus promotes significant lipid-independent phosphorylation (see below).

Arginine-rich peptides and proteins have the unusual ability of releasing the pseudosubstrate of protein kinase C and thus maximally activating the enzyme.[11] Thus, protamine sulfate is an excellent cofactor-independent substrate of protein kinase C.

The substrate specificity of protein kinase C μ is significantly different from that of other protein kinase C family members, and a peptide based on an optimal sequence obtained from an oriented peptide library LVRQMSVAFFF is recommended for assaying this isozyme.[8]

In Vitro Assay of Protein Kinase C Activity

Standard in Vitro Assay

Table II outlines a typical protein kinase C assay that can be used for measuring activity in crude cell lysates or in fractions of pure protein. Protein kinase C activity is assayed by measuring the initial rate of [32P]phosphate incorporation from [γ-32P]ATP into saturating amounts of substrate. Each sample consists of (1) protein kinase C [on the order of 2 nM if using pure protein; diluted into 20 mM HEPES (pH 7.4), 2 mM dithiothreitol (DTT)]; (2) lipid (10-fold dilution of 1.4 mM PS,

---

[8] K. Nishikawa, A. Toker, F. J. Johannes, Z. Songyang, and L. C. Cantley, *J. Biol. Chem.* **272**, 952 (1997).

[9] B. R. Chakravarthy, A. Bussey, J. F. Whitfield, M. Sikorska, R. E. Williams, and J. P. Durkin, *Anal. Biochem.* **196**, 144 (1991).

[10] A. S. Edwards and A. C. Newton, *J. Biol. Chem.* **272**, 18382 (1997).

[11] J. W. Orr and A. C. Newton, *J. Biol. Chem.* **269**, 8383 (1994).

TABLE II
STANDARD ASSAY FOR MEASURING PROTEIN KINASE C ACTIVITY

Sample[a] (48 μl)	Lipid[b] (8 μl)	Ca^{2+c} (8 μl)	Chelator[d] (8 μl)	GO[e] (6 μl)	STOP[f] (25 μl)
1–3	+	+	−	+	+
4–6	−	−	+	+	+

[a] Triplicate samples of 1–5 μl of protein kinase C-containing solution diluted into 20 mM HEPES (pH 7.4 at 22°) containing a final concentration of 1 mM DTT. Note that, when assaying cell lysates, the amount of Triton X-100 brought into the assay should not exceed 0.05%, otherwise the mole fraction of PS and DG in micelles drops to subactivating levels. If pure protein kinase C is used, 2 nM enzyme in the assay is recommended. Glycerol levels should not exceed 10% for optimal activity.
[b] Sonicated suspension of 1.4 mM PS and 38 μM DG dispersed in 20 mM HEPES, pH 7.4.
[c] CaCl$_2$, 2 mM in H$_2$O.
[d] EGTA (1 mM) and 1 mM EDTA in H$_2$O.
[e] ATP (500 μM) containing [γ-^{32}P]ATP (75 μCi ml^{-1}), 25 mM MgCl$_2$, 20 mM HEPES (pH 7.4), peptide substrate (500 μg ml^{-1}).
[f] 0.1 M ATP, 0.1 M EGTA (pH 8).

38 μM DG stock) or water; and (3) CaCl$_2$ (10-fold dilution of 2 mM stock) or chelator (10-fold dilution of 1 mM EDTA, 1 mM EGTA stock). The phosphorylation reaction is initiated by addition of 16 μl of GO solution [500 μM ATP containing [γ-^{32}P]ATP (75 μCi ml^{-1}), 25 mM MgCl$_2$, 20 mM HEPES (pH 7.4), peptide substrat (500 μg ml^{-1})]. The phosphorylation reaction is allowed to proceed for 6 min at 30° (to measure initial rate) and then quenched by addition of 25 μl of STOP solution containing 0.1 M ATP and 0.1 M EDTA, pH 8. Aliquots (85 μl) are spotted on Whatman (Clifton, NJ) P81 ion-exchange chromatography paper and washed four times (500 ml/wash) with 0.4% (v/v) phosphoric acid, followed by a rinse in 95% (v/v) ethanol. Papers are added to 5 ml of scintillation fluid and ^{32}P is detected by liquid scintillation counting. Protein kinase C activity is taken as the net Ca^{2+}/lipid-dependent activity. Results from a typical protein kinase C assay are shown in Fig. 2.

Autophosphorylation of pure or partially pure protein kinase C can be measured by the same method as substrate phosphorylation, except that substrate is omitted from the GO solution and samples are analyzed by sodium dodecyl sulfate–polyacrylamide gel electrophoresis (SDS–PAGE) and autoradiography. Autophosphorylation is an intramolecular reaction and thus does not depend on enzyme concentration; for conventional protein kinase Cs, the rate of autophosphorylation is on the order of 0.1 mol of phosphate per mole of protein kinase C per minute.[12]

[12] A. C. Newton and D. E. Koshland, Jr., *J. Biol. Chem.* **262,** 10185 (1987).

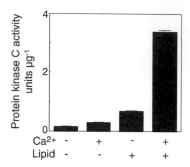

FIG. 2. Ca^{2+} and lipid dependence of conventional protein kinase C measured by monitoring the phosphorylation of the protein kinase C selective peptide under the conditions outlined in Table II. Protein kinase C βII was expressed in insect cells, using the baculovirus expression system. Cells were lysed and protein kinase C activity was determined directly from the cell lysate. One unit is defined as 1 nmol of phosphate incorporated per minute onto the peptide. The amount of protein kinase C in the cell extract was determined by Western blot analysis relative to known amounts of protein kinase C βII. Data are expressed as the mean ± SD of a triplicate experiment.

Autophosphorylation is a less sensitive assay than measuring substrate phosphorylation, but this assay avoids substrate-induced lipid aggregation and is a true measure of the intrinsic catalytic activity of protein kinase C.

Assay of Specific Isozymes after Immunoprecipitation

The standard assay described above detects total lipid-stimulated activity and does not discriminate between isozymes. Such discrimination can be achieved by immunoprecipitating specific isozymes with commercially available isozyme-specific antibodies and performing the above-described kinase assay with the immune complexes, as outlined below.

Cells are lysed in 1 ml of buffer composed of 20 mM Tris (pH 7.5), 10% (v/v) glycerol, 1% (v/v) Triton X-100, 10 mM EDTA, 150 mM NaCl, 20 mM sodium fluoride, 5 mM sodium pyrophosphate, 1 mM sodium vanadate, leupeptin (1 mg ml^{-1}), pepstatin A (1 mg ml^{-1}), and 1 mM phenylmethylsulfonyl fluoride (PMSF) at 4°. Ten percent of the total lysate is analyzed by SDS–PAGE to normalize for total protein kinase C content. Protein kinase C isozymes are immunoprecipitated either with isoform-specific antibodies or with antibodies against epitope tags [e.g., hemagglutinin (HA), Myc, hexahistidine (His$_6$)] and incubated with a 25 μl ml^{-1} concentration of a 50% mix of protein A/G beads. For immunocomplex kinase assays, immunoprecipitates are washed stringently as follows: twice in phosphate-buffered saline (PBS), 1% (v/v) Nonidet P-40 (NP-40); twice in 10 mM Tris (pH 7.5), 0.5 M LiCl; and twice in 10 mM Tris (pH 7.5), 100 mM NaCl, 1 mM EDTA. Immunocomplex protein kinase assays are carried out by the procedure for the standard assay described above. Care should be taken to ensure that the Triton

X-100 concentration is below 0.05% (v/v) to avoid dilution of the PS and DG to subactivating levels. For greater sensitivity, myelin basic protein (170 μg ml^{-1}) or histone H2B (170 μg ml^{-1}) can be used as substrate, reactions can be resolved by 12.5% (w/v) SDS–PAGE, and substrate phosphorylation quantitated by molecular imaging of radioactivity.

Phosphorylation of Transmembrane Substrates by Protein Kinase C

Protein kinase C phosphorylates a number of transmembrane substrates, including G-protein-coupled receptors such as the β-adrenergic receptor and rhodopsin. Such substrates can be presented to protein kinase C in their native membrane environment (e.g., membrane fraction of cell lysate) or in reconstituted membranes containing PS and DG. If the native environment is used, it is more effective to activate protein kinase C by adding phorbol ester [e.g., 1 nM phorbol dibutyrate (PDBu) for 1 μM lipid membrane] to the membrane sample. Addition of PS : DG will sequester protein kinase C on the artificial membranes and may reduce accessibility and hence phosphorylation of the substrate in the native membrane.

Assay of Activation State of Protein Kinase C in Cells

Because the activity of conventional and novel protein kinase Cs is allosterically regulated by second messengers (Ca^{2+} and DG) compartmentalized in intact cells, assaying protein kinase C activity in cell lysates does not indicate the previous activity state of these protein kinase Cs in the cell. Although agonist stimulation may increase the phosphorylation state of novel protein kinase Cs, such increases are subtle relative to the order of magnitude increase in activity mediated by cofactors. This is not true, however, for atypical protein kinase Cs. These isozymes are directly activated by phosphorylation so that probing the phosphorylation state or activity is a sensitive readout of the activation state of the enzyme in the cell. For conventional and novel protein kinase Cs, measuring stimulus-dependent membrane translocation or stimulus dependent substrate phosphorylation is the most effective measure of activity in the cell.

Atypical Protein Kinase Cs

The basal (lipid-independent) activity of atypical protein kinase Cs is significantly higher than that of other protein kinase C family members.[13] As a result, substantial activity can be detected in the complete absence of added PS. These protein kinase Cs are not significantly phosphorylated in unstimulated cells but become phosphorylated as a result of activation of the PI3 kinase signaling pathway. Thus, the intrinsic activity of protein kinase C ζ immunoprecipitated from

---

[13] A. Toker, *Front. Biosci.* **3**, D1134 (1998).

cell lysates increases in cells stimulated with agonists that activate PI3 kinase. It should be stressed that this activity results from regulation by phosphorylation. For other isozymes, phosphorylation is not sufficient for activation: novel and conventional protein kinase Cs require allosteric regulation by cofactors in order to be active.

Membrane Translocation

A hallmark of protein kinase C activation is its translocation to discrete intracellular locations. Consequently, measuring the subcellular distribution of specific isozymes of protein kinase C is a standard approach for determining whether a particular stimulus has activated a specific isozyme. This approach works well for phorbol ester-treated cells, where the membrane interaction induced by phorbol esters is so strong that cell lysis does not significantly perturb it. However, for agonists that activate protein kinase C by production of DG, the membrane interaction is considerably weaker and thus disrupted on lysis. In particular, changes in Ca^{2+} levels on lysis, coupled with the rapid metabolism of DG, affect the fraction of membrane-bound protein kinase C. For this, *in vivo* analysis of the subcellular distribution of protein kinase C, using fluorescence imaging of green fluorescent protein (GFP)-tagged constructs, is more informative.[14–16] This method has the advantage of allowing real-time monitoring of the subcellular location of protein kinase C after agonist stimulation, but suffers from the fact that overexpressed protein may not always reflect the localization of endogenous protein.

In Vivo Labeling

The *in vivo* activation of protein kinase C can also be assessed by monitoring the stimulus-dependent phosphorylation of major known substrates such as the MARCKS protein or, in hematopoietic cells, pleckstrin.[17,18] Cells are *in vivo* labeled with [^{32}P]phosphate according to standard procedures, stimulated with agonist, and the phosphorylation of these proteins can be directly analyzed in lysates by SDS–PAGE followed by autoradiography. Greater sensitivity can be achieved by extracting MARCKS protein with 40% (v/v) acetone, as described by Robinson and co-workers.[19] The advantage of this method is that it provides a direct readout for the activity of protein kinase C in cells.

[14] N. Sakai, K. Sasaki, N. Ikegaki, Y. Shirai, Y. Ono, and N. Saito, *J. Cell Biol.* **139,** 1465 (1997).

[15] E. Oancea and T. Meyer, *Cell* **95,** 307 (1998).

[16] Q. J. Wang, D. Bhattacharyya, S. Garfield, K. Nacro, V. E. Marquez, and P. M. Blumberg, *J. Biol. Chem.* **274,** 37233 (1999).

[17] B. R. Chakravarthy, R. J. Isaacs, P. Morley, and J. F. Whitfield, *J. Biol. Chem.* **270,** 24911 (1995).

[18] A. Toker, C. Bachelot, C.-S. Chen, J. R. Falck, J. H. Hartwig, L. C. Cantley, and T. Kovacsovics, *J. Biol. Chem.* **270,** 29525 (1995).

[19] P. J. Robinson, J. P. Liu, W. Chen, and T. Wenzel, *Anal. Biochem.* **210,** 172 (1993).

[40] Assays for Protein-Tyrosine Phosphatases

By DANIEL F. MCCAIN and ZHONG-YIN ZHANG

Many studies have implicated protein-tyrosine phosphatases (PTPases) as important regulators of cellular functions and as potential drug targets. This has led to an increased interest in the development of precise assays for detailed mechanistic studies of PTPases, rapid assays for high-throughput PTPase inhibitor screening, and more biologically relevant assays using physiological protein substrates.

The first assays aimed specifically at PTPases utilized tyrosine-phosphorylated proteins such as myosin light chain LC_{20},[1] insulin receptor, epidermal growth factor receptor,[2] RCM-lysozyme,[3,4] or myelin basic protein[5,6] as substrates while $^{32}P_i$ release was monitored as the readout of PTPase activity. The tyrosine-phosphorylated forms of other proteins and peptides such as Raytide, angiotensin II,[7] $p56^{lck}$,[8] glutamine synthase,[9] angiotensin, RR src,[10] and angiotensin II peptides[11,12] have also been used as PTPase substrates. Studies involving phosphorylated, physiological substrates have also been carried out in which phosphoamino acid analysis,[13–18] phosphopeptide mapping,[18–20] and sodium dodecyl sulfate–polyacrylamide gel electrophoresis (SDS–PAGE) with phosphoimaging[18] were used to monitor PTPase activity.

[1] C. P. Chan, B. Gallis, D. K. Blumenthal, C. J. Pallen, J. H. Wang, and E. G. Krebs, *J. Biol. Chem.* **261,** 9890 (1986).

[2] M. J. King and G. J. Sale, *Biochem. J.* **256,** 893 (1988).

[3] N. K. Tonks, C. D. Diltz, and E. H. Fischer, *J. Biol. Chem.* **263,** 6722 (1988).

[4] N. K. Tonks, C. D. Diltz, and E. H. Fischer, *Methods Enzymol.* **201,** 427 (1991).

[5] M. Streuli, N. X. Krueger, T. Thai, M. Tang, and H. Saito, *EMBO J.* **9,** 2399 (1990).

[6] N. K. Tonks, C. D. Diltz, and E. H. Fischer, *Methods Enzymol.* **201,** 442 (1991).

[7] M. Streuli, N. X. Krueger, A. Y. M. Tsai, and H. Saito, *Proc. Natl. Acad. Sci. U.S.A.* **86,** 8698 (1989).

[8] C. J. Pallen, D. S. Y. Lai, H. P. Chia, I. Boulet, and P. H. Tong, *Biochem. J.* **276,** 315 (1991).

[9] K. Mashima, Y. Okajima, J. Usui, T. Shimizu, and K. Kimura, *J. Biochem.* **115,** 333 (1994).

[10] J. Lu, A. L. Notkins, and M. S. Lan, *Biochem. Biophys. Res. Commun.* **204,** 930 (1994).

[11] Y. Minami, F. J. Stafford, J. Lippincott-Schwartz, L. C. Yuan, and R. D. Klausner, *J. Biol. Chem.* **266,** 9222 (1991).

[12] F. M. Uckun, L. Tuel-Ahlgren, C. W. Song, K. Waddick, D. E. Myers, J. Kirihara, J. A. Ledbetter, and G. L. Schieven, *Proc. Natl. Acad. Sci. U.S.A.* **89,** 9005 (1992).

[13] J. A. Cooper, B. M. Sefton, and T. Hunter, *Methods Enzymol.* **99,** 387 (1983).

[14] W. J. Boyle, P. van der Geer, and T. Hunter, *Methods Enzymol.* **201,** 110 (1991).

[15] M. P. Kamps, *Methods Enzymol.* **201,** 21 (1991).

[16] H. Sun, *Methods Mol. Biol.* **84,** 307 (1998).

[17] S. Dowd, A. A. Sneddon, and S. M Keyse, *J. Cell Sci.* **111,** 3389 (1998).

[18] M. L. Sohasky and J. E. Ferrel, Jr., *Mol. Biol. Cell* **10,** 3729 (1999).

[19] K. Luo, T. R. Hurley, and B. M. Sefton, *Methods Enzymol.* **201,** 149 (1991).

[20] E. D. Cahir McFarland, T. R. Hurley, J. T. Pingel, B. M. Sefton, A. Shaw, and M. L. Thomas, *Proc. Natl. Acad. Sci. U.S.A.* **90,** 1402 (1993).

For the sake of safety, convenience, and ease in waste disposal, a variety of nonradioactive PTPase assays have been developed. In this respect phosphotyrosine antibodies have proved to be extremely useful.[20-25] The technical difficulties and time requirements associated with obtaining phosphorylated proteins in sufficient quantities led to the use of phosphotyrosine, synthetic phosphopeptides, or other phosphoylated small molecules in favor of large proteins as substrates for detailed mechanistic studies of PTPases. In some cases, the PTPase activity is measured by separating and quantifying the phosphorylated and dephosphorylated peptides via high-performance liquid chromatography (HPLC)[21,26] or even capillary zone electrophoresis.[27] Continuous spectrophotometric assays using phosphotyrosine or phosphopeptides have also been developed that are particularly useful because they are rapid and precise.[26,28-30] In addition, there are many examples of studies employing synthetic, small molecules as substrates for PTPases.[31-33] Interestingly, however, as the number of known physiological substrates of PTPases has continued to grow, an increased emphasis has been placed on the use of these proteins as substrates in detailed mechanistic studies.[17,18]

This chapter describes several assays well suited for detailed biochemical analysis of PTPases. Figure 1 shows the reactions catalyzed by PTPases, where X represents an artificial substrate. As seen from Fig. 1, PTPase activity can be assayed by monitoring the appearance of inorganic phosphate or dephosphorylated product, or by monitoring the disappearance of phosphorylated substrate. Protein-tyrosine phosphatase assays, as well as those of most other enzymes, generally fall into two categories: quenched, end-point assays or continuous assays. This chapter includes several assay procedures under each of these categories.

[21] J. A. Madden, M. I. Bird, Y. Man, T. Raven, and D. D. Myles, *Anal. Biochem.* **199,** 210 (1991).

[22] J. Babcook, J. Watts, R. Aebersold, and H. J. Ziltener, *Anal. Biochem.* **196,** 245 (1991).

[23] T. Ishibashi, D. P. Bottaro, A. Chan, T. Miki, and S. A. Aaronson, *Proc. Natl. Acad. Sci. U.S.A.* **89,** 12170 (1992).

[24] S. Singh, B. G. Darnay, and B. B. Aggarwal, *J. Biol. Chem.* **271,** 31049 (1996).

[25] S. Walchli, M.-L. Curchod, R. P. Gobert, S. Arkinstall, and R. H. van Huijsduijnen, *J. Biol. Chem.* **275,** 9792 (2000).

[26] K. Nash, M. Feldmuller, J. de Jersey, P. Alewood, and S. Hamilton, *Anal. Biochem.* **213,** 303 (1993).

[27] T. N. Gamble, C. Ramachandran, and K. P. Bateman, *Anal. Chem.* **71,** 3469 (1999).

[28] Z. Zhao, N. F. Zander, D. A. Malencik, S. R. Anderson, and E. H. Fischer, *Anal. Biochem.* **202,** 361 (1992).

[29] Z.-Y. Zhang, D. Maclean, A. M. Thieme-Sefler, R. W. Roeske, and J. E. Dixon, *Anal. Biochem.* **211,** 7 (1993).

[30] G. Daum, F. Solca, C. D. Diltz, Z. Zhao, D. E. Cool, and E. H. Fischer, *Anal. Biochem.* **211,** 50 (1993).

[31] E. B. Gottlin, X. Xu, D. M. Epstein, S. P. Burke, J. W. Eckstein, D. P. Ballou, and J. E. Dixon, *J. Biol. Chem.* **271,** 27445 (1996).

[32] B. Zhou and Z.-Y. Zhang, *J. Biol. Chem.* **274,** 35526 (1999).

[33] C. C. Fjeld, A. E. Rice, Y. Kim, K. R. Gee, and J. M. Denu, *J. Biol. Chem.* **275,** 6749 (2000).

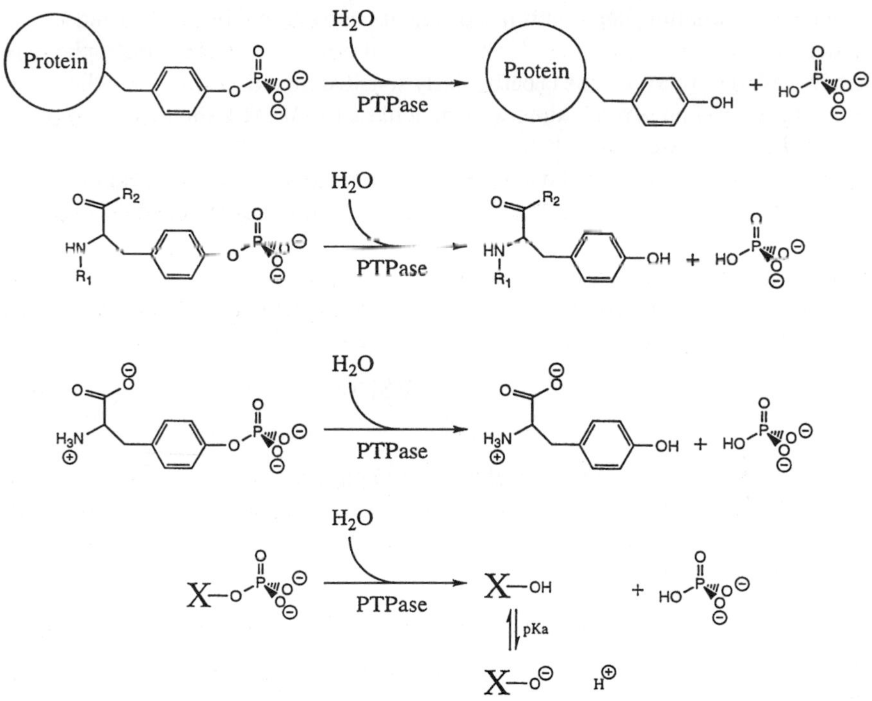

FIG. 1. Reactions catalyzed by protein-tyrosine phosphatases. Four reactions are shown to illustrate that PTPase assays have been developed employing proteins, peptides, phosphotyrosine, or small molecules (X) as substrates.

General Procedures

As the kinetic parameters of any enzyme can be affected by pH, ionic strength, and even choice of buffer, these things must be considered carefully.[34,35] The following buffers are typically used in PTPase assays: from pH 4.0 to 5.6, 100 mM acetate (sodium acetate, enzyme grade; Fisher, Pittsburgh, PA); from pH 5.6 to 6.5, 50 mM succinate (succinic acid, A.C.S. reagent; Aldrich, Milwaukee, WI); from pH 6.6 to 7.3, 50 mM 3,3-dimethylglutarate; from pH 7.3 to 8.8, 100 mM Tris (tris(hydroxymethyl)aminomethane, molecular biology grade; Fisher); and from pH 8.8 to 10.0, 100 mM glycine (Fisher). In addition, each buffer should also contain 1 mM EDTA (ethylenediaminetetraacetic acid, A.C.S. reagent; Fisher), and the ionic strength should be adjusted to 0.15 M with sodium chloride (A.C.S. reagent; Fisher). All PTPases have an active site cysteine that must be in the reduced, thiolate form for the enzyme to be active. This makes all PTPases somewhat

[34] K. J. Ellis and J. F. Morrison, *Methods Enzymol.* **87,** 405 (1982).
[35] J. S. Blanchard, *Methods Enzymol.* **104,** 404 (1984).

sensitive to oxidation and for this reason all enzymes should be purified and stored under reducing conditions—at least 1 mM dithiothreitol (DTT, electrophoresis grade; Fisher). If the enzyme is particularly sensitive to oxidation and/or the assay time is greater than several minutes, then at least 1 mM DTT should be present in the final assay mixture as well.

In all assays described, the kinetic parameters are obtained by fitting the data to some form of the Michaelis–Menten equation. The standard Michaelis–Menten equation [Eq. (1)] is used directly to fit initial velocity, v_0, versus substrate concentration, [S], data by nonlinear regression, using programs such as Enzyme Kinetics (Trinity Software, Plymouth, NH), KinetAsyst (IntelliKinetics, State College, PA), or KaleidaGraph (Synergy Software, Reading, PA).

$$d[P]/dt = v_0 = V_{max}[S]/(K_m + [S]) \tag{1}$$

$$V_{max} = k_{cat}[E] \tag{2}$$

The integrated form of the Michaelis–Menten equation, [Eq. (3)], is used to obtain values of k_{cat} and K_m for certain continuous reactions that are monitored to completion. In this case, $[S]_0$ is the initial substrate concentration while [P] is the product concentration at time t.[29,36]

$$t = [P]/k_{cat}[E] + (K_m/k_{cat}[E])\ln\{[S]_0/([S]_0 - [P])\} \tag{3}$$

When substrate concentration is much less than K_m, the limiting case of the Michaelis–Menten equation [Eq. (4)], is used to obtain V_{max}/K_m.

$$v = V_{max}[S]/K_m \tag{4}$$

For continuous assays, in which $S_0 < K_m$, the integrated form [Eq. (5)] is used to determine V_{max}/K_m.

$$[P] = S_0[1 - \exp(-V_{max}t/K_m)] \tag{5}$$

where [P] is the product concentration at time t, and S_0 is the initial substrate concentration.

Quenched Protein-Tyrosine Phosphatase

All quenched assays described below involve measuring the initial rate of the PTPase reaction at various substrate concentrations under steady state conditions and using the Michaelis–Menten equation to obtain kinetic parameters. To do this, several conditions must be met. The enzyme concentration must be much lower than the substrate concentration so that the steady state assumption is valid. And so that the assay is done under initial rate conditions, the reaction should

[36] I. H. Segel, in "Enzyme Kinetics," p. 54. John Wiley & Sons, New York, 1975.

be allowed to proceed less than 10% closer to equilibrium. Because phosphatase reactions are essentially irreversible, this means that the substrate concentration can drop only 10% or less during the course of the assay. For example, if the initial substrate concentration is 20 mM, the reaction should be stopped before the substrate concentration drops below 18 mM. The general procedure for a quenched assay follows.

1. A stock of an appropriate buffer solution is prepared as described under General Procedures (above).

2. A substrate stock solution is prepared in the chosen buffer. As phosphorylated substrates have ionizable phosphate groups, it may be necessary to adjust the pH of the substrate solution to the desired pH. The substrate will also contribute to the ionic strength of the solution; however, this contribution is often ignored at low substrate concentration (<50 mM). Once prepared, substrate solutions are usually kept on ice and stored long term at −20° to minimize hydrolysis.

3. A series of reaction mixtures is prepared in test tubes (usually 5–10 in duplicate or triplicate for each assay), each containing 200 μl minus the amount of enzyme solution to be added. These reaction mixtures are of various concentrations of substrate prepared with appropriate amounts of the substrate stock solution and the buffer solution. If a sufficient amount of substrate is available and exact values of both the k_{cat} and K_m for the PTPase reaction are desired, substrate concentrations spanning 0.2 to 5 K_m (e.g., 0.2, 0.4, 0.7, 1, 1.5, 2, 3, and 5 K_m) are chosen. If substrate concentrations above the K_m are not attainable or only the k_{cat}/K_m value is desired, then a series of substrate concentrations much lower than the K_m is used (<0.1 K_m; e.g., 0.02, 0.04, 0.06, 0.8, and 0.1 K_m). One set of blank tubes with substrate concentrations identical to those of the assay tubes, each with a total volume of 200 μl, is also prepared.

4. The assay tubes and the blank tubes are placed into a water bath of the appropriate temperature for at least 10 min before starting the assay so that they can equilibrate to that temperature. Commonly employed temperatures are 25, 30, and 37°.

5. The reaction is started by adding enzyme solution (2–20 μl) to the first assay tube, vortexing, and placing the tube back into the water bath. A stopwatch is started at this time and enzyme solution is sequentially added to each assay tube at appropriate time intervals (15–30 sec between tubes) just as it was added to the first.

6. After a specified amount of time has elapsed, quenching solution is added to each reaction in the same order and at the same time intervals in which enzyme was added. Quenching solution is added to each of the blank tubes as well.

7. When all reactions have been quenched, the tubes are removed from the water bath and each is analyzed for PTPase activity as specified below depending on the specific assay type.

8. The PTPase activity versus substrate concentration is now plotted. If the dependence of the PTPase activity on substrate concentration is hyperbolic, the data can be fit directly to the Michaelis–Menten equation [Eq. (1)] or Lineweaver–Burke equation to obtain both V_{max} and K_m. If low substrate concentrations ([S] <0.1 K_m) are used and the data show a linear dependence, then linear regression can be used to fit the data and the limitting case of the Michaelis–Menten equation [Eq. (4)] can be used in which the slope of the line is V_{max}/K_m. The V_{max} and V_{max}/K_m values are now in terms of the signal that was measured in the experiment and appropriate conversions must be made to obtain meaningful values for k_{cat} and k_{cat}/K_m, respectively. When PTPase activity versus [S] is hyperbolic V_{max} is converted to k_{cat} as follows [Eq. (6)]:

$$k_{cat} = \frac{V_{max}}{\varepsilon[\text{E}]\text{assay time}} \left(\frac{\text{quenched volume}}{\text{initial volume}} \right) (\text{conversion factors}) \quad (6)$$

where ε is an assay-dependent constant that relates the signal obtained to the product concentration such as a molar absorption coefficient, and "conversion factors" represents the fact that variable unit conversion factors may be needed 50 that k_{cat} can be obtained solely with units of time in the denominator (i.e., $\sec^{-1}$). In the case of low substrate concentrations, k_{cat}/K_m can be obtained from the linear regression of the PTPase activity versus [S] data as follows [Eq. (7)]:

$$\frac{k_{cat}}{K_m} = \frac{\text{slope}}{\varepsilon[\text{E}]\text{assay time}} \left(\frac{\text{quenched volume}}{\text{initial volume}} \right) (\text{conversion factors}) \quad (7)$$

The slope of the linear regression line is obtained with units of PTPase activity/[S]. Equation (7) converts this into appropriate units of k_{cat}/K_m (i.e., $\sec^{-1} M^{-1}$).

Assays for Inorganic Phosphate

The detection of inorganic phosphate has long been of great interest in the biochemical field. Some of the earliest PTPase assays involved the use of ^{32}P-phosphorylated proteins as substrates and quantitating the released ^{32}P$_i$. These kinds of assays are particularly useful for characterizing the interaction of a PTPase with its physiological substrate as described elsewhere.[16–18] Several inorganic phosphate assays have also been developed that can be used to monitor inorganic phosphate release in both quenched or continuous assays.[37,38] Here we describe two nonradioactive, colorimetric procedures for the quantitation of inorganic phosphate released during a PTPase assay. Method 1 was originally developed by Baginski *et al.*[39] and modified by Black and Jones.[40] Method 2 was described by

[37] M. R. Webb, *Proc. Natl. Acad. Sci. U.S.A.* **89**, 4884 (1992).
[38] M. Brune, J. L. Hunter, J. E. T. Corrie, and M. R. Webb, *Biochemistry* **33**, 8262 (1994).
[39] E. S. Baginski, P. P. Foa, and B. Zak, *Clin. Chim. Acta* **15**, 155 (1967).
[40] M. J. Black and M. E. Jones, *Anal. Biochem.* **135**, 233 (1983).

Lanzetta et al.[41] The utility of these assays lies in the fact that virtually any phosphorylated peptide or small molecule can be employed as a substrate. In general, they should be used to assay purified proteins with synthetic substrates and not crude lysates, as these usually have high background inorganic phosphate levels.

Method 1: Ammonium Molybdate Method

SOLUTIONS

Solution A: 2% (w/v) Ammonium molybdate tetrahydrate (A.C.S. reagent; Aldrich)

Solution B: 14% (w/v) Ascorbic acid (L-ascorbic acid, A.C.S. reagent; Aldrich) in 50% (w/v) trichloroacetic acid (A.C.S. reagent; Fisher). The trichloroacetic acid solution is made first. Ascorbic acid, 0.42 g, in 3 ml of 50% (w/v) trichloroacetic acid is enough for 20 reaction vessels

Solution C: 2% (w/v) Trisodium citrate dihydrate (Aldrich) and 2% (w/v) sodium arsenite (Sigma, St. Louis, MO) in 2% (v/v) acetic acid (glacial acetic acid, A.C.S. Plus reagent; Fisher). The acetic acid is added after the other reagents have dissolved

Quenching solution: 10% (w/v) Trichloroacetic acid

Solution B is usually made freshly 1 hr before it is used and stored on ice. It is stable for approximately 5 days at 4°. The other solutions are stable for months.

PROCEDURE

1. The general procedure for quenched assays is followed, using 200-μl reaction volumes and quenching each reaction and blank with 100 μl of 10% (w/v) trichloroacetic acid.

2. Two milliliters of solution A is added to 3 ml of solution B and vortexed. The resulting solution must be yellow, not green.

3. Of this A/B mixture, 250 μl is added to each quenched reaction mixture and vortexed.

4. Solution C (500 μl) is added 1–3 min after the addition of the A/B mixture.

5. After at least 5 minutes, the absorption of each mixture at 700 nm is measured, using the appropriate blank.

6. Absorption versus substrate concentration is plotted and analyzed by a graphing, equation-fitting program such as one of those described under General Procedures (above). The V_{max} obtained is in terms of absorbance units at 700 nm. To convert this into concentration units, a standard curve is constructed by executing the assay with a series of standard sodium phosphate solutions. Depending on the conditions used Eq. (6) or (7) converts these values into standard kinetic parameters.

[41] P. A. Lanzetta, L. J. Alvarez, P. S. Reinach, and A. O. Candia, *Anal. Biochem.* **100,** 95 (1979).

Method 2: Malachite Green–Ammonium Molybdate Method

SOLUTIONS

MG: 0.045% (w/v) Malachite green hydrochloride (Sigma)
AM: 4.2% (w/v) Ammonium molybdate tetrahydrate (A.C.S. reagent; Aldrich)
   in 4 N hydrochloric acid (trace metal grade; Fisher)
Tw: 10% (w/v) Tween 20 (enzyme grade; Fisher)
MG/AM/Tw: Add 1 part AM to 3 parts MG. Stir for 30 min and filter with
   Whatman (Clifton, NJ) grade 5 filter paper. Add 200 μl of Tw for every
   100 ml of MG/AM.

PROCEDURE

1. The general procedure for quenched assays is followed, using 200-μl reaction volumes and quenching with 800 μl of MG/AM/Tw reagent.

2. The color is allowed to develop for 30 min at room temperature.

3. The absorbance of each reaction mixture is measured at 650 nm, using the appropriate blank.

4. The kinetic parameters are obtained as in step 6 of method 1—constructing an appropriate calibration curve with standard sodium phosphate solutions and using Eq. (6) or (7).

Assays Using Artificial, Small-Molecule Substrates That Yield
*   Chromogenic Products*

Small-molecule substrates such as 4-nitrophenyl phosphate (pNPP) have been used extensively to study alkaline phosphatases, acid phosphatases, purple acid phosphatases, and the serine/threonine phosphatases. These same assays have been adapted to the study of PTPases. Further, as phosphatase-conjugated antibodies are extensively employed in immunological assays, much work has been done in the development of small-molecule phosphatase substrates with even greater sensitivity than pNPP, the substrate that is described below. It should be noted that this procedure can be generalized to any substrate that yields a highly chromogenic product. As shown in Fig. 1, the product will contain an alcohol oxygen that may be protonated or deprotonated depending on the pH. In general it is the deprotonated form of the product that is highly chromogenic. For this reason the reactions are quenched with a strong base to give maximum signal.

Method 3: Quenched 4-Nitrophenyl Phosphate Assay

SOLUTIONS

Quenching solution: 1 M sodium hydroxide (A.C.S. reagent; Fisher)
Substrate solution: 10–50 mM 4-nitrophenyl phosphate (pNPP, >99%; Fluka,

Rokonkoma, NY) prepared in the appropriate buffer and adjusted to the appropriate pH as described under General Procedures.

PROCEDURE

1. The general procedure for quenched assays is followed, using 200-μl reaction volumes and quenching with 1000 μl of 1 M sodium hydroxide.

2. The absorbance of each reaction mixture is measured at 405 nm, using the apropriate blank.

3. The absorbance versus pNPP concentration is plotted and the kinetic parameters are obtained in terms of A_{405} units, using Eq. (1) or (4).

4. The k_{cat} or k_{cat}/K_m is obtained with the absorption coefficient $\varepsilon = 18,000\ M^{-1}$ cm^{-1} for pNPP and Eq. (6) or (7).

Continuous Protein-Tyrosine Phosphatase Assays

Because they are rapid, efficient, and precise, continuous assays are of particular interest in the PTPase field. Under appropriate conditions, they allow the explicit determination of both k_{cat} and K_m in a single experiment that requires only one reaction vessel and one blank. In inhibition studies, they can reveal time-dependent inhibition or inactivation of PTPases that may go undetected in quenched assays. For these reasons they are of particular value in high-throughput PTPase inhibitor screening as well as in mechanistic analysis. All of the continuous assays described follow the same general procedure. They all involve monitoring the reaction progress spectrophotomerically over time, and analyzing the resulting data using some form of the Michaelis–Menten equation [Eqs. (1)–(5)]. The general continuous assay procedure follows.

1. Buffer and substrate solutions are prepared as in steps 1 and 2 under Quenched Assays.

2. A substrate solution of appropriate concentration is prepared in a spectrophotometric cuvette. The volume used will depend on the size of the cuvette, but will typically be 500 or 1000 μl. A blank cuvette such that the final substrate concentration will be identical to that of the reaction cuvette is also prepared. Both cuvettes are placed in the appropriate cell of a temperature-controlled spectrophotometer or fluorimeter depending on the assay. They are allowed to equilibrate to the appropriate temperature for 15 min. Typically PTPase assays are done at 25, 30, or 37°.

3. A catalytic amount of the PTPase, usually 2–20 μl, is added to the reaction cuvette and mixed by inverting it several times. It is returned to the spectrophotometer and data acquisition is begun.

4. After the progress curve has been recorded, it is analyzed according to the specific assay procedures described below to obtain the kinetic parameters.

Assays Using Small, Phosphorylated Peptides or Phosphotyrosine

When phosphotyrosine is hydrolyzed into tyrosine, its absorbance and fluorescence change significantly and this is the basis for method 4.[28,29]

Method 4: Using Tyrosine-Phosphorylated Peptides to Determine Both k_{cat} and K_m

SOLUTIONS

Substrate solution: Tyrosine-phosphorylated peptide dissolved in the appropriate buffer

PROCEDURE

1. The general procedure for continuous assays is followed, with the reaction mixture prepared such that the final concentration of phosphorylated peptide will be at least 2 K_m (typically 10–500 μM).
2. The cuvette containing the reaction mixture is placed into the temperature-controlled spectrophotometer and allowed to equilibrate to the appropriate temperature.
3. Rather than using a blank cuvette, the absorbance of the reaction mixture can be set to zero before enzyme is added.
4. A catalytic amount of PTPase, 2–20 μl, is added and mixed by inversion.
5. The absorbance at 282 nm or the fluorescence at 305 nm with excitation at 280 nm is monitored until the signal no longer increases.
6. The integrated form of the Michaelis–Menten equation [Eq. (3)] is used to obtain the Kinetic parameters k_{cat} and K_m. A curve-fitting program is used to fit the data, using Eq. (3) optimizing for $\sum(t_{exp} - t_{calc})^2$. Figure 2 shows such a progress curve monitoring the dephosphorylation of a phosphopeptide by a mutant PTP1B.

Method 5: Using Tyrosine-Phosphorylated Peptides to Determine k_{cat}/K_m Only. Method 5 is performed in the same way as method 4 except that a low substrate concentration is used ($\ll K_m$). Also, Eq. (5) is used to fit the data. Under these conditions the reaction is first order with respect to [S], where V_{max}/K_m is the apparent first-order rate constant. The k_{cat}/K_m is determined by dividing this first-order rate constant by [E]. Figure 3 shows a progress curve generated in this way, monitoring the dephosphorylation of a phosphopeptide by PTP1B where the peptide concentration is $\ll K_m$.

Assays Using Artificial, Small-Molecule Substrates

Method 6: Using Continuous 4-Nitrophenyl Phosphate Assay to Determine Both k_{cat} and K_m. Method 6 is done in the same way as the previous continuous assays except that the reaction is monitored via absorbance at 405 nm. Also, instead

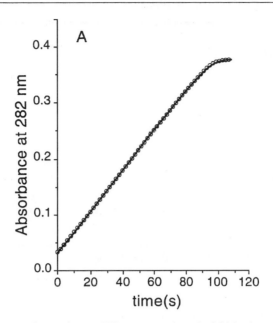

FIG. 2. Time course of a continuous PTPase assay where the initial substrate concentration is greater than the K_m. This shows the PTP1B/R47A-catalyzed hydrolysis of the phosphopeptide DADEpYLIPQQG at pH 7.0 and 30°. The enzyme and peptide concentrations were 156 nM and 560 μM, respectively. The theoretical curve (solid line) was obtained through a nonlinear least-squares fitting algorithm to the experimental data points, using Eq. (3).

of analyzing the entire progress curve, only the initial, linear portion of the curve will be used. The slope of this line gives the initial rate as the change in absorbance per second. The kinetic parameters k_{cat} and K_m are obtained via the method of initial rates, where continuous assays are repeated at different pNPP concentrations spanning 0.2–5 K_m. In this way the analysis is the same as is employed in quenched assays. However, corrections must be made for the change in the molar absorptivity coefficient (ε) of the product, p-nitrophenol, at different pH values. At high pH (>9) the absorptivity coefficient of p-nitrophenol is 18,000 M^{-1} cm^{-1}. However, the pK_a of the phenolic proton is 7.1 for pNPP. As stated earlier, the chromogenic species is the deprotonated form. Therefore, at pH values below 9 or 8, a standard curve for p-nitrophenol should be constructed in the buffer and at the pH of interest. For example, the absorptivity coefficient for pNPP at pH 7.0 is 7800 M^{-1} cm^{-1}. Phosphatase substrates with lower pK_a values have been synthesized for greater sensitivity at neutral and acidic pHs and are particularly useful for continuous assays at these pH values.[42]

[42] K. R. Gee, W.-C. Sun, M. K. Bhalgat, R. H. Upson, D. H. Klaubert, K. A. Latham, and R. P. Haugland, *Anal. Biochem.* **273**, 41 (1999).

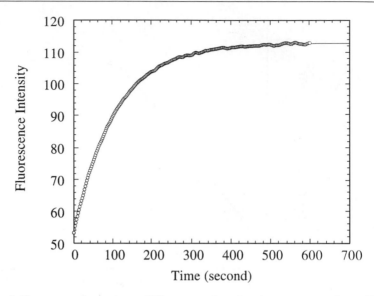

FIG. 3. Time course of a continuous PTPase assay where the substrate concentration $\ll K_m$. This shows the PTP1B-catalyzed hydrolysis of Ac-AAAApYAAAA-NH$_2$. The peptide concentration was 10 μM and the enzyme concentration was 6.7 nM. The reaction was monitored continuously at 305 nm for the increase in tyrosine fluorescence with excitation at 280 nm. The theoretical curve (solid line) was obtained through a nonlinear least-squares fit to the experimental data points, using Eq. (5).

Several rapid, nonradioactive assays for PTPases have been described here. These are useful for the detailed biochemical characterization of PTPases as well as high-throughput screening for PTPase inhibitors and activators.

Acknowledgments

This work was supported by NIH Grants CA69202 (Z.-Y.Z.) and ST32GM07260 (D.F.M.). We thank Mauro Sarmiento and Yen-Fang Keng for the preparation of Figs. 2 and 3, respectively, and Nicole McCain for help in preparing this manuscript.

Section VIII

Regulation of Gene Expression

[41] Differential Display of mRNAs Regulated by G-Protein Signaling

By HIRONORI EDAMATSU, YOSHITO KAZIRO, and HIROSHI ITOH

Introduction

Heterotrimeric GTP-binding proteins (G proteins) play important roles in transducing signals from seven-transmembrane receptors to intracellular signaling cascades.[1,2] On stimulation of a G-protein-coupled receptor, G proteins are activated, and in turn stimulate intracellular signaling pathways that regulate transcription. For instance, lysophosphatidic acid (LPA) stimulates transcription of early response genes through G-protein-mediated signals.[3] In neuronal cells, neurotransmitters stimulate transcription, which might contribute to memory and learning. It has been reported that glucagon-like peptide 2 activates transcription of early response genes, such as c-*fos,* c-*jun, junB,* and *zif218,* in baby hamster kidney cells.[4] Overexpression of the G protein $\beta\gamma$ subunits stimulates mitogen-activated protein (MAP) kinase cascades, including extracellular signal-regulated kinase (ERK), c-Jun N-terminal kinase (JNK), and p38MAPK cascades, which regulate promoter activity of early response genes.[5-10] In addition to the $\beta\gamma$ subunits, expression of constitutively active mutants of the G protein α subunit can also activate these pathways.[11-13] In Rat-1 fibroblast cells, expression of a constitutively active mutant of $G\alpha_{12}$ induces activation of the Ras/ERK/c-*fos* promoter and the JNK pathways that result in induction of their transformation.[12,14-17] Moreover, somatic mutations of *Gαi2* and *Gαs* genes, which result in constitutive activation of these

[1] Y. Kaziro, H. Itoh, T. Kozasa, M. Nakafuku, and T. Satoh, *Annu. Rev. Biochem.* **60,** 349 (1991).

[2] J. R. Hepler and A. G. Gilman, *Trends Biochem. Sci.* **17,** 383 (1992).

[3] J. K. Chuprun, J. R. Raymond, and P. J. Blackshear, *J. Biol. Chem.* **272,** 773 (1997).

[4] B. Yusta, R. Somwar, F. Wang, D. Munroe, S. Grinstein, A. Klip, and D. J. Drucker, *J. Biol. Chem.* **274,** 30459 (1999).

[5] P. Crespo, N. Xu, W. F. Simonds, and J. S. Gutkind, *Nature (London)* **369,** 418 (1994).

[6] M. Faure, T. A. Voyono-Yasenetskaya, and H. R. Bourne, *J. Biol. Chem.* **269,** 7851 (1994).

[7] A. Ito, T. Satoh, Y. Kaziro, and H. Itoh, *FEBS Lett.* **368,** 183 (1995).

[8] J. Yamauchi, Y. Kaziro, and H. Itoh, *J. Biol. Chem.* **272,** 7602 (1997).

[9] J. Yamauchi, M. Nagao, Y. Kaziro, and H. Itoh, *J. Biol. Chem.* **272,** 27771 (1997).

[10] Y. Sun, J. Yamauchi, Y. Kaziro, and H. Itoh, *J. Biochem. (Tokyo)* **125,** 515 (1999).

[11] M. Nagao, J. Yamauchi, Y. Kaziro, and H. Itoh, *J. Biol. Chem.* **273,** 22892 (1998).

[12] H. Edamatsu, Y. Kaziro, and H. Itoh, *FEBS Lett.* **440,** 231 (1998).

[13] M. Nagao, Y. Kaziro, and H. Itoh, *Oncogene* **18,** 4425 (1999).

[14] A. M. Pace, Y. H. Wong, and H. R. Bourne, *Proc. Natl. Acad. Sci. U.S.A.* **88,** 7031 (1991).

[15] S. K. Gupta, C. Gallego, J. M. Lowndes, C. M. Pleiman, C. Sable, B. J. Eisffelder, and G. L. Johnson, *Mol. Cell. Biol.* **12,** 190 (1992).

[16] S. K. Gupta, C. Gallego, G. L. Johnson, and L. E. Heasley, *J. Biol. Chem.* **267,** 7987 (1992).

0076-6879/02 $35.00

gene products, have been found in certain human tumor tissues.[18,19] However, the detail of the effect of the activation of G protein signaling on the gene expression remains unclear.

mRNA differential display is a powerful technique to identify the genes whose expression is regulated by G protein signaling. The method of mRNA differential display is based on reverse transcription (RT) polymerase chain reaction (PCR), and was originally developed by Liang and co-workers.[20,21] The method consists of two parts: (1) synthesis of cDNA with oligo(dT) primers anchored with one or two additional bases at the 3' end, which enables the grouping of the cDNA population into three or more groups, and (2) PCR with low stringency. In the original method, short oligonucleotide primers (~10-mer) were used for differential display. However, short oligonucleotide primers sometimes result in low reproducibility of each experiment. To obtain good reproducibility, several methods using long oligonucleotide primers (~20-mer) have been reported.[22,23] As illustrated in Fig. 1, PCR of the modified method consists of low-stringency cycles followed by high-stringency cycles. In the low-stringency cycles with an annealing temperature of 37°, the long oligonucleotide primers can anneal with cDNA species as short oligonucleotide primers. In the high-stringency cycles with an annealing temperature of 55°, the long oligonucleotide primers strictly anneal with and amplify cDNAs that possess the primer sequences at their 5' and 3' ends. In this article, we describe the use of differential display with long oligonucleotide primers for the analysis of G-protein signaling-regulated genes.

Procedures for Identification of Genes Whose Expression Is Regulated by G-Protein Signaling

The method for identification of genes whose expression is regulated by G-protein signaling consists of (1) preparation of total RNA from the cells before and after stimulation, (2) cDNA synthesis using one base-anchored oligo(dT)

[17] T. Ikezu, T. Okamoto, Y. Murayama, T. Okamoto, Y. Homma, E. Ogata, and I. Nishimoto, *J. Biol. Chem.* **269,** 31955 (1994).

[18] C. A. Landis, S. B. Masters, A. Spada, A. M. Pace, H. R. Bourne, and L. Vallar, *Nature (London)* **340,** 692 (1989).

[19] J. Lyons, C. A. Landis, G. Harsh, L. Vallar, K. Grünewald, H. Feichtinger, Q.-Y. Duh, O. H. Clark, E. Kawasaki, H. R. Bourne, and F. McCormick, *Science* **249,** 655 (1990).

[20] P. Liang and A. B. Pardee, *Science* **257,** 967 (1992).

[21] P. Liang, W. Zhu, X. Zhang, Z. Guo, R. P. O'Connell, A. Lidia, F. Wang, and A. B. Pardee, *Nucleic Acids Res.* **22,** 5763 (1994).

[22] M. H. K. Linskens, J. Feng, W. H. Andrews, B. E. Enlow, S. M. Saati, L. A. Tonkin, W. D. Funk, and B. Villeponteau, *Nucleic Acids Res.* **23,** 3244 (1995).

[23] T. Ito, K. Kito, N. Adati, Y. Mitsui, H. Hagiwara, and Y. Sasaki, *FEBS Lett.* **351,** 231 (1994).

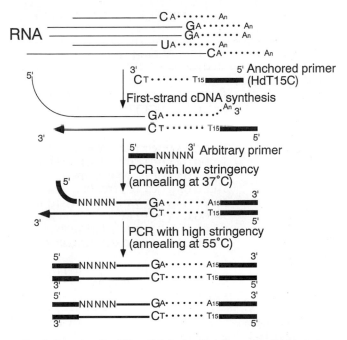

FIG. 1. Procedure for differential display. Details are described in text.

primers (anchored primers), (3) PCR using an arbitrary primer and an anchored primer with low stringency, (4) determination of the sequence of the candidate cDNAs, and (5) confirmation of the upregulation or downregulation of the expression of the candidates by Northern blot or RT-PCR analysis.

Materials

Arbitrary primer, 10 μM (see Table I)[22]

Anchored primer, 50 μM (see Table I)[22]

SuperScript II reverse transcriptase (250 U/μl; GIBCO-BRL, Gaithersburg, MD)

dNTP mix, 2.5 mM: 2.5 mM dATP, 2.5 mM dCTP, 2.5 mM dGTP, 2.5 mM dTTP

0.1 mM dNTP mix, 0.1 mM: 0.1 mM dATP, 0.1 mM dCTP, 0.1 mM dGTP, 0.1 mM dTTP

$[\alpha\text{-}^{32}\text{P}]$dCTP, 370 MBq/ml ($>$111 TBq/mmol)

Taq DNA polymerase, 5 U/μl

First-strand synthesis buffer, 5$\times$: 250 mM Tris-HCl (pH 8.3), 375 mM KCl, 15 mM MgCl$_2$

TABLE I
PRIMERS FOR DIFFERENTIAL DISPLAY[a]

Oligo nucleotide	Sequence (5' → 3')
Arbitrary primers	
22A00	5'-CGG GAA GCT TAT CGA CTC CAA G-3'
22A01	5'-CGG GAA GCT TTA GCT AGC ATG G-3'
22A02	5'-CGG GAA GCT TGC TAA GAC TAG C-3'
22A03	5'-CGG GAA GCT TTG CAG TGT GTG A-3'
22A04	5'-CGG GAA GCT TGT GAC CAT TGC A-3'
22A05	5'-CGG GAA GCT TGT CTG CTA GGT A-3'
22A06	5'-CGG GAA GCT TGC ATG GTA GTC T-3'
22A07	5'-CGG GAA GCT TGT GTT GCA CCA T-3'
22A08	5'-CGG GAA GCT TAG ACG CTA GTG T-3'
22A09	5'-CGG GAA GCT TTA GCT AGC AGA C-3'
22A10	5'-CGG GAA GCT TCA TGA TGC TAC C-3'
22A14	5'-CGG GAA GCT TAT CTT TCT ACC C-3'
Anchored primers	
HdT15A	5'-GCG CAA GCT TTT TTT TTT TTT TTA-3'
HdT15C	5'-GCG CAA GCT TTT TTT TTT TTT TTC-3'
HdT15G	5'-GCG CAA GCT TTT TTT TTT TTT TTG-3'

[a] These primers were originally designed by Linskens *et al.* [M. H. K. Linskens, J. Feng, W. H. Andrews, B. E. Enlow, S. M. Saati, L. A. Tonkin, W. D. Funk, and B. Villeponteau, *Nucleic Acids Res.* **23,** 3244 (1995)]. The underlines indicate *Hin*dIII restriction sites.

PCR buffer with Mg^{2+}, 10× : 100 mM Tris-HCl (pH 8.3), 500 mM KCl, 15 mM MgCl$_2$

RNase-free DNase (GIBCO-BRL)

Diethyl pyrocarbonate (DEPC)-treated H_2O[24]

Sterilized H_2O (not treated with DEPC)

TE: 10 mM Tris-HCl (pH 8.0), 1 mM EDTA

Stop solution: 95% (v/v) formamide, 20 mM EDTA, 0.05% (w/v) bromphenol blue, 0.05% (w/v) xylene cyanol

Programmable thermal controller [Perkin-Elmer, MJ Research (Norwalk, CT); (Watertown, MA)]

DNA sequencing gel [6% (w/v) urea-denatured gel]

Whatman (Clifton, NJ) 3 MM filter paper

Plastic film (e.g., Saran Wrap)

Cloning vector (e.g., pBluescript II; Stratagene, La Jolla, CA)

[24] F. M. Ausubel, R. Brent, R. E. Kingston, D. D. Moore, J. G. Seidman, J. A. Smith, and K. Struhl, "Short Protocols in Molecular Biology." John Wiley & Sons, New York, 1999.

Preparation of RNA

For differential display, total cellular RNA is sufficient, and enrichment of poly(A)$^+$ RNA is not required. However, the RNA should be free from chromosomal DNA because contamination of chromosomal DNA will seriously affect the results of differential display. Therefore, total cellular RNAs are prepared by CsCl step gradient after extraction of RNA with guanidine thiocyanate. When RNA is prepared by a "single-step" purification method such as the acid–phenol method, the RNA should be treated with RNase-free DNase.

Synthesis of Complementary DNA

To anneal an anchored primer to RNA, 50 pmol of an anchored primer is mixed with 2.5 μg of total cellular RNA in a clean tube, and the final volume is adjusted to 10 μl by adding DEPC-treated H_2O. The tube is incubated with a programmable thermal controller at 70° for 10–20 min to denature RNA. During denaturing RNA, 10 μl of 2× RT mix is prepared for each tube as follows.

First-strand synthesis buffer, 5×	4 μl
Dithiothreitol (DTT), 0.1 M	2 μl
dNTP mix, 2.5 mM	1 μl
SuperScript II reverse transcriptase, 250 U/μl	1 μl
DEPC-treated H_2O	2 μl

After denaturing the RNA, the reaction tube is chilled on ice, and centrifuged briefly. Then, 10 μl of 2× RT mix is added to the RNA–primer mixture and mixed well by gentle pipetting. The tube is set on the programmable thermal controller, and incubated at 25° for 10 min and then at 42° for 50 min. To inactivate reverse transcriptase, the tube is incubated at 70° for 15 min. After brief centrifugation, 80 μl of TE is added and mixed well. This first-strand cDNA solution is ready for PCR and can be stored at −20°.

Polymerase Chain Reaction for Differential Display

For PCR, 19 μl of PCR premix is prepared for each tube as follows.

Sterilized H_2O	12.9 μl
First-strand cDNA solution	2 μl
PCR buffer with Mg^{2+}, 10×	2 μl
dNTP mix, 0.1 mM	1.5 μl
Anchored primer 50 μM	0.2 μl
[α-^{32}P]dCTP, 370 MBq/ml	0.2 μl
Taq DNA polymerase, 5 U/μl	0.2 μl

A 1-μl (10-pmol) aliquot of arbitrary primer is added to each PCR tube, and mixed with 19 μl of PCR premix. The tube is set on the programmable thermal controller, and incubated as follows.

No. of cycles	Time	Temperature (°C)
1	3 min	94
	1 min	37
	2 min	72
3 (low stringency)	45 sec	94
	1 min	37
	2 min	72
21 (high stringency)	45 sec	94
	45 sec	55
	2 min	72
1	5 min	72

The resulting PCR sample (3 μl) is mixed with 2.5 μl of stop solution, and incubated at 90° for 3 min. About 4 μl of the sample is electrophoresed on a 6% (w/v) urea-denatured sequencing gel. After electrophresis, the gel is transferred onto Whatman filter paper, covered with plastic film (e.g., Saran Wrap), and dried without fixing by methanol. By using radioactive ink or a luminescent label, at least three points on the gel or the attached filter paper are marked. The gel is exposed to X-ray film [New England Nuclear (Boston, MA) or Eastman Kodak (Rochester, NY)] overnight at room temperature without an intensifier screen.

Isolation and Reamplification of cDNAs of Interest

After developing the film, the film is aligned with the gel. The bands whose density has changed after stimulation are selected, and both edges of the bands of interest are punched by a needle. The band of interest is cut out with a razor blade, and the gel slice along with the attached filter paper and plastic film is transferred into a clean PCR tube. To elute DNA fragments from the gel slice, 50 μl of sterilized water is added and the tube is incubated at 60° for 10 to 20 min. For reamplification of the isolated DNA fragment, 100 μl of mixture is prepared and PCR is carried out as follows.

Sterilized H$_2$O	73.5 μl
Eluate (described above)	10 μl
dNTP mix, 2.5 mM	10 μl
Anchored primer, 50 μM	1 μl
Arbitrary primer, 10 μM	5 μl
Taq DNA polymerase, 5 U/μl	0.5 μl

No. of cycles	Time	Temperature (°C)
25	1	94
	1	55
	2	72
1	7	72

Reamplification is checked by 3% (w/v) agarose gel electrophoresis. The reamplified DNA is isolated and recovered from the agarose gel according to standard procedures. The DNA is digested with *Hin*dIII restriction enzyme, and subcloned into the *Hin*dIII site of pBluescript II SK(−). DNA is introduced into *Escherichia coli* and transformants are plated on LB agar plates.

Characterization of cDNA

Because each band of differential display sometimes contains two or more cDNA species, at least three independent clones of *E. coli* should be characterized with sequencing or restriction mapping. The obtained sequences are subject to a homology search through DNA databases such as GenBank by the Basic Local Alignment Search Tool (BLAST).[25] Confirmation of the changes in mRNA level of the isolated candidates by Northern blot or RT-PCR analysis is required.

Example: Detection and Isolation of Genes Whose Expression Is Regulated by Gα_{i2}- and Ras-Regulated Signals in Rat-1 Fibroblast Cells

In Rat-1 fibroblast cells, a constitutively active mutant of Gα_{i2} [Gα_{i2}(Q205L)] stimulates the Ras/ERK/c-*fos* promoter and the JNK pathways that induce transformation.[12,14-17] To identify the genes whose expression is regulated through Gα_{i2}-mediated signaling, we utilized differential display. We prepared a Rat-1 transfectant that expresses the constitutively active mutant of Gα_{i2} [Gα_{i2}(Q205L)] or Ras [Ras(G12V)] in an isopropyl-β-D-thiogalactopyranoside (IPTG)-dependent manner.[12,26] Total cellular RNAs were prepared from these cell lines before and after IPTG-induced expression of these mutant proteins, and differential display was performed to analyze the changes in the patterns of the mRNA expression.

An example of the differential display of Gα_{i2}- or Ras-dependent genes is shown in Fig. 2A. The density of the bands indicated by the arrows was increased after IPTG-induced expression of Gα_{i2}(Q205L) or Ras(G12V). These bands were isolated and their DNA sequences were determined. BLAST searches revealed that clone 1 was the 3′ noncoding region of a rat homolog of matrix metalloprotease

[25] S. F. Altschul, W. Gish, W. Miller, E. W. Myers, and D. J. Lipman, *J. Mol. Biol.* **215,** 403 (1990).
[26] H. Edamatsu, Y. Kaziro, and H. Itoh, *Gene* **187,** 289 (1997).

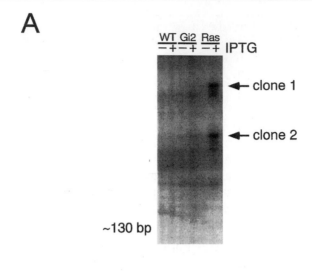

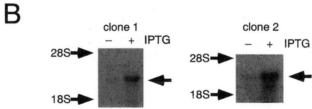

FIG. 2. Activation of Ras or Gα$_{i2}$ stimulates gene expression. (A) Differential display of Rat-1 cells and their transfectants. Parental Rat-1 (WT), and the transfectants with the Gα$_{i2}$(Q205L) (G$_{i2}$) or Ras(G12V) (Ras) gene under the control of the *lac* operator repressor system, were treated with or without IPTG, and total cellular RNAs were prepared. Complement DNAs were synthesized with the anchored primer HdT15C, and PCR was performed with the primers HdT15C and 22A00. The arrows indicate the positions of clones 1 and 2. (B) Northern analysis of changes in the expression of clones 1 and 2. Rat-1 transfectants harboring the Ras(G12V) gene were treated with or without IPTG, and total cellular RNAs were prepared. Each RNA (20 μg) was analyzed by Northern blot, using digoxigenin (DIG)-labeled probes generated from the DNA fragment of clone 1 or 2.

13 (MMP13),[27] and that clone 2 was the 3' noncoding region of a rat homolog of the cell surface receptor for gibbon ape leukemia virus (Glvr-1).[28] Northern blot analysis using these cDNA fragments as probes demonstrated that these transcriptions are upregulated on stimulation of the Ras pathway (Fig. 2B). It has been shown that the expression of MMP13 is upregulated in human tumor tissues, such

[27] C. O. Quinn, D. K. Scott, C. E. Brinckerhoff, L. M. Matrisian, J. J. Jeffrey, and N. C. Partridge, *J. Biol. Chem.* **265,** 22342 (1990).

[28] G. Palmer, J. P. Bonjour, and J. Caverzasio, *Endocrinology* **138,** 5202 (1997).

as breast carcinomas,[29] and regulated by AP-1 activity.[30] Expression of Glvr-1 is stimulated by insulin-like growth factor I in human SaOS-2 osteoblast-like cells.[28]

[29] J. M. Freije, I. Diez-Itza, M. Balbin, L. M. Sanchez, R. Blasco, J. Tolivia, and C. Lopez-Otin, *J. Biol. Chem.* **269**, 16766 (1994).

[30] A. M. Pendas, M. Balbin, E. Llano, M. G. Jimenez, and C. Lopez-Otin, *Genomics* **40**, 222 (1997).

[42] Gene Profiling of Transgenic Mice with Targeted Expression of Activated Heterotrimeric G Protein α Subunits Using DNA Microarray

By HSIEN-YU WANG, XIAOSONG SONG, XI-PING HUANG, and JIANGCHUAN TAO

Introduction

The study of gene expression is fundamental to research in development, metabolic regulation, and human disease. Genetics has made great use of a variety of molecular techniques to measure the expression of genes relevant to development. Traditional methods include DNA–excess solution hybridization, riboprobe hybridization, and Northern analysis to ascertain the steady state levels of specific RNA molecules, products of genes of interest. Although each of the technologies has specific advantages, all of the traditional approaches are labor intensive and designed to explore in great detail a rather limited number of RNAs. The pace of technological advancement in the analysis of gene expression has been unparalleled, affording a rich set of tools with which development research can be accelerated. The Human Genome Project and efforts to identify unique fragments of gene [expressed sequence tags (ESTs), etc.] have resulted in an impressive expansion in the amount and quality of molecular probes amenable to aid research in the study of development specifically, and human disease in general. cDNA spotting and DNA oligonucleotide arraying offer powerful, high-throughput capabilities for gene profiling and is highlighted herein. Differential display, serial analysis of gene expression (SAGE), and subtractive cloning technologies offer alternative approaches, but are not featured in this work.

Genome-wide expression profiling using DNA microarray and cDNA spotting enables a comprehensive, high-throughput screening of genes regulated by development, metabolic regulation, and disease.[1–3] Techniques for cDNA spotting

[1] L. M. Staudt and P. O. Brown, *Annu. Rev. Immunol.* **18**, 829 (2000).

enable the creation of arrays with >1000 cDNA clones/cm^2 of area.[4,5] The oligonu-cleotide-based arrays are densely organized, offer specificity, high sensitivity, and the ability to define false-positive hybridizations.[6,7] Both approaches are limited to analysis of known genes (or ESTs) and require microgram quantities of start-ing RNA from target cells or tissues. From a practical standpoint, these high-throughput strategies require a mass of 1000 to 10,000 cells for adequate amounts of RNA. The technology is likely to advance to higher levels of sensitivity, enabling reliable analysis from <1000 cells in the future.

These techniques can assist in identifying the roles of both well-known and newly discovered genes in development, metabolic regulation, and human disease. This broad coverage of large numbers (10,000–80,000) of genes also benefits from relative ease of use, readily exportable to routine use in laboratories. Analysis of such large arrays of interesting gene products by DNA microarray and cDNA spot-ting, for example, can provide a molecular "signature" enabling the description and differentiation of various cancers, as reported for two human leukemias.[8] Furthermore, many genes associated with signaling pathways in development (e.g., Wnt-Frizzled signaling via the β-catenin pathway) have been shown to be associated, when mutated, with human cancers. The gene expression profiles not only provide "signatures" and elucidate specific biomarkers associated with dis-eases, but also can be used in tandem with therapies to predict outcomes and pre-ferred therapeutic interventions based on the signatures. All of this great promise remains to be fulfilled, as the technologies transit from the drawing board to the research laboratory and ultimately to the bedside for diagnosis and improved treat-ments of human diseases, such as diabetes, cancer, and developmental defects.

For heterotrimeric G proteins, both *loss-of-function* and *gain-of-function* mut-nts have been identified as the basis of human disease.[9] Gain-of-function mutations in the stimulatory G protein α subunit (Gα_s) can provoke endocrine neoplasias as

[2] N. L. van Hal, O. Vorst, A. M. van Houwelingen, E. J. Kok, A. Peijnenburg, A. Aharoni, A. J. van Tunen, and J. Keijer, *J. Biotechnol.* **78,** 271 (2000).

[3] J. Khan, M. L. Bittner, Y. Chen, P. S. Meltzer, and J. M. Trent, *Biochim. Biophy. Acta* **1423,** M17 (1999).

[4] U. Scherf, D. T. Ross, M. Waltham, L. H. Smith, J. K. Lee, L. Tanabe, K. W. Kohn, W. C. Reinhold, T. G. Myers, D. T. Andrews, D. A. Scudiero, M. B. Eisen, E. A. Sausville, Y. Pommier, D. Botstein, P. O. Brown, and J. N. Weinstein, *Nat. Genet.* **24,** 236 (2000).

[5] P. T. Spellman, G. Sherlock, M. Q. Zhang, V. R. Iyer, K. Anders, M. B. Eisen, P. O. Brown, D. Botstein, and B. Futcher, *Mol. Biol. Cell* **9,** 3273 (1998).

[6] U. Alon, N. Barkai, D. A. Notterman, K. Gish, S. Ybarra, D. Mack, and A. J. Levine, *Proc. Natl. Acad. Sci. U.S.A.* **96,** 6745 (1999).

[7] M. E. Nau, L. R. Emerson, R. K. Martin, D. E. Kyle, D. F. Wirth, and M. Vahey, *J. Clin. Microbiol.* **38,** 1901 (2000).

[8] T. R. Golub, D. K. Slonim, P. Tamayo, C. Huard, M. Gaasenbeek, J. P. Mesirov, H. Coller, M. L. Loh, J. R. Downing, M. A. Caligiuri, C. D. Bloomfield, and E. S. Lander, *Science* **286,** 531 (1999).

[9] A. J. Morris and C. C. Malbon, *Physiol. Rev.* **79,** 1373 (1999).

well as the constellation of developmental and endocrine defects associated with McCune–Albright syndrome (MAS).[10] Loss-of-function mutation of $G\alpha_s$ in contrast, is associated with Albright hereditary osteodystrophy (AHO).[11] Although the role of mutations of $G\alpha_s$ in human diseases such as MAS and AHO has been well documented, how these mutations lead to sustained changes in gene expression is not known. Application of DNA microarray technology to analysis of the genes regulated by the changes in $G\alpha_s$ seems ideally suited to address this central question. In an earlier article we discussed the creation of transgenic mice that harbor gain-of-function or loss-of-function mutant forms of heterotrimeric G protein α subunits, expressed in targeted tissues under the influence of inducible promoters. In this work we outline the experimental strategy and technical issues that are encountered in the application of DNA microarray technology to the analysis of tissues in mice harboring gain-of-function, activating mutations of heterotrimeric G protein α subunits. In particular, we discuss experimental data collected in an analysis of mice that harbor an inducible, tissue-specific expression vector transgene that expresses the GTPase-deficient Q205L $G\alpha_{i2}$ protein in liver, adipose, and skeletal muscle.[12]

Methods

Experimental Considerations for Gene Profiling in Transgenic Mice

Much of the data derived from DNA microarray have been obtained from two basic experimental approaches: analysis of cells in culture in response to a stimulus expected to change gene expression (e.g., steroid hormone); and analysis of tissue samples from pathological states (e.g., cancer) and their nonpathological controls. The first situation is almost ideal. Theoretically, the response being monitored is acute and the time course of the analysis minimizes the issues of compensating responses to the changes induced by the stimulus. Activation of cells in culture with a hormone capable of stimulating a sharp increase in intracellular cyclic AMP can be subjected to gene profiling over the short term in a search for the genes sensitive to increased cyclic AMP (early event) as well as those genes whose expression then is regulated by the truly cyclic AMP-sensitive genes (later event). One can expect compensatory responses (late event), in which the cell may be undergoing desensitization of the transducing receptor while elevating the levels of cyclic AMP-dependent phosphodiesterases (PDEs) in order to rectify the initial stimulus. Sampling at each of these phases will yield differing profiles of gene activity.

[10] A. M. Spiegel, *J. Inherit. Metab. Dis.* **20,** 113 (1997).

[11] A. M. Spiegel, *Annu. Rev. Physiol.* **58,** 143 (1996).

[12] J. F. Chen, J. H. Guo, C. M. Moxham, H. Y. Wang, and C. C. Malbon, *J. Mol. Med.* **75,** 283 (1997).

The second situation, in which tissues from pathological conditions are subjected to gene profiling and compared with profiles from nonpathological controls, is best exemplified in the search for gene profiles of specific cancers. In this case, the analysis cannot be performed during the transition from normal to malignant, but rather well after the malignant state has been achieved and documented by other analyses. The goal of these studies is to obtain a molecular signature of the genes whose activity has been stably altered in the malignant state. This approach is limited by the ability to discern the key genes whose activity may effect global changes in other downstream genes, but benefits from a survey of the widest array of gene targets and unknown ESTs that can provide a molecular signature of the disease, a "fingerprint" of the disease state, against which future samples can be compared. When one considers most human diseases, which by virtue of their onset and progression are already well underway before gene profiling can be performed, the analysis of gene activity is really confined to those genes that escape all of the compensatory processes available to the cell/organism. Thus, for the study of mutations of specific G protein α subunits in transgenic mice or humans, gene profiling will study the genes that are activated in a sustained fashion and are operating outside of the normal compensatory processes that we know to operate.

Based on these basic parameters, it becomes clear that for analysis of cells in culture far more flexibility is available that may enable the study of key early events in gene activity that control a cascade of downstream groups of genes. For samples from transgenic mice or patient material, a DNA microarray will not likely reveal the temporally related genes of interest, but only the subset of genes required to support a steady state pathological state.

Design of Inducible, Targeted Expression of G Protein Subunit Generation of Transgenic Mice

The methodologies employed to create transgenic mice that harbor activated, GTPase-deficient versions of G protein α subunits and the use of inducible, antisense RNA expression to suppress the expression of specific G protein α subunits are described elsewhere (please consult the original reports[12–15]).

Isolation of mRNA from Targeted Tissues of Transgenic Mice

The overall procedure for gene profiling of samples from tissues has multiple steps (Fig. 1): mRNA extraction and validation, cDNA synthesis and validation, cRNA synthesis and labeling, test chip validation, and expression arraying. High-quality preparation of total RNA is crucial to the overall success of the entire

[13] C. M. Moxham, Y. Hod, and C. C. Malbon, *Science* **260,** 991 (1993).
[14] C. M. Moxham and C. C. Malbon, *Nature (London)* **379,** 840 (1996).
[15] H. Y. Wang, F. Lin, and C. C. Malbon, *Methods Mol. Biol.* **126,** 241 (2000).

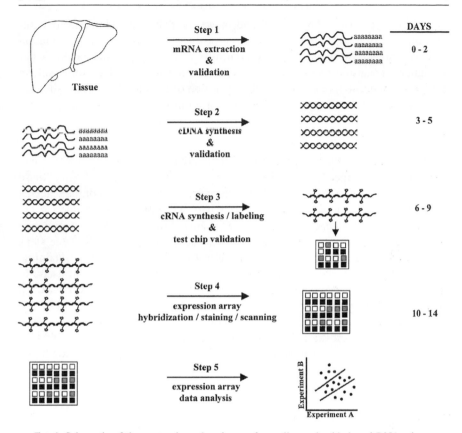

FIG. 1. Schematic of the protocol employed to perform oligonucleotide-based DNA microarray starting from RNA isolated from cells or tissues. See the text for details.

procedure. There exist many experimental approaches to the isolation of total cellular RNA. In the laboratory, we have employed virtually all of the available methodologies over the past 15 years, but routinely use only the one-step, RNA STAT-60 (Tel-Test, Friendswood, TX) methodology. This methodology is highly reliable and makes use of reagent solutions ideal for RNA isolation from animal cells and tissues. Total cellular RNA prepared via this method is ready to use as the starting material for the cDNA synthesis that is the first step in the DNA microarray strategy.

For tissues, 50 mg (wet weight)/ml of RNA STAT-60 is optimal. The tissue is rapidly excised, placed in the RNA STAT-60, and immediately homogenized with a Polytron (Brinkmann, Westbury, NY) homogenizer. The protocol for the remaining steps is as outlined by the commercial supplier (Tel-Test), with a few modifications as outlined below. The homogenization and RNA extraction steps are followed precisely. We have found, however, that three additional phenol-chloroform extractions are optimal before proceeding to the RNA precipitation and wash steps. The yield of RNA from liver is about 5 μg of total RNA per milligram of starting tissue, and somewhat lower when isolated from skeletal muscle and adipose tissues. The final preparation of total RNA is essentially free of DNA and protein and displays an absorbance ratio (260 nm/280 nm) of >1.8.

Preparation of Labeled cRNAs

The second and third steps of the protocol create the necessary intermediate cDNA from which the cRNA can be synthesized and labeled (Fig. 1). Each of these steps has a detailed protocol provided by the commercial suppliers of the reagents to be employed. In most cases, the researchers will not be responsible for the production of cDNAs and labeled cRNAs for the hybridization. More commonly, these steps are performed by "core" research facilities in which all of the steps from cDNA synthesis through cRNA synthesis (and on to hybridization, staining, and scanning) are performed within the core laboratory for individual users. In our own studies we have made use of the Affymetrix (Santa Clara, CA) GeneChip system. All of the protocols and reagents used in the core are those recommended by the commercial supplier. Our selection of the Affymetrix platform was based simply on the state-of-the-art technology and the advantages offered by this supplier. Affymetrix has pioneered in the development of the oligonucleotide-based platform, which converged from technologies such as oligonucleotide synthesis, photolithographic masking, and combinatorial chemistry.

DNA Microarraying on Affymetrix Platform

Some brief background is required in order to appreciate the technological advantages of the Affymetrix platform. The array for an individual gene sequence is composed of a 2 × 10 boxed array in which 16 to 20, 25-oligomer probes have been created that span the entire 5′ → 3′ sequence and another 16 to 20, 25-oligomer probes have been designed to be intentional mismatches for hybridization. The boxes are 24 μm per side and contain several million copies of a specific probe, tethered to the chip surface. Probe selection is based on several criteria, in addition to the sequence itself, such as empirical observations of probe performance versus mismatched controls, overrepresentation of the 3′ region of the sequence, and deletion of those potential probes that may already be represented in other gene sequences on a given array. Thus, for each gene sequence there are theoretically 10–20 positive matches for hybridization and 10–20 corresponding negative, mismatch controls from which to measure signal. A representative 2 × 10 array of several candidate genes is displayed to illuminate the organization of the chips (Fig. 2). The positive probe arrays are labeled "P," whereas the mismatch probe arrays are labeled "M" for abitrary genes A–E. The potential shortcomings of the

Gene

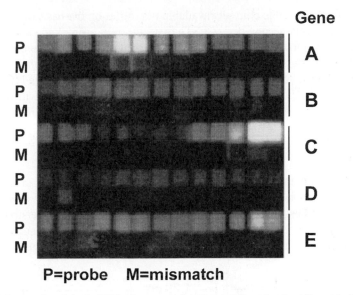

P=probe   M=mismatch

Fig. 2. Image of section of a DNA microarray of gene expression using the Affymetrix GeneChip system: details of the post-hybridization arrays of five arbitrary gene arrays. Note that for each of the genes (A to E), a positive probe (P) array of 14 oligomers is arranged above a companion control set of probes, each with an intentional mismatch (M). See the text for details.

cDNA spotting approach, in which false positives would be difficult to identify, are made convincingly clear when arrays from a large number of genes are examined. The occasional "false" positive is apparent in the Affymetrix system, but would likely go unnoticed in a cDNA spot in which this region of sequence was represented.

Selection of Gene Chips

The selection of gene chips to be employed in the hybridization is dependent on three criteria: source (species) of the RNA to be analyzed; general targets sought for analysis by the investigator; and availability of special arrays for more individualized uses. There are two basic types of chips required for an arraying experiment. The first is the "test" chip that is employed to ensure that the labeling reactions were successful. The test chips are analyzed in advance of hybridizations with the second type of chip, "gene chips," actually used for the arraying. The test chip costs a fraction of the gene chips and successful use of the test chip validates the sample preparation and signals that the samples now are ready for arraying on the gene chips.

Whole genome arrays are available currently in human, rat, mouse, and yeast, as well as in *Escherichia coli*. Genome arrays for *Drosophila* and *Arabidopsis* have recently been released. For arraying of transgenic mouse tissues/cells, there are several murine gene chip sets available that differ on the basis of complexity. Many include a large number of ESTs that currently are unknown, but will be revealed as the mouse genome is completed. In addition, new targeted arrays are available for more specialized purposes. A cancer gene chip is based on human genes well known to be regulated in cancers. Others, more specialized for analysis of metabolic, apoptotic, and developmental gene profiles, likely will be created in response to consumer demand. Custom arraying in the Affymetrix platform is available, although cost-prohibitive for most laboratories.

Gene Profiles

The density of the data yield by the array of a single chip is shown in a magnified version of an actual gene chip, examining the profile of a sample of adipose tissue taken from a transgenic mouse harboring the Q205L-activated version of $G\alpha_{i2}$ (Fig. 3). The surface engaged in the hybridization for this chip is a square, 1.28 cm on each side. The image of the hybridized probe array reveals the signal from

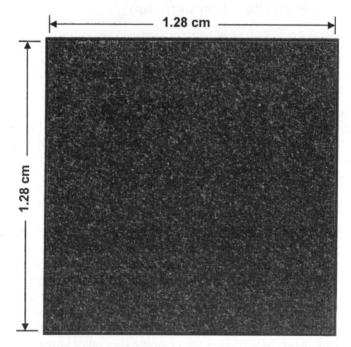

FIG. 3. Image of entire Affymetrix GeneChip array for adipose tissue sampled from a transgenic mouse harboring the GTPase-deficient, Q205L mutant form of $G\alpha_{i2}$. Total RNA was harvested from 500 mg of fresh adipose tissue from a transgenic Q205L mouse. The samples were prepared as outlined in Fig. 1 and the hybridization was performed. The actual image shown has been magnified from its actual 1.28×1.28 cm size. See text for details.

more than 6000 genes, each with an individual set of 2×10 custom-designed probe sequences and corresponding mismatched sequences. An arraying often requires the use of several chips to cover the available genome. For murine chips, a full array of more than 36,000 genes and ESTs is available currently.

It would be a disservice to avoid discussion of the issue of cost of DNA microarray analysis, because this one issue often dissuades researchers from considering the technology early. Establishing a platform for DNA microarray analysis incurs significant expense, no matter what system is selected. For the Affymetrix platform an initial investment in equipment of ~\$ 200,000 is required. On-going, recurring expense for the technical support and consumable supplies is about \$70,000 per year to operate one system at near capacity. The price to the investigator will reflect the extent to which a core facility is supported by the institution and the ability of the institution to secure a favorable discount for the test and gene chips. Even with considerable discounts and institutional support, the cost of testing two test chips and arraying two experimental conditions (e.g., transgenic tissue vs. tissue from nontransgenic littermate) ranges from \$2000 to \$2500. At first glance this would seem a considerable cost. But when this cost is compared with that of many molecular/immunological kits that yield 100 data points, the rich data yield of the chips takes on a different light. Discussions comparing the cost of cDNA-spotted slides versus gene chips are not relevant. cDNA spotting requires that the laboratory first amplify and validate cDNAs for several hundred or thousand targets at a large initial cost of human resource and reagents before any indication that the analysis will yield important new information. Thus, the power of high-throughput analysis of tens of thousands of genes must be evaluated carefully, but much like transgenic technology, DNA microarray analysis is emerging as an indispensable tool for much of modern biology.

Data Analysis

Without doubt, the greatest challenge in the application of DNA microarray to gene profiling of important medical and biological questions is in the data analysis itself. The data sets from these experiments are rich and cannot be adequately addressed by routine methods. To gain some appreciation of this challenge, merely arranging the output of a three-chip array set onto a Microsoft Excel spreadsheet with a single line devoted to each ID of the gene and/or EST would produce a printout of several hundred pages of dense type. The allure of these dense data sets rapidly dissipates and the investigator is overcome by the sheer enormity of the task of comparing even two data sets, let alone many more if the analysis concerns creating molecular signatures of tumors from >50 patients.

Standard analysis packages of software are available that will rank order the signals from the genes posting the greatest positive change relative to those with the most negative change from control (Table I). The analyses of a few of the

Table I

Increase (A) or Decrease (B) of RNA Expression of Q205L Transgenic vs wild - type FVB mice: analysis by DNA Microarray

A. Increase

Probe set	Abs Call	Diff Call	Avg Diff Change	B=A	Fold Change	Hyperlink	Description
92546_r_a	P	I	18122	*	~24.0	AB006361	Cluster Incl AB006361; mus musculus mRNA for prostaglandin D synthetase, complete cds/cds = (0,629) /gb = AB006361/gi =2317285/ug=Mm.1008/len = 789/STRA = for
100440_f_a	P	I	23427.1		6.5	U76758	Cluster Incl U76758:Ankyrin 1, erythroid /cds=(0.164) /gb=U76758 /gi=1698953 /ug=Mm.4789 /len=191 /STRA=for
102040_at	P	I	3537.1	*	~5.5	Y15798	Cluster Incl Y15798:Mus musculus mRNA for G-protein coupled receptor kinase 6-B /cds=(22,1791) /gb=Y15798 /gi=3341457 /ug=Mm.10193 /len=2841 /STRA=for
98602_at	P	I	2398.9	*	~4.0	U08110	Cluster Incl U08110:RAN GTPase activating protein 1 /cds=(274,2043) /gb= U08110 /gi=472851 /ug=Mm. 3833/len=2981 /STRA=for
103839	P	I	2334.9	*	~4.0	AF068748	Cluster Incl AF068748:Mus musculus sphingosine kinase (SPHK1a) mRNA, partial cds /cds(0,1515) /gb = AF068748 /gi=AF068748 /gi=3659691 ug=Mn.20944/len=1815 /STRA=for
97525_at	P	I	10805.2		3.8	U48403	Cluster Incl U48403:Glycerol kinase /cds=(93,1667) /gb=U48403 /gi=1480468 /ug= Mn.3242 /len = 2493 /STRA = for
97844_at	P	I	2080.1	*	~3.6	U67187	Cluster Incl U67187:Mus musculus G protein signaling regulator RGS2 (rgs2) mRNA, complete cds / cds = (20,655) /gb = U67187 /gi =1911236/ug = Mn.28262 /len =1240 /STRA = for
104580_at	P	I	1561.3	*	~3.0	U85711	Cluster Incl U85711: mus musculus phospholipase C delta-1 mRNA, complete cds /cds = (105,2375) gb= U85711 /gi = 4099290 /ug = Mm.23963 /len = 2665 /STRA = for
100967_at	P	I	1368.5	*	~2.7	AF072757	Cluster Incl AF072757:Mus musculus fatty acid transport protein 2mRNA, complete cds / cds=(0,1871) /gb=AD072757 /gi=3335564 /ug=Mm.6611 /len=1872 /STRA=for
104174_at	P	I	1334.1	*	~2.7	J02700	Cluster Incl J02700:Phosphodiesterase I/nucleotide pyrophosphatase 1cds/ = (111,2828) /gb = J02700 /gi = 200236 /ug = Mm.27254 /len = 3224/STRA=rev
92356_at	P	I	1325.4	*	~2.7	M90388	Cluster Incl M90388:Mouse protein tyrosine phosphatase (70zped) mRNA, complete cds /gi =200522 /ug = Mm.395 /len = 2734 /STRA = rev
101947_at	P	I	1637.2		2.4	AB028921	Cluster Incl AB028921:Mus musculus mRNA for NAKAP95, complete dcs /cds = (71,1999) /gb = AB028921 /gi=5931619 /ug = Mm.9590 /len =2068 /len =2068 /STRA = rev
103392_at	P	I	1050.1	*	~2.3	U12919	Cluster Incl U12919:Adenylate cyclase 7 /cds = (678,3977) /gb = U12919 /gi = 602411 /ug = Mm.18658 /len = 5199 /STRA = rev

B. Decrease

Probe Set	Abs Call	Diff Call	Avg Diff Change	Fold Chan	Hyperlink	Description
94057_g_a	P	D	-126799	-2.4	M21285	Cluster Incl M21285:Stearoyl-coenzyme A desaturase 1 /cds=(0,1067) /gb=M21285 /gi=200949 /ug=Mm.22026 /len=1068 /STRA=for
99671_at	P	D	-107028	-2.4	X04673	Cluster Incl X04673:Adipsin /cds = (19,798) /gb = X04673 /gi = 49883 / ug = Mm.4407 /len=867 /STRA=for
95356_at	P	D	-117830	-2.3	D00466	Cluster Incl D00466:Apolipoprotein E/cds=(23,958)/gb=D00466/gi=220334/ug=Mm.21417/len=1068 /STRA=for

genes that are either activated (positive relative to control) or suppressed (negative relative to control) are displayed. More detailed mining of these data requires more sophisticated software that can perform shadowing of data and cluster analysis. Many leading laboratories are struggling with these formidable problems, but as DNA microarray analysis is only at the beginning of the road, not the end, rapid development of algorithms and statistical tools for better data mining most certainly will emerge.

Concluding Remarks

Although rapidly assimilated since its early development in 1997,[16] DNA microarrays have provided a powerful new tool capable of gene profiling on a genome-wide scale. The application of this technology to assay of the genes controlled by the mutation of specific G protein α subunits is only in its infancy, full of promise and many formidable problems, not the least of which is data mining. Understanding how signaling pathways control gene expression is a central goal of biology and medicine. DNA microarray technology puts this exciting new capability within reach.

[16] J. L. DeRisi, V. R. Iyer, and P. O. Brown, *Science* **278,** 680 (1997).

[43] Retroviral Vectors Applied to Gene Regulation Studies

By T. J. Murphy, Grace K. Pavlath, Xiaofei Wang, Valerie Boss, Karen L. Abbott, Aaron M. Robida, Jim Nichols, Kaiming Xu, Michelle L. Ellington, and James R. Loss II

There are instances in G-protein-coupled receptor studies when it is desirable to express a transgene in some specialized cell phenotype to understand a process better. Most such cells are more resistant to plasmid transfection than are the transformed fibroblasts often used in G protein signaling studies [e.g., COS, 293, and chinese hamster ovary (CHO)]. This often presents a substantial barrier against understanding context-specific mechanisms of Gα protein-coupled receptor signaling and cellular regulation. Highly efficient gene transfer methods applicable to broader ranges of cell phenotypes can surmount this problem.

Retroviruses use naturally evolved mechanisms to gain entry into cells, release their transgene cargoes, and permanently tag the new host. Their most significant advantages over plasmid transfection include high gene transfer efficiencies in

Copyright © 2002 by Academic Press
All rights of reproduction in any form reserved.
0076-6879/02 $35.00

many cell types, with the genes integrating into host cell chromatin permanently. One limitation of the Moloney murine leukemia-based retroviruses, which is the basis of the system and vectors described in this article, is that cell division is necessary to take advantage of these properties.[1] A more detailed description of Moloney-based viruses, their tropism, and other features is available in a previous volume of this series.[2] For additional background, retroviruses have been compared with several other virus-based gene delivery approaches.[3] The newer lentivirus-based vectors, which appear to share the simplicity and efficiency of Moloney-based systems, should prove a popular alternative for transduction of nondividing cells.[4]

Relatively simple and rapid approaches to produce infectious recombinant retroviral particles have been described.[5-7] These transient retrovirus producer cell lines express genes necessary for the viral life cycle, subverting the need for *trans*-delivery using a helper virus. As such, they avoid delays associated with isolating and characterizing producer cell subclones, while producing among the highest titers possible. An additional advantage is that they are all based on the 293 cell background, so that transfection with retroviral vector plasmids is highly efficient and reproducible. Last, foreign gene insertion in retroviral vectors is as simple as any other standard plasmid. Altogether, these attributes make retrovirus-mediated expression no less difficult than plasmid transfection-based approaches for transgene delivery. For cells that transfect poorly, retroviral methods are more reproducible, easier, and less costly over the long haul.

There are limitations to retrovirus-based expression systems, however. In plasmid transfection, it is increasingly common to transfect cells with two or more experimental plasmids simultaneously. Such an experiment is ill-advised with retroviruses, because mixing retroviral stocks dilutes each virus, and high viral titer is essential for efficient transduction. Such multitiered transgene experiments are best performed by sequential viral infections, but at a cost requiring more time and effort in maintaining various transgenic cell lines.

Perhaps the most significant limitation of retrovirus-mediated transgenesis rests in the inherent peculiarity of their biology. Transcription of the viral mRNA begins in the 5' long terminal repeat (LTR) and ends by polyadenylation within the 3' LTR. Cryptic transcriptional termination between the two LTRs will render a proviral mRNA that is either not packaged efficiently or will fail to integrate

[1] D. G. Miller, M. A. Adam, and A. D. Miller, *Mol. Cell. Biol.* **10**, 4239 (1990).

[2] A. D. Miller, D. G. Miller, J. V. Garcia, and C. M. Lynch, *Methods Enzymol.* **217**, 581 (1993).

[3] D. Stone, A. David, F. Bolognani, P. R. Lowenstein, and M. G. Castro, *J. Endocrinol.* **164**, 103 (2000).

[4] T. Kafri, H. van Praag, L. Ouyang, F. H. Gage, and I. M. Verma, *J. Virol.* **73**, 576 (1999).

[5] W. S. Pear, G. P. Nolan, M. L. Scott, and D. Baltimore, *Proc. Natl. Acad. Sci. U.S.A.* **90**, 8392 (1993).

[6] T. M. Kinsella and G. P. Nolan, *Hum. Gene Ther.* **7**, 1405 (1996).

[7] F. Grignani, T. Kinsella, A. Mencarelli, M. Valtieri, D. Riganelli, L. Lanfrancone, C. Peschle, G. P. Nolan, and P. G. Pelicci, *Cancer Res.* **58**, 14 (1998).

properly into target cell chromatin. This is extremely rare in our experience, but emphasizes the need for carefully designing sequence insertions. Viral RNA packaging efficiency also diminishes markedly as the sequence interval between the two LTRs approaches 8 to 9 kilobases (kb).

A well-known limitation, but often not obvious to new practitioners, is associated with expression from internal promoters lying between the flanking LTRs. These internal promoters tend to function in only 30 to 50% of cells actually transduced by such retroviruses, through poorly understood epigenetic suppression mechanisms.[8] In many instances, this is acceptable; for example, this level of expression efficiency may vastly exceed the best results of plasmid DNA transfection in some cells. On the other hand, alternatives are necessary if higher efficiencies of transgene expression are necessary. The system and newer vectors we describe below, when used judiciously, can overcome many of these inherent difficulties and allow for transgene expression in as many or as few cells as desired in a culture.

Maintenance of Retroviral Plasmids in Bacteria

Transgenes are cloned into retroviral vectors between their essential 5' LTR, ψ packaging signals and the 3' LTR sequences, which are necessary components in viral production. The remainders of these circular plasmids are typical pUC-based sequences for bacterial selection and amplification. Even so, retroviral plasmids grow less efficiently in bacteria then other pUC-based plasmids, likely because of recombinations triggered by the virtually identical LTR regions. The following protocol provides optimal conditions to limit spurious plasmids, enhancing yields of the desired vector. Given additional efficiencies after the plasmid growth stage, a single 500-μg plasmid DNA yield typically far exceeds the amount necessary to perform a complete series of experiments with one vector.

1. Add 1–30 μl of retroviral plasmid (resuspended in either water or within a ligation reaction mixture) to the bottom of a sterile 13 × 100 mm polypropylene tube and place on ice. Add 50–100 μl of chemically competent Top10F' Escherichia coli (Invitrogen, Carlsbad, CA). This bacterial strain is superior compared with several others tested for growing these plasmids. After incubating the tubes on ice for 30 min, place the tubes in a 42° water bath for 45–60 sec. Add 2× yeast–tryptone (YT) bacterial growth medium to a final volume of 500 μl before incubating for 45 min at 37° with agitation.

2. Spread 200 μl of the recovered bacteria on a Luria–Bertani (LB) agar plate containing ampicillin (100 μg/ml) and tetracycline (12.5 μg/ml). Incubate the plates at 37° for 12–18 hr. The β-Lactamase resistance gene is carried on the retroviral plasmid whereas tetracycline resistance is conferred from the resident F'

---

[8] M. Emerman and H. M. Temin, Cell 39, 459 (1984).

plasmid in the Top10F′ bacteria. For unclear reasons, coselection with tetracycline dramatically improves plasmid quality compared with selection with ampicillin alone.

3. Pick colonies and amplify by liquid cultures in 2× YT with ampicillin (100 μg/ml) and tetracycline (12.5 μg/ml). The best possible aeration is crucial for efficient plasmid yield.

4. Extract plasmid DNA with Qiagen (Valencia, CA) Miniprep (for 2.5-ml cultures) or Maxiprep (for 250-ml cultures) kits. Typical plasmid yield is between 1 and 2 μg of DNA per milliliter of culture medium; as much as 5- to 10-fold less than for other pUC plasmids grown under identical conditions.

Transient, Helper Virus-Free Retroviral Production

The following protocols are adopted from others.[9] We currently use the 293 cell-derived Phoenix amphotropic and ecotropic cell lines[7] obtained through the American Type Culture Collection (ATCC, Manassas, VA) which express the Moloney virus structural genes (*gag, pol, pro,* and *env*). Briefly, viral mRNA is transcribed from the retroviral plasmid template within transfected producer cells. The viral mRNA is encapsulated and shed nonlytically into the cell culture medium, collected, and then added to target cells. Standard biosafety level 2 precautions are used in working with the producer cells and any material derived from them. This is relatively safe material to work with, but because the viruses are infectious, heightened awareness of potential dangers is necessary. Sterile, single-use aliquots of 25 mM chloroquine, 2 M CaCl$_2$, and 2× N,N-bis-(2-hydroxyethyl)-2-aminoethanesulfonic acid (BES) solution should be stored frozen at $-20°$ for consistency.

1. Producer cells received from the ATCC require expansion and storage as frozen stocks for subsequent use. Plate the contents of the original cell vial in a 100-mm dish in bicarbonate-buffered Dulbecco's modified Eagle's medium (DMEM) with 10% heat-inactivated fetal bovine serum (FBS) and 1× penicillin and streptomycin, at 37° in a 5% CO$_2$ incubator. After growing to ~80% confluence, remove by gentle trypsinization (0.05% trypsin–Versene) and distribute onto 15 new 100-mm plates. Repeating this expansion one or two additional times provides sufficient stocks for an extended period.

2. After the cells have reached ~70–80% confluence, freeze two aliquots per plate in 1 ml of heat-inactivated FBS plus 10% (v/v) dimethyl sulfoxide (DMSO). Slowly freezing the vials overnight at $-80°$ in a 2-propanol-buffered container before long-term storage in liquid nitrogen improves subsequent viability.

[9] H. Kotani, P. B. Newton III, S. Zhang, Y. L. Chiang, E. Otto, L. Weaver, R. M. Blaese, W. F. Anderson, and G. J. McGarrity, *Hum. Gene Ther.* **5,** 19 (1994).

3. To prepare cells for retroviral production, thaw a freezer vial and place its contents in 25 ml of growth medium on a 150-mm plate. From this point on, the use of heat-inactivated serum is no longer necessary. After reaching ~80% confluence, trypsinize and inoculate a 100-mm plate for each planned transfection, using 7×10^6 cells in 15 ml of medium and incubate for 16 hr before transfection. Rather than maintaining constant selection pressure on the retroviral structural genes, we typically subculture the contents of a frozen vial for five to eight passages before thawing earlier passages of frozen stock.

4. Just before transfection, replace growth medium with 8 ml of fresh medium and add chloroquine to a final concentration of 25 μM.

5. Bring 25 to 40 μg of retroviral plasmid DNA to a final volume of 875 μl in deionized H_2O, using capped, sterile 13×75 mm polypropylene tubes. Mix in 125 μl of 2 M CaCl$_2$, and then while vortexing at a medium setting rapidly add 1.0 ml of 2× BES solution [50 mM BES (Sigma, St. Louis, MO), 280 mM NaCl, 1.5 mM Na$_2$HPO$_4$; optimal pH ranges between pH 6.95 and 7.05]. Immediately transfer the slightly clouded 2-ml mixture to the producer cell plate by dropwise addition before returning the plate to a 37°, 5% CO$_2$ incubator. When performing multiple transfections, we find that working with one plate at a time is better. This, and vortexing, prevent large clumps from forming, and yield a finer precipitate and more efficient transfection.

6. After 6 to 8 hr, aspirate the transfection medium and add 25 ml of fresh growth medium to the plates.

7. Beginning 24 hr after initiating transfection, aspirate the growth medium and replace with 9 ml of fresh medium; then place the plates in a 5% CO$_2$ incubator set at 32°. Twenty-four hours after this (i.e., 48 hr after starting a producer cell transfection), collect the conditioned medium and pass it through a 0.45-μm pore size cellulose acetate syringe tip filter. Either freeze aliquots immediately by immersion in liquid nitrogen before storage at −80°, or use fresh as described below. Add 9 ml of fresh growth medium to the producer cells and incubate again at 32° for 12 hr before the next collection. We typically collect two to four supernatants at 12-hr intervals after the first harvest. Unused frozen supernatants are discarded 3–6 months later, after autoclaving.

Retroviral Infection

Incubate target cells with the conditioned producer cell supernatants in the presence of Polybrene by a plate centrifugation method as described below, which improves transduction efficiency markedly. Infection of the target cells can begin as early as 48 hr after starting producer cell transfections.

1. Twenty-four hours before beginning infection, seed wells in a 6×35 mm multiwell plate with sufficient target cells such that they are no more than 40%

confluent the following day when infection begins. For our vascular smooth muscle cells, we typically inoculate each well with ~150,000 cells.

2. For infection, either thaw a frozen retroviral supernatant or filter a freshly collected aliquot. Exchange the target cell growth medium by adding 2 ml of retroviral supernatant to each well. Mix in 20 μl of a 100× Polybrene stock solution per well [100× stock: hexadimethrine bromide (0.8 mg/ml; Sigma) dissolved in phosphate-buffered saline (PBS), filter sterilized, and stored at −20° in aliquots].

3. Incubate the plates at 32° for 15 min. During this time, equilibrate a refrigerated bench top centrifuge to 32° by running the rotor with buckets at 2500 rpm to generate frictional heat, setting the temperature dial to 40°. We use a Beckman (Fullerton, CA) model GS-6R centrifuge equipped with a GH 3.8 swing bucket rotor.

4. Place the cell culture plates in suitable microplate carrier buckets and centrifuge at 2500 rpm for 30 min at 32°. Most cells seem to tolerate this force, but are damaged if less then 2 ml of medium per well is used.

5. After the spin, aspirate the retroviral medium and replace with fresh growth medium. Incubate the cells at 37° until the next infection cycle.

6. Infections can be repeated successively at 12-hr intervals, timed to fresh retroviral supernatant harvesting. Some cells may require only one round of infection, whereas others may require three or more rounds to achieve the highest possible infection efficiency.

7. Typically, the cells can be used directly for experiments if desired 24 to 48 hr after the last infection cycle. However, they likely are undergoing cell stress due to the foreign RNA sequence, which can complicate many parameters. To avoid this, allow several days time for chromatin integration, after which the stress should subside. More often, they are expanded by subculture, and treated appropriately for any selection strategies before performing experiments.

Retroviral Vectors

The following section describes features of several basic derivatives of the LNCX retroviral vector.[2] Each derivative confers individual advantages for different types of experiments. These multiple vector options take fuller advantage of the high infection efficiency and stable chromatin integration associated with retroviruses, while also surmounting most of the limitations outlined above. Many of these vectors have been used in several types of cells, including primary cultures of endothelial cells, astrocytes, or skeletal and smooth muscle cells, and also smooth muscle, PC12, lymphocyte, and glioma cell lines.

Long Terminal Repeat-Driven Expression

pCL1. pCL1 is used for straightforward constitutive expression driven by the LTR promoter and is most likely to yield quantitative gene transfer and expression efficiencies, or for gain-of-function experiments that do not depend on quantitative infection efficiency if not possible. The LNCX neomycin resistance and cytomegalovirus (CMV) promoter sequences have been exchanged in pCL1 for a short multicloning site. We find that stable chromatin integration outweighs any need for a neomycin resistance marker, which if selected for, still cannot correct the epigenetic suppression of internal CMV promoter activity. Thus, even with 100% infection efficiency by an LNCX vector, expression efficiency from the internal CMV promoter is only 30 to 50%. The noncohesive *Bst*XI restriction sites in pCL1 provide a universal cloning site. Inserts are derived in any way desired, blunt ended with the Klenow enzyme, and ligated with a molar excess of *Bst*XI adapters (formed using 5'-CTGGCGCG and 5'-CGCGCCAGCACA) before cloning into the vector. Inserting a marker in pCL1, such as Zeo : eGFP, which simultaneously fluoresces and confers resistance to bleomycin analogs (vector pTJ84), provides a positive control to establish and monitor efficiencies of both producer cell transfection and target cell infection in new systems (Fig. 1).

pTJ66. pTJ66 is designed to ensure transgene expression in all cells that may not infect quantitatively (Fig. 2). This vector produces a chimeric mRNA that simultaneously expresses an experimental protein and the Zeo : eGFP reporter/selection marker. The cDNA for the experimental protein is cloned into the noncohesive universal cloning *Sfi*I sites (after ligation with *Sfi*I adapters as described above formed by 5'-AGGCCTAG and 5'-CTAGGCCTACA) and is expressed by normal 5' cap-dependent translation. Zeo : eGFP is expressed by cap-independent translation from an internal ribosome entry site (IRES) positioned downstream of the experimental cDNA. Cells that express these mRNAs can be visualized, quantified by fluorescence-activated cell sorting (FACS), and selected for by using either bleomycin or even FACS if the target cell is bleomycin resistant.

Transcriptional Reporter Vectors

pKA9. pKA9 has a luciferase cDNA driven by a minimal human interleukin 2 (IL-2) promoter. Enhancer sequences cloned between the *Sal*I and *Bam*HI sites are tested for transcriptional function, for example, to measure how endogenous $G\alpha$-coupled receptors activate specific transcription factors. This minimal promoter is quiescent alone, but supports a variety of different *cis*-acting enhancer elements. For unclear reasons, enhancer induction is more robust if transcription is directed from the bottom strand toward the 5' LTR. Several functional derivatives of pKA9 have been created, including specific reporters for NF-κB, NFAT, CREB, AP1, the c-*fos* serum response element (SRE), STAT, and the γ-interferon activated sequence (GAS).[10]

[10] K. L. Abbott, A. M. Robida, M. E. Davis, G. K. Pavlath, J. M. Camden, J. T. Turner, and T. J. Murphy, *J. Mol. Cell. Cardiol.* **32**, 391 (2000).

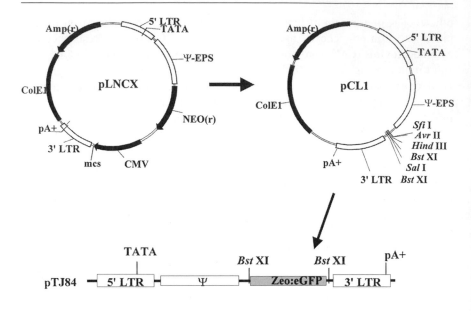

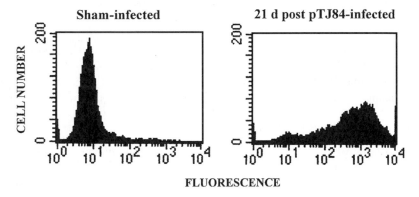

FIG. 1. Conversion of LNCX. Approximate locations of TATA boxes, packaging signals, restriction sites, and poly(A)⁺ signals are indicated. PTJ84 was cloned into pCL1 by the *Bst*XI adapter strategy. Vascular smooth muscle cells were sham infected, or infected with pTJ84. Three weeks after infection, the cells were analyzed by FACS for the Zeo : eGFP marker. The peak in sham cells represents autofluorescence whereas ~90% of the pTJ84 cells have a greater fluorescence due to the marker. The 3-week interval from selection to analysis is used to illustrates stability of expression.

pCAT/CUL. pCAT/CUL builds on the pKA9 concept by replacement of the neomycin resistance cassette with a chloramphenicol acetyltransferase (CAT) cDNA. This vector is designed for promoter deletion analysis. Measures of constitutive expression of CAT activity from the 5′ LTR quantitatively normalize luciferase activity, thereby controlling for gene dosage across a panel of different

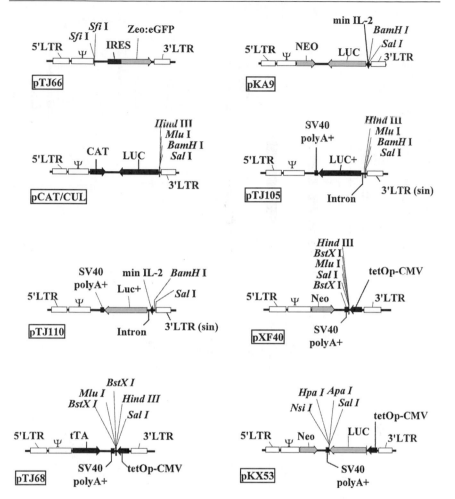

FIG. 2. Various retroviral vectors described in text. The invariant elements are open, whereas distinct functional elements particular to each vector are solid.

promoters. Promoter fragments are cloned upstream of luciferase using unique *Sal*I, *Bam*HI, *Mlu*I, and *Hind*III restriction sites. In a proof of principle test for pCAT/CUL, deletion analysis of the well-described human IL-8 promoter established the same obligate role for a proximal promoter region NF-κB-like enhancer element as identified with regular plasmids[11] (Fig. 3).

pTJ105. pTJ105 has some important differences and potential advantages over those described above. First, a mutation in the 3′ LTR TATA box renders it

[11] A. R. Brasier, M. Jamaluddin, A. Casola, W. Duan, Q. Shen, and R. P. Garofalo, *J. Biol. Chem.* **273,** 3551 (1998).

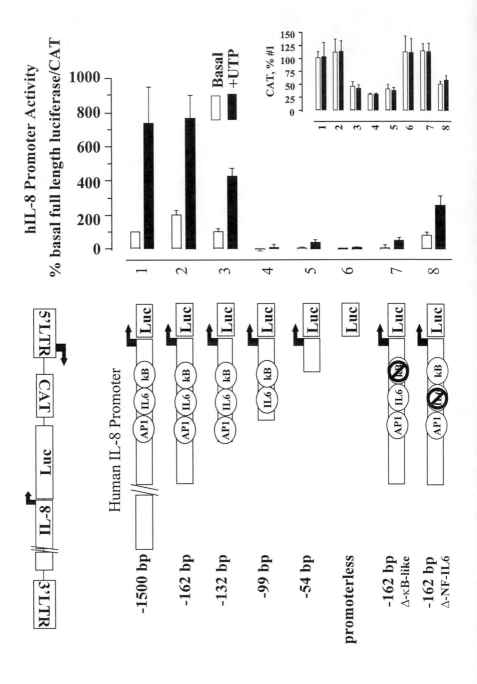

self-inactivating (SIN). Expression from chromatin-integrated viral DNA is actually driven by these 3′ LTR sequences, which replace the 5′ LTR sequences through recombination. Certain mutations in the 3′ LTR can thus inactivate integrated LTR promoter activity without diminishing the capacity for efficient retroviral production. In concept, internal promoters escape LTR-mediated epigenetic suppression in SIN vectors. pTJ105 has *Sal*I, *Bam*HI, *Mlu*I, and *Hind*III sites for insertion of promoters to drive luciferase expression. Because the LTR promoter is functionally deficient, CAT-based normalization as in pCAT/CUL is unavailable, and so other ways for gene dosage correction are necessary if used for deletion analysis. Other modifications in pTJ105 include the use of codon-optimized luciferase cDNA derived from pGL-3 (Promega, Madison, WI), a small intron from the simian virus 40 (SV40) genome placed in the 5′ UTR, and SV40-derived poly(A)$^+$ signals inserted downstream of luciferase.

pTJ110. pTJ110 is identical to pTJ105 except that it has a minimal human IL-2 promoter sequence with upstream *Sal*I and *Bam*HI restriction sites to accept enhancer elements. Limited experience so far with pTJ110 and pTJ105 is promising; absolute values for basal and induced luciferase levels from these are as much as ∼100 times greater than from the earlier gereration, yet fold responsiveness to agonist stimulation is retained. These will be useful for conducting luciferase activity measurements that previously had been less sensitive than possible, owing to equipment or assay conditions.

Tetracycline-Regulated Expression

pXF40. pXF40 allows conditional expression from the minimal CMV promoter/*tet* operon (tetOp-CMV) enhancer sequences first described by Gossen and Bujard.[12] This vector functions in cells that expresses the tetracycline *trans*-activator (tTA) protein ("tet-off"), which is delivered by another retroviral vector. This vector has been used to drive a host of inserts including luciferase, eGFP, Zeo : eGFP, several eGFP fusion proteins, and full-length mRNA constructs for posttranscriptional gene regulation studies. Similar to the LNCX platform, the tetOp-regulated promoter is subject to epigenetic suppression. Fusing *trans*-dominant protein sequences to the carboxyl terminus of a Zeo : eGFP module, or to either end of eGFP, is an effective strategy, to enrich for operating cells carrying such moieties. As a cautionary note, we generally find that the maximal expression

---

FIG. 3. A promoter deletion analysis result, using pCAT/CUL. Studies were performed in rat vascular smooth muscle cells activated for 5 hr with UTP to stimulate P2Y nucleotide receptors. Luciferase activity from various segments of the human IL-8 promoter were examined within sublines prepared independently, and normalized by parallel measures of CAT activity in each sample. Note how basal and inducible transcriptional function is lost by deletions or mutations within the proximal enhancersome. *Inset:* CAT activity is unchanged by stimulus, but varying across cell lines. Each column represents the mean of four experiments performed in duplicate. A. Brasier (Galveston, TX) kindly provided plasmids containing the various promoter inserts.

amplitude is ~30-fold over basal levels. The tetOp-mCMV promoter in our experience with smooth muscle cells is simply weaker than the LTR or a full-length CMV promoter. Based on estimates of cellular eGFP fusion protein concentrations, *trans*-dominant proteins most likely to work with this vector are those that have biological activities in concentrations ranges of ~100 nM or lower. Last, if used to induce protein expression, effective clearance of maintenance anhydrotetracycline requires doses no more than 10-fold higher than are maximally suppressive.

pTJ68. pTJ68 represents an all-in-one tetracycline-regulated expression system. Expression of the tTA is driven by the LTR promoter, and a tetOp-driven promoter is on the bottom strand of the same vector. Packaging the regulatory system into a single vector simplifies several types of experiments, and the expression dynamics differ little from the dual-vector pXF40-based system described above. One current strategy to create tTA-enriched cell cultures is by expressing Zeo : eGFP from pTJ68. Selecting for the positive cells that escape epigenetic suppression ensures tTA activity in all cells indirectly. These can subsequently be infected with retroviruses driving other tetOp-regulated sequences.

Posttranscriptional Reporter Vector

pKX53. pKX53 has a tetOp-driven luciferase cDNA that terminates in downstream SV40-derived poly(A)$^+$ signals. Between these latter two elements is a multicloning site for insertion of heterologous mRNA sequences. Thus, chimeric mRNAs are expressed to a steady state and by rapidly suppressing transcription with anhydrotetracycline, the chimeric mRNA decay can be measured. This avoids the toxicity associated with actinomycin D mRNA chase experiments, providing more carefully controlled conditions to examine the regulatory characteristics of *cis*-acting heterologous mRNA sequences. As for pXF40 described above, the cells must first be prepared to express the tTA, using a second vector.

Conclusions

Retroviral vector-mediated expression involves simple protocols. Any limitations imposed by their biology are outweighed by problems with less efficient methods of transgene delivery, and/or are solvable through optimized vector design. Most of our vector development has been along the lines of creating platforms for studying gene expression regulation, to measure both transcriptional and posttranscriptional processes.[10,13–17] A major advantage these confer is the

[12] M. Gossen and H. Bujard, *Proc. Natl. Acad. Sci. U.S.A.* **89,** 5547 (1992).

[13] V. Boss, X. Wang, L. F. Koppelman, K. Xu, and T. J. Murphy, *Mol. Pharmacol.* **54,** 264 (1998).

[14] K. L. Abbott, B. B. Friday, D. Thaloor, T. J. Murphy, and G. K. Pavlath, *Mol. Biol. Cell* **9,** 2905 (1998).

ability to measure larger numbers of parameters than possible through repeated plasmid transfection paradigms. For example, multidata point dose responses for several agonists or antagonists are readily acquired and replicated through repeated passages of a cell subline generated through a single infection protocol. This advantage of the stable chromatin integration feature limits the experimental variation associated otherwise with repeating transfections. Thus, the researcher more rapidly approaches confidence in a larger data set than is otherwise possible. Other retroviral vectors described here are as useful to express recombinant proteins, but which vector is best depends on experimental objectives. The researcher should not expect extremely high expression levels from these vectors. Rather, efficiency of expression is their principal advantage. However, both constitutive and tet-regulated vectors are suitable for expressing *trans*-dominant recombinant molecules so long as they are capable of function within cells in the submicromolar concentration range.[18] Whether a particular protein can reach such levels will depend on its intrinsic stability characteristics. As always, solutions likely exist for instances when even this might prove problematic.

[15] V. Boss, K. L. Abbott, X. Wang, G. K. Pavlath, and T. J. Murphy, *J. Biol. Chem.* **273**, 19664 (1998).
[16] B. Adams, T. S. Obertone, X. Wang, and T. J. Murphy, *Mol. Pharmacol.* **55**, 1028 (1999).
[17] K. Xu and T. J. Murphy, *J. Biol. Chem.* **275**, 7604 (2000).
[18] X. Wang and T. J. Murphy, *Mol. Pharmacol.* **54**, 514 (1998).

[44] Overexpression of Tightly Regulated Proteins: Protein Phosphatase 2A Overexpression in NIH 3T3 Cells

By PRAHLAD T. RAM

Introduction

Normal physiological function is dependent on the systemic and cellular homeostatic properties of the organism. Regulation of protein levels plays an important role in maintaining cellular homeostasis. A number of proteins regulate their own levels, either negatively or positively, and these regulatory mechanisms can occur at both at the transcriptional and translational level. Tightly regulated protein levels present a problem to investigators who may want to overexpress the protein in cell lines in order to study the effect of overexpression of the protein on a particular physiological event or end point. An example of a tightly regulated protein is the catalytic subunit of protein phosphatase 2A (PP2A).[1] PP2A is a protein

Fig. 1. Scheme to overexpress tightly regulated proteins. This schematic representation shows the strategy used to transfect NIH 3T3 cells for a short time and then sort for cells that are expressing a marker protein (EGFP). The cells are cotransfected with the protein of choice and EGFP for 8 hr, following which the cells are sorted for EGFP expression. This method allows selection of cells that overexpress PP2A. This method yields about 0.1% of cells that are initially transfected.

phosphatase that dephosphorylates serine and threonine residues on a number of enzymes and molecules and regulates the activities of multiple signal transduction pathways.[2] PP2A activity has been shown to regulate several cellular functions such as DNA replication, RNA splicing, transcription and translation, signal transduction, and also several physiological processes such as cell cycle progression, development, and transformation.[2] Efforts to overexpress PP2A have not been successful thus far because PP2A tightly regulates its own levels at the translational level.[1]

Several avenues of research depend on overexpression of proteins to study a particular aspect of protein function. In these instances stable or transient expression of the proteins in tissue culture cells allows for overexpression of the protein of interest. Although this method is feasible for a great many proteins there are several that cannot be overexpressed by simple transient or stable transfection. Another useful method is the use of inducible expression vectors that can be used to express proteins that may have a deleterious effect on the cells, but even this method may not work for proteins whose levels are tightly regulated. The method presented in this article utilizes transient transfection for a short time period followed by sorting and selection of the transfected cells and in this way collects a population of cells that do overexpress the protein in question. This is schematically shown in Fig. 1. This method makes use of enhanced green fluorescent protein (EGFP) expression to select only transfected cells. By cotransfecting EGFP and the protein of choice we are able to overexpress a protein whose levels are tightly regulated, and in this way we are able to overexpress the protein phosphatase 2A catalytic subunit in NIH 3T3 cells.

Method

NIH 3T3 cells are cultured in Dulbecco's modified Eagle's medium (DMEM; GIBCO-BRL, (Gaithersburg, MD) supplemented with 5% (v/v) calf serum

[1] Z. Baharians and A. H. Schonthal, *J. Biol. Chem.* **273,** 19019 (1998).
[2] T. A. Millward, S. Zolnierowicz, and B. A. Hemmings, *Trends Biochem. Sci.* **24,** 186 (1999).

(HyClone, Logan, UT) in a humidified $37°$, 5% CO_2 incubator. Cells are plated in 10-cm dishes at 1×10^6 cells/plate. The following evening the cells are serum starved by changing the medium to DMEM containing 0.1% (w/v) cell culture-grade bovine serum albumin (BSA; Sigma, St. Louis, MO) and the cells are cultured overnight. The following day the cells are transfected with an equal amount of PP2A catalytic subunit (epiPP2A/pcDNA neo; kind gift of B. E. Wadzinski, Vanderbilt University Medical Center, Nashville, TN)[3] and EGFP (Clontech, Palo Alto, CA) cDNAs, using LipofectAMINE 2000 (LF2000; GIBCO-BRL). For the transfection 15 μg of EGFP cDNA plus 15 μg of PP2A cDNA are added to 1 ml of Opti-MEM (GIBCO-BRL) and the tube is mixed gently. In another tube 42 μl of LF2000 is added to 1 ml of Opti-MEM, and the tube is mixed gently and allowed to sit at room temperature for 5 min. After this the LF2000–Opti-MEM mixture is added to the tube containing the cDNAs. The cDNA–LF2000 solution is gently mixed and incubated at room temperature for 15 min. The transfection mixture is then added to the cells and the cells are placed back in the incubator for 8 hr.

After the 8 hr incubation the medium is aspirated and the cells are rinsed with phosphate-buffered saline (PBS). The cells are then detached from the plate by incubating them in 5 ml of trypsin (GIBCO-BRL) at $37°$ in the incubator. The detached cells are collected in a 15-ml conical tube and centrifuged in a tabletop centrifuge for 5 min at 2500 rpm at room temperature. The supernatant is aspirated and the cells are resuspended in 1 ml of PBS. The resuspended cells are passed through a 35-μm pore size nylon mesh (Falcon; Becton Dickinson, Lincoln Park, NJ) and placed on ice. The cells are passed through a MoFlo cell sorter (Cytomation; Fort Collins, CO) and sorted for GFP expression. GFP-positive cells are collected in a 1.5-ml Eppendorf tube containing 20 μl of lysis buffer [50 mM Tris-HCl (pH 7.4), 1% (v/v) Nonidet P-40 (NP-40), 0.25% (w/v) sodium deoxycholate, 150 mM NaCl, 1 mM EGTA, 1 mM phenylmethylsulfonyl fluoride (PMSF), aprotinin (1 μg/ml), and leupeptin (1 μg/ml)]. The walls of the collecting Eppendorf tube are coated with the lysis buffer and an effort is made to ensure that the collecting stream of cells is directed to the bottom of the Eppendorf tube and not allowed to hit the sides of the tube, as this may cause drying of the cells before they reach the lysis buffer. After the cells are collected the tube is spun down to ensure that all the cells are collected into the lysis buffer. The walls of the tube are also washed down with the lysis buffer and the tube is spun down again. An appropriate amount of 6× SDS sample buffer [62 mM Tris-HCl (pH 6.8), 2% (w/v) sodium dodecyl sulfate (SDS), 10% (v/v) glycerol, 50 mM dithiothreitol (DTT), 0.1% (w/v) bromphenol blue] is added to the tube, boiled for 5 min, and resolved on a 10% (w/v) SDS–polyacrylamide gel. Using this method levels of PP2A are increased 2- to 4-fold in NIH 3T3 cells (Fig. 2).

---

[3] B. E. Wadzinski, B. J. Eisfelder, L. F. Peruski, Jr., M. C. Mumby, and G. L. Johnson, *J. Biol. Chem.* **267**, 16883 (1992).

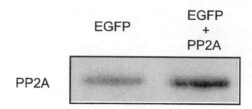

FIG. 2. Overexpression of PP2A in NIH 3T3 cells. The immunoblot using a PP2A catalytical subunit-specific antibody (Upstate Biotechnology) shows overexpression in NIH 3T3 cells that were cotransfected with EGFP and PP2A cDNAs and sorted. The cells show a 2- to 4-fold increase in PP2A levels as compared with cells that were transfected only with the EGFP vector. A total of 7000 cells were lysed in lysis buffer (see text) and resolved by 10% (w/v) SDS–PAGE. The proteins were transferred to Hybond-C and blocked in 5% (w/v) nonfat dry milk in PBS for 30 min. The blot was incubated with 1 μg of anti-PP2A antibody (clone 1D6; Upstate Biotechnology) overnight. The blot was washed twice with water and incubated in goat anti-mouse HRP-conjugated antibody (1 : 5000; Roche, Nutley, NJ) for 1 hr. The blot was washed twice with water and once with PBS–0.05% (v/v) Tween 20 for 5 min. The blot was rinsed with water and visualized enhanced chemiluminescence (ECL) kit (Amersham).

Experimental Concerns

The optimum expression of GFP is 36–48 hr after transfection, and in the protocol described here the cells are sorted for GFP expression after only 8 hr. Because of this exceedingly short time frame the expression levels of GFP and the percentage of cells that can be collected are low. In a typical experiment 3×10^6 cells are collected from the 10-cm dish after trypsinization. After the 5-min spin and resuspension about 2.3×10^6 cells are left. After the filtering step 2.1×10^6 cells are collected and used for sorting. Sorting for GFP expression cells 8 hr after transfection typically yields between 3000 and 4000 cells per 10-cm plate that is transfected. The transfections are performed in duplicate each time, and so typically the sorting yields about 7000 to 8000 cells. Protein from 7000 cells is sufficient for immunoblot analysis when the lysate is run on a minigel (Hoefer, San Francisco, CA). For our purpose, which is to examine the effects of PP2Ac overexpression on mitogen-activated protein kinase 1/2 (MAPK1,2) activity, the cell lysate is loaded on a 10% SDS–polyacrylamide gel and transferred onto Hybond-C membrane (Amersham, Arlington Heights, IL). The blot is probed with a phospho-MAPK1,2-specific antibody (New England BioLabs, Beverly, MA) and subsequently stripped and reprobed for MAPK1,2 (New England BioLabs) and restripped and reprobed for PP2Ac (Upstate Biotechnology, Lake Placid, NY). For this experimental protocol the method described here yields a sufficient number of cells to perform the assay.

Conclusion

The method described in this article allows for overexpression of a tightly regulated protein in tissue culture cells. Although this method does work there are

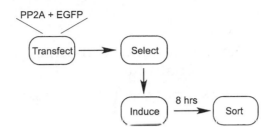

FIG. 3. Alternative scheme to enhance overexpression of tightly regulated proteins. This alternative approach is a modification of the method that was used and could potentially increase the yield of cells that overexpress the protein of choice and at the same time reduce the cost of the experiment. Although parts of this have not been experimentally tested, the basic principle of a short expression time frame has been shown to work here.

several limitations that need to be noted. The main drawback is the low output of overexpressing cells compared with the amount of cells that was used at the start of the experimental protocol.

An additional drawback is the cost of lipid transfection reagents that are used for the protocol. Although this protocol is successful it does have a relatively high cost-to-output ratio, and the low output of cells may not be suitable for all end-point assays. Although we have demonstrated that as few as 7000 cells is sufficient for immunoblot analysis, this amount of cells may not be sufficient for other assays. While keeping the same basis of the assay we can make some additional improvement that may yield larger quantities of cells that overexpress tightly regulated proteins. This can be achieved by transfecting cells with a bicistronic inducible expression vector that expresses EGFP and the protein of choice on the same vector. In this approach the cells will be stably transfected with the bicistronic inducible vector and then the expression of the protein will be induced and the cells sorted after 6 to 8 hr. In this way, by stable expression we will have an enriched pool of transfected cells to start off with and the induction will allow for almost all cells to express the protein, whereas in the method presented here only about 30% of the cells are transfected, out of which only 3% of the cells express sufficient levels of GFP to be able to be selected by the sorter. Therefore, if we were to start off with a pool of cells, all of which were transfected, then by the inducible selection method we would theoretically end up with a much larger number of cells. This is schematically shown in Fig. 3. This approach would be much more cost-effective because the transfection could also be done by less expensive means and with far fewer cells. With these additional changes short-term expression and sorting appears to be a feasible method to overexpress proteins whose levels might otherwise be tightly controlled.

[45] Monitoring G-Protein-Coupled Receptor Signaling with DNA Microarrays and Real-Time Polymerase Chain Reaction

By Tony Yuen, Wen Zhang, Barbara J. Ebersole, and Stuart C. Sealfon

Introduction

The physiological effects achieved through activation of G-protein-coupled receptors (GPCRs) by agonists may be highly sensitive both to the precise pattern of receptor stimulation and to specific properties of the agonist. In the gonadotropin-releasing hormone (GnRH) receptor system, different concentrations or patterns of release of GnRH elicit distinct hormone secretion and biosynthetic responses.[1,2] Many receptors activate multiple signaling pathways and the relative intrinsic activities of specific agonists may depend on which pathway is assayed. For several GPCRs, the relative activation of distinct signal transduction pathways by various agonists acting at the same receptor can differ, a phenomenon referred to as agonist-dependent "stimulus trafficking."[3] For example, differential stimulation of the inositol phosphate and arachidonic acid second-messenger system has been reported for a series of partial agonists at the human 5-hydroxytryptamine (serotonin) 2A and 2C (5-HT$_{2A}$ and 5-HT$_{2C}$) receptors.[4]

Ideally, an assay of cellular responses after receptor activation should reflect the changes that occur in the various cellular signaling pathways. Biochemical signal transduction assays do not easily lend themselves to the simultaneous assay of multiple signal transduction pathways. To sample the overall signaling response, we quantify the gene response pattern by microarray and quantitative real-time polymerase chain reaction (PCR) techniques.

This approach can be understood by analogy with gene reporter constructs. A variety of gene reporters have been used to measure GPCR responses. In many of these assays, a promoter sequence that is activated by signal transduction drives the synthesis of a reporter molecule. For example, we have used an NF-κB–luciferase reporter gene to monitor GPCR activation.[5] These constructs can be used to generate satisfactory concentration–response curves. While

[1] L. Wildt, A. Hausler, G. Marshall, J. S. Hutchison, T. M. Plant, P. E. Belchetz, and E. Knobil, *Endocrinology* **109**, 376 (1981).

[2] K. Ishizaka, S. Kitahara, H. Oshima, P. Troen, B. Attardi, and S. J. Winters, *Endocrinology* **130**, 1467 (1992).

[3] T. Kenakin, *Trends Pharmacol. Sci.* **16**, 232 (1995).

[4] K. A. Berg, S. Maayani, J. Goldfarb, C. Scaramellini, P. Leff, and W. P. Clarke, *Mol. Pharmacol.* **54**, 94 (1998).

individual reporter constructs may respond to more than one signaling pathway, constructs with distinct response elements, such as those for cAMP or NF-κB, can differentiate among signal transduction pathways. Therefore, the measurement of multiple reporter constructs would allow sampling of the activation of different signaling pathways. A specific gene reporter does not monitor a single G protein or signal transduction pathway. A panel of reporters, however, could provide a measurement that reflects the pattern of second-messenger activation in the various signaling pathways in the cell.

Although gene reporter constructs can be transfected into cells in culture and even introduced *in vivo* by transgenic approaches,[6] it is not practical to monitor a large group of reporter constructs. A related approach, which we have developed, is to use the endogenous promoters of the cells as the equivalent of signal transduction reporter constructs. Rather than determine the change in a reporter gene, such as luciferase, the activity readout becomes the level of the specific mRNAs driven by these promoters. We call these markers intrinsic reporters of cell signaling (IRC). The IRCs we monitor are genes (such as *Arc, fos,* and *egr-1*) that are modulated in response to perturbations of intracellular signaling. Selection of a large number of IRCs can provide a sampling that reflects the relative activation of different signal transduction pathways.

The evolution of gene profiling techniques makes it feasible to monitor IRC levels after receptor activation. The techniques for monitoring global patterns of gene expression have advanced rapidly. Microarrays allow comparison of the level of expression of thousands of mRNAs among different RNA samples. Several approaches to microarrays have been developed.[7,8] Microarrays rely on the detection of hybridization of a labeled cRNA or cDNA to a complementary sequence that is deposited at a specific location on a solid support. In general, microarray techniques have the advantage of measuring the comparative expression of a large number of genes and the disadvantages of lack of absolute quantitation of expression. In addition, the relatively high cost per RNA sample analyzed introduces a practical limit on the number of samples that can be studied. Real-time PCR allows more precise measurement of a limited number of genes in a large number of samples. The development of the network signal transduction assay we describe, based on profiling the expression of multiple IRCs, requires an integration of microarray screening, bioinformatics approaches, and detailed concentration-dependence studies using real-time PCR.

[5] C. Klein, N. Gurvich, M. Sena-Esteves, S. Bressman, M. F. Brin, B. J. Ebersole, S. Fink, L. Forsgren, J. Friedman, D. Grimes, G. Holmgren, M. Kyllerman, A. E. Lang, D. de Leon, J. Leung, C. Prioleau, D. Raymond, G. Sanner, R. Saunders-Pullman, *Ann. Neurol.* **47,** 369 (2000).

[6] S. Impey, M. Mark, E. C. Villacres, S. Poser, C. Chavkin, and D. R. Storm, *Neuron* **16,** 973 (1996).

[7] V. G. Cheung, M. Morley, F. Aguilar, A. Massimi, R. Kucherlapati, and G. Childs, *Nat. Genet.* **21,** 15 (1999).

[8] R. J. Lipshutz, S. P. Fodor, T. R. Gingeras, and D. J. Lockhart, *Nat. Genet.* **21,** 20 (1999).

TABLE I
OVERVIEW OF INTRINSIC REPORTER OF CELL
SIGNALING ASSAY

A. Selection of candidate IRCs
  1. RNA preparation for microarray screening
  2. Sample validation
  3. Microarray labeling and hybridization
  4. Data validation
  5. Bioinformatics and cluster analysis
B. Determination of response of IRCs
  1. Oligonucleotide design for real-time PCR assays
  2. RNA extraction and reverse transcription for PCR
  3. Real-time PCR

Overview of Experimental Design

The overall approach to the study of GPCR signaling is summarized in Table I. We have used cell lines that either endogenously express or are stably transfected with the receptor to be studied.

The panel of IRCs is initially identified by microarray screening. Microarray experiments must be carefully designed if meaningful data are to be obtained and unnecessary expense avoided. As discussed below, there are several different experimental platforms that can be used for microarray studies. Microarray experiments require access to specialized instrumentation for data acquisition. Because of the large number of RNA measurements made in these experiments, planning at the early stages will save considerable effort in following up false-positive results later. Human and mouse commercial oligonucleotide arrays of the current generation contain approximately 12,500 genes on each chip. The sheer number of genes assayed leads to a sizeable number of genes in any given sample (hundreds to thousands, depending on the cutoff values utilized) that show variation due to biological noise and measurement variability. The sample-to-sample variations in expression levels that are not due to the treatment studied should be randomly distributed and decrease dramatically with the analysis of multiple samples from each treatment group. The number of samples required for analysis depends on the sample variation in the system under study and the size of the gene changes that occur. In our cell line experiments, we find that high-quality array data sets (see below) obtained from three independent samples in treatment and control groups are sufficient when screening 25,000 mRNAs. Microarrays can be used to study responses in more complex tissues, but a larger number of samples in each treatment group would be required in order to obtain an adequate signal-to-noise ratio. Although analysis of multiple samples from each treatment group may seem profligate, it is in fact more economical than a more limited study because it simplifies the data analysis and eliminates extremely time-consuming analysis of false-positive responses. The importance of assaying an adequate number of independent samples from each treatment group cannot be overemphasized. As described below, bioinformatics approaches, such as clustering algorithms, can assist in the analysis of the data.

After identification of candidate genes by microarray screening, the responses are confirmed by real-time PCR. The PCR measurements allow detailed and precise study of the responses of a selected population of genes. Whereas the microarray screening is useful for testing the responses of large numbers of genes, the PCR assays allow more precise measurements and are much better adapted to the analysis of a large number of samples.

RNA Preparation for Microarray Screening

Starting material sufficient to yield at least 100 μg of total RNA will allow at least one labeled RNA sample for array hybridization. The RNA yield from specific cell lines varies greatly and should be determined from pilot experiments. For example, the HEK293 cells used to generate the regression data in Fig. 1 yield approximately 1 mg of total RNA from one 15-cm plate grown to 90% confluence (30 million cells). In contrast, the mouse gonadotrope cell line, LβT2 yields less than 200 μg of RNA from 30 million cells.

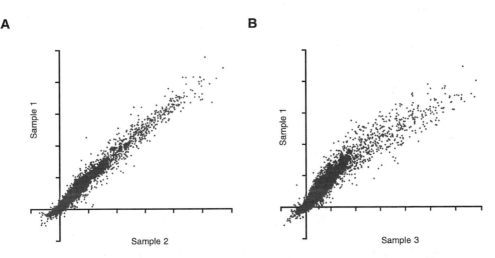

FIG. 1. Regression analysis of microarray data. All gene cluster measurements are compared in two samples. The dispersion of the plot reflects the quality of particular samples and assay. (A) Representative "good" data samples. The plot is linear and scatter is limited ($r = 0.98$) (B) Poor regression. The plot is less linear and scatter is greater ($r = 0.95$). Note that the data for sample 1 are included in both regression plots, indicating that the excessive scatter results from the data from sample 3. Data were generated with RNA samples from HEK293 cells and Affymetrix U95A oligonucleotide arrays.

Cells are treated with agonist and harvested at selected time points for RNA extraction. Total RNA is isolated by the guanidinium thiocyanate method described by Chomczynski and Sacchi.[9] All plasticware and solutions should be RNase free.[10] The protocol is as follows.

1. Remove the culture medium and add 10 ml of lysis buffer per 15-cm plate. Lysis buffer consists of 4 M guanidinium thiocyanate, 25 mM sodium citrate (pH 7.0), 0.5% (w/v) sarcosyl (N-lauroylsarcosine), and 0.1 M 2-mercaptoethanol. The method for preparation of the lysis buffer is described in Chomczynski and Sacchi.[9]

2. Triturate the sample several times. It may be viscous initially. Briskly swirling the plates on a laboratory rotator helps to decrease viscosity.

3. Collect the lysate into a 50-ml centrifuge tube (Falcon; Becton Dickinson Labware, Lincoln Park, NJ).

4. Add 1 ml of 2 M sodium acetate (pH 4.0). Mix thoroughly by inversion.

5. Add 10 ml of water-saturated phenol. Mix thoroughly by inversion.

6. Add 2 ml of chloroform–isoamyl alcohol (49 : 1, v/v). Shake vigorously for 15 sec.

7. Centrifuge at 2000g for 30 min at 4° [e.g., Sorvall (Newtown, CT) RT-6000B benchtop centrifuge fitted with an H1000B rotor].

8. Collect the upper aqueous phase in a new 50-ml centrifuge tube. Avoid taking the interphase or the lower phase.

9. Add 10 ml of 2-propanol and mix thoroughly by inversion. Incubate at −20° for at least 1 hr to precipitate the RNA.

10. Centrifuge at 2000g for 30 min at 4°.

11. Remove the supernatant.

12. Resuspend the pellet in 0.75 ml of lysis buffer. Transfer to a 1.5-ml microcentrifuge tube.

13. Add 0.75 ml of 2-propanol and mix thoroughly by inversion. Incubate at −20° for at least 1 hr to precipitate the RNA.

14. Centrifuge at 16,000g for 10 min at 4° in a microcentrifuge (Eppendorf, Hamburg, Germany).

15. Remove the supernatant.

16. Wash with 1 ml of 70% (v/v) ethanol. Centrifuge at 16,000g for 5 min at 4°.

17. Remove the supernatant and air dry the pellet for 15 min.

18. Resuspend the RNA in at least 100 μl of diethyl pyrocarbonate (DEPC)-treated water. If a large quantity of RNA is prepared, the pellet can be difficult to dissolve. If needed, increase the volume of water, heat for 10 min at 65°, and vortex vigorously.

---

[9] P. Chomczynski and N. Sacchi, *Anal. Biochem.* **162,** 156 (1987).

[10] D. D. Blumberg, *Methods Enzymol.* **152,** 20 (1987).

19. The concentration of the RNA is determined by absorbance at 260 nm and is then adjusted to 0.5 $\mu g/\mu l$.

For isolating poly(A)$^+$ RNA from total RNA, we presently utilize a commercial magnetic separation kit (PolyATtract mRNA isolation system; Promega, Madison, WI) according to the manufacturer instructions. Although it is possible to label total RNA for microarray screening, we have used only poly(A)$^+$ RNA. Poly(A)$^+$ isolation takes 1–2 hr.

Sample Validation

Before performing the microarray labeling and hybridization, we determine the quality of the mRNA samples and, if possible, confirm the presence of consistent regulatory responses. The mRNA samples are tested by real-time PCR (see details below). The expression of several control genes and of genes that are known from the published literature or previous experience to be regulated by receptor activation is quantified. We find that *egr-1*, for example, is activated by stimulation of many receptors. These assays allow confirmation that the mRNA samples to be labeled are of good quality. These validation studies also confirm that the level of control genes, such as actin, shows an expected stability and that at least some regulated genes show a consistent increase of a reasonable magnitude. If the PCR analysis of these samples reveals significant variation among samples in the same experimental group, the treatment and RNA isolation are repeated.

Microarray Labeling and Hybridization

A variety of array formats, both commercial and custom printed, are available. On the basis of our experience, we now avoid commercial filter arrays that are hybridized with radioactively labeled probe. Custom-printed dual-fluorescent label cDNA microarrays are also suitable, if available. Many academic institutions, including ours, have established shared facilities to print and distribute high-density cDNA arrays. Custom-printed arrays, while requiring more expertise to utilize, are considerably less expensive.[7,11] In dual-labeled cDNA microarrays, the experimental samples are labeled with one fluorophore and are mixed with a pooled reference sample labeled with a second fluorophore. Thus the results obtained for each experimental sample are compared with the internal standard for each microarray. High-density oligonucleotide arrays are available from Affymetrix (Santa Clara, CA). See Lipshutz *et al.*[8] for a description of the design and manufacture of

[11] V. R. Iyer, M. B. Eisen, D. T. Ross, G. Schuler, T. Moore, J. C. F. Lee, J. M. Trent, L. M. Staudt, J. Hudson, Jr., M. S. Boguski, D. Lashkari, D. Shalon, D. Botstein, and P. O. Brown, *Science* **283**, 83 (1999).

commercial high-density oligonucleotide arrays. In this system, each array is hybridized with biotinylated cRNA probe that is subsequently stained with a secondary fluorescent antibody. Hybridization and data acquisition require access to hybridization and data acquisition instruments dedicated to the Affymetrix arrays. The regression data generated in Fig. 1 were obtained with this system. The probe is generated according to the instructions of the supplier and its suitability for high-density arrays is assessed with a manufacturer-supplied test array. The test array hybridization is evaluated to judge probe length (ratio of 3′ to 5′ probes), the presence of spiked bacterial controls, and the level of noise and background. The samples showing satisfactory test hybridization are then hybridized with the high-density arrays.

Data Validation

There are a variety of academic and commercial software programs available for the acquisition and analysis of microarray data.[12] We perform an initial analysis of fluorescence intensity oligonucleotide hybridization data with Microarray Suite 4.0 (Affymetrix). After initial data extraction, we export a value reflecting the average expression (average perfect match/mismatch difference, as explained in software documentation) for each gene to an Excel spreadsheet.

We first estimate the quality of the data in each sample by examining the regression pattern obtained by comparing the data set of samples with that of other samples. In a comparison of any two samples after receptor stimulation, the majority of genes are expected to be unaltered. Therefore an ideal regression plot comparing the level of expression of all genes in two sample should approach a straight line through the origin with a uniform slope of unity. The deviation from a straight line predominantly represents sample variation and measurement error. Two types of patterns are shown in Fig. 1. In Fig. 1A, the linear regression is straight and the scatter relatively small ($r = 0.98$). In Fig. 1B, the curve is slightly S-shaped and the scatter is greater ($r = 0.95$). As sample 3 is unique to Fig. 1B, this pattern suggests that the data obtained from sample 3 is noisier. A poor regression pattern obtained with a given sample appears the same when that sample is compared with other samples from both experimental and control groups. Only sample data sets showing acceptable regression patterns are used for subsequent analysis. The initial analysis of satisfactory data sets can be performed by determining the mean and standard error of the changes in expression and evaluation of the levels of the previously documented control and regulated genes (see above). Data sets containing multiple time points or conditions are analyzed by clustering algorithms (see below). All regulated genes of interest identified by microarray are confirmed and studied in full concentration–response experiments, using real-time PCR.

[12] D. E. Bassett, Jr., M. B. Eisen, and M. S. Boguski, *Nat. Genet.* **21,** 51 (1999).

Bioinformatics and Cluster Analysis

Multiple time point microarray studies[11] and multigene PCR experiments[13] can generate extremely large data sets. In studying responses to GPCR stimulation, the treatments can vary in terms of concentration of agonist and duration of exposure before harvesting. One experiment performed in triplicate at six time points and a single agonist concentration on high-density arrays can generate several hundred thousand gene measurements. Gene clustering techniques facilitate the identification of groups of genes that show similar patterns of expression in the different treatment groups. Appropriate application of the various clustering approaches may benefit from expert bioinformatics assistance.

Although the specific algorithms utilized can be complex, the objective of this type of analysis is straightforward. In an experiment studying time course, clustering techniques can be utilized to determine which genes show similar changes across all samples. The analysis is performed by representing the expression data in a conceptual space, where each axis represents a different experimental group. In an experiment with only three time points, all the expression data could be represented in three-dimensional space, where each axis corresponds to a different time point and the distance along each axis the relative expression of each gene. The points (vectors) that are located near each other represent genes that show a similar pattern of change in these samples. More treatment groups require more dimensions that, while difficult to visualize, can be represented in a matrix and subjected to a similar mathematical treatment.

There are two general approaches to clustering: supervised and unsupervised. In supervised clustering, the researcher selects the type of expression pattern sought. In unsupervised clustering, the data are grouped without preconceptions about their pattern. In a time course experiment, one might seek early, middle, and late expressed gene populations by a supervised clustering approach. However, groups of genes with unusual patterns, such as genes that are induced early, down-regulate at intermediate time, and elevate again at late time points would more likely be identified by unsupervised clustering approaches.

The commonly used unsupervised methods are hierarchical clustering[14] and self-organizing map clustering.[15] In the hierarchical clustering method, the similarity of expression patterns of any two genes or elements (groups of genes) across multiple microarray experiments is calculated in the form of a correlation coefficient. Branches are used to link two genes or elements to assemble all the genes into a phylogenetic tree. The shorter the branch, the more similar the expression pattern of the two linked elements is.[16]

[13] X. Wen, S. Fuhrman, G. S. Michaels, D. B. Carr, S. Smith, J. L. Barker, and R. Somogyi, *Proc. Natl. Acad. Sci. U.S.A.* **95,** 334 (1998).

[14] R. R. Sokal and C. D. Michener, *Univ. Kans. Sci. Bull.* **38,** 1409 (1958).

[15] T. Kohoen, "Self-Organizing Maps." Springer-Verlag, Berlin, 1997.

Self-organizing maps, another unsupervised clustering method, was originally developed by Kohonen for artificial neural networks. When this method is applied to microarray data analysis, each gene is considered as a vector in N-dimensional space, where N is the number of experimental conditions. For each vector, the values of the N coordinates are the gene expression profile across all experiments.[17] In this method, the number of clusters (nodes) and the geometry of the nodes (e.g., 3×4 grid) are predetermined by the user. These nodes are then mapped into the N-dimensional space with randomly initialized positions. There are no relationships among these nodes, and thus the map is unorganized. For each gene vector, its distances to the nodes in the N-dimensional space are calculated. The information is used to move the closest node and its neighboring nodes in the original grid in the N-dimensional space. This procedure, called a learning process, is iterated many times. Eventually, each node with values of N coordinates represents the expression pattern of one cluster of genes and each gene is assigned to a cluster to which it has the shortest distance.

To identify a cluster of genes with a particular expression pattern, supervised clustering methods are preferred.[18,19] A "model" expression pattern is first selected. A similarity score is then assigned to each gene in the microarray experiment according to how similar its expression profile is, compared with the "model" gene in the form of correlation coefficient. A cluster of genes most similar to the model gene is selected. In a multiple time point experiment, a model early gene expression profile can be used to identify other early genes induced by an agonist.

Several free clustering software programs can be downloaded from the web (Cluster and TreeView by M. Eisen and Xcluster by G. Sherlock: http://rana. Stanford.EDU/software/, GeneCluster by P. Tamayo: http://www.genome.wi.mit. edu/MPR/software.html). Self-organizing map analysis can be performed via a user-friendly web server (EZSOM by W. Zhang: http://transport.physiobio.mssm. edu/services/ezsom). For data analysis by the hierarchical method, the Cluster program accepts tab-delimited text as input (usually gene expression data with genes in rows and array experiments in columns). It generates a text description of the clustering results and an input file for color-coded graphical representation of the hierarchy in TreeView. EZSOM accepts the same type of input file and, in addition, requires the user to select the number of clusters sought.

[16] M. B. Eisen, P. T. Spellman, P. O. Brown, and D. Botstein, *Proc. Natl. Acad. Sci. U.S.A.* **95,** 14863 (1998).

[17] P. Tamayo, D. Slonim, J. Mesirov, Q. Zhu, S. Kitareewan, E. Dmitrovsky, E. S. Lander, and T. R. Golub, *Proc. Natl. Acad. Sci. U.S.A.* **96,** 2907 (1999).

[18] T. R. Golub, D. K. Slonim, P. Tamayo, C. Huard, M. Gaasenbeek, J. P. Mesirov, H. Coller, M. L. Loh, J. R. Downing, M. A. Caligiuri, C. D. Bloomfield, and E. S. Lander, *Science* **286,** 531 (1999).

[19] S. Chu, J. DeRisi, M. Eisen, J. Mulholland, D. Botstein, P. O. Brown, and I. Herskowitz, *Science* **282,** 699 (1998).

Several commercial software packages support cluster analysis, including S-Plus (MathSoft, Cambridge, MA) and LifeArray (Incyte Genomics, Palo Alto, CA).

Oligonucleotide Design for Real-Time Polymerase Chain Reaction Assays

Real-time PCR provides rapid and precise confirmation and follow-up studies of gene changes identified by microarray studies.[20,21] There are several different approaches to real-time PCR. For studies of IRCs, we utilize SYBR Green (Molecular Probes, Eugene, OR) detection real-time PCR because multiple reactions can be set up rapidly and inexpensively with standard oligonucleotides. Real-time PCR relies on the fluorescent quantification of PCR product during each cycle of amplification. Specific detection systems, such as molecular beacons[21] and Taqman assays,[20] rely on the synthesis of a fluorescently labeled detection oligonucleotide. These specific assays have the advantage of specificity, but the disadvantage of added expense and a delay in obtaining the fluorescently labeled detection oligonucleotides. Assay of PCR product through the use of the fluorescent dye SYBR Green allows the reaction to be based on standard oligonucleotides. Because SYBR Green will detect any PCR product, including nonspecific products and primer–dimers, careful oligonucleotide design for the reaction is required.

Primers should be designed, if possible, within 1 kb of the polyadenylation site. Amplicons of 100–200 bp are ideal for real-time applications. It is advantageous to design the primers to have the same melting temperature so that PCR with different primer sets can be performed in the same run. We design primers that are 20-mers with 55% GC content and a single 3'-G or -C. Candidate primers are tested for specificity by BLAST (http://www.ncbi.nlm.nih.gov/BLAST/) and for folding and self-annealing by standard DNA analysis software. Primer pairs are first tested for specificity and absence of primer–dimer formation (low molecular weight products) by PCR followed by gel electrophoresis. Designing each primer pair takes about 1 hr.

RNA Extraction and Reverse Transcription for Polymerase Chain Reaction

PCR is reliable with relatively small quantities of the RNA target. Therefore only small numbers of cells are needed for sample generation. We typically utilize cells grown in multiwell plates (6, 12, or 24 wells) and isolate RNA with a commercial 96-well plate format RNA isolation column (StrataPrep 96; Stratagene, La Jolla, CA). Results from most cell lines are more reliable if lysed cells

[20] C. A. Heid, J. Stevens,, K. J. Livak, and P. M. Williams, *Genome Res.* **6,** 986 (1996).

[21] S. Tyagi and F. R. Kramer, *Nat. Biotechnol.* **14,** 303 (1996).

are first prefiltered through 0.45-μm pore size acetate filters (Spin-X; Costar, Cambridge, MA). Two hundred samples can easily be treated and processed within 1 day, using this approach. This allows the analysis of full concentration–response curves with triplicate samples.

We utilize 5 μg of total RNA in the first-strand cDNA synthesis. To a sterile, nuclease-free, 0.5-ml microcentrifuge tube, add the following:

Diethylpyrocarbonate (DEPC)-treated water	1 μl
Total RNA (0.5 μg/μl)	10 μl
Oligo(dT)$_{18}$ (0.5 μg/μl)	1 μl

Incubate the mixture at 65° for 5 min. Quick-chill on ice. Centrifuge at 16000g for 10 sec at 4° to collect the contents of the tube. Add the following.

First-strand buffer (5×): 250 mM Tris-HCl (pH 8.3 at room temperature), 375 mM KCl, 15 mM MgCl$_2$)	4 μl
Dithiothreitol (DTT), 0.1 M	2 μl
Superscript II (200 units/μl) (Life Technologies, Rockville, MD)	1 μl
dNTP mix (10 mM each dNTP)	1 μl

Mix gently and incubate at 42° for 50 min. The reaction is then stopped by heating at 70° for 15 min.

To prepare the samples for PCR, 2 μl of the first-strand cDNA is diluted with deionized water to 100 μl. Typically 5 μl of this dilution is used in each PCR. Assuming that the poly(A)$^+$ RNA population is 2% of the total RNA, approximately 500 pg of the first-strand cDNA is used as template in the real-time PCR.

Real-Time Polymerase Chain Reaction

Real-time PCR requires a specialized thermocycler with fluorescence detection. A variety of commercial instruments are available. We utilize the ABI Prism 7700 (Applied Biosystems, Foster City, CA), which allows assays to be performed in the 96-well plate format. Good PCR technique is required to avoid contamination of subsequent reactions.[22] This includes isolating PCR products and plasmids from RNA preparation and reaction setup. A dedicated bench for RNA isolation and PCR setup and dedicated pipettors should be maintained. Aerosol-resistant pipette tips should be used.

Commercial kits for SYBR Green-based PCRs are available from Applied Biosystems and perform reliably (SYBR Green PCR core reagents; SYBR Green PCR master mix). We assemble our own reaction components as follows:

---

[22] S. Kwok and R. Higuchi, *Nature (London)* **339,** 237 (1989).

	Volume per 20-μl reaction (μl)	Volume per 100 reactions (μl)
Deionized water	10.4	1040
PCR buffer (10×): 200 mM Tris-HCl (pH 8.4), 500 mM KCl	2	200
MgCl$_2$, 50 mM	1.2	120
SYBR Green I, 10×	0.1	10
dNTPs, 10 mM each	0.4	40
Primer mix, 5 μM each	0.8	80
Taq DNA polymerase, 5 U/μl	0.1	10

"Hot start" Taq polymerase should be used. Platinum Taq (Life Technologies) and AmpliTaq Gold (Applied Biosystems) both perform well. The 10× SYBR Green I is prepared by diluting 10 μl of the stock 10,000× concentrate (Molecular Probes) into 10 ml of Tris-HCl, pH 8.0, and is stored in 0.5-ml aliquots at $-20°$.

Fifteen microliters of the master mix is aliquoted into 0.2-ml MicroAmp optical tubes (Applied Biosystems). Alternatively, a 96-well optical reaction plate (Applied Biosystems) can be used. Five microliters of the first-strand cDNA is then added to the tube and the solution is mixed by repeat pipetting. This achieves a final concentration reaction containing 20 mM Tris-HCl, 50 mM KCl, 3 mM MgCl$_2$, 0.5× SYBR Green I, 200 μM dNTPs, 0.2 μM each of forward and reverse primers, approximately 500 pg of first strand cDNA, and 0.5 unit of Taq polymerase.

The reaction tubes are covered with MicroAmp optical caps (Applied Biosystems) using a cap-installing tool (Applied Biosystems). The contents are collected to the bottom of the tube by brief centrifugation in an RT-6000B benchtop centrifuge fitted with a microplate carrier (Sorvall, Newtown, CT). The tubes are then placed in the ABI 7700 thermocycler and incubated at 95° for 2 min (10 min if using AmpliTaq Gold) to activate the enzyme and denature the DNA template. Forty cycles of PCR amplification are then performed as follows: denature at 95° for 15 sec, anneal at 55° for 20 sec, extend at 72° for 30 sec.

This protocol works well for amplicons up to 500 base pairs (bp). For longer amplicons, the extension step should be adjusted accordingly (approximately 1 min/kb). Either the FAM or SYBR channel can be used for fluorescence detection of SYBR Green I. Fluorescent emission values are collected every 7 sec during the extension step. Data are analyzed with Sequence Detector version 1.7 software (Applied Biosystems). To obtain the threshold cycle (C_T) values, the threshold is set in the linear range of a semilog amplification plot of ΔRn against cycle number. This ensures that C_T is within the log phase of the amplification. Here ΔRn is the fluorescence emission value minus baseline fluorescence value. When the PCR is at 100% efficiency, the C_T decreases by one cycle as the concentration of DNA template doubles.

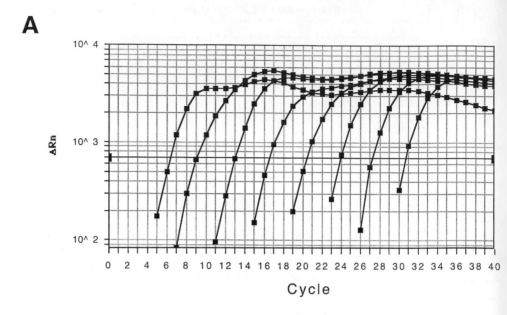

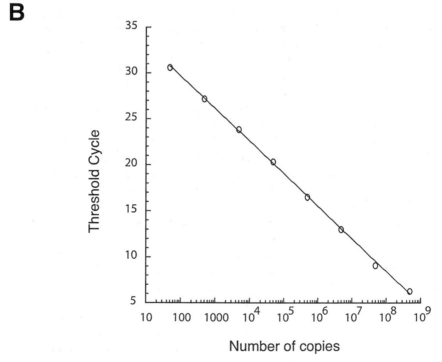

FIG. 2. Standard curve generated by real-time PCR. PCR for an *egr-3* amplicon was performed with serial 10-fold dilutions of *egr-3* cDNA as a template. (A) Amplification plots obtained. (B) Plot of threshold cycle against cDNA copy number. The large linear range of the assay is evident.

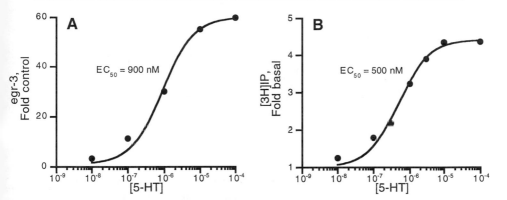

FIG. 3. Intrinsic reporter concentration–response curve. 5-HT$_{2A}$ receptor-expressing cells were treated in triplicate with varying concentrations of serotonin. (A) The level of *egr-3* gene expression after 1 hr of treatment was determined by real-time PCR. The data were analyzed by nonlinear regression. (B) For comparison, an inositol phosphate concentration–response curve produced with the same cell line is shown. Accumulation of [^{3}H]inositol phosphates was determined as previously described. (See Ref. 23.)

To confirm that the correct amplicon is made, the amplified products are analyzed by agarose gel electrophoresis and visualized by ethidium bromide staining. A good reaction yields a single band of the expected size and has no smearing or primer–dimer formation.

To generate a standard curve for each primer pair, 10-fold serial dilutions are made from a plasmid with a known number of copies of the gene. The C_T of each dilution is determined, and is plotted against the log value of the copy number. Amplification efficiency of each primer pair is obtained by the slope of regression (Fig. 2). A 100% efficient PCR has a slope of -3.32. The number of copies in the samples is intrapolated by its C_T value using the respective standard curve.

The potential for IRC assays to monitor signal transduction is illustrated in Fig. 3, which shows a serotonin concentration–response curve of *egr-3* levels generated with 5-HT$_{2A}$ receptor-expressing HEK293 cells. The *egr-3* response curve closely resembles that obtained by a classic assay of accumulation of inositol phosphates.[23]

Acknowledgments

The work described was supported by NIH Grants DA12923 and DK46943. We are grateful to Drs. Rene Hen and Etienne Sibille for advice and assistance with the microarray experiments and to Dr. Sanjay Tyagi for assistance in developing the PCR assays. We also thank Drs. Venu Nair and Elisa Wurmbach for helpful comments.

[23] N. Almaula, B. J. Ebersole, D. Zhang, H. Weinstein, and S. C. Sealfon, *J. Biol. Chem.* **271,** 14672 (1996).

[46] cAMP Response Element-Mediated Gene Expression in Transgenic Reporter Gene Mouse Strain

By KARL OBRIETAN, SOREN IMPEY, and DANIEL R. STORM

Introduction

Inducible gene expression plays a critical role in virtually every aspect of nervous system physiology. Alterations in gene expression patterns are required for such diverse phenomena as long-term memory formation, drug dependence, and circadian timing. Inducible gene expression can be initiated by cell surface receptor-mediated activation of intracellular kinase pathways. In turn these kinases phosphorylate transcription factors, thereby leading to transcriptional activation.

Transcription factors facilitate (or repress) transcription by binding to enhancer (or repressor) elements that may be located up to several thousand of bases from the transcription initiation site. Transcription factor *trans*-activation stimulates transcription by facilitating the formation and stability of the basal transcriptional complex. Much work aimed at characterizing the transcriptional basis of neural plasticity in the CNS has focused on the transcription factor CREB (cyclic AMP response element-binding protein) and on the cyclic AMP response element (CRE).

The CRE consists of a palindromic eight-base pair sequence (TGACGTCA). It binds members of the leucine zipper family of transcription factors, including CREB, CREM (cAMP response element modulation protein), and ATF-1 (activating transcription factor 1). These transcription factors can form homodimers or heterodimers that, once activated, facilitate CRE-dependent transcription. The transcriptional activation of a variety of genes is regulated by CREs within their enhancer regions. Many of these genes, including the immediate-early genes c-*fos* and *jun*-B, and genes encoding brain-derived neurotrophic factor (BDNF) and the neuropeptides somatostatin, vasopressin, and vasoactive intestinal peptide, play important roles in CNS physiology.

The CRE transcriptional pathway can integrate signals from multiple kinase pathways into marked variations in downstream gene transcription. The *trans*-activation potential of CRE-binding proteins is regulated by the phosphorylation of a set of amino acid residues within their kinase-inducible domains. For example, the phosphorylation of CREB at Ser-133 is required for CREB-dependent transcription.[1] A wide variety of kinase cascades trigger CREB phosphorylation at

[1] P. K. Brindle and M. R. Montminy, *Curr. Opin. Genet. Dev.* **2**, 199 (1992).

Ser-133. These signal pathways, including protein kinase A (PKA), Ca^{2+}/calmodulin kinase IV, and mitogen-activated protein kinase (MAPK)/RSK,[2-5] are activated by a diverse array of transmitters, ranging from fast excitatory neurotransmitters to modulatory peptides and neurotrophic factors. CRE-mediated transcription can also be potently repressed via phosphatase activation. For example, Ser/Thr phosphatase type 2A (PP2A) can rapidly inactivate CREB by dephosphorylating Ser-133; blockade of PP2A facilitates CRE-dependent transcription by maintaining the phospho-Ser-133-activated form of CREB.[6] In addition, CRE-mediated gene expression can be repressed as a result of CREB heterodimerization with inhibitory transcription factors, or as a result of ICER (inducible cAMP early repressor) induction. CREB-mediated transcription may also be inhibited by stimulus-induced phosphorylation. For example, phosphorylation at Ser-142 disrupts the ability of the Ser-133 phosphorylated form of CREB to stimulate transcription.[7]

Transcriptional activation via the CRE pathway depends on the overall strength and duration of kinase pathway activation. For example, CRE-dependent gene expression is induced only by long-term CREB phosphorylation at Ser-133; brief CREB phosphorylation at Ser-133 does not stimulate transcription.[8-12] The CRE pathway also integrates activation of multiple signaling pathways into alterations in downstream gene expression. For example, pairing increased cAMP with increased cytosolic Ca^{2+} leads to synergistic increases in CRE-mediated gene expression.[3] It is important to note that synergism does not occur at the level of CREB phosphorylation at Ser-133. Along these lines, synergistic activation of CRE-mediated gene expression does not result in a parallel level of potentiated CREB phosphorylation. Together, these results reveal that monitoring CREB phosphorylation at Ser-133 is an imperfect tool for assessing CRE-dependent gene expression and underscore the importance of monitoring CRE-dependent transcription. To assess the regulation of the CRE pathway *in vivo,* we have generated

[2] M. Sheng, G. McFadden, and M. E. Greenberg, *Neuron* **4,** 571 (1990).

[3] S. Impey, K. Obrietan, S. T. Wong, S. Poser, S. Yano, G. Wayman, J. C. Deloulme, G. Chan, and D. R. Storm, *Neuron* **21,** 869 (1998).

[4] G. A. Gonzalez and M. R. Montminy, *Cell* **59,** 675 (1989).

[5] J. Xing, D. D. Ginty, and M. E. Greenberg, *Science* **273,** 959 (1996).

[6] B. E. Wadzinski, W. H. Wheat, S. Jaspers, L. F. J. Peruski, R. L. Lickteig, G. L. Johnson, and D. J. Klemm, *Mol. Cell. Biol.* **13,** 2822 (1993).

[7] Y. Sun, J. Yamauchi, Y. Kaziro, and H. Itoh, *J. Biochem.* **1999,** 515 (1999).

[8] F. C. Liu and A. M. Graybiel, *Neuron* **17,** 1133 (1996).

[9] P. Brindle, T. Nakajima, and M. Montminy, *Proc. Natl. Acad. Sci. U.S.A.* **92,** 10521 (1995).

[10] H. Enslen, P. Sun, D. Brickey, S. H. Soderling, E. Klamo, and T. R. Soderling, *J. Biol. Chem.* **269,** 15520 (1994).

[11] M. Schwaninger, R. Blume, M. Kruger, G. Lux, E. Oetjen, and W. Knepel, *J. Biol. Chem.* **270,** 8860 (1995).

[12] M. A. Thompson, D. D. Ginty, A. Bonni, and M. E. Greenberg, *J. Biol. Chem.* **270,** 4224 (1995).

a transgenic mouse strain in which six CREs drive the expression of a reporter gene, β-galactosidase. Thus, by monitoring β-galactosidase expression, one is able to characterize CRE-dependent transcription *in vivo*. To facilitate the examination of CRE-mediated gene expression using this mouse strain, we present a set of techniques optimized for β-galactosidase protein detection in CRE–*lacZ* mice. Because CRE–*lacZ* mice express low levels of β-galactosidase we strongly recommend the use of immunochemical detection methods and have found that 5-bromo-4-chloro-3-indolyl-β-D-galactopyranoside (X-Gal) histochemistry is useful only for paradigms that induce exceptionally high levels of LacZ expression. Furthermore, we have found that adult mouse CNS tissue contains significant β-galactosidase-like activity and recommend the use of Western analysis of LacZ expression rather than enzymatic assays when quantitative analysis is necessary.

Methods

Construct Design

Expression of full-length β-galactosidase is driven by six CREs (CTCGAGCT-GCGTCATATGTAACTTC) in tandem repeat upstream of a Rous sarcoma virus (RSV) minimal promoter. Transcriptional background is silenced by placing a thymidine kinase (TK)-driven *neo*^{r} gene with a simian virus 40 (SV40) poly(A) region in the reverse orientation upstream of the CRE–*lacZ* transgene. C57BL/6a × SJL F$_2$ mouse blastocytes are microinjected with the linearized construct to generate the transgenic mouse strain. Founders are bred to the C57BL/6 mouse strain.

Polymerase Chain Reaction-Based Genotyping

A polymerase chain reaction (PCR)-based approach to genotyping is used to identify transgenic mice from heterozygous/wild-type matings. Initially, genomic DNA is purified from a 0.5-cm tail snip. The tail is digested in 750 μl of digestion buffer overnight at 56°. Phenol–chloroform (500 μl) is then added to each tube. Next, the tubes are vortexed and centrifuged at 14,000 rpm for 10 min. The aqueous phase is then removed and 500 μl of 2-propanol is added to precipitate the DNA. Next, the samples are inverted several times, then spun at 14,000 rpm for 10 min. After removing the supernatant, the pelleted DNA is washed in 70% (v/v) ethanol and then briefly dried, followed by resuspension in 200 μl of TE buffer, and heated to 85° for 1 hr. Heating facilitates DNA resuspension and inactivates DNases. The DNA concentration should be 0.25–0.50 μg/μl.

The primer pair ACCAGAAGCCGTGCCGGAAA and CCCGTAGGTAGT-CACGCAAC is used to amplify a 572-base pair fragment of the *Escherichia coli* β-galactosidase gene. Genomic DNA (0.25 μg/μl) is combined with a 0.2 μM concentration of each primer and PCR reagents. PCR should be performed at the following temperatures and cycle times: 95° denaturation for 45 sec, 59° annealing for 30 sec, and 72° extension for 30 sec. Thirty cycles are run. Samples are then electrophoresed on a 1% (w/v) agarose gel.

Tail digestion buffer
  EDTA, 5 mM
  NaCl, 200 mM
  Tris (pH 8.0), 100 mM
  Sodium dodecyl sulfate (SDS), 0.2% (w/v)
  Proteinase K (Sigma, St. Louis, MO), 10 mg/ml
TE buffer
  Tris-HCl (pH 8.0), 10 mM
  EDTA (pH 8.0), 1 mM

Sacrifice and Sectioning

For the examination of CRE-dependent gene expression, mice are killed either by CO_2-mediated asphyxiation, or by cervical dislocation. Mice are then rapidly decapitated and their brains removed and immediately immersed in ice-cold oxygenated artificial cerebral spinal fluid (ACSF). Brains are then cut into 400-μm sections using a vibratome. These conditions were found to be optimal for maintaining the overall health of the neural tissue. Tissue destined for electrophysiological recording is maintained in ACSF buffer. Tissue destined for immediate immunocytochemical examination of β-galactosidase is placed in a 6% (v/v) formaldehyde–phosphate-buffered saline (PBS) solution containing 50 mM HEPES and gently rocked for 2–4 hr at room temperature. Tissue is then cryoprotected with 30% (w/v) sucrose for at least 6 hr. Thin (35- to 40-μm) sections are then cut through the tissue, using a sliding microtome or cryostat. Alternatively, the brain is rapidly dissected and then immediately frozen in O.C.T. (Tissue-Tek, Elkhart, IN) embedding medium in a plastic mold. These are kept at −80° until the time of sectioning. Fifteen 20-μm sections are obtained with a cryostat and mounted on Superfrost slides (Fisher, Pittsburgh, PA). The slides are typically stored at −80° until further processing or used immediately. Slides are briefly air dried and then fixed in 4% (w/v) paraformaldehyde for 30 min and washed twice with PBS. For Western analysis of reporter expression, 400-μm coronal brain sections are quick-frozen onto glass coverslips and the region of interest is excised with the use of a dissecting microscope. Tissue should be stored at −70° or in liquid nitrogen.

ACSF: pH 7.4
  NaCl, 120 mM
  KCl, 3.5 mM

> MgCl$_2$, 1.3 mM
> CaCl$_2$, 2.5 mM
> NaH$_2$PO$_4$, 1.25 mM
> NaHCO$_3$, 25.6 mM
> Glucose, 10 mM

PBS: pH 7.4

> NaCl, 137 mM
> KCl, 2.7 mM
> Na$_2$KPO$_4$ · 7H$_2$O, 4.3 mM
> KH$_2$PO$_4$, 1.3 mM

Formaldehyde (6%, v/v)–PBS solution

> Heat 100 ml of PBS to 100° in a fume hood. Add 6 g of paraformalde-hyde with continuous stirring. Once in solution, cool to room temperature and add 50 mM HEPES, pH 7.4. Solution should be prepared weekly or aliquoted and stored frozen

Immunocytochemistry

For β-galactosidase immunocytochemistry, free-floating sections or mounted slides are blocked for 2 hr in 10% bovine serum albumin (BSA) and 1% (v/v) normal goat serum in phosphate-buffered saline with 0.1% (v/v) Triton X-100 (PBST), and 0.02% (w/v) azide. After blocking, sections or slides are incubated overnight at 4° with an affinity-purified rabbit antibody against β-galactosidase [1 : 500–1000, final dilution (5 Prime → 3 Prime, Boulder, CO) or 1 : 2000 final dilution (ICN, Costa Mesa, CA)]. The antibody is diluted in PBST containing 2.5% (w/v) BSA and 0.05% (w/v) azide. The tissue is then incubated with a rhodamine X-conjugated anti-rabbit IgG (2 μg/ml, final dilution; Jackson Immuno Research, West Grove, PA. or Alexa 594-conjugated secondary anti-rabbit IgG (1 μg/ml, final dilution; Molecular Probes, Eugene, OR). The antibody is diluted in PBST containing 2.5% (w/v) BSA and 0.05% (w/v) azide. This immunolabeling protocol works well for visualization of relatively high β-galactosidase expression. However, if relatively weak or subtle changes in expression are to be studied a more sensitive method of detecting β-galactosidase expression should be employed. Toward this end, we have used a signal amplification system that relies on the Alexa line of fluorophores from Molecular Probes. As described above, tissue is incubated with an antibody against β-galactosidase, and then with a fluorescein-conjugated secondary antibody (2 μg/ml, final dilution; Jackson Immuno Research). To amplify the fluorescent signal, the tissue is then incubated with a tertiary Alexa Fluor 488- labeled anti-fluorescein antibody raised in rabbit (2 μg/ml, final dilution). This step dramatically boosts the fluorescent signal, thereby allowing better resolution of variations in signal intensity. Because the absorption and emission characteristics of Alexa Fluor

488 are similar to those of fluorescein, a fluorescein filter set is used for visualization of Alexa Fluor 488 fluorescein. The immunofluorescent signal can also be amplified by the administration of an Alexa secondary antibody in conjunction with an Alexa Fluor 488-conjugated tertiary antibody directed against the IgG domain of the secondary antibody. After the blocking step and between each antibody treatment, tissue is washed at least six times (5 min per wash with PBST). Free-floating sections are mounted with Gelmount (Biomedia, Foster City, CA). If another mounting medium is used, care should be taken to ensure that it does not interfere with the fluorescent properties of Alexa Fluor 488. Molecular Probes provides a compatibility chart assessing Alex immunofluorescence in tissues mounted with a variety of media. To prevent spurious immunolabeling variability within an experiment, all samples should be processed concurrently.

Western Analysis of β-Galactosidase

Another approach to monitor CRE-dependent gene expression is to immunoblot for expression of the β-galactosidase reporter. To characterize expression in a particular brain region, such as the hippocampus or suprachiasmatic nuclei (SCN), tissue is cut into 400-μm sections, as described above, and then quick frozen onto glass coverslips and the region of interest is excised from the slice. Excised brain tissue is then be resuspended in lysis buffer and sonicated or homogenized. An aliquot of the homogenate is taken to determine total protein by the bicinchoninic acid method (Pierce, Rockford, IL). The remaining sample is resuspended in an equal volume of 6× sample buffer. The samples are then heated to 90° for 10 min and vortexed (20 sec). Next, tissue homogenates are centrifuged (8 min at 13,000g). Extracts are then loaded onto an 8% (w/v) SDS–polyacrylamide gel and subjected to electrophoresis, using procedures recommended by the manufacturer. Protein is then transblotted onto polyvinylidenedifluoride (PVDF) membrane (Immobilon P; Millipore, Bedford, MA). After blocking nonspecific binding with 10% (w/v) powdered milk in PBST (1 hr at room temperature), membranes are incubated overnight at 4° with an antibody against β-galactosidase (1 : 500; 5 Prime → 3 Prime) diluted in PBST with 5% (w/v) BSA and 0.02% (w/v) azide. A goat anti-rabbit IgG alkaline phosphatase-conjugated secondary antibody (1 : 2000; ICN-Cappel) diluted in PBST with 5% (w/v) BSA and 0.02% (w/v) azide is then applied to the membrane for 4 hr at room temperature. Immunoreactivity is developed with a luminescent AP detection system (New England Nuclear, Boston, MA). After each antibody treatment, membranes are washed at least six times (5 min per wash) with a 5% (w/v) milk–PBST solution. β-Galactosidase runs as a 120-kDa band. Care should be taken to ensure that the molecular weight standards do not include β-galactosidase as a marker protein. If β-galactosidase is used, the lane containing the makers should be excised from the membrane before immunolabeling.

Lysis buffer: pH 7.4
β-Glycerophosphate, 50 mM
EGTA, 1.5 mM
Na$_3$VO$_4$, 0.1 mM
Dithiothreitol (DTT), 1 mM
Aprotinin, 10 μg/ml
Pepstatin, 0.2 μg/ml
Leupeptin, 10 μg/ml
Phenylmethylsulfonyl fluoride (PMSF), 1 mM
Sample buffer (6×)
Tris-HCl (pH 6.8), 0.3 M
Glycerol, 30% (v/v)
DTT, 0.1 g/ml
SDS, 0.15 g/ml
Bromphenol blue, 2.0 mg/ml

Neuronal Cell Culture

Initially, neonatal mice are decapitated and their brains are removed and placed in sterile dissociation medium (DM). The brain region of interest is then excised, washed three times, and finely chopped. Next, the tissue is placed in a mild proteolytic solution containing papain latex (100 units/ml; Worthington, Freehold, NJ) and cysteine (0.45 mg/ml; Sigma, St. Louis, MO) diluted in DM. Incubation times vary depending on the age of the tissue. For example, tissue from embryonic day 18 mice is incubated for 20 min at 37°, whereas postnatal day 1 brain tissue is incubated for 30 min at 37°. Tissue is then washed three times in DM to remove the proteolytic solution. The tissue is then transferred to tissue culture medium. The tissue is then slowly triturated 20 to 30 times into a single-cell suspension, using a 10-ml serological pipette. The cell suspension is allowed to sit for 5 min to precipitate tissue that did not dissociate. Next, the cell suspension is removed and plated onto coverslips or tissue culture dishes coated with poly-D-lysine. Cell cultures are maintained at 37° and 5% CO_2 in an incubator. Ninety minutes after plating, the tissue culture medium is replaced. After 2 days in culture, medium is supplemented with 5 μM cytosine arabinofuranoside to reduce glial cell proliferation. Tissue culture medium is replaced twice weekly.

DM, pH 7.7
Na$_2$SO$_4$, 90 mM
K$_2$SO$_4$, 30 mM
MgCl$_2$, 16 mM
CaCl$_2$, 0.25 mM

HEPES, 32 mM
Phenol red, 0.01% (w/v)
Tissue culture medium
Neurobasal (GIBCO, Grand Island, NY) supplemented with:
B27 (GIBCO), 1×
Glutamine (GIBCO), 0.32 mM
Fetal bovine serum (GIBCO), 1% (v/v)
Penicillin–streptomycin (GIBCO), 100 units/ml
Poly-D-lysine solution
Poly-D-lysine should be high molecular mass (>300 kDa; Sigma). To coat glass coverslips, dilute poly-D-lysine to 0.5 mg/ml. To coat tissue culture-treated polystyrene, dilute poly-D-lysine to 0.1 mg/ml. Incubate the surface for at least 15 min with the poly-D-lysine solution. Wash the dishes/coverslips twice with sterile distilled water.

Slice Stimulation

Impey *et al.*[3,13] have used electrical stimulation paradigms to study the relationship between induction of CRE-dependent transcription and induction of long-lasting forms of long-term potentiation (L-LTP).[3,14] For L-LTP in area CA1 of the hippocampus, transverse brain slices of hippocampus are collected and continuously perfused in oxygenated ACSF. Brain slices are maintained at a constant temperature of 34° for 1–2 hr before the beginning of the experiment. A tungsten bipolar electrode (100-μm diameter) is used to stimulate the slice and a glass microelectrode (1- to 2-μm tip) filled with 3 M NaCl is used for recording. Three 100-Hz (0.2-msec square pulse) tetani delivered at 5-min intervals to the stratum radiatum are used to generate L-LTP. Decremental LTP is generated by delivering one 100-Hz stimulus with a pulse duration of 0.1 msec. After stimulation, slices are perfused in ACSF for a minimum of 4 hr to allow for expression of the CRE-regulated reporter. Slices are then transferred to a 6% (v/v) formaldehyde–PBS solution and immunocytochemically processed, as described above, for β-galactosidase expression.

Quantitation of β-Galactosidase

Images of immunofluorescently labeled sections are collected with an MRC-600 scanning laser confocal microscope (Bio-Rad, Hercules, CA) equipped with a krypton–argon laser. There are several reasons for using a confocal microscope to

[13] S. Impey, D. M. Smith, K. Obrietan, R. Donahue, C. Wade, and D. R. Storm, *Nat. Neurosci.* **1**, 595 (1998).
[14] S. Impey, M. Mark, E. C. Villacres, S. Poser, C. Chavkin, and D. R. Storm, *Neuron* **16**, 973 (1996).

collect images. For example, relative to a conventional microscope, the confocal microscope offers superior resolution, greater sensitivity, and, most importantly, the ability to resolve signal from a single-cell layer, thereby diminishing signal from both out-of-focus cells and background fluorescence. Typically, images are collected with either a ×20 or ×10 objective. Kalman filtering (four frames) is used to reduce noise levels. For each experiment the dynamic range of the signal is optimized. Toward this end, the intensity of the weakest fluorescent sample is set to the minimum intensity that gives a visible signal ($\sim$20 on a 0–255 intensity scale). Attenuation of the signal is accomplished by placing neutral density filters in the light path and by decreasing gain and background levels. Data from all samples are collected using these settings. Without maximizing the dynamic range it may be difficult to characterize variations in reporter expression with the methods described above. Image collection is performed "blind." Immunofluorescence densitometry is used to quantify CRE-regulated reporter gene expression. Various data analysis programs are suitable for fluorescent signal quantitation. We utilized Metamorph image analysis software (Universal Imaging, West Chester, PA).

An example of fluorescence quantitation is described below. For quantitation of CRE-mediated transcription-associated contextual learning, the integrated fluorescent pixel intensity from a 40×40-pixel square was measured in hippocampal areas CA1, CA3, the dentate gyrus, and the molecular cell layer. Quantitation was performed on hippocampal regions between bregma -1.46 and bregma -1.94. Measurements were taken from three different areas of the same structure/region and the mean signal was determined. Signal from the region of interest is then normalized to the signal from another cell layer or another anatomical region within the same slice that does not vary in response to stimulus conditions. Thus, the integrated pixel signals from the above-described regions were normalized to the integrated pixel signal from the area between the stratum radiatum and stratum lacunosum. The Student t test is used to determine significance between two groups. One-way analysis of variance (ANOVA) followed by the Tukey–Kramer multiple comparison posttest is used to determine significance for greater than two experimental groups. In instances in which only a subset of cells exhibits CRE-dependent transcription, conventional counting of immunopositive cells is a good alternative to densitometric analysis.

Results and Discussion

CREB/CRE in CNS Plasticity

We have utilized the CRE–*lacZ* transgenic mouse strain to investigate the contribution of the CRE transcriptional pathway to hippocampal-dependent contextual learning and assess its contribution to the endogenous circadian timing properties of the SCN. Below, we present a synopsis of these results.

Long-Term Memory and CRE

Memory is divided into a short-term and long-term form. Short-term memory is defined by its relatively brief duration (persisting no longer than several hours). Long-term memory is defined by its persistence (lasting days or more) and by its dependence on new protein synthesis; administration of transcription inhibitors blocks long-term but not short-term memory formation.

A variety of model systems have been used to show that increases in both Ca^{2+} and cAMP are required for memory formation. For example, associative learning is blocked in *Drosophila,* which lacks Ca^{2+}-sensitive adenylyl cyclase.[15] Given the transcriptional basis of long-term memory (LTM) formation, it is logical to hypothesize that increased cAMP and Ca^{2+} bring about LTM formation by initiating transcriptional activation. One likely transcription pathway targeted by cAMP and Ca^{2+} is the CREB/CRE pathway. In support of this possibility, CRE-dependent transcription is synergistically regulated by coactivatation of Ca^{2+}- and cAMP-dependent signaling.[3] Additional data supporting a role for activation of the CREB/CRE signaling pathway during learning come from CREB disruption studies. For example, CREB mutant mice show impaired learning, and CREB antisense injection into the amygdala blocks cued fear conditioning.[16,17] Furthermore, disruption of CREB-dependent signaling in *Aplysia* attenuates long-term facilitation,[18,19] and in *Drosophila,* overexpression of transcriptionally inactive CREB blocks long-term memory.[20] Although these studies suggest that CREB/CRE-dependent gene expression is required for memory formation, it is unclear whether activation of this transcriptional pathway occurs during long-term memory formation. Thus, the CRE reporter gene mouse strain affords a unique opportunity to examine transcriptional regulation via the CRE in an intact animal.

To assess the effects of hippocampus-dependent forms of learning on CRE-dependent gene expression, mice were trained under either contextual conditioning or passive avoidance learning paradigms.[13] Contextual conditioning is a form of learning in which an association is made between an innocuous environment and

[15] M. S. Livingstone, P. P. Sziber, and W. G. Quinn, *Cell* **37,** 205 (1984).

[16] R. Bourtchuladze, B. Frenguelli, J. Blendy, D. Cioffi, G. Schutz, and A. J. Silva, *Cell* **79,** 59 (1994).

[17] R. Lamprecht, S. Hazvi, and Y. Dudai, *J. Neurosci.* **17,** 8443 (1997).

[18] K. C. Martin, A. Casadio, E. Y. Zhu H, J. C. Rose, M. Chen, C. H. Bailey, and E. R. Kandel, *Cell* **91,** 927 (1997).

[19] P. K. Dash, B. Hochner, and E. R. Kandel, *Nature (London)* **21,** 718 (1990).

[20] J. C. Yin, J. S. Wallach, M. Del Vecchio, E. L. Wilder, H. Zhou, W. G. Quinn, and T. Tully, *Cell* **79,** 49 (1994).

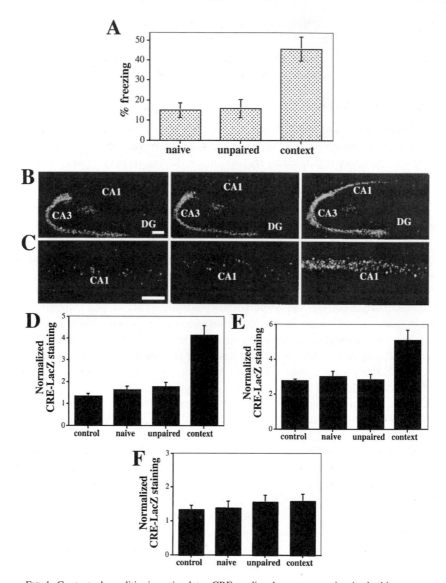

FIG. 1. Contextual conditioning stimulates CRE-mediated gene expression in the hippocampus. (A) Associative learning measured as a percentage of time spent "freezing" after reintroduction to the conditioning chamber (8 hr after training). Contextually trained mice exhibit significantly ($p < 0.0005$ vs naive and unpaired) greater amounts of time freezing. (B) Representative images of β-galactosidase expression in the hippocampus from naive (*leftmost*), unpaired (*center*), and context-trained (*far right*) mice. Bar: 500 μm. (C) Magnification of CA1 cell layer from naive (*leftmost*), unpaired (*center*), and context-trained (*far right*) mice. Bar: 100 μm. (D) Quantitated immunoexpression of β-galactosidase reporter in area CA1. Significance increases ($p < 0.001$, Student t test, relative to unhandled control mice) were observed only for the context-trained group. (E) Similar results were observed in area CA3. (F) No significant changes (relative to expression in unhandled control mice) were observed in the dentate gyrus. [Reprinted from Impey *et al.* (1998) with permission from Nature America; see Ref. 13.]

an aversive stimulus. In this case, an association was formed between a conditioning chamber and a mild foot shock. Associative learning was scored by measuring freezing behavior during reexposure to the conditioning chamber. Data presented in Fig. 1 reveal that contextually conditioned mice showed significant increases in CRE-mediated transcription in areas CA1 and CA3 of the hippocampus compared with "control" mice (no shock or exposure to the chamber), "unpaired" mice (exposure to the chamber was disassociated from the shock) and "naive" mice (placed in the conditioning chamber, but did not receive a shock). Auditory cue fear conditioning, an amygdala-dependent form of associative learning, triggered CRE-mediated gene expression in the amygdala, whereas reporter expression in the hippocampus was not altered. These data reveal that CRE-mediated gene expression is discretely induced in brain regions involved in associative learning.

As mentioned above, long-lasting long-term potentiation (LTP) is used as a model paradigm to examine the cellular neurophysiological events that may underlie memory formation. As with memory formation, LTP can be broken into temporally and mechanistically distinct components; long-lasting LTP (L-LTP) requires *de novo* gene expression and persists for greater than 3 hr, whereas decremental LTP (D-LTP) is not dependent on new gene expression and only persists for 1–3 hr. Thus, if there is a causal link between L-LTP and long-term memory formation, one may hypothesize that signaling events that elicit L-LTP also stimulate CRE-mediated transcription. To test this hypothesis, hippocampal area CA1 from CRE–*lacZ* mice were stimulated under conditions that would lead to either L-LTP or D-LTP. Data presented in Fig. 2 reveal that CRE-mediated gene expression was increased only by stimulus paradigms that elicit L-LTP; induction of D-LTP did not stimulate CRE-mediated gene expression. These results reveal the utility of the CRE–*lacZ* mouse strain, and add further support to the growing body of evidence implicating the CREB/CRE transcriptional pathway in long-term memory formation.

Circadian Timing

Rhythmic gene expression is required for endogenous circadian timekeeping. Although work has identified an important role for the E-box, the role of other enhancer elements has not been characterized. Given the crucial role the CREB/CRE transcriptional pathway plays in hippocampus-dependent forms of learning, we were interested in examining whether the CREB/CRE transcriptional pathway contributes to endogenous circadian timing and to light entrainment of the clock. Previous work in the SCN revealed that light triggers CREB phosphorylation at Ser-133[21]; however, it was not known whether CRE-mediated gene expression is

---

[21] D. D. Ginty, J. M. Kornhauser, M. A. Thompson, H. Bading, K. E. Mayo, J. S. Takahashi, and M. E. Greenberg, *Science* **260,** 238 (1993).

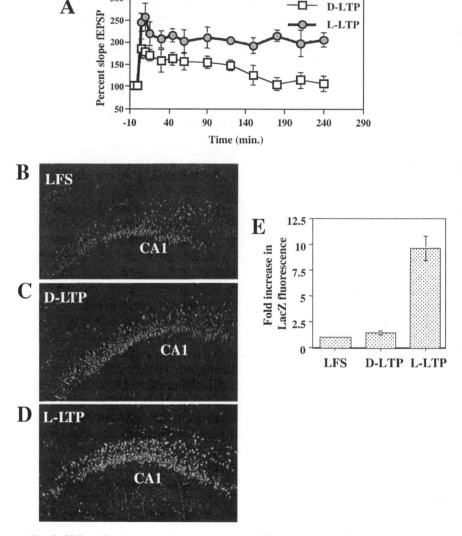

Fig. 2. CRE-mediated transcription is increased by L-LTP but not by D-LTP in hippocampal area CA1. (A) L-LTP was generated by three 100-Hz, 1-sec tetani spaced 5 min apart. D-LTP was generated by one 100-Hz, 1-sec tetanus. Inset shows representative EPSPs from a slice before and after potentiation. Horizontal scale bar represents 5 msec, vertical scale bar represents 0.3 mV. (B–D) Representative β-galactosidase expression 4 hr after low-frequency stimulation, D-LTP and L- LTP. (E) Quantitation of β-galactosidase signal from slices that generated L-LTP, D-LTP and from control, low-frequency-stimulated (LFS) slices. Error bars denote the SEM. Fold increase in β-galactosidase signal intensity initiated by L-LTP was statistically greater ($p < 0.001$, Student t test) than the signal intensity associated with LFS and D-LTP. [Reprinted from Impey *et al.* (1998) with permission from Cell Press; see Ref. 3.]

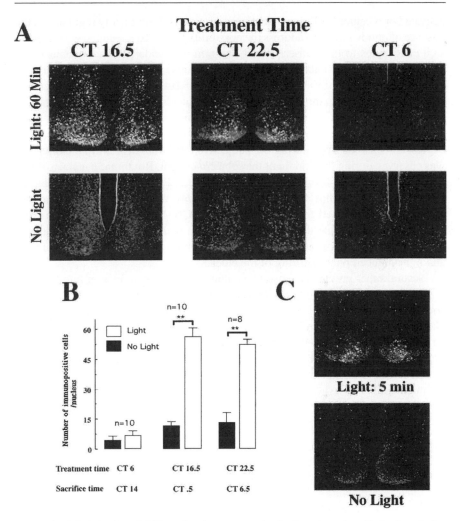

FIG. 3. Light induction of CRE-mediated gene expression is phase dependent. Dark-adapted mice were exposed to light (60 min, 400 lux) at different circadian times: early subjective night (CT 16.5), late subjective night (CT 22.5), or midsubjective day (CT 6). Mice were then returned to total darkness. Eight hours later mice were killed and their brains were processed for β-galactosidase expression. Relative to control animals not exposed to light, photic stimulation at CT 16.5 and CT 22.5 initiated robust reporter expression in the SCN, whereas no induction was observed at CT 6. (B) Quantitation of CRE-mediated gene expression. Error bars denote SEM. **$p < 0.0001$, Student t test. (C) A 5-min light treatment at CT 16.5 induced CRE-regulated reporter expression only in a discrete band of cells in the ventrolateral SCN. [Reprinted from Obrietan *et al.* (1999) with permission from the American Society for Biochemistry and Molecular Biology; see Ref. 22.]

activated or whether CRE-dependent transcription is rhythmically regulated in the absence of photic input. To address the possibility that CRE-mediated gene expression contributes to the transcriptional program that underlies endogenous clock timing, the CRE–β-galactosidase transgenic mouse strain was used to monitor CRE-mediated transcription in the SCN *in vivo*.[22] By using immunocytochemical detection techniques to monitor β-galactosidase levels at different time points over a 24-hr period, we were able to show that CRE-mediated gene expression is regulated in a circadian manner under conditions of total darkness. Maximal reporter expression was observed during the subjective day, whereas minimal expression was observed during the middle of the subjective night. Rhythmic β-galactosidase expression was also observed by Western analysis. Photic stimulation during the subjective night, but not during the subjective day, also stimulates a marked increase in CRE-mediated gene expression in the SCN (Fig. 3). Together, these result raise the possibility that the CRE transcriptional pathway plays an essential role in facilitating transcriptional events required for both circadian rhythm generation and light-induced phase shifting of the clock.

Several other groups have used this mouse strain to gain insight into the transcription basis of a variety of neurophysiological events. For example, Pham *et al.*[23] and Barth *et al.*[24] examined CRE-regulated gene expression during developmental plasticity of the visual and barrel cortices, respectively. Thome *et al.*[25] characterized variations in CRE-dependent transcription associated with chronic antidepressant treatment. In conclusion, these studies reveal the CRE–*lacZ* mouse strain as an invaluable tool for the parsimonious examination of CRE-dependent transcription.

[22] K. Obrietan, S. Impey, D. Smith, J. Athos, and D. R. Storm, *J. Biol. Chem.* **274,** 17748 (1999).

[23] T. A. Pham, S. Impey, D. R. Storm, and M. P. Stryker, *Neuron* **22,** 63 (1999).

[24] A. L. Barth, M. McKenna, S. Glazewski, P. Hill, S. Impey, D. Storm, and K. Fox, *J. Neurosci.* **20,** 4206 (2000).

[25] J. Thome, N. Sakai, K.-H. Shin, C. Steffen, Y.-J. Zhang, S. Impey, D. Storm, and R. S. Duman, *J. Neurosci.* **20,** 4030 (2000).

[47] Functional Genomic Search of G-Protein-Coupled Receptors Using Microarrays with Normalized cDNA Library

By Susumu Katsuma, Satoshi Shiojima, Akira Hirasawa,
Yasuhito Suzuki, Hiroshi Ikawa, Kazuchika Takagaki,
Yoshinori Kaminishi, Masatoshi Murai, Tadaaki Ohgi,
Junichi Yano, and Gozoh Tsujimoto

Introduction

G-protein-coupled receptors (GPCRs) represent the single most important drug targets for medical therapy, and information from genome sequencing and genomic databases has substantially accelerated their discovery. Despite its present large size, the GPCR superfamily continues to expand rapidly as new receptors are discovered through automated sequencing of cDNA libraries and bioinformatics. It is estimated that several thousand GPCRs may exist in the human genome, and at present with most of the genome sequenced, as many as 250 GPCRs have been cloned. However, a systematic approach to identify the function of newly discovered GPCRs is lacking.

Large-scale monitoring of gene expression is a powerful approach to clarify cellular events. DNA microarray[1,2] technologies permit us to recognize genome-wide expression profiling, and have a profound impact to biological research, especially pharmacology. This technology can also be applied to drug discovery and molecular classification of diseases.[3,4] We introduced microarray to discover novel functions of GPCRs involved in disease states. In analyzing tissues of an animal disease model, we first attempted to use microarrays with cDNAs randomly selected from Unigene clones (Research Genetics, Huntsville, AL) and found that few clones successfully hybridized presumably because of the tissue-specific gene expression. To resolve this problem, we aimed to fabricate microarrays with the cDNA library of the tissue to be analyzed; however, ordinary cDNA libraries contain a high frequency of undesirable clones because of the redundancy of mRNA species in the cell.[5] To get rid of the redundancy from ordinary cDNA

[1] M. Schena, D. Shalon, R. W. Davis, and P. O. Brown, *Science* **270,** 467 (1995).
[2] J. L. DeRisi, V. R. Iyer, and P. O. Brown, *Science* **278,** 680 (1997).
[3] D. T. Ross, U. Scherf, M. B. Eisen, C. M. Perou, C. Rees, P. Spellman, V. Iyer, S. S. Jeffrey, M. Van de Rijn, M. Waltham, A. Pergamenschikov, J. C. Lee, D. Lashkari, D. Shalon, T. G. Myers, J. N. Weinstein, D. Botstein, and P. O. Brown, *Nat. Genet.* **24,** 227 (2000).
[4] U. Scherf, D. T. Ross, M. Waltham, L. H. Smith, J. K. Lee, L. Tanabe, K. W. Kohn, W. C. Reinhold, T. G. Myers, D. T. Andrews, D. A. Scudiero, M. B. Eisen, E. A. Sausville, Y. Pommier, D. Botstein, P. O. Brown, and J. N. Weinstein, *Nat. Genet.* **24,** 236 (2000).

libraries ("normalization"), we chose two approaches, the "subtractive library" approach and "hit-picking." The subtractive library approach is a conventional method reported by Bonaldo et al.,[6] while the hit-picking approach is a novel one we developed by combining macroarray and robotic systems; we named the latter method "hit-picking" because desirable clones are collected after selection by filter hybridization.

This article describes microarray analysis in combination with two normalization methods of cDNA libraries, and also presents preliminary comparative results of both methods. We propose that microarray analysis with a normalized cDNA library is a powerful approach to identify gene function related to disease.

Preparation of Normalized Libraries

Construction of Normalized Library by in Vitro Subtraction

We constructed a normalized library as previously described by Bonaldo et al.[6] with minor modifications (Fig. 1). We developed two procedures to normalize cDNA libraries based on hybridization of in vitro synthesized DNA or RNA (driver) from an entire library with the library itself in the form of single-stranded circles (tracer). The normalization of rat kidney and liver is described below.

Construction of Normalized Rat Kidney cDNA Library:
Double-Stranded DNA as Driver

CONSTRUCTION OF RAT KIDNEY cDNA LIBRARY. Total RNA is isolated from rat kidney with a Sepasol-RNA I RNA extraction kit (Nacalai, Kyoto, Japan). Poly(A)$^+$ RNA prepared from total RNA by Oligotex-dT30 (TaKaRa, Kyoto, Japan) is used for construction of a phage cDNA library, using a ZAP Express vector kit (Stratagene, San Diego, CA). Size fractionation of cDNAs is performed by 1% (w/v) agarose gel electrophoresis and cDNAs ranging from 500 to 1500 bp are recovered by the GeneClean (BIO 101, Vista, CA) method. Phagemid cDNA library is excised in vivo from the phage library with helper phage (ExAssist; Stratagene).

PREPARATION OF SINGLE-STRANDED LIBRARY DNA. Double-stranded phagemid DNA is converted to single-stranded circles by the combined action of gene II (phage F1 endonuclease) and Escherichia coli exonuclease III enzymes.

1. The gene II reaction should contain 4 μg of plasmid DNA from the rat kidney cDNA library and 1 μl of Gene II (GIBCO-BRL, Gaithersburg, MD) for a total volume of 20 μl. The reaction is performed at 30° for 1 hr.
2. Gene II is heat inactivated at 65° for 5 min.

[5] M. D. Adams, M. Dubnick, A. R. Kerlavage, R. Moreno, J. M. Kelly, T. R. Utterback, J. W. Nagle, C. Fields, and J. C. Venter, Nature (London) 355, 632 (1992).
[6] M. F. Bonaldo, G. Lennon, and M. B. Soares, Genome Res. 6, 791 (1996).

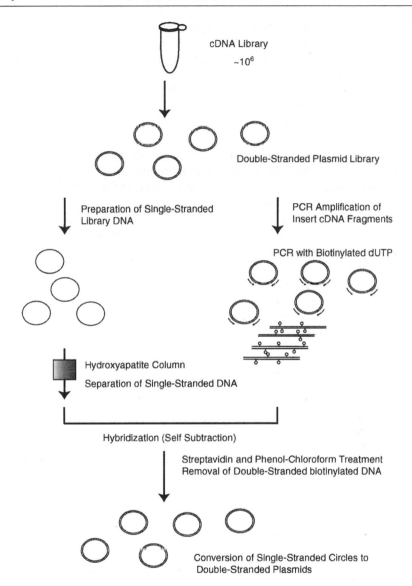

FIG. 1. Diagram of the normalization method with a double-stranded DNA driver.

3. The reaction mixture is chilled on ice and 2 μl of exonuclease III (GIBCO-BRL) is added.

4. The reaction mixture is incubated at 37° for 30 min.

5. Gene II and exonuclease III are digested with proteinase K (Roche Diagnostics, Tokyo, Japan) at 50° for 15 min in a 100-μl reaction mixture containing

10 mM Tris-HCl (pH 7.8), 5 mM EDTA, 0.5% (w/v) sodium dodecyl sulfate (SDS), and 136 μg of proteinase K.

6. After extraction with an equal volume of phenol–chloroform, the DNA is recovered by ethanol precipitation.

7. The library DNA is digested with PvuII at 37° for 2 hr. Note that PvuII restriction enzyme does not cleave single-stranded DNA.

8. The reaction mixture is diluted with 2 ml of loading buffer [0.12 M sodium phosphate, 10 mM EDTA, 1% (w/v) SDS] and purified by hydroxyapatite chromatography at 60° using a column preequilibrated with the same buffer [1-ml bed volume; 0.4 g of hydroxyapatite (Bio-Rad, Hercules, CA)]. After washing with loading buffer, 6 ml of the wash fraction is combined with the flowthrough fraction and concentrated by freeze–drying.

9. The freeze-dried sample is dissolved in 2 ml of distilled H_2O. The sample solution is divided into two tubes, and each sample is desalted by gel filtration, using NAP-10 columns (Amersham Pharmacia Biotech, Uppsala, Sweden). Each sample is collected up and extracted with 2-butanol, and ethanol precipitated.

10. To measure the amount of single-stranded DNA, DNA pellets are resuspended in distilled H_2O and quantified spectrophotometrically using the equation: 1 OD_{260} = 40 μg/ml. DNA is ethanol precipitated again, and each 50 μg of single-strand DNA is resolved in deionized formamide and stored at −20°.

PREPARATION OF DOUBLE-STRANDED BIOTINYLATED DNA. The DNA biotinylation step is performed by incorporation of biotinylated 11-dUTP during polymerase chain reaction (PCR) amplification, using the Expand High Fidelity PCR system (Roche Diagnostics).

1. As a template, 5 ng of double-stranded plasmid cDNA library is mixed with 2 μl of dNTP stock (the final concentration of each dNTP except for dTTP is 200 μM, and that of dTTP is 140 μM), 5 μl of a 20 μM solution of T3 (5′-AATTAACCCTCACTAAAGGG-3′) and T7 primer (5′-GTAATACGACTCA-CTATAGGGC-3′), 10 μl of 10× Expand High Fidelity buffer, 0.75 μl of Expand High Fidelity enzyme mix (2.6 units), and 70.25 μl of distilled H_2O.

2. Then 50 μl mineral oil is added and the reaction mixture is subjected to the following amplification cycle conditions in a thermal cycler: After 1 min of incubation at 94°, 20 cycles of 1 min at 94°, 2 min at 55°, 3 min at 72°, and 7 min at 72°.

3. The excess primer and substrate are removed from PCR-amplified fragments by gel filtration, using MicroSpin S-300HR columns (Amersham Pharmacia Biotech). Sample volume is adjusted to 200 μl each and samples are applied to MicroSpin columns and gel filtrated by centrifugation (3000g, 2 min).

4. The PCR product is ethanol precipitated and dissolved in 5 μl of TE buffer.

5. The amount of biotinylated DNA product is estimated by comparison between absorbance at 260 nm of biotinylated PCR product and that of nonbiotinylated control reaction.

HYBRIDIZATION

1. Biotinylated double-stranded DNA (1.5 μl, 0.5 μg) is mixed with 5 μl (50 ng) of covalently closed single stranded library DNA in deionized formamide, 0.5 μl (10 μg) of 5′ blocking oligonucleotide (GTGGGGTACCAGGTAAGTGTACC-CAATTCGCCCTATAGTGAGTCGTATTACAATTT), and 0.5 μl (10 ng) of 3′ blocking oligonucleotide (GAAATTAACCCTCACTAAAGGGAACAAAAGC-TGGAGCTCGCGCGCCTGCAGGTCG).

2. This mixture is heated at 80° for 3 min. Then 2.5 μl of buffer B containing 0.48 M NaCl, 0.04 M Tris-HCl (pH 8.0), 20 mM EDTA, and 4% (w/v) SDS is added.

3. The reaction mixture is incubated at 30° for 24 h.

SEPARATION OF SINGLE-STRANDED DNA

1. The hybridization reaction is transferred to 10 μl of TE100 (10 mM Tris-HCl, 1 mM EDTA, 100 mM NaCl).

2. Ten microliters of a 0.1% (w/v) solution of streptavidin (dissolved in 10 mM sodium phosphate buffer, pH 7.0) is added and mixed at room temperature for 10 min.

3. The mixture is extracted with 30 μl of phenol–chloroform, and the aqueous phase containing unhybridized cDNA is removed into a new tube. This streptavidin and phenol–chloroform treatment is repeated on the pooled aqueous phase once more.

4. The aqueous phase is ethanol precipitated.

CONVERSION OF SINGLE-STRANDED CIRCLES TO
DOUBLE-STRANDED PLASMIDS

1. Ethanol-precipitated single-stranded DNA is resuspended in 11 μl of distilled H$_2$O. Then 4 μl of 5× Sequenase buffer (Amersham Pharmacia Biotech) and 1 μl (1 μg) of F23 primer (5′-CCCAGTCACGACGTTGTAAAACG-3′) are added, and the mixture is incubated at 65° for 5 min and then at 37° for 3 min.

2. Then 1 μl of Sequenase, version 2.0 (Amersham Pharmacia Biotech), 1 μl of 0.1 M dithiothreitol (DTT), and 2 μl of dNTP stock (the final concentration of each dNTP is 10 mM) are added, and the reaction is incubated at 37° for 30 min.

3. The total volume is taken up to 100 μl with TE and the reaction is extracted once with phenol–chloroform. Plasmid DNA is ethanol precipitated and dissolved in 3 μl of distilled H$_2$O.

4. Double-stranded plasmid solution (1.5 μl) is electroporated into DH10B bacteria (GIBCO-BRL), and propagated under kanamycin selection. To determine the total number of colonies, some transformants are diluted and inoculated on a kanamycin plate, and colonies are counted. The number of independent clones in the normalized library is estimated.

Using this method, we constructed a rat kidney normalized library that consisted of 2.4×10^4 independent clones.

Construction of Normalized Rat Liver cDNA Library:
Single-Stranded RNA as Driver

The procedure for the normalization of rat liver cDNA library is essentially identical to that for construction of a normalized rat kidney cDNA library, except for the use of biotinylated cRNA in hybridization (Fig. 2).

PREPARATION OF SINGLE-STRANDED LIBRARY DNA. Construction of a rat liver cDNA library from poly(A)$^+$ RNA and preparation of single-stranded DNA are performed as described above.

PREPARATION OF BIOTINYLATED cRNA. Biotinylated RNA is synthesized by an *in vitro* transcription reaction from the plasmid cDNA library, using a DIG DNA labeling kit (Roche Diagnostics).

1. For obtaining runoff transcripts, 1 μg of plasmid DNA containing cDNA inserts is linearized by digestion with *Eco*RI, which digests plasmid DNAs at the end of cDNA insert on the other side of the T7 promoter.

2. After phenol–chloroform treatment and ethanol precipitation, DNA pellets are resuspended in 13 μl of distilled H_2O. Then 2 μl of 10× transcription buffer, 1 μl of RNase inhibitor, 2 μl of RNA polymerase T7, and 2 μl of biotin RNA labeling mix (Roche Diagnostics) are added, and the mixture is incubated at 37° for 2 hr.

3. DNase I (2 μl) is added and the reaction is incubated at 37° for 15 min. The reaction is stopped by adding 2 μl of 0.2 M EDTA (pH 8.0).

4. The RNA transcripts are analyzed by 1% (w/v) agarose gel electrophoresis and it is confirmed that 10 μg of full-length biotin-labeled RNA is obtained from 1 μg of template DNA.

5. After RNA transcripts are ethanol precipitated, each 1 μg of the biotinylated RNA is resuspended in 1.5 μl of TE buffer.

HYBRIDIZATION

1. Biotinylated single-stranded RNA (1.5 μl, 1 μg) is mixed with 5 μl (50 ng) of covalently closed single-stranded library DNA in deionized formamide and 1 μl of distilled H_2O.

2. This mixture is heated at 80° for 3 min. Then 2.5 μl of buffer B [0.48 M NaCl, 0.04 M Tris-HCl (pH 8.0), 20 mM EDTA, 4% (w/v) SDS] is added.

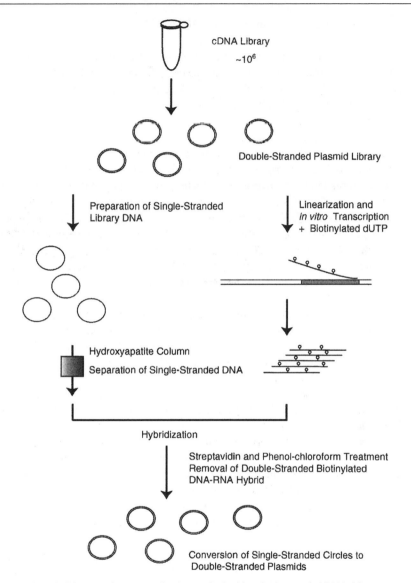

FIG. 2. Diagram of the normalization method with a single-stranded RNA driver.

3. The reaction mixture is incubated at 30° for 24 hr.

After hybridization, the remaining single-stranded DNA is recovered, converted to double-stranded plasmids, and electroporated into DH10B bacteria as described above. By this method, we constructed a rat liver normalized library that consisted of 1.6×10^4 independent clones.

Efficiency of Normalization

To confirm the efficiency of normalization, we first identified abundant genes in the library, and used those genes as a measure of normalization.

The cDNA inserts of original libraries are excised by *Eco*RI–*Xho*I digestion, and separated by 1% (w/v) agarose gel electrophoresis. The cDNA fragments are recovered by the GeneClean (BIO 101) method and labeled with $[\alpha\text{-}^{32}P]dCTP$, using a random primer DNA-labeling kit (TaKaRa). Comparative analysis of gene expression in cDNA libraries is performed by Southern hybridization, using an Atlas RAT cDNA expression array (Clontech, Palo Alto, CA). After hybridization with labeled cDNA probes and several washes of the membrane, hybridization signals are detected by the BAS system (Fuji Film, Tokyo, Japan) and several genes that have intense signals are selected.

Using the cDNA library as a template, polymerase chain reaction (PCR) amplification of these selected genes is performed with specific primers in a reaction mixture containing Platinum PCR mix (GIBCO-BRL). The conditions for gene-specific PCR amplification of cDNA mixtures are preheating (at 94° for 2 min), and 25 cycles of denaturation (at 94° for 1 min), annealing (at 55° for 1 min), and extension (at 72° for 2 min). The PCR products are run on 6% (w/v) polyacrylamide gels and stained with ethidium bromide. By comparing the intensity of gene-specific bands, two genes are selected as abundant genes.

The efficiency of normalization is examined by dilution ratio of highly abundant genes between nontreated cDNA libraries and normalized cDNA libraries. DNAs from nontreated and normalized cDNA libraries are PCR amplified for 12, 15, 18, 21, 24, and 27 cycles with each gene-specific primer in a platinum PCR mix reaction mixture containing 0.2 μl of $[\alpha\text{-}^{32}P]dCTP$ (Amersham Pharmacia Biotech; ~110 TBq/mmol). PCR products are analyzed on a 6% (w/v) polyacrylamide gel, and determination of normalization efficiency is performed by measurement of intensity of gene-specific bands by the BAS system. The dilution ratio of highly abundant genes is estimated in terms of the logarithmic radioactivity of the band increase in proportion to PCR cycle number.

In a normalized rat kidney cDNA library, the frequency of abundant clones cyclophilin and ferritin H are reduced to 50 and 20%, respectively, of that in the nontreated library (Fig. 3). Similar to the normalized kidney library, in a normalized rat liver cDNA library, the frequencies of serum albumin and α_1-protease inhibitor are reduced to 20 and 5%, respectively (Fig. 3). These results indicate that normalization in both normalized cDNA libraries was efficient.

(A)

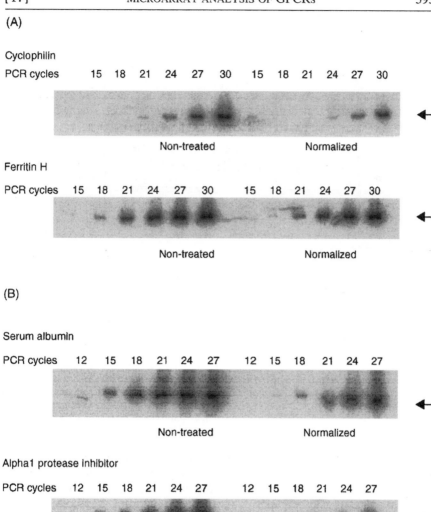

FIG. 3. (A) Comparative analysis of original and normalized rat kidney cDNA libraries. The normalization process is carried out by the double-stranded DNA driver method. Reduction of highly abundant genes is estimated in terms of the logarithmic radioactivity of the band increase in proportion to PCR cycle number. (B) Comparative analysis of original and normalized rat liver cDNA libraries. The normalization process is carried out by the single-stranded RNA driver method.

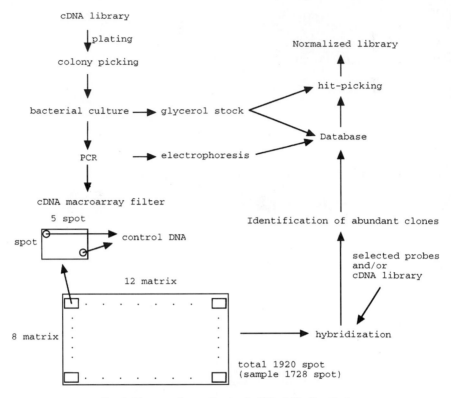

FIG. 4. Diagram of normalization by "hit-picking" method.

To further confirm the efficiency of normalization in a rat liver library, we randomly selected 200 clones from the normalized and original libraries, and sequenced these. The frequency of the serum albumin gene, the most frequently expressed gene in the rat liver cDNA library, is reduced from 38 clones to 1 clone, and all 200 clones are unique in the normalized library, suggesting that this normalization process works well.

Construction of Normalized Library by Hit-Picking

As shown in Fig. 4, the hit-picking method is dependent on robotics, and each process is performed in the 96-well or 384-well format. It is a complicated system compared with a conventional *in vitro* normalization method (Figs. 1 and 4); however, we can remove highly expressed genes visually by this method, hence genes expressed at low abundance can certainly be retained in the library after the normalization process (Fig. 5). Details of the process of hit-picking are given below.

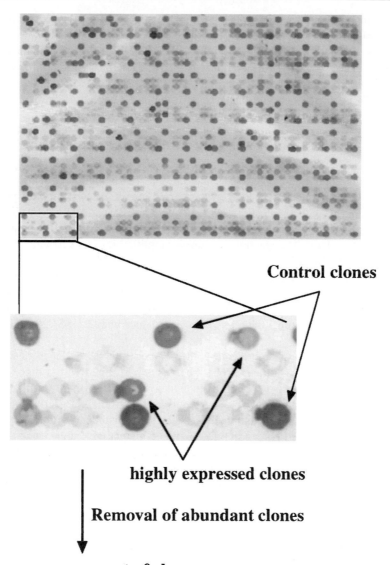

Control clones

highly expressed clones

Removal of abundant clones

rearrangement of clones

FIG. 5. Removal of abundant clones by the hit-picking method. Shown are macroarray filters on which 1920 genes (including 192 control genes) are prepared by robotics. The filters are hybridized with DIG-labeled probes. After color development, images are scanned, and the signal intensity of each clone is measured. Abundant clones are removed from an original library by robotics. Rearranged clones are used as the normalized library.

1. Construction of a cDNA plasmid library is carried out as described above. Transformation of the plasmids is performed by a conventional method. Bacterial colonies are picked at random and inoculated in 96-well deep-well plates by a colony picker. After 20–24 hr of culture, 100 μl of bacterial solution is mixed with 20 μl of 80% (v/v) glycerol, and stored at $-80°$ as bacterial stocks.

2. PCR is carried out with bacterial solution as a template to prepare cDNA fragments for macroarray filters. PCRs are performed in 96-well format. A stock mixture of sufficient volume for the number of reactions is prepared. For each reaction, the stock should contain 5 μl of 10× reaction buffer, 4 μl of 2.5 mM dNTP mix, 2.5 μl of 10 μM forward primer, 2.5 μl of 10 μM reverse primer, 0.25 μl of Taq polymerase (5 U/μl), and 35.75 μl of distilled H_2O in a total volume of 50 μl. Sequences of PCR primers used are as follows: forward primer, 5′-GTTTTCCCAGTCACGAC; reverse primer, 5′-CAGGAAACAGCTATGAC. A small amount of bacterial solution is transferred to the PCR mixture by a replicater (96-pin plate). The reaction mixture is amplified for 30 cycles in a thermal cycler. The amplification conditions are as follows: denaturation at 94° for 40 sec, annealing at 58° for 40 sec, and extension at 72° for 90 sec. We routinely use a Biomek2000 HDRT (high-density replicating tool, 96-pin plate) system (Beckman Coulter, Fullerton, CA) to make macroarray filters on which PCR products are spotted. After spotting, the filters are washed once with 2× SSC, followed by UV cross-linking.

3. Labeled probes are hybridized with prepared filters in order to determine abundant clones in the library. Probe labeling is performed with a digoxigenin (DIG) DNA-labeling kit (Roche Diagnostics). Probe DNAs are purified from cDNA libraries or synthesized by reverse transcription from tissue-derived mRNA. Hybridization and color development are carried out under conditions recommended by the supplier. After color development, images are scanned, and the signal intensity of each clone is measured. After identifying abundant clones, they are removed from an original library by robotics, such as a Biomek2000 workstation (Beckman Coulter). Rearranged clones are stored at $-80°$ as a normalized library.

We constructed a rat kidney normalized cDNA library by this hit-picking method (Fig. 5). Preliminary experiments indicated that redundancy of mRNA species in kidney was quite high; thus more than 20 genes represented more than 1% of the total mRNA mass. To assess the efficiency of removal of the abundant clones, we sequenced more than 2000 clones isolated from the library constructed by this method, and observed that only 2 genes (aldolase B and mitochondrial cytochrome oxidase c) represented more than 1% of total mRNA species. These results show that abundant species can be reduced drastically by this hit-picking method. In addition, we compared the efficiency of this method with the above-mentioned subtractive library method. Sequencing analysis revealed that the library normalized by the conventional DNA–DNA subtractive library method contained three genes that represented more than 1% of total clones, indicating that similar results were obtained by both methods. In the case of CDK109, however, the library normalized by the subtractive library method contained 1.7% of total clones, whereas the library constructed by hit-picking contained only 0.4% of total clones. This comparison suggests that the efficiency of normalization may differ for the individual clone, and in some cases the hit-picking method might be more efficient than the subtractive library method.

Fabrication of Microarrays with Normalized cDNA Library

A flow chart for the fabrication process of microarrays with a normalized cDNA library is shown in Fig. 6. Detailed methodology is explained below.

1. Bacterial clones from the normalized library are cultured in 96-well deep-well plates. Plasmid preparation is carried out by the alkaline–SDS method by a Biomek2000 workstation. cDNA inserts of plasmids are amplified by PCR, using 5'-amino-linked primers. Portions of PCR products are electrophoresed in agarose gels to confirm accurate amplification.

2. Before spotting of probe DNA, PCR products in 96-well plates are transferred to 384-well format plates. Probe DNA is dried, dissolved in 15 μl of 3× SSC, and soaked at room temperature for 12 hr. Spotting probe DNA to slide glasses is performed by a GMS417 arrayer (Affimetrix, Santa Clara, CA) under conditions recommended by the manufacturer.

3. Immobilization of the probe DNA is performed: glass slides on which probe DNA has been spotted are incubated at 80° for 1 hr. After incubation, UV cross-linking is carried out by UV Stratalinker 1800 (Stratagene) at an energy of 60 mJ.

4. Immobilized glass slides are washed in 0.2% (w/v) SDS for 2 min. After this wash, glass slides are washed twice in distilled H_2O for 2 min, followed by soaking the glass slides in blocking solution for 20 min. Blocking solution is made up as follows: dissolve 2.7 g of succinic anhydride in 164.7 ml of 1-methyl-2-pyrrolidinone, add 15.3 ml of 1 M boric acid, and mix well. After blocking, glass slides are washed three times in distilled H_2O, followed by soaking them in boiling water for 2 min. Glass slides are soaked in 95% (v/v) ethanol for 1 min, and dried by centrifugation at low speed. Prepared microarrays are kept shaded and desiccated until hybridization.

Hybridization and Analysis

1. Isolation of total RNA is carried out as described above. mRNA is prepared from total RNA by Oligotex-dT30 (TaKaRa).

2. Two micrograms of each poly(A)$^+$ RNA is used to produce the first-strand cDNA. Each reaction is carried out as follows: each sample contains 14.8 μl of

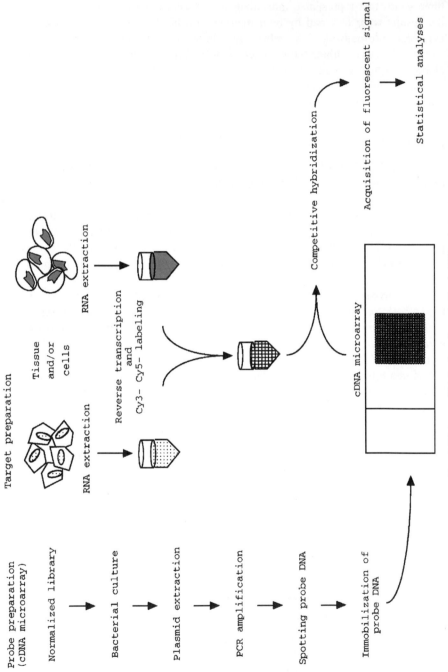

FIG. 6. Diagram of fabrication of a normalized cDNA microarray.

diethyl pyrocarbonate (DEPC)-treated distilled H_2O containing 2 μg of mRNA, 0.6 μl of random hexamer (1 μg/μl; TaKara), 6 μl of 5× reaction buffer (as supplied with reverse transcriptase), 3 μl of 0.1 M dithiothreitol (DTT, as supplied with reverse transcriptase), 0.6 μl of dNTPs (a mix containing 25 mM concentrations of dATP, dGTP, and dCTP, and a 10 mM concentration of dTTP), 3 μl of Cy3- or Cy5-conjugated dUTP (Amersham Pharmacia Biotech), and 2 μl of Superscript II reverse transcriptase (GIBCO-BRL) in a final volume of 30 μl. To run the reaction, samples are incubated at 42° for 1 hr. One hour later, 1 μl of Superscript II reverse transcriptase is further added to the mixture, and incubated at 42° for 1 hr. After incubation, 1.5 μl of alkaline solution (1 N NaOH, 0.5 M EDTA) is added and incubated 10 min at 65° to degrade the rest of the RNA. Neutralization is performed by adding 1.5 μl of 1 N HCl.

3. Reaction mixtures labeled with both dyes are combined and concentrated to 10 μl, using a Microcon filter device (Millipore, Bedford, MA) after adding 270 μl of TE buffer. At this time, if cDNA labeling is efficient, the concentrated cDNA appears purple. After adding 500 μl of TE buffer and 20 μg of salmon sperm DNA to the labeled cDNA, this solution is concentrated to 6 μl by centrifugation, followed by collection of the labeled DNA. When 22 × 22 mm glass coverslips are used, 1 μl of tRNA (20 μg/μl; Sigma, St. Louis, MO), 1 μl of poly(A) (20 μg/μl; Roche Diagnostics), 1.7 μl of 20× SSC, and 0.3 μl of 10% (w/v) SDS are added to the labeled DNA in a total volume of 10 μl. When larger glass coverslips are used, it is recommended that the volume of 20× SSC and 10% (w/v) SDS be increased in proportion to total volume.

4. Labeled cDNA is incubated at 100° for 2 min and cooled at room temperature for 30 min. Denatured cDNA is applied to the microarray, and a glass coverslip is put on it without introducing air bubbles. The glass coverslip is attached to the microarray with a paper bond to avoid being dried out during hybridization. Hybridization is performed at 65° for 12–16 hr of incubation. After removal of the glass coverslips in 2× SSC–0.5% (w/v) SDS, glass slides are washed twice in 2× SSC–0.5% (w/v) SDS for 5 min at room temperature, followed by a wash with 0.2× SSC–0.5% (w/v) SDS at 40° for 3 min. A final wash is done with 0.2× SSC for 3 min. Glass slides in slide cases are centrifuged at low speed to dry them. Hybridization images are scanned by a scanner, such as a ScanArray 5000 (GSI Lumonics, Billerica, MA), and analyzed by software such as Array Gauge (Fuji Film).

By assembling microarrays with a normalized kidney cDNA library as described above, we analyzed gene expression of the diseased kidney in HIGA mouse, an animal model of immunoglobulin A nephropathy.[7] We observed several

---

[7] S. Miyawaki, E. Muso, E. Takeuchi, H. Matsushima, Y. Shibata, S. Sasayama, and H. Yoshida, *Nephron* **76**, 201 (1997).

GPCRs to be upregulated in the diseased kidney. Among the GPCRs, leukotriene B_4 (LTB$_4$) and serotonin 5HT-2 receptors were included. LTB$_4$ is a chemoattractive lipid mediator, and an antagonist against LTB$_4$ receptor has been shown to inhibit nephrotoxic serum nephritis in rats.[8] 5HT-2 is known to stimulate proliferation and prostaglandin synthesis in mesangial cells by activating phospholipase and protein kinase C.[9] Taken together with these previous results, our microarray analysis strongly indicates that these GPCRs may play a role in the pathological phenotypes of this disease model. Furthermore, our microarray analysis showed that some GPCRs, which have never been noted to be involved in renal disorders, were markedly altered in their expression levels, suggesting that those GPCRs may play important roles in the diseased kidney. Also, as exemplified in this study, microarray analysis with a normalized library is a powerful approach to monitor changes in expression of the low-abundance genes, such as GPCRs. It may be due to the fact that the normalized library, constructed by either the subtractive library or hit-picking method, could enrich the low-abundance genes by reducing the high frequency of undesirable ("junky") clones. In conclusion, microarrays with the normalized cDNA library would be a powerful tool for functional genomics of GPCRs.

---

[8] S. Suzuki, T. Kuroda, J. I. Kazama, N. Imai, H. Kimura, M. Arakawa, and F. Gejyo, *J. Am. Soc. Nephrol.* **10,** 264 (1999).

[9] P. Mene, F. Pugliese, and G. A. Cinotti, *Hypertension* **17,** 151 (1991).

Author Index

A

Aaronson, S. A., 508
Abbondanzo, S. J., 209, 210(23)
Abbott, K. L., 539, 545, 550, 550(10; 15), 551
Abdel-Majidl, R. M., 227, 229
Abe, J., 443
Abel, T., 426
Abney, C. C., 477
Abramovitch, E., 472, 473(19)
Acker, J., 402
Ackerman, L., 68, 70(14)
Adam, A. C., 146
Adam, M. A., 540
Adamik, R., 397, 399, 401(6), 403, 403(6), 404(6)
Adamo, S., 105
Adams, B., 550(16), 551
Adams, M., 488
Adams, M. D., 585(5), 586
Adams, M. E., 67, 365
Adams, S. P., 175
Adati, N., 522
Aebersold, R., 508
Aelst, L. V., 396(53), 397
Afar, D. E., 466
Aggarwal, B. B., 508
Aguiar, J. Q., 276(33), 278
Aguilar, F., 557, 561(7)
Aharoni, A., 529(2), 530
Ahlijanian, M. K., 67
Ahmed, F. A., 37, 41(1)
Ahn, N. G., 413, 418(3)
Ahn, S., 88(26), 89, 90(26)
Aitken, A., 471, 471(14; 15), 472, 472(12), 481(12; 14), 482, 491
Akasaki, T., 355
Akkerman, J.-W., 384, 389(26), 396(26)
Alberts, A. S., 406
Albouz, S., 294
Aldrich, R. W., 188
Alessandrini, A., 490

Alessi, D. R., 89, 448, 448(9), 449, 450, 455(7), 458(16; 17), 459, 490, 499
Alewood, P., 508
Alexandropoulos, K., 383
Alexinsky, T., 427
Allen, D. G., 106
Allende, G., 37
Almaula, N., 569
Almiron, M., 235
Alnemri, E. S., 276(23), 278
Alon, U., 530
Althoefer, H., 404
Altschul, S. F., 527
Altschuler, D. L., 384, 387(24), 388(24)
Altshuller, Y. A., 266, 271(5)
Altshuller, Y. M., 255, 259, 266, 267(6), 271(6), 330, 331(6)
Alvarez, L. J., 513
Alvarez, R., 42, 95, 108, 111(28)
Amano, M., 371
Amess, B., 471(15), 472
Amherdt, M., 359, 360, 360(8), 363(8)
Amrein, K. E., 465
An, S., 299
Anantharam, A., 51, 53, 60
Anders, K., 530
Anderson, D. J., 230
Anderson, S. R., 508, 516(28)
Anderson, W. F., 542
Andjelkovic, M., 448, 448(9), 449, 450, 451, 455(7), 458(16; 17), 459, 460, 462, 462(19), 463
Andreotti, A. H., 469
Andrews, D. T., 530, 585
Andrews, L., 102
Andrews, R. K., 232(8), 233
Andrews, W. H., 522, 524
Anjard, C., 150
Anneren, C., 387(36), 388, 393(36)
Antonio, L., 67
Antonny, B., 28, 37(9), 402
Aosaki, T., 54
Applebury, M. L., 38(9), 39

601

Appleman, M. M., 29
Arakawa, M., 600
Arbabian, M., 188
Arden, S., 107
Arkinstall, S., 508
Arnold, B. S., 361
Arnold, R. S., 255
Arshavsky, V. Y., 30
Artemyev, N. O., 27, 28, 29, 30, 31, 31(12), 33(12; 13), 36, 36(25), 37, 37(9), 44
Arvindsson, A., 89
Ashworth, A., 490, 496
Aston-Jones, G., 427
Atherton, D., 102
Athos, J., 207, 220(15), 224(15), 226(15), 229, 583(22), 584
Attardi, B., 556
Ausubel, F. M., 248, 524
Avaeva, S. M., 341
Avruch, J., 384, 393, 396(25), 491
Aylwin, M. L., 189

B

Baba, K., 464
Babcook, J., 508
Bach, M. E., 230
Bachelot, C., 506
Bader, M.-F., 270
Bading, H., 581
Bae, C. D., 255, 256(7), 260(7), 261, 261(7), 264(17)
Bae, Y. S., 338
Baehr, W., 38(9), 39
Baginski, E. S., 512
Bagrodia, S., 396(53), 397, 491
Baharians, Z., 552
Bailey, C. H., 579
Baird, A. M., 383, 392(11), 395(11)
Baird, S., 425
Bakalyar, H. A., 127, 206
Baker, L. P., 207, 224, 225(42)
Bakre, M. M., 37, 41, 43(16), 44(16)
Balbin, M., 529
Balch, W. E., 360
Balla, T., 88(27), 89
Ballou, D. P., 508
Baltimore, D., 540
Bankaitis, V. A., 336

Banker, G. A., 385
Barad, M., 427, 431(14)
Barbar, D., 404
Barber, D. L., 404
Barbier, A. J., 128, 129(10), 130(10), 162(9), 163, 165, 166, 167(17), 169(17), 170(17), 172(13; 17), 173(13; 17), 175(13), 179(17), 232(17; 18), 233
Barkai, N., 530
Barker, J. L., 563
Barnard, D., 491
Barrett, T., 420, 421(19; 20)
Barria, A., 426
Bar-Sagi, D., 255, 415
Barth, A. L., 584
Bassett, D. E., Jr., 562
Bastianutto, C., 108, 111(28)
Basu, S., 276
Bate, M., 70
Bateman, K. P., 508
Bauerfeind, R., 340
Baumann, N., 294
Baykov, A. A., 341
Bayoumy, S., 276
Beauve, A., 164
Beavo, J. A., 29, 41, 43(16), 44(16), 207
Beckers, J. M., 366
Beeman, D., 126
Belchetz, P. E., 556
Bell, R. M., 286, 491, 501
Bellacosa, A., 448
Bellairs, R., 63
Benard, V., 349, 354, 355(5; 6), 396(52), 397
Bence, K., 464, 466, 466(3)
Bennett, N., 27
Berg, K. A., 556
Berg, L. J., 469
Berger, A. J., 154
Berk, B. C., 443
Berman, D. M., 181, 185, 407
Bernert, J. T., Jr., 294
Berridge, C. W., 426
Berridge, M. J., 107
Berthold, H., 477
Bertics, P. J., 169, 170(21), 182(21), 183(21)
Bevan, A. P., 452
Bezprozvanny, I., 76
Bhala, U. S., 3, 6(5)
Bhalgat, M. K., 517
Bhalla, U. S., 126, 426

Bhat, N. R., 276(26), 278
Bhattacharyya, D., 506
Bianchi, G., 112
Bielawska, A., 275(7), 276, 276(25; 32), 278, 286, 293(38)
Bilan, P. J., 450
Bird, M. I., 508
Birnbaumer, L., 56(19), 57, 184, 185(39), 404, 406(2), 409(2)
Birnbaumer, M., 184, 185(39), 404, 406(2), 409(2)
Birrell, G. B., 341
Bischoff, J. R., 389(39), 390, 393(39), 396(39)
Bisgaier, C. L., 294
Bishop, A. L., 371
Bishop, W. R., 286
Bitensky, M. W., 29
Bitter, I., 88(28), 89
Bittner, M. L., 529(3), 530
Bkalyar, H. A., 230
Black, M. J., 512
Blackshear, P. J., 521
Blackstock, W. P., 427
Blaese, R. M., 542
Blanchard, J. S., 509
Blasco, R., 529
Blendy, J., 579
Bligh, E. G., 279, 280(35)
Blinks, J. R., 106
Bliss, T. V., 426
Blitzer, R. D., 430
Blobe, G. C., 276
Bloomfield, C. D., 530, 564
Blumberg, D. D., 560
Blumberg, P. M., 506
Blume, R., 571
Blumenthal, D. K., 507
Bodmer, R., 70
Boeke, J. D., 141
Boeston, B., 164
Boguski, M. S., 561, 562, 563(11)
Bohl, B. P., 354, 355(5), 396(52), 397
Bohm, A., 318
Bokoch, G. M., 349, 354, 355(5), 387(35), 388, 396(52), 397
Bolen, J. B., 464
Bollag, G., 371, 371(10; 11), 372, 381(11), 382(10), 405, 407, 407(14; 15), 410(14; 15)
Bollag, R. J., 255

Bollag, W. B., 255
Bolognani, F., 540
Bolshakov, V. Y., 221
Boman, A. L., 359, 361
Bonaldo, M. F., 586
Bonifacino, J. S., 360
Bonjour, J. P., 528
Bonni, A., 571
Boring, D. L., 189
Borleis, J., 151
Bornfeldt, K. E., 41
Bos, J. L., 383, 383(18; 36), 384, 387(9; 18; 36), 388, 389(26), 391(6), 392, 393, 393(1), 396(26; 44)
Boss, G. R., 397
Boss, V., 539, 550, 550(15), 551
Bossy-Wetzel, E., 276(19), 278
Bost, F., 425
Botstein, D., 359, 530, 561, 563(11; 16), 564, 585
Bottaro, D. P., 508
Bouchard, D., 62
Boucher, P. D., 212, 220
Boudou, V., 115, 122(25)
Boulet, I., 507
Boulikas, T., 40
Bourne, H. R., 226, 241, 242(3), 404, 521, 522, 527(14)
Bourtchuladze, R., 579
Boutry, J. M., 294
Bowden, G. T., 276(30), 278
Bower, J. M., 126
Bownds, M. D., 30
Boyar, W. C., 255
Boyle, W. J., 507
Boynton, J. D., 295
Brachmann, R. K., 141
Bradford, M. M., 390, 475
Bradley, A., 207, 212, 214(19), 216(30), 219
Branch, K. D., 255
Brasier, A. R., 547, 549
Braun, P., 141
Braun, T., 95, 96(6), 105(1)
Bredesen, D. E., 276(22), 278
Breiner, M., 365
Brendel, S., 29, 30(24)
Brenner, B., 276(31), 278
Brenner, D. A., 276(25), 278
Brent, R., 248, 524
Bressman, S., 556(5), 557

Brewster, R., 70
Brickey, D., 571
Bright, N., 107, 360
Brin, M. F., 556(5), 557
Brinckerhoff, C. E., 528
Brindle, P. K., 570, 571
Brini, M., 107, 108, 111(28)
Broach, J. T., 276
Brodbeck, D., 448, 450
Broek, D., 241
Brose, N., 482
Brothers, C. A., 62
Brothers, G. M., 60, 62
Brown, A. M., 56(19), 57
Brown, H. A., 258, 261(12), 264(12), 328(3; 5),
    329, 331(5; 10), 332, 360, 363
Brown, P. O., 529, 530, 539, 561, 563(11; 16),
    564, 585
Brown, R. L., 31, 36(28), 39
Brown, S., 337
Brownstein, M., 107
Bruder, J. T., 384, 396(25)
Brugge, J. S., 464
Brune, M., 512
Brunet, A., 448
Brunissen, A., 487
Brunner, M., 359, 360(8), 363(8)
Brunton, L. L., 16
Buck, J., 95, 100, 102(2), 104, 105(2), 127,
    150, 231
Buckley, C. D., 383, 387(9)
Buensucuso, C. S., 383
Buhl, A. M., 404
Bujard, H., 550
Bunnell, S. C., 469
Bunz, F., 424
Burakoff, S. J., 89
Burbelo, P. D., 350
Burczak, J. D., 245
Burgering, B. M., 450
Burgess, J. W., 452
Burke, S. P., 508
Burn, P., 465
Burnay, M. M., 106
Burnett, P., 107
Buser, C. A., 273
Bushfield, M., 112
Bussey, A., 502
Bussione, R., 397
Button, D., 107

C

Cadwallader, K., 255, 257(3), 260(3), 264(3),
    265, 267, 267(4), 490(14), 491
Cahir McFarland, E. D., 507, 508(20)
Caldwell, J., 476, 478(28)
Cali, J. J., 112, 219
Caligiuri, M. A., 530, 564
Callahan, M., 244
Calvet, S., 487
Camden, J. M., 545, 550(10)
Campbell, A. K., 107
Campbell, D. G., 89, 420, 490, 496
Campbell, S., 491
Candia, A. O., 513
Canfield, D. R., 60
Cann, M. J., 95, 100, 102(2), 104, 105(2), 127,
    150, 151, 152, 152(6; 7), 158(6), 231
Cantley, L. C., 70, 72(6), 74, 89, 335, 502, 506
Cantor, C. R., 122, 306, 320(1)
Capecchi, M. R., 207
Capper, A. B., 241
Capponi, A. M., 106
Carafoli, E., 173, 176(24)
Carey, K. D., 141, 147(5), 383, 392(11), 395(11)
Carpenter, C. L., 335
Carpenter, G., 482
Carr, D. B., 563
Carr, D. W., 383
Carraway, K. L., 89
Casadio, A., 427, 431(14), 579
Casanova, J. E., 398
Casey, G., 295
Casey, P. J., 149
Casola, A., 547
Casperson, G. F., 241, 242(3)
Castro, M. G., 540
Catterall, W. A., 51, 57, 67
Caudwell, B., 448, 450, 455(7)
Caudwell, F. B., 448(9), 449
Caumont, A.-S., 270
Cavanagh, J., 420, 421, 421(19; 20)
Cavenagh, M. M., 359, 365
Caverzasio, J., 528
Cerasi, E., 472, 473(19)
Cericola, C., 397
Cerione, R. A., 27, 28, 371, 396(53), 397
Chabre, M., 27, 28, 37(9)
Chakravarthy, B. R., 502, 506
Chalfant, C. E., 275(7), 276

Chalk, P., 350
Chan, A., 508
Chan, C. P., 507
Chan, G., 571, 577(3), 579(3), 582(3)
Chan, G. C. K., 207
Chan, K. W., 71, 75, 80(1)
Chan, M., 224, 225(42)
Chan, T. A., 424
Chan, T. O., 448
Chang, P. P., 399
Chan-Hui, P. Y., 276
Chap, H., 304, 305(14), 337
Chapline, C., 476
Chardin, P., 28, 37(9)
Chassaing, G., 487
Chasserot-Golaz, S., 270
Chavkin, C., 207, 220, 220(15), 224(15),
     226(15), 227, 229, 557, 577
Chen, C. A., 449
Chen, C. H., 470, 472, 473(11; 22; 24), 476,
     478(11; 28; 29), 479(29), 483(11; 24),
     484, 484(24), 485(22; 24; 51), 486(11),
     487(22; 24), 488(24)
Chen, C. S., 89, 499, 506
Chen, D. F., 230
Chen, E., 141, 245, 470, 471(5)
Chen, H.-C., 399
Chen, J., 6, 7(10), 72(7), 74, 76(9), 206, 359
Chen, J. F., 531, 532(12)
Chen, L., 129, 149, 162(8), 163, 242, 472,
     473(18)
Chen, M., 579
Chen, S. S.-Y., 312, 314(6), 316(6)
Chen, W., 90, 506
Chen, Y., 166, 167(18), 172(18), 529(3), 530
Chen, Z., 162(9), 163, 169, 170(21), 182(21),
     183(21), 189
Cheng, Y.-C., 36
Chereson, A. R., 440
Cherfils, J., 402
Cheung, R., 318
Cheung, V. G., 557, 561(7)
Chia, H. P., 507
Chiang, Y. L., 542
Chiariello, M., 437, 438, 440, 440(6)
Chien, C. T., 70
Childs, G., 557, 561(7)
Chiono, M., 106
Choi, E. J., 206
Chomczynski, P., 560

Chou, M. M., 499
Choudhary, J. S., 427
Chousterman, S., 212
Chruscinski, A., 426, 427(9)
Chu, S., 564
Chu, Y., 437(4), 438
Chun, J., 299
Chung, E., 151, 152(6), 158(6)
Chung, J. K., 338
Chung, S., 62
Chung, U., 276(20), 278
Chuprun, J. K., 521
Cinotti, G. A., 600
Cioffi, C. L., 425
Cioffi, D., 579
Clapham, D. E., 51, 60
Clapp, P., 241
Clark, J., 360, 361(18), 367(18)
Clark, K. L., 255
Clark, O. H., 522
Clark, R. B., 188, 387(35), 388
Clark, S. F., 460
Clarke, W. P., 556
Clark-Lewis, I., 341
Clerc, A., 27
Cleveland, D. W., 480
Cleves, A. E., 336
Coadwell, J., 267
Cobb, M. H., 437
Cobbold, P. H., 106, 107
Cockcroft, S., 265
Codina, J., 56(19), 57
Coffer, P. J., 355, 448
Cogan, E. B., 341
Cohen, P., 305, 448, 448(9), 449, 450, 455(7),
     458(16), 459, 490, 496, 499
Coker, G., 312, 314(6), 316(6)
Colanzi, A., 397
Colhere-Garapin, F., 212
Coleman, D. E., 318
Collard, J. G., 355
Coller, H., 530, 564
Colley, W. C., 255
Collingridge, G. L., 426
Colowick, S. P., 117
Communi, D., 335
Conklin, B. R., 404
Connolly, M. L., 440
Conti, M., 105
Contos, J. J., 299

Cook, G. H., 115
Cook, S., 255, 257(3), 260(3), 264(3), 265, 267(4)
Cooke, F. T., 267, 336, 337, 337(10; 17), 341(10), 342(10)
Cool, D. E., 508
Cool, R. H., 383(18), 384, 387(18)
Cooper, D. M. F., 105, 106, 107, 108, 109(23), 112, 206, 239, 241
Cooper, J. A., 490, 491, 507
Cooper, R. G., 491
Copeland, N. G., 255
Corbin, J. D., 42
Corda, D., 397
Cormier, M. J., 107
Corrie, J. E. T., 512
Coso, O. A., 276(27), 278, 404, 438
Costa, T., 6
Costantini, F., 207, 213(20)
Cotecchia, S., 6
Cottom, J., 476
Couet, J., 206
Courtneidge, S. A., 338, 339(22)
Coussen, F., 127, 206
Coussens, L., 470, 471(5)
Couvillon, A. D., 89
Cowley, S., 490
Cox, D. H., 51
Crabb, J., 476
Craig Venter, J., 585(5), 586
Crasnier, M., 164
Cremona, O., 336
Crespo, P., 440, 521
Crews, C. M., 490
Crompton, A., 89
Cron, P., 448, 450, 451, 455(7), 458(16; 17), 459, 460, 462, 462(19)
Cross, D. A., 450
Csukai, M., 472, 473(24), 476, 478(29), 479(29), 483(24), 484(24), 485(24), 487(24), 488(24)
Cubbold, P. H., 110(29), 111
Cuffe, J. E., 188
Cui, Y., 294
Cunningham, C. C., 74, 76(9)
Curchod, M.-L., 508
Currie, R. G., 89
Cuthbertson, K. S., 106, 107
Cutler, R. E., 384, 490
Cutler, R. E., Jr., 491

Cuvillier, O., 276(27), 278
Cvejic, S., 464
Czarny, M., 265
Czech, M. P., 89

D

Daar, I. O., 491
Dadgar, J., 472, 473(22), 485(22), 487(22)
Daly, J. W., 188, 189
Damak, S., 43
Danchin, A., 164
Danho, W., 38(10), 39, 41(10), 44(10)
Daniel, L., 491
Daniels, D. V., 42
Darnay, B. G., 508
Dascher, C., 360
Dash, P. K., 579
Datta, S. R., 448
Daum, G., 508
David, A., 540
David, C., 340
Davis, A. C., 207, 214(19)
Davis, L. J., 395
Davis, M. E., 545, 550(10)
Davis, R. J., 413, 418(1), 419(1), 420, 421, 421(19–22), 422, 422(22)
Davis, R. L., 97, 151
Davis, R. W., 585
Dbaibo, G. S., 276, 276(21; 25), 278
De, C. P., 340
Deak, M., 89
Dean, N. M., 425
Debnath, J., 464
de Bruyn, K. M., 450
De Cammilli, P., 335, 336
DeClue, J. E., 384
Defer, N., 220
de Jersey, J., 508
De Lean, A., 6
de Leon, D., 556(5), 557
Delgado, P. S., 276(33), 278
Delikat, S., 276
Dell'Angelica, E. C., 360
Deloulme, J. C., 571, 577(3), 579(3), 582(3)
Del Vecchio, M., 579
DeMatteis, M. A., 397, 476, 478(29), 479(29)
Demott, M., 102
Deniell, L., 336

Dennerly, D. A., 304, 305(16)
Dennis, P. B., 450
Dent, P., 490
Denton, R. M., 107
Denu, J. M., 508
Der, C. J., 387(35), 388
Derijard, B., 413, 415(4), 420, 420(4)
DeRisi, J. L., 539, 564, 585
Derkach, V., 426
Derksen, A., 305
Dermott, J. M., 438
de Rooij, J., 383(18), 384, 387(18)
Derossi, D., 487
Desaubry, L., 112, 114, 114(8; 9), 115(9; 17),
    199
Desiderio, D. M., 305
Desnick, R. J., 294
de Souza, N. J., 189
Dessauer, C. W., 112, 113, 114, 114(11; 16),
    115, 115(11; 16; 22), 116(11; 22), 117(19;
    21), 120(11; 16; 22), 121(11), 122(16), 127,
    127(11), 128, 129(8; 9), 140(11), 160, 173,
    173(1), 178(1), 179(36), 180, 198, 199,
    199(8), 205(4), 231, 232(2; 10–16), 233
Deterre, P., 27
Devreotes, P. N., 151
de Vries, L., 60, 371
de Vries-Smits, A. M., 450
Dhanasekaran, N., 404, 438
Dhand, R., 338, 339(22)
Diaz, B., 491
Dickens, M., 421
Diehl, S. R., 384
Diez-Itza, I., 529
DiGirolamo, M., 397
Dignard, D., 242
DiJulio, D. H., 221, 226(38)
Dillehay, D. L., 276, 294(8)
Dillon, T. J., 383, 392(11), 395(11; 12)
Diltz, C. D., 507, 508
Ding, M., 276(30), 278
Di Paolo, G., 336
Disatnik, M. H., 482
Disteche, C. M., 220, 224(35)
Dittman, A. H., 206
Diversé-Pierluissi, M. A., 51, 53, 56, 57, 60,
    67(4), 69(4)
Dixon, J. E., 508, 510(29), 516(29)
Dmitrovsky, E., 564
Dobrowsky, R. T., 276(32), 278

Dobson, M. J., 227, 229
Dods, R. F., 95, 105(1)
Dodson, G. G., 482
Doe, C. Q., 70
Doi-Yoshioka, H., 294
Dolphin, A. C., 54
Donahue, R., 577, 580(13)
Donaldson, J. G., 397, 398
Dong, C., 415, 424(9)
Dong, Z., 276(30), 278
Doolittle, R. F., 456
Dormer, R. L., 107
Dorn, G. W. II, 472, 473(21; 24), 483(24),
    484(24), 485(21; 24), 486(21), 487(24),
    488(21; 24), 489(21)
Doronin, S., 232(11), 233
Døskeland, S. O., 4, 7(6)
dos Santos, M., 72(5), 74
Dousteblazy, L., 304, 305(14)
Dove, S. K., 336, 337, 337(10; 17), 341(10),
    342(10)
Dowd, S., 507, 508(17), 512(17)
Dowdy, S. F., 487
Dower, W. J., 245
Dowler, S. A., 89
Downes, C. P., 89, 448
Downing, J. R., 530, 564
Downward, J., 384, 490
Doyle, J., 220
Drayer, A. L., 337
Drechsel, D., 350
Dreyer, W. J., 38
Dreyfuss, G., 276(23), 278
Drucker, D. J., 521
Drugan, J. K., 491
D'Sa-Eipper, C., 221
Du, G., 265, 266, 267(6), 270, 271(6)
Duan, W., 547
Dubnick, M., 585(5), 586
Dudai, Y., 579
Duddy, S., 476
Dufner, A., 450
Duh, Q.-Y., 522
Duman, R. S., 584
Dumke, C. L., 30
Dumler, I. L., 31
Dunlap, K., 51, 56, 57, 60, 67(4), 69(4)
Durkin, J. P., 502
Duronio, R. J., 175
Dutil, E. M., 499

Dyer, E. L., 137
Dyer, W. J., 279, 280(35)

E

Ebersole, B. J., 556, 556(5), 557, 569
Ebina, T., 96
Ebisu, S., 276(20), 278
Eck, M. J., 469
Eckert, S. P., 383, 395(12)
Eckstein, F., 29
Eckstein, J. W., 508
Edamatsu, H., 521, 527, 527(12)
Edelhoff, S., 220, 224(35)
Edwards, A. S., 502
Eguinoa, A., 267
Eichholtz, T., 299
Eisen, M. B., 530, 561, 562, 563(11; 16), 564, 585
Eisfelder, B. J., 521, 527(15), 553
Elenko, E., 371
Ella, K. M., 295
Elledge, S. J., 384, 396(25)
Ellig, C. L., 383, 395(12)
Ellington, M. L., 539
Ellis, C. A., 471, 471(14), 472, 472(12), 481(12; 14)
Ellis, K. J., 509
elMasry, N., 371
El-Sherif, N., 472, 473(18)
Emerman, M., 541
Emerson, L. R., 530
Emr, S. D., 335, 339
End, P., 338
Engebrecht, J. A., 255, 266, 269(8), 271(7), 273(8), 296, 330, 331(6), 359
English, D., 294, 328
English, J. D., 427, 430(12; 13)
Enkvetchakul, D., 77
Enlow, B. E., 522, 524
Enslen, H., 571
Epstein, D. M., 508
Erdjument-Bromage, H., 267
Erickson, J. W., 28
Erikson, R. L., 490
Erlich, S., 294
Erneux, C., 335, 337
Essen, L. O., 318
Etingof, R. N., 31

Eusebi, F., 189
Evanczuk, A. T., 29
Evangelopoulos, A. E., 341
Evans, R. M., 230
Evans, T., 226, 387(34), 388
Evans, W. H., 107
Eversole-Cire, P., 404
Evtushenko, O. A., 341
Ewald, D. A., 53
Exton, J. H., 255, 256, 256(7), 259, 259(11), 260, 260(7), 261, 261(7), 262(15), 264(13; 15; 17), 296, 328(2), 329, 331, 404

F

Fabian, J. R., 491
Fagan, K. A., 106, 107, 108, 109(23)
Fahrenfort, I., 299
Fakhrai, H., 425
Falck, J. R., 76, 463, 506
Fales, H. M., 361
Falke, J. J., 482
Fallon, J. R., 427
Fancy, D. A., 129, 130(22), 198
Fang, X.-J., 295
Fanjul, L. F., 276(33), 278
Fantus, I. G., 452
Farquhar, M. G., 371
Farr, G. H. III, 407(23), 408
Fattaey, A., 141
Faure, M., 521
Faure, R., 452
Feder, J. N., 70
Federman, A., 226
Feichtinger, H., 522
Feig, L. A., 396(51), 397
Feinstein, P. G., 97, 127, 151, 206, 230
Feldmuller, M., 508
Feller, S. M., 390
Felsenstein, J., 155
Feng, J., 522, 524
Feng, S., 71, 81(4), 82(4), 83(4), 87(4), 469
Fenton, A. A., 427
Ferguson, K. M., 185
Ferlinz, K., 294
Fernandez, A., 450, 458(16), 459
Fernandez, J., 102
Ferrans, V. J., 403
Ferrell, J. E., Jr., 507, 508(18), 512(18)

Ferrige, A. G., 361
Fest, W., 301
Fields, C., 585(5), 586
Fields, P. E., 389(38), 390
Fields, S., 141, 171, 183(22)
Figueroa, X. A., 207, 220(15), 224(15), 226(15), 229
Finan, P. M., 337, 340(14)
Finazzi, D., 398
Find, A., 227, 229
Fink, S., 556(5), 557
Fischer, E. H., 507, 508, 516(28)
Fischer, S. G., 480
Fischer, T., 371
Fishman, P. A., 163
Fitch, F. W., 389(38), 390
Fiucci, G., 265
Fjeld, C. C., 508
Flanagan, L. A., 74, 76(9)
Flati, S., 397
Flavell, R. A., 221, 336, 415
Fleming, I. N., 261, 264(17)
Fletcher, J. A., 384
Fleurdelys, B., 395
Flint, N. A., 465
Florio, V. A., 112, 113
Foa, F. F., 512
Fodor, S. P., 557, 561(8)
Folch, J., 279
Foote, S. L., 426
Forman-Kay, J., 70
Forsgren, L., 556(5), 557
Foster, L. M., 276(22), 278
Fox, K., 584
Francis, S. H., 42
Frank, H., 95
Franke, B., 384, 389(26), 393, 396(26)
Frankel, A., 295
Frech, M., 450, 458(16), 459, 463
Freije, J. M., 529
Freissmuth, M., 135, 232(9), 233, 385
Frenguelli, B., 579
Friday, B. B., 550
Friedman, J., 556(5), 557
Fritsch, E. F., 244, 446(13), 447
Frohman, A. A., 296
Frohman, M. A., 255, 257(3), 259, 260(3), 264(3), 265, 266, 267(4; 6), 269, 269(8), 270, 271(5–7), 272(9), 273(8), 295, 328(4), 329, 330, 331(6), 398, 491

Frohman, M. J., 255
Fromm, C., 404
Frorath, B., 477
Fruman, D. A., 72(6), 74
Fry, D., 38(10), 39, 41(10), 44(10), 46
Fry, M. J., 338, 339(22)
Fuhrman, S., 563
Fujli, M., 488
Fujisawa, H., 415
Fujisawa, K., 350
Fujita, A., 350
Fujita, T., 294
Fuks, Z., 276(29), 278
Fukuda, Y., 141
Fulgham, D. L., 169, 170(21), 182(21), 183(21)
Fuller, D., 150
Fung, B. K.-K., 27, 28
Funk, W. D., 522, 524
Furth, M. E., 395
Furuya, S., 274(9), 276
Furuyashiki, T., 350
Fusella, A., 397
Futcher, B., 530

G

Gaasenbeek, M., 530, 564
Gadi, I., 209, 210(23)
Gaffney, P. R., 448
Gage, F. H., 540
Gajewski, T. F., 389(38), 390
Galarreta, C. M. R., 276(33), 278
Galetic, I., 448
Gallego, C., 521, 527(15; 16)
Gallis, B., 507
Gamard, C. J., 276(21), 278
Gamble, T. N., 508
Gamblin, S. J., 482
Gao, B. N., 230
Garapin, A. C., 212
Garber, S. S., 188, 486
Garbers, D. L., 137
Garcia, D. W., 51
Garcia, E. P., 340
Garcia, J. V., 540, 544(2)
Garcia, P., 307
Garcia-Sainz, J. A., 37
Garfield, S., 506
Garofalo, R. P., 547

Garotta, G., 200
Garrison, J. C., 7, 56, 180, 182(37), 185(37), 186(37), 187(37), 232(19), 233
Gaudette, D. C., 295
Gautam, M., 404
Ge, C., 383
Gee, K. R., 508, 517
Geijsen, N., 355
Gejyo, F., 600
Geladopoulos, T. P., 341
Gelinson, J., 299
Geneste, O., 406
Gennis, R. B., 318
George, C. H., 107
Geppert, M., 221
Gerber, D., 230
Geremia, R., 105
Gerrard, J. M., 304
Geske, R., 212
Gether, U., 6
Ghosh, S., 491
Gibbs, J. B., 384
Gibbs, T. C., 295, 296
Gibson, S., 437, 438(3)
Gietz, R. D., 242, 246(9)
Gilbert, R. L., 404
Gilbertson, T. A., 43
Gillespie, P. G., 29, 41
Gilman, A. G., 40, 54, 56(19), 57, 62, 112, 113, 114, 114(11–13; 16), 115, 115(11; 13; 16; 22), 116(11; 22), 117(19; 21), 120(11; 16; 22), 121(11), 122(13; 16), 127, 127(11), 128, 128(6), 129, 129(8; 9), 130(22), 135, 137(14), 139, 140, 140(11), 149, 150, 160, 162(4; 6; 7), 163, 166, 173, 173(1; 6; 19), 175, 178, 178(1; 6), 179(36), 180, 181, 183(19), 185, 192, 197(22), 198, 199, 199(3; 6; 8), 200, 201(13a), 202(13), 205(2; 4), 206, 230, 231, 232, 232(2; 4; 10; 12–16; 21; 22), 233, 236, 239, 239(27), 249, 318, 359, 371, 371(10; 11), 372, 375, 381(11), 382(10), 404, 405, 407, 407(14; 15), 410(14; 15), 464, 521
Gingeras, T. R., 557, 561(8)
Ginty, D. D., 571, 581
Giordano, J., 402, 403(22)
Giovannini, M. G., 426
Gish, K., 530
Gish, W., 527
Giuriato, S., 337

Glazewski, S., 584
Glick, J. G., 41, 43(16), 44(16), 149
Gobert, R. P., 508
Goda, Y., 221
Goeddel, D. V., 245
Goetzl, E. J., 299
Goi, T., 396(51), 397
Gold, M. R., 383, 391(6)
Goldfarb, J., 556
Goldin, A. L., 75
Goldsmith, P. K., 51
Golub, T. R., 530, 564
Gomez, N., 496
Gonzales, E., 107, 108, 109(23)
Gonzalez, G. A., 571
Gonzalez, I. H., 276(33), 278
Goodbody, A., 295
Gordeladze, J. O., 95, 96(5)
Gordon, J., 175, 458, 475, 477(27), 480(27)
Gosselin, G., 113, 114(13), 115, 115(13), 122(13; 25), 128, 137(14), 232(22), 233
Gossen, M., 550
Gotoh, T., 384
Gotoh, Y., 415, 418(13)
Gottlin, E. B., 508
Gout, I., 338, 339(22)
Govaerts, C., 335
Goy, M. F., 56
Goyns, M. H., 106
Grabs, D., 340
Graf, R., 56(19), 57
Graham, L. K., 499
Grandea, A. G. d., 476
Granovsky, A. E., 27, 36, 37(9)
Grant, S. G., 427
Grassi, F., 189
Graves, B., 38(10), 39, 41(10), 44(10)
Gray, M. O., 470, 472, 473(11; 19; 20; 24), 478(11), 483(11; 20; 24), 484, 484(24), 485(24; 51), 486(11; 20), 487(24), 488(24)
Graybiel, A. M., 571
Graziano, M. P., 56(19), 57, 135
Greeley, D., 38(10), 39, 41(10), 44(10)
Green, D. R., 276(19), 278
Greenberg, M. E., 448, 571, 581
Greener, A., 244
Greengard, P., 117
Grewal, S. S., 383(17), 384, 385, 387(30)
Griffith, L. C., 426
Griffith, O. H., 341

Griffiths, G., 364
Grignani, F., 540, 542(7)
Grimes, D., 556(5), 557
Grinstein, S., 521
Grips, M., 490
Groffen, J., 354, 355(6)
Grünewald, K., 522
Gu, C., 106
Gu, Q.-M., 255, 257(3), 260(3), 264(3)
Guan, C. D., 476
Guan, Z.-W., 305
Gudz, T. I., 276
Guernsey, D. L., 227, 229
Guillou, J. L., 107, 108, 109(23)
Gulbins, E., 276(31), 278
Guo, J. H., 531, 532(12)
Guo, M., 70
Guo, S., 336, 340(9), 342(9)
Guo, Z., 522
Gupta, R., 307
Gupta, S. K., 414, 415(7), 420, 422, 424(7), 521, 527(15; 16)
Gurtner, G., 212, 220(27), 227
Gurvich, N., 556(5), 557
Gutkind, J. S., 437, 438, 438(1), 440, 440(6), 521
Gutkind, S., 276(27), 278, 404
Gutowski, S., 258, 261(12), 264(12), 328, 328(5), 329, 330, 331(5; 7), 363, 371
Gutstein, H. B., 67

H

Hacker, B. M., 224, 225(42), 227, 229
Hagiwara, H., 522
Hahn, D., 114, 117(18), 127, 128(7), 129(7), 232, 233(5), 236
Hahn, H., 472, 473(21), 485(21), 486(21), 488(21), 489(21)
Haimovitz-Friedman, A., 276(29), 278
Hall, A., 350, 354, 371, 379, 380(14)
Hall, D. A., 452
Hall, J. P., 413
Hall, M. N., 337, 337(17)
Hall, R. A., 62
Hallberg, B., 490
Hallbrink, M., 487
Hallett, M. B., 107
Halnon, N. J., 129, 162(8), 163, 242

Hamaguchi, M., 391, 396(43)
Hamer, D. H., 242
Hamill, O. P., 53, 79
Hamilton, S., 508
Hamm, H. E., 28, 29, 30, 31(12), 33(12), 36(25), 44, 318
Hammer, R. E., 221
Hammes, G. G., 136
Hammond, S. A., 266, 271(5)
Hammond, S. M., 255, 257(3), 260(3), 264(3), 265, 266, 267(4), 271(7), 330, 331(6)
Han, J. M., 266, 267(6), 271(6)
Han, J.-S., 338, 398
Han, J. W., 450
Han, P. L., 97, 151
Hanahan, D. J., 304, 305
Hancock, J. F., 490(14), 491
Hancock, R., 40
Haneda, M., 437(5), 438
Hanks, S. K., 495
Hannun, Y. A., 275, 275(6; 7), 276, 276(4; 18; 21; 24; 25; 32), 278, 286, 293, 293(38), 501
Hanoune, J., 220
Hansel, C., 227, 229
Hansson, V., 95, 96(5)
Harder, K. W., 341
Harding, A., 491
Hardman, J. G., 137
Harlow, E., 40, 141
Harris, A., 471, 472(12), 481(12)
Harrison, S. C., 469
Harry, A., 166, 167(18), 172(18)
Harsh, G., 522
Hart, A. C., 154
Hart, M. J., 371, 371(10), 372, 382(10), 405, 407, 407(14; 15), 410(14; 15)
Hartman, A. R. E., 227
Hartwieg, E., 482
Hartwig, J. H., 506
Harvey, B. J., 188
Hashimoto, Y., 391, 396(43)
Hasty, P., 216(30), 219
Hata, A., 466
Hatase, O., 384
Hatley, M. E., 127, 150, 200, 201(13a)
Hattori, M., 383
Hattori, S., 384, 464, 469(4)
Haugland, R. P., 517
Haun, R. S., 399

Hauri, H. P., 364
Hausler, A., 556
Hauw, J. J., 294
Hawkins, P. T., 267
Hawkins, R. D., 230
Hayaishi, O., 117
Hayashi, Y., 184, 185(40), 187(40)
Haydar, T. F., 221
Hayes, F., 266, 271(5)
Haystead, C. M., 490
Haystead, T. A., 490
Hazvi, S., 579
He, C., 75, 81(16), 82(16), 83(16), 87(16)
Heasley, L. E., 438, 521, 527(16)
Heid, C. A., 565
Heinicke, T., 491
Hellevuo, K., 220, 239
Hellriegel, M., 490
Helms, J. B., 398
Hemeking, H., 424
Hemmings, B. A., 448, 448(9), 449, 450, 451,
    455(7), 458(16; 17), 459, 460, 462,
    462(19), 463, 552
Henderson, H. E., 216(29), 219
Hepler, J. R., 239, 371, 521
Her, J. H., 490
Herlitze, S., 51, 57
Hernandez-Sotomayor, S. M., 482
Herrmann, C., 383, 389(40), 390, 393
Herskowitz, I., 564
Hescheler, J., 54
Hess, P., 54, 415
Heuschneider, G., 188
Hibi, M., 414, 415(6), 417(6), 422(6)
Higashijima, T., 185
Higuchi, R., 249, 566
Hilal-Dandan, R., 16
Hiles, I., 338
Hilgemann, D. W., 71, 72(2), 75, 81(4), 82(4),
    83(4), 86, 87(4)
Hill, C. S., 405
Hill, M. M., 448, 460
Hill, P., 584
Hille, B., 51, 60
Hinds, T. R., 206
Hingorani, V. N., 28
Hinzmann, B., 490
Hirabayashi, Y., 274(9), 276
Hirasawa, A., 585
Hirose, K., 89

Hirst, J., 360
Ho, A., 487
Ho, I. H. M., 72(8), 74, 81(8), 83(8)
Ho, P. S., 306, 320(2)
Ho, W.-T., 255, 256, 256(7), 259(11), 260(7),
    261(7)
Ho, Y.-K., 28
Hochner, B., 579
Hockerman, G. H., 57
Hod, Y., 532
Hoffman, P. L., 239
Hofmann, F., 173, 176(24)
Hogan, B., 207, 213(20)
Hohenegger, M., 232(9), 233
Hokin, L. E., 286
Hokin, M. R., 286
Hokin-Neaverson, M., 189
Holden, J., 261, 264(16)
Holdorf, A. D., 383, 392(11), 395(11)
Hollenberg, S. M., 490
Holler, C., 385
Holmes, A. B., 337, 337(17), 448
Holmgren, G., 556(5), 557
Holub, B. J., 295
Homcy, C., 129, 162(5; 8), 163, 242
Homma, Y., 521(17), 522, 527(17)
Honda, A., 398
Hooley, R., 404
Hope, M. J., 317
Hoppel, C. L., 276
Horgan, A. M., 385
Horinouchi, K., 294
Horn, G., 393
Horodniceanu, F., 212
Horowitz, J. M., 276
Horstmann, H., 364
Horvitz, H., 482
Hoshi, T., 188
Hou, W., 499
Houghten, R. A., 440
House, C., 484
Howard, P., 482
Howell, S., 471(15), 472
Hoyt, M. A., 359
Hsuan, J. J., 72(5), 74, 337, 338, 340(14)
Hu, C. D., 384, 387(29), 393(29), 394(29)
Huang, C.-L., 71, 81, 81(4), 82(4), 83(4; 24),
    87(4), 276(30), 278
Huang, W., 490
Huang, X., 410

Huang, X.-P., 529
Huang, X. Y., 387(33), 388, 464, 466, 466(3), 469, 469(4)
Huang, Y. Y., 230
Huang, Z.-H., 114, 117(18), 127, 128(7), 129(7), 232, 232(6–8), 233, 233(5), 236
Huard, C., 530, 564
Hudson, J., Jr., 561, 563(11)
Hughes, K. T., 335
Hughes, W. E., 336, 337(10), 341(10), 342(10)
Hulmes, J., 476
Hundle, B., 472, 473(22), 485(22), 487(22)
Hunt, D. F., 490
Hunter, J. L., 512
Hunter, T., 495, 507
Hurley, J. B., 27, 127, 128, 192, 197(21)
Hurley, J. H., 198, 231, 232(1; 20), 233
Hurley, T. R., 507, 508(20)
Hurteau, J., 295
Husi, H., 427
Hutchison, J. S., 556
Huttner, W. B., 301
Hutton, J. C., 107
Hwang, P. M., 97, 151

Iourgenko, V., 152, 158(9)
Ip, Y. T., 413, 418(1), 419(1)
Iranfar, N., 150
Irie, S., 383
Irvine, R. F., 335
Isaacs, R. J., 506
Ishakawa, Y., 96
Ishibashi, T., 508
Ishikawa, Y., 129, 162(5; 8), 163, 206, 242
Ishisaki, T., 350
Ishizaka, K., 556
Ishizaki, T., 350
Ito, A., 521
Ito, H., 141
Ito, T., 522
Itoh, H., 521, 527, 527(12), 571
Itoh, T., 336
Iwanaga, S., 107
Iyengar, R., 6, 7(10), 112, 126, 127, 140, 141, 147(5), 160, 166, 167(18), 172(18), 173(2), 178(2), 206, 383, 426, 430
Iyer, V. R., 530, 539, 561, 563(11), 585

I

Idzerda, R., 221, 226(38)
Igarashi, T., 276(20), 278
Iino, M., 89
Ikawa, H., 585
Ikawa, Y., 383
Ikeda, H., 239
Ikeda, S. R., 51, 53, 60
Ikegaki, N., 488, 506
Ikehara, Y., 397
Ikezu, T., 521(17), 522, 527(17)
Imai, N., 600
Imamura, A., 212, 220(27)
Impey, S., 206, 207, 224, 225(42), 557, 570, 571, 577, 577(3), 579(3), 580(13), 582(3), 583(22), 584
Inagaki, M., 184, 185(40), 187(40)
Ingham, R. J., 383, 391(6)
Ingley, E., 463
Iniguez-Lluhi, J. A., 239, 318
Inouye, S., 107
Insel, P. A., 16, 226

J

Jackson, C. L., 398, 402
Jackson, S. P., 337
Jacobowitz, O., 6, 7(10), 206
Jacobson, K. L., 106, 221, 226(38)
Jahn, R., 482
Jahr, C. E., 37
Jain, V. K., 169
Jaken, S., 476
Jakobs, K. H., 112
Jakubowicz, T., 448, 450
Jalink, K., 299
Jamaluddin, M., 547
James, D. E., 460
James, S. R., 448
Jamieson, L., 476, 478(28)
Jan, L. Y., 68, 70, 70(14), 75
Jan, Y. N., 68, 70, 70(14)
Janmey, P. A., 74, 76(9)
Janz, R., 221
Jarpe, M. B., 437, 438(3)
Jascur, T., 465
Jaspers, S., 571
Jayadev, S., 276

Jeffrey, J. J., 528
Jeffrey, S. S., 585
Jenco, J. M., 255, 257(3), 260(3), 264(3), 265, 267(4), 307
Jenkins, G. M., 276(18), 278
Jenkins, N. A., 255
Jensen, O. N., 427(21), 428
Jeong, S. W., 53
Jiang, H., 404
Jiang, M. M., 70
Jiang, X., 328, 330, 331(7; 10), 332, 371, 371(10; 11), 372, 381(11), 382(10), 405, 407(14; 15), 410(14; 15)
Jiang, Y., 410, 464, 469(4)
Jimenez, M. G., 529
Jin, J., 450
Jin, T., 71, 75, 89(18), 90(18), 92(18)
Jin, T. G., 384, 387(29), 393(29), 394(29)
Jin, Y., 482
Jirik, F., 341
Johannes, F. J., 502
Johnson, F. H., 106
Johnson, G. L., 404, 437, 438, 438(3), 521, 527(15; 16), 553, 571
Johnson, J. A., 470, 472, 473(11; 18; 19), 478(11), 483(11), 484, 485(51), 486(11), 499
Johnson, K. S., 416, 476
Johnson, N. L., 404
Johnson, P., 341
Johnson, R. A., 95, 112, 113, 114, 114(8; 9; 13; 16), 115, 115(9; 13; 16; 17), 116, 120(16), 122(13; 16; 25), 128, 137, 137(14), 198, 199, 205(2), 232(11; 22), 233
Johnson, W. C., 306, 320(2)
Joliot, A. H., 487
Jones, D., 265, 471(15), 472, 482, 491
Jones, G., 482
Jones, L. P., 53
Jones, M. E., 512
Jones, P. F., 448
Jones, S. N., 415
Jonkers, J., 407(23), 408
Joost, H. G., 365
Jordan, D. J., 60, 383
Jordan, J. D., 140, 141, 147(5)
Jordan, K. A., 396(53), 397
Jorgensen, E., 482
Joyner, A., 208, 213(21)
Juni, N., 464

K

Kadowaki, S., 89
Kafri, T., 540
Kahn, R. A., 261, 264(16), 331, 359, 360, 360(8), 361, 361(18), 363, 363(8), 365, 367(18)
Kaibuchi, K., 371
Kain, S. R., 488
Kamath, R., 359
Kameshita, I., 415
Kamibayashi, C., 276
Kaminishi, Y., 585
Kamps, M. P., 507
Kanaho, Y., 266, 272(9), 398
Kandel, E. R., 230, 426, 427, 427(9), 431(14), 579
Kang, H. S., 338
Kaplan, J. M., 154
Kaplan, M. W., 29
Karagenic, N., 107
Karin, M., 414, 415(6), 417(6), 422(6)
Kariya, K., 384, 387(29), 393(29), 394(29)
Karliner, J. S., 472, 473(20), 483(20), 484, 485(51), 486(20)
Karolak, L. A., 275, 276(4)
Karpen, J. W., 105
Kasai, H., 54
Kashiwagi, K., 488
Kassis, S., 163
Katagiri, K., 383
Katan, M., 318
Kataoka, T., 241, 384, 387(29), 393(29), 394(29)
Katayama, K., 255
Katbuchi, K., 384
Katsuma, S., 585
Katsushika, S., 162(5), 163, 242
Kawabe, J.-I., 96, 129, 162(5; 8), 163, 242
Kawamoto, K., 398
Kawasaki, E., 522
Kawasaki, T., 294
Kay, L. E., 70
Kayalar, C., 276(22), 278
Kazama, J. I., 600
Kaziro, Y., 384, 521, 527, 527(12), 571
Kaznacheyeva, E., 76
Keane, R. W., 276(22), 278
Keijer, J., 529(2), 530
Kelly, J. M., 585(5), 586

Kelly, K., 437(4), 438
Kemp, B. E., 484
Kenakin, T., 556
Kendall, J. M., 107
Kennedy, E. P., 294
Kennedy, M. B., 427, 427(21), 428
Kent, R. S., 6
Kerlavage, A. R., 585(5), 586
Keyse, S. M., 507, 508(17), 512(17)
Khan, J., 529(3), 530
Khaner, H., 470, 471(9; 13; 16), 472, 475(13; 16), 480(13), 481(13), 482(9)
Khorana, H. G., 39
Kikkawa, R., 437(5), 438
Kikkawa, U., 488
Kiley, S., 476
Kim, E.-G., 259, 264(13)
Kim, H. D., 188
Kim, H. K., 255
Kim, S. R., 338
Kim, W. T., 336
Kim, Y., 266, 267(6), 271(6), 508
Kimelman, D., 407(23), 408
Kimura, A., 141, 507
Kimura, H., 600
Kinashi, T., 383
Kincaid, R. L., 29
Kinet, J. P., 466
King, A. J., 491
King, M. J., 507
Kingston, R. E., 248, 524
Kinsella, T. M., 540, 542(7)
Kinzler, K. W., 424
Kiosses, W. B., 358, 396(54), 397
Kirihara, J., 507
Kirschner, M. W., 480
Kishikawa, K., 275(7), 276
Kitabatake, A., 149, 383
Kitahara, S., 556
Kitareewan, S., 564
Kitayama, H., 383, 384
Kito, K., 522
Kitts, P., 488
Kiyokawa, E., 149, 383
Kjoller, L., 371
Klamo, E., 571
Klaubert, D. H., 517
Klauck, T., 476
Klausner, R. D., 397, 398, 507
Klein, C., 556(5), 557

Klemm, D. J., 571
Kleuser, B., 276(27), 278
Kleuss, C. K., 29, 30(24), 114, 117(21), 128, 129(8), 173, 199, 232(16), 233
Klinger, M., 385
Kliot, B., 152
Klip, A., 450, 521
Kluck, R. M., 276(19), 278
Knaus, U. G., 354, 355(6)
Knepel, W., 571
Knoepp, S. M., 295
Knopfel, L., 173, 176(24)
Kobilka, B. K., 6, 426, 427(9)
Kobrinsky, E., 75, 89(18), 90(18), 92(18)
Kodaki, T., 255, 338, 339
Koenderman, L., 355
Koga, H., 355
Koheon, T., 563
Kohn, K. W., 530, 585
Kok, E. J., 529(2), 530
Kolesnick, R. N., 275, 276, 276(2; 29), 278
Kolter, R., 235
Kondepudi, A., 488
Kondo, S., 464
Konishi, H., 488
Köntgen, F., 210
Koop, E. A., 383, 387(9)
Koppelman, L. F., 550
Koppenhoefer, U., 276(31), 278
Koretzky, G. A., 276(28), 278
Kornhauser, J. M., 581
Korswagen, H. C., 154
Koshland, D. E., Jr., 502, 503
Kosik, K. S., 74, 76(9)
Kosutsumi, Y., 294
Kotani, H., 542
Kourilsky, P., 212
Kovacsovics, T., 506
Kozasa, T., 40, 149, 371(10; 11), 372, 375, 381(11), 382(10), 405, 407(14; 15), 410(14; 15), 464, 466(3), 469(4), 521
Kozma, S. C., 450
Kramer, F. R., 565
Kramp, W., 305
Kratzin, H., 301
Krebs, E. G., 443, 490, 507
Kreiman, G., 427
Kreis, T. E., 364
Krishna, U. M., 76
Krueger, N. X., 507

Kruger, M., 571
Krumins, A. M., 62
Krupinski, J., 127, 206, 219, 230, 236, 239, 239(27)
Krupouski, J., 112
Ktistakis, N. T., 360
Kuai, J., 359, 361
Kuan, C. Y., 221, 424
Kubiak, P., 427
Kucherlapati, R., 557, 561(7)
Kuhn, H., 38
Kuida, K., 221
Kular, G., 89
Kumakura, S., 294
Kumar, S., 276(23), 278
Kurashina, Y., 117
Kurata, T., 149, 383, 384, 391, 396(43)
Kurihara, P., 43
Kuriyama, R., 359
Kuriyan, J., 469
Kuroda, S., 371, 394
Kuroda, T., 600
Kurosaki, T., 383, 391(6), 465, 466
Kuruga, M., 212, 220
Kusuhara, M., 443
Kuzmic, P., 126
Kwak, K. S., 41
Kwok, S., 566
Kyle, D. E., 530
Kyllerman, M., 556(5), 557
Kyriakis, J. M., 384, 393, 396(25)

L

Lachance, D., 452
Lacy, E., 207, 213(20)
Laemmli, U. K., 62, 458, 480
Lai, D. S. Y., 507
Lakowicz, J., 306, 307(3), 309(3), 320(3)
Lamb, N. J., 450, 458(16), 459
Lambeth, J. D., 255
Lambright, D. G., 318
Lammers, J.-W. J., 355
Lamprecht, R., 579
Lan, M. S., 507
Landau, E. M., 141, 430
Lander, E. S., 530, 564
Landis, C. A., 522

Lane, A., 491
Lane, D., 40
Lanfrancone, L., 540, 542(7)
Lang, A. E., 556(5), 557
Lang, F., 276(31), 278
Langel, U., 487
Lanzetta, P. A., 513
Lapetina, E. G., 384, 387(24), 388(24)
Lasala, D. J., 255
Lashkari, D., 561, 563(11), 585
Latham, K. A., 517
Latinis, K. M., 276(28), 278
Laurenza, A., 189, 191
Lavin, M., 276(23), 278
Ledbetter, J. A., 507
Lee, C., 338
Lee, E., 129, 200, 202(13), 318
Lee, H., 299
Lee, J. A. C., 110(29), 111
Lee, J. C., 585
Lee, J. C. F., 561, 563(11)
Lee, J. D., 443
Lee, J. K., 530, 585
Lee, J. Y., 276
Lee, M. H., 499
Lees, M., 279
Leevers, S. J., 490(15), 491
Leff, P., 556
Lefkowitz, R. J., 6, 62
Le Good, J. A., 499
Leinders-Zufall, T., 207
Lemrow, S. M., 336, 340(9), 342(9)
Lenardo, M. J., 464
Lengauer, C., 424
Lennon, G., 586
Leong, W. L., 227, 229
Lerner, R. A., 440
Le Saux, F., 294
Leung, D. W., 245
Leung, J., 556(5), 557
Levin, L. R., 95, 97, 100, 102(2), 104, 105(2), 127, 150, 151, 152, 152(6; 7), 158(4; 6; 9), 173, 176(25), 231
Levine, A. J., 530
Lewis, T. S., 413, 418(3)
Li, L., 407(23; 24), 408
Li, Q., 27
Li, S. C., 70
Li, W., 389(37), 390
Li, Y., 476

Liang, P., 522
Liao, X. C., 464
Lichtenstein, H., 276
Lickteig, R. L., 571
Lidia, A., 522
Liebman, P. A., 29
Light, Y., 384, 393, 393(27)
Liliom, K., 305
Lin, A., 414, 415(6), 417(6), 422(6)
Lin, F., 532
Lin, K. C., 151
Lin, S., 6
Lin, X. H., 276
Linardic, C. M., 275, 276(4)
Linden, D., 227, 229
Linden, J., 36
Linder, M. E., 129, 175, 200, 202(13)
Linderkamp, O., 276(31), 278
Lindgren, M., 487
Lindorfer, M. A., 7, 56, 180, 182(37), 185(37), 186(37), 187(37), 232(19), 233
Lin-Goerke, J. L., 245
Linskens, M. H. K., 522, 524
Liou, H.-H., 81, 83(24)
Lipkin, V. M., 31
Lipman, D. J., 527
Lipp, H. P., 219, 225(32)
Lippinscott-Schwartz, J., 397, 507
Lipshutz, R. J., 557, 561(8)
Liron, T., 472, 473(21; 24), 483(24), 484(24), 485(21; 24), 486(21), 487(24), 488(21; 24), 489(21)
Lisanti, M. P., 206
Liscovitch, M., 265, 275, 276(3)
Lisman, J. E., 426, 427
Litman, B. J., 38
Littman, D. R., 464
Litwack, G., 276(23), 278
Liu, B., 275(6), 276, 276(18), 278
Liu, F. C., 571
Liu, J. P., 506
Liu, P., 207
Liu, Y., 128, 188, 192, 197(21), 198, 232(20), 233
Liu, Z., 450
Livak, K. J., 565
Livingstone, M. S., 579
Liyanage, M., 398
Llano, E., 529
Lockhart, D. J., 557, 561(8)

Logothetis, D. E., 54, 71, 72(7), 74, 75, 79(3), 80(1), 81(3; 16), 82(16), 83(3; 16), 87, 87(16), 89(18), 90(18), 92(18)
Loh, M. L., 530, 564
Lok, J. M., 175
Londos, C., 112, 115, 115(3), 116, 134, 166, 236
Londos, D., 224, 226(43)
Loomis, C., 501
Loomis, W. F., 150
Lopatin, A. N., 79
Lopez, I., 255
Lopez, J., 470, 471(9; 13; 16), 472, 475(13; 16), 480(13), 481(13), 482(9)
Lopez-Otin, C., 529
Lorenz, J. N., 472, 473(21; 24), 483(24), 484(24), 485(21; 24), 486(21), 487(24), 488(21; 24), 489(21)
Loss, J. R. II, 539
Loussouarn, G., 77
Lovric, J., 491
Lowe, P. N., 350
Lowenstein, P. R., 540
Lowndes, J. M., 189, 521, 527(15)
Lowry, W., 464
Lowy, D. R., 384
Lu, B., 70
Lu, C. C., 75
Lu, J., 507
Lu, T., 84, 86(25)
Luberto, C., 275
Lucocq, J. M., 450, 458(16), 459
Luedke, C. E., 227
Lui, M., 267
Luini, A., 397
Lum, B. S., 452
Luo, J., 472, 473(17), 482(17)
Luo, K., 507
Luo, Z., 491
Lupu, V. D., 76
Luthi, A., 336
Lux, G., 571
Lynch, C. M., 540, 544(2)
Lyons, E., Jr., 112
Lyons, J., 522

M

Ma, M., 207
Ma, W., 276(30), 278, 410, 464, 466(3), 469(4)

Ma, Y.-C., 464, 469
Maayani, S., 556
Macdonald, S. G., 490(14), 491
MacGregor, R. J., 17
Mack, D., 530
Mackie, K., 51
Maclean, D., 508, 510(29), 516(29)
Madaule, P., 350
Madden, J. A., 508
Madison, V., 38(10), 39, 41(10), 44(10)
Magnusson, R., 166, 167(18), 172(18)
Magrath, I. T., 169
Mahey, R., 106
Maina, C. V., 476
Maira, S. M., 450, 451, 460, 462, 462(19)
Majerus, P. W., 335
Majima, K., 464
Makhina, E. N., 77, 79
Makhinson, M., 426, 430(8)
Makino, A., 274(9), 276
Makowske, M., 484
Malbon, C. C., 464, 530, 531, 532, 532(12)
Malencik, D. A., 508, 516(28)
Malinski, J. A., 27
Man, Y., 508
Mancino, V., 405
Manganiello, V. C., 29
Mangion, J., 62
Maniatis, T., 244, 446(13), 447
Mann, M., 427(21), 428
Mano, H., 405, 407, 407(16), 410(21), 469
Mansour, S. J., 402, 403(22)
Mao, J., 404, 405, 407, 407(16; 22–24), 408, 410(21), 469
Marais, R., 384, 393, 393(27), 491
Margolskee, R. F., 37, 38(10), 39, 41, 41(1; 10), 43, 43(1; 16), 44(1; 10; 16), 46, 46(1)
Marinissen, M. J., 438, 440, 440(6)
Mark, M., 557, 577
Marquez, V. E., 506
Marsault, R., 107, 108, 111(28)
Marshall, C. J., 384, 393, 393(27), 490, 490(15), 491, 496
Marshall, G., 556
Marshall, M. S., 384, 396(25), 491
Martin, G., 389(40), 390
Martin, H., 471(15), 472
Martin, K. C., 426, 427, 427(9), 431(14), 579
Martin, R. K., 530
Martino, P., 490

Marty, A., 53, 79
Mashima, K., 507
Maskevitch, G., 383
Mason, C. S., 393, 491
Masry, N., 407
Massenburg, D., 398
Massimi, A., 557, 561(7)
Masters, S. B., 522
Masuda-Inoue, S., 294
Mathes, K. D., 491
Mathias, S., 275, 276(2)
Matile, H., 200
Matrisian, L. M., 528
Matsuda, M., 149, 383, 384, 391, 396(43)
Matsui, H., 384
Matsumoto, K., 241
Matsuo, K., 391, 396(43)
Matsuoka, I., 220
Matsushima, H., 599
Matsuzaki, T., 383
Mattera, R., 56(19), 57
Matzaris, M., 337
Mauco, G., 304, 305(14)
Maurer, F., 448
Max, M., 37, 41, 41(1), 43(16), 44(16), 46
Mayer, B. J., 383
Mayford, M., 230
Mayne, G. C., 472, 473(23)
Mayo, K. E., 581
McCabe, P., 371, 407
McCain, D. F., 507
McCann, R. O., 107
McCleskey, E. W., 383, 385, 387(30), 395(12)
McCormick, D. A., 336
McCormick, F., 387(34; 35), 388, 522
McDonald, N. Q., 336, 337(10), 341(10), 342(10)
McDowell, J. H., 38
McEnery, M. W., 67
McEntaffer, R., 36
McEwen, R. K., 337, 337(17)
McFadden, G., 571
McGarrity, G. J., 542
McGee, T. P., 336
McGinley, M., 276
McHugh-Sutkowski, E., 192
McIntire, W. E., 56
McKay, R., 425
McKenna, M., 584
McKinnon, M., 338

McKinnon, P. J., 37, 41(1), 43(1), 44(1), 46(1)
McLaughlin, S. K., 37, 41(1), 43(1), 44(1), 46(1), 273, 307
McLeod, S. J., 383, 391(6)
McMahon, E., 407(24), 408
McMahon, T., 472, 473(22), 485(22), 487(22)
McNamara, B., 188
McPherson, P. S., 335, 340
McPherson, R. A., 491
McReynolds, L. A., 476
McVicar, D., 464
Meacci, E., 399, 400(18)
Medkova, M., 482
Mehler, E., 46
Meier, K. E., 294, 295, 296
Meier, R., 448, 450, 458(16), 459
Melançon, P., 402, 403(22)
Mellors, A., 295
Meltzer, P. S., 529(3), 530
Mencarelli, A., 540, 542(7)
Mene, P., 600
Meng, J., 149
Menon, T., 137, 206
Mercer, E. H., 230
Mercer, J. A., 427(21), 428
Mercola, D., 425
Merrill, A. H., Jr., 276, 294(8)
Mesirov, J. P., 530, 564
Messing, R. O., 472, 473(22), 485(22), 487(22)
Methfessel, C., 79
Mettei, M. G., 220
Metzger, H., 128, 189
Meyer, T., 88(26), 89, 90, 90(26), 488, 506
Meyers, R. E., 72(6), 74
Miao, W., 491
Michael, D., 427, 431(14)
Michaels, G. S., 563
Michaud, N. R., 191
Michel, I. M., 115
Michell, R. H., 337, 337(17)
Michener, C. D., 563
Miki, K., 397
Miki, T., 508
Miller, A. D., 540, 544(2)
Miller, B. S., 276(26), 278
Miller, D. G., 540, 544(2)
Miller, J. F., 245
Miller, K. G., 470, 471(9), 482(9)
Miller, R. J., 53

Miller, R. T., 380
Miller, W., 527
Mills, G. B., 295
Mills, J. S., 28, 31(12), 33(12)
Millward, T. A., 552
Milona, N., 151
Min, D. S., 259, 260, 261, 262(15), 264(13; 15; 17), 331
Minami, Y., 507
Minato, N., 383
Minden, A., 414, 415(6), 417(6), 422(6)
Ming, X. F., 450
Minh Vuong, T., 28, 37(9)
Minogue, S., 72(5), 74
Mira, J.-P., 354, 355(6)
Mironov, A., 397
Mirshahi, T., 71, 75, 81(16), 82(16), 83(16), 87(16), 89(18), 90(18), 92(18)
Mische, S. M., 102
Mishina, M., 79
Mishiro, S., 276(20), 278
Misra-Press, A., 383
Missler, M., 221
Misumi, K., 397
Misumi, Y., 397
Mitchell, C. A., 337
Mitoma, J., 274(9), 276
Mitsui, Y., 522
Mittal, R., 28
Mitterauer, T., 232(9), 233
Miura, K., 336
Miyake, Y., 294
Miyata, T., 107
Miyawaki, S., 599
Mizuno, T., 380
Moarefi, I., 469
Mochizuki, N., 149, 383, 391, 396(43)
Mochly-Rosen, D., 470, 471(9; 13; 16), 472, 472(3), 473, 473(11; 17–22; 24), 475(13; 16), 476, 478(11; 28; 29), 479(29), 480(13), 481, 481(13), 482, 482(9; 17; 38), 483(11; 20; 24), 484, 484(24), 485(21; 22; 24; 51), 486, 486(11; 20; 21), 487(22; 24), 488(21; 24), 489(21)
Moelling, K., 491
Molliver, D. C., 385, 387(30)
Mollner, S., 173, 176(24)
Molnos, J., 465
Monaco, L., 189

Mondino, A., 389(37), 390
Monesi, V., 105
Monia, B. P., 425
Mons, N., 105, 106, 239
Montaner, S., 404
Montero, M., 108
Montgomery, C., 212
Montminy, M. R., 570, 571
Moodie, S. A., 490
Moody, T. D., 426, 430(8)
Moolenaar, W. H., 294, 299
Moomaw, C. R., 258, 261(12), 264(12), 328(5), 329, 331(5), 363
Moore, D. D., 524
Moore, O. D., 248
Moore, T., 561, 563(11)
Moos, M., Jr., 192
Moran, L. S., 476
Moreno, R., 585(5), 586
Morgan, C., 265
Morgan, S. J., 338
Morii, N., 350
Morinaga, N., 397, 399, 401, 401(6), 402(4), 403(4–7), 404(6)
Morita, E. A., 38(9), 39
Moriyama, K., 37, 46
Morley, M., 506, 557, 561(7)
Morrice, N., 448, 450, 455(7)
Morris, A. J., 255, 257(3), 259, 260(3), 264(3), 265, 266, 267(4; 6), 269, 269(8), 270, 271(5–7), 272(9), 273(8), 295, 296, 307, 328(4), 329, 330, 331(6), 338, 359, 398, 464, 530
Morris, D. I., 188, 192
Morrison, D. K., 384, 490, 491
Morrison, J. F., 509
Morton, R. E., 299
Mosior, M., 501
Moss, B., 266, 271(7)
Moss, J., 359, 397, 398, 399, 401, 401(6; 9), 402(4), 403, 403(4–7), 404(6)
Mousseau, B., 476
Moutjdir, M., 472, 473(18)
Moxham, C. M., 531, 532, 532(12)
M'Rabet, L., 393
Mueller, D. L., 389(37), 390
Muglia, L. J., 207, 212, 220(15; 27), 224(15), 226(15), 229
Muglia, L. M., 212, 220(27), 227
Mulholland, J., 564

Mullenix, J. B., 7, 128, 129(10), 130(10), 165, 166, 167(17), 169, 169(17), 170(17; 21), 172(13; 17), 173(13; 17), 175(13), 179(17), 180, 182(21; 37), 183(21), 185(37), 186(37), 187(37), 232(17–19), 233
Muller, D., 426
Müller, J., 490
Muller, R. U., 427
Mumby, M. C., 276, 553
Mumby, S. M., 175
Munroe, D., 299, 521
Muradov, K. G., 31, 36
Murai, M., 585
Murakami, T., 149, 383
Muramatsu, M., 360
Murata, K., 141
Murayama, Y., 184, 185(40), 187(40), 521(17), 522, 527(17)
Murgia, M., 107
Murphy, P. J., 212, 220
Murphy, T. J., 539, 545, 550, 550(10; 15–17), 551
Murray, L. D., 113, 114(16), 115(16), 120(16), 122(16), 198, 205(4)
Murrell-Lagnado, R., 72(8), 74, 81(8), 83(8)
Muso, E., 599
Muzzio, I. A., 426, 427(9)
Myers, D. E., 507
Myers, E. W., 527
Myers, T. G., 530, 585
Myles, D. D., 508
Myrray, A. W., 472, 473(23)
Myung, C. S., 56

N

Nacro, K., 506
Nagamachi, Y., 255
Nagao, M., 521
Nagashima, K., 149, 383, 391, 396(43)
Nagle, J. W., 585(5), 586
Nair, B. G., 166
Nakafuku, M., 521
Nakahashi, Y., 107, 108, 109(23)
Nakajima, T., 571
Nakamura, H., 398
Nakamura, S., 384
Nakanishi, H., 380
Nakanishi, S., 219, 225(32)

Nakano, A., 360
Nakashima, S., 255, 257(3), 260(3), 264(3), 265, 267(4)
Nakaya, M., 391, 396(43)
Nakayama, K., 398
Nalefski, E. A., 482
Nanott, C., 232(9), 233
Nara, F., 294
Nardi, A., 227
Narumiya, S., 350
Narvey, M., 304
Nash, K., 508
Natochin, M., 27, 36, 37, 37(9)
Nau, M. E., 530
Nawy, S., 37
Nazaire, F., 276
Neer, E. J., 95, 175, 404
Neher, E., 53, 79
Nelson, E., 107, 108, 109(23)
Nemoto, Y., 336, 340
Nesher, R., 472, 473(19)
Neubig, R., 51, 57, 60, 67(4), 69(4)
Neumann, P. E., 227, 229
Newmeyer, D. D., 276(19), 278
Newton, A. C., 470, 471(6), 482(6), 499, 501, 502, 503
Newton, P. B. III, 542
Ng, D. H., 341
Ng, J. B., 452
Nguyen, B., 84, 86(25)
Nguyen, O., 338, 339(22)
Nguyen, P. V., 426
Nichols, C. G., 77, 79
Nichols, J., 539
Nickels, J. T., 276
Nield, H. S., 162(9), 163
Nielson, M. D., 224, 225(42)
Nijman, S. M., 383(18), 384, 387(18)
Niman, H. L., 440
Nimnual, A., 415
Nishikawa, K., 502
Nishimoto, I., 184, 185(40; 41), 187(40), 521(17), 522, 527(17)
Nishishita, T., 276(20), 278
Nishizuka, Y., 470, 471(4), 476(10), 499
No, D., 230
Noda, M., 383, 384, 399
Noeldeke, J., 491
Nogami, M., 398
Noguchi, M., 107

Nolan, G. P., 540, 542(7)
Nomura, H., 43
Nonet, M., 482
Nonner, D., 276(22), 278
Northemann, W., 477
Notkins, A. L., 507
Notterman, D. A., 530
Nouranifar, R., 430
Novick, P., 335
Nozawa, Y., 255, 257(3), 260(3), 264(3), 265, 267(4)
Numa, S., 79
Nurse, P., 339

O

Oancea, E., 488, 506
Obeid, L. M., 275, 275(6), 276, 276(4; 18; 21; 24; 32), 278
Obertone, T. S., 550(16), 551
Obrietan, K., 570, 571, 577, 577(3), 579(3), 580(13), 582(3), 583(22), 584
O'Connell, R. P., 522
O'Dell, T. J., 426, 430(8)
Oetjen, E., 571
Offermanns, S., 405
O'Gara, F., 164
Ogasawara, M., 397, 399, 401, 403(7)
Ogata, E., 184, 185(40; 41), 187(40), 276(20), 278, 521(17), 522, 527(17)
Ogita, T., 294
Øgreid, D., 4, 7(6)
Ohba, Y., 149, 383, 391, 396(43)
Ohgi, T., 585
Ohmori, S., 488
Ohtsuka, T., 394
Oka, N., 96
Okada, T., 384, 387(29), 393(29), 394(29)
Okajima, Y., 507
Okamoto, H., 117
Okamoto, T., 184, 185(40), 187(40), 521(17), 522, 527(17)
Okayama, H., 449
Okazaki, T., 276(20), 278
Olofsson, B., 376(13), 377
Olson, M. F., 393
Ono, Y., 488, 506
Ooi, C. E., 360
Orci, L., 359, 360, 360(8), 363(8)

Ord, T., 276(22), 278
O'Rourke, E. C., 387(35), 388
Orr, E., 476, 478(28)
Orr, J. W., 502
Orth, K., 127, 206
Ortiz, D. F., 60
Oshima, H., 556
Oshima, Y., 241
Osinska, H., 472, 473(21), 485(21), 486(21),
    488(21), 489(21)
Ostrom, R. S., 226
Ostrowski, M. C., 384, 387(24), 388(24)
O'Sullivan, G. C., 188
Otmakhova, N. A., 426
O'Toole, T. E., 383
Otsu, M., 338
Ott, S. M., 106, 221, 226(38)
Otterbach, B., 294
Otto, E., 542
Otto-Bruc, A., 28, 37(9)
Ouyang, L., 540
Owen, D., 350
Ozaki, T., 149, 383

P

Pace, A. M., 226, 404, 521, 522, 527(14)
Pacheco-Rodriguez, G., 397, 399, 400(18),
    403
Page, R. D., 157
Painter, G., 337, 337(17)
Palczewski, K., 29
Pallante, M., 438, 440, 440(6)
Pallen, C. J., 507
Palmer, D. J., 360
Palmer, G., 528
Palmiter, R. D., 220, 229
Pan, M.-G., 383, 385(4), 387(4), 388(4), 394(4),
    395(4)
Panayotou, G., 338, 339(22)
Pang, I.-H., 53, 175
Papageorge, A. G., 384
Papermaster, D. S., 38
Pardee, A. B., 522
Parekh, D. B., 499
Parent, A., 227, 229
Park, H., 466
Park, J., 448
Park, M., 70

Park, S.-K., 255, 256(7), 260, 260(7), 261(7),
    262(15), 264(15), 331, 404
Parker, P. J., 335, 336, 337, 337(7; 10; 17), 338,
    338(7), 339, 339(7), 340(14), 341(10),
    342(3; 7; 10), 450, 451, 460, 462,
    462(19), 470, 471(5), 499
Partridge, N. C., 528
Passast, M., 29, 30(24)
Pasti, L., 107
Patel, T. B., 7, 128, 129(10), 130(10), 160,
    162(9), 163, 165, 166, 167(17), 169,
    169(17), 170(17; 21), 172(13; 17), 173,
    173(1; 13; 17), 175(13), 177(26), 178(1),
    179(17; 26), 180, 182(21; 37), 183(21),
    184(16), 185(37), 186(37), 187, 187(16;
    37), 232(17–19), 233
Patel, Y., 471(15), 472
Paterson, H. F., 384, 393, 393(27), 490(15), 491
Patil, P. G., 53
Patrick, R. S., 336, 337(10), 341(10), 342(10)
Patton, W. A., 398
Pavlath, G. K., 539, 545, 550, 550(10; 15), 551
Pawson, T., 70, 470
Paylor, R., 427
Payne, D. M., 490
Payrestre, B., 337
Pear, W. S., 540
Pearson, R. B., 484
Peijnenburg, A., 529(2), 530
Pelicci, P. G., 540, 542(7)
Pena, L. A., 275, 276(2)
Pendas, A. M., 529
Pentyala, S. N., 88(28), 89, 320
Pepperkok, R., 364
Pergamenschikov, A., 585
Perisic, O., 318
Perl, D. P., 294
Perou, C. M., 585
Perrelet, A., 360
Perry, D. K., 275(7), 276, 276(21; 24), 278, 286,
    293, 293(38)
Peruski, L. F., Jr., 553, 571
Peschle, C., 540, 542(7)
Peseese, X., 335
Peterson, S. N., 384, 387(24), 388(24)
Petit Jacques, J., 71, 79(3), 81(3), 83(3)
Petitou, M., 342
Petrenko, A. G., 482
Petrova, V., 307
Peyroche, 402

Pfeuffer, T., 128, 173, 176(24)
Pham, T. A., 584
Picard, D., 107
Pieroni, J. P., 6, 7(10), 112
Pierra, C., 115, 122(25)
Pierre, S. C., 7, 180, 182(37), 185(37), 186(37), 187(37), 232(19), 233
Pilli, G., 188
Pilz, R. B., 397
Pina, E., 37
Pineda, V., 207, 220(15), 224(15), 226(15), 229
Pingel, J. T., 507, 508(20)
Pirianov, G., 276(27), 278
Pitossi, F. J., 448
Pitt, G. S., 151
Plant, P. J., 482
Plant, T. M., 556
Plantavid, M., 337
Plasterk, R. H., 154
Platt, R., 276(21), 278
Pleiman, C. M., 521, 527(15)
Plonk, S. G., 404
Plummer, M. R., 54
Podtelejnikov, A. V., 301
Poirier, G. G., 276(21; 24), 278
Polaina, J., 146
Polakis, P., 149, 371, 387(34), 388, 407
Pommier, Y., 530, 585
Poppleton, H. M., 7, 169, 170(21), 180, 182(21; 37), 183(21), 185(37), 186(37), 187(37), 232(19), 233
Poser, S. W., 224, 225(42), 557, 571, 577, 577(3), 579(3), 582(3)
Posern, G., 390
Posner, B. A., 318
Posner, B. I., 452
Post, S. R., 16, 226
Potapova, O., 425
Poullle, Y., 220
Pouli, A. E., 107
Pouyssegur, J., 226
Pozzan, T., 107, 108, 111(28)
Prasad, M. V., 438
Prasher, D., 107
Premont, R. T., 150, 151(3), 206, 207, 220, 225(16)
Prendergast, F. G., 106
Preston, M. S., 112, 115(3)
Prestwich, G. D., 72(7), 74, 76(9), 255, 257(3), 260(3), 264(3), 266, 269(8), 273(8)

Price, S., 43
Priess, J., 286, 293(37)
Prioleau, C., 556(5), 557
Prochiantz, A., 487
Provance, D. W., Jr., 427(21), 428
Provost, J. J., 255, 256(7), 260(7), 261(7)
Prusoff, W. H., 36
Pugliese, F., 600
Puille, Y., 220
Pullen, N., 450
Pushkareva, M. Y., 276
Putney, J. W., Jr., 105

Q

Qin, J., 192
Qiu, R. G., 371, 407
Quarmby, L. M., 128, 178, 239
Quest, A. F., 488
Quilliam, L. A., 387(35), 388
Quinn, C. O., 528
Quinn, W. G., 579

R

Raaijmakers, J. A. M., 355
Radley, E., 337, 340(14)
Radziwill, G., 491
Ragsdale, C. W., 245
Raingeaud, J., 413, 415(4), 420, 420(4)
Rajkowski, J., 427
Rakic, P., 221
Rall, T. W., 137, 206
Ram, P. T., 551
Ramachandran, C., 508
Rameh, L. E., 89
Ramfrez-Solis, R., 207, 214(19)
Ramsay, K., 476
Randazzo, P. A., 331, 361, 365
Rao, V. D., 232(7), 233
Rapp, U. R., 384, 390, 396(25), 490
Rarick, H. M., 28, 31(12), 33(12), 44
Rashed, H. M., 166
Rathbun, G., 89
Ratner, N., 384
Raucher, D., 90
Ravazzola, M., 360
Raven, T., 508

Ravichandran, K. S., 89
Rawlings, D. J., 466
Raymond, D., 556(5), 557
Raymond, J. R., 521
Rayter, S. I., 490
Rebecchi, M. J., 88(28), 89, 307, 320
Reddy, K. K., 463
Reden, J., 189
Reed, J. C., 276(22), 278
Reed, R. R., 97, 127, 151, 173, 176(25), 206, 230
Reedquist, K. A., 383, 387(9; 36), 388, 392, 393, 393(36), 396(44)
Rees, C., 585
Reese, C. B., 448
Reid, T., 350
Reinach, P. S., 513
Reinhold, W. C., 530, 585
Remmers, A. E., 51, 57, 60, 67(4), 69(4)
Ren, X.-D., 358, 396(54), 397
Retief, J. D., 155, 156(14)
Reuveny, E., 75
Revel, J. P., 405
Reyes, J. G., 276(33), 278
Rhee, L., 470, 471(5)
Rhee, S. G., 338, 398
Rice, A. E., 508
Rice, V., 106
Riganelli, D., 540, 542(7)
Riggs, P. D., 476
Riley, R. T., 276, 294(8)
Rim, C., 383, 385(4), 387(4), 388(4), 394(4), 395(4)
Rincon, M., 420
Rittenhouse, S. E., 448
Rivera-Perez, J., 216(30), 219
Rizzo, M. A., 491
Rizzuto, R., 107, 108, 111(28)
Robayna, I. G., 276(33), 278
Robbins, D. J., 245
Robbins, J., 472, 473(21), 485(21), 486(21), 488(21), 489(21)
Robbins, J. D., 189
Robertson, E. J., 208(22), 209, 213(22)
Robichon, A., 37, 38(10), 39, 41(1; 10), 43(1), 44(1; 10), 46(1)
Robida, A. M., 539, 545, 550(10)
Robineau, S., 402
Robinson, K., 471(15), 472
Robinson, M. J., 437
Robinson, M. S., 360

Robinson, P., 304
Robinson, P. J., 506
Robles-Flores, M., 37
Rodbell, M., 116, 134, 136, 166, 224, 226(43), 236
Roeske, R. W., 508, 510(29), 516(29)
Rohács, T., 71, 72(7), 74
Rohrer, D., 426, 427(9)
Roll, R., 255
Romero, G., 491
Rommel, C., 491
Ron, D., 472, 473(17; 19), 476, 478(28), 481, 482(17; 38), 484
Rong, Y., 188
Roper, R., 266, 271(7)
Rosas, F. E., 276(33), 278
Roscoe, W., 371(11), 372, 381(11), 405, 407(15), 410(15)
Rose, H. J. C., 579
Rose, J. C., 427, 431(14)
Rose, K., 330, 331(6)
Rosen, O. M., 484
Rosenfeld, C., 342
Rosenstein, A., 427
Rosenthal, A., 490
Rosenthal, W., 29, 30(24)
Rosenwald, A. G., 360, 361(18), 365, 367(18)
Rosenzweig, S. A., 276(26), 278
Roskoski, R., Jr., 454
Ross, D. T., 530, 561, 563(11), 585
Ross, E., 383, 387(9)
Ross, E. M., 112
Rossier, M. F., 106
Rossomando, A. J., 490
Roth, K. A., 221
Roth, M. G., 360
Rothenberg, M., 70
Rothman, J. E., 359, 360, 360(8), 363(8), 364, 366, 398
Rotin, D., 482
Rous, B., 360
Roy, A., 164
Roy, S., 491
Rubinfeld, B., 387(34), 388
Rubsam, L. Z., 212, 220
Rudge, S. A., 255, 266, 269(8), 271(7), 273(8), 307, 330, 331(6), 359
Rudnick, D. A., 175
Ruiz, G. M., 70
Ruiz-Avila, L., 37, 41(1), 43(1), 44(1), 46(1)

Ruiz-Larrea, F., 338, 339(22)
Rullea, C., 261, 264(16)
Runnels, L. W., 307, 313, 314
Ruoho, A. E., 128, 188, 189, 192, 193, 197(21), 232(20), 233
Ruoho, Y. A. E., 198
Rusanescu, G., 396(51), 397
Rutter, G. A., 107
Rybalkin, S., 41, 43(16), 44(16)
Rybalkina, I. G., 41
Ryu, S. H., 266, 267(6), 271(6)

S

Saati, S. M., 522, 524
Sabacan, L., 30
Sable, C., 521, 527(15)
Sacchi, N., 560
Sahyoun, N., 474
Saiga, Y., 106
Saito, H., 507
Saito, N., 488, 506
Saito, Y., 350, 490, 496
Sakai, N., 488, 506, 584
Sakaki, Y., 107
Sakisaka, T., 336
Sakmann, B., 53, 79
Sakmar, T. P., 39
Sale, G. J., 507
Salmon, M., 383, 387(9)
Saloman, Y., 134
Salomon, Y., 95, 116, 166, 224, 226(43), 236
Samama, P., 6
Sambrook, J., 244, 446(13), 447
Sanchez, L. M., 529
Sanders, L. C., 354, 355(6)
Sandhoff, K., 294
Sanner, G., 556(5), 557
Santini, G., 397
Saracino, M. R., 491
Sasaki, K., 488, 506
Sasaki, Y., 522
Sasayama, S., 599
Satoh, T., 384, 521
Satterthwaite, A., 464
Saunders-Pullman, R., 556(5), 557
Saur, W., 112
Sausville, E. A., 530, 585
Scanarini, M., 477

Scaramellini, C., 556
Scarlata, S., 306, 307, 313, 314, 320
Schaber, M. D., 384
Schaefer, M. L., 212, 220(27), 227
Schaeffer, E. M., 464
Schäfer, R., 490
Schaffhausen, B. S., 499
Schalwyk, L. C., 227, 229
Schapal, L., 95, 100, 102(2), 104, 105(2), 127, 150, 231
Scharenberg, A. M., 466
Schechtman, D., 470, 473
Scheel, J., 364
Scheller, R. H., 470, 471(9), 482(9)
Schena, M., 585
Scherer, S. W., 402, 403(22)
Scherf, U., 530, 585
Scheuer, T., 51, 57
Schewach, D. S., 220
Schiestl, R. H., 242, 246(9)
Schieven, G. L., 507
Schiff, M. L., 60
Schimmel, P. R., 122, 306, 320(1)
Schmelz, E. M., 276, 294(8)
Schmidt, A., 301
Schmitt, J. M., 383, 392(11), 395(11)
Schmitz, A. C., 490
Schneider, P. B., 294
Schoenwaelder, S. M., 337
Schofield, J. G., 107
Scholich, K., 7, 128, 129(10), 130(10), 165, 166, 167(17), 169, 169(17), 170(17; 21), 172(13; 17), 173, 173(13; 17), 175(13), 177(26), 179(17; 26), 180, 182(37; 21), 183(21), 185(37), 186(37), 187, 187(37), 232(17–19), 233
Schonthal, A. H., 552
Schrader, K. A., 230
Schreiber, S. L., 469
Schroeder, J. J., 276, 294(8)
Schuchman, E. H., 294
Schuler, G., 561, 563(11)
Schultz, G., 29, 30(24)
Schuman, E. M., 427
Schurmann, A., 365
Schuske, K., 482
Schutz, G., 579
Schwaninger, M., 571
Schwartz, J. H., 470, 471(7)
Schwartz, M. A., 358, 396(54), 397

Schwartz, R. D., 188
Schwartzberg, P. L., 464
Schwarz, J. K., 276
Schwarze, S. R., 487
Schwencke, C., 96, 206
Sciorra, V. A., 265, 266, 269(8), 270, 273(8), 359, 491
Sciulli, M. G., 397
Scolnick, E. M., 384, 395
Scott, D. K., 528
Scott, J. D., 470
Scott, M. L., 540
Scott, R. H., 54
Scudiero, D. A., 530, 585
Scully, T. T., 114, 115(22), 116(22), 120(22), 198, 232(14), 233
Sealfon, S. C., 556, 569
Seamon, K. B., 188, 189, 191, 192, 193
Sefton, B. M., 507, 508(20)
Segel, I. H., 115, 117(23), 125(23), 510
Seger, R., 443, 490
Seidel, M. G., 385
Seidman, J. G., 524
Seidman, S., 75, 248
Sekiya, F., 338
Self, A. J., 354, 379, 380(14)
Selleres, L. A., 471(14), 472, 481(14)
Sellers, L. A., 471, 472(12), 481(12)
Sena-Esteves, M., 556(5), 557
Sepsenwol, S., 95
Serafini, T., 359, 360(8), 363(8), 364
Sergeant, S., 188
Sers, C., 490
Seth, A., 422
Settleman, J., 384, 396(25)
Seyer, J. M., 166, 184(16), 187(16)
Shabanowitz, J., 490
Shaefer, M. L., 207, 220(15), 224(15), 226(15), 229
Shah, S., 307
Shalon, D., 561, 563(11), 585
Shanahan, M. F., 193
Shank, R., 189
Shaprio, P. S., 413, 418(3)
Sharma, A., 295
Sharma, S., 371, 407
Sharp, A. H., 67
Sharrocks, A. D., 420, 421(22), 422(22)
Shaw, A. S., 383, 392(11), 395(11), 507, 508(20)

Shaw, R. S., 232(6), 233
Shayman, J. A., 276, 294(8)
Shears, S. B., 336
Sheetz, M. P., 90
Shen, K., 90
Shen, Q., 547
Shen, Z., 299
Sheng, M., 571
Sheperd, S., 68, 70(14)
Sher, A., 464
Sherlock, G., 530
Shewach, D. S., 212
Shibata, Y., 599
Shimizu, K., 394
Shimizu, T., 507
Shimomura, O., 106
Shin, K.-H., 584
Shindler, K. S., 221
Shiojima, S., 585
Shirai, Y., 488, 506
Shome, K., 491
Shore, P., 420
Shoshani, I., 112, 114, 114(8), 115, 115(17), 122(25), 199
Shuichi, K., 129, 162(8), 163
Shuttleworth, T. J., 106
Shyng, S. L., 77
Sicheri, F., 469
Siddiqui, R. A., 294
Siderovski, D. P., 60, 62, 67
Siebold, A., 188
Siegele, D. A., 235
Sieglebaum, S. A., 221
Sievert, M. K., 188
Sigal, I. S., 384
Sigler, P. B., 318
Signorelli, P., 275
Sigworth, F. J., 53, 79
Sikorska, M., 502
Silva, A. J., 219, 225(32), 427, 579
Silver, L., 219, 221(33), 225(33)
Simmons, M. L., 220, 229
Simon, M.-F., 304, 305(14)
Simon, M. I., 404, 405, 407, 407(16), 410(21), 469
Simonds, W. F., 192, 438, 440, 521
Simpson, A. W., 107
Simpson, E. M., 219, 225(32)
Sinclair, M. L., 95, 100, 102(2), 104, 105, 105(2), 127, 150, 231

Singer, W. D., 328, 328(3), 329, 331(10), 332,
    371(10; 11), 372, 380, 381(11), 382(10),
    405, 407(14; 15), 410(14; 15)
Singh, J. C., 106, 221, 226(38)
Singh, S., 508
Sithanandam, G., 490
Sitia, R., 108
Skaug, J., 402, 403(22)
Skiba, N. P., 28, 30, 31(12), 33(12), 36(25)
Skinner, H. B., 336
Skuntz, S., 360, 361(18), 367(18)
Slatko, B. E., 476
Slaughter, C., 206, 258, 261(12), 264(12),
    328(5), 329, 331(5), 363
Slazas, M. M., 482
Slepak, V. Z., 30
Slepnev, V. I., 340
Slonim, D. K., 530, 564
Smallman, D. S., 227, 229
Smeal, T., 414, 415(6), 417(6), 422(6)
Smerdon, S. J., 482
Smigel, M. D., 134, 185
Smit, M. J., 127
Smith, A. D., 338
Smith, B. L., 470, 471(9; 16), 472, 475(16),
    482(9), 486
Smith, D., 583(22), 584
Smith, D. B., 416, 476
Smith, D. M., 577, 580(13)
Smith, G. E., 249
Smith, H. G., Jr., 38
Smith, J. A., 248, 524
Smith, K. E., 106
Smith, L. H., 530, 585
Smrcka, A. S., 267
Smyth, M. J., 276(24), 278
Sneddon, A. A., 507, 508(17), 512(17)
Snow, B. E., 60, 62, 67
Snowman, A. M., 67
Snutch, T. P., 53
Snyder, S. H., 67
Soares, M. B., 586
Soderbom, F., 150
Soderling, S. H., 571
Soderling, T. R., 426, 571
Sohasky, M. L., 507, 508(18), 512(18)
Sokal, R. R., 563
Solca, F., 508
Solessio, E. C., 37
Soling, H.-D., 301

Somogyi, R., 563
Somwar, R., 450, 521
Sondek, J., 30, 318
Soneji, Y., 482
Song, C. W., 507
Song, O., 141
Song, Q., 276(23), 278
Song, X., 529
Songyang, Z., 70, 502
Sonnenburg, W. K., 41
Soreq, H., 75
Soriano, P., 212
Sossin, W. S., 340, 470, 471(7)
Sotiroudis, T. G., 341
Souroujon, M. C., 470, 472, 472(3), 473(24),
    483(24), 484(24), 485(24), 487(24),
    488(24)
Spaargaren, M., 389(39), 390, 393, 393(39),
    396(39)
Spada, A., 522
Spana, E. P., 70
Sparrow, L., 276(23), 278
Spector, I., 88(28), 89
Speicher, L. A., 188
Spellman, P. T., 530, 563(16), 564, 585
Spickofsky, N., 37, 38(10), 39, 41(1; 10), 43(1),
    44(1; 10), 46(1)
Spiegel, A. M., 192, 531
Spiegel, S., 276, 276(27), 278, 294(8)
Sprang, S. R., 112, 113, 114, 114(12; 13; 16),
    115(13; 16), 120(16), 122(13; 16), 127(11),
    128, 129, 130(22), 137(14), 166, 173(19),
    179(36), 180, 183(19), 192, 197(22), 198,
    199(3), 205(2; 4), 232(15; 21; 22), 233,
    318, 407, 470, 471(8), 483(8), 485(8)
Springer, C. J., 491
Srinivasan, A., 276(22), 278
Staal, S. P., 448
Stabel, S., 470, 471(5)
Staehelin, T., 458, 475, 477(27), 480(27)
Stafford, F. J., 507
Stamnes, M. A., 360
Stanley, G. H. S., 279
Stanzel, M., 139, 162(4), 163, 239
Staub, O., 482
Staudt, L. M., 529, 561, 563(11)
Stauffer, T. P., 88(26), 89, 90, 90(26)
Stearns, T., 359
Steed, P. M., 255
Steffen, C., 584

Steitz, T. A., 128
Stenberg, P. E., 385, 387(30)
Stephens, L. R., 267, 335
Stephens, R. M., 491
Sterling, A. E., 337, 340(14)
Sternglanz, R., 171, 183(22)
Sternweis, P. C., 53, 175, 258, 261(12), 264(12), 328, 328(3; 5), 329, 330, 331(5; 7; 10), 332, 360, 363, 371, 371(10; 11), 372, 380, 381(11), 382(10), 405, 407(14; 15), 410(14; 15)
Sternweis, P. J., 175
Sternweis, P. M., 405, 407(14), 410(14)
Stevens, C. F., 427
Stevens, J., 565
Stevens, R. C., 240
Stewart, C. L., 209, 210, 210(23), 294
Stieger, M., 465
Stoffel, W., 294
Stokoe, D., 490(14), 491
Stolz, D. B., 491
Stolz, L. E., 336, 340(9), 342(9)
Stone, D., 540
Stork, P. J. S., 141, 147(5), 383, 383(17), 384, 385, 385(4), 387(4; 30), 388(4), 392(11), 394(4), 395(4; 11; 12)
Storm, D. R., 106, 150, 173, 176(23), 206, 207, 220, 220(15), 221, 224, 224(15; 35), 225(42), 226(15; 38), 227, 229, 557, 570, 571, 577, 577(3), 579(3), 580(13), 582(3), 583(22), 584
Strathman, M. P., 404
Streuli, M., 507
Striessnig, J., 67
Strike, P., 276(23), 278
Stringfield, T., 173, 177(26), 179(26)
Struhl, K., 248, 524
Strum, J. C., 491
Stryer, L., 27, 29, 39
Stryker, M. P., 584
Stubbs, G. W., 38
Stubbs, L., 220
Stuber, D., 200
Stübner, D., 112
Sturch, S., 361
Sturgill, T. W., 490
Su, M. S., 420, 421(22), 422(22)
Sudhof, T. C., 221, 482
Sue-Ling, C. K., 255
Suggs, S., 67

Sugimoto, T., 437(5), 438
Sui, J. L., 71, 75, 79(3), 80(1), 81(3), 83(3)
Sumikawa, K., 75
Sumimoto, H., 355
Summers, M. D., 249
Sun, H., 162(9), 163, 166, 169, 170(21), 182(21), 183(21), 184(16), 187(16), 491, 507, 512(16)
Sun, P., 571
Sun, W.-C., 517
Sun, Y., 521, 571
Sunahara, R. K., 113, 114, 114(12; 13; 16), 115(13; 16), 117(21), 120(16), 122(13; 16), 127, 128, 129, 129(8), 130(22), 137(14), 150, 166, 173, 173(19), 183(19), 192, 197(22), 198, 199, 199(3), 200, 201(13a), 205(2; 4), 231, 232(2; 16; 21; 22), 233
Sung, T.-C., 255, 259, 265, 266, 269, 271(7), 295, 328(4), 329, 330, 331(6), 491
Superti-Furga, G., 491
Sussman, D. J., 407(23; 24), 408
Sutherland, E. W., 137, 206
Sutkowski, E. M., 188, 189
Sutton, R. B., 470, 471(8), 483(8), 485(8)
Suzuki, C., 117
Suzuki, S., 600
Suzuki, Y., 585
Suzuki-Konagi, K., 294
Swanson, R. J., 38(9), 39
Sweatt, J. D., 427, 430(12; 13)
Symons, M., 404, 490(14), 491
Sziber, P. P., 579

T

Tabakoff, B., 239
Tagliamonte, J. A., 476
Takacs, B., 465
Takagaki, K., 585
Takagi, Y., 107
Takahashi, H., 384
Takahashi, J. S., 219, 225(32), 581
Takahashi, T., 79
Takai, K., 117
Takai, Y., 380, 384, 394
Takata, M., 465
Takatsu, K., 383
Takatsuki, A., 397
Takeda, K., 403

Takei, K., 336
Takenawa, T., 336
Takeshima, H., 89
Takeshita, A, 461
Takeuchi, E., 599
Takeuchi-Suzuki, E., 384, 396(25)
Tall, E. G., 88(28), 89
Tam, C., 466
Tamayo, P., 530, 564
Tamura, G., 397
Tanabe, L., 530, 585
Tanabe, M., 89
Tanaka, M., 294
Tang, W.-J., 114, 117(18), 127, 128(6; 7),
    129(7), 139, 160, 162(4; 6), 163, 173(6),
    178, 178(6), 189, 192, 198, 199(6), 206,
    230, 231, 232, 232(1; 6–9), 233, 233(4; 5),
    236, 239, 239(27), 249
Tang, X., 265
Tao, J., 529
Tate, G., 106
Tatsumi, M., 29
Taussig, R., 112, 128, 137, 178, 239, 241,
    242(6), 246, 249, 250
Tavare, J. M., 107
Taylor, S. J., 396(53), 397
Tchernitsa, O. I., 490
Tembleau, A., 487
Temel, R., 266, 271(7)
Temin, H. M., 541
Tempst, P., 267
Teramoto, H., 438
Teruel, M. N., 488
Terui, T., 361, 365
Tesmer, J. J. G., 112, 113, 114, 114(12; 13; 16),
    115(13, 16), 120(16), 122(13; 16), 127(11),
    128, 129, 130(22), 137(14), 166, 173(19),
    179(36), 180, 183(19), 192, 197(22), 198,
    199(3), 205(2; 4), 232(15; 21; 22), 233, 407
Testa, J. R., 448
Testa, M. P., 276(22), 278
Tew, K. D., 188
Thai, T., 507
Thaloor, D., 550
Thelen, M., 267
Thiele, C., 301
Thiele, D. J., 242
Thieme-Sefler, A. M., 508, 510(29), 516(29)
Thiriot, D. S., 188
Thomas, D. Y., 242

Thomas, G., 450
Thomas, M. J., 426, 430(8)
Thomas, M. L., 507, 508(20)
Thomas, S. A., 220, 229
Thome, J., 584
Thompson, A., 338
Thompson, G., 350
Thompson, J. L., 106
Thompson, M. A., 571, 581
Thompson, W. J., 29
Tibbles, L. A., 413, 418(2)
Tigyi, G., 305
Tobin, H., 476
Togawa, A., 397, 399, 401, 403, 403(7)
Toh-e, A., 241
Toker, A., 471, 471(14; 15), 472, 472(12),
    481(12; 14), 499, 502, 505, 506
Tokumura, A., 305
Tolivia, J., 529
Tom, C., 230
Tonegawa, S., 230, 427
Tong, P. H., 507
Tonkin, L. A., 522, 524
Tonks, N. K., 507
Tormo, A., 235
Totty, N. F., 337, 338, 340(14), 470, 471(5)
Tournier, C., 415, 420, 421, 421(19; 20)
Towbin, H., 458, 475, 477(27), 480(27)
Towler, D. A., 175
Toya, Y., 96, 206
Treisman, R., 405, 406
Trent, J. M., 529(3), 530, 561, 563(11)
Troen, P., 556
Tsai, A. Y. M., 507
Tsai, S.-C., 397, 399, 402(4), 403(4)
Tseng, J.-L., 305
Tserng, K. Y., 276
Tsichlis, P. N., 448
Tsien, J. Z., 230
Tsirka, S. E., 255
Tsuji, F. I., 107
Tsujimoto, G., 585
Tuel-Ahlgren, L., 507
Tully, T., 579
Turck, C. W., 466
Turner, J. T., 545, 550(10)
Turner, T. K., 415
Tuy, F., 342
Tyagi, S., 565
Tzivion, G., 491

U

Ucki, A., 117
Uckun, F. M., 507
Uemura, T., 68, 70(14)
Ulevitch, R. J., 443
Ullman, M. D., 294
Ullrich, A., 470, 471(5)
Unni, V., 221
Upson, R. H., 517
Urano, T., 396(51), 397
Urusawa, K., 16
Usuda, S., 276(20), 278
Usui, J., 507
Utterback, T. R., 585(5), 586

V

Vahey, M., 530
Vallar, L., 522
Vallotton, M. B., 106
Valtieri, M., 540, 542(7)
van Delf, S., 355
van der Geer, P., 507
Van de Rijn, M., 585
van der Linden, A. M., 154
van der Meer, W., 312, 314(6), 316(6)
Vanek, P. G., 276(27), 278
Vanhaesebroeck, B., 335
van Hal, N. L., 529(2), 530
van Holde, K., 306, 320(2)
van Houwelingen, A. M., 529(2), 530
van Huijsduijnen, R. H., 508
Vanier, M. T., 294
van Kooyk, Y., 383, 387(9)
van Lint, J., 450
van Praag, H., 540
Van Renterghem, B., 89
van Tunen, A. J., 529(2), 530
VanValkenburgh, H., 361
Van Veldhoven, P. P., 286
van Weeren, P. C., 450
Varmess, H. E., 464
Varnai, P., 88(27), 89
Vass, W. C., 384
Vasudevan, C., 491
Vaughan, M., 359, 397, 398, 399, 400(18), 401, 401(6; 9), 402(4), 403, 403(4–7), 404(6)
Veda, R., 464

Veillete, A., 466
Venable, M. E., 276
Verheij, M., 276(29), 278
Verheijen, M. H., 383(18), 384, 387(18)
Verma, I. M., 540
Vernet, T., 242
Vexler, Z. S., 404
Vicendo, P., 338
Vidal, M., 141
Villacres, E. C., 220, 224(35), 227, 229, 557, 577
Villeponteau, B., 522, 524
Vincent, S. J., 70
Vitale, N., 270, 399, 400(18)
Vivaudou, M., 75
Vo, B., 192
Vocero-Akbani, A., 487
Vogelstein, B., 424
Vogt, S. K., 212, 220(27), 227
Vojtek, A. B., 490
von Lintig, F. C., 397
Vorherr, T., 173, 176(24)
Vorst, O., 529(2), 530
Voss, K. A., 276, 294(8)
Vossler, M., 383, 385(4), 387(4), 388(4), 394(4), 395(4)
Voyono-Yasenetskaya, T. A., 404, 521

W

Waddick, K., 507
Wade, C., 577, 580(13)
Wadzinski, B. E., 193, 553, 571
Wahl, M. I., 466
Walchli, S., 508
Walikonis, R. S., 427(21), 428
Walker, N., 241, 242(3)
Wall, M. A., 318
Wallach, J. S., 579
Walling, M. J., 242
Waltham, M., 530, 585
Wan, Y., 387(33), 388, 464, 466, 469(4)
Wang, D. S., 89
Wang, E., 276, 294(8)
Wang, F., 521, 522
Wang, H. G., 276(22), 278
Wang, H. Y., 529, 531, 532, 532(12)
Wang, J. H., 507
Wang, L., 230

Wang, Q. J., 450, 506
Wang, S., 70
Wang, T., 320
Wang, X , 539, 550, 550(15; 16), 551
Wang, Y., 427
Ward, M. A., 427
Ward, W. W., 488
Wasmeier, C., 107
Watanabe, G., 350
Watanabe, H., 398
Watanabe, N., 350
Waterfield, M. D., 335, 337, 337(7), 338, 338(7), 339, 339(7; 22), 340(14), 342(7), 470, 471(5)
Waterhouse, N., 276(23), 278
Waters, M. G., 360
Watkins, S. C., 491
Watsky, M. A., 305
Watson, E. L., 106, 221, 226(38)
Watson, P. A., 112
Watters, D., 276(23), 278
Watts, J., 508
Wayman, G. A., 206, 207, 571, 577(3), 579(3), 582(3)
Weaver, C., 407(23), 408
Weaver, L., 542
Webb, M. R., 512
Weber, C. K., 390
Weber, G., 320
Weber, M. J., 490
Weber, P., 202
Wehner, J. A., 219, 225(32)
Wehner, J. M., 427
Wei, J., 206, 207
Weigert, R., 397
Weinryb, I , 115
Weinstein, H., 46, 569
Weinstein, J. N., 530, 585
Weinstock, C., 276(31), 278
Weipz, G. J., 169, 170(21), 182(21), 183(21)
Weiss, O., 261, 264(16), 331, 361
Wells, C., 330, 331(7), 371
Welsh, M., 387(36), 388, 393(36)
Wen, X., 563
Weng, G., 126, 166, 167(18), 172(18)
Wenk, M. R., 336
Wensel, T. G., 27
Wenzel, T., 506
West, M. A., 360
Westenbroek, R. E., 66

Westwick, J. K., 276(25), 278
Whaley, C. D., 389(37), 390
Wheat, W. H., 571
Whisnant, R. E., 114, 117(19; 21), 128, 129 (8; 9), 173, 198, 199, 199(8), 232(10; 16), 233
White, M. M , 189
Whitfield, J. F., 502, 506
Whitmarsh, A. J., 420, 421, 421(19; 20; 22), 422(22)
Whitters, E. A., 336
Wickman, K., 51, 60
Widmann, C., 437, 438(3)
Wieland, F. T., 398
Wigler, M., 241
Wilder, E. L., 579
Wildman, D., 37, 41(1), 43(1), 44(1), 46(1)
Wildt, L., 556
Wiley, S., 70
Wilkie, T. M., 181, 185
Williams, P. M., 565
Williams, R. E., 502
Williams, R. L., 318
Willingham, M. C., 360, 361(18), 367(18)
Willumsen, B. M., 490
Wilson, I. A., 440
Winder, D. G., 426, 427(9)
Winkler, D. G., 491
Winter, D. C., 188
Winters, S. J., 556
Wirth, D. F., 530
Withers, G. S., 385
Witke, W., 301
Witte, O. N., 464, 466
Wittinghofer, A., 383, 383(18), 384, 387(18), 393, 490
Wittinghofer, Q., 389(40), 390
Wittpoth, C., 7, 166, 167(17), 169(17), 170(17), 172(17), 173, 173(17), 177(26), 179 (17; 26), 180, 182(37), 185(37), 186(37), 187, 187(37), 232(18; 19), 233
Witzemann, V., 79
Wolder, M., 301
Wolf, L. G., 175
Wolf, M., 474
Wolff, J., 112, 115
Wolff, R. A., 276(32), 278
Wolfman, A., 490
Wolin, M. S., 113
Wolthuis, R. M., 383, 387(9)

Wong, S. T., 150, 173, 176(23), 206, 207, 220(15), 221, 224(15), 226(15; 38), 227, 229, 571, 577(3), 579(3), 582(3)
Wong, T., 430
Wong, Y. H., 226, 521, 527(14)
Wonhwa, C., 482
Woodgett, J. R., 413, 418(2), 448, 450
Woscholski, R., 335, 336, 337, 337(7; 10), 338, 338(7), 339, 339(7), 340(14), 341(10), 342(3; 7; 10)
Wozniak, D. F., 227
Wray, V. P., 40
Wray, W., 40
Wright, C. F., 242
Wu, D., 404, 405, 407, 407(16; 22–24), 408, 410(21), 469
Wu, G., 472, 473(21; 24), 483(24), 484(24), 485(21; 24), 486(21), 487(24), 488(21; 24), 489(21)
Wu, J., 490
Wu, S., 270
Wu, Z. L., 106, 173, 176(23), 188, 207, 220, 229
Wynne, J., 405

X

Xanthoudakis, S., 276(20), 278
Xia, Z., 206, 220, 229
Xiao, B., 482
Xie, W., 405, 407, 407(16; 22; 24), 408, 410(21), 469
Xie, Y., 255, 294
Xie, Z., 256, 259(11)
Xing, J., 571
Xing, L., 383
Xiong, W. H., 37
Xu, J., 415
Xu, K., 539, 550, 550(17), 551
Xu, N., 404, 440, 521
Xu, W., 469
Xu, X., 508
Xu, Y., 295, 299

Y

Yagi, K., 488
Yaich, L. E., 70
Yaish, P., 338

Yamaji, R., 403
Yamamori, B., 394
Yamamoto, D., 464
Yamanaka, G., 29
Yamashita, S., 255, 391, 396(43)
Yamauchi, J., 521, 571
Yamawaki-Kataoka, Y., 384, 387(29), 393(29), 394(29)
Yamazaki, A., 29
Yamazaki, M., 398
Yan, S.-Z., 114, 117(18), 127, 128(7), 129(7), 231, 232, 232(6–8), 233, 233(5), 236
Yan, X., 75, 81(16), 82(16), 83(16), 87(16)
Yang, D. D., 414, 415, 419, 420(8; 10), 421(8), 424(8; 10)
Yang, J., 84, 86(25)
Yang, S. H., 420, 421(22), 422(22)
Yang, T. T., 488
Yano, J., 585
Yano, S., 571, 577(3), 579(3), 582(3)
Yao, B., 276
Yao, H., 383, 385(4), 387(4), 388(4), 394(4), 395(4; 12)
Yao, T. P., 230
Yap, G., 464
Yarfitz, S., 27
Yasuda, J., 421
Yatani, A., 56(19), 57, 472, 473(21), 485(21), 486(21), 488(21), 489(21)
Yau, K. W., 37
Ybarra, S., 530
Yedovitzky, M., 472, 473(19)
Yeger, H., 482
Yeung, S.-M. H., 112
Yigzaw, Y., 173, 177(26), 179(26)
Yin, J. C., 579
Yokozeki, T., 398
York, J. D., 90, 336, 340(9), 342(9), 383
York, R. D., 383, 383(17), 384, 385, 385(4), 387(4; 30), 388(4), 394(4), 395(4; 12)
Yoshida, H., 599
Yoshimura, M., 206, 239
Young, S., 470, 471(5)
Yuan, H., 405, 407, 407(16; 22–24), 408, 410(21), 469
Yuan, L. C., 507
Yue, D. T., 53
Yuen, T., 556
Yurivich, D. A., 286
Yusta, B., 521

Z

Zachary, I., 226
Zak, B., 512
Zambrano, M. M., 235
Zander, N. F., 508, 516(28)
Zeltser, R., 383
Zhang, C. J., 359, 360, 361(18), 365, 367(18)
Zhang, D., 569
Zhang, G., 128, 192, 197(21), 198, 232(20), 233
Zhang, H., 71, 72(8a), 74, 75, 81(16), 82(16), 83(16), 87, 87(16), 89(18), 90(18), 92(18), 216(29), 219
Zhang, J., 276(24), 278
Zhang, K., 387(35), 388
Zhang, M. Q., 530
Zhang, P., 276(18; 26), 278
Zhang, S., 542
Zhang, W., 556
Zhang, X.-F., 84, 86(25), 335, 340, 384, 393, 396(25), 522
Zhang, Y.-J., 266, 269, 271(5; 7), 272(9), 276, 584
Zhang, Z.-H., 472, 473(18)
Zhang, Z.-Y., 507, 508, 510(29), 516(29)
Zhang-Sun, G., 452
Zhao, A. Z., 207

Zhao, X.-H., 402, 403(22)
Zhao, Z., 508, 516(28)
Zheng, B., 207, 371
Zheng, T. S., 221
Zheng, Y., 371
Zhong, W., 70
Zhou, B., 508
Zhou, D., 137, 239, 241, 242(6), 246, 250
Zhou, H. Z., 472, 473(24), 483(24), 484(24), 485(24), 487(24), 488(24), 579
Zhou, S.-S., 81, 83(24)
Zhou, Z., 56(19), 57
Zhu, E. Y., 579
Zhu, H., 427, 431(14)
Zhu, Q., 564
Zhu, W., 522
Zhu, X. J., 359, 361
Zhu, Y., 30
Ziegler, W. H., 499
Ziltener, H. J., 508
Zimmermann, G., 137, 239, 241, 242(6), 246, 250
Zolnierowicz, S., 552
Zuber, J., 490
Zufall, F., 207
Zwahlen, C., 70
Zwartkruis, F. J., 383, 383(18), 384, 387(9; 18), 393, 393(1)

Subject Index

A

AC, *see* Adenylyl cyclase

Adenylyl cyclase
calcium-sensitive enzymes, *see*
  Calcium-sensitive adenylyl cyclase
catalytic domains of membrane-bound
  enzymes
  activity assay
    $[\alpha\text{-}^{32}P]ATP$ as substrate, 134–135
    concentration of domains, 135–136
    incubation conditions, 136–137, 139
  applications, 140
  crystal structures, 127–128, 140
  expression vector construction
    type I enzyme C1 domain, 128
    type II enzyme C2 domain, 128–130
    type V enzyme C1 domain, 129
  kinetic parameters, 139
  purification from recombinant *Escherichia coli*
    overview, 131
    type II enzyme C2 domain, 132–134
    type V enzyme C1 domain, 131–132
  topology, 127
domain structure, 198, 232
gene cloning of transmembrane enzymes from
  Drosophila, 151–152
genomic resources for transmembrane
  enzyme identification, 152–154
isoforms, 95, 127, 150, 207, 231–232, 241
knockout mice, *see* Knockout mouse,
  adenylyl cyclase
model organisms for transmembrane isoform
  studies, 150–151
module for signaling pathway modeling, 7
photoaffinity labeling
  enzyme fragments
    expression and purification, 192
    photoaffinity labeling, 195–196
  enzyme purification, 191–192
  forskolin labels
    applications, 188–189

binding affinity, 191, 197
synthesis, 189–191, 196–197
materials, 189
membranes
  isolation, 189
  photoaffinity labeling, 192–193
pure enzyme labeling, 193–195
P-site inhibitors
  cytoplasmic domains in binding, 114
  definition, 112
  forward reaction inhibition kinetics
    assays, 116–117
    inhibition patterns, 115–116
    multiple inhibitor analysis, 117
    reagents, 116
  kinetic modeling of mechanisms, 125–126
  mechanisms of action, 113–114
  potency of types, 112
  reverse reaction inhibition kinetics
    coupled enzyme assay for ATP synthesis,
      117–119
    dead-end inhibition, 117
  reversible binding measurements
    equilibrium dialysis, 120–122
    filter binding, 122–123, 125
    overview, 119–120
phylogenetic analysis of transmembrane
  enzymes
  distance analysis, 157
  maximum likelihood analysis, 157
  overview, 154–155
  parsimony analysis, 155–157
  prospects, 159
  tree construction, 157–159
regulatory diversity, 206–207, 241
regulatory mutant selection in yeast
  applications of system, 250–251
  baculovirus–Sf9 expression of mutants for
    characterization, 249–250
  CYR1 gene deletion, 241
  $G_s\alpha$-insensitive mutant screening, 247–248
  mutant library generation, 243–246
  sequencing, 248–249

transformation, 246–247
type V adenylyl cyclase expression,
    242–243
yeast strains and media, 242
soluble enzyme production from
    membrane-bound enzyme, *see* Type V
    adenylyl cyclase; Type 7 adenylyl
    cyclase
soluble enzyme purification from rat testis
    anion-exchange chromatography, 97,
        100–101
    dye affinity chromatography, 98, 100–101
    extract preparation, 97
    gel electrophoresis, 102
    gel filtration, 97–98
    gene cloning, 105
    high-performance liquid chromatography,
        101–102
    pilot studies, 95–96
    yield, 102, 104
type I enzyme–G protein interactions,
    173–179
type V enzyme, *see* Type V adenylyl cyclase
type 7 enzyme, *see* Type 7 adenylyl cyclase
ADP-ribosylation factor
    brefeldin A-inhibited guanine nucleotide
        exchange protein activation
        assays
            ADP-ribosylation factor activation of
                cholera toxin-catalyzed ADP-
                ribosylagmatine synthesis, 400
            GTPγS binding by ADP-ribosylation
                factor, 400–402
        overview, 397–398
    coatomer
        coat protein recruitment assay
            bovine brain cytosol preparation,
                364–365
            Golgi-enriched membrane preparation,
                366
            immunoblot analysis, 367
            incubation conditions, 366–367
            overview, 363–364
        effector overview, 359–360
    Golgi vesiculation by transient expression
        immunofluorescence microscopy, 369
        microinjection of plasmids, 368–369
        normal rat kidney cell preparation, 368
        overview, 367–368
        plasmid concentration effects, 369–370

plasmids, 368
time response, 369–370
function, 359–360, 398
phospholipase D
    assay, 361–363, 370
    effector overview, 359–360
purification of recombinant protein, 331, 361
Q71L dominant activating mutant, 360–361
Aequorin
    calcium-sensitive adenylyl cyclase chimeras
        constructs, 107–108
        intracellular calcium flux measurement,
            111–112
        luminescence assay, 110–111
        rationale for construction, 107
        synthesis, 108–109
        validation
            aequorin sensitivity to calcium, 109
            cyclase sensitivity to calcium, 109
            targeting to regions of capacitive calcium
                entry, 109–110
    intracellular calcium measurement, 106–107
    intracellular targeting, 107
Akt, *see* Protein kinase B
Antisense oligonucleotide, c-Jun N-terminal
    kinase inhibition, 424–425
ARF, *see* ADP-ribosylation factor

B

Baculovirus–Sf9 cell expression system
    adenylyl cyclase
        mutants 249–250
        type V enzyme, 162–163
    brefeldin A-inhibited guanine nucleotide
        exchange proteins, 403–404
    Gα13, 375–376
    p115 RhoGEF
        protease inhibitors, 372
        detergents, 383
        virus stock preparation, 374
        cell growth and infection, 374
        purification, 374–375
    phospholipase D
        human PLD1
            cell growth and infection, 267
            Glu–Glu tagging, 266–267
            histidine-tagged protein, 361–362
            immunoaffinity chromatography,
                267–268

materials, 268
protein assay, 268
ultrafiltration for concentrating and
buffer exchange, 270–271
vectors, 267
yield, 268–269
rat brain enzyme
activation studies, 264–265
purification, 262–264
recombinant virus preparation, 262
transfection, 262
virus infection, 262
RhoA
nonprenylated protein, 377–378
prenylated protein, 376–377
transducin α subunit, 39–40
BIG1, see Brefeldin A-inhibited guanine
nucleotide exchange proteins
BIG2, see Brefeldin A-inhibited guanine
nucleotide exchange proteins
Brefeldin A-inhibited guanine nucleotide
exchange proteins
ADP-ribosylation factor activation, 397–398
discovery, 398
guanine nucleotide exchange assays
ADP-ribosylation factor activation of
cholera toxin-catalyzed ADP-
ribosylagmatine synthesis, 400
GTPγS binding by ADP-ribosylation
factor, 400–402
materials, 399–400
purification of BIG1 and BIG2
baculovirus–Sf9 cell expression, 403–404
native proteins, 402–403
subcellular localization, 403
Bruton's tyrosine kinase
defects in disease, 464
domains, 464
function, 464, 469
G protein regulation
activation assay
buffer, 466
gel electrophoresis, 467–468
incubation conditions, 467
substrate, 466
Gα12, 464, 469
Gαq, 464, 468–469
purification of histidine-tagged human
enzyme from Escherichia coli
anion-exchange chromatography, 466

cell induction and lysis, 465
nickel affinity chromatography, 465
vectors, 464–465
Btk, see Bruton's tyrosine kinase

C

Calcium/calmodulin kinase II, long-term
potentiation
confocal microscopy immunohistochemistry
of activated kinases
antibody incubation, 432–433
imaging, 433–435
localization of kinases, 436
materials, 428–429
slicing of slices, 432
role, 427
Calcium channel
patch clamp, see Whole-cell patch clamp
receptor–G protein pathway inhibition, 51
RGS12 interactions
antibodies, 62–63, 67
chick dorsal root ganglion neurons
dissection, 60–61, 63
feeding, 64
neurotransmitter application, 61–62,
64–64
plating, 61, 64
trituration, 61, 63–64
function, 60
immunoprecipitation
calcium channel α subunit, 66, 68
tyrosine-phosphorylated proteins, 62,
65–67
materials, 60–63
Numb control overlay assay, 69–70
Western blot
far Western analysis, 68–69
phosphorylated proteins, 62, 66–67
Calcium-sensitive adenylyl cyclase
aequorin chimeras
constructs, 107–108
intracellular calcium flux measurement,
111–112
luminescence assay, 110–111
rationale for construction, 107
synthesis, 108–109
validation
aequorin sensitivity to calcium, 109
cyclase sensitivity to calcium, 109

targeting to regions of capacitive calcium
entry, 109–110
capacitive calcium entry
sensitivity of enzyme, 105–106
stimulation of release, 106
knockout mice, *see* Knockout mouse,
adenylyl cyclase
Cdc42
activation
GTP:GDP ratio measurement, 349
indirect assays, 349
inhibitor protein dissociation, 349
p21-activated kinase immunoassay
advantages, 358–359
affinity precipitation assay comparison,
358
cell extract preparation, 354
controls, 353
glutathione *S*-transferase–binding
domain fusion protein preparation,
350–352
GTPase-binding domain, 350
immunobead precipitation, 355
incubation with fusion protein, 354–355
interaction time course and stability,
355–357
principle, 350
function, 349
Ceramide
degradation, 275
mass assays
diacyglycerol kinase assay
applications, 293
cell culture, 286
extraction of lipids, 286–287
incubation conditions, 287–288
mixed micelle preparation, 287
principle, 286
standard curve, 286
thin-layer chromatography, 288–290
radiolabeling of cells
advantages, 293
cell culture and precursor uptake,
279–280
extraction of lipids, 280–281
hydrolysis of lipids, 282
lipid phosphate colorimetric assay,
281–282
principles, 278–279
thin-layer chromatography, 282–285

sphingolipid metabolism
inhibitors, 293–294
overview, 277
stress signaling, 275
synthesis
enzyme activation, 276
overview, 275–276
precursor radiolabeling assay, 290–291
Coatomer, ADP-ribosylation factor effects
coat protein recruitment assay
bovine brain cytosol preparation, 364–365
Golgi-enriched membrane
preparation, 366
immunoblot analysis, 367
incubation conditions, 366–367
overview, 363–364
effector, 359–360
Golgi vesiculation by transient expression
immunofluorescence microscopy, 369
microinjection of plasmids, 368–369
normal rat kidney cell preparation, 368
overview, 367–368
plasmid concentration effects, 369–370
plasmids, 368
time response, 369–370
Confocal microscopy
long-term potentiation-activated kinases
antibody incubation, 432–433
imaging, 433–435
localization of kinases, 436
materials, 428–429
slicing of slices, 432
phosphatidylinositol bisphosphate hydrolysis
assay
marker translocation analysis, 90, 92
pleckstrin homology domain–green
fluorescent protein fusion protein
marker and expression, 88–90
CRE, *see* Cyclic AMP response element
Cyclic AMP response element
binding protein regulation, 570–571
gene expression in β-galactosidase reporter
transgenic mice
animal sacrifice and tissue sectioning,
573–574
circadian rhythm stidies, 581, 584
construct design, 572
genotyping with polymerase chain reaction,
572–573
immunocytochemistry, 574–575

long-term potentiation studies
long-term memory studies, 579, 581
neuronal cell culture, 576–577
reporter immunofluorescence
quantification, 577–578
stimulation of slices, 577
overview, 571–572
Western blot analysis, 575–576
sequence, 570
transcription factors, 570

D

Diacyglycerol kinase, ceramide mass assay
applications, 293
cell culture, 286
extraction of lipids, 286–287
incubation conditions, 287–288
mixed micelle preparation, 287
principle, 286
standard curve, 286
thin-layer chromatography, 288–290
Differential display, see Messenger RNA
differential display
DNA microarray
G protein-coupled receptor gene searching
hybridization of microarrays, 597, 599
kidney disease genes in HIGA mouse,
599–600
microarray fabrication, 597
normalized library construction by
hit-picking
complementary DNA library
preparation, 596
efficiency of normalization, 596–597
hybridization, 596
polymerase chain reaction, 596
robotics, 594
normalized library construction by
subtraction
biotinylation of DNA, 588–589
complementary DNA library
construction, 586
efficiency of normalization, 592, 594
hybridization, 589
plasmid construction, 589–590
RNA as driver, 590, 592
single-stranded library DNA preparation,
586–588
overview, 585–586

G protein-coupled receptor signaling
monitoring
data analysis
cluster analysis, 563–565
validation, 562
intrinsic reporters of cell signaling, 557
messenger RNA sample validation, 561
microarray labeling and hybridization,
561–562
principles and overview, 556–559
real-time polymerase chain reaction
amplification reaction, 567
gel electrophoresis of products, 569
materials, 566–567
primer design, 565
RNA isolation and reverse transcription,
565–566
standard curve generation, 569
RNA isolation, 559–561
transgenic mouse profiling of Gα_{i2}
subunit-activated genes
advantages, 529–530, 539
cell culture analysis, 531–532
complementary RNA labeling, 534
data analysis, 537, 539
DNA microarraying, 534–535
gene chip selection, 535–536
gene profiling, 536–537
messenger RNA isolation, 532–534
throughput and sensitivity, 530
tissue sample analysis, 531–532
Dorsal root ganglion neuron
calcium channel–RGS12 interactions
antibodies, 62–63, 67
chick dorsal root ganglion neurons
dissection, 60–61, 63
feeding, 64
neurotransmitter application, 61–62,
64–64
plating, 61, 64
trituration, 61, 63–64
function, 60
immunoprecipitation
calcium channel α subunit, 66, 68
tyrosine-phosphorylated proteins, 62,
65–67
materials, 60–63
Numb control overlay assay, 69–70
Western blot
far Western analysis, 68–69

phosphorylated proteins, 62, 66–67
whole-cell patch clamp studies, *see*
    Whole-cell patch clamp, chick dorsal
    root ganglion neurons

E

Embryonic stem cell, *see* Knockout mouse,
    adenylyl cyclase
ERK, *see* Extracellular signal-regulated kinase
Extracellular signal-regulated kinase, *see also*
    EMitogen-activated protein kinase
    G protein-coupled receptor regulation,
        437–438
    long-term potentiation
        confocal microscopy
            immunohistochemistry of activated
                kinases
            antibody incubation, 432–433
            imaging, 433–435
            localization of kinases, 436
            materials, 428–429
            slicing of slices, 432
        role, 427

F

Fluorescence resonance energy transfer,
    phospholipase C–G_q interactions
    acceptor selection, 312–313
    complex systems and antibody tagging,
        315–316
    distance of energy transfer, 311–312, 316
    donor selection, 313–314
    homotransfer, 314–315
    orientation factor, 316
Forskolin
    crystal structure of adenylyl
        cyclase : forskolin : $G_s\alpha$: GTPγS
        complex formation and crystallization,
        202–205
    photoaffinity labels, *see* Photoaffinity labeling
FRET, *see* Fluorescence resonance energy
    transfer

G

$G\alpha_0$
    guanosine nucleotide exchange and activation,
        140–141

yeast two-hybrid analysis
    bait construction, 143
    confirmation of interactions, 147
    confirmation of interactions, 148
    controls, 143–144
    library screening, 144–145
    materials, 141
    plasmid isolation for interacting proteins,
        146
    positive clone characterization, 145–146
    protein types in interaction, 149
    rationale, 140–141
    sequencing of clones, 147–148
    specificity of interactions, 147
    transformation, small-scale, 141–143
$G\alpha_{12}$, Bruton's tyrosine kinase regulation,
    464, 469
$G\alpha_{13}$
    baculovirus–Sf9 cell expression and
        purification, 375–376
    function, 404–405
    GTPase assay, 381–382
    p115 RhoGEF interactions
        Dbl homology domain deletion
            effects, 407
        luciferase reporter gene assay
            controls, 406
            criteria for interaction, 406–407
            c-fos serum response element utilization,
                405–406
            luciferase assay, 409
            normalization with green fluorescent
                protein, 407–408, 410
            transfection, 406, 409–410
        overview, 405
    RhoA activation pathways, 410
GENESIS, *see* Signaling pathway modeling
$G\alpha$i1, type V adenylyl cyclase interaction with
    α subunit
    calcium/calmodulin stimulation, 176
    expression of enzyme
        construct preparation, 173–175
        Escherichia coli expression, 175
        reconstitution, 175–176
    $G_i\alpha$ interacting sites, 178–179
$G\alpha_{i2}$
    differential display of regulated genes,
        527–529
    DNA microarray profiling of $G\alpha_{i2}$
        subunit-activated genes

advantages, 529–530, 539
cell culture analysis, 531–532
complementary RNA labeling, 534
data analysis, 537, 539
DNA microarraying, 534–535
gene chip selection, 535–536
gene profiling, 536–537
messenger RNA isolation, 532–534
throughput and sensitivity, 530
tissue sample analysis, 531–532
GIRK, *see* Inwardly rectifying potassium
  channel
G protein-coupled receptor
  DNA microarray analysis, *see* DNA
    microarray
  mitogen-activated protein kinase regulation,
    437–438, 521
  phospholipase D signaling, *see*
    Phospholipase D
G protein-coupled receptor kinase, whole-cell
  patch clamp and microinjection of
  recombinant protein, 58–59
G_q
  Bruton's tyrosine kinase regulation, 464,
    468–469
  phospholipase C interactions, fluorescence
    assay
    association rate determination, 323–324
    concentrating factor, 327
    controls, 319–320
    degree of association determination, 323
    dissociation constant determination, 321,
      323, 325–327
    dissociation rate determination, 323–324
    emission spectroscopy
      advantages and disadvantages,
        310–311
      center of spectral mass, 309
      dansyl-labeled $G\beta\gamma$, 309–310
      titration with acrylodan-labeled $G\beta\gamma$,
        307–309
    fluorescence resonance energy transfer
      acceptor selection, 312–313
      complex systems and antibody tagging,
        315–316
      distance of energy transfer,
        311–312, 316
      donor selection, 313–314
      homotransfer, 314–315
      orientation factor, 316

instrumentation, 319
labeling of enzyme, 317
membrane concentration in titration,
  321–322
membrane crowding, 318
overview of approaches, 306–307
photobleaching, 320
probe handling, 316
reconstitution of membrane proteins,
  317–318
titration conditions, 318–320
GRIN3, $G_0 \alpha$ subunit interactions, 149
GRK, *see* G protein-coupled receptor kinase
G_s
  gain-of-function mutants, 530–531
  loss-of-function mutants, 530–531
  type V adenylyl cyclase interactions with
    α subunit
    activity stimulation by constitutively active
      mutant $G_s\alpha$, 167–168
    C1b interactiions with C2 domain, 166,
      170–173
    C2–$G_s\alpha$ interactions, 181–182
    crystal structures
      C1 histidine-tagged domain purification,
        199–200
      C1 : C2 : forskolin:$G_s\alpha$: GTPγS complex
        formation and crystallization,
        202–205
      C1a–C2a domain reconstitution,
        198–199
      C2 domain purification from type II
        enzyme, 201
      $G_s\alpha$ histidine-tagged protein purification,
        201–202
      prospects, 205–206
    GTPase-activating protein activity
      assay, 180–181
      C2 subdomains, 182–184
      enzyme domain expression and
        purification, 180
    peptide competition studies, 171–173
    regulatory mutant selection in yeast
      applications of system, 250–251
      baculovirus–Sf9 expression of mutants
        for characterization, 249–250
      CYR1 gene deletion, 241
      $G_s\alpha$-insensitive mutant screening,
        247–248
      mutant library generation, 243–246

sequencing, 248–249
transformation, 246–247
type V adenylyl cyclase expression,
    242–243
yeast strains and media, 242
yeast two-hybrid assay
    C1b intramolecular interactions, 168
    C2–C1b interactions, 170–171
    controls, 169
    plasmids, 168–169
    transformation, 169
Guanine nucleotide exchange factor, *see*
    Brefeldin A-inhibited guanine nucleotide
    exchange proteins; p115 RhoGEF
GzGAP, $G_0 \alpha$ subunit interactions, 149

H

Hippocampal brain slice, *see* Long-term
    potentiation

I

Immunoprecipitation, kinase assays
    mitogen-activated protein kinases
        antibodies, 439
        cell culture, 441
        epitope-tagged kinases, 439–440
        glutathione *S*-transferase fusion proteins as
            substrates, 443–445
        immuniprecipitation, 442–443
        kinase assay conditions, 443–444
        lysate preparation, 441–442
        principles, 439
    protein kinase B
        buffers, 452–453
        controls, 455
        immunoprecipitation, 453–454
        kinase assay conditions, 454–455
        lysate preparation, 453
        troubleshooting, 455
Inwardly rectifying potassium channel,
    phosphatidylinositol bisphosphate
    interactions
    GIRK1/GIRK4 single-channel kinetics
        channel mean opening time, 77, 80–81
        data analysis, 79–81
        open channel probability, 77
        opening frequency, 77, 80–81

patch total activity, 77, 80–81
recording technique, 78–79
handing of lipids, 76
hydrolysis of phospholipids
    calcium-activated chloride current
        measurement, 87
    channel inhibition studies, 87–88
    confocal microscopy and marker
        translocation analysis, 90, 92
    pleckstrin homology domain–green
        fluorescent protein fusion protein
        marker and expression, 88–90
IRK1 dose–response, 74
macropatch recording in *Xenopus* oocytes,
    75–76
phospholipid antibody inhibition studies
    IRK1, 81–84
    reversibility, 83–84
    rundown prevention, 83
phospholipid specificity, 72, 74
polylysine block studies, 84, 86
rundown studies, 71
scanning cysteine acessibility mutagenesis
    mapping of interaction sites, 86–87
IRK, *see* Inwardly rectifying potassium
    channel

J

JNK, *see* c-Jun N-terminal kinase
c-Jun N-terminal kinase, *see also*
    Mitogen-activated protein kinase
    activation
        kinases in activation, 420
        overview, 418–419
        phosphoprotein Western blot analysis,
            418–419, 421, 423
        reporter gene assay, 421–423
    assays
        in gel assay for isoform detection,
            415, 418
        immune complex kinase assay, 413–414,
            416–417
        materials, 415–416
        solid-phase kinase assay, 414, 417–418
    dominant negative mutants, 420–421
    G protein-coupled receptor regulation,
        437–438
    inhibition
        antisense oligonucleotides, 424–425

knockout mouse, 424
isoforms, 414–415, 424
c-Jun transcriptional activation assay,
419–420
stress response, 413
subcellular localization, 419

K

Kinetikit, *see* Signaling pathway modeling
Kir channel, *see* Inwardly rectifying potassium
channel
Knockout mouse, adenylyl cyclase
behavioral studies, 227, 229
CRE–*lox* system, 230–231
disruption strategy for *ac1*, *ac3*, and *ac8*, 210,
212
embryonic stem cell culture, 209–210
enzyme assays
calcium-stimulated activity, 226–227
cell accumulation, 226
membrane assay, 226–227
radioimmunoassay, 226
gene compensation, 225–226
gene loci for isoforms, 220
genotyping
polymerase chain reaction, 216, 218
Southern blot, 214, 216
inducible gene expression systems,
230–231
isoform selection, 207
materials, 207–209
mouse breeding
chimeras, 218
double knockout mouse production
AC1/AC8 knockout, 221–222
crosses, 223, 224
genetic background issues, 224–225
probability analysis, 221–222
rationale, 219–220
strategies, 220–221, 223
F_1 generation, 218
knockout mouse production, 219
phenotyping, 225, 227, 229
recombinant embryonic stem cell
generation
antibiotic selection, 214
cell culture, 213
freezing, 214
overview, 212–213, 227, 230

transfection, 213–214
vector consruction, 212

L

Long-term potentiation
calcium/calmodulin kinase II role, 427
calcium signaling, 426
confocal microscopy immunohistochemlsuy
of activated kinases
antibody incubation, 432–433
imaging, 433–435
localization of kinases, 436
materials, 428–429
slicing of slices, 432
cyclic AMP response element studies in
transgenic mice
long-term memory studies, 579, 581
neuronal cell culture, 576–577
reporter immunofluorescence
quantification, 577–578
stimulation of slices, 577
electrophysiology of hippocamplal slices
brain slice advantages, 429–430
dissection, 430
induction of long-term potentiation,
430–431
monosynaptic excitatory postsynaptic
potentials, 430
recording, 430
solutions, 428
extracellular signal-regulated kinase role, 427
induction by theta pulse stimulation with
β-receptor stimulation, 426–427, 436
LPA, *see* Lysophosphatidic acid
LTP, *see* Long-term potentiation
Lysophosphatidic acid
assay
agonist incubation of cells, 301
extraction of lipids, 301–302, 304
materials, 300
metabolic labeling of cells with tritiated
lyso-platelet-activating factor, 301,
304
radioactive precursor selection, 303–305
radioactivity quantification, 302–303
thin-layer chromatography, 302, 305
cellular responses, 299–300
metabolism, 299
transcription stimulation, 521

M

Membrane protein interaction fluorescence
assay, *see* Phospholipase C
Messenger RNA differential display, G protein
signaling
complementary DNA
characterization, 527
isolation and reamplification, 526–527
synthesis, 525
$G\alpha_{i2}$-regulated genes, 527–529
materials, 523–524
polymerase chain reaction, 525–526
principles, 522–523
RNA isolation, 525
Mitogen-activated protein kinase, *see also*
Extracellular signal-regulated kinase; c-Jun
N-terminal kinase
activated phosphoprotein analysis
immunofluorescence, 447
Western blot, 445–447
G protein-coupled receptor regulation,
437–438, 521
immunoprecipitated enzyme assay
antibodies, 439
cell culture, 441
epitope-tagged kinases, 439–440
glutathione *S*-transferase fusion proteins as
substrates, 443–445
immuniprecipitation, 442–443
kinase assay conditions, 443–444
lysate preparation, 441–442
principles, 439
kinases in activation, 437, 445
types, 437

N

4-Nitrophenyl phosphate, protein-tyrosine
phosphatase assays, 514–517

P

p115 RhoGEF
assays
exchange activity
dissociation assay conditions, 380
GTPγS binding, 380
principle, 378
Rho prenylation effects, 380–381

RhoA loading with tritiated GDP,
378–379
GTPase
assay conditions, 382
$G\alpha_{13}$ loading with GTP, 381–382
baculovirus–Sf9 cell expression of proteins
cell growth and infection, 374
detergents, 383
G_{13} α-subunit purification, 375–376
p115 RhoGEF purification, 374–375
protease inhibitors, 372
RhoA preparation
nonprenylated protein, 377–378
prenylated protein, 376–377
virus stock preparation, 374
function in G protein signaling, 371
$G\alpha_{13}$ interactions
Dbl homology domain deletion effects, 407
luciferase reporter gene assay
controls, 406
criteria for interaction, 406–407
c-fos serum response element utilization,
405–406
luciferase assay, 409
normalization with green fluorescent
protein, 407–408, 410
transfection, 406, 409–410
overview, 405
Paladin, G_0 α subunit interactions, 149
Patch clamp, *see* Inwardly rectifying potassium
channel; Whole-cell patch clamp
PCR, *see* Polymerase chain reaction
PDE, *see* Phosphodiesterase
Phosphatidylinositol bisphosphate
availability and sources, 72
forms, 71
inwardly rectifying potassium channel
interactions
GIRK1/GIRK4 single-channel kinetics
channel mean opening time, 77, 80–81
data analysis, 79–81
open channel probability, 77
opening frequency, 77, 80–81
patch total activity, 77, 80–81
recording technique, 78–79
handing of lipids, 76
hydrolysis of phospholipids
calcium-activated chloride current
measurement, 87
channel inhibition studies, 87–88

confocal microscopy and marker
translocation analysis, 90, 92
pleckstrin homology domain–green
fluorescent protein fusion protein
marker and expression, 88–90
IRK1 dose–response, 74
macropatch recording in *Xenopus* oocytes,
75–76
phospholipid antibody inhibition studies
IRK1, 81–84
reversibility, 83–84
rundown prevention, 83
phospholipid specificity, 72, 74
polylysine block studies, 84, 86
rundown studies, 71
scanning cysteine acessibility mutagenesis
mapping of interaction sites, 86–87
metabolism, 71–72
phospholipase D assay, phospholipid vesicle
delivery of substrate
activator studies, 332
assay, 331–332
comparison of vesicle formation methods,
332–333
gel-filtered vesicle preparation, 330
materials, 328–329
phosphatidylinositol bisphosphate
inclusion rationale, 328
sonicated vesicle preparation, 329
troubleshooting, 333–334
Phosphatidylinositol trisphosphate
5-phosphatase
assays
colorimetric assay, 341–342
incubation conditions, 340–341
radioactive substrate preparation with
phosphatidylinositol 3-kinase
enzyme sources, 337–338
extraction of lipids, 338–339
incubation conditions, 339
thin-layer chromatography, 339–340
radioactivity quantification, 340–341
classification, 335–336
synaptojanin
function, 336–337
inhibitor profile, 345
purification from rat brain, 342–343
Sac1 domain, 336
substrate specificity, 340, 344
Phosphodiesterase

purification
bovine retinal PDE6, 38
chicken pineal enzyme, 39
gustatory enzymes, 38
PDE6γ subunit from baculovirus–Sf9 cell
system, 39–40
signaling pathway modeling, 9
transducin α subunit binding to inhibitory
γ subunit
activation function, 27, 37
activity assay of activation
acid alumina column assay, 42
difficulty, 27–28
reconstitution, 28
tritiated cyclic GMP as substrate,
29–30, 42
fluorescence assay of binding to Pγ subunit
competition assay, 35–36
dissociation constant determination,
35–36
fluorescence measurement, 35
labeling with 3-(bromoacetyl)-
7-diethylaminocoumarin, 33
mapping of subunit interface, 36
recombinant subunit expression and
purification, 32–33
site-directed mutagenesis, 30–32
transducin effector residue identification,
36–37
gustatory enzyme activation and
fractionation, 43
mutant studies of α-transducin, 46–48
peptide stimulation assays
peptide selection, 41
phosphate release assay, 41
reagents, 41
tritiated cyclic GMP assay, 41
proton evolution assay, 28–29
subunit interaction residue identification,
36–37, 44, 46
tryptophan fluorescence assay, 28
Phospholipase C, fluorescence assay of G$_q$
interactions
association rate determination, 323–324
concentrating factor, 327
controls, 319–320
degree of association determination, 323
dissociation constant determination, 321, 323,
325–327
dissociation rate determination, 323–324

emission spectroscopy
  advantages and disadvantages, 310–311
  center of spectral mass, 309
  dansyl-labeled G$\beta\gamma$, 309–310
  titration with acrylodan-labeled G$\beta\gamma$,
    307–309
fluorescence resonance energy transfer
  acceptor selection, 312–313
  complex systems and antibody tagging,
    315–316
  distance of energy transfer, 311–312, 316
  donor selection, 313–314
  homotransfer, 314–315
  orientation factor, 316
instrumentation, 319
labeling of enzyme, 317
membrane concentration in titration, 321–322
membrane crowding, 318
overview of approaches, 306–307
photobleaching, 320
probe handling, 316
reconstitution of membrane proteins, 317–318
titration conditions, 318–320
Phospholipase D
ADP-ribosylation factor
  assay, 361–363, 370
  effector overview, 359–360
assays
  BODIPY-phosphatidylcholine as substrate
    for membrane assays, 298–299
  phospholipid vesicle delivery of substrate
    activator studies, 332
    assay, 331–332
    comparison of vesicle formation
      methods, 332–333
    gel-filtered vesicle preparation, 330
    materials, 328–329
    phosphatidylinositol bisphosphate
      inclusion rationale, 328
    sonicated vesicle preparation, 329
    troubleshooting, 333–334
  transphosphatidylation in cells,
    phosphatidylethanol detection,
    296–297
baculovirus–Sf9 expression, purification of
    human PLD1
  cell growth and infection, 267
  Glu–Glu tagging, 266–267
  histidine-tagged protein, 361–362
  immunoaffinity chromatography, 267–268

materials, 268
protein assay, 268
ultrafiltration for concentrating and buffer
    exchange, 270–271
vectors, 267
yield, 268–269
baculovirus–Sf9 expression, purification of
    rat brain PLD1
  activation studies, 264–265
  purification, 262–264
  recombinant virus preparation, 262
  transfection, 262
  virus infection, 262
cation-exchange chromatography of
    recombinant protein, 330–331
COS 7 cell expression of rat brain PLD1
  activation studies, 260–261
  assays
    membrane preparation and assay, 258
    transphosphatidylation assay *in vivo*, 257
  immunoprecipitation, 259–260
  transfection
    FuGENE 6, 256–257
    Lipofectamine, 256
  Western blot analysis, 259
enriched membrane preparation from Sf9
    cells, 331
family member homology, 295
G protein-coupled receptor signaling assay
  agonist preparation, 272–273
  mammalian cell culture, 271
  materials, 271
  sucrose-loaded vesicle binding assay,
    273–274
  transfection, 271–272
  tritiated palmitic acid incorporation, 272
isozymes, 255, 265, 295, 328
lysophosphatidic acid production, *see*
    Lysophosphatidic acid
product metabolism, 294–295
regulatory diversity, 255, 296, 328
signaling function, 265
Photoaffinity labeling
adenylyl cyclase
  enzyme fragments
    expression and purification, 192
    photoaffinity labeling, 195–196
  enzyme purification, 191–192
  forskolin labels
    applications, 188–189

binding affinity, 191, 197
  synthesis, 189–191, 196–197
  materials, 189
  membrance
    isolation, 189
  photoaffinity labeling, 192–193
  pure enzyme labeling, 193–195
  specificity determinants, 188
Phylogenetic analysis, adenylyl cyclase
    transmembrane enzymes
  distance analysis, 157
  maximum likelihood analysis, 157
  overview, 154–155
  parsimony analysis, 155–157
  prospects, 159
  tree construction, 157–159
PIP₂, see Phosphatidylinositol bisphosphate
PKA, see Protein kinase A
PKB, see Protein kinase B
PKC, see Protein kinase C
PLC, see Phospholipase C
PLD, see Phospholipase D
Polymerase chain reaction
  knockout mouse genotyping, 216, 218
  real-time DNA microarray analysis of G
    protein-coupled receptor signaling
  amplification reaction, 567
  gel electrophoresis of products, 569
  materials, 566–567
  primer design, 565
  RNA isolation and reverse transcription,
    565–566
  standard curve generation, 569
Potassium channel, see Inwardly rectifying
  potassium channel
PP2A, see Protein phosphatase 2A
Protein kinase A, module for signaling pathway
  modeling, 7, 9
Protein kinase B
  activation signaling, 448
  immunoprecipitated enzyme assay
    buffers, 452–453
    controls, 455
    immunoprecipitation, 453–454
    kinase assay conditions, 454–455
    lysate preparation, 453
    troubleshooting, 455
  isoforms, 448
  phosphoprotein analysis
    antibody production, 456, 458

dephospho-peptide column, 456–457
dephosphorylation with γ-phosphatase, 457
phospho-peptide column, 457
protein A–Sepharose chromatography, 456
sites of phosphorylation, 455–456
Western blot analysis, 458
purification of hemagglutinin-tagged enzyme,
  463
subcellular localization
  immunofluorescence microscopy, 459–460
  rationale, 458–459
  subcellular fractionation
    analysis of fractions, 462–463
    buffers, 460
    crude plasma membrane fraction
      preparation, 461–462
    crude total membrane fraction
      preparation, 460
substrates, 448–449
transient transfection
  calcium phosphate transfection, 449
  constructs, 450
  treatment of cells
    growth factors, 452
    inhibitors, 452
    pervanadate, 452
  troubleshooting, 449
Protein kinase C
  activation
    assays
      atypical protein kinases, 505–506
      labeling in vivo, 506
      membrane translocation, 506
    pathways, 499
    phosphorylation, 499–500
  assay
    autophosphorylation assay, 503–504
    isozyme immunoprecipitation and assay,
      504–505
    lipid cofactor inclusion, 500–501
    phosphorous-32 incorporation assay,
      502–503
    substrates, 502
    transmembrane substrates, 505
  cofactors, 500
  isozymes, 470–471
  peptide modulators
    applications, 470
    coimmunoprecipitation, 479–480
    expression–interaction cloning assay, 476

far Western blot
  gel electrophoresis and overlay, 475–476
  lysate fractionation, 474–475
  materials, 474–475
immunofluorescence colocalization
  cell culture and fixation, 478
  staining, 479
introduction in cells
  carrier peptides, 487
  microinjection, 485–486
  overview, 484–485
  transfection, 487–488
  transgenic mice, 488–489
  transient permeabilization, 486–487
peptide competition assays, 480–481
pulldown assay
  bead washes, 477–478
  materials, 477
  principle, 476–477
  Western blot analysis, 477
sequence identification
  common sequences in nonrelated
    proteins, 481–482
  homology between binding proteins,
    471–472, 482–483
  importance of evolutionarily conserved
    sequences, 483
  pseudo-RACK sites, 484
  types and isoform specificity, 473
purification of recombinant protein, 331
receptor for activated C-kinases, 471
Protein phosphatase 2A
  overexpression in NIH 3T3 cells
    cell culture, 552–553
    green fluorescent protein expression
      sorting, 553–554
    immunoblotting, 553
    limitations, 554–555
    optimization, 555
    transfection, 553
  regulation, 551–552
Protein-tyrosine phosphatase
  assay
    buffers, 509
    continuous assays
      4-nitrophenyl phosphate assay, 516–517
      reaction conditions, 515
      tyrosine-phosphorylated peptides as
        substrates, 516
    inorganic phosphate assays

ammonium molybdate assay, 513
malachite green–ammonium molybdate
  assay, 514
overview, 512–513
nonradioactive assays, overview, 508, 518
quenched end-point assays
  data analysis, 512
  4-nitrophenyl phosphate assay, 514–515
  reaction conditions, 510–511
substrates, 507–508
kinetic parameter analysis, 510
oxidation sensitivity, 509–510
PTPase, see Protein-tyrosine phosphatase

R

Rac
  activation
    GTP : GDP ratio measurement, 349
    indirect assays, 349
    inhibitor protein dissociation, 349
    p21-activated kinase immunoassay
      advantages, 358–359
      affinity precipitation assay
        comparison, 358
      cell extract preparation, 354
      controls, 353
      glutathione S-transferase–binding
        domain fusion protein preparation,
        350–352
      GTPase-binding domain, 350
      immunobead precipitation, 355
      incubation with fusion protein, 354–355
      interaction time course and stability,
        355–357
      principle, 350
  function, 349
Raf-1
  activation
    membrane targeting, 490–491
    phosphorylation, 491
    Ras role, 490–491
  assays
    coupled kinase assay, 496–497
    extract preparation, 493–494
    immunoprecipitation, 494–496
    materials, 491–492
    MEK1 gel electrophoresis, 496
    phosphoprotein-specific antibodies,
      497–498

Western blot, 497
substrate specificity, 498
RalGDS
  G protein specificity, 393
  Rap1
    glutathione S-transferase fusion protein
      pull-down assay of activation
      FLAG–Rap1 assay, 392–393
      fusion protein preparation, 390–391
      immunobloting, 391–392
      incubation conditions, 391
      lysate preparation, 390
      principles, 388–389
      time course of activation, 393
    protein–protein interactions, 385,
      387–388
Rap1
  activation assays
    activators, 387–388
    GTP : GDP ratio, 384
    nonisotopic assays, 384–386
    RalGDS–glutathione S-transferase
      pull-down assay
      FLAG–Rap1 assay, 392–393
      fusion protein preparation, 390–391
      immunobloting, 391–392
      incubation conditions, 391
      lysate preparation, 390
      principles, 388–389
      time course of activation, 393
    Rap1 GAP1 activation, 387–388
  antibodies, 384
  protein–protein interactions
    binding partners, 383, 393–394
    pull-down of effectors
      advantages, 395–396
      immunoblotting, 395
      lysate preparation, 395
      nickel affinity chromatography of
        histidine-tagged Rap1 and
        associated proteins, 395
      transfection, 394–395
    RalGDS, 385, 387–388
  signaling function, 383
Rap1GAP, G_0 α subunit interactions, 149
Regulators of G protein signaling, *see specific*
  RGS proteins
Retrovirus-mediated transgenesis
  advantages, 550–551
  infection, 5453–544

limitations, 540–541
Moloney-based viruses, 540
plasmid maintenance in bacteria, 541–542
transient, helper virus-free retroviral
  production, 542–543
vectors
  long terminal repeat-driven expression
    pCL1, 544–545
    pTJ66, 545
  posttranscriptional reporter vector
    pKX53, 550
  tetracycline-regulated expression
    pTJ68, 550
    pXF40, 549–550
  transcriptional reporter vectors
    pCAT/CUL, 545–546
    pKA9, 545
    pTJ105, 547, 549
    pTJ110, 549
RGS12, calcium channel interactions
  antibodies, 62–63, 67
  chick dorsal root ganglion neurons
    dissection, 60–61, 63
    feeding, 64
    neurotransmitter application, 61–62,
      64–64
    plating, 61, 64
    trituration, 61, 63–64
  function, 60
  immunoprecipitation
    calcium channel α subunit, 66, 68
    tyrosine-phosphorylated proteins,
      62, 65–67
  materials, 60–63
  Numb control overlay assay, 69–70
  Western blot
    far Western analysis, 68–69
    phosphorylated proteins, 62, 66–67
RGS17, G_0 α subunit interactions, 149
Rho
  activation
    GTP : GDP ratio measurement, 349
    indirect assays, 349
    inhibitor protein dissociation, 349
    p21-activated kinase immunoassay
      advantages, 358–359
      affinity precipitation assay
        comparison, 358
      cell extract preparation, 354
      controls, 353

glutathione *S*-transferase–binding
    domain fusion protein preparation,
    350–352
GTPase-binding domain, 350
immunobead precipitation, 355
incubation with fusion protein, 354–355
interaction time course and stability,
    355–357
principle, 350
function, 349, 371
guanine nucleotide exchange factor, *see* p115
    RhoGEF
purification of recombinant protein, 331
RhoA
    baculovirus–Sf9 cell expression
        nonprenylated protein, 377–378
        prenylated protein, 376–377
    p115 RhoGEF exchange activity assay
        dissociation assay conditions, 380
        GTPγS binding, 380
        principle, 378
        Rho prenylation effects, 380–381
        RhoA loading with tritiated GDP,
            378–379

S

SCAM, *see* Scanning cysteine acessibility
    mutagenesis
Scanning cysteine acessibility mutagenesis,
    inwardly rectifying potassium
    channel–phosphatidylinositol bisphosphate
    interaction studies, 86–87
Signaling pathway modeling
    complexity, 3, 22
    detail, 20–21
    emergent properties, 22
    GENESIS/Kinetikit
        availability, 6
        enzyme objects, 10, 12
        enzyme regulation modeling, 12–13
        error correction, 13
        features and functions, 19–20
        file management, 18–19
        groups, 12
        hardware requirements, 6
        molecular pools, 9–10
        output display, 17
        reaction objects, 9–10
        reaction rates, 10–11

simulation running, 16
        time steps and accuracy, 17–18
        volumes and units, 18
    parameter specification
        concentration units, 13
        enzyme constant setting, 15–16
        mechanisms versus parameters, 16
        pool concentration setting, 14
        reaction kinetics setting, 14–15
        volume units, 13
    prospects, 22–23
    rationale, 3
    reaction mechanisms in pathways
        data sources, 4, 6
        modules for simulation
            adenylyl cyclase, 7
            protein kinase A, 7, 9
            receptor–G-protein, 6–7
        phosphodiesterase, 9
    reliability, 21
    robustness, 21
Southern blot, knockout mouse genotyping,
    214, 216
Sphingomyelin
    assay
        extraction, 291–292
        precursor radiolabeling of cells, 291
        principle, 291
        thin-layer chromatography, 292–293
    sphingolipid metabolism
        inhibitors, 293–294
        overview, 277
Synaptojanin
    function, 336–337
    inhibitor profile, 345
    purification from rat brain, 342–343
    Sac1 domain, 336
    substrate specificity, 340, 344

T

Thin-layer chromatography
    ceramide mass assays, 282–285, 288–290
    lysophosphatidic acid, 302, 305
    phosphoinositides, 339–340
    sphingomyelin, 292–293
TLC, *see* Thin-layer chromatography
Transducin
    conformational assay of α subunit by limited
        proteolysis, 42–43

GTPase assay, 43
GTPγS-binding assay, 43
phosphodiesterase inhibitory γ subunit
  binding
  activation function, 27, 37
  activity assay of activation
    acid alumina column assay, 42
    difficulty, 27–28
    reconstitution, 28
    tritiated cyclic GMP as substrate,
      29–30, 42
  fluorescence assay of binding to Pγ subunit
    competition assay, 35–36
    dissociation constant determination,
      35–36
    fluorescence measurement, 35
    labeling with 3-(bromoacetyl)-
      7-diethylaminocoumarin, 33
    mapping of subunit interface, 36
    recombinant subunit expression and
      purification, 32–33
    site-directed mutagenesis, 30–32
    transducin effector residue identification,
      36–37
  gustatory enzyme activation and
    fractionation, 43
  mutant studies of α-transducin, 46–48
  peptide stimulation assays
    peptide selection, 41
    phosphate release assay, 41
    reagents, 41
    tritiated cyclic GMP assay, 41
  proton evolution assay, 28–29
  subunit interaction residue identification,
    36–37, 44, 46
  tryptophan fluorescence assay, 28
  purification of α subunit from
    baculovirus–Sf9 cell system, 39–40
  subunits, 27
Transgenic mouse
  cyclic AMP response element-mediated gene
    expression in β-galactosidase reporter
    strain
    animal sacrifice and tissue sectioning,
      573–574
    circadian rhythm stidies, 581, 584
    construct design, 572
    genotyping with polymerase chain reaction,
      572–573
    immunocytochemistry, 574–575

    long-term potentiation studies
      long-term memory studies, 579, 581
      neuronal cell culture, 576–577
      reporter immunofluorescence
        quantification, 577–578
      stimulation of slices, 577
      overview, 571–572
      Western blot analysis, 575–576
DNA microarray profiling of Gα$_{i2}$
    subunit-activated genes
    advantages, 529–530, 539
    cell culture analysis, 531–532
    complementary RNA labeling, 534
    data analysis, 537, 539
    DNA microarraying, 534–535
    gene chip selection, 535–536
    gene profiling, 536–537
    messenger RNA isolation, 532–534
    throughput and sensitivity, 530
    tissue sample analysis, 531–532
  protein kinase C peptide modulators, 488–489
Type V adenylyl cyclase
  activity assay, 166
  crystal structures
    C1 histidine-tagged domain purification,
      199–200
    C1 : C2 : forskolin : Gsα : GTPγS complex
      formation and crystallization, 202–205
    C1a–C2a domain reconstitution, 198–199
    C2 domain purification from type II
      enzyme, 201
    G$_s$α histidine-tagged protein purification,
      201–202
    prospects, 205–206
  expression
    C1a–C2 enzyme
      Escherichia coli expression, 164–165
      plasmid construction, 163–164
      purification, 165–166
    C1–C2 enzyme
      Escherichia coli expression, 164–165
      plasmid construction, 163–164
      purification, 165–166
    cytosolic domain structure, 160, 162
    full-length enzyme in baculovirus–Sf9 cell
      system, 162–163
  G$_i$α and Gβγ interacting site identification
    calcium/calmodulin stimulation, 176
    expression of enzyme
      construct preparation, 173–175

Escherichia coli expression, 175
reconstitution, 175–176
$G_i\alpha$ interacting sites, 178–179
$G_s\alpha$ interaction studies
activity stimulation by constitutively active
mutant $G_s\alpha$, 167–168
C1b interactiions with C2 domain, 166,
170–173
peptide competition studies, 171–173
yeast two-hybrid assay
C1b intramolecular interactions, 168
C2–C1b interactions, 170–171
controls, 169
plasmids, 168–169
transformation, 169
GTPase-activating protein activity for $G_s\alpha$
assay, 180–181
C2 subdomains
activating protein activity, 183–184
interactions, 182–183
enzyme domain expression and
purification, 180
yeast two-hybrid assay for C2–$G_s\alpha$
interactions, 181–182
guanine nucleotide exchange
factor-enhancing activity
GTPγS binding to G_s, 186–187
overview, 184–185
receptor peptide-mediated activity studies,
185, 187
regulatory mutant selection in yeast
applications of system, 250–251
baculovirus–Sf9 expression of mutants for
characterization, 249–250
CYR1 gene deletion, 241
$G_s\alpha$-insensitive mutant screening, 247–248
mutant library generation, 243–246
sequencing, 248–249
transformation, 246–247
type V adenylyl cyclase expression,
242–243
yeast strains and media, 242
type VI enzyme homology, 160
Type 7 adenylyl cyclase, soluble enzyme
production
C1a domain preparation
chromatography, 238
Escherichia coli expression, 237–238, 240
overview, 237–238
C2a domain preparation

anion-exchange chromatography, 235–236
comparison with other isoforms, 233–234
Escherichia coli expression, 235, 240
extract preparation, 236
materials, 234
nickel affinity chromatography, 235–236
domains, 232–233
membrane-bound enzyme comparison of
properties, 239
proteolysis sensitivity of domains, 239

W

Western blot
β-galactosidase reporter analysis in transgenic
mice, 575–576
c-Jun N-terminal kinase phosphoproteins,
418–419, 421, 423
mitogen-activated protein kinase
phosphoproteins, 445–447
phospholipase D, 259
protein kinase B phosphoprotein, 458, 462
protein kinase C overlay assay of peptide
modulators
gel electrophoresis and overlay, 475–476
lysate fractionation, 474–475
materials, 474–475
protein phosphatase 2A, 553
Raf-1, 497
Rap1
associated protein assay, 395
RalGDS–glutathione *S*-transferase
pull-down assay, 391–392
small G protein detection, 396–397
Whole-cell patch clamp, chick dorsal root
ganglion neurons
data analysis, 54
G protein-mediated pathways
overview, 51
tools for elucidation
antibodies against G proteins, 55–56
cholera toxin, 55
G protein α subunit peptides, 57
GTP analogs, 54
pertussis toxin, 55
recombinant G proteins, 56–57
neuron culture, 53
recording pipette introduction studies
guanine nucleotide analogs, 58
peptide inhibitors of protein kinases, 58

recombinant G protein-coupled receptor kinases, 58–59
technique, 57–58
recordings, 53–54
solutions, 53

Y

Yeast two-hybrid system
  G₀ α subunit effectors
    bait construction, 143
    confirmation of interactions, 147
    confirmation of interactions, 148
    controls, 143–144
    library screening, 144–145
    materials, 141

plasmid isolation for interacting proteins, 146
positive clone characterization, 145–146
protein types in interaction, 149
rationale, 140–141
sequencing of clones, 147–148
specificity of interactions, 147
transformation, small-scale, 141–143
$G_s\alpha$ interaction with type V adenylyl cyclase
  C1b intramolecular interactions, 168
  C2–C1b interactions, 170–171
  C2–$G_s\alpha$ interactions, 181–182
  controls, 169
  plasmids, 168–169
  transformation, 169

ISBN 0-12-182246-X

90051

9 780121 822460